Introduction to Modern Cryptography

Chapman & Hall/CRC Cryptography and Network Security Series

Introduction to Modern Cryptography Jonathan Katz and Yehuda Lindell

Series Editors: Douglas R. Stinson and Jonathan Katz

Secret History: The Story of Cryptology, Second Edition
Craig P. Bauer

Data Science for Mathematicians
Nathan Carter

Discrete Explorations
Craig P. Bauer

Cryptography: Theory and Practice, Fourth Edition
Douglas R. Stinson and Mary P. Paterson

Cryptology: Classical and Modern, Second Edition
Richard Klima and Neil Sigmon

Group Theoretic Cryptography
Maria Isabel Gonzalez Vasco and Rainer Steinwandt

Advances of DNA Computing in Cryptography
Suyel Namasudra and Ganesh Chandra Deka

Mathematical Foundations of Public Key Cryptography
Xiaoyun Wang, Guangwu Xu, Minggiang Wang, Xianmeng Meng

Guide to Pairing-Based Cryptography
Nadia El Mrabet and Marc Joye

https://www.crcpress.com/Chapman--HallCRC-Cryptography-and-Network-Security-Series/book-series/CHCRYNETSEC

Introduction to Modern Cryptography

Third Edition

Jonathan Katz and Yehuda Lindell

CRC Press
Taylor & Francis Group
Boca Raton London New York

CRC Press is an imprint of the
Taylor & Francis Group, an **informa** business

A CHAPMAN & HALL BOOK

Third edition published 2021
by CRC Press
6000 Broken Sound Parkway NW, Suite 300, Boca Raton, FL 33487-2742

and by CRC Press
2 Park Square, Milton Park, Abingdon, Oxon, OX14 4RN

First edition published by Taylor and Francis 2007 Second edition published by Taylor and Francis 2014

CRC Press is an imprint of Taylor & Francis Group, LLC

ISBN: 780815354369 (hbk)
ISBN: 9781351133036 (ebk)

Typeset in Computer Modern font
by KnowledgeWorks Global Ltd.

Visit the companion website/eResources:[insert CW/eResources URL

To Jill, Abigail, and Rena
– JK

To Yael, Yehonatan, Itamar,
Orel, Shirel, and Noam
– YL

Contents

Preface

The goal of our book remains the same as in the first edition: to present the core paradigms and principles of modern cryptography to a general audience with a basic mathematics background. We have designed this book to serve as a textbook for undergraduate- or graduate-level courses in cryptography (in computer science, electrical engineering, or mathematics departments), as a general introduction suitable for self-study (especially for beginning graduate students), and as a reference for students, researchers, and practitioners.

There are numerous other cryptography textbooks available today, and the reader may rightly ask whether another book on the subject is needed. We would not have written this book—nor worked on revising it for the second and third editions—if the answer to that question were anything other than an unequivocal *yes*. What, in our opinion, distinguishes our book from others is that it provides a *rigorous* treatment of modern cryptography in an *accessible* and *introductory* manner.

Our focus is on *modern* (post-1980s) cryptography, which is distinguished from classical cryptography by its emphasis on definitions, precise assumptions, and rigorous proofs of security. We briefly discuss each of these in turn (these principles are explored in greater detail in Chapter 1):

- **The central role of definitions:** A key intellectual contribution of modern cryptography has been the recognition that *formal definitions of security are an essential first step in the design of any cryptographic primitive or protocol.* The reason, in retrospect, is simple: if you don't know what it is you are trying to achieve, how can you hope to know when you have achieved it? As we will see in this book, cryptographic definitions of security are quite strong and—at first glance—may appear impossible to achieve. One of the most amazing aspects of cryptography is that efficient constructions satisfying such strong definitions can be proven to exist (under rather mild assumptions).

- **The importance of precise assumptions:** As will be explained in Chapters 2 and 3, many cryptographic constructions cannot currently be proven secure unconditionally. Security, instead, generally relies on some widely believed (though unproven) assumption(s). The modern cryptographic approach dictates that *any such assumptions must be clearly stated and unambiguously defined.* This not only allows for objective evaluation of the assumptions but, more importantly, enables rigorous proofs of security (as described next).

- **The possibility of proofs of security:** The previous two principles serve as the basis for the idea that *cryptographic constructions can be proven secure* with respect to clearly stated definitions of security and relative to well-defined cryptographic assumptions. This concept is the essence of modern cryptography, and is what has transformed the field from an art to a science.

  The importance of this idea cannot be overemphasized. Historically, cryptographic schemes were designed in a largely heuristic fashion, and were deemed to be secure if the designers themselves could not find any attacks. In contrast, modern cryptography advocates the design of schemes with formal, mathematical proofs of security in well-defined models. Such schemes are *guaranteed* to be secure (with respect to a certain security definition) unless the underlying assumption is false. By relying on long-standing assumptions, it is thus possible to obtain schemes that are extremely unlikely to be broken.

A unified approach. The above principles of modern cryptography are relevant not only to the "theory of cryptography" community. The importance of precise definitions is, by now, widely understood and appreciated by developers and security engineers who use cryptographic tools to build secure systems, and rigorous proofs of security have become one of the requirements for cryptographic schemes to be standardized.

Changes in the Third Edition

In preparing the third edition, we have continued to integrate a more practical perspective without sacrificing a rigorous approach. This is reflected in a number of changes and additions as compared to the second edition:

- We have divided our treatment of symmetric-key encryption into two parts: Chapter 3 deals with security against "passive" attacks (i.e., CPA-security), while Chapter 5 addresses "active" attacks (i.e., CCA-security and authenticated encryption). Besides breaking up what was previously a long chapter, this also allows us to introduce message authentication codes before discussing active attacks against encryption schemes.

- With an eye toward symmetric-key schemes used in practice, we have improved our coverage of stream ciphers and stream-cipher modes of operation (Sections 3.6.1 and 3.6.2); added a treatment of nonce-based encryption (Section 3.6.4); and incorporated material about standardized schemes such as GMAC and Poly1305 (Section 4.5) as well as GCM, CCM, and ChaCha20-Poly1305 (Section 5.3.2).

- With similar motivation, we have added sections on the ChaCha20 stream cipher and SHA-3 to Chapter 7. As part of our discussion about SHA-3, we also describe the sponge construction.

- We have further increased our coverage of elliptic-curve cryptography (Section 9.3.4), including a discussion of elliptic curves used in practice.

- Our treatment of TLS in Section 13.7 has been updated to reflect the latest version (TLS 1.3).

- Reflecting recent trends, we have added a chapter (Chapter 14) describing the impact of quantum computers on cryptography, and providing examples of "post-quantum" encryption and signature schemes.

For those currently using the first edition of our book, as well as for reference, we also summarize the changes/additions we have already made in the second edition (all of which remain here):

- We have increased our coverage of *stream ciphers*, including stream-cipher modes of operation as well as stream-cipher design principles and examples of stream ciphers used in practice.

- We have emphasized the importance of authenticated encryption and secure communication sessions in Sections 5.2–5.4.

- We have moved our treatment of hash functions into its own chapter (Chapter 6), and have added a section on hash-function design principles and widely used constructions (Section 7.3). We have also improved our treatment of generic attacks on hash functions, including a discussion of rainbow tables (Section 6.4.3).

- We have included several important attacks on cryptographic *implementations* that arise in practice, including chosen-plaintext attacks on chained-CBC encryption (Section 3.6.3), timing attacks on MAC verification (Section 4.2), and padding-oracle attacks on CBC-mode encryption (Section 5.1.1).

- After much deliberation, we have decided to introduce the random-oracle model earlier in the book (Section 6.5). This has several benefits, including allowing for an integrated treatment of standardized public-key encryption and signature schemes in Chapters 12 and 13.

- We have strengthened our coverage of elliptic-curve cryptography (Section 9.3.4) and have added a discussion of its impact on recommended key lengths (Section 10.4).

- In the chapter on public-key encryption, we introduce the KEM/DEM paradigm as a form of hybrid encryption (see Section 12.3). We also cover DHIES/ECIES in addition to the RSA PKCS #1 standards.

- In the chapter on digital signatures, we now describe the construction of signatures from identification schemes using the Fiat–Shamir transform, with the Schnorr signature scheme as a prototypical example. We have also improved our coverage of DSA/ECDSA. We include brief discussions of SSL/TLS and signcryption, both of which serve as culminations of material covered up to that point.

- In the "advanced topics" chapter, we have amplified our treatment of homomorphic encryption, and have added sections on secret sharing and threshold encryption.

Beyond the above, we have also edited the entire book to make extensive corrections as well as smaller adjustments, including more worked examples, to improve the exposition. Several additional exercises have also been added.

Guide to Using This Book

This section is intended primarily for instructors seeking to adopt this book for their course, though the student picking up this book on his or her own may also find it a useful overview.

Required background. We have structured the book so the only formal prerequisite is a course on discrete mathematics. Even here we rely on very little: we only assume familiarity with basic (discrete) probability and modular arithmetic. Students reading this book are also expected to have had some exposure to algorithms, mainly to be comfortable reading pseudocode and to be familiar with big-$\mathcal{O}$ notation. Many of these concepts are reviewed in Appendix A and/or when first used in the book.

Notwithstanding the above, the book does use definitions, proofs, and abstract mathematical concepts, and therefore requires some mathematical maturity. In particular, the reader is assumed to have had some exposure to proofs, whether in an upper-level mathematics course or a course on discrete mathematics, algorithms, or computability theory.

Suggestions for course organization. The core material of this book, which we recommend should be covered in any introductory course on cryptography, consists of the following (in all cases, starred sections are excluded; more on this below):

- *Introduction and Classical Cryptography:* Chapters 1 and 2 discuss classical cryptography and set the stage for modern cryptography.

- *Private-Key (Symmetric) Cryptography:* Chapter 3–5 provide a thorough treatment of private-key encryption and message authentication, and Chapter 6 covers hash functions and their applications. (Section 6.6 could be skipped if that material will not be used later.)

  We also highly recommend covering at least part of Chapter 7, which deals with symmetric-key primitives used in practice; in our experience students really enjoy this material, and it makes the abstract ideas they have learned in previous chapters more concrete. Although we do consider this core material, it is not used in the remainder of the book and so can be safely skipped if desired.

- *Public-Key Cryptography:* Chapter 9 gives a self-contained introduction to all the number theory needed for the remainder of the book. The material in The public-key revolution, including Diffie–Hellman key exchange, is described in Chapter 11. Chapters 12 and 13 go into detail about public-key encryption and digital signatures; those pressed for time can pick and choose what to cover appropriately.

We are typically able to cover most of the above in a one-semester (35-hour) undergraduate or Masters-level course (omitting some proofs and skipping some topics, as needed) or, with some changes to add more material on theoretical foundations, in the first three-quarters of a one-semester PhD-level course. Instructors with more time available can proceed at a more leisurely pace or incorporate additional topics, as discussed below.

Those wishing to cover additional material, in either a longer course or a faster-paced graduate course, will find that the book is structured to allow flexible incorporation of other topics as time permits (and depending on the interests of the instructor). Specifically, the starred (*) sections and chapters may be covered in any order, or skipped entirely, without affecting the overall flow of the book. We have taken care to ensure that none of the core (i.e., unstarred) material depends on any of the starred material and, for the most part, the starred sections do not depend on each other. (When they do, this dependence is explicitly noted.)

We suggest the following from among the starred topics for those wishing to give their course a particular flavor:

- *Theory:* A more theoretically inclined course could include material from Section 3.2.2 (semantic security); Chapter 8 (one-way functions and hard-core predicates, and constructing pseudorandom generators, functions, and permutations from one-way permutations); Section 9.4 (one-way functions and collision-resistant hash functions from number-theoretic assumptions); Section 12.5.3 (RSA encryption without random oracles); and Section 15.3 (cryptographic protocols).

- *Mathematics:* A course directed at students with a strong mathematics background—or being taught by someone who enjoys this aspect of cryptography—could incorporate Section 4.6 (information-theoretic MACs in finite fields); some of the more advanced number theory from Chapter 9 (e.g., the Chinese remainder theorem, the Miller–Rabin primality test, and more on elliptic curves); and all of Chapter 10 (algorithms for factoring and computing discrete logarithms).

In either case, a selection of advanced public-key schemes from Chapters 14 and 15 could also be included.

xx

Feedback and Errata

Our goal in writing this book was to make modern cryptography accessible to a wide audience beyond the "theoretical computer science" community. We hope you will let us know if we have succeeded! The many enthusiastic emails we have received in response to our first and second editions have made the whole process of writing this book worthwhile.

We are always happy to receive feedback. We hope there are no errors or typos in the book; if you do find any, however, we would greatly appreciate it if you let us know. You can email your comments and errata to jkatz2@gmail.com and lindell@biu.ac.il; please put "Introduction to Modern Cryptography" in the subject line. A list of known errata will be maintained at http://www.cs.umd.edu/~jkatz/imc.html.

Acknowledgments

We continue to be grateful to all those who have sent us comments, suggestions, and corrections for the book. We would like to thank, in particular, Jack Aaron, Rounak Agarwal, Ionut Ambrosie, Dan Bernstein, Jeremiah Blocki, David Cash, Claude Crépeau, Dana Dachman-Soled, Daniel Escudero, Pooya Farshim, Rolf Haenni, Imededdine Jerbi, Ali El Kaafarani, Zach Kissel, Angelique Faye Loe, Wilde Luo, Tal Malkin, Alejandro Mardones, Kurt Pan, Greg Plaxton, Kyle Andrew Porter, Christian Schaffner, Jim Tallent, Hanh Tang, Markus Triska, and Rui Xue for their feedback on the second edition.

Finally, we thank our wives and children for all their support during the now over a decade(!) we have spent working on this project.

Part I

Introduction and Classical Cryptography

Chapter 1

Introduction

1.1 Cryptography and Modern Cryptography

The *Concise Oxford English Dictionary (9th ed.)* defines cryptography as "the art of writing or solving codes." This is historically accurate, but does not capture the current breadth of the field or its modern scientific foundations. The definition focuses solely on the *codes* that have been used for centuries to enable secret communication. But cryptography nowadays encompasses much more than this: it deals with mechanisms for ensuring integrity, techniques for exchanging secret keys, protocols for authenticating users, electronic voting, cryptocurrency, and more. Without attempting to provide a complete characterization, we would say that modern cryptography involves *the study of mathematical techniques for securing digital information, systems, and distributed computations against adversarial attacks.*

The dictionary definition also refers to cryptography as an *art*. Until late in the 20th century cryptography was, indeed, largely an art. Constructing good codes, or breaking existing ones, relied on creativity and a developed sense of how codes work. There was little theory to rely on and, for a long time, no working definition of what constitutes a good code. Beginning in the 1970s and 1980s, this picture of cryptography radically changed. A rich theory began to emerge, enabling the rigorous study of cryptography as a *science* and a mathematical discipline. This perspective has, in turn, influenced how researchers think about the broader field of computer security.

Another very important difference between classical cryptography (say, before the 1980s) and modern cryptography relates to its adoption. Historically, the major consumers of cryptography were military organizations and governments. Today, cryptography is everywhere! If you have ever authenticated yourself by typing a password, purchased something by credit card over the Internet, or downloaded a verified update for your operating system, you have used cryptography. And, more and more, programmers with relatively little experience are being asked to "secure" the applications they write by incorporating cryptographic mechanisms.

In short, cryptography has gone from a heuristic set of techniques for ensuring secret communication for a few niche applications to a science that helps secure systems more generally for ordinary people around the world.

Goals of this book. Our goal is to make the basic principles of modern cryptography accessible to students of computer science, electrical engineering, or mathematics; to professionals who want to incorporate cryptography in systems or software they are developing; and to anyone with a basic level of mathematical maturity who is interested in understanding this fascinating field. After completing this book, the reader should appreciate the security guarantees common cryptographic primitives are intended to provide; be aware of standard (secure) constructions of such primitives; and be able to perform a basic evaluation of new schemes based on their proofs of security (or lack thereof) and the mathematical assumptions underlying those proofs. It is not our intention for readers to become experts—or to be able to design new cryptosystems—after finishing this book, but we have attempted to provide the terminology and foundational material needed for the interested reader to subsequently study the more advanced literature in this field.

This chapter. The focus of this book is the formal study of *modern* cryptography, but we begin in this chapter with a more informal discussion of "classical" cryptography. Besides allowing us to ease into the material, our treatment in this chapter will also serve to motivate the more rigorous approach we will be taking in the rest of the book. Our intention here is not to be exhaustive and, as such, this chapter should not be taken as a representative historical account. The reader interested in the history of cryptography is invited to consult the references at the end of this chapter.

1.2   The Setting of Private-Key Encryption

Classical cryptography was concerned with designing and using *codes* (or *ciphers*) that enable two parties to send messages while keeping those messages hidden from an eavesdropper who can monitor all communication between them. In modern parlance, codes are called *encryption schemes* and that is the terminology we will use here. Security of all classical encryption schemes relies on a secret—a *key*—shared by the communicating parties in advance and unknown to the eavesdropper. This scenario, in which the communicating parties share some secret information in advance, is known as the *private-key* (or *shared-/secret-key*) setting, and *private-key encryption* is one example of a cryptographic primitive used in this setting. Before describing some historical encryption schemes, we discuss private-key encryption more generally.

In the context of private-key encryption, two parties share a key and use that key when they want to communicate secretly. One party can send a message, or *plaintext*, to the other by using the shared key to *encrypt* (or "scramble") the message and thus obtain a *ciphertext* that is transmitted to the receiver. The receiver uses the same key to *decrypt* (or "unscramble") the

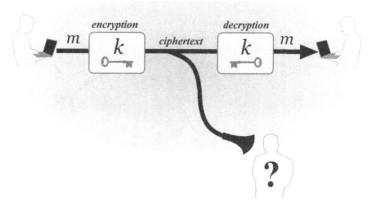

FIGURE 1.1: One common use case for private-key cryptography (here, encryption): two parties share a key that they use to communicate securely.

ciphertext and recover the original message. The same key is used to convert the plaintext into a ciphertext and back; that is why this setting is also known as the *symmetric-key* setting, where the symmetry lies in the fact that both parties hold the same key that is used for encryption and decryption. This is in contrast to *asymmetric*, or *public-key*, encryption (introduced in Chapter 11), where encryption and decryption use different keys.

As already noted, the goal of encryption is to keep the plaintext hidden from an eavesdropper who can monitor the communication channel and observe the ciphertext. We discuss this in more detail later in this chapter, and spend a great deal of time in Chapters 2, 3, and 5 formally defining this goal.

There are two canonical applications of private-key cryptography. In the first (cf. Figure 1.1), the two communication parties are separated in *space*, e.g., a worker in New York communicating with her colleague in California. These two users are assumed to have been able to securely share a key in advance of their communication. (Note that if one party simply sends the key to the other over the public communication channel, then the eavesdropper obtains the key also!) This could be accomplished, for example, by having the parties physically meet in a secure location to share a key before they separate; in the example just given, the co-workers might arrange to share a key when they are both in the New York office. In other cases, sharing a key securely is more difficult. For the next several chapters we simply assume that sharing a key is possible; we revisit this issue in Chapter 11.

The second widespread application of private-key cryptography involves the same party communicating with itself over *time*. (See Figure 1.2.) Consider, e.g., disk encryption, where a user encrypts some plaintext and stores the resulting ciphertext on his hard drive; the same user will return at a later point in time to decrypt the ciphertext and recover the original data. The hard drive here serves as the communication channel on which an attacker

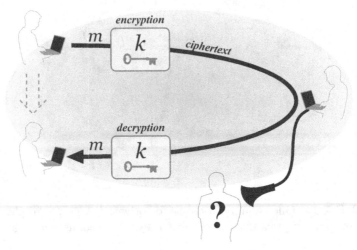

FIGURE 1.2: Another common use case of private-key cryptography (again, encryption): a single user stores data securely over time.

might eavesdrop if it can gain access to the hard drive and read its contents. "Sharing" the key is now trivial, though the user still needs a secure and reliable way to remember/store the key for use at a later point in time.

The syntax of encryption. Formally, a private-key encryption scheme is defined by specifying a *message space* $\mathcal{M}$ along with three algorithms: a procedure for generating keys (Gen), a procedure for encrypting (Enc), and a procedure for decrypting (Dec). The message space $\mathcal{M}$ defines the set of "legal" messages, i.e., those supported by the scheme. The algorithms of the scheme have the following functionality:

1. The *key-generation algorithm* Gen is a probabilistic algorithm that outputs a key k chosen according to some distribution.

2. The *encryption algorithm* Enc takes as input a key k and a message m and outputs a ciphertext c. We denote by $\mathsf{Enc}_k(m)$ the encryption of the plaintext m using the key k.

3. The *decryption algorithm* Dec takes as input a key k and a ciphertext c and outputs a plaintext m. We denote the decryption of the ciphertext c using the key k by $\mathsf{Dec}_k(c)$.

An encryption scheme must satisfy the following correctness requirement: for every key k output by Gen and every message $m \in \mathcal{M}$, it holds that

$$\mathsf{Dec}_k(\mathsf{Enc}_k(m)) = m.$$

In words: encrypting a message and then decrypting the resulting ciphertext using the same key yields the original message.

The set of all possible keys output by the key-generation algorithm is called the *key space* and is denoted by $\mathcal{K}$. Almost always, Gen simply chooses a key uniformly from the key space; in fact, one can assume without loss of generality that this is the case (see Exercise 2.1).

Reviewing our earlier discussion, an encryption scheme can be used by two parties who wish to communicate secretly as follows. First, Gen is run to obtain a key k that the parties share. Later, when one party wants to send a plaintext m to the other, she computes $c := \mathsf{Enc}_k(m)$ and sends the resulting ciphertext c over the public channel to the other party.[1] Upon receiving c, the other party computes $m := \mathsf{Dec}_k(c)$ to recover the original plaintext.

Keys and Kerckhoffs' principle. As should be clear from the above, if an eavesdropping adversary knows the algorithm Dec as well as the key k shared by the two communicating parties, then that adversary will be able to decrypt any ciphertexts transmitted by those parties. It is for this reason that the communicating parties must share the key k securely and keep k completely hidden from everyone else. Perhaps they should keep the decryption algorithm Dec secret, also? For that matter, would it not be better for them to keep all the details of the encryption scheme secret?

Auguste Kerckhoffs [114, 115] argued the opposite in the late 19th century when elucidating several design principles for military ciphers. One of the most important of these, now known simply as *Kerckhoffs' principle*, was:

> *The cipher method must not be required to be secret, and it must be able to fall into the hands of the enemy without inconvenience.*

That is, an encryption scheme should be designed to be secure *even if* an eavesdropper knows all the details of the scheme, so long as the attacker doesn't know the key being used. Stated differently, security should not rely on the encryption scheme being secret; instead, Kerckhoffs' principle demands that *security rely solely on secrecy of the key*.

There are three primary arguments in favor of Kerckhoffs' principle. The first is that it is significantly easier to maintain secrecy of a short key than to keep secret a (more complicated) encryption scheme. This is especially true when encryption is used widely. For example, consider the case where encryption is used for communication between all pairs of employees in some organization. Unless each pair of parties use their own, unique scheme, some parties will know the scheme being used by others. Moreover, information about the scheme might be leaked by one of those employees (say, after being fired), or obtained by an attacker using reverse engineering. In short, it is simply unrealistic to assume that the encryption scheme will remain secret.

Second, in case the honest parties' shared, secret information *is* ever exposed, it will be much easier for them to change the key than to replace the

---

[1]We use ":=" to denote deterministic assignment, and assume for now that Enc is deterministic. A list of common notation can be found in the back of the book.

encryption scheme. (Consider updating a file versus installing a new program.) Moreover, it is relatively trivial to generate a new random secret, whereas it would be a huge undertaking to design a new encryption scheme.

Finally, prior to widespread deployment of an encryption scheme, there is a significant benefit to encouraging public review of that scheme in order to check for possible weaknesses. Going further, it is desirable to *standardize* encryption schemes so that (1) compatibility between different users is ensured and (2) the general public will use strong schemes that have undergone public scrutiny. Overall, perhaps counter-intuitively, it is advantageous to have broad, public dissemination of the full details of an encryption scheme—the exact opposite of keeping the scheme secret.

Nowadays Kerckhoffs' principle is understood as advocating that the entire cryptographic design process be made completely public, in stark contrast to the notion of "security by obscurity" that suggests keeping algorithms secret improves security. In fact, it is very dangerous to use a proprietary, "homebrewed" algorithm (i.e., a non-standardized algorithm designed in secret) since published designs undergo public peer review and are therefore likely to be stronger. Many years of experience have demonstrated that it is very difficult to construct good cryptographic schemes. Our confidence in the security of a scheme is much higher if it has been extensively studied by experts (beyond the designers of the scheme) and no flaws have been found. As simple and obvious as it may sound, the principle of open cryptographic design (i.e., Kerckhoffs' principle) has been ignored over and over again with disastrous results. Fortunately, today there are several secure, standardized, and widely available cryptosystems and no reason to use anything else.

1.3   Historical Ciphers and Their Cryptanalysis

In our study of "classical" cryptography we will examine some historical encryption schemes and show that they are insecure. Our main aims in presenting this material are (1) to highlight the weaknesses of heuristic approaches to cryptography, and thus motivate the modern, rigorous approach that will be taken in the rest of the book, and (2) to demonstrate that simple approaches to achieving secure encryption are unlikely to succeed. Along the way, we will present some central principles of cryptography inspired by the weaknesses of these historical schemes.

In this section, plaintext characters are written in `lower case` and ciphertext characters are written in `UPPER CASE` for clarity.

Caesar's cipher. One of the oldest recorded ciphers, known as *Caesar's cipher*, is described in *De Vita Caesarum, Divus Iulius* ("The Lives of the Caesars, the Deified Julius"), written in approximately 110 CE:

> *There are also letters of his to Cicero, as well as to his intimates*
> *on private affairs, and in the latter, if he had anything confidential*
> *to say, he wrote it in cipher, that is, by so changing the order of*
> *the letters of the alphabet, that not a word could be made out...*

Julius Caesar encrypted by shifting the letters of the alphabet 3 places forward: a was replaced with D, b with E, and so on. At the very end of the alphabet, the letters wrap around and so z was replaced with C, y with B, and x with A. For example, encryption of the message begin the attack now, with spaces removed, gives:

EHJLQWKHDWWDFNQRZ.

An immediate problem with this cipher is that the encryption method is *fixed*; there is no key. Thus, anyone learning how Caesar encrypted his messages would be able to decrypt effortlessly.

Interestingly, a variant of this cipher called ROT-13 (where the shift is 13 places instead of 3) is still used nowadays in various online forums. It is understood that this does not provide any cryptographic security; it is used merely to ensure that the text (say, a movie spoiler) is unintelligible unless the reader of a message makes the conscious decision to decrypt it.

The shift cipher and the sufficient key-space principle. The *shift cipher* can be viewed as a keyed variant of Caesar's cipher.[2] Specifically, in the shift cipher the key k is a number between 0 and 25. To encrypt, letters are shifted as in Caesar's cipher, but now by k places. Mapping this to the syntax of encryption described earlier, the message space consists of arbitrary length strings of English letters with punctuation, spaces, and numerals removed, and with no distinction between upper and lower case. Algorithm Gen outputs a uniform key $k \in \{0, \ldots, 25\}$; algorithm Enc takes a key k and a plaintext and shifts each letter of the plaintext forward k positions (wrapping around at the end of the alphabet); and algorithm Dec takes a key k and a ciphertext and shifts every letter of the ciphertext *backward* k positions.

A more mathematical description is obtained by equating the English alphabet with the set $\{0, \ldots, 25\}$ (so a = 0, b = 1, etc.). The message space $\mathcal{M}$ is then any finite sequence of integers from this set. Encryption of the message $m = m_1 \cdots m_\ell$ (where $m_i \in \{0, \ldots, 25\}$) using key k is given by

$$\mathsf{Enc}_k(m_1 \cdots m_\ell) = c_1 \cdots c_\ell, \quad \text{where } c_i = [(m_i + k) \bmod 26].$$

(The notation $[a \bmod N]$ denotes the remainder of a upon division by N, with $0 \leq [a \bmod N] < N$. We refer to the process mapping a to $[a \bmod N]$ as *reduction modulo N*; see also Chapter 9.) Decryption of a ciphertext $c = c_1 \cdots c_\ell$ using key k is given by

$$\mathsf{Dec}_k(c_1 \cdots c_\ell) = m_1 \cdots m_\ell, \quad \text{where } m_i = [(c_i - k) \bmod 26].$$

---

[2] In some books, "Caesar's cipher" and "shift cipher" are used interchangeably.

Is the shift cipher secure? Before reading on, try to decrypt the following ciphertext that was generated using the shift cipher and a secret key k:

OVDTHUFWVZZPISLRLFZHYLAOLYL.

Is it possible to recover the message without knowing k? Actually, it is trivial! The reason is that there are only 26 possible keys. So one can try to decrypt the ciphertext using every possible key and thereby obtain a list of 26 candidate plaintexts. The correct plaintext will certainly be on this list; moreover, if the ciphertext is "long enough" then the correct plaintext will likely be the only candidate on the list that "makes sense." By scanning the list of candidates it is easy to recover the original plaintext. (This is not necessarily true, but will be true most of the time. Even when it is not, this attack narrows down the set of potential plaintexts to at most 26 possibilities.)

An attack that involves trying every possible key is called a *brute-force* or *exhaustive-search* attack. Clearly, for an encryption scheme to be secure it must not be vulnerable to such an attack.[3] This observation is known as the *sufficient key-space principle*:

> *Any secure encryption scheme must have a key space that is sufficiently large to make an exhaustive-search attack infeasible.*

One can debate what amount of effort makes a task "infeasible," and an exact determination of feasibility depends on both the resources of a potential attacker and the length of time for which the sender and receiver want to ensure secrecy of their communication. Nowadays, attackers can use supercomputers, thousands of cloud servers, or graphics processing units (GPUs) to speed up brute-force attacks. To protect against such attacks the key space must be very large—say, of size at least 2^{80}, and even larger in many settings.

The sufficient key-space principle gives a *necessary* condition for security, but not a *sufficient* one. The next example demonstrates this.

The mono-alphabetic substitution cipher. In the shift cipher, the key defines a map from each letter of the (plaintext) alphabet to some letter of the (ciphertext) alphabet, where the map is a fixed shift determined by the key. In the *mono-alphabetic substitution cipher* the key also defines a map on the alphabet, but the map is now allowed to be *arbitrary* subject only to the constraint that it be one-to-one (so that decryption is possible). The key space thus consists of all *bijections*, or *permutations*, of the alphabet. So, for example, the key that defines the following permutation

a	b	c	d	e	f	g	h	i	j	k	l	m	n	o	p	q	r	s	t	u	v	w	x	y	z
X	E	U	A	D	N	B	K	V	M	R	O	C	Q	F	S	Y	H	W	G	L	Z	I	J	P	T

---

[3]Technically, this is only true if the message space is larger than the key space; we will return to this point in Chapter 2. Encryption schemes used in practice have this property.

(in which a maps to X, etc.) would encrypt the message `tellhimaboutme` to `GDOOKVCXEFLGCD`. The name of this cipher comes from the fact that the key defines a (fixed) substitution for individual characters of the plaintext.

Assuming the English alphabet is being used, the key space is of size $26! = 26 \cdot 25 \cdot 24 \cdots 2 \cdot 1$, or approximately 2^{88}, and a brute-force attack is infeasible. This, however, does not mean the cipher is secure! In fact, as we will show next, it is easy to break this scheme even though it has a large key space.

Assume English-language text is being encrypted (i.e., the text is grammatically correct English writing, not just text written using characters of the English alphabet). The mono-alphabetic substitution cipher can then be attacked by utilizing statistical properties of the English language. (Of course, the same idea works for any language.) The attack relies on the facts that:

1. For any key, the mapping of each letter is fixed, and so if e is mapped to D, then every appearance of e in the plaintext will result in the appearance of D in the ciphertext.

2. The frequency distribution of individual letters in English-language text is known. (See Figure 1.3.) Of course, very short texts may deviate from this distribution, but even texts consisting of only a few sentences tend to have distributions that are very close to it.

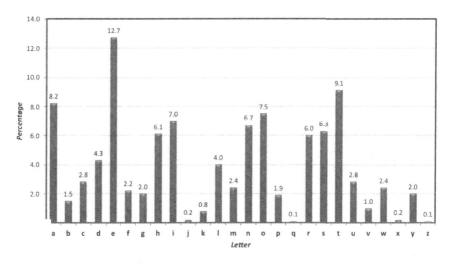

FIGURE 1.3: Average letter frequencies for English-language text.

The attack works by tabulating the frequency distribution of characters in the ciphertext, i.e., recording that A appeared 12% of the time, B appeared 3% of the time, and so on. These frequencies are then compared to the known letter frequencies of normal English text. One can then guess parts of the mapping defined by the key based on the observed frequencies. For example, since **e** is the most frequent letter in English, one can guess that the most frequent character in the ciphertext corresponds to the plaintext character **e**, and so on. Some of the guesses may be wrong, but enough of the guesses will be correct to enable relatively quick decryption (especially utilizing other knowledge of English, such as the fact that **q** is generally followed by **u**, and that **h** is likely to appear between **t** and **e**). We conclude that although the mono-alphabetic substitution cipher has a large key space, it is still insecure.

It should not be surprising that the mono-alphabetic substitution cipher can be quickly broken, since puzzles based on this cipher are common (and are solved by some people before their morning coffee!). We recommend that you try to decipher the following ciphertext—this should convince you how easy it is to carry out the attack. (Use Figure 1.3.)

```
JGRMQOYGHMVBJWRWQFPWHGFFDQGFPFZRKBEEBJIZQQOCIBZKLFAFGQVFZFWWE
OGWOPFGFHWOLPHLRLOLFDMFGQWBLWBWQOLKFWBYLBLYLFSFLJGRMQBOLWJVFP
FWQVHQWFFPQOQVFPQOCFPOGFWFJIGFQVHLHLROQVFGWJVFPFOLFHGQVQVFILE
OGQILHQFQGIQVVOSFAFGBWQVHQWIJVWJVFPFWHGFIWIHZZRQGBABHZQOCGFHX
```

An improved attack on the shift cipher. We can use letter-frequency tables to give an improved attack on the shift cipher. Our previous attack on the shift cipher required decrypting the ciphertext using each possible key, and then checking which key results in a plaintext that "makes sense." A drawback of this approach is that it is somewhat difficult to automate, since it is difficult for a computer to check whether a given plaintext "makes sense." (We do not claim that it would be impossible, as the attack could be automated using a dictionary of valid English words.) More importantly, there may be cases—we will see one later—where the plaintext characters follow the same distribution as English-language text even though the plaintext itself is not valid English, in which case checking for a plaintext that "makes sense" will not work.

We now describe an attack that does not suffer from these drawbacks. As before, associate the letters of the English alphabet with $0, \ldots, 25$. Let p_i, with $0 \leq p_i \leq 1$, denote the frequency of the ith letter in normal English text (i.e., $p_0 = 0.082$ using Figure 1.3). Calculation using Figure 1.3 gives

$$\sum_{i=0}^{25} p_i^2 \approx 0.065. \tag{1.1}$$

Now, say we are given some ciphertext and let q_i denote the frequency of the ith letter of the alphabet in this ciphertext; i.e., q_i is simply the number of occurrences of the ith letter of the alphabet in the ciphertext divided by the length of the ciphertext. If the key is k, then p_i should be roughly equal to q_{i+k} for all i because the ith letter is mapped to the $(i + k)$th letter. (We use $i+k$ instead of the more cumbersome $[i+k \bmod 26]$.) Thus, if we compute

$$I_j \stackrel{\text{def}}{=} \sum_{i=0}^{25} p_i \cdot q_{i+j} \tag{1.2}$$

for each value of $j \in \{0, \ldots, 25\}$, then we expect to find that $I_k \approx 0.065$ (where k is the actual key), whereas I_j for $j \neq k$ will be different from 0.065. This leads to a key-recovery attack that is easy to automate: compute I_j for all j, and then output the value k for which I_k is closest to 0.065.

The Vigenère (poly-alphabetic shift) cipher. The statistical attack on the mono-alphabetic substitution cipher can be carried out because the key defines a fixed mapping that is applied letter-by-letter to the plaintext. Such an attack could be thwarted by using a *poly-alphabetic substitution cipher* where the key instead defines a mapping that is applied on *blocks* of plaintext characters. Here, for example, a key might map the 2-character block ab to DZ while mapping ac to TY; note that the plaintext character a does not get mapped to a fixed ciphertext character. Poly-alphabetic substitution ciphers "smooth out" the frequency distribution of characters in the ciphertext and make it harder to perform statistical analysis.

The *Vigenère cipher*, a poly-alphabetic shift cipher that is a special case of the above, can be viewed as applying different instances of the shift cipher to different parts of the plaintext. The key is now viewed as a *string* of letters; encryption is done by shifting each plaintext character by the amount indicated by the next character of the key, wrapping around in the key when necessary. (This degenerates to the shift cipher if the key has length 1.) For example, encryption of the message tellhimaboutme using the key cafe would work as follows:

Plaintext:	tellhimaboutme
Key (repeated):	cafecafecafeca
Ciphertext:	VEQPJIREDOZXOE

(The key need not be an English word.) This is exactly the same as encrypting the first, fifth, ninth, . . . characters with the shift cipher and key c; the second, sixth, tenth, . . . characters with key a; the third, seventh, . . . characters with f; and the fourth, eighth, . . . characters with e. Notice that in the above example l is mapped once to Q and once to P. Furthermore, the ciphertext character E is sometimes obtained from e and sometimes from a. Thus, the character frequencies of the ciphertext are "smoothed out," as desired.

If the key is sufficiently long, cracking this cipher appears daunting. Indeed, it had been considered by many to be "unbreakable," and although it was invented in the 16th century, a systematic attack on the scheme was only devised hundreds of years later.

Attacking the Vigenère cipher. A first observation in attacking the Vigenère cipher is that *if the length of the key is known* then attacking the cipher is relatively easy. Specifically, say the length of the key, also called the *period*, is t. Write the key k as $k = k_1 \cdots k_t$ where each k_i is a letter of the alphabet. An observed ciphertext $c = c_1 c_2 \cdots$ can be divided into t parts where each part can be viewed as having been encrypted using the shift cipher. Specifically, for all $j \in \{1, \ldots, t\}$ the ciphertext characters

$$c_j, c_{j+t}, c_{j+2t}, \ldots$$

all resulted by shifting the corresponding characters of the plaintext by k_j positions. We refer to the above sequence of characters as the jth *stream*. All that remains is to determine, for each of the t streams, which of the 26 possible shifts was used. This is not as trivial as in the case of the shift cipher, because it is no longer possible to simply try different shifts in an attempt to determine when decryption of a stream "makes sense." (Recall that a stream does not correspond to consecutive letters in the plaintext.) Furthermore, trying to guess the entire key k at once would require a brute-force search through 26^t different possibilities, which is infeasible for large t. Nevertheless, we can still use letter-frequency analysis to analyze each stream independently. Namely, for each stream we tabulate the frequency of each ciphertext character and then check which of the 26 possible shifts yields the "right" probability distribution for that stream. Since this can be carried out independently for each stream (i.e., for each character of the key), this attack takes time $26 \cdot t$ rather than time 26^t.

A more principled, easier-to-automate approach is to apply the improved attack on the shift cipher (discussed earlier) to each stream. That attack did not rely on checking for a plaintext that "made sense," but only relied on the underlying frequency distribution of characters in the plaintext.

Either of the above approaches gives a successful attack when the key length is known. What if the key length is unknown?

Note first that as long as the maximum length T of the key is not too large, we can simply repeat the above attack T times (once for each possible value $t \in \{1, \ldots, T\}$). This leads to at most T different candidate plaintexts, among which the true plaintext will likely be easy to identify. So an unknown key length is not a serious obstacle.

There are also more efficient ways to determine the key length from an observed ciphertext. One is to use *Kasiski's method*, published in the mid-19th century. The first step here is to identify repeated patterns of length 2 or 3 in the ciphertext. These are likely the result of certain bigrams or trigrams that appear frequently in the plaintext. For example, consider the common word "`the`." This word will be mapped to different ciphertext characters, depending on its position in the plaintext. However, if it appears twice in the same relative position, then it will be mapped to the same ciphertext characters. For a sufficiently long plaintext, there is thus a good chance that "`the`" will be mapped repeatedly to the same ciphertext characters.

Consider the following concrete example with the key `beads` (spaces have been added for clarity):

Plaintext:	the	man	and	the	woman	retrieved	the	letter	from	the	post	office
Key:	bea	dsb	ead	sbe	adsbe	adsbeadsb	ead	sbeads	bead	sbe	adsb	eadsbe
Ciphertext:	ULE	PSO	ENG	LII	WREBR	RHLSMEYWE	XHH	DFXTHJ	GVOP	LII	PRKU	SFIADI

The word `the` is mapped sometimes to `ULE`, sometimes to `LII`, and sometimes to `XHH`. However, it is mapped *twice* to `LII`, and in a long enough text it is likely that it would be mapped multiple times to each possibility. Kasiski's observation was that the distance between such repeated appearances (assuming they are not coincidental) is a multiple of the period. (In the above example, the period is 5 and the distance between the two appearances of `LII` is 30, which is 6 times the period.) Therefore, the greatest common divisor of the distances between repeated sequences (assuming they are not coincidental) will yield the key length t or a multiple thereof.

An alternative approach, called the *index of coincidence method*, is more methodical and hence easier to automate. Recall that if the key length is t, then the ciphertext characters

$$c_1, c_{1+t}, c_{1+2t}, \ldots$$

in the first stream all resulted from encryption using the same shift. This means that the frequencies of the characters in this sequence are expected to be identical to the character frequencies of standard English text *in some shifted order*. In more detail: let q_i denote the observed frequency of the ith letter in this stream; this is simply the number of occurrences of the ith letter of the alphabet divided by the total number of letters in the stream. If the shift used here is j (i.e., if the first character k_1 of the key is equal to j), then for all i we expect $q_{i+j} \approx p_i$, where p_i is the frequency of the ith letter of the alphabet in standard English text. (Once again, we use q_{i+j} in place of $q_{[i+j \bmod 26]}$.) But this means that the sequence $q_0, \ldots, q_{25}$ is just the sequence $p_0, \ldots, p_{25}$ shifted j places. As a consequence (cf. Equation (1.1)):

$$\sum_{i=0}^{25} q_i^2 \approx \sum_{i=0}^{25} p_i^2 \approx 0.065.$$

This leads to a nice way to determine the key length t. For $\tau = 1, 2, \ldots, T$, look at the sequence of ciphertext characters $c_1, c_{1+\tau}, c_{1+2\tau}, \ldots$ and tabulate $q_0, \ldots, q_{25}$ for this sequence. Then compute

$$S_\tau \stackrel{\text{def}}{=} \sum_{i=0}^{25} q_i^2. \tag{1.3}$$

When $\tau = t$ we expect $S_\tau \approx 0.065$, as discussed above. On the other hand, if τ is not a multiple of t we expect that all characters will occur with roughly equal probability in the sequence $c_1, c_{1+\tau}, c_{1+2\tau}, \ldots$, and so we expect $q_i \approx 1/26$ for all i. In this case we will obtain

$$S_\tau \approx \sum_{i=0}^{25} \left(\frac{1}{26}\right)^2 \approx 0.038.$$

The smallest value of τ for which $S_\tau \approx 0.065$ is thus likely the key length. One can further validate a guess τ by carrying out a similar calculation using the second stream $c_2, c_{2+\tau}, c_{2+2\tau}, \ldots$, etc.

Ciphertext length and cryptanalytic attacks. The above attacks on the Vigenère cipher require a longer ciphertext than the attacks on previous schemes. For example, the index of coincidence method requires c_1, c_{1+t}, c_{1+2t} (where t is the actual key length) to be sufficiently long in order to ensure that the observed frequencies are close to what is expected; the ciphertext itself must then be roughly t times larger. Similarly, the attack we showed on the mono-alphabetic substitution cipher requires a longer ciphertext than the attack on the shift cipher (which can work for encryptions of even a single word). This illustrates that a longer key can, in general, require the cryptanalyst to obtain more ciphertext in order to carry out an attack. (Indeed, the Vigenère cipher can be shown to be secure if the key is as long as what is being encrypted. We will see a related phenomenon in the next chapter.)

Conclusions. We have presented only a few historical ciphers. Beyond their historical interest, our aim in presenting them was to illustrate some important lessons. Perhaps the most important is that *designing secure ciphers is hard.* The Vigenère cipher remained unbroken for a long time. Far more complex schemes have also been used. But a complex scheme is not necessarily secure, and all historical schemes have been broken.

1.4   Principles of Modern Cryptography

As should be clear from the previous section, cryptography was historically more of an art than a science. Schemes were designed in an heuristic manner and evaluated based on their perceived complexity or cleverness. A scheme would be analyzed to see if any attacks could be found; if so, the scheme would be "patched" to thwart that attack, and the process repeated. Although there may have been agreement that some schemes were *not* secure (as evidenced by an especially damaging attack), there was no agreed-upon notion of what requirements a "secure" scheme should satisfy, and no way to give evidence that any specific scheme *was* secure.

Over the past several decades, cryptography has developed into more of a science. Schemes are now developed and analyzed in a more systematic manner, with the ultimate goal being to give a rigorous *proof* that a given construction is secure. In order to articulate such proofs, we first need *formal definitions* that pin down exactly what "secure" means; such definitions are useful and interesting in their own right. As it turns out, most cryptographic proofs rely on currently unproven *assumptions* about the algorithmic hardness of certain mathematical problems; any such assumptions must be made explicit and be stated precisely. An emphasis on definitions, assumptions, and proofs distinguishes *modern* cryptography from classical cryptography; we now discuss these three principles in greater detail.

1.4.1 Principle 1 – Formal Definitions

One of the key contributions of modern cryptography has been the recognition that formal definitions of security are *essential* for the proper design, study, evaluation, and usage of cryptographic primitives. Put bluntly:

> *If you don't understand what you want to achieve, how can you possibly know when (or if) you have achieved it?*

Formal definitions provide such understanding by giving a clear description of what threats are in scope and what security guarantees are desired. As such, definitions can help guide the design of cryptographic schemes. Indeed, it is much better to formalize what is required *before* the design process begins, rather than to come up with a definition *post facto* once the design is complete. The latter approach risks having the design phase end when the designers' patience is exhausted (rather than when the goal has been met), or may result in a construction achieving *more* than is needed at the expense of efficiency.

Definitions also offer a way to evaluate and analyze constructions. With a definition in place, one can study a proposed scheme to see if it achieves the desired guarantees; in some cases, one can even *prove* a given construction secure (see Section 1.4.3) by showing that it meets the definition. On the flip side, definitions can be used to conclusively show that a given scheme is *not* secure, insofar as the scheme does not satisfy the definition. In particular, observe that the attacks in the previous section do not conclusively demonstrate that any of the schemes shown there is "insecure." For example, the attack on the Vigenère cipher assumed that sufficiently long English text was being encrypted, but perhaps the Vigenère cipher is "secure" if short English text, or compressed text (which will have roughly uniform letter frequencies), is encrypted? It is hard to say without a formal definition in place.

Definitions enable a meaningful comparison of schemes. As we will see, there can be *multiple* (valid) ways to define security; the "right" one depends on the context in which a scheme is used. A scheme satisfying a weaker definition may be more efficient than another scheme satisfying a stronger definition; with precise definitions we can properly evaluate the trade-offs between the two schemes. Along the same lines, definitions enable secure usage of schemes. Consider deciding which encryption scheme to use for some larger application. A sound way to approach the problem is to first understand what notion of security is required for that application, and then find an encryption scheme satisfying that notion. A side benefit of this approach is *modularity*: a designer can "swap out" one encryption scheme and replace it with another (that also satisfies the necessary definition of security) without having to worry about affecting security of the overall application.

Writing a formal definition forces one to think about what is essential to the problem at hand and what properties are extraneous. Going through the process often reveals subtleties of the problem that were not obvious at first glance. We illustrate this next for the case of encryption.

An example: secure encryption. A common mistake is to think that formal definitions are not needed, or are trivial to come up with, because "everyone has an intuitive idea of what security means." This is not the case. As an example, we consider the case of encryption. (The reader may want to pause here to think about how they would formally define what it means for an encryption scheme to be secure.) Although we postpone a formal definition of secure encryption to subsequent chapters, we describe here informally what such a definition should capture.

In general, a security definition has two components: a security guarantee (or, from the attacker's point of view, what constitutes a successful attack) and a threat model. The security guarantee defines what the scheme is intended to prevent the attacker from doing, while the threat model describes the power of the adversary, i.e., what actions the attacker is assumed able to carry out.

Let's start with the first of these. What should a secure encryption scheme guarantee? Here are some thoughts:

- *It should be impossible for an attacker to recover the key.* We have previously observed that if an attacker can determine the key shared by two parties using some scheme, then that scheme cannot be secure. However, it is easy to come up with schemes for which key recovery is impossible, yet the scheme is blatantly insecure. Consider, e.g., the scheme where $\mathsf{Enc}_k(m) = m$. The ciphertext leaks no information about the key (and so the key cannot be recovered if it is long enough) yet the message is sent in the clear! We thus see that inability to recover the key is necessary but not sufficient for security. This makes sense: the aim of encryption is to protect the *message*; secrecy of the key is a means for achieving this goal, but is not itself the objective.

- *It should be impossible for an attacker to recover the plaintext from the ciphertext.* This definition is better, but is still far from satisfactory. In particular, this definition would consider an encryption scheme secure if its ciphertexts revealed 90% of the plaintext, as long as 10% of the plaintext remained hard to figure out. This is clearly unacceptable in most common applications of encryption; for example, when encrypting a salary database, we would be justifiably upset if 90% of employees' salaries were revealed!

- *It should be impossible for an attacker to recover any character of the plaintext from the ciphertext.* This looks like a good definition, yet is still not sufficient. Going back to the example of encrypting a salary database, we would not consider an encryption scheme secure if it reveals whether an employee's salary is more than or less than $100,000, even if it does not reveal any particular digit of that employee's salary. Similarly, we would not want an encryption scheme to reveal whether one particular employee makes more than another.

Another issue is how to formalize what it means for an adversary to "recover a character of the plaintext." What if an attacker correctly guesses, through sheer luck or external information, that the least significant digit of someone's salary is 0? Clearly that should not render an encryption scheme insecure, and so any viable definition must somehow rule out such behavior from qualifying as a successful attack.

- *The "right" answer: regardless of any information an attacker already has, a ciphertext should leak no* additional *information about the underlying plaintext.* This informal definition captures all the concerns outlined above. Note in particular that it does not try to define what information about the plaintext is "meaningful"; it simply requires that *no* information be leaked. This is important, as it means that a secure encryption scheme is suitable for all potential applications in which secrecy is required.

  What is missing here is a precise, mathematical formulation of the definition. How should we capture an attacker's prior knowledge about the plaintext? And what does it mean to (not) leak information? We will return to these questions in the next two chapters; see especially Definitions 2.3 and 3.12.

Now that we have fixed a security *goal*, it remains to specify a *threat model*. This specifies what "power" the attacker is assumed to have, but does not place any restrictions on the adversary's *strategy*. This is an important distinction: we specify what we assume about the adversary's abilities, but we do *not* assume anything about *how it uses* those abilities. It is impossible to foresee what strategies might be used in an attack, and history has proven that attempts to do so are doomed to failure.

There are several plausible options for the threat model in the context of encryption; standard ones, in order of increasing power of the attacker, are:

- **Ciphertext-only attack:** This is the most basic attack, where the adversary just observes a ciphertext (or multiple ciphertexts) and attempts to determine information about the underlying plaintext (or plaintexts). This is the threat model we have been implicitly assuming when discussing classical encryption schemes in the previous section.

- **Known-plaintext attack:** Here, the adversary is able to learn one or more plaintext/ciphertext pairs generated using some key. The aim of the adversary is then to deduce information about the underlying plaintext of some *other* ciphertext produced using the same key.

  All the classical encryption schemes we have seen are trivial to break using a known-plaintext attack; we leave a demonstration as an exercise.

- **Chosen-plaintext attack:** In this attack, the adversary can obtain plaintext/ciphertext pairs, as above, for plaintexts *of its choice.*

- **Chosen-ciphertext attack:** The final type of attack is one where the adversary is additionally able to obtain (some information about) the *decryption* of ciphertexts of its choice, e.g., whether the decryption of some ciphertext chosen by the attacker yields a valid English message. The adversary's aim, once again, is to learn information about the underlying plaintext of some *other* ciphertext (whose decryption the adversary is unable to obtain directly) generated using the same key.

Although the threat models are listed in order of increasing strength, none of them is inherently better than any other; the right one to use depends on the environment in which an encryption scheme is deployed.

The first two types of attack are the easiest to carry out. In a ciphertext-only attack, the only thing the adversary needs to do is eavesdrop on the communication channel over which encrypted messages are sent. In a known-plaintext attack it is assumed the adversary also obtains ciphertexts corresponding to known plaintexts. This is often easy to accomplish because not all encrypted messages are secret, at least not indefinitely. As a trivial example, two parties may always encrypt a "hello" message whenever they begin communicating. As a more complex example, encryption may be used to keep quarterly-earnings reports secret until their release date; in this case, anyone eavesdropping on the ciphertext will later obtain the corresponding plaintext.

In the latter two attacks the adversary is assumed to be able to obtain encryptions and/or decryptions of plaintexts/ciphertexts of its choice. This may at first seem strange, and we defer a more detailed discussion of these attacks, and their practicality, to Section 3.4.2 (for chosen-plaintext attacks) and Section 5.1 (for chosen-ciphertext attacks).

1.4.2   Principle 2 – Precise Assumptions

Most modern cryptographic constructions cannot be proven secure unconditionally; such proofs would require resolving questions in the theory of computational complexity that seem far from being answered today.[4] The result of this unfortunate state of affairs is that proofs of security typically rely on *assumptions.* Modern cryptography requires any such assumptions to be made explicit and mathematically precise. At the most basic level, this is because proofs of security require this. But there are other reasons as well:

1. *Validation of assumptions:* By their very nature, assumptions are statements that are not proven but are instead conjectured to be true. In order to strengthen our belief in some assumption, it is necessary to

---

[4]In particular, most of cryptography requires the unproven assumption that $\mathcal{P} \neq \mathcal{NP}$.

study it: The more the assumption is examined and tested without being refuted, the more confident we are that the assumption is true. Furthermore, study of an assumption can provide evidence of its validity by showing that it is implied by some other assumption that is also widely believed.

If the assumption being relied upon is not precisely stated, it cannot be effectively studied and (potentially) refuted. Thus, a precondition to increasing our confidence in an assumption is having a precise statement of what exactly is being assumed.

2. *Comparison of assumptions:* Often in cryptography we are presented with two schemes that can both be proven to satisfy some definition, each based on a different assumption. Assuming all else is equal, which scheme should be preferred? If the assumption on which the first scheme is based is *weaker* than the assumption on which the second scheme is based (i.e., if the second assumption implies the first), then the first scheme is preferable since it may turn out that the second assumption is false while the first assumption is true. If the assumptions used by the two schemes are not comparable, then the general rule is to prefer the scheme that is based on the better-studied assumption in which there is presumably greater confidence.

3. *Understanding the necessary assumptions:* An encryption scheme may be based on some underlying building block. If some weaknesses are later found in the building block, how can we tell whether the encryption scheme is still secure? If the underlying assumptions regarding the building block are made clear as part of proving security of the scheme, then we need only check whether the required assumptions are affected by the new weaknesses that were found.

A question that sometimes arises is: rather than prove a scheme secure based on some other assumption, why not simply assume that the scheme *itself* is secure? In some cases—e.g., when the definition is simple and a scheme has successfully resisted attack for many years—this may be an acceptable approach. But this approach is not preferred, and is downright dangerous when a new construction is being introduced. The reasons above help explain why. First, an assumption that has been studied for several years is preferable to a new, arbitrary assumption that is introduced along with a new construction. Second, there is a general preference for "simpler" assumptions—i.e., an assumption about the hardness of a clean mathematical problem vs. an assumption that a complex scheme satisfies an elaborate security definition—since simpler assumptions are in general easier to understand and study. Another advantage of relying on "lower-level" assumptions (rather than just assuming a scheme is secure) is that these low-level assumptions can typically be used in other constructions. Finally, low-level assumptions enable *modularity.* Consider an encryption scheme whose security relies on some assumed property of one of its building blocks. If the underlying building block turns out *not* to satisfy the stated assumption, the encryption scheme can be instantiated using a different component that satisfies the necessary requirements.

1.4.3   Principle 3 – Proofs of Security

The two principles just described allow us to achieve our goal of providing rigorous *proof* that a construction satisfies a given definition under certain assumptions. Such proofs are especially important in the context of cryptography where there is an attacker who is actively trying to "break" some scheme. Proofs of security give an iron-clad guarantee—relative to the definition and assumptions—that no attacker will succeed; this is much better than taking an unprincipled or heuristic approach to the problem. Without a proof that no adversary with the specified resources can break some scheme, we are left only with our intuition that this is the case. Experience has shown that intuition in cryptography and computer security is disastrous. There are countless examples of unproven schemes that were broken, sometimes immediately and sometimes years after being developed.

Summary: Rigorous vs. Heuristic Approaches to Security

Reliance on definitions, assumptions, and proofs constitutes a rigorous approach to cryptography that is distinct from the informal approach of classical cryptography. Unfortunately, unprincipled, "off-the-cuff" solutions are still designed and deployed by those wishing to obtain a quick solution to a problem, or by those who are simply unknowledgable. We hope this book will contribute to an awareness of the rigorous approach and its importance in developing provably secure schemes.

1.4.4   Provable Security and Real-World Security

Much of modern cryptography now rests on sound mathematical foundations. But this does not mean that the field is no longer partly an *art* as well. The rigorous approach leaves room for creativity in developing definitions suited to contemporary applications and environments, in proposing new mathematical assumptions and designing new primitives, and in constructing novel schemes and proving them secure. There will also always be the art of *attacking* deployed cryptosystems, even when they are proven secure. We expand on this point next.

The approach taken by modern cryptography has revolutionized the field, and helps provide confidence in the security of cryptographic schemes deployed in the real world. But it is important not to overstate what a proof of security implies. A proof of security is always relative to the *definition* being considered and the *assumption(s)* being used. If the security guarantee does not match what is needed, or the threat model does not capture the adversary's true abilities, then the proof may be irrelevant. Similarly, if the assumption that is relied upon turns out to be false, then the proof of security is meaningless.

The take-away point is that provable security of a scheme does not necessarily imply security of that scheme in the real world.[5] While some have viewed this as a drawback of provable security, we view this optimistically as illustrating the *strength* of the approach. To attack a provably secure scheme in the real world, the attacker is forced to focus attention on the definition (i.e., to explore how the idealized definition differs from the real-world requirements) or the underlying assumptions (i.e., to see whether they hold). In turn, it is the job of cryptographers to continually refine their definitions to more closely match the real world, and to investigate their assumptions to test their validity. Provable security does not end the age-old battle between attacker and defender, but it does provide a framework that helps shift the odds in the defender's favor.

References and Additional Reading

In this chapter, we have studied just a few of the known historical ciphers. There are many others of both historical and mathematical interest, and we refer the reader to textbooks by Stinson [195] or Trappe and Washington [196] for further details. The important role cryptography has played throughout history is a fascinating subject covered in books by Kahn [106] and Singh [188].

Shannon [177] was the first to pursue a rigorous approach to cryptography based on precise definitions and mathematical proofs; we explore his work in the next chapter.

Exercises

1.1 Decrypt the ciphertext provided at the end of the section on mono-alphabetic substitution ciphers.

1.2 Provide a formal definition of the Gen, Enc, and Dec algorithms for the mono-alphabetic substitution cipher.

1.3 Provide a formal definition of the Gen, Enc, and Dec algorithms for the Vigenère cipher. (Note: there are several plausible choices for Gen; choose one.)

---

[5]Here we are not even considering the possibility of an incorrect *implementation* of the scheme. Poorly implemented cryptography is a serious problem in the real world, but this problem is largely outside the scope of this book.

1.4 Say you are given a ciphertext that corresponds to English-language text that was encrypted using either the shift cipher or the Vigenère cipher with period greater than 1. How could you tell which was the case?

1.5 Implement the attacks described in this chapter for the shift cipher and the Vigenère cipher.

1.6 The shift and Vigenère ciphers can also be defined on the 256-character alphabet consisting of all possible bytes (8-bit strings), and using XOR instead of modular addition.

   (a) Provide a formal definition of both schemes in this case.

   (b) Discuss how the attacks we have shown in this chapter can be modified to break these schemes.

1.7 The index of coincidence method relies on a known value for the sum of the *squares* of plaintext-letter frequencies (cf. Equation (1.1)). Why would it not work using the sum $\sum_i p_i$ itself?

1.8 Show that the shift, substitution, and Vigenère ciphers are all trivial to break using a chosen-plaintext attack. How much chosen plaintext is needed to recover the key for each of the ciphers?

1.9 Assume an attacker knows that a user's password is either abcd or bedg. Say the user encrypts his password using the shift cipher, and the attacker sees the resulting ciphertext. Show how the attacker can determine the user's password, or explain why this is not possible.

1.10 Repeat the previous exercise for the Vigenère cipher using period 2, using period 3, and using period 4.

1.11 The attack on the Vigenère cipher has two steps: (a) find the key length by identifying τ with $S_\tau \approx 0.065$ (cf. Equation (1.3)) and (b) for each character of the key, find j maximizing I_j (cf. Equation (1.2)), using $\{p_i\}$ corresponding to English text. What happens in each case if the underlying plaintext is in a language other than English?

Chapter 2

Perfectly Secret Encryption

In the previous chapter we presented some historical encryption schemes and showed that they can be broken easily. In this chapter, we look at the other extreme and study encryption schemes that are *provably* secure even against an adversary with unbounded computational power. Such schemes are called *perfectly secret*. We rigorously define this notion, and explore conditions under which perfect secrecy can be achieved.

The material in this chapter belongs, in some sense, more to the world of "classical" cryptography than to the world of "modern" cryptography. Besides the fact that all the material introduced here was developed before the revolution in cryptography that took place in the mid-1970s and 1980s, the constructions we study in this chapter rely only on the first and third principles outlined in Section 1.4. That is, precise mathematical definitions are used and rigorous proofs are given, but it will not be necessary to rely on any unproven computational assumptions. It is clearly advantageous to avoid such assumptions; we will see, however, that doing so has inherent limitations. Thus, in addition to serving as a good basis for understanding the principles underlying modern cryptography, the results of this chapter also justify our later adoption of all three of the aforementioned principles.

Beginning with this chapter, we will define security and analyze schemes using probabilistic experiments involving randomized algorithms. (We assume familiarity with basic probability theory. The relevant notions are reviewed in Appendix A.3.) A simple example is given by the "experiment" in which the parties who wish to communicate using a private-key encryption scheme generate a random key. Since randomness is so essential, we briefly discuss the issue of generating randomness suitable for cryptographic applications before returning to a discussion of cryptography *per se*.

Generating randomness. Throughout the book, we will assume for simplicity that parties have access to an unlimited supply of independent, unbiased (i.e., uniform) bits. Where do these random bits come from? Since classical computation is deterministic, it is not at all clear how computers can be used to generate random bits. In principle, one could generate a small number of uniform bits by hand, e.g., by flipping a fair coin. But that approach is not very convenient, nor does it scale.

Modern *random-number generation* proceeds in two steps. First, a "pool" of high-entropy data is collected. (For our purposes a formal definition of entropy is not needed, and it suffices to think of entropy as a measure of unpredictability.) Next, this high-entropy data is processed to yield a sequence of nearly independent and unbiased bits. This second step is necessary since high-entropy data is not necessarily uniform.

For the first step, some source of unpredictable data is needed. This can come from external inputs, for example, delays between network events, hard-disk access times, keystrokes or mouse movements made by the user, and so on. More sophisticated approaches—which, by design, incorporate random-number generation more tightly into the system at the hardware level—can also be used. These rely on physical phenomena such as thermal/shot noise or radioactive decay; for example, certain Intel processors use thermal noise to generate high-entropy data on-chip. Hardware random-number generators of this sort generally produce high-entropy data at a faster rate than techniques relying on external sources.

The processing needed to "smooth" the high-entropy data to obtain (nearly) independent and uniform bits is non-trivial, and is discussed briefly in Section 6.6.4. Here, we consider a simple example to give an idea of what can be done. Imagine that our high-entropy pool contains a sequence of *biased* bits, where 1 occurs with probability p and 0 occurs with probability $1 - p$. (We do assume, however, that the bits are all *independent*. In practice this assumption is typically not valid and so more-complex processing must be done.) Thousands of such bits have lots of entropy, but are not close to uniform. We can obtain a uniform sequence of bits by taking the original bits in pairs: if we see a 1 followed by a 0 then we output 0, and if we see a 0 followed by a 1 then we output 1. (If we see two 0s or two 1s in a row we output nothing, and simply move on to the next pair.) The probability that any pair results in a 0 is $p \cdot (1 - p)$, which is exactly equal to the probability that any pair results in a 1. (Note that we do not even need to know the value of p!) We thus obtain a uniformly distributed output from our initial high-entropy pool.

Care must be taken in how random bits are produced, and using poor random-number generators can often leave a good cryptosystem vulnerable to attack. One should use a random-number generator that is *designed for cryptographic use*, rather than a "general-purpose" random-number generator that is generally not suitable for cryptographic applications. In particular, the rand() function in the C stdlib.h library is *not* cryptographically secure, and using it in cryptographic settings can have disastrous consequences.

2.1 Definitions

We begin by recalling and expanding upon the syntax of encryption, as introduced in the previous chapter. An encryption scheme is defined by three algorithms Gen, Enc, and Dec, as well as a specification of a *message space* $\mathcal{M}$ with $|\mathcal{M}| > 1$.[1] The key-generation algorithm Gen is a probabilistic algorithm that outputs a key k chosen according to some distribution. We denote by $\mathcal{K}$ the (finite) *key space*, i.e., the set of all possible keys that can be output by Gen. The encryption algorithm Enc takes as input a key $k \in \mathcal{K}$ and a message $m \in \mathcal{M}$, and outputs a ciphertext c. We now explicitly allow the encryption algorithm to be probabilistic (so $\mathsf{Enc}_k(m)$ might output a different ciphertext when run multiple times), and we write $c \leftarrow \mathsf{Enc}_k(m)$ to denote the possibly probabilistic process by which message m is encrypted using key k to give ciphertext c. (Looking ahead, we also sometimes use the notation $x \leftarrow S$ to denote uniform selection of x from a set S. In case Enc is deterministic, we may emphasize this by writing $c := \mathsf{Enc}_k(m)$.) We let $\mathcal{C}$ denote the set of all possible ciphertexts that can be output by $\mathsf{Enc}_k(m)$, for all possible choices of $k \in \mathcal{K}$ and $m \in \mathcal{M}$ (and for all random choices of Enc in case it is randomized). The decryption algorithm Dec takes as input a key $k \in \mathcal{K}$ and a ciphertext $c \in \mathcal{C}$ and outputs a message $m \in \mathcal{M}$. We assume *perfect correctness*, meaning that for all $k \in \mathcal{K}$, $m \in \mathcal{M}$, and any ciphertext c output by $\mathsf{Enc}_k(m)$, it holds that $\mathsf{Dec}_k(c) = m$ with probability 1. Perfect correctness implies that we may assume Dec is deterministic without loss of generality, since $\mathsf{Dec}_k(c)$ must give the same output every time it is run. We will thus write $m := \mathsf{Dec}_k(c)$ to denote the (deterministic) process of decrypting ciphertext c using key k to yield the message m.

In the definitions and theorems below, we refer to probability distributions over $\mathcal{K}$, $\mathcal{M}$, and $\mathcal{C}$. The distribution over $\mathcal{K}$ is the one defined by running Gen and taking the output. (It is almost always the case that Gen chooses a key uniformly from $\mathcal{K}$ and, in fact, we may assume this without loss of generality; see Exercise 2.1.) We let K be the random variable denoting the value of the key output by Gen; thus, for any $k \in \mathcal{K}$, $\Pr[K = k]$ denotes the probability that the key output by Gen is equal to k. Similarly, we let M be the random variable denoting the message being encrypted, so $\Pr[M = m]$ denotes the probability that the message takes on the value $m \in \mathcal{M}$. The probability distribution of the message is not determined by the encryption scheme itself, but instead reflects the likelihood of different messages being sent by the parties using the scheme, as well as an adversary's uncertainty about what will be sent. As an example, an adversary may know that the message will either be `attack today` or `don't attack`. The adversary may even know (by other means) that with probability 0.7 the message will be a command to attack and with probability 0.3 the message will be a command not to attack. In this case, we have $\Pr[M = \texttt{attack today}] = 0.7$ and $\Pr[M = \texttt{don't attack}] = 0.3$.

K and M are required to be independent, i.e., what is being communicated by the parties must be independent of the key they share. This makes sense,

---

[1] If $|\mathcal{M}| = 1$ there is only one message and no point in communicating, let alone encrypting.

among other reasons, because the distribution over $\mathcal{K}$ is determined by the encryption scheme itself (since it is defined by Gen), while the distribution over $\mathcal{M}$ depends on the context in which the encryption scheme is being used.

Fixing an encryption scheme and a distribution over $\mathcal{M}$ determines a distribution over the space of ciphertexts $\mathcal{C}$ given by choosing a key $k \in \mathcal{K}$ (according to Gen) and a message $m \in \mathcal{M}$ (according to the given distribution), and then computing the ciphertext $c \leftarrow \mathsf{Enc}_k(m)$. We let C be the random variable denoting the resulting ciphertext and so, for $c \in \mathcal{C}$, write $\Pr[C = c]$ to denote the probability that the ciphertext is equal to the fixed value c.

Example 2.1

We work through a simple example for the shift cipher (cf. Section 1.3). Here, by definition, we have $\mathcal{K} = \{0, \ldots, 25\}$ with $\Pr[K = k] = 1/26$ for each $k \in \mathcal{K}$.

Say we are given the following distribution over $\mathcal{M}$:

$$\Pr[M = \mathsf{a}] = 0.7 \quad \text{and} \quad \Pr[M = \mathsf{z}] = 0.3.$$

What is the probability that the ciphertext is B? There are only two ways this can occur: either $M = \mathsf{a}$ and $K = 1$, or $M = \mathsf{z}$ and $K = 2$. By independence of M and K, we have

$$\Pr[M = \mathsf{a} \wedge K = 1] = \Pr[M = \mathsf{a}] \cdot \Pr[K = 1]$$

$$= 0.7 \cdot \left(\frac{1}{26}\right).$$

Similarly, $\Pr[M = \mathsf{z} \wedge K = 2] = 0.3 \cdot \left(\frac{1}{26}\right)$. Therefore,

$$\Pr[C = \mathsf{B}] = \Pr[M = \mathsf{a} \wedge K = 1] + \Pr[M = \mathsf{z} \wedge K = 2]$$

$$= 0.7 \cdot \left(\frac{1}{26}\right) + 0.3 \cdot \left(\frac{1}{26}\right) = 1/26.$$

We can calculate conditional probabilities as well. For example, what is the probability that the message a was encrypted, given that we observe ciphertext B? Using Bayes' Theorem (Theorem A.8) we have

$$\Pr[M = \mathsf{a} \mid C = \mathsf{B}] = \frac{\Pr[C = \mathsf{B} \mid M = \mathsf{a}] \cdot \Pr[M = \mathsf{a}]}{\Pr[C = \mathsf{B}]}$$

$$= \frac{\Pr[C = \mathsf{B} \mid M = \mathsf{a}] \cdot 0.7}{1/26}.$$

Note that $\Pr[C = \mathsf{B} \mid M = \mathsf{a}] = 1/26$, since if $M = \mathsf{a}$ then the only way $C = \mathsf{B}$ can occur is if $K = 1$ (which occurs with probability 1/26). We conclude that $\Pr[M = \mathsf{a} \mid C = \mathsf{B}] = 0.7$.   ◇

Example 2.2

Consider the shift cipher again, but with the following distribution over $\mathcal{M}$:

$$\Pr[M = \mathsf{kim}] = 0.5, \ \Pr[M = \mathsf{ann}] = 0.2, \ \Pr[M = \mathsf{boo}] = 0.3.$$

What is the probability that $C = \mathtt{DQQ}$? The only way this ciphertext can occur is if $M = \mathtt{ann}$ and $K = 3$, or $M = \mathtt{boo}$ and $K = 2$, which happens with probability $0.2 \cdot 1/26 + 0.3 \cdot 1/26 = 1/52$.

We can also compute the probability that $\mathtt{ann}$ was encrypted, conditioned on observing the ciphertext $\mathtt{DQQ}$? A calculation as above using Bayes' Theorem gives $\Pr[M = \mathtt{ann} \mid C = \mathtt{DQQ}] = 0.4$. $\diamondsuit$

Perfect secrecy. We are now ready to define the notion of *perfect secrecy*. We imagine an adversary who knows the probability distribution of M; that is, the adversary knows the likelihood that different messages will be sent. The adversary also knows the encryption scheme being used. The only thing unknown to the adversary is the key shared by the parties. A message is chosen by one of the honest parties and encrypted, and the resulting ciphertext is transmitted to the other party. The adversary can *eavesdrop* on the parties' communication, and thus observe this ciphertext. (That is, this is a ciphertext-only attack, where the attacker sees only a single ciphertext.) For a scheme to be perfectly secret, observing this ciphertext should have *no effect* on the adversary's knowledge regarding the actual message that was sent; in other words, the *a posteriori* probability that some message $m \in \mathcal{M}$ was sent, conditioned on the ciphertext that was observed, should be no different from the *a priori* probability that m would be sent. This means that the ciphertext reveals nothing about the underlying plaintext, and the adversary learns absolutely nothing about the plaintext that was encrypted. Formally:

DEFINITION 2.3 *An encryption scheme* (Gen, Enc, Dec) *with message space $\mathcal{M}$ is* perfectly secret *if for every probability distribution for M, every message $m \in \mathcal{M}$, and every ciphertext $c \in \mathcal{C}$ for which $\Pr[C = c] > 0$:*

$$\Pr[M = m \mid C = c] = \Pr[M = m].$$

(The requirement that $\Pr[C = c] > 0$ is a technical one needed to prevent conditioning on a zero-probability event.)

Example 2.4
We show that the shift cipher is *not* perfectly secret when used with the message space $\mathcal{M}$ consisting of all two-character plaintexts. To do so, we work with Definition 2.3, and show a probability distribution over $\mathcal{M}$ for which, for some message m and ciphertext c,

$$\Pr[M = m \mid C = c] \neq \Pr[M = m].$$

Many such distributions are possible, but we pick a simple one: say the message is either $\mathtt{aa}$ or $\mathtt{ab}$, each with half probability. Set $m = \mathtt{ab}$ and $c = \mathtt{XX}$. Then clearly $\Pr[M = \mathtt{ab} \mid C = \mathtt{XX}] = 0$, as there is no way that $\mathtt{XX}$ can ever result from the encryption of $\mathtt{ab}$. But $\Pr[M = \mathtt{ab}] = 1/2$. $\diamondsuit$

We now give an equivalent formulation of perfect secrecy. This formulation defines perfect secrecy by requiring that the distribution of the ciphertext does not depend on the plaintext, i.e., for any two messages $m, m' \in \mathcal{M}$ the distribution of the ciphertext when m is encrypted should be identical to the distribution of the ciphertext when m' is encrypted. That is, for every $m, m' \in \mathcal{M}$, and every $c \in \mathcal{C}$, we have

$$\Pr[\mathsf{Enc}_K(m) = c] = \Pr[\mathsf{Enc}_K(m') = c] \qquad (2.1)$$

(where the probabilities are over choice of K and any randomness of Enc). Note that the above probabilities depend *only* on the encryption scheme, and make no reference to any underlying distribution on $\mathcal{M}$. The above condition implies that a ciphertext contains no information about the plaintext, and that it is impossible to distinguish an encryption of m from an encryption of m', since the distributions of the ciphertext are the same in each case.

LEMMA 2.5   *An encryption scheme* (Gen, Enc, Dec) *with message space* $\mathcal{M}$ *is perfectly secret if and only if Equation (2.1) holds for every* $m, m' \in \mathcal{M}$ *and every* $c \in \mathcal{C}$.

PROOF   The proof is straightforward, but we go through it in detail. The key observation is that for any scheme, any distribution on $\mathcal{M}$, any $m \in \mathcal{M}$ for which $\Pr[M = m] > 0$, and any $c \in \mathcal{C}$, we have

$$
\begin{aligned}
\Pr[C = c \mid M = m] &= \Pr[\mathsf{Enc}_K(M) = c \mid M = m] \\
&= \Pr[\mathsf{Enc}_K(m) = c \mid M = m] \\
&= \Pr[\mathsf{Enc}_K(m) = c], \qquad (2.2)
\end{aligned}
$$

where the first equality is by definition of the random variable C, the second is because we are conditioning on the event that $M = m$, and the third is because K is independent of M. We also use the fact that for any $c \in \mathcal{C}$ with $\Pr[C = c] > 0$, we have

$$\Pr[M = m \mid C = c] \cdot \Pr[C = c] = \Pr[C = c \mid M = m] \cdot \Pr[M = m]. \quad (2.3)$$

Take the uniform distribution over $\mathcal{M}$. If the scheme is perfectly secret then $\Pr[M = m \mid C = c] = \Pr[M = m]$, and so Equation (2.3) implies that $\Pr[C = c \mid M = m] = \Pr[C = c]$. Since m and c were arbitrary, this shows that for every $m, m' \in \mathcal{M}$ and every $c \in \mathcal{C}$,

$$
\begin{aligned}
\Pr[\mathsf{Enc}_K(m) = c] &= \Pr[C = c \mid M = m] \\
&= \Pr[C = c] \\
&= \Pr[C = c \mid M = m'] = \Pr[\mathsf{Enc}_K(m') = c]
\end{aligned}
$$

(using Equation (2.2)), proving one direction of the lemma.

Conversely, say Equation (2.1) holds for every $m, m' \in \mathcal{M}$ and every $c \in \mathcal{C}$. Fix some distribution over $\mathcal{M}$, a message $m \in \mathcal{M}$, and a ciphertext $c \in \mathcal{C}$ with $\Pr[C = c] > 0$. If $\Pr[M = m] = 0$ then we trivially have

$$\Pr[M = m \mid C = c] = \Pr[M = m] = 0.$$

So, assume $\Pr[M = m] > 0$. For $c \in \mathcal{C}$, define $p_c \overset{\text{def}}{=} \Pr[\mathsf{Enc}_K(m) = c]$. Equations (2.1) and (2.2) imply that $\Pr[C = c \mid M = m'] = p_c$ for every $m' \in \mathcal{M}$. So,

$$
\begin{aligned}
\Pr[C = c] &= \sum_{m' \in \mathcal{M}} \Pr[C = c \mid M = m'] \cdot \Pr[M = m'] \\
&= \sum_{m' \in \mathcal{M}} p_c \cdot \Pr[M = m'] = p_c = \Pr[C = c \mid M = m],
\end{aligned}
$$

where the sum is over m' with $\Pr[M = m'] > 0$. Equation (2.3) implies that $\Pr[M = m \mid C = c] = \Pr[M = m]$, so the scheme is perfectly secret. ■

Perfect (adversarial) indistinguishability. We conclude this section by presenting another equivalent definition of perfect secrecy. This definition is based on an *experiment* involving an adversary passively observing a ciphertext and then trying to guess which of two possible messages was encrypted. We introduce this notion since it will serve as our starting point for defining computational security in the next chapter; throughout the rest of the book we will often use experiments like this one to define security.

In the present context, we consider the following experiment: An adversary $\mathcal{A}$ first specifies two arbitrary messages $m_0, m_1 \in \mathcal{M}$. Next, a key k is generated using Gen. Then, one of the two messages specified by $\mathcal{A}$ is chosen (each with probability $1/2$) and encrypted using k; the resulting ciphertext is given to $\mathcal{A}$. Finally, $\mathcal{A}$ outputs a "guess" as to which of the two messages was encrypted; $\mathcal{A}$ *succeeds* if it guesses correctly. An encryption scheme is *perfectly indistinguishable* if *no* adversary $\mathcal{A}$ can succeed with probability better than $1/2$. (Note that, for any encryption scheme, $\mathcal{A}$ can succeed with probability $1/2$ by outputting a uniform guess; the requirement is simply that no attacker can do any better than this.) We stress that no limitations are placed on the computational power of $\mathcal{A}$.

Formally, let $\Pi = (\mathsf{Gen}, \mathsf{Enc}, \mathsf{Dec})$ be an encryption scheme with message space $\mathcal{M}$. Let $\mathcal{A}$ be an adversary, which is formally just a (stateful) algorithm that we may assume is deterministic without loss of generality. We define an experiment $\mathsf{PrivK}^{\mathsf{eav}}_{\mathcal{A},\Pi}$, based, on $\mathcal{A}$ and Π, as follows:

The adversarial indistinguishability experiment $\mathsf{PrivK}^{\mathsf{eav}}_{\mathcal{A},\Pi}$:

1. *The adversary $\mathcal{A}$ outputs a pair of messages $m_0, m_1 \in \mathcal{M}$.*

2. *A key k is generated using Gen, and a uniform bit $b \in \{0, 1\}$ is chosen. Ciphertext $c \leftarrow \mathsf{Enc}_k(m_b)$ is computed and given to $\mathcal{A}$. We refer to c as the challenge ciphertext.*

3. $\mathcal{A}$ *outputs a bit* b'.

4. *The output of the experiment is defined to be 1 if* $b' = b$, *and 0 otherwise. We write* $\mathsf{PrivK}^{\mathsf{eav}}_{\mathcal{A},\Pi} = 1$ *if the output of the experiment is 1 and in this case we say that* $\mathcal{A}$ succeeds.

As noted earlier, it is trivial for $\mathcal{A}$ to succeed with probability $1/2$ by outputting a random guess. Perfect indistinguishability requires that it is impossible for any $\mathcal{A}$ to do better.

DEFINITION 2.6   *Encryption scheme* $\Pi = (\mathsf{Gen}, \mathsf{Enc}, \mathsf{Dec})$ *with message space* $\mathcal{M}$ is perfectly indistinguishable *if for every* $\mathcal{A}$ *it holds that*

$$\Pr\left[\mathsf{PrivK}^{\mathsf{eav}}_{\mathcal{A},\Pi} = 1\right] = \frac{1}{2}.$$

The following lemma states that Definition 2.6 is equivalent to Definition 2.3. We leave the proof of the lemma as Exercise 2.6.

LEMMA 2.7   *Encryption scheme* Π *is perfectly secret if and only if it is perfectly indistinguishable.*

Example 2.8
We show that the Vigenère cipher is *not* perfectly indistinguishable, at least for certain parameters. Concretely, let Π denote the Vigenère cipher for the message space of two-character strings, and where the period is chosen uniformly in $\{1, 2\}$. To show that Π is not perfectly indistinguishable, we exhibit an adversary $\mathcal{A}$ for which $\Pr\left[\mathsf{PrivK}^{\mathsf{eav}}_{\mathcal{A},\Pi} = 1\right] > \frac{1}{2}$.
   Adversary $\mathcal{A}$ does:

1. Output $m_0 = \mathsf{aa}$ and $m_1 = \mathsf{ab}$.

2. Upon receiving the challenge ciphertext $c = c_1 c_2$, do the following: if $c_1 = c_2$ output 0; else output 1.

Computation of $\Pr\left[\mathsf{PrivK}^{\mathsf{eav}}_{\mathcal{A},\Pi} = 1\right]$ is tedious but straightforward.

$$\Pr\left[\mathsf{PrivK}^{\mathsf{eav}}_{\mathcal{A},\Pi} = 1\right]$$
$$= \frac{1}{2} \cdot \Pr\left[\mathsf{PrivK}^{\mathsf{eav}}_{\mathcal{A},\Pi} = 1 \mid b = 0\right] + \frac{1}{2} \cdot \Pr\left[\mathsf{PrivK}^{\mathsf{eav}}_{\mathcal{A},\Pi} = 1 \mid b = 1\right]$$
$$= \frac{1}{2} \cdot \Pr[\mathcal{A} \text{ outputs } 0 \mid b = 0] + \frac{1}{2} \cdot \Pr[\mathcal{A} \text{ outputs } 1 \mid b = 1], \qquad (2.4)$$

where b is the uniform bit determining which message gets encrypted. As defined, $\mathcal{A}$ outputs 0 if and only if the two characters of the ciphertext $c = c_1 c_2$ are equal. When $b = 0$ (so $m_0 = \mathsf{aa}$ is encrypted) then $c_1 = c_2$ if either (1) a key of period 1 is chosen, or (2) a key of period 2 is chosen and both characters of the key are equal. The former occurs with probability $\frac{1}{2}$, and the latter occurs with probability $\frac{1}{2} \cdot \frac{1}{26}$. So

$$\Pr[\mathcal{A} \text{ outputs } 0 \mid b = 0] = \frac{1}{2} + \frac{1}{2} \cdot \frac{1}{26} \approx 0.52.$$

When $b = 1$ then $c_1 = c_2$ only if a key of period 2 is chosen and the first character of the key is one more than the second character of the key, which happens with probability $\frac{1}{2} \cdot \frac{1}{26}$. So

$$\Pr[\mathcal{A} \text{ outputs } 1 \mid b = 1] = 1 - \Pr[\mathcal{A} \text{ outputs } 0 \mid b = 1] = 1 - \frac{1}{2} \cdot \frac{1}{26} \approx 0.98.$$

Plugging into Equation (2.4) then gives

$$\Pr\left[\mathsf{PrivK}^{\mathsf{eav}}_{\mathcal{A},\Pi} = 1\right] = \frac{1}{2} \cdot \left(\frac{1}{2} + \frac{1}{2} \cdot \frac{1}{26} + 1 - \frac{1}{2} \cdot \frac{1}{26}\right) = 0.75 > \frac{1}{2},$$

and the scheme is not perfectly indistinguishable. $\diamondsuit$

2.2 The One-Time Pad

In 1917, Vernam patented a perfectly secret encryption scheme now called the *one-time pad*. At the time Vernam proposed the scheme, there was no proof that it was perfectly secret; in fact, the notion of perfect secrecy was not yet defined. Approximately 25 years later, however, Shannon introduced the definition of perfect secrecy and demonstrated that the one-time pad satisfied that definition.

In describing the scheme we let $a \oplus b$ denote the *bitwise exclusive-or* (XOR) of two equal-length binary strings a and b. (i.e., if $a = a_1 \cdots a_\ell$ and $b = b_1 \cdots b_\ell$ are ℓ-bit strings, then $a \oplus b$ is the ℓ-bit string given by $(a_1 \oplus b_1) \cdots (a_\ell \oplus b_\ell)$.) In the one-time pad encryption scheme the key is a uniform string of the same length as the message, and the ciphertext is computed by simply XORing the key and the message; a formal definition is given as Construction 2.9. Before discussing security, we first verify correctness: For every key k and every message m it holds that $\mathsf{Dec}_k(\mathsf{Enc}_k(m)) = k \oplus k \oplus m = m$, and so the one-time pad constitutes a valid encryption scheme.

One can easily prove perfect secrecy of the one-time pad using Lemma 2.5 because the ciphertext is uniformly distributed regardless of what message is encrypted. We give a direct proof based on the original definition.

THEOREM 2.10 *The one-time pad encryption scheme is perfectly secret.*

CONSTRUCTION 2.9

Fix an integer $\ell > 0$. The message space $\mathcal{M}$, key space $\mathcal{K}$, and ciphertext space $\mathcal{C}$ are all equal to $\{0,1\}^\ell$ (the set of all binary strings of length ℓ).

- **Gen:** the key-generation algorithm chooses a key from $\mathcal{K} = \{0,1\}^\ell$ according to the uniform distribution (i.e., each of the 2^ℓ strings in the space is chosen as the key with probability exactly $2^{-\ell}$).

- **Enc:** given a key $k \in \{0,1\}^\ell$ and a message $m \in \{0,1\}^\ell$, the encryption algorithm outputs the ciphertext $c := k \oplus m$.

- **Dec:** given a key $k \in \{0,1\}^\ell$ and a ciphertext $c \in \{0,1\}^\ell$, the decryption algorithm outputs the message $m := k \oplus c$.

The one-time pad encryption scheme.

PROOF   We first compute $\Pr[C = c \mid M = m]$ for arbitrary $c \in \mathcal{C}$ and $m \in \mathcal{M}$ with $\Pr[M = m] > 0$. For the one-time pad, we have

$$
\begin{aligned}
\Pr[C = c \mid M = m] &= \Pr[K \oplus m = c \mid M = m] \\
&= \Pr[K = m \oplus c \mid M = m] \\
&= 2^{-\ell},
\end{aligned}
$$

where the first equality is by definition of the scheme and the fact that we condition on the event $M = m$, and the final equality holds because the key K is a uniform ℓ-bit string that is independent of M. Fix any distribution over $\mathcal{M}$. Using the above result, we see that for any $c \in \mathcal{C}$ we have

$$
\begin{aligned}
\Pr[C = c] &= \sum_{m \in \mathcal{M}} \Pr[C = c \mid M = m] \cdot \Pr[M = m] \\
&= 2^{-\ell} \cdot \sum_{m \in \mathcal{M}} \Pr[M = m] \\
&= 2^{-\ell},
\end{aligned}
$$

where the sum is over $m \in \mathcal{M}$ with $\Pr[M = m] \neq 0$. Bayes' Theorem gives:

$$
\begin{aligned}
\Pr[M = m \mid C = c] &= \frac{\Pr[C = c \mid M = m] \cdot \Pr[M = m]}{\Pr[C = c]} \\
&= \frac{2^{-\ell} \cdot \Pr[M = m]}{2^{-\ell}} \\
&= \Pr[M = m].
\end{aligned}
$$

We conclude that the one-time pad is perfectly secret. ∎

The one-time pad was used by several national-intelligence agencies in the mid-20th century to encrypt sensitive traffic. Perhaps most famously, the "red phone" linking the White House and the Kremlin during the Cold War was protected using one-time pad encryption, where the governments of the US and the USSR would exchange extremely long keys using trusted couriers carrying briefcases of paper on which random characters were written.

Notwithstanding the above, one-time pad encryption is rarely used nowadays because it has a number of drawbacks. Most prominent is that *the key is as long as the message*.[2] This limits the usefulness of the scheme for sending very long messages (as it may be difficult to securely share and store a very long key), and is problematic when the parties cannot predict in advance (an upper bound on) how long the message will be.

Moreover, the one-time pad—as the name indicates—*is only secure if used once* (with a given key). Although we did not yet define a notion of secrecy when multiple messages are encrypted, it is easy to see that encrypting more than one message with the same key leaks a lot of information. In particular, say two messages m, m' are encrypted using the same (unknown) key k. An adversary who obtains $c = m \oplus k$ and $c' = m' \oplus k$ can compute

$$c \oplus c' = (m \oplus k) \oplus (m' \oplus k) = m \oplus m'$$

and thus learn the XOR of the two messages or, equivalently, exactly where the two messages differ. This attack extends to more than two messages as well, where it enables the attacker to learn the XOR of all pairs of messages. While this may not seem very significant, it is enough to rule out any claims of perfect secrecy for encrypting more than one message using the same key. Moreover, if the messages correspond to natural-language text, then given the XOR of sufficiently many pairs of messages—or even two sufficiently long messages—it is possible to perform frequency analysis (as in the previous chapter, though more complex) and recover the messages themselves. (See Exercise 2.16 for an example.) An interesting historical example of this is given by the *VENONA project*, as part of which the US and UK were able to decrypt ciphertexts sent by the Soviet Union that were mistakenly encrypted with repeated portions of a one-time pad over several decades.

2.3 Limitations of Perfect Secrecy

We ended the previous section by noting some drawbacks of the one-time pad encryption scheme. Here, we show that these drawbacks are not specific to that scheme, but are instead *inherent* limitations of perfect secrecy. Specifically, we prove that *any* perfectly secret encryption scheme must have a key space that is at least as large as the message space. If all keys are the same

---

[2]This does not make the one-time pad useless, since it may be easier for two parties to share a key at some point in time before the message to be communicated is known.

length, and the message space consists of all strings of some fixed length, this implies that the key is at least as long as the message. In particular, the key length of the one-time pad is optimal. (The other limitation—namely, that a key can be used only once—is also inherent; see Exercise 2.19.)

THEOREM 2.11 If (Gen, Enc, Dec) *is a perfectly secret encryption scheme with message space* $\mathcal{M}$ *and key space* $\mathcal{K}$*, then* $|\mathcal{K}| \geq |\mathcal{M}|$*.*

PROOF We show that if $|\mathcal{K}| < |\mathcal{M}|$ then the scheme cannot be perfectly secret. Assume $|\mathcal{K}| < |\mathcal{M}|$. Consider the uniform distribution over $\mathcal{M}$ and let $c \in \mathcal{C}$ be a ciphertext that occurs with nonzero probability. Let $\mathcal{M}(c)$ be the set of all possible messages that are possible decryptions of c; that is

$$\mathcal{M}(c) \stackrel{\text{def}}{=} \{m \mid m = \mathsf{Dec}_k(c) \text{ for some } k \in \mathcal{K}\}.$$

Clearly $|\mathcal{M}(c)| \leq |\mathcal{K}|$. (Recall that we may assume Dec is deterministic.) If $|\mathcal{K}| < |\mathcal{M}|$, there is some $m' \in \mathcal{M}$ such that $m' \notin \mathcal{M}(c)$. But then

$$\Pr[M = m' \mid C = c] = 0 \neq \Pr[M = m'],$$

and so the scheme is not perfectly secret. ∎

Perfect secrecy with shorter keys? The above theorem shows an inherent limitation of schemes that achieve perfect secrecy. Even so, individuals occasionally claim they have developed a radically new encryption scheme that is "unbreakable" and achieves the security of the one-time pad without using keys as long as what is being encrypted. The above proof demonstrates that such claims *cannot* be true; anyone making such claims either knows very little about cryptography or is blatantly lying.

2.4 *Shannon's Theorem

In his work on perfect secrecy, Shannon also provided a characterization of perfectly secret encryption schemes. This characterization says that, under certain conditions, the key-generation algorithm Gen must choose the key *uniformly* from the set of all possible keys (as in the one-time pad); moreover, for every message m and ciphertext c there is a *unique* key mapping m to c (again, as in the one-time pad). Beyond being interesting in its own right, this theorem is a useful tool for proving (or disproving) perfect secrecy of schemes. We discuss this further after the proof.

The theorem as stated here assumes $|\mathcal{M}| = |\mathcal{K}| = |\mathcal{C}|$, meaning that the sets of plaintexts, keys, and ciphertexts all have the same size. We have already seen that for perfect secrecy we must have $|\mathcal{K}| \geq |\mathcal{M}|$. It is easy to see that correct decryption requires $|\mathcal{C}| \geq |\mathcal{M}|$. Therefore, in some sense, perfectly secret encryption schemes with $|\mathcal{M}| = |\mathcal{K}| = |\mathcal{C}|$ are "optimal."

THEOREM 2.12 (Shannon's theorem) *Let* $(\mathsf{Gen}, \mathsf{Enc}, \mathsf{Dec})$ *be an encryption scheme with message space* $\mathcal{M}$*, for which* $|\mathcal{M}| = |\mathcal{K}| = |\mathcal{C}|$*. The scheme is perfectly secret if and only if:*

1. *Every key* $k \in \mathcal{K}$ *is chosen with (equal) probability* $1/|\mathcal{K}|$ *by* Gen*.*

2. *For every* $m \in \mathcal{M}$ *and every* $c \in \mathcal{C}$*, there is a unique key* $k \in \mathcal{K}$ *such that* $\mathsf{Enc}_k(m)$ *outputs* c*.*

PROOF The intuition behind the proof is as follows. To see that the stated conditions imply perfect secrecy, note that condition 2 means that any ciphertext c could be the result of encrypting any possible plaintext m, because there is some key k mapping m to c. Since there is a *unique* such key, and each key is chosen with equal probability, perfect secrecy follows as for the one-time pad. For the other direction, perfect secrecy immediately implies that for every m and c there is at least one key mapping m to c. The fact that $|\mathcal{M}| = |\mathcal{K}| = |\mathcal{C}|$ means, moreover, that for every m and c there is exactly *one* such key. Given this, each key must be chosen with equal probability or else perfect secrecy would fail to hold. A formal proof follows.

We assume for simplicity that Enc is deterministic. (One can show that this is without loss of generality here.) We first prove that if the encryption scheme satisfies conditions 1 and 2, then it is perfectly secret. The proof is essentially the same as the proof of perfect secrecy for the one-time pad, so we will be relatively brief. Fix arbitrary $c \in \mathcal{C}$ and $m \in \mathcal{M}$. Let k be the unique key, guaranteed by condition 2, for which $\mathsf{Enc}_k(m) = c$. Then,

$$\Pr[\mathsf{Enc}_K(m) = c] = \Pr[K = k] = 1/|\mathcal{K}|,$$

where the final equality holds by condition 1. Since this holds for arbitrary m and c, Lemma 2.5 implies that the scheme is perfectly secret.

For the second direction, assume the encryption scheme is perfectly secret; we show that conditions 1 and 2 hold. Fix arbitrary $c \in \mathcal{C}$. There must be some message m^* for which $\Pr[\mathsf{Enc}_K(m^*) = c] \neq 0$. Lemma 2.5 then implies that $\Pr[\mathsf{Enc}_K(m) = c] \neq 0$ for every $m \in \mathcal{M}$. In other words, if we let $\mathcal{M} = \{m_1, m_2, \ldots\}$, then for each $m_i \in \mathcal{M}$ we have a nonempty set of keys $\mathcal{K}_i \subset \mathcal{K}$ such that $\mathsf{Enc}_k(m_i) = c$ if and only if $k \in \mathcal{K}_i$. Moreover, when $i \neq j$ then $\mathcal{K}_i$ and $\mathcal{K}_j$ must be disjoint or else correctness fails to hold. Since $|\mathcal{K}| = |\mathcal{M}|$, we see that each $\mathcal{K}_i$ contains only a single key k_i, as required by condition 2. Now, Lemma 2.5 shows that for any $m_i, m_j \in \mathcal{M}$ we have

$$\Pr[K = k_i] = \Pr[\mathsf{Enc}_K(m_i) = c] = \Pr[\mathsf{Enc}_K(m_j) = c] = \Pr[K = k_j].$$

Since this holds for all $1 \leq i, j \leq |\mathcal{M}| = |\mathcal{K}|$, and $k_i \neq k_j$ for $i \neq j$, this means each key is chosen with probability $1/|\mathcal{K}|$, as required by condition 1. ∎

Shannon's theorem is useful for deciding whether a given scheme is perfectly secret. Condition 1 is easy to check, and condition 2 can be demonstrated (or contradicted) without having to compute any probabilities (in contrast to working with Definition 2.3 directly). As an example, perfect secrecy of the one-time pad is trivial to prove using Shannon's theorem. We stress, however, that the theorem only applies when $|\mathcal{M}| = |\mathcal{K}| = |\mathcal{C}|$.

References and Additional Reading

The one-time pad is popularly credited to Vernam [200], who filed a patent on it, but recent historical research [28] shows that it was invented some 35 years earlier. Analysis of the one-time pad had to await the ground-breaking work of Shannon [177], who introduced the notion of perfect secrecy.

In this chapter we studied perfectly secret *encryption*. Some other cryptographic problems can also be solved with "perfect" security. A notable example is the problem of message authentication where the aim is to prevent an adversary from (undetectably) modifying a message sent from one party to another. We study this problem in depth in Chapter 4, discussing "perfectly secure" message authentication in Section 4.6.

Exercises

2.1 Prove that, by redefining the key space, we may assume that the key-generation algorithm Gen chooses a uniform key from the key space, without changing $\Pr[C = c \mid M = m]$ for any m, c.

> **Hint:** Define the key space to be the set of all possible random bits used by the randomized algorithm Gen.

2.2 Prove that, by redefining the key space as well as the encryption algorithm, we may assume that encryption is deterministic without changing $\Pr[C = c \mid M = m]$ for any m, c.

2.3 Prove or refute: An encryption scheme with message space $\mathcal{M}$ is perfectly secret if and only if for every probability distribution on $\mathcal{M}$ and every $c_0, c_1 \in \mathcal{C}$ we have $\Pr[C = c_0] = \Pr[C = c_1]$.

2.4 Prove or refute: For every perfectly secret encryption scheme it holds that for every distribution on the message space $\mathcal{M}$, every $m, m' \in \mathcal{M}$, and every $c \in \mathcal{C}$:

$$\Pr[M = m \mid C = c] = \Pr[M = m' \mid C = c].$$

2.5 Prove that in Definition 2.6 we may assume $\mathcal{A}$ is deterministic without loss of generality.

2.6 Prove Lemma 2.7.

2.7 What is the ciphertext that results when the plaintext 0x012345 (written in hex) is encrypted using the one-time pad with key 0xFFEEDD?

2.8 For each of the following encryption schemes, state whether the scheme is perfectly secret. Justify your answer in each case.

(a) The message space is $\mathcal{M} = \{0, \ldots, 4\}$, and Gen chooses a uniform key from the key space $\mathcal{K} = \{0, \ldots, 5\}$. $\text{Enc}_k(m)$ returns $[m + k \bmod 5]$, and $\text{Dec}_k(c)$ returns $[c - k \bmod 5]$.

(b) The message space is $\mathcal{M} = \{m \in \{0, 1\}^\ell \mid$ the last bit of m is $0\}$. Gen chooses a uniform key from $\{0, 1\}^{\ell-1}$. $\text{Enc}_k(m)$ returns ciphertext $m \oplus (k \| 0)$, and $\text{Dec}_k(c)$ returns $c \oplus (k \| 0)$.

2.9 In each of the following schemes, $\text{Enc}_k(m) = [m + k \bmod 3]$. State in each case whether the scheme is perfectly secret, and justify your answers.

(a) The message space is $\mathcal{M} = \{0, 1\}$, and Gen chooses a uniform key from the key space $\mathcal{K} = \{0, 1\}$.

(b) The message space is $\mathcal{M} = \{0, 1, 2\}$, and Gen chooses a uniform key from the key space $\mathcal{K} = \{0, 1, 2\}$.

(c) The message space is $\mathcal{M} = \{0, 1\}$, and Gen chooses a uniform key from the key space $\mathcal{K} = \{0, 1, 2\}$.

2.10 The following questions concern the message space $\mathcal{M} = \{0, 1\}^{\leq \ell}$, the set of all nonempty binary strings of length at most ℓ.

(a) Consider the encryption scheme in which Gen chooses a uniform key from $\mathcal{K} = \{0, 1\}^\ell$, and $\text{Enc}_k(m)$ outputs $k_{|m|} \oplus m$, where k_t denotes the first t bits of k. Show that this scheme is not perfectly secret for message space $\mathcal{M}$.

(b) Design a perfectly secret encryption scheme for message space $\mathcal{M}$.

2.11 When using the one-time pad with the key $k = 0^\ell$, we have $\text{Enc}_k(m) = k \oplus m = m$ and the message is sent in the clear! It has therefore been suggested to modify the one-time pad by only encrypting with $k \neq 0^\ell$ (i.e., to have Gen choose k uniformly from the set of *nonzero* keys of length ℓ). Is this modified scheme still perfectly secret? Explain.

2.12 Let Π denote the Vigenère cipher where the message space consists of all 3-character strings (over the English alphabet), and the period t is fixed to 2 (and so the key is a uniform string of length 2). Define $\mathcal{A}$ as follows: $\mathcal{A}$ outputs $m_0 = $ aaa and $m_1 = $ aab. When given a ciphertext c, it outputs 0 if the first character of c is the same as the third character of c, and outputs 1 otherwise. Compute $\Pr[\mathsf{PrivK}^{\mathsf{eav}}_{\mathcal{A},\Pi} = 1]$.

2.13 Let Π denote the Vigenère cipher where the message space consists of all 3-character strings (over the English alphabet), and the key is generated by first choosing the period t uniformly from $\{1, 2, 3\}$ and then letting the key be a uniform string of length t.

   (a) Define $\mathcal{A}$ as follows: $\mathcal{A}$ outputs $m_0 = $ aab and $m_1 = $ abb. When given a ciphertext c, it outputs 0 if the first character of c is the same as the second character of c, and outputs 1 otherwise. Compute $\Pr[\mathsf{PrivK}^{\mathsf{eav}}_{\mathcal{A},\Pi} = 1]$.

   (b) Construct and analyze an adversary $\mathcal{A}'$ for which $\Pr[\mathsf{PrivK}^{\mathsf{eav}}_{\mathcal{A}',\Pi} = 1]$ is greater than your answer from part (a).

2.14 In this exercise, we look at different conditions under which the shift, mono-alphabetic substitution, and Vigenère ciphers are perfectly secret:

   (a) Prove that if only a single character is encrypted, then the shift cipher is perfectly secret.

   (b) What is the largest message space $\mathcal{M}$ for which the mono-alphabetic substitution cipher provides perfect secrecy?

   (c) Prove that the Vigenère cipher using (fixed) period t is perfectly secret when used to encrypt messages of length t.

   Reconcile this with the attacks shown in the previous chapter.

2.15 Give a direct proof that a scheme satisfying Definition 2.6 must have $|\mathcal{K}| \geq |\mathcal{M}|$. Specifically, let Π be an arbitrary encryption scheme with $|\mathcal{K}| < |\mathcal{M}|$. Show an $\mathcal{A}$ for which $\Pr\left[\mathsf{PrivK}^{\mathsf{eav}}_{\mathcal{A},\Pi} = 1\right] > \frac{1}{2}$.

   Hint: It may be easier to let $\mathcal{A}$ be randomized.

2.16 The following questions concern multiple encryptions of single-character ASCII plaintexts with the one-time pad using the same 8-bit key. You may assume that the plaintexts are either (upper- or lower-case) English letters or the space character.

   (a) Say you see the ciphertexts 1011 0111 and 1110 0111. What can you deduce about the plaintext characters these correspond to?

   (b) Say you see the three ciphertexts 0110 0110, 0011 0010, and 0010 0011. What can you deduce about the plaintext characters these correspond to?

   Hint: Focus on the second bit of the ciphertexts.

2.17 Assume we require only that an encryption scheme (Gen, Enc, Dec) with message space $\mathcal{M}$ satisfy the following: For all $m \in \mathcal{M}$, we have $\Pr[\mathsf{Dec}_K(\mathsf{Enc}_K(m)) = m] \geq 2^{-t}$. (This probability is taken over choice of the key as well as any randomness used during encryption/decryption.) Show that perfect secrecy can be achieved with $|\mathcal{K}| < |\mathcal{M}|$ when $t \geq 1$. Prove a lower bound on the size of $\mathcal{K}$ in terms of t.

2.18 Let $\varepsilon > 0$ be a constant. Say an encryption scheme is ε-*perfectly secret* if for every adversary $\mathcal{A}$ it holds that

$$\Pr\left[\mathsf{PrivK}^{\mathsf{eav}}_{\mathcal{A},\Pi} = 1\right] \leq \frac{1}{2} + \varepsilon.$$

(Compare to Definition 2.6.) Consider a variant of the one-time pad where $\mathcal{M} = \{0,1\}^\ell$ and the key is chosen uniformly from an arbitrary set $\mathcal{K} \subseteq \{0,1\}^\ell$ with $|\mathcal{K}| = (1 - \varepsilon) \cdot 2^\ell$; encryption and decryption are otherwise the same.

(a) Prove that this scheme is ε-perfectly secret.

(b) Prove that this scheme is $\left(\frac{\varepsilon}{2(1-\varepsilon)}\right)$-perfectly secret when $\varepsilon \leq 1/2$. (Note that $\frac{\varepsilon}{2(1-\varepsilon)} \leq \varepsilon$ here, so this is an improvement over part (a).)

(c) Prove that any deterministic scheme that is ε-perfectly secret must have $|\mathcal{K}| \geq (1 - 2\varepsilon) \cdot |\mathcal{M}|$. (Note: It is an open question to prove a tight lower bound that also holds for randomized schemes.)

2.19 In this problem we consider definitions of perfect secrecy for the encryption of *two* messages (using the same key). Here we consider distributions on *pairs* of messages from the message space $\mathcal{M}$; we let M_1, M_2 be random variables denoting the first and second message, respectively. (These random variables are not assumed to be independent.) We generate a (single) key k, sample a pair of messages (m_1, m_2) according to the given distribution, and then compute ciphertexts $c_1 \leftarrow \mathsf{Enc}_k(m_1)$ and $c_2 \leftarrow \mathsf{Enc}_k(m_2)$; this induces a distribution on pairs of ciphertexts and we let C_1, C_2 be the corresponding random variables.

(a) Say encryption scheme (Gen, Enc, Dec) is *perfectly secret for two messages* if for all distributions on $\mathcal{M} \times \mathcal{M}$, all $m_1, m_2 \in \mathcal{M}$, and all ciphertexts $c_1, c_2 \in \mathcal{C}$ with $\Pr[C_1 = c_1 \wedge C_2 = c_2] > 0$:

$$\Pr[M_1 = m_1 \wedge M_2 = m_2 \mid C_1 = c_1 \wedge C_2 = c_2]$$
$$= \Pr[M_1 = m_1 \wedge M_2 = m_2].$$

Prove that *no* encryption scheme can satisfy this definition.

Hint: Take $c_1 = c_2$.

(b) Say encryption scheme (Gen, Enc, Dec) is *perfectly secret for two distinct messages* if for all distributions on $\mathcal{M} \times \mathcal{M}$ where the first and second messages are guaranteed to be different (i.e., distributions on pairs of *distinct* messages), all $m_1, m_2 \in \mathcal{M}$, and all $c_1, c_2 \in \mathcal{C}$ with $\Pr[C_1 = c_1 \wedge C_2 = c_2] > 0$:

$$\Pr[M_1 = m_1 \wedge M_2 = m_2 \mid C_1 = c_1 \wedge C_2 = c_2]$$
$$= \Pr[M_1 = m_1 \wedge M_2 = m_2].$$

Show an encryption scheme that provably satisfies this definition.

Hint: The encryption scheme you propose need not be efficient, although an efficient solution is possible.

Part II

Private-Key (Symmetric) Cryptography

Chapter 3

Private-Key Encryption

In the previous chapter we saw some fundamental limitations of perfect secrecy. In this chapter we begin our study of modern cryptography by introducing the weaker (but sufficient) notion of *computational* secrecy. We then show how this definition can be used to bypass the impossibility results shown previously for perfect secrecy and, in particular, how a short key (say, 128 bits long) can be used to encrypt many long messages (say, gigabytes in total).

Along the way we will study the fundamental notion of *pseudorandomness*, which captures the idea that something can "look" completely random even though it is not. This powerful concept underlies much of modern cryptography, and has applications and implications beyond the field as well.

3.1 Computational Security

In Chapter 2 we introduced the notion of perfect secrecy. While perfect secrecy is a worthwhile goal, it is also unnecessarily strong. Perfect secrecy requires that *absolutely no information* about an encrypted message is leaked, even to an eavesdropper *with unlimited computational power*. For all practical purposes, however, an encryption scheme would still be considered secure if it leaked information with some *tiny probability* to eavesdroppers with *bounded computational power*. For example, a scheme that leaks information with probability at most 2^{-60} to eavesdroppers investing up to 200 years of computational effort on the fastest available supercomputer (or cluster of computers) would be more than adequate for real-world applications. *Computational* security definitions take into account computational limits on an attacker, and allow for a small probability that security is violated, in contrast to notions (such as perfect secrecy) that are *information-theoretic* in nature. Computational security is now the *de facto* way in which security is defined for almost all cryptographic applications.

We stress that although we give up on obtaining perfect secrecy, this does not mean we do away with the rigorous mathematical approach we have been taking so far. Definitions and proofs are still essential; the only difference is that we now consider a weaker (but still meaningful) notion of security.

As discussed, computational security definitions incorporate two relaxations relative to information-theoretic notions of security:

1. *Security is only guaranteed against **efficient** adversaries that run for some feasible amount of time.* This means that given enough time (or sufficient computational resources) an attacker may be able to violate security. If we can make the resources required to break the scheme larger than those available to any realistic attacker, however, then for all practical purposes the scheme is unbreakable.

2. *Adversaries can potentially succeed (i.e., security can potentially fail) with some **very small probability**.* If we can make this probability sufficiently small, we need not worry about it.

(As we will see, both these relaxations are necessary in order to overcome the limitations of perfect secrecy shown in the last chapter.) To obtain a meaningful theory, we need to precisely define what we mean by the above relaxations. There are two general approaches for doing so: the *concrete approach* and the *asymptotic approach*. These are described next.

3.1.1   The Concrete Approach

The concrete approach to computational security quantifies the security of a cryptographic scheme by explicitly bounding the maximum success probability of a (randomized) adversary running for some specified amount of time or, more precisely, investing some specified amount of computational effort. Thus, a concrete definition of security takes the following form:

> *A scheme is (t, ε)-secure if any adversary running for time at most t succeeds in breaking the scheme with probability at most ε.*

(Of course, the above serves only as a general template, and for it to make sense we need to define exactly what it means to "break" the scheme in question.) As an example, one might have a scheme with the guarantee that no adversary running for at most 200 years using the fastest available supercomputer can succeed in breaking the scheme with probability better than 2^{-60}. Or, it may be more convenient to measure running time in terms of CPU cycles, and to construct a scheme such that no adversary using at most 2^{80} cycles can break the scheme with probability better than 2^{-60}.

It is instructive to get a feel for the large values of t and the small values of ε that are typical of modern cryptosystems.

Example 3.1
Modern private-key encryption schemes are generally assumed to give almost optimal security in the following sense: when the key has length n—and so the key space has size 2^n—an adversary running for time t (measured in, say,

CPU cycles) succeeds in breaking the scheme with probability at most $ct/2^n$ for some fixed constant c. (This simply corresponds to a brute-force search of the key space.)

Assuming $c = 1$ for simplicity, a key of length $n = 64$ provides adequate security against an adversary using a standard desktop computer. Indeed, on a 4 GHz processor with 16 cores that executes 4×10^9 cycles per second per core, 2^{64} CPU cycles require $2^{64}/(4 \times 10^9 \times 16)$ seconds, or about 9 years. (The above numbers are for illustrative purposes only; in practice $c \neq 1$, and several other factors—including the time required for accessing memory—can significantly affect the performance of brute-force attacks.)

However, there is no reason to assume that an adversary is limited to a desktop computer, and powerful adversaries are able to carry out computations orders of magnitude faster. Today, the minimum recommended key length is $n = 128$. The difference between 2^{64} and 2^{128} is a *multiplicative factor* of 2^{64}. To get a feeling for how big this is, note that according to physicists' estimates the number of seconds since the Big Bang is on the order of 2^{58}.

If the probability that an attacker can successfully recover an encrypted message in one year is at most 2^{-60}, then it is much more likely that the sender and receiver will both be hit by lightning in that time period than that the attacker will recover the message! Something that occurs with probability 2^{-60} each second is expected to occur roughly once every 10 billion years. ◇

The concrete approach is important in practice, since concrete guarantees are what users of a cryptographic scheme are ultimately interested in. However, precise concrete guarantees are difficult to provide. Furthermore, one must be careful in interpreting concrete-security claims. For example, a claim that no adversary running for 5 years can break a given scheme with probability better than ε begs the questions: what type of computing power (e.g., desktop PC, supercomputer, network of hundreds of computers) does this assume? Does this take into account expected future advances in computing power (which, by Moore's Law, roughly doubles every 18 months)? Does the estimate assume the use of "off-the-shelf" algorithms, or dedicated hardware optimized for the attack? Furthermore, such a guarantee says little about the success probability of an adversary running for 2 years (other than the fact that it can be at most ε) and says nothing about the success probability of an adversary running for 10 years.

3.1.2 The Asymptotic Approach

As partly noted above, there are some technical and theoretical difficulties in using the concrete-security approach. These issues must be dealt with in practice but when describing schemes abstractly (as we do in this book) it is convenient instead to use an *asymptotic* approach. This approach, rooted in complexity theory, introduces an integer-valued *security parameter* (denoted by n) that parameterizes both cryptographic schemes as well as all involved parties (i.e., the honest parties as well as the attacker). When honest parties use a scheme (e.g., when they generate a key), they choose some value for the security parameter; for the purposes of this discussion, one can view the security parameter as corresponding to the length of the key. We also view the running time of the adversary, as well as its success probability, as functions of the security parameter rather than as fixed, concrete values. Then:

1. We equate "efficient adversaries" with randomized (i.e., probabilistic) algorithms running in time *polynomial in n*. This means there is some polynomial p such that the adversary runs for time at most $p(n)$ when the security parameter is n. We also require—for real-world efficiency— that honest parties run in polynomial time, although we stress that the adversary may be much more powerful (and run much longer) than the honest parties.

2. We equate the notion of "small probabilities of success" with success probabilities *smaller than any inverse polynomial in n*. (See Definition 3.4.) Such probabilities are called *negligible*.

Let PPT stand for "probabilistic polynomial-time." A definition of asymptotic security then takes the following general form:

> *A scheme is* secure *if any* PPT *adversary succeeds in breaking the scheme with at most negligible probability.*

This notion of security is *asymptotic* since it depends on a scheme's behavior for sufficiently large values of n. The following example illustrates this.

Example 3.2
Say we have a scheme that is asymptotically secure. Then it may be the case that an adversary running for n^3 minutes can succeed in "breaking the scheme" with probability $2^{40}/2^n$ (which is a negligible function of n). When $n \leq 40$ this means that an adversary running for 40^3 minutes (about 6 weeks) can break the scheme with probability 1, so such values of n are not very useful. Even for $n = 50$ an adversary running for 50^3 minutes (about 3 months) can break the scheme with probability roughly $1/1000$, which may not be acceptable. On the other hand, when $n = 500$ an adversary running for 200 years breaks the scheme only with probability roughly 2^{-500}.   ◇

As indicated by the previous example, we can view the security parameter as a mechanism that allows the honest parties to "tune" the security of a scheme to some desired level. (Increasing the security parameter also increases the time required to run the scheme, as well as the length of the key, so the honest parties will want to set the security parameter as small as possible subject to defending against the class of attacks they are concerned about.) Viewing the security parameter as the key length, this corresponds to the fact that the time required for an exhaustive-search attack grows exponentially in the length of the key. The ability to "increase security" by increasing the security parameter has important practical ramifications, since it enables honest parties to defend against increases in computing power. The following example gives a sense of how this might play out in practice.

Example 3.3

Let us see the effect that the availability of faster computers might have on security in practice. Say we have a cryptographic scheme in which the honest parties run for $10^6 \cdot n^2$ cycles, and for which an adversary running for $10^8 \cdot n^4$ cycles can succeed in "breaking" the scheme with probability at most $2^{-n/2}$. (The numbers are intended to make calculations easier, and are not meant to correspond to any existing cryptographic scheme.)

Assume all parties are using 2 GHz computers and the honest parties set $n = 80$. Then the honest parties run for $10^6 \cdot 6400$ cycles, or 3.2 seconds, and an adversary running for $10^8 \cdot (80)^4$ cycles, or roughly 3 weeks, can break the scheme with probability only 2^{-40}.

Say 8 GHz computers become available, and all parties upgrade. Honest parties can increase n to 160 (which requires generating a fresh key) and maintain a running time of 3.2 seconds (i.e., $10^6 \cdot 160^2$ cycles at $8 \cdot 10^9$ cycles/second). In contrast, the adversary now has to run for over 8 million seconds, or more than 13 weeks, to achieve a success probability of 2^{-80}. The effect of a faster computer has been to make the adversary's job *harder*.   ◇

Even when using the asymptotic approach it is important to remember that when a cryptosystem is ultimately deployed a concrete security guarantee will be needed. (After all, some value of n must be chosen, and it is important to understand what level of security is being provided.) As the above examples indicate, however, an asymptotic security claim can typically be translated into a concrete security bound for any desired value of n.

The Asymptotic Approach in Detail

We now discuss more formally the notions of "polynomial-time algorithms" and "negligible success probabilities."

Efficient algorithms. A function f from the natural numbers to the non-negative real numbers is *polynomially bounded* (or simply *polynomial*) if there is a constant c such that $f(n) < n^c$ for all n. An algorithm A runs in polynomial time if there exists a polynomial p such that, for every input $x \in \{0,1\}^*$, the computation of $A(x)$ terminates within at most $p(|x|)$ steps. (Here, $|x|$ denotes the length of the string x.) As mentioned earlier, we equate efficient adversaries with those whose running time is polynomial in the security parameter n. When it is necessary to explicitly indicate this, we provide the security parameter in unary (i.e., the string 1^n consisting of n ones) as input to an algorithm. An algorithm may take other inputs besides the security parameter—for example, a message to be encrypted—and in that case we allow its running time to be polynomial in the total length of its inputs.

A technical advantage of working with polynomials is that they obey certain closure properties. In particular, if p_1, p_2 are two polynomials, then the function $p(n) = p_1(p_2(n))$ is also polynomial.

By default, we allow all algorithms to be probabilistic (i.e., randomized). Any such algorithm is assumed to have access to a sequence of unbiased, independent random bits. Equivalently, a randomized algorithm is given (in addition to its input) a uniformly distributed *random tape* of sufficient length whose bits it can use, as needed, throughout its execution.

We consider randomized algorithms by default for two reasons. First, randomness is essential to cryptography (e.g., in order to choose random keys and so on) and so honest parties must be probabilistic; given this, it is natural to allow adversaries to be probabilistic as well. Second, randomization is practical and—as far as we know—gives attackers additional power. Since our goal is to model *all* realistic attacks, we prefer a more liberal definition of efficient computation.

Negligible success probability. A negligible function is one that is asymptotically smaller than any inverse polynomial function. Formally:

DEFINITION 3.4 *A function f from the natural numbers to the nonnegative real numbers is* negligible *if for every polynomial p there is an N such that for all $n > N$ it holds that $f(n) < \frac{1}{p(n)}$.*

The above is equivalently stated as follows: for every polynomial p and *all sufficiently large values of n* it holds that $f(n) < \frac{1}{p(n)}$. Or, in other words, for all constants c there exists an N such that for all $n > N$ it holds that $f(n) < n^{-c}$. We typically denote an arbitrary negligible function by negl.

Example 3.5
The functions $2^{-n}, 2^{-\sqrt{n}}$, and $n^{-\log n}$ are all negligible. However, they approach zero at very different rates. For example, we can look at the minimum value of n for which each function is smaller than $1/n^5$:

1. Solving $2^{-n} < n^{-5}$ we get $n > 5 \log n$. The smallest integer value of $n > 1$ for which this holds is $n = 23$.

2. Solving $2^{-\sqrt{n}} < n^{-5}$ we get $n > 25 \log^2 n$. The smallest integer value of $n > 1$ for which this holds is $n \approx 3500$.

3. Solving $n^{-\log n} < n^{-5}$ we get $\log n > 5$. The smallest integer value of n for which this holds is $n = 33$.

From the above you may have the impression that $n^{-\log n}$ is smaller than $2^{-\sqrt{n}}$. However, this is incorrect; for all $n > 65536$ it holds that $2^{-\sqrt{n}} < n^{-\log n}$. Nevertheless, this does show that for values of n in the hundreds or thousands, an adversarial success probability of $n^{-\log n}$ is preferable to an adversarial success probability of $2^{-\sqrt{n}}$.   ◇

A technical advantage of working with negligible success probabilities is that they obey certain closure properties. The following is an easy exercise.

PROPOSITION 3.6   *Let* negl_1 *and* negl_2 *be negligible functions. Then,*

1. *The function* $\mathsf{negl}_3(n) = \mathsf{negl}_1(n) + \mathsf{negl}_2(n)$ *is negligible.*

2. *For any polynomial* p, *the function* $\mathsf{negl}_4(n) = p(n)\cdot\mathsf{negl}_1(n)$ *is negligible.*

The second part of the above proposition implies that if a certain event occurs with only negligible probability in some experiment, then the event occurs with negligible probability even if that experiment is repeated polynomially many times. (This relies on the union bound; see Proposition A.7.) For example, the probability that n fair coin flips all come up "heads" is 2^{-n}, which is negligible. This means that even if we repeat the experiment of flipping n coins polynomially many times, the probability that *even one* of those experiments results in n heads is still negligible.

A corollary of the second part of the above proposition is that if a function g is *not* negligible, then neither is the function $f(n) \stackrel{\text{def}}{=} g(n)/p(n)$ for any polynomial p.

Asymptotic Security: A Summary

Any security definition consists of two components: a definition of what is considered a "break" of the scheme, and a specification of the power of the adversary. The power of the adversary can relate to many issues (e.g., in the case of encryption, whether we assume a ciphertext-only attack or a chosen-plaintext attack). However, when it comes to the *computational* power of the adversary, we will from now on model the adversary as efficient and thus only consider adversarial strategies that can be implemented in probabilistic polynomial time. (The only exceptions are Section 4.6, where we revisit information-theoretic security, and Chapter 14, where we consider *quantum* polynomial-time attackers.) Definitions will also be formulated so that a break that occurs with negligible probability is not considered significant. Thus, the general framework of any security definition will be:

> A scheme is *secure* if for every *probabilistic polynomial-time* adversary $\mathcal{A}$ carrying out an attack (of some formally specified type), the probability that $\mathcal{A}$ succeeds in the attack (where success is also formally specified) is *negligible*.

Such a definition is *asymptotic* because it is possible that for small values of n an adversary can succeed with high probability. In order to see this more clearly, we expand the term "negligible" in the above statement:

> A scheme is *secure* if for every PPT adversary $\mathcal{A}$ carrying out an attack, and every polynomial p, there is an integer N such that when $n > N$ the probability that $\mathcal{A}$ succeeds in the attack is less than $\frac{1}{p(n)}$.

Note that nothing is guaranteed for values $n \leq N$.

On the Choices Made in Defining Asymptotic Security

In defining the general notion of asymptotic security, we have made two choices: we have identified efficient adversarial strategies with the class of *probabilistic polynomial-time algorithms*, and have equated small chances of success with *negligible probabilities*. Both these choices are—to some extent—arbitrary, and one could build a perfectly reasonable theory by defining, say, efficient strategies as those running in time $2^{o(n)}$, or small success probabilities as those bounded by 2^{-n}. Nevertheless, we briefly justify the choices we have made (which are the standard ones).

Those familiar with complexity theory or algorithms will recognize that the idea of equating efficient computation with (probabilistic) polynomial-time algorithms is not unique to cryptography. One advantage of using (probabilistic) polynomial time as our notion of efficiency is that this frees us from having to specify our model of computation precisely, since the extended Church–Turing thesis states that all "reasonable" models of computation are polynomially equivalent.[1] Thus, we need not specify whether we use Turing machines, boolean circuits, or random-access machines; we can present algorithms in high-level pseudocode and be confident that if our analysis shows that an algorithm runs in polynomial time, then any reasonable implementation of that algorithm will run in polynomial time.

Another advantage of (probabilistic) polynomial-time algorithms is that they satisfy desirable closure properties: in particular, an algorithm that does only polynomial computation and makes polynomially many calls to polynomial-time subroutines will itself run in polynomial time.

The most important feature of negligible probabilities is the closure property we have already seen in Proposition 3.6(2): a polynomial multiplied by a negligible function is still negligible. This means, in particular, that if a polynomial-time algorithm makes polynomially many calls to some subroutine that "fails" with negligible probability each time it is called, then the probability that *any* call to that subroutine fails is still negligible.

---

[1]Note, however, that *quantum* computers may may give super-polynomial speedup (for some problems) relative to classical computers. We defer further discussion to Chapter 14.

Necessity of the Relaxations

Computational secrecy introduces two relaxations of perfect secrecy: first, secrecy is guaranteed only against efficient adversaries; second, secrecy may "fail" with small probability. Both these relaxations are essential for achieving practical encryption schemes, and in particular for bypassing the negative results for perfectly secret encryption. We informally discuss why this is the case. Assume we have an encryption scheme where the size of the key space $\mathcal{K}$ is smaller than the size of the message space $\mathcal{M}$. (As shown in the previous chapter, this means the scheme cannot be perfectly secret.) Two attacks apply regardless of how the encryption scheme is constructed:

- Given a ciphertext c, an adversary can decrypt c using all keys $k \in \mathcal{K}$. This gives a list of all the messages to which c can possibly correspond. Since this list cannot contain all of $\mathcal{M}$ (because $|\mathcal{K}| < |\mathcal{M}|$), this attack leaks *some* information about the message that was encrypted.

  Moreover, say the adversary carries out a known-plaintext attack and learns that ciphertexts $c_1, \ldots, c_\ell$ correspond to the messages $m_1, \ldots, m_\ell$, respectively. The adversary can again try decrypting each of these ciphertexts with all possible keys until it finds a key k for which $\mathsf{Dec}_k(c_i) = m_i$ for all i. Later, given a ciphertext c that is the encryption of an unknown message m, it is almost surely the case that $\mathsf{Dec}_k(c) = m$.

  Brute-force attacks like the above allow an adversary to "succeed" with probability ≈ 1 in time $\mathcal{O}(|\mathcal{K}|)$.

- Consider again the case where the adversary learns that ciphertexts $c_1, \ldots, c_\ell$ correspond to messages $m_1, \ldots, m_\ell$. The adversary can *guess* a uniform key $k \in \mathcal{K}$ and check whether $\mathsf{Dec}_k(c_i) = m_i$ for all i. If so, then, as above, the attacker can use k to decrypt anything subsequently encrypted by the honest parties.

  Here the adversary runs in constant time and "succeeds" with nonzero probability $1/|\mathcal{K}|$.

Nevertheless, by setting $|\mathcal{K}|$ large enough we can hope to achieve meaningful secrecy against attackers running in time much less than $|\mathcal{K}|$ (so the attacker does not have sufficient time to carry out a brute-force attack), except possibly with small probability on the order of $1/|\mathcal{K}|$.

3.2 Defining Computationally Secure Encryption

Given the background of the previous section, we are ready to present a definition of computational security for private-key encryption. First, we redefine the *syntax* of private-key encryption; this will be largely the same as the syntax introduced in Chapter 2 except that we now explicitly take into account the security parameter n. We also make two other changes: we allow the decryption algorithm to output an error (e.g., in case it is presented with an invalid ciphertext), and let the message space be the set $\{0,1\}^*$ of all (finite-length) binary strings by default.

DEFINITION 3.7 *A* private-key encryption scheme *consists of three probabilistic polynomial-time algorithms* (Gen, Enc, Dec) *such that:*

1. *The* key-generation algorithm Gen *takes as input 1^n (i.e., the security parameter written in unary) and outputs a key k; we write $k \leftarrow$ Gen(1^n) (emphasizing that* Gen *is a randomized algorithm). We assume without loss of generality that any key k output by* Gen(1^n) *satisfies $|k| \geq n$.*

2. *The* encryption algorithm Enc *takes as input a key k and a plaintext message $m \in \{0,1\}^*$, and outputs a ciphertext c. Since* Enc *may be randomized, we write this as $c \leftarrow$ Enc$_k(m)$.*

3. *The* decryption algorithm Dec *takes as input a key k and a ciphertext c, and outputs a message $m \in \{0,1\}^*$ or an error. We denote a generic error by the symbol $\perp$.*

It is required that for every n, every key k output by Gen(1^n)*, and every $m \in \{0,1\}^*$, it holds that* Dec$_k($Enc$_k(m)) = m$.

If (Gen, Enc, Dec) *is such that for k output by* Gen(1^n)*, algorithm* Enc$_k$ *is only defined for messages $m \in \{0,1\}^{\ell(n)}$, then we say that* (Gen, Enc, Dec) *is a* fixed-length private-key encryption scheme for messages of length $\ell(n)$.

Almost always, Gen(1^n) simply outputs a uniform n-bit string as the key. When this is the case, we omit Gen and define a private-key encryption scheme to be a *pair* of algorithms (Enc, Dec). Without significant loss of generality, we assume Dec is deterministic throughout this book, and so write $m := $ Dec$_k(c)$.

The above definition considers *stateless* schemes, in which each invocation of Enc is independent of all prior invocations (and similarly for Dec). Later in this chapter, we will discuss *stateful* schemes in which parties may maintain local state that is updated after each invocation of Enc and/or Dec. We assume encryption schemes are stateless (as in the above definition) unless explicitly noted otherwise.

3.2.1 The Basic Definition of Security (EAV-Security)

We begin by presenting the most basic notion of computational security for private-key encryption: security against a ciphertext-only attack where the adversary observes only a *single* ciphertext or, equivalently, security when a given key is used to encrypt just a *single* message. We consider stronger definitions of security later.

Motivating the definition. As we have already discussed, any definition of security consists of two distinct components: a threat model (i.e., a specification of the assumed power of the adversary) and a security goal (usually specified by describing what constitutes a "break" of the scheme). We begin our definitional treatment by considering the simplest threat model, where there is an *eavesdropping adversary* who observes the encryption of a single message. This is exactly the threat model we considered in the previous chapter. This only difference here is that, as explained in the previous section, we are now interested only in computationally bounded adversaries that are limited to running in polynomial time.

Although we have made two assumptions about the adversary's capabilities (namely, that it eavesdrops on one ciphertext, and that it runs in polynomial time), we make no assumptions whatsoever about the adversary's *strategy* in trying to decipher the ciphertext it observes. This is crucial for obtaining meaningful notions of security: the definition ensures protection against *any* computationally bounded eavesdropper, regardless of the algorithm it uses.

Correctly defining the security goal for encryption is not trivial, but we have already discussed this issue at length in Section 1.4.1 and in the previous chapter. We therefore just recall that the idea behind the definition is that the adversary should be unable to learn *any partial information* about the plaintext from the ciphertext. The definition of *semantic security* (cf. Section 3.2.2) exactly formalizes this notion, and was the first definition of computationally secure encryption to be proposed. Semantic security is complex and difficult to work with. Fortunately, there is an equivalent definition called *indistinguishability* that is much simpler.

The definition of indistinguishability is patterned on the alternative definition of perfect secrecy given as Definition 2.6. (This serves as further justification that the definition of indistinguishability is a good one.) Recall that Definition 2.6 considers an experiment $\mathsf{PrivK}^{\mathsf{eav}}_{\mathcal{A},\Pi}$ in which an adversary $\mathcal{A}$ outputs two messages m_0 and m_1, and is then given an encryption of one of those messages using a randomly generated key. The definition states that a scheme Π is secure if no adversary $\mathcal{A}$ can determine which of the messages m_0, m_1 was encrypted with probability any different from $1/2$ (which is the probability that $\mathcal{A}$ is correct if it just makes a random guess).

Here, we keep the experiment $\mathsf{PrivK}^{\mathsf{eav}}_{\mathcal{A},\Pi}$ almost exactly the same (except for some technical differences discussed below), but introduce two important modifications in the definition itself:

1. We now consider only adversaries running in *polynomial time*, whereas Definition 2.6 considered even adversaries with unbounded running time.

2. We now concede that the adversary might determine the encrypted message with probability *negligibly better than* $1/2$.

As discussed extensively in the previous section, the above relaxations constitute the core elements of computational security.

As for the other differences, the most prominent is that we now parameterize the experiment by a security parameter n. The running time of the adversary $\mathcal{A}$, as well as its success probability, are then both viewed as functions of n. We write $\mathsf{PrivK}^{\mathsf{eav}}_{\mathcal{A},\Pi}(n)$ to denote the experiment being run with security parameter n, and write

$$\Pr[\mathsf{PrivK}^{\mathsf{eav}}_{\mathcal{A},\Pi}(n) = 1] \tag{3.1}$$

to denote the probability that the output of experiment $\mathsf{PrivK}^{\mathsf{eav}}_{\mathcal{A},\Pi}(n)$ is 1. Note that with $\mathcal{A}, \Pi$ fixed, the expression in Equation (3.1) is a function of n.

A second difference is that we now explicitly require the adversary to output two messages m_0, m_1 *of equal length*. (In Definition 2.6 this requirement is implicit if the message space $\mathcal{M}$ only contains messages of some fixed length, as is the case for the one-time pad encryption scheme.) This means that, by default, we do not require a secure encryption scheme to hide the length of the plaintext. We revisit this point at the end of this section; see also Exercises 3.2 and 3.3.

Indistinguishability in the presence of an eavesdropper. We now give the formal definition, beginning with the experiment outlined above. The experiment is defined for a private-key encryption scheme $\Pi = (\mathsf{Gen}, \mathsf{Enc}, \mathsf{Dec})$, an adversary $\mathcal{A}$, and a value n for the security parameter:

The adversarial indistinguishability experiment $\mathsf{PrivK}^{\mathsf{eav}}_{\mathcal{A},\Pi}(n)$:

1. *The adversary $\mathcal{A}$ is given input 1^n, and outputs a pair of messages m_0, m_1 with $|m_0| = |m_1|$.*

2. *A key k is generated by running $\mathsf{Gen}(1^n)$, and a uniform bit $b \in \{0,1\}$ is chosen. Ciphertext $c \leftarrow \mathsf{Enc}_k(m_b)$ is computed and given to $\mathcal{A}$. We refer to c as the* challenge *ciphertext.*

3. *$\mathcal{A}$ outputs a bit b'.*

4. *The output of the experiment is defined to be 1 if $b' = b$, and 0 otherwise. If $\mathsf{PrivK}^{\mathsf{eav}}_{\mathcal{A},\Pi}(n) = 1$, we say that $\mathcal{A}$ succeeds.*

There is no limitation on the lengths of m_0 and m_1, as long as they are the same. (Of course, if $\mathcal{A}$ runs in polynomial time, then m_0 and m_1 have length polynomial in n.) If Π is a fixed-length scheme for messages of length $\ell(n)$, the above experiment is modified by requiring $m_0, m_1 \in \{0,1\}^{\ell(n)}$.

The fact that the adversary can only eavesdrop is implicit in the fact that the adversary is given only a (single) ciphertext, and does not have any further interaction with the sender or the receiver. (As we will see later, allowing additional interaction makes the adversary significantly stronger.)

The definition of indistinguishability states that an encryption scheme is secure if no PPT adversary $\mathcal{A}$ succeeds in guessing which message was encrypted in the above experiment with probability significantly better than random guessing (which is correct with probability $1/2$):

DEFINITION 3.8 *A private-key encryption scheme* $\Pi = (\mathsf{Gen}, \mathsf{Enc}, \mathsf{Dec})$ *has* indistinguishable encryptions in the presence of an eavesdropper, *or is* EAV-secure, *if for all probabilistic polynomial-time adversaries* $\mathcal{A}$ *there is a negligible function* negl *such that, for all* n,

$$\Pr\left[\mathsf{PrivK}^{\mathsf{eav}}_{\mathcal{A},\Pi}(n) = 1\right] \leq \frac{1}{2} + \mathsf{negl}(n).$$

The probability above is taken over the randomness used by $\mathcal{A}$ *and the randomness used in the experiment (for choosing the key and the bit* b, *as well as any randomness used by* Enc).

It should be clear that Definition 3.8 is *weaker* than Definition 2.6, which is equivalent to perfect secrecy. Thus, any perfectly secret encryption scheme is also EAV-secure. Our goal, therefore, is to show that there exist encryption schemes satisfying Definition 3.8 that can circumvent the limitations of perfect secrecy, and in particular for which the key is shorter than the message. (Note that this must be the case if the scheme can handle arbitrary length messages.) That is, we will show schemes that satisfy Definition 3.8 but cannot satisfy Definition 2.6.

An equivalent formulation. Definition 3.8 requires that no PPT adversary can determine which of two messages was encrypted with probability significantly better than $1/2$. An equivalent formulation is that every PPT adversary *behaves the same* whether it observes an encryption of m_0 or of m_1. Since $\mathcal{A}$ outputs a bit, "behaving the same" means it outputs 1 with almost the same probability in each case. To formalize this, define $\mathsf{PrivK}^{\mathsf{eav}}_{\mathcal{A},\Pi}(n, b)$ as above except that the fixed bit $b \in \{0,1\}$ is used (rather than being chosen at random). Let $\mathsf{out}_{\mathcal{A}}(\mathsf{PrivK}^{\mathsf{eav}}_{\mathcal{A},\Pi}(n, b))$ denote the output bit b' of $\mathcal{A}$ in this experiment. The following states that the output distribution of $\mathcal{A}$ is not significantly affected by whether it is running in experiment $\mathsf{PrivK}^{\mathsf{eav}}_{\mathcal{A},\Pi}(n, 0)$ or experiment $\mathsf{PrivK}^{\mathsf{eav}}_{\mathcal{A},\Pi}(n, 1)$.

DEFINITION 3.9 *A private-key encryption scheme* $\Pi = (\mathsf{Gen}, \mathsf{Enc}, \mathsf{Dec})$ *has* indistinguishable encryptions in the presence of an eavesdropper *if for all* PPT *adversaries* $\mathcal{A}$ *there is a negligible function* negl *such that*

$$\left| \Pr[\mathsf{out}_{\mathcal{A}}(\mathsf{PrivK}^{\mathsf{eav}}_{\mathcal{A},\Pi}(n, 0)) = 1] - \Pr[\mathsf{out}_{\mathcal{A}}(\mathsf{PrivK}^{\mathsf{eav}}_{\mathcal{A},\Pi}(n, 1)) = 1] \right| \leq \mathsf{negl}(n).$$

The fact that this is equivalent to Definition 3.8 is left as an exercise.

On Revealing the Plaintext Length

The default notion of secure encryption does not require the encryption scheme to hide the plaintext length and, in fact, all commonly used encryption schemes reveal the plaintext length (or a close approximation thereof). The main reason for this is that it is *impossible* to support arbitrary length messages while hiding all information about the plaintext length (cf. Exercise 3.2). In many cases this is inconsequential since the plaintext length is already public or is not sensitive. This is not always the case, however, and sometimes leaking the plaintext length is problematic. As examples:

- *Simple numeric/text data:* Say the encryption scheme being used reveals the plaintext length exactly. Then encrypted salary information would reveal whether someone makes a 5-figure or a 6-figure salary. Similarly, encryption of "yes"/"no" responses would leak the answer exactly.

- *Auto-suggestions:* Websites often include an "auto-complete" or "auto-suggestion" functionality by which the website suggests a list of potential words or phrases based on partial information the user has already typed. The *size* of this list can reveal information about the letters the user has typed so far. (For example, the number of auto-completions returned for "th" is far greater than the number for "zo.")

- *Database searches:* Consider a user querying a database for all records matching some search term. The *number of records returned* can reveal a lot of information about what the user was searching for. This can be particularly damaging if the user is searching for medical information and the query reveals information about a disease the user has.

- *Compressed data:* If the plaintext is compressed before being encrypted, then information about the plaintext might be revealed even if only fixed-length plaintext is being encrypted. For example, a short compressed plaintext would indicate that the original (uncompressed) plaintext has a lot of redundancy. If an adversary can control a portion of what gets encrypted, this vulnerability can enable an adversary to learn additional information about the rest of the plaintext; it has been shown possible to use an attack of exactly this sort (called the *CRIME attack*) to reveal secret session cookies from encrypted HTTPS traffic.

When using encryption one should determine whether leaking the plaintext length is a concern and, if so, take steps to mitigate or prevent such leakage by padding all messages to some pre-determined length before encrypting them.

3.2.2   *Semantic Security

We motivated the definition of secure encryption by saying that it implies the inability of an adversary to learn any partial information about the plaintext from the ciphertext. At first glance, however, Definition 3.8 looks very different. As we have mentioned, though, that definition is equivalent to a definition called *semantic security* that formalizes exactly the notion we want.

We build up to that definition by first introducing two weaker notions and showing that they are implied by EAV-security.

We begin by showing that EAV-security implies that ciphertexts leak no information about individual bits of the plaintext. Formally, we show that if an EAV-secure encryption scheme (Enc, Dec) (recall that if Gen is omitted, the key is a uniform n-bit string) is used to encrypt a uniform message $m \in \{0,1\}^{\ell}$, then for any i it is infeasible for an attacker given the ciphertext to guess the ith bit of m (here denoted by m^i) with probability much better than $1/2$.

THEOREM 3.10 *Let $\Pi = (\mathsf{Enc}, \mathsf{Dec})$ be a fixed-length private-key encryption scheme for messages of length ℓ that is EAV-secure. Then for all PPT adversaries $\mathcal{A}$ and $i \in \{1, \ldots, \ell\}$, there is a negligible function negl such that*

$$\Pr\left[\mathcal{A}(1^n, \mathsf{Enc}_k(m)) = m^i\right] \leq \frac{1}{2} + \mathsf{negl}(n), \tag{3.2}$$

where the probability is taken over uniform $m \in \{0,1\}^{\ell}$ and $k \in \{0,1\}^n$, the randomness of $\mathcal{A}$, and the randomness of Enc.

PROOF The idea behind the proof of this theorem is that if it were possible to determine the ith bit of m from $\mathsf{Enc}_k(m)$, then it would also be possible to distinguish between encryptions of messages m_0 and m_1 whose ith bits differ. We formalize this via a *proof by reduction*, in which we show how to use any efficient adversary $\mathcal{A}$ to construct an efficient adversary $\mathcal{A}'$ such that if $\mathcal{A}$ violates Equation (3.2), then $\mathcal{A}'$ violates EAV-security of Π. (See Section 3.3.2 for more discussion of proofs by reduction.) Since Π is EAV-secure, this implies that no such $\mathcal{A}$ can exist.

Fix an arbitrary PPT adversary $\mathcal{A}$ and $i \in \{1, \ldots, \ell\}$. Let $I_0 \subset \{0,1\}^{\ell}$ be the set of all strings whose ith bit is 0, and let $I_1 \subset \{0,1\}^{\ell}$ be the set of all strings whose ith bit is 1. We have

$$\Pr\left[\mathcal{A}(1^n, \mathsf{Enc}_k(m)) = m^i\right]$$
$$= \frac{1}{2} \cdot \Pr_{m_0 \leftarrow I_0}\left[\mathcal{A}(1^n, \mathsf{Enc}_k(m_0)) = 0\right] + \frac{1}{2} \cdot \Pr_{m_1 \leftarrow I_1}\left[\mathcal{A}(1^n, \mathsf{Enc}_k(m_1)) = 1\right].$$

Construct the following eavesdropping adversary $\mathcal{A}'$:

Adversary $\mathcal{A}'(1^n)$:

1. Choose uniform $m_0 \in I_0$ and $m_1 \in I_1$. Output m_0, m_1.

2. Upon observing a ciphertext c, invoke $\mathcal{A}(1^n, c)$. If $\mathcal{A}$ outputs 0, output $b' = 0$; otherwise, output $b' = 1$.

Note that $\mathcal{A}'$ runs in polynomial time since $\mathcal{A}$ does.

By the definition of experiment $\mathsf{PrivK}^{\mathsf{eav}}_{\mathcal{A}',\Pi}(n)$, we have that $\mathcal{A}'$ succeeds if and only if $\mathcal{A}$ outputs b upon receiving $\mathsf{Enc}_k(m_b)$. So

$$\Pr\left[\mathsf{PrivK}^{\mathsf{eav}}_{\mathcal{A}',\Pi}(n) = 1\right]$$
$$= \Pr\left[\mathcal{A}(1^n, \mathsf{Enc}_k(m_b)) = b\right]$$
$$= \frac{1}{2} \cdot \Pr_{m_0 \leftarrow I_0}\left[\mathcal{A}(1^n, \mathsf{Enc}_k(m_0)) = 0\right] + \frac{1}{2} \cdot \Pr_{m_1 \leftarrow I_1}\left[\mathcal{A}(1^n, \mathsf{Enc}_k(m_1)) = 1\right]$$
$$= \Pr\left[\mathcal{A}(1^n, \mathsf{Enc}_k(m)) = m^i\right].$$

Since $(\mathsf{Enc}, \mathsf{Dec})$ is EAV-secure, there is a negligible function negl such that $\Pr\left[\mathsf{PrivK}^{\mathsf{eav}}_{\mathcal{A}',\Pi}(n) = 1\right] \leq \frac{1}{2} + \mathsf{negl}(n)$. We conclude that

$$\Pr\left[\mathcal{A}(1^n, \mathsf{Enc}_k(m)) = m^i\right] \leq \frac{1}{2} + \mathsf{negl}(n),$$

completing the proof. ∎

We next argue that EAV-security implies that no PPT adversary can learn *any* function f of the plaintext m from the ciphertext, regardless of the distribution $\mathcal{D}$ of m. This requirement is not trivial to define formally, because it needs to distinguish information that the attacker knows about the message due to $\mathcal{D}$ from information the attacker learns about the message from the ciphertext. (For example, if $\mathcal{D}$ is only over messages for which f evaluates to 1, then it is easy for an attacker to determine $f(m)$. But in this case the attacker is not *learning* $f(m)$ from the ciphertext.) This is taken into account in the definition by requiring that if there exists an adversary who, with some probability, correctly computes $f(m)$ when given $\mathsf{Enc}_k(m)$, then there exists an adversary that can correctly compute $f(m)$ with almost the same probability *without being given the ciphertext at all* (and knowing only the distribution $\mathcal{D}$ of m).

THEOREM 3.11 *Let $(\mathsf{Enc}, \mathsf{Dec})$ be a fixed-length private-key encryption scheme for messages of length ℓ that is EAV-secure. Then for any PPT algorithm $\mathcal{A}$ there is a PPT algorithm $\mathcal{A}'$ such that for any distribution $\mathcal{D}$ over $\{0,1\}^\ell$ and any function $f : \{0,1\}^\ell \rightarrow \{0,1\}$, there is a negligible function negl such that:*

$$\left| \Pr\left[\mathcal{A}(1^n, \mathsf{Enc}_k(m)) = f(m)\right] - \Pr\left[\mathcal{A}'(1^n) = f(m)\right] \right| \leq \mathsf{negl}(n),$$

where the first probability is taken over choice of m according to $\mathcal{D}$, uniform choice of $k \in \{0,1\}^n$, and the randomness of $\mathcal{A}$ and Enc, and the second probability is taken over choice of m according to $\mathcal{D}$ and the randomness of $\mathcal{A}'$.

PROOF (Sketch) The fact that $(\mathsf{Enc}, \mathsf{Dec})$ is EAV-secure means that, for any $\mathcal{D}$, no PPT adversary can distinguish between $\mathsf{Enc}_k(m)$ for m chosen according to $\mathcal{D}$, and $\mathsf{Enc}_k(1^\ell)$ (i.e., an encryption of the all-1 string). (We leave a proof of this claim to the reader.) Consider now the probability that $\mathcal{A}$ successfully computes $f(m)$ given $\mathsf{Enc}_k(m)$. We claim that $\mathcal{A}$ should successfully compute $f(m)$ given $\mathsf{Enc}_k(1^\ell)$ with almost the same probability; otherwise, $\mathcal{A}$ could be used to distinguish between $\mathsf{Enc}_k(m)$ and $\mathsf{Enc}_k(1^\ell)$. The distinguisher is easily constructed: choose m according to $\mathcal{D}$, and output $m_0 = m$, $m_1 = 1^\ell$. When given a ciphertext c that is an encryption of either m_0 or m_1, invoke $\mathcal{A}(1^n, c)$ and output 0 if and only if $\mathcal{A}$ outputs $f(m)$. If $\mathcal{A}$ outputs $f(m)$ when given an encryption of m with probability that is significantly different from the probability that it outputs $f(m)$ when given an encryption of 1^ℓ, then the described distinguisher violates Definition 3.9.

The above suggests the following algorithm $\mathcal{A}'$ that does not receive an encryption of m, yet computes $f(m)$ almost as well as $\mathcal{A}$ does: $\mathcal{A}'(1^n)$ chooses a uniform key $k \in \{0,1\}^n$, invokes $\mathcal{A}$ on $c \leftarrow \mathsf{Enc}_k(1^\ell)$, and outputs whatever $\mathcal{A}$ does. By the above, we have that $\mathcal{A}$ outputs $f(m)$ when run as a subroutine by $\mathcal{A}'$ with almost the same probability as when it receives $\mathsf{Enc}_k(m)$. Thus, $\mathcal{A}'$ fulfills the property required by the theorem. ∎

Semantic security. The full definition of semantic security guarantees considerably more than what is considered in Theorem 3.11. The definition allows arbitrary (efficiently sampleable) distributions over messages, generated by some polynomial-time sampling algorithm Samp. The definition also takes into account arbitrary "external" information $h(m)$ about the message m that may be available to the adversary via other means (e.g., because the message is used for some other purpose as well). It also allows messages of varying lengths, although—as discussed at the end of the previous section—it assumes the message length is revealed.

DEFINITION 3.12 *A private-key encryption scheme* (Enc, Dec) *is semantically secure in the presence of an eavesdropper if for every* PPT *algorithm* $\mathcal{A}$ *there exists a* PPT *algorithm* $\mathcal{A}'$ *such that for any* PPT *algorithm* Samp *and polynomial-time computable functions* f *and* h, *the following is negligible:*

$$\left| \Pr[\mathcal{A}(1^n, \mathsf{Enc}_k(m), h(m)) = f(m)] - \Pr[\mathcal{A}'(1^n, |m|, h(m)) = f(m)] \right|,$$

where the first probability is taken over m *output by* Samp(1^n), *uniform choice of* $k \in \{0,1\}^n$, *and the randomness of* Enc *and* $\mathcal{A}$, *and the second probability is taken over* m *output by* Samp(1^n) *and the randomness of* $\mathcal{A}'$.

The adversary $\mathcal{A}$ is given the ciphertext $\mathsf{Enc}_k(m)$ as well as the external information $h(m)$, and attempts to guess the value of $f(m)$. Algorithm $\mathcal{A}'$ also attempts to guess the value of $f(m)$, but is given *only* the length of m and $h(m)$. The security requirement states that $\mathcal{A}$'s probability of correctly guessing $f(m)$ is about the same as that of $\mathcal{A}'$. Intuitively, then, this means that the ciphertext $\mathsf{Enc}_k(m)$ does not reveal any information about $f(m)$ except for $|m|$.

Definition 3.12 is a very strong and convincing formulation of the security guarantees that should be provided by an encryption scheme. Definition 3.8 is much easier to work with. Fortunately, the definitions are *equivalent*:

THEOREM 3.13   *A private-key encryption scheme has indistinguishable encryptions in the presence of an eavesdropper (i.e., is EAV-secure) if and only if it is semantically secure in the presence of an eavesdropper.*

Looking ahead, a similar equivalence to a "semantic security"-based definition is known for all the definitions we present in this chapter and Chapter 5. We can therefore use a simpler notion as our working definition, while being assured that it implies the strong guarantees of semantic security.

3.3   Constructing an EAV-Secure Encryption Scheme

Having defined what it means for an encryption scheme to be secure, the reader may expect us to turn immediately to constructions of secure encryption schemes. Before doing so, however, we need to introduce the notion of *pseudorandom generators* (PRGs), which are important building blocks for private-key encryption. This, in turn, will lead to a discussion of *pseudorandomness*, which plays a fundamental role in cryptography in general and private-key encryption in particular.

3.3.1   Pseudorandom Generators

A pseudorandom generator G is an efficient, deterministic algorithm for transforming a short, uniform string (called a *seed*) into a longer, "uniform-looking" (or "pseudorandom") output string. Stated differently, a pseudorandom generator uses a small amount of true randomness in order to generate a large amount of pseudorandomness. This is useful whenever a large number of random(-looking) bits are needed, since generating true random bits is often difficult and slow. (See the discussion at the beginning of Chapter 2.) Pseudorandom generators have been studied since at least the 1940s when they were used for running statistical simulations. In that context, researchers proposed various statistical tests that a pseudorandom generator should pass in order to be considered "good." As a simple example, one could require that the first bit of the output of a pseudorandom generator should be equal to 1 with probability very close to $1/2$ (where the probability is taken over uniform choice of the seed), since the first bit of a uniform string is equal to 1 with probability exactly 1/2. As another example, the parity of any fixed subset of the output bits should also be 1 with probability very close to 1/2. More complex statistical tests can also be considered.

This historical approach to determining the quality of some candidate pseudorandom generator is unsatisfying, as it is not clear when passing some set of statistical tests is sufficient to guarantee the soundness of using a candidate pseudorandom generator for some application. (In particular, there may be another statistical test that *does* successfully distinguish the output of the generator from true random bits.) The historical approach is even more problematic when using pseudorandom generators for cryptographic applications; there, security may be compromised if any attacker is able to distinguish the output of a pseudorandom generator from uniform, and we do not know in advance what strategy an attacker might use.

The above considerations motivated a cryptographic approach to defining pseudorandom generators in the 1980s. The basic realization was that a good pseudorandom generator should pass *all* (efficient) statistical tests. That is, for *any* efficient statistical test (or *distinguisher*) D, the probability that D returns 1 when given the output of the pseudorandom generator should be close to the probability that D returns 1 when given a uniform string of the same length. Informally, then, this means the output of a pseudorandom generator "looks like" a uniformly generated string to *any* efficient observer.

We begin by defining what it means for a distribution to be pseudorandom. Let Dist be a distribution on ℓ-bit strings. (This means that Dist assigns some probability to every string in $\{0,1\}^\ell$; sampling from Dist means that we choose an ℓ-bit string according to this probability distribution.) Informally, Dist is *pseudorandom* if the experiment in which a string is sampled from Dist is indistinguishable from the experiment in which a uniform string of length ℓ is sampled. (Strictly speaking, since we are in an asymptotic setting we need to speak of the pseudorandomness of a *sequence* of distributions Dist $= \{\text{Dist}_n\}$, where distribution Dist_n is used for security parameter n. We ignore this point in our current discussion.) More precisely, it should be infeasible for any polynomial-time algorithm to determine (better than guessing) whether it is given a string sampled according to Dist, or whether it is given a uniform ℓ-bit string. This means that *a pseudorandom string is just as good as a uniform string*, as long as we consider only polynomial-time observers. We stress that it does not make sense to say that any fixed string is "pseudorandom," in the same way that it is meaningless to refer to any fixed string as "uniform." Rather, pseudorandomness is a property of a *distribution* on strings. (Nevertheless, we sometimes informally call a string output by a pseudorandom generator a "pseudorandom string" in the same way we might say that a string sampled according to the uniform distribution is a "uniform string.") Just as indistinguishability is a computational relaxation of perfect secrecy, pseudorandomness is a computational relaxation of true randomness.

Let G be an efficiently computable function that maps strings of length n to outputs of length $\ell(n) > n$, and define Dist_n to be the distribution on $\ell(n)$-

bit strings obtained by choosing a uniform $s \in \{0,1\}^n$ and outputting $G(s)$. Then G is a pseudorandom generator if and only if the distribution Dist_n (technically, the sequence of distributions $\{\mathsf{Dist}_n\}$) is pseudorandom.

The formal definition. As discussed above, G is a pseudorandom generator if no efficient distinguisher can detect whether it is given a string output by G or a string chosen uniformly at random. As in Definition 3.9, this is formalized by requiring that every efficient algorithm outputs 1 with almost the same probability when given $G(s)$ (for uniform seed s) or a uniform string. (For an equivalent definition analogous to Definition 3.8, see Exercise 3.7.) We obtain a definition in the asymptotic setting by letting the security parameter n determine the length of the seed, and insisting that G be computable by an efficient algorithm. As a technicality, we also require that G's output be longer than its input; otherwise, G is not very useful or interesting.

DEFINITION 3.14 *Let G be a deterministic polynomial-time algorithm such that for any n and any input $s \in \{0,1\}^n$, the result $G(s)$ is a string of length $\ell(n)$. G is a pseudorandom generator if the following conditions hold:*

1. **(Expansion.)** *For every n it holds that $\ell(n) > n$.*

2. **(Pseudorandomness.)** *For any PPT algorithm D, there is a negligible function* negl *such that*

$$\big| \Pr[D(G(s)) = 1] - \Pr[D(r) = 1] \big| \leq \mathsf{negl}(n),$$

   where the first probability is taken over uniform choice of $s \in \{0,1\}^n$ and the randomness of D, and the second probability is taken over uniform choice of $r \in \{0,1\}^{\ell(n)}$ and the randomness of D.

We call $\ell(n)$ the expansion factor *of G.*

We give an example of an insecure pseudorandom generator to gain familiarity with the definition.

Example 3.15
Define $G(s)$ to output s followed by $\oplus_{i=1}^n s_i$ (i.e., the XOR of all the bits of s), so the expansion factor of G is $\ell(n) = n + 1$. The output of G can be distinguished easily from uniform. Consider the following efficient distinguisher D: on input a string w, output 1 if and only if the final bit of w is equal to the XOR of all the preceding bits of w. Since this holds for all strings output by G, we have $\Pr[D(G(s)) = 1] = 1$. On the other hand, if r is uniform, the final bit of r is uniform and so $\Pr[D(r) = 1] = \frac{1}{2}$. The quantity $|\frac{1}{2} - 1|$ is constant, not negligible, and so G is not a pseudorandom generator. (Note that D is not always "correct," since it sometimes outputs 1 even when given a uniform string. But D is still a good distinguisher.)   ◇

Discussion. The distribution of the output of a pseudorandom generator G is far from uniform. To see this, consider the case that $\ell(n) = 2n$ and so G doubles the length of its input. Under the uniform distribution on $\{0,1\}^{2n}$, each of the 2^{2n} possible strings is chosen with probability exactly 2^{-2n}. In contrast, consider the distribution of the output of G when it is run on a uniform n-bit seed. The number of different strings in the range of G is at most 2^n. The fraction of strings of length $2n$ that are in the range of G is thus at most $2^n/2^{2n} = 2^{-n}$, and we see that the vast majority of strings of length $2n$ have probability 0 of being output by G.

This in particular means that it is trivial to distinguish between a random string and a pseudorandom string *given an unlimited amount of time*. Let G be as above and consider the exponential-time distinguisher D that works as follows: $D(w)$ outputs 1 if and only if there exists an $s \in \{0,1\}^n$ such that $G(s) = w$. (This computation is carried out in exponential time by exhaustively computing $G(s)$ for every $s \in \{0,1\}^n$. Recall that by Kerckhoffs' principle, the specification of G is known to D.) Now, if w were output by G, then D outputs 1 with probability 1. In contrast, if w is uniformly distributed in $\{0,1\}^{2n}$, then the probability that there exists an s with $G(s) = w$ is at most 2^{-n}, and so D outputs 1 in this case with probability at most 2^{-n}. So

$$\big| \Pr[D(r) = 1] - \Pr[D(G(s)) = 1] \big| \geq 1 - 2^{-n},$$

which is large. This is just another example of a *brute-force attack*, and does not contradict the pseudorandomness of G since the attack is not efficient.

The seed and its length. The seed for a pseudorandom generator is analogous to the key used by an encryption scheme, and—just as in the case of a cryptographic key—the seed s must be chosen uniformly and be kept secret from any adversary if we want $G(s)$ to look random. Another important point, evident from the above discussion of brute-force attacks, is that s must be long enough so that it is not feasible to enumerate all possible seeds. In an asymptotic sense this is taken care of by setting the length of the seed equal to the security parameter, so exhaustive search over all possible seeds requires exponential time. In practice, the seed length n must at least be large enough so that a brute-force attack running in time 2^n is infeasible.

On the existence of pseudorandom generators. Do pseudorandom generators exist? They certainly seem difficult to construct, and one may rightly ask whether any algorithm G satisfies Definition 3.14. Although we do not know how to unconditionally prove the existence of pseudorandom generators, we have strong reasons to believe they exist for any (polynomial) expansion factor. For one, they can be constructed under the rather weak assumption that *one-way functions* exist (which is true if certain problems like factoring are hard); this is discussed in detail in Chapter 8. We also have several practical constructions of candidate pseudorandom generators called *stream ciphers* for which no efficient distinguishers are known; see Section 3.6.1 for details and Section 7.1 for concrete examples. In this chapter, we simply assume pseudorandom generators exist for any polynomial expansion factor, and explore how they can be used to build secure encryption schemes. Doing so in a sound way relies on the idea of *proofs by reduction*, which we describe next.

3.3.2 Proofs by Reduction

If we wish to prove that a given construction (e.g., encryption scheme) is computationally secure, then—unless the scheme is information-theoretically secure—we must rely on unproven assumptions. Our strategy will be to assume that some mathematical problem is hard, or that some *low-level* cryptographic primitive is secure, and then to *prove* that the given construction based on that problem/primitive is secure as long as our initial assumption is correct. In Section 1.4.2 we have already explained in great detail the advantages of this approach, so we do not repeat those arguments here.

A proof that some cryptographic construction Π is secure as long as some underlying problem X is hard generally proceeds by presenting an explicit *reduction* showing how to transform any efficient adversary $\mathcal{A}$ that succeeds in "breaking" Π into an efficient algorithm $\mathcal{A}'$ that solves X. Since this is so important, we walk through a high-level outline of the steps of such a proof in detail. (We will see numerous concrete examples throughout the book, beginning with the proof of Theorem 3.16 in the next section.) We start with the assumption that some problem X cannot be solved (in some precisely defined sense) by any polynomial-time algorithm, except with negligible probability. We then want to prove that some cryptographic construction Π is secure (again, in some sense that is precisely defined). A proof by reduction proceeds via the following steps (see also Figure 3.1):

1. Fix some efficient (i.e., probabilistic polynomial-time) adversary $\mathcal{A}$ attacking Π. Denote this adversary's success probability by $\varepsilon(n)$.

2. Construct an efficient algorithm $\mathcal{A}'$ that attempts to solve problem X by using adversary $\mathcal{A}$ as a subroutine. An important point here is that $\mathcal{A}'$ knows nothing about how $\mathcal{A}$ works; the only thing $\mathcal{A}'$ knows is that $\mathcal{A}$ is expecting to attack Π. So, given some input instance x of problem X, our algorithm $\mathcal{A}'$ will *simulate* for $\mathcal{A}$ an instance of Π such that:

   (a) As far as $\mathcal{A}$ can tell, it is interacting with Π. That is, the view of $\mathcal{A}$ when run as a subroutine by $\mathcal{A}'$ should be distributed identically to (or at least close to) the view of $\mathcal{A}$ when it interacts with Π itself.

   (b) When $\mathcal{A}$ succeeds in "breaking" the instance of Π that is being simulated by $\mathcal{A}'$, this should allow $\mathcal{A}'$ to solve the instance x it was given, at least with inverse polynomial probability $1/p(n)$.

   I.e., we attempt to *reduce* the problem of solving X to the problem of breaking Π.

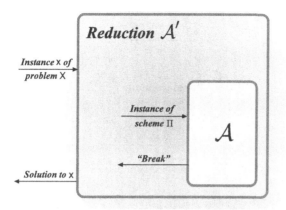

FIGURE 3.1: A high-level overview of a proof by reduction.

3. Taken together, the above imply that $\mathcal{A}'$ solves X with probability $\varepsilon(n)/p(n)$. Now, if $\varepsilon(n)$ is not negligible then neither is $\varepsilon(n)/p(n)$. Moreover, if $\mathcal{A}$ is efficient then we obtain an efficient algorithm $\mathcal{A}'$ solving X with non-negligible probability, contradicting our initial assumption.

4. Given our assumption regarding X, we conclude that *no* efficient adversary $\mathcal{A}$ can succeed in breaking Π with non-negligible probability. Stated differently, Π is computationally secure.

As an illustration of the above idea, we show in the following section how to use a pseudorandom generator G to construct an encryption scheme, and we prove the encryption scheme secure by showing that any attacker who can "break" the encryption scheme can be used to distinguish the output of G from a uniform string. Under the assumption that G is a pseudorandom generator, then, the encryption scheme is secure.

3.3.3 EAV-Security from a Pseudorandom Generator

A pseudorandom generator provides a natural way to construct a secure, fixed-length encryption scheme with a key shorter than the message. Recall that in the one-time pad (see Section 2.2), encryption is done by XORing a *random* pad with the message. The crucial insight is that we can use a *pseudorandom* pad instead. Rather than sharing this long, pseudorandom pad, however, the sender and receiver can instead share a uniform *seed* that is used to generate the pad when needed (see Figure 3.2); this seed will be shorter than the pad and hence shorter than the message. As for security, the intuition is that a pseudorandom string "looks random" to any polynomial-time adversary and so a computationally bounded eavesdropper cannot distinguish between a message encrypted using the one-time pad or a message encrypted using this "pseudo-"one-time pad encryption scheme.

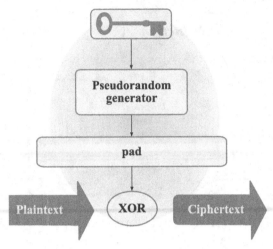

FIGURE 3.2: Encryption with a pseudorandom generator.

The encryption scheme. Fix some message length $\ell(n)$ and let G be a pseudorandom generator with expansion factor $\ell(n)$ (that is, $|G(s)| = \ell(|s|)$). Recall that an encryption scheme is defined by three algorithms: a key-generation algorithm Gen, an encryption algorithm Enc, and a decryption algorithm Dec. The key-generation algorithm is the trivial one: $\mathsf{Gen}(1^n)$ simply outputs a uniform key $k \in \{0,1\}^n$. Encryption works by applying G to the key (which serves as a seed) in order to obtain a pad that is then XORed with the plaintext. Decryption applies G to the key and XORs the resulting pad with the ciphertext to recover the message. The scheme is described formally in Construction 3.17. In Section 3.6.2, we describe how stream ciphers are used to implement a variant of this scheme in practice.

THEOREM 3.16 *If G is a pseudorandom generator, then Construction 3.17 is an EAV-secure, fixed-length private-key encryption scheme for messages of length $\ell(n)$.*

PROOF   Let Π denote Construction 3.17. We show that Π satisfies Definition 3.8 (under the assumption that G is a pseudorandom generator). Namely, we show that for any probabilistic polynomial-time adversary $\mathcal{A}$ there is a negligible function negl such that

$$\Pr\left[\mathsf{PrivK}^{\mathsf{eav}}_{\mathcal{A},\Pi}(n) = 1\right] \leq \frac{1}{2} + \mathsf{negl}(n). \tag{3.3}$$

The intuition is that if Π used a uniform pad in place of the pseudorandom pad $G(k)$, then the resulting scheme would be identical to the one-time pad encryption scheme and $\mathcal{A}$ would be unable to correctly guess which message was encrypted with probability any better than $1/2$. Thus, if Equation (3.3)

CONSTRUCTION 3.17

Let G be a pseudorandom generator with expansion factor $\ell(n)$. Define a fixed-length private-key encryption scheme for messages of length $\ell(n)$ as follows:

- **Gen**: on input 1^n, choose uniform $k \in \{0,1\}^n$ and output it as the key.

- **Enc**: on input a key $k \in \{0,1\}^n$ and a message $m \in \{0,1\}^{\ell(n)}$, output the ciphertext
$$c := G(k) \oplus m.$$

- **Dec**: on input a key $k \in \{0,1\}^n$ and a ciphertext $c \in \{0,1\}^{\ell(n)}$, output the message
$$m := G(k) \oplus c.$$

A private-key encryption scheme based on any pseudorandom generator.

does *not* hold then $\mathcal{A}$ must implicitly be distinguishing the output of G from a random string. We make this explicit by showing a *reduction*; namely, by showing how to use $\mathcal{A}$ to construct an efficient distinguisher D, with the property that D's ability to distinguish the output of G from a uniform string is directly related to $\mathcal{A}$'s ability to determine which message was encrypted by Π. Security of G then implies security of Π.

Let $\mathcal{A}$ be an arbitrary PPT adversary. We construct a distinguisher D that takes a string w as input, and whose goal is to determine whether w was chosen uniformly (i.e., w is a "random string") or whether w was generated by choosing a uniform k and computing $w := G(k)$ (i.e., w is a "pseudorandom string"). We construct D so that it emulates the eavesdropping experiment for $\mathcal{A}$, as described below, and observes whether $\mathcal{A}$ succeeds or not. If $\mathcal{A}$ succeeds then D guesses that w must be a pseudorandom string, while if $\mathcal{A}$ does not succeed then D guesses that w is a random string. In detail:

Distinguisher D:
D is given as input a string $w \in \{0,1\}^{\ell(n)}$. (We assume that n can be determined from $\ell(n)$.)

1. Run $\mathcal{A}(1^n)$ to obtain a pair of messages $m_0, m_1 \in \{0,1\}^{\ell(n)}$.

2. Choose a uniform bit $b \in \{0,1\}$. Set $c := w \oplus m_b$.

3. Give c to $\mathcal{A}$ and obtain output b'. Output 1 if $b' = b$, and output 0 otherwise.

D clearly runs in polynomial time (assuming $\mathcal{A}$ does).

Before analyzing the behavior of D, we define a modified encryption scheme $\widetilde{\Pi} = (\widetilde{\mathsf{Gen}}, \widetilde{\mathsf{Enc}}, \widetilde{\mathsf{Dec}})$ that is exactly the one-time pad encryption scheme, except that we now incorporate a security parameter that determines the length of the message to be encrypted. That is, $\widetilde{\mathsf{Gen}}(1^n)$ outputs a uniform key k of length $\ell(n)$, and the encryption of message $m \in \{0,1\}^{\ell(n)}$ using key

$k \in \{0,1\}^{\ell(n)}$ is the ciphertext $c := k \oplus m$. (Decryption can be done as usual, but is inessential to what follows.) Perfect secrecy of the one-time pad implies

$$\Pr\left[\mathsf{PrivK}^{\mathsf{eav}}_{\mathcal{A},\widetilde{\Pi}}(n) = 1\right] = \frac{1}{2}. \tag{3.4}$$

To analyze the behavior of D, the main observations are:

1. If w is chosen uniformly from $\{0,1\}^{\ell(n)}$, then the view of $\mathcal{A}$ when run as a subroutine by D is distributed identically to the view of $\mathcal{A}$ in experiment $\mathsf{PrivK}^{\mathsf{eav}}_{\mathcal{A},\widetilde{\Pi}}(n)$. This is because when $\mathcal{A}$ is run as a subroutine by $D(w)$ in this case, $\mathcal{A}$ is given a ciphertext $c = w \oplus m_b$ where $w \in \{0,1\}^{\ell(n)}$ is uniform. Since D outputs 1 exactly when $\mathcal{A}$ succeeds in its eavesdropping experiment, we therefore have (using Equation (3.4))

$$\Pr_{w \leftarrow \{0,1\}^{\ell(n)}}[D(w) = 1] = \Pr\left[\mathsf{PrivK}^{\mathsf{eav}}_{\mathcal{A},\widetilde{\Pi}}(n) = 1\right] = \frac{1}{2}. \tag{3.5}$$

   (The subscript on the first probability just makes explicit that w is chosen uniformly from $\{0,1\}^{\ell(n)}$ there.)

2. If w is instead generated by choosing uniform $k \in \{0,1\}^n$ and then setting $w := G(k)$, the view of $\mathcal{A}$ when run as a subroutine by D is distributed identically to the view of $\mathcal{A}$ in experiment $\mathsf{PrivK}^{\mathsf{eav}}_{\mathcal{A},\Pi}(n)$. This is because $\mathcal{A}$, when run as a subroutine by D, is now given a ciphertext $c = w \oplus m_b$ where $w = G(k)$ for a uniform $k \in \{0,1\}^n$. Thus,

$$\Pr_{k \leftarrow \{0,1\}^n}[D(G(k)) = 1] = \Pr\left[\mathsf{PrivK}^{\mathsf{eav}}_{\mathcal{A},\Pi}(n) = 1\right]. \tag{3.6}$$

Since G is a pseudorandom generator (and since D runs in polynomial time), we know there is a negligible function negl such that

$$\left|\Pr_{k \leftarrow \{0,1\}^n}[D(G(k)) = 1] - \Pr_{w \leftarrow \{0,1\}^{\ell(n)}}[D(w) = 1]\right| \leq \mathsf{negl}(n).$$

Using Equations (3.5) and (3.6), we thus see that

$$\left|\Pr\left[\mathsf{PrivK}^{\mathsf{eav}}_{\mathcal{A},\Pi}(n) = 1\right] - \frac{1}{2}\right| \leq \mathsf{negl}(n),$$

which implies $\Pr\left[\mathsf{PrivK}^{\mathsf{eav}}_{\mathcal{A},\Pi}(n) = 1\right] \leq \frac{1}{2} + \mathsf{negl}(n)$. Since $\mathcal{A}$ was an arbitrary PPT adversary, this completes the proof that Π is EAV-secure. ∎

It is easy to get lost in the details of the proof and wonder whether anything has been gained as compared to the one-time pad; after all, the one-time pad also encrypts an ℓ-bit message by XORing it with an ℓ-bit string! The point of the construction, of course, is that the shared key k can be *much shorter* than the ℓ-bit string $G(k)$. In particular, using the above scheme it may be possible to securely encrypt a 1 Mb file using only a 128-bit key. By relying on computational secrecy we have thus circumvented the impossibility result of Theorem 2.11, which states that any *perfectly* secret encryption scheme must use a key at least as long as the message.

Reductions—a discussion. We do *not* prove unconditionally that Construction 3.17 is secure. Rather, we prove that it is secure *under the assumption* that G is a pseudorandom generator. This approach of *reducing* the security of a higher-level construction to a lower-level primitive has a number of advantages (as discussed in Section 1.4.2). One of these advantages is that, in general, it is easier to design a lower-level primitive than a higher-level one; it is also easier, in general, to directly analyze an algorithm G with respect to a lower-level definition than to analyze a more complex scheme Π with respect to a higher-level definition. This does not mean that constructing a pseudorandom generator is "easy," only that it is easier than constructing an encryption scheme from scratch. (In the present case the encryption scheme does nothing except XOR the output of a pseudorandom generator with the message and so this isn't quite true. Soon, however, we will see more complex constructions and in those cases the ability to reduce the task to a simpler one is very useful.) Another advantage is that the construction can be instantiated with any pseudorandom generator G, providing some flexibility to the users of the scheme.

Concrete security. Although Theorem 3.16 and its proof are in an asymptotic setting, we can readily adapt the proof to bound the *concrete* security of the encryption scheme in terms of the concrete security of G. Fix some value of n for the remainder of this discussion, and let Π now denote Construction 3.17 using this value of n. Assume G is (t, ε)-pseudorandom (for the given value of n), in the sense that for all distinguishers D running in time at most t we have

$$\left| \Pr[D(G(s)) = 1] - \Pr[D(r) = 1] \right| \leq \varepsilon. \tag{3.7}$$

(Think of $t \approx 2^{80}$ CPU cycles and $\varepsilon \approx 2^{-60}$, though precise values are irrelevant for our discussion.) We claim that Π is $(t - t', \varepsilon)$-secure for some (small) constant t', in the sense that for all $\mathcal{A}$ running in time at most $t - t'$ we have

$$\Pr\left[\mathsf{PrivK}^{\mathsf{eav}}_{\mathcal{A},\Pi} = 1 \right] \leq \frac{1}{2} + \varepsilon. \tag{3.8}$$

(Note that the above are now fixed numbers, not functions of n, since we have fixed n and are no longer in an asymptotic setting.) To see this, let $\mathcal{A}$ be an arbitrary adversary running in time at most $t - t'$. Distinguisher D, as constructed in the proof of Theorem 3.16, has very little overhead besides running $\mathcal{A}$; setting t' appropriately ensures that D runs in time at most t. Our assumption on the concrete security of G then implies Equation (3.7); proceeding exactly as in the proof of Theorem 3.16, we obtain Equation (3.8).

3.4   Stronger Security Notions

Until now we have considered a relatively weak definition of security in which the adversary only passively eavesdrops on a single ciphertext sent between the honest parties. Here we consider stronger security notions.

3.4.1   Security for Multiple Encryptions

Definition 3.8 deals with the case where the communicating parties transmit a single ciphertext that is observed by an eavesdropper. It would be convenient, however, if the communicating parties could securely send multiple ciphertexts to each other—all generated using the same key—even if an eavesdropper might observe all of them. For such applications we need an encryption scheme secure for the encryption of multiple messages.

We begin with an appropriate definition of security for this setting. As in the case of Definition 3.8, we first introduce an appropriate experiment defined for any encryption scheme Π, adversary $\mathcal{A}$, and security parameter n:

The multiple-message eavesdropping experiment $\mathsf{PrivK}^{\mathsf{mult}}_{\mathcal{A},\Pi}(n)$:

1. *The adversary $\mathcal{A}$ is given input 1^n, and outputs a pair of equal-length lists of messages $\vec{M}_0 = (m_{0,1}, \ldots, m_{0,t})$ and $\vec{M}_1 = (m_{1,1}, \ldots, m_{1,t})$, with $|m_{0,i}| = |m_{1,i}|$ for all i.*

2. *A key k is generated by running $\mathsf{Gen}(1^n)$, and a uniform bit $b \in \{0,1\}$ is chosen. For all i, the ciphertext $c_i \leftarrow \mathsf{Enc}_k(m_{b,i})$ is computed and the list $\vec{C} = (c_1, \ldots, c_t)$ is given to $\mathcal{A}$.*

3. *$\mathcal{A}$ outputs a bit b'.*

4. *The output of the experiment is defined to be 1 if $b' = b$, and 0 otherwise.*

The definition of security is the same as before, except that it now refers to the above experiment.

DEFINITION 3.18   *A private-key encryption scheme $\Pi = (\mathsf{Gen}, \mathsf{Enc}, \mathsf{Dec})$ has indistinguishable multiple encryptions in the presence of an eavesdropper if for all probabilistic polynomial-time adversaries $\mathcal{A}$ there is a negligible function negl such that*

$$\Pr\left[\mathsf{PrivK}^{\mathsf{mult}}_{\mathcal{A},\Pi}(n) = 1\right] \leq \frac{1}{2} + \mathsf{negl}(n).$$

Any scheme that has indistinguishable multiple encryptions in the presence of an eavesdropper clearly also satisfies Definition 3.8, since experiment $\mathsf{PrivK}^{\mathsf{eav}}_{\mathcal{A},\Pi}(n)$ corresponds to the special case of $\mathsf{PrivK}^{\mathsf{mult}}_{\mathcal{A},\Pi}(n)$ where the adversary outputs two lists containing only a single message each. In fact, our new definition is strictly stronger than Definition 3.8, as the following shows.

PROPOSITION 3.19 *There is a private-key encryption scheme that has indistinguishable encryptions in the presence of an eavesdropper, but not in-distinguishable* multiple *encryptions in the presence of an eavesdropper.*

PROOF We do not have to look far to find an example of an encryption scheme satisfying the proposition. The one-time pad is perfectly secret, and so also has indistinguishable encryptions in the presence of an eavesdropper. We show that it is not secure in the sense of Definition 3.18. (We have discussed this attack in Chapter 2 already; here, we merely analyze the attack with respect to Definition 3.18.)

Concretely, consider the following adversary $\mathcal{A}$ attacking the scheme Π (in the sense defined by experiment $\mathsf{PrivK}^{\mathsf{mult}}_{\mathcal{A},\Pi}(n)$): $\mathcal{A}$ outputs $\vec{M}_0 = (0^\ell, 0^\ell)$ and $\vec{M}_1 = (0^\ell, 1^\ell)$. (The first contains the same message twice, while the second contains two different messages.) Let $\vec{C} = (c_1, c_2)$ be the list of ciphertexts that $\mathcal{A}$ receives. If $c_1 = c_2$, then $\mathcal{A}$ outputs $b' = 0$; otherwise, $\mathcal{A}$ outputs $b' = 1$.

We now analyze the probability that $b' = b$. The crucial point is that the one-time pad is *deterministic*, so encrypting the same message twice (using the same key) yields the same ciphertext. Thus, if $b = 0$ then we must have $c_1 = c_2$ and $\mathcal{A}$ outputs 0 in this case. On the other hand, if $b = 1$ then a different message is encrypted each time; hence $c_1 \neq c_2$ and $\mathcal{A}$ outputs 1. We conclude that $\mathcal{A}$ correctly outputs $b' = b$ with probability 1, and so the encryption scheme is not secure with respect to Definition 3.18. ∎

Necessity of probabilistic encryption. The above might appear to show that Definition 3.18 is impossible to achieve using *any* encryption scheme. This is true as long as the encryption scheme is (stateless[2] and) *deterministic*, and so encrypting the same message multiple times using the same key always yields the same result. This is important enough to state as a theorem.

THEOREM 3.20 *If Π is a encryption scheme in which* Enc *is a deterministic function of the key and the message, then Π cannot have indistinguishable multiple encryptions in the presence of an eavesdropper.*

This should not be interpreted to mean that Definition 3.18 is too strong. Indeed, leaking to an eavesdropper the fact that two encrypted messages are the same can be a significant security breach. (Consider, e.g., a scenario in which someone encrypts a series of yes/no answers!)

---

[2]Theorem 3.20 refers *only* to encryption schemes satisfying the syntax of Definition 3.7. We will see in Section 3.6.2 that if an encryption scheme is stateful (something that is not covered by Definition 3.7), then it is possible to securely encrypt multiple messages even if Enc is deterministic. In Section 3.6.4, we consider a third syntax for encryption that also enables deterministic encryption to be used to securely encrypt multiple messages.

To construct a scheme secure for encrypting multiple messages, we must design a scheme in which encryption is *randomized,* so that when the same message is encrypted multiple times different ciphertexts can be produced. This may seem impossible since decryption must always be able to recover the message. However, we will soon see how to achieve it.

While achieving security for the encryption of multiple messages is important, we do not extensively consider Definition 3.18 itself but instead focus on the stronger definition that we introduce in the following section.

3.4.2 Chosen-Plaintext Attacks and CPA-Security

Chosen-plaintext attacks capture the ability of an adversary to exercise (partial) control over what the honest parties encrypt. Imagine a scenario in which two honest parties share a key k, and the attacker can influence those parties to encrypt messages $m_1, m_2, \ldots$ (using k) and send the resulting ciphertexts over a channel that the attacker can observe. At some later point in time, the attacker observes a ciphertext corresponding to some *unknown* message m encrypted using the same key k; let us even assume that the attacker knows that m is one of two possibilities m_0, m_1. Security against chosen-plaintext attacks means that even in this case the attacker cannot tell which of those two messages was encrypted with probability significantly better than random guessing. (For now we revert back to the case where the eavesdropper is given only a single encryption of an unknown message. Shortly, we will return to consideration of the multiple-message case.)

Chosen-plaintext attacks in the real world. Are chosen-plaintext attacks a realistic concern? For starters, note that chosen-plaintext attacks also encompass *known-plaintext attacks*—in which the attacker knows some of the messages being encrypted, even if it does not get to choose them—as a special case. Moreover, there are several real-world scenarios in which an adversary might have significant influence over what messages get encrypted. A simple example is given by an attacker typing on a terminal, which in turn encrypts everything the adversary types using a key (unknown to the attacker) shared with a remote server. Here the attacker exactly controls what gets encrypted, and the encryption scheme should still reveal nothing when it is used—with the same key—to encrypt data typed by another user.

Interestingly, chosen-plaintext attacks have also been used successfully as part of historical efforts to break military encryption schemes. For example, during World War II the British placed mines at certain locations, knowing that the Germans—when finding those mines—would encrypt the locations and send them back to headquarters. Those encrypted messages were used by cryptanalysts at Bletchley Park to break the German encryption scheme.

Another example is given by the famous story involving the Battle of Midway. In May 1942, US Navy cryptanalysts intercepted an encrypted message from the Japanese that they were able to partially decode. The result indicated that the Japanese were planning an attack on AF, where AF was a ciphertext fragment that the US was unable to decode. For other reasons, the US believed that Midway Island was the target. Unfortunately, their attempts to convince planners in Washington that this was the case were futile; the general belief was that Midway could not possibly be the target. The Navy cryptanalysts devised the following plan: They instructed US forces at Midway to send a fake message that their freshwater supplies were low. The Japanese intercepted this message and immediately sent an encrypted message to their superiors that "AF is low on water." The Navy cryptanalysts now had their proof that AF corresponded to Midway, and the US dispatched three aircraft carriers to that location. The result was that Midway was saved, and the Japanese incurred significant losses. This battle was a turning point in the war between the US and Japan in the Pacific.

The Navy cryptanalysts here carried out a chosen-plaintext attack, as they were able to influence the Japanese (albeit in a roundabout way) to encrypt the word "Midway." If the Japanese encryption scheme had been secure against chosen-plaintext attacks, this strategy by the US cryptanalysts would not have worked (and history may have turned out very differently)!

CPA-security. In the formal definition we model chosen-plaintext attacks by giving the adversary $\mathcal{A}$ access to an *encryption oracle*, viewed as a "black box" that encrypts messages of $\mathcal{A}$'s choice using a key k that is unknown to $\mathcal{A}$. That is, we imagine $\mathcal{A}$ has access to an "oracle" $\mathsf{Enc}_k(\cdot)$; when $\mathcal{A}$ *queries* this oracle by providing it with a message m as input, the oracle returns a ciphertext $c \leftarrow \mathsf{Enc}_k(m)$ as the reply. (If Enc is randomized, the oracle uses fresh randomness each time it answers a query.) The adversary can interact with the encryption oracle adaptively, as many times as it likes.

Consider the following experiment defined for any encryption scheme $\Pi =$ (Gen, Enc, Dec), adversary $\mathcal{A}$, and value n for the security parameter:

The CPA indistinguishability experiment $\mathsf{PrivK}^{\mathsf{cpa}}_{\mathcal{A},\Pi}(n)$:

1. *A key k is generated by running* $\mathsf{Gen}(1^n)$.

2. *The adversary $\mathcal{A}$ is given input 1^n and oracle access to $\mathsf{Enc}_k(\cdot)$, and outputs a pair of messages m_0, m_1 of the same length.*

3. *A uniform bit $b \in \{0, 1\}$ is chosen, and then a ciphertext $c \leftarrow \mathsf{Enc}_k(m_b)$ is computed and given to $\mathcal{A}$.*

4. *The adversary $\mathcal{A}$ continues to have oracle access to $\mathsf{Enc}_k(\cdot)$, and outputs a bit b'.*

5. *The output of the experiment is defined to be 1 if $b' = b$, and 0 otherwise. In the former case, we say that $\mathcal{A}$ succeeds.*

DEFINITION 3.21 *A private-key encryption scheme* $\Pi =$ (Gen, Enc, Dec) *has* indistinguishable encryptions under a chosen-plaintext attack, *or is* CPA-secure, *if for all probabilistic polynomial-time adversaries $\mathcal{A}$ there is a negligible function* negl *such that*

$$\Pr\left[\mathsf{PrivK}^{\mathsf{cpa}}_{\mathcal{A},\Pi}(n) = 1\right] \leq \frac{1}{2} + \mathsf{negl}(n),$$

where the probability is taken over the randomness used by $\mathcal{A}$, as well as the randomness used in the experiment.

CPA-security is nowadays the minimal notion of security an encryption scheme should satisfy, though it is becoming more common to require even stronger security notions that we will discuss in Chapter 5.

3.4.3 CPA-Security for Multiple Encryptions

Definition 3.21 can be extended to the case of multiple encryptions in the same way that Definition 3.8 is extended to give Definition 3.18, i.e., by using lists of plaintexts. Here, we take a different approach that is somewhat simpler and has the advantage of modeling attackers that can *adaptively* choose pairs of plaintexts to be encrypted. Specifically, we now give the attacker access to a "left-or-right" oracle $\mathsf{LR}_{k,b}$ that, on input a pair of equal-length messages m_0, m_1, computes the ciphertext $c \leftarrow \mathsf{Enc}_k(m_b)$ and returns c. That is, if $b = 0$ then the adversary always receives an encryption of the "left" plaintext, and if $b = 1$ then it always receives an encryption of the "right" plaintext. The bit b is a uniform bit chosen at the beginning of the experiment, and as in previous definitions the goal of the attacker is to guess b.

Consider the following experiment defined for any encryption scheme $\Pi = (\mathsf{Gen}, \mathsf{Enc}, \mathsf{Dec})$, adversary $\mathcal{A}$, and value n for the security parameter:

The LR-oracle experiment $\mathsf{PrivK}^{\mathsf{LR\text{-}cpa}}_{\mathcal{A},\Pi}(n)$:

1. *A key k is generated by running $\mathsf{Gen}(1^n)$.*
2. *A uniform bit $b \in \{0,1\}$ is chosen.*
3. *The adversary $\mathcal{A}$ is given input 1^n and oracle access to $\mathsf{LR}_{k,b}(\cdot, \cdot)$, as defined above.*
4. *The adversary $\mathcal{A}$ outputs a bit b'.*
5. *The output of the experiment is defined to be 1 if $b' = b$, and 0 otherwise. In the former case, we say that $\mathcal{A}$ succeeds.*

DEFINITION 3.22 *Private-key encryption scheme Π has* indistinguishable multiple encryptions under a chosen-plaintext attack, *or is* CPA-secure for multiple encryptions, *if for all probabilistic polynomial-time adversaries $\mathcal{A}$ there is a negligible function* negl *such that*

$$\Pr\left[\mathsf{PrivK}^{\mathsf{LR\text{-}cpa}}_{\mathcal{A},\Pi}(n) = 1\right] \leq \frac{1}{2} + \mathsf{negl}(n),$$

where the probability is taken over the randomness used by $\mathcal{A}$ and the randomness used in the experiment.

Note that an attacker given access to $\mathsf{LR}_{k,b}$ can simulate access to an encryption oracle: to obtain the encryption of a message m, the attacker simply queries $\mathsf{LR}_{k,b}(m, m)$. Given this observation, it is immediate that if Π is CPA-secure for multiple encryptions then it is also CPA-secure. It should also be clear that if Π is CPA-secure for multiple encryptions then it has indistinguishable multiple encryptions in the presence of an eavesdropper. In other words, Definition 3.22 is at least as strong as Definitions 3.18 and 3.21.

It turns out that CPA-security is *equivalent* to CPA-security for multiple encryptions. (This stands in contrast to the case of eavesdropping adversaries; cf. Proposition 3.19.) We state the following without proof; an analogous result in the public-key setting is proved in Section 12.2.2.

THEOREM 3.23 *Any private-key encryption scheme that is CPA-secure is also CPA-secure for* multiple *encryptions.*

Thus, it suffices to prove that a scheme is CPA-secure (for a single encryption), and we may then conclude that it is CPA-secure for multiple encryptions as well.

Fixed-length vs. arbitrary-length messages. An advantage of working with the notion of CPA-security for multiple messages (or, equivalently, CPA-security) is that it allows us to treat fixed-length encryption schemes without loss of generality. In particular, given any CPA-secure *fixed-length* encryption scheme $\Pi = (\mathsf{Gen}, \mathsf{Enc}, \mathsf{Dec})$, it is possible to construct a CPA-secure encryption scheme $\Pi' = (\mathsf{Gen}', \mathsf{Enc}', \mathsf{Dec}')$ for *arbitrary-length* messages quite easily. For simplicity, say Π encrypts messages that are 1-bit long. Leave Gen' the same as Gen. Define Enc'_k for any message m (having some arbitrary length ℓ) as $\mathsf{Enc}'_k(m) = \mathsf{Enc}_k(m_1), \ldots, \mathsf{Enc}_k(m_\ell)$, where m_i denotes the ith bit of m. Decryption is done in the natural way. It follows from Theorem 3.23 that if Π is CPA-secure then so is Π'.

There are more efficient ways to encrypt messages of arbitrary length than by adapting a fixed-length encryption scheme in this manner. We explore this further in Section 3.6.

3.5 Constructing a CPA-Secure Encryption Scheme

Before constructing encryption schemes secure against chosen-plaintext attacks, we first introduce the important notion of *pseudorandom functions*.

3.5.1   Pseudorandom Functions and Permutations

Pseudorandom functions (PRFs) generalize the notion of pseudorandom generators. Now, instead of considering "random-looking" *strings* we consider "random-looking" *functions*. As in our earlier discussion of pseudorandomness, it does not make much sense to say that any *fixed* function $f : \{0,1\}^* \rightarrow \{0,1\}^*$ is pseudorandom (in the same way it makes little sense to say that any fixed function is random). Instead, we must consider the pseudorandomness of a *distribution* on functions. Such a distribution is induced naturally by considering *keyed functions*, defined next.

A keyed function $F : \{0,1\}^* \times \{0,1\}^* \rightarrow \{0,1\}^*$ is a two-input function, where the first input is called the *key* and typically denoted by k. We say F is *efficient* if there is a polynomial-time algorithm that computes $F(k, x)$ given k and x. (We will only be interested in efficient keyed functions.) The security parameter n dictates the key length, input length, and output length. That is, we associate with F three functions ℓ_{key}, ℓ_{in}, and ℓ_{out}; for any key $k \in \{0,1\}^{\ell_{key}(n)}$, the function F_k is only defined for inputs $x \in \{0,1\}^{\ell_{in}(n)}$, in which case $F_k(x) \in \{0,1\}^{\ell_{out}(n)}$. Unless stated otherwise, we assume for simplicity that F is *length preserving*, meaning $\ell_{key}(n) = \ell_{in}(n) = \ell_{out}(n) = n$. (Note, however, that this is only to reduce notational clutter, and it is not uncommon to have pseudorandom functions that are not length preserving.) Let Func_n denote the set of all functions mapping n-bit strings to n-bit strings.

In typical usage a key $k \in \{0,1\}^n$ is chosen and fixed, and we are then interested in the single-input function $F_k : \{0,1\}^n \rightarrow \{0,1\}^n$ defined by $F_k(x) \stackrel{\text{def}}{=} F(k, x)$ mapping n-bit input strings to n-bit output strings. A keyed function F thus induces a distribution on functions in Func_n, where the distribution is given by choosing a uniform key $k \in \{0,1\}^n$ and then considering the resulting single-input function F_k. We call F *pseudorandom* if the function F_k (for a uniform key k) is indistinguishable from a function chosen uniformly at random from the set Func_n of all functions having the same domain and range; that is, if no efficient adversary can distinguish—in a sense we more carefully define below—whether it is interacting with F_k (for uniform k) or f (where f is chosen uniformly from Func_n).

Since choosing a uniform function is less intuitive than choosing a uniform string, it is worth spending a bit more time on this idea. The set Func_n is finite, and selecting a uniform function mapping n-bit strings to n-bit strings simply means choosing a function uniformly from this set. How large is Func_n? A function f is specified by giving its value on each point in its domain. We can view any function (over a finite domain) as a large look-up table that stores $f(x)$ in the row of the table labeled by x. For $f \in \mathsf{Func}_n$, the look-up table for f has 2^n rows (one for each string in the domain $\{0,1\}^n$), with each row containing an n-bit string (since the range of f is $\{0,1\}^n$). Concatenating all the entries of this table, we see that any function in Func_n can be represented by a string of length $2^n \cdot n$. Moreover, this correspondence is one-to-one, as each string of length $2^n \cdot n$ (i.e., each table containing 2^n entries of length n)

defines a unique function in Func_n. Thus, the size of Func_n is exactly the number of strings of length $n \cdot 2^n$, i.e., $|\mathsf{Func}_n| = 2^{n \cdot 2^n}$.

Viewing a function as a look-up table provides another useful way to think about selecting a uniform function $f \in \mathsf{Func}_n$: It is exactly equivalent to choosing each row in the look-up table of f uniformly. This means, in particular, that the values $f(x)$ and $f(y)$, for any two inputs $x \neq y$, are uniform and independent. We can view this look-up table as being populated by uniform entries in advance, before f is evaluated on any input, or we can view entries of the table as being chosen uniformly "on-the-fly," as needed, whenever f is evaluated on a new input on which it was never evaluated before.

A pseudorandom function is a keyed function F such that F_k (for uniform $k \in \{0,1\}^n$) is indistinguishable from f (for uniform $f \in \mathsf{Func}_n$). The former is chosen from a distribution over (at most) 2^n distinct functions, whereas the latter is chosen from all $2^{n \cdot 2^n}$ functions in Func_n. Despite this, the "behavior" of those functions must look the same to any polynomial-time distinguisher.

A first attempt at formalizing the notion of a pseudorandom function would be to proceed as in Definition 3.14. That is, we could require that every polynomial-time distinguisher D that receives a description of F_k outputs 1 with "almost" the same probability as when it receives a description of a random function f. However, this definition is inappropriate since the description of a random function has *exponential length* (given by its look-up table of length $n \cdot 2^n$), while D is limited to running in polynomial time. So, D would not even have sufficient time to examine its entire input.

Instead, we allow D to probe the input/output behavior of the function by giving D access to an *oracle* $\mathcal{O}$ which is either equal to F_k or f. The distinguisher D may query its oracle at any point x, in response to which the oracle returns $\mathcal{O}(x)$. We treat the oracle as a black box in the same way as when we provided the adversary with oracle access to the encryption algorithm in the definition of a chosen-plaintext attack. Here, however, the oracle computes a deterministic function and so returns the same result if queried twice on the same input. D may interact freely with its oracle, choosing its queries adaptively based on all previous outputs. Since D runs in polynomial time, however, it can ask only polynomially many queries.

We now present the formal definition. (The definition assumes F is length preserving for simplicity.)

DEFINITION 3.24   *An efficient, length preserving, keyed function F : $\{0,1\}^* \times \{0,1\}^* \to \{0,1\}^*$ is a* pseudorandom *function if for all probabilistic polynomial-time distinguishers D, there is a negligible function negl such that:*

$$\left| \Pr[D^{F_k(\cdot)}(1^n) = 1] - \Pr[D^{f(\cdot)}(1^n) = 1] \right| \leq \mathsf{negl}(n),$$

where the first probability is taken over uniform choice of $k \in \{0,1\}^n$ and the randomness of D, and the second probability is taken over uniform choice of $f \in \mathsf{Func}_n$ and the randomness of D.

We stress that D is *not* given the key k (in the same way that D is not given the seed when defining a pseudorandom generator). It is meaningless to require that F_k "look random" if k is known, since given k it is trivial to distinguish an oracle for F_k from an oracle for f. (All the distinguisher has to do is query the oracle at any point x to obtain the answer y, and compare this to the result $y' := F_k(x)$ that it computes itself using the known value k. An oracle for F_k will return $y = y'$, while an oracle for a random function will return $y = y'$ only with probability 2^{-n}.) This means that if k is revealed, any claims about pseudorandomness no longer hold.

Example 3.25

We can gain familiarity with the definition by considering an insecure example. Define the keyed, length-preserving function F by $F(k, x) = k \oplus x$. For any input x, the value of $F_k(x)$ is uniformly distributed (when k is uniform). Nevertheless, F is not pseudorandom since its values on any *two* points are correlated. Consider the distinguisher D that queries its oracle $\mathcal{O}$ on distinct points x_1, x_2 to obtain values $y_1 = \mathcal{O}(x_1)$ and $y_2 = \mathcal{O}(x_2)$, and outputs 1 if and only if $y_1 \oplus y_2 = x_1 \oplus x_2$. If $\mathcal{O} = F_k$, for any k, then D outputs 1. On the other hand, if $\mathcal{O} = f$ for f chosen uniformly from Func_n, then

$$\Pr[f(x_1) \oplus f(x_2) = x_1 \oplus x_2] = \Pr[f(x_2) = x_1 \oplus x_2 \oplus f(x_1)] = 2^{-n},$$

since $f(x_2)$ is uniform and independent of x_1, x_2, and $f(x_1)$. We thus have $\Pr[D^{F_k(\cdot)}(1^n) = 1] = 1$ and $\Pr[D^{f(\cdot)}(1^n) = 1] = 2^{-n}$, and the difference between these two is not negligible. $\diamondsuit$

Pseudorandom functions and pseudorandom generators.   As one might expect, there is a close relationship between pseudorandom functions and pseudorandom generators. It is fairly easy to construct a pseudorandom generator G from a pseudorandom function F by simply evaluating F on a series of distinct inputs; e.g., we can define $G(s) \stackrel{\text{def}}{=} F_s(1) \| F_s(2) \| \cdots \| F_s(\ell)$ for any desired ℓ (where $\|$ denotes concatenation). If F_s were replaced by a uniform function f, the output of G would be uniform; when using F, the output is pseudorandom. You are asked to prove this formally in Exercise 3.16.

Considering the other direction, a pseudorandom generator G immediately gives a pseudorandom function F *with small input length*. Specifically, say G has expansion factor $\ell(n) = n \cdot 2^{t(n)}$. We can define the keyed function $F : \{0, 1\}^n \times \{0, 1\}^{t(n)} \to \{0, 1\}^n$ as follows: to compute $F_k(i)$, first compute $G(k)$ and interpret the result as a look-up table with $2^{t(n)}$ rows each containing n bits; output the ith row. (We leave the proof that F is pseudorandom to the reader.) Note, however, that F is efficient only if $t(n) = \mathcal{O}(\log n)$. It is possible, though more difficult, to construct pseudorandom functions *with large input length* from pseudorandom generators; see Section 8.5. Since pseudorandom generators can be constructed based on certain mathematical problems conjectured to be hard, we conclude that pseudorandom functions (for long inputs) can be constructed based on those same problems. The fact that pseudorandom functions can based on hard mathematical problems represents one of the amazing contributions of modern cryptography.

Pseudorandom Permutations

Let $\mathsf{Perm}_n \subset \mathsf{Func}_n$ be the set of all permutations (i.e., bijections) on $\{0,1\}^n$. Viewing any $f \in \mathsf{Perm}_n$ as a look-up table as before, we now have the added constraint that the entries in any two distinct rows must be different. We have 2^n different choices for the entry in the first row of the table; once we fix that entry, we are left with only $2^n - 1$ choices for the second row, and so on. We thus see that the size of Perm_n is $(2^n)!$.

Let F be a keyed function where, for the moment, ℓ_{key}, ℓ_{in}, and ℓ_{out} can be arbitrary. We call F a *keyed permutation* if $\ell_{in} = \ell_{out}$, and furthermore for all $k \in \{0,1\}^{\ell_{key}(n)}$ the function $F_k : \{0,1\}^{\ell_{in}(n)} \to \{0,1\}^{\ell_{in}(n)}$ is one-to-one (i.e., F_k is a permutation). We call ℓ_{in} the *block length* of F in this case. A keyed permutation is *efficient* if there is a polynomial-time algorithm for computing $F_k(x)$ given k and x, as well as a polynomial-time algorithm for computing $F_k^{-1}(y)$ given k and y. That is, F_k should be both efficiently computable and *efficiently invertible* given k. As before, unless stated otherwise we assume F is *length preserving* for simplicity and so $\ell_{key}(n) = \ell_{in}(n) = n$.

The definition of what it means for an efficient, keyed permutation F to be a *pseudorandom permutation* is exactly analogous to Definition 3.24, with the only difference being that now we require F_k to be indistinguishable from a uniform *permutation* rather than a uniform function. That is, we require that no efficient algorithm can distinguish between access to F_k (for uniform key k) and access to f (for uniform $f \in \mathsf{Perm}_n$). We remark that whenever the block length is sufficiently long (as is usually the case in practice), a random permutation is indistinguishable from a random function with the same domain and range; thus, we can equally well define a pseudorandom permutation by requiring that no efficient algorithm can distinguish between access to F_k (for uniform key k) and access to f (for uniform $f \in \mathsf{Func}_n$). This is a consequence of the following proposition, proven formally in Appendix A.4.

PROPOSITION 3.26   *If F is a pseudorandom permutation for which $\ell_{in}(n) \geq n$, then F is also a pseudorandom function.*

While the above is true asymptotically, concrete security may be impacted when a pseudorandom permutation is viewed as a pseudorandom function.

Strong pseudorandom permutations. If F is a keyed permutation then cryptographic schemes based on F might require the honest parties to compute the inverse F_k^{-1} in addition to computing F_k itself. This potentially introduces new security concerns. In particular, it may now be necessary to impose the stronger requirement that F_k be indistinguishable from a uniform permutation *even if the distinguisher is additionally given oracle access to the inverse of the permutation.* If F has this property, we call it a *strong* pseudorandom permutation.

DEFINITION 3.27   *Let $F : \{0,1\}^* \times \{0,1\}^* \to \{0,1\}^*$ be an efficient, length preserving, keyed permutation. F is a* strong pseudorandom permutation *if for all probabilistic polynomial-time distinguishers D, there exists a negligible function* negl *such that:*

$$\left| \Pr[D^{F_k(\cdot), F_k^{-1}(\cdot)}(1^n) = 1] - \Pr[D^{f(\cdot), f^{-1}(\cdot)}(1^n) = 1] \right| \leq \mathsf{negl}(n),$$

where the first probability is taken over uniform choice of $k \in \{0,1\}^n$ and the randomness of D, and the second probability is taken over uniform choice of $f \in \mathsf{Perm}_n$ and the randomness of D.

Of course, any strong pseudorandom permutation is also a pseudorandom permutation. However, the converse is not true.

3.5.2 CPA-Security from a Pseudorandom Function

We focus here on constructing a CPA-secure fixed-length encryption scheme. By what we have said at the end of Section 3.4.3, this implies the existence of a CPA-secure encryption scheme for arbitrary-length messages. In Section 3.6 we will discuss more efficient ways of encrypting messages of arbitrary length.

A naive attempt at constructing an encryption scheme from a pseudorandom permutation is to define $\mathsf{Enc}_k(m) = F_k(m)$. Although we expect that this "reveals no information about m" (since, if f is a uniform permutation, then $f(m)$ is a uniform n-bit string for any m), this method of encryption is *deterministic* and so cannot possibly be CPA-secure since encrypting the same plaintext twice will yield the same ciphertext.

Our CPA-secure construction uses *randomized* encryption. Specifically, we encrypt by applying a pseudorandom function to a *random value* $r \in \{0,1\}^n$ and XORing the output with the plaintext; the ciphertext includes both the result as well as r (to enable the receiver to decrypt). See Figure 3.3 and Construction 3.28. Encryption can again be viewed as XORing a pseudorandom pad with the plaintext (just like in the "pseudo-"one-time pad), with the major difference being the fact that here a *fresh* pseudorandom pad—that depends on r—is used each time a message is encrypted. (The pseudorandom pad is only "fresh" if the pseudorandom function is applied to a "fresh" value r on which it has never been evaluated before. The proof below shows that with overwhelming probability this is always the case.)

Note that for any key k, every message m has 2^n corresponding ciphertexts. Nevertheless, the receiver is able to decrypt correctly. (Check for yourself that decryption always returns the correct result!) This scheme also has the

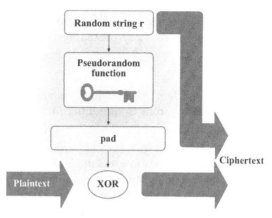

FIGURE 3.3: Encryption with a pseudorandom function.

property that *the ciphertext is longer than the plaintext*. This is the first encryption scheme we have seen that has either of these properties.

Before turning to the proof that the above construction is CPA-secure, we highlight a common template that is used by most proofs of security (even outside the context of encryption) for constructions based on pseudorandom functions. The first step of such proofs is to consider a hypothetical version of the construction in which the pseudorandom function is replaced with a random function. It is then argued—using a proof by reduction—that this modification does not significantly affect the attacker's success probability. We are then left with analyzing a scheme that uses a completely random function. The rest of the proof typically relies on probabilistic analysis and does not rely on any computational assumptions. We will utilize this proof template several times in this and the next two chapters.

CONSTRUCTION 3.28

Let F be a pseudorandom function. Define a fixed-length, private-key encryption scheme for messages of length n as follows:

- Gen: on input 1^n, choose uniform $k \in \{0,1\}^n$ and output it.

- Enc: on input a key $k \in \{0,1\}^n$ and a message $m \in \{0,1\}^n$, choose uniform $r \in \{0,1\}^n$ and output the ciphertext
$$c := \langle r, \, F_k(r) \oplus m \rangle.$$

- Dec: on input a key $k \in \{0,1\}^n$ and a ciphertext $c = \langle r, s \rangle$, output the message
$$m := F_k(r) \oplus s.$$

A CPA-secure encryption scheme from any pseudorandom function.

THEOREM 3.29   *If F is a pseudorandom function, then Construction 3.28 is a CPA-secure, fixed-length private-key encryption scheme for messages of length n.*

PROOF   Let $\widetilde{\Pi} = (\widetilde{\mathsf{Gen}}, \widetilde{\mathsf{Enc}}, \widetilde{\mathsf{Dec}})$ be an encryption scheme that is exactly the same as $\Pi = (\mathsf{Gen}, \mathsf{Enc}, \mathsf{Dec})$ from Construction 3.28, except that a truly random function f is used in place of F_k. That is, $\widetilde{\mathsf{Gen}}(1^n)$ chooses a uniform function $f \in \mathsf{Func}_n$, and $\widetilde{\mathsf{Enc}}$ encrypts just like Enc except that f is used instead of F_k. (This modified encryption scheme is not efficient. But we can still define it as a hypothetical encryption scheme for the sake of the proof.)

Fix an arbitrary PPT adversary $\mathcal{A}$, and let $q(n)$ be an upper bound on the number of queries that $\mathcal{A}(1^n)$ makes to its encryption oracle. (Note that q must be upper-bounded by some polynomial.) As the first step of the proof, we show that there is a negligible function negl such that

$$\left| \Pr\left[\mathsf{PrivK}_{\mathcal{A},\Pi}^{\mathsf{cpa}}(n) = 1 \right] - \Pr\left[\mathsf{PrivK}_{\mathcal{A},\widetilde{\Pi}}^{\mathsf{cpa}}(n) = 1 \right] \right| \leq \mathsf{negl}(n). \qquad (3.9)$$

We prove this by reduction. We use $\mathcal{A}$ to construct a distinguisher D for the pseudorandom function F. The distinguisher D is given oracle access to a function $\mathcal{O}$, and its goal is to determine whether $\mathcal{O}$ is "pseudorandom" (i.e., equal to F_k for uniform $k \in \{0,1\}^n$) or "random" (i.e., equal to f for uniform $f \in \mathsf{Func}_n$). To do this, D simulates experiment $\mathsf{PrivK}^{\mathsf{cpa}}$ for $\mathcal{A}$ in the manner described below, and observes whether $\mathcal{A}$ succeeds or not. If $\mathcal{A}$ succeeds then D guesses that its oracle must be a pseudorandom function, whereas if $\mathcal{A}$ does not succeed then D guesses that its oracle must be a random function. In detail:

> **Distinguisher D:**
> D is given input 1^n and access to an oracle $\mathcal{O}: \{0,1\}^n \to \{0,1\}^n$.
>
> 1. Run $\mathcal{A}(1^n)$. Whenever $\mathcal{A}$ queries its encryption oracle on a message $m \in \{0,1\}^n$, answer this query in the following way:
>   (a) Choose uniform $r \in \{0,1\}^n$.
>   (b) Query $\mathcal{O}(r)$ and obtain response y.
>   (c) Return the ciphertext $\langle r, y \oplus m \rangle$ to $\mathcal{A}$.
>
> 2. When $\mathcal{A}$ outputs messages $m_0, m_1 \in \{0,1\}^n$, choose a uniform bit $b \in \{0,1\}$ and then:
>   (a) Choose uniform $r \in \{0,1\}^n$.
>   (b) Query $\mathcal{O}(r)$ and obtain response y.
>   (c) Return the challenge ciphertext $\langle r, y \oplus m_b \rangle$ to $\mathcal{A}$.
>
> 3. Continue answering encryption-oracle queries of $\mathcal{A}$ as before until $\mathcal{A}$ outputs a bit b'. Output 1 if $b' = b$, and 0 otherwise.

D runs in polynomial time since $\mathcal{A}$ does. The key points are as follows:

1. If D's oracle is a pseudorandom function, then the view of $\mathcal{A}$ when run as a subroutine by D is distributed identically to the view of $\mathcal{A}$ in experiment $\mathsf{PrivK}^{\mathsf{cpa}}_{\mathcal{A},\Pi}(n)$. This is because, in this case, a uniform key k is chosen and then every encryption is carried out by choosing a uniform r, computing $y := F_k(r)$, and setting the ciphertext equal to $\langle r, y \oplus m \rangle$, exactly as in Construction 3.28. Thus,

$$\Pr_{k \leftarrow \{0,1\}^n} \left[D^{F_k(\cdot)}(1^n) = 1 \right] = \Pr \left[\mathsf{PrivK}^{\mathsf{cpa}}_{\mathcal{A},\Pi}(n) = 1 \right], \quad (3.10)$$

where we emphasize on the left-hand side that k is chosen uniformly.

2. If D's oracle is a random function, then the view of $\mathcal{A}$ when run as a subroutine by D is distributed identically to the view of $\mathcal{A}$ in experiment $\mathsf{PrivK}^{\mathsf{cpa}}_{\mathcal{A},\widetilde{\Pi}}(n)$. This can be seen exactly as above, with the only difference being that a uniform function $f \in \mathsf{Func}_n$ is used instead of F_k. Thus,

$$\Pr_{f \leftarrow \mathsf{Func}_n} \left[D^{f(\cdot)}(1^n) = 1 \right] = \Pr \left[\mathsf{PrivK}^{\mathsf{cpa}}_{\mathcal{A},\widetilde{\Pi}}(n) = 1 \right], \quad (3.11)$$

where f is chosen uniformly from Func_n on the left-hand side.

By the assumption that F is a pseudorandom function (and since D is efficient), there exists a negligible function negl for which

$$\left| \Pr \left[D^{F_k(\cdot)}(1^n) = 1 \right] - \Pr \left[D^{f(\cdot)}(1^n) = 1 \right] \right| \leq \mathsf{negl}(n).$$

Combining the above with Equations (3.10) and (3.11) gives Equation (3.9). For the second part of the proof, we show that

$$\Pr \left[\mathsf{PrivK}^{\mathsf{cpa}}_{\mathcal{A},\widetilde{\Pi}}(n) = 1 \right] \leq \frac{1}{2} + \frac{q(n)}{2^n}. \quad (3.12)$$

(Recall that $q(n)$ is a bound on the number of encryption queries made by $\mathcal{A}$.) The above holds even if we place no computational restrictions on $\mathcal{A}$. To see this, observe that every time a message m is encrypted in $\mathsf{PrivK}^{\mathsf{cpa}}_{\mathcal{A},\widetilde{\Pi}}(n)$ (either by the encryption oracle or when the challenge ciphertext is computed), a uniform $r \in \{0,1\}^n$ is chosen and the ciphertext is set equal to $\langle r, f(r) \oplus m \rangle$. Let r^* denote the random string used when generating the challenge ciphertext $\langle r^*, f(r^*) \oplus m_b \rangle$. There are two possibilities:

1. *The value r^* is never used when answering any of $\mathcal{A}$'s encryption-oracle queries:* In this case, $\mathcal{A}$ learns nothing about $f(r^*)$ from its interaction with the encryption oracle (since f is a truly random function). This means that, from the perspective of $\mathcal{A}$, the value $f(r^*)$ that is XORed with m_b is uniformly distributed and independent of the rest of the experiment, and so the probability that $\mathcal{A}$ outputs $b' = b$ in this case is exactly $1/2$ (as in the case of the one-time pad).

2. *The value r^* is used when answering at least one of $\mathcal{A}$'s encryption-oracle queries:* In this case, $\mathcal{A}$ may easily determine whether m_0 or m_1 was encrypted. This is so because if the encryption oracle ever returns a ciphertext $\langle r^*, s \rangle$ in response to a request to encrypt the message m, the adversary learns that $f(r^*) = s \oplus m$.

However, since $\mathcal{A}$ makes at most $q(n)$ queries to its encryption oracle (and thus at most $q(n)$ values of r are used when answering $\mathcal{A}$'s encryption-oracle queries), and since r^* is chosen uniformly from $\{0,1\}^n$, the probability of this event is at most $q(n)/2^n$.

Let repeat denote the event that r^* is used by the encryption oracle when answering at least one of $\mathcal{A}$'s queries. As just discussed, the probability of repeat is at most $q(n)/2^n$, and the probability that $\mathcal{A}$ succeeds in $\mathsf{PrivK}^{\mathsf{cpa}}_{\mathcal{A},\widetilde{\Pi}}$ if repeat does not occur is exactly $1/2$. Therefore:

$$\Pr[\mathsf{PrivK}^{\mathsf{cpa}}_{\mathcal{A},\widetilde{\Pi}}(n) = 1]$$
$$= \Pr[\mathsf{PrivK}^{\mathsf{cpa}}_{\mathcal{A},\widetilde{\Pi}}(n) = 1 \wedge \mathsf{repeat}] + \Pr[\mathsf{PrivK}^{\mathsf{cpa}}_{\mathcal{A},\widetilde{\Pi}}(n) = 1 \wedge \overline{\mathsf{repeat}}]$$
$$\leq \Pr[\mathsf{repeat}] + \Pr[\mathsf{PrivK}^{\mathsf{cpa}}_{\mathcal{A},\widetilde{\Pi}}(n) = 1 \mid \overline{\mathsf{repeat}}] \ \leq \ \frac{q(n)}{2^n} + \frac{1}{2}.$$

Combining the above with Equation (3.9), we see that there is a negligible function negl such that $\Pr[\mathsf{PrivK}^{\mathsf{cpa}}_{\mathcal{A},\Pi}(n) = 1] \leq \frac{1}{2} + \frac{q(n)}{2^n} + \mathsf{negl}(n)$. Since q is polynomial, $\frac{q(n)}{2^n}$ is negligible. In addition, the sum of two negligible functions is negligible, and thus there exists a negligible function negl' such that $\Pr[\mathsf{PrivK}^{\mathsf{cpa}}_{\mathcal{A},\Pi}(n) = 1] \leq \frac{1}{2} + \mathsf{negl}'(n)$. $\blacksquare$

Concrete security. The above proof shows that

$$\Pr[\mathsf{PrivK}^{\mathsf{cpa}}_{\mathcal{A},\Pi}(n) = 1] \leq \frac{1}{2} + \frac{q(n)}{2^n} + \mathsf{negl}(n)$$

for some negligible function negl. The final term depends on the security of F as a pseudorandom function; it is a bound on the distinguishing advantage of algorithm D (which has roughly the same running time as the adversary $\mathcal{A}$). The term $\frac{q(n)}{2^n}$ represents a bound on the probability that the value r^* used to encrypt the challenge ciphertext was used to encrypt some other message, and depends on the number of encryption-oracle queries the attacker makes.

3.6 Modes of Operation and Encryption in Practice

The encryption schemes described in Sections 3.3.3 and 3.5.2 (namely, Constructions 3.17 and 3.28) have a number of drawbacks that make them ill-suited for practical applications. For starters, Construction 3.17 is only EAV-secure. In addition, both constructions are defined only for the encryption of *fixed-length* messages. While Construction 3.28 could be used to encrypt arbitrary-length messages using the approach discussed at the end of Section 3.4.3, this would result in a scheme in which the ciphertext length is a constant multiple of the plaintext length, which is rather inefficient. In this section, we show how to overcome these drawbacks.

While we are dealing with practical considerations, we also begin to discuss how the underlying building blocks of secure encryption schemes—namely, pseudorandom generators and pseudorandom permutations—are instantiated in the real world using *stream ciphers* and *block ciphers*, respectively. Our goal here is mainly to introduce the appropriate terminology and syntax; we defer an in-depth discussion of how stream ciphers and block ciphers are designed, and some popular candidates for those primitives, to Chapter 7.

3.6.1 Stream Ciphers

A pseudorandom generator G as in Definition 3.14 is rather inflexible since its output length is fixed. This makes G a poor fit for adapting Construction 3.17 to handle arbitrary-length messages. Specifically, say G has expansion factor ℓ. We cannot easily use G to encrypt messages of length $\ell' > \ell$ using a single n-bit key. And, although we can encrypt messages of length $\ell' < \ell$ by truncating the output of G, doing so is wasteful since it involves generating ℓ pseudorandom bits and then discarding $\ell - \ell'$ of them.

Stream ciphers, used in practice to instantiate pseudorandom generators, provide greater flexibility. The output bits of a stream cipher are produced gradually and on demand, so that an application can request exactly as many pseudorandom bits as it needs. This extends their usefulness (since there is no upper bound on the number of bits that can be generated) and improves efficiency (since no extraneous pseudorandom bits are generated).

Formally, a stream cipher is a pair of deterministic algorithms (Init, Next) where:

- Init takes as input a seed s and an optional *initialization vector IV*, and outputs some initial state st.

- Next takes as input a current state st and outputs a bit[3] y along with updated state st'.

Starting from some initial state st_0, we can generate any desired number of bits by repeatedly calling Next as many times as needed. As shorthand for this, we define an algorithm GetBits that takes as input an initial state st_0

---

[3]In practice, Next might output a byte or even a larger number of random bits, rather than just outputting a single bit at a time. We assume it outputs a bit for simplicity here.

and a desired output length 1^ℓ (specified in unary, since GetBits runs in time linear in ℓ) and then does:

1. For $i = 1$ to ℓ, compute $(y_i, \mathsf{st}_i) := \mathsf{Next}(\mathsf{st}_{i-1})$.

2. Return the ℓ-bit string $y = y_1 \cdots y_\ell$ as well as the final state st_ℓ.

We let $\mathsf{GetBits}_1$ be the algorithm that runs GetBits and only returns its initial output (namely, the ℓ-bit string y).

A secure stream cipher *without* an IV is just a pseudorandom generator with a more flexible interface. That is, we require that when we run Init on a uniform seed s to obtain st_0, and then generate *any* (polynomial) number of bits using $\mathsf{GetBits}_1$, the resulting output is pseudorandom. Formally, given a stream cipher (Init, Next) and a parameter $\ell = \ell(n) > n$, we may define the deterministic function G^ℓ as

$$G^\ell(s) \overset{\text{def}}{=} \mathsf{GetBits}_1(\mathsf{Init}(s), 1^\ell).$$

Then the stream cipher is *secure* if G^ℓ is a pseudorandom generator for any polynomial ℓ.

Security for a stream cipher that *does* take an IV can be defined in multiple ways. We define security in this case to be akin to that of a pseudorandom function. Specifically, here we consider the setting where a uniform seed s is chosen and then $\mathsf{Init}(s, \cdot)$ is run repeatedly using different values for the IV; the requirement is that running $\mathsf{GetBits}_1$ using the different initial states should produce output streams that appear *independently* uniform. Formally, given a stream cipher (Init, Next) (where Init takes an n-bit IV) and a parameter $\ell = \ell(n)$, we may define the keyed function $F^\ell : \{0,1\}^n \times \{0,1\}^n \to \{0,1\}^\ell$ as

$$F^\ell_s(IV) \overset{\text{def}}{=} \mathsf{GetBits}_1(\mathsf{Init}(s, IV), 1^\ell).$$

Then the stream cipher is *secure* if F^ℓ is a pseudorandom function for any polynomial ℓ.

Practical stream ciphers typically do not support arbitrary values of n (which determines the length of the seed and the IV), but instead work only for some fixed values of n. Concrete-security definitions are thus more appropriate than the asymptotic definitions given above.

Constructing stream ciphers from pseudorandom functions. A pseudorandom function F can be used to construct a stream cipher (Init, Next) that takes an IV. (This is very similar to the construction of pseudorandom generators from pseudorandom functions discussed briefly in Section 3.5.1.) The basic idea is to use the seed s for the stream cipher as a key for F, and to evaluate F_s on a sequence of consecutive inputs starting from a value determined by IV. Concretely, if we set the length of the initialization vector to $3n/4$, then the output of the stream cipher will be

$$F_s(IV \parallel \langle 0 \rangle), F_s(IV \parallel \langle 1 \rangle), \ldots$$

(see Construction 3.30), where $\langle i \rangle$ denotes the binary encoding of integer i as an $n/4$-bit string. Informally, this will be secure (assuming F is a pseudorandom function) as long as no more than $2^{n/4}$ output blocks are generated for any IV, since in that case F_s is evaluated at distinct inputs when the stream cipher is used with different IVs.

CONSTRUCTION 3.30

Let F be a pseudorandom function. Define a stream cipher (Init, Next) as follows, where Init accepts a $3n/4$-bit initialization vector and Next outputs n bits in each call:

- Init: on input $s \in \{0,1\}^n$ and $IV \in \{0,1\}^{3n/4}$, output st $= (s, IV, 0)$.

- Next: on input st $= (s, IV, i)$, output $y := F_s(IV \| \langle i \rangle)$ and updated state st$' = (s, IV, i+1)$.

A stream cipher from a pseudorandom function.

Although stream ciphers can be constructed from pseudorandom functions in this way, dedicated constructions of stream ciphers used in practice typically have better performance, especially in resource-constrained environments.

3.6.2 Stream-Cipher Modes of Operation

We discuss two modes of operation for encrypting arbitrary-length messages using a stream cipher (Init, Next): *synchronized mode* and *unsynchronized mode*.

Synchronized mode. Stream ciphers are often used to encrypt an online communication session between two parties. In that case, a fresh key k is generated by the parties (e.g., using methods described in Chapter 11) and then that key is used to encrypt the messages sent during the session. Assuming that the communication between the parties is such that all messages arrive in order and no messages are lost (as is the case, e.g., when communicating over TCP), the two parties are *synchronized* and the following method can be used to encrypt a series of messages from a sender S to a receiver R:

1. Both parties call Init(k) to obtain the same initial state st$_0$.

2. Let st$_S$ be the current state of S. If S wants to encrypt a message m, it computes $(y, \text{st}'_S) := \text{GetBits}(\text{st}_S, 1^{|m|})$, sends $c := m \oplus y$ to the receiver, and updates its local state to st$'_S$.

3. Let st$_R$ be the current state of R. When R receives a ciphertext c from the sender, it computes $(y, \text{st}'_R) := \text{GetBits}(\text{st}_R, 1^{|c|})$, outputs the message $m := c \oplus y$, and updates its own local state to st$'_R$.

In the above description, the same party always acts as a sender. But by sharing a second key the parties can support bidirectional communication.

Let ℓ denote the total combined length of all messages encrypted during the course of a session. Conceptually, synchronized mode encryption can be viewed as a counterpart to Construction 3.17 where (1) ℓ need not be fixed in advance, and (2) the entire "message" need not be encrypted at once.

The above is an example of *stateful* encryption where the sender and receiver are required to maintain state between the encryption/decryption of different messages. One can define an appropriate notion of CPA-security suitable for stateful encryption, and prove that the above scheme meets that definition if the underlying stream cipher is secure.

Observe that for synchronized mode, the stream cipher does not need to use an IV. Note also that *there is no ciphertext expansion*, since the total communication from the sender to the receiver is exactly equal to the total length of the messages being encrypted.

Unsynchronized mode. When a stream cipher does take an IV, it can be used to construct a stateless encryption scheme that is exactly analogous to Construction 3.28; see Construction 3.31. CPA-security of this scheme follows as in the proof of Theorem 3.29. We stress that the main advantage here is that the encryption scheme directly handles arbitrary-length messages.

CONSTRUCTION 3.31

Let (Init, Next) be a stream cipher that takes an n-bit IV. Define a private-key encryption scheme for arbitrary-length messages as follows:

- Gen: on input 1^n, choose a uniform $k \in \{0,1\}^n$ and output it.

- Enc: on input a key $k \in \{0,1\}^n$ and a message $m \in \{0,1\}^*$, choose uniform $IV \in \{0,1\}^n$, and output the ciphertext
$$\langle IV, \mathsf{GetBits}_1(\mathsf{Init}(k, IV), 1^{|m|}) \oplus m \rangle.$$

- Dec: on input a key $k \in \{0,1\}^n$ and a ciphertext $\langle IV, c \rangle$, output the message
$$m := \mathsf{GetBits}_1(\mathsf{Init}(k, IV), 1^{|c|}) \oplus c.$$

Unsynchronized mode encryption from a stream cipher that takes an IV.

3.6.3   Block Ciphers and Block-Cipher Modes of Operation

A *block cipher* is simply another name for a (strong) pseudorandom permutation. That is, a block cipher $F : \{0,1\}^n \times \{0,1\}^\ell \to \{0,1\}^\ell$ is a keyed function such that, for all k, the function F_k defined by $F_k(x) \stackrel{\text{def}}{=} F(k, x)$ is a bijection (i.e., a permutation). Recall that n is the *key length* of F, and ℓ

is its *block length*. The main distinction between block ciphers and pseudo-random permutations is that the former typically only support a specific set of key/block lengths, and in particular do not support arbitrary-length keys. For simplicity, we will assume in this section that $\ell = n$.

As shown earlier (cf. Construction 3.30), a block cipher can be used to construct a stream cipher that accepts an IV; this means we can use any block cipher F to implement the stream-cipher modes of operation discussed in Section 3.6.2. Several other block-cipher modes of operation are also possible; here, we present four of the most common ones and discuss their security. In our discussion, we assume for simplicity that all messages m being encrypted have length a multiple of n (the block length of F), and write $m = m_1, m_2, \ldots, m_\ell$ where each $m_i \in \{0,1\}^n$ represents a block of the plaintext. (Messages whose length is not a multiple of n can be unambiguously padded to have length a multiple of n by appending a 1 followed by sufficiently many 0s, and so this assumption is without much loss of generality.)

Electronic Code Book (ECB) mode. This is a naive mode of operation in which the ciphertext is obtained by direct application of the block cipher to each plaintext block. That is, $c := F_k(m_1), F_k(m_2), \ldots, F_k(m_\ell)$; see Figure 3.4. Decryption is done in the obvious way, using the fact that F_k^{-1} is efficiently computable.

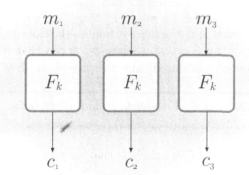

FIGURE 3.4: Electronic Code Book (ECB) mode.

ECB mode is *deterministic* and therefore cannot be CPA-secure. Worse, ECB-mode encryption is not even EAV-secure. This is because if a block is repeated in the plaintext, it will result in a repeating block in the ciphertext. Thus, for example, it is easy to distinguish the encryption of a plaintext that consists of two identical blocks from the encryption of a plaintext that consists of two different blocks. This is not just a theoretical problem. Consider encrypting an image in which small groups of pixels correspond to a plaintext block. Encrypting using ECB mode may reveal a significant amount of information about patterns in the image, something that should not happen when using a secure encryption scheme. (Figure 3.5 demonstrates this.) For these reasons, ECB mode should never be used.

FIGURE 3.5: An illustration of the dangers of using ECB mode. The middle figure is an encryption of the image on the left using ECB mode; the figure on the right is an encryption of the same image using a secure mode. (Taken from `http://en.wikipedia.org` and derived from images created by Larry Ewing (`lewing@isc.tamu.edu`) using The GIMP.)

Cipher Block Chaining (CBC) mode. To encrypt here, a uniform initialization vector (IV) of length n is first chosen as the initial ciphertext block. Then, ciphertext blocks are generated by applying the block cipher to the XOR of the current plaintext block and the previous ciphertext block. That is, set $c_0 := IV$ and then, for $i = 1$ to ℓ, set $c_i := F_k(c_{i-1} \oplus m_i)$. The final ciphertext is $c_0, c_1, \ldots, c_\ell$. (See Figure 3.6.) Decryption of a ciphertext $c_0, \ldots, c_\ell$ is done by computing $m_i := F_k^{-1}(c_i) \oplus c_{i-1}$ for $i = 1, \ldots, \ell$. Note that the IV is included in the ciphertext (and so the ciphertext is n bits longer than the plaintext); this is crucial so decryption can be done.

CBC encryption is randomized, and it is possible to show:

THEOREM 3.32   *If F is a pseudorandom permutation, then CBC mode is CPA-secure.*

The main drawback of CBC mode is that encryption must be carried out

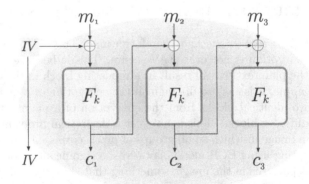

FIGURE 3.6: Cipher Block Chaining (CBC) mode.

sequentially because the previous ciphertext block c_{i-1} is needed in order to process the next plaintext block m_i. Thus, if parallel processing is available, CBC-mode encryption may not be the most efficient choice.

There is a stateful variant of CBC-mode encryption—called *chained CBC mode*—in which the last block of the previous ciphertext is used as the *IV* when encrypting the next message. This reduces the bandwidth, as a new *IV* need not be sent each time. See Figure 3.7, where an initial message m_1, m_2, m_3 is encrypted using a uniform *IV*, and then subsequently a second message m_4, m_5 is encrypted using the final ciphertext block of the previous ciphertext (i.e., c_3) as the *IV*. (In contrast, encryption using standard CBC mode would generate a fresh, random *IV* when encrypting the second message.) Chained CBC mode was used in SSL 3.0 and TLS 1.0.

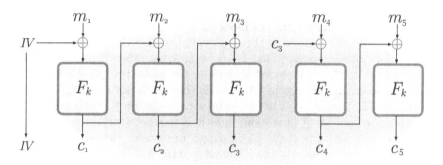

FIGURE 3.7: Chained CBC mode.

It may appear that chained CBC mode is as secure as CBC mode, since the chained-CBC encryption of m_1, m_2, m_3 followed by encryption of m_4, m_5 yields the same ciphertext blocks as CBC-mode encryption of the (single) message m_1, m_2, m_3, m_4, m_5. Nevertheless, chained CBC mode is vulnerable to a chosen-plaintext attack. The basis of the attack is that the adversary knows *in advance* the "initialization vector" c_3 that will be used for the second encrypted message. We describe the attack informally, based on Figure 3.7. Assume the attacker knows that $m_1 \in \{m_1^0, m_1^1\}$, and observes the first ciphertext IV, c_1, c_2, c_3. The attacker then requests an encryption of a second message m_4, m_5 with $m_4 = IV \oplus m_1^0 \oplus c_3$, and observes a second ciphertext c_4, c_5. One can verify that $m_1 = m_1^0$ if and only if $c_4 = c_1$, and so the attacker learns m_1. This example serves as a warning against making any modifications to cryptographic schemes, even if those modifications seem benign.

Output Feedback (OFB) mode. The third mode we present can be viewed as an unsynchronized stream-cipher mode, where the stream cipher is constructed in a specific way from the underlying block cipher. We describe the mode directly. To encrypt a message m, first a uniform $IV \in \{0,1\}^n$ is chosen. Then, a pseudorandom stream is generated from IV in the following way:

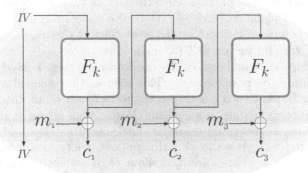

FIGURE 3.8: Output Feedback (OFB) mode.

Define $y_0 := IV$, and set the ith block y_i of the stream to be $y_i := F_k(y_{i-1})$. Each block of the plaintext is then encrypted by XORing it with the appropriate block of the stream; that is, $c_i := y_i \oplus m_i$. (See Figure 3.8.) As in CBC mode, the IV is included as part of the ciphertext to enable decryption. However, in contrast to CBC mode, here it is not required that F be invertible. (In fact, it need not even be a permutation.) Furthermore, as in stream-cipher modes of operation, here it is not necessary for the plaintext length to be a multiple of the block length n; instead, the generated stream can be truncated to exactly the plaintext length. Another advantage of OFB mode is that its stateful variant (in which the final value y_ℓ is used as the IV for encrypting the next message, and is not sent) *is* secure. This stateful variant is equivalent to a *synchronized* stream-cipher mode, with the stream cipher constructed from the block cipher in a specific way.

OFB mode can be shown to be CPA-secure if F is a pseudorandom function. Although encryption must be carried out sequentially, this mode has the advantage relative to CBC mode that the bulk of the computation (namely, computation of the pseudorandom stream) can be done independently of the actual message to be encrypted. That is, it is possible to generate a pseudorandom stream ahead of time using preprocessing, after which encryption of the plaintext (once it is known) is incredibly fast.

Counter (CTR) mode. Counter mode can also be viewed as an unsynchronized stream-cipher mode, where the stream cipher is constructed from the block cipher in a way that is analogous to Construction 3.30. We give a self-contained description here. To encrypt a message with $\ell < 2^{n/4}$ blocks using CTR mode, a uniform $IV \in \{0,1\}^{3n/4}$ is first chosen. Then, a pseudorandom stream is generated by computing $y_i := F_k(IV \parallel \langle i \rangle)$ for $i = 1, 2, \ldots$, where the counter i is encoded as an $n/4$-bit string. (The lengths of the IV and the counter are somewhat arbitrary, as long as they sum to n. A longer IV leads to better concrete security—cf. the proof of Theorem 3.33—but reduces the maximum length of messages that can be encrypted.) The ith ciphertext block

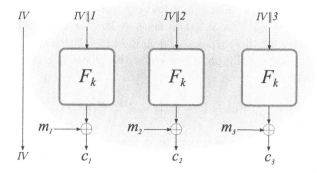

FIGURE 3.9: Counter (CTR) mode.

is computed as $c_i := y_i \oplus m_i$. As in CBC and OFB modes, the IV is included as part of the ciphertext to enable decryption; see Figure 3.9. Note again that decryption does not require F to be invertible, or even a permutation. As with OFB mode—another "stream-cipher" mode—the generated stream can be truncated to exactly the plaintext length, and preprocessing can be used to generate the pseudorandom stream before the message is known.

In contrast to all the secure modes discussed previously, CTR mode has the advantage that encryption and decryption can be fully *parallelized*, since all the blocks of the pseudorandom stream can be computed independently of each other. It is also possible to recover the ith block of the plaintext from the ciphertext using only a single evaluation of F. These features make CTR mode an attractive choice in practice.

We provide a proof that CTR mode is CPA-secure, since the proof of security in this case is relatively straightforward. We directly prove CPA-security for multiple encryptions (cf. Definition 3.22), rather than relying on Theorem 3.23, since the proof is equally simple and a direct proof yields a better concrete-security bound.

THEOREM 3.33 *If F is a pseudorandom function, then CTR mode is CPA-secure for multiple encryptions.*

PROOF We follow the same template as in the proof of Theorem 3.29: We first replace F with a random function and then analyze the resulting scheme.

Fix an arbitrary PPT adversary $\mathcal{A}$, and let $q(n)$ be a polynomial upper-bound on the number of queries made by $\mathcal{A}(1^n)$ to its left-or-right oracle. We assume for simplicity that the messages $\mathcal{A}$ submits to its oracle always contain fewer than $2^{n/4}$ blocks. (This must be true for large enough n since $\mathcal{A}$ runs in polynomial time.) Let $\Pi = (\mathsf{Gen}, \mathsf{Enc}, \mathsf{Dec})$ be the CTR-mode encryption scheme, and let $\widetilde{\Pi} = (\widetilde{\mathsf{Gen}}, \widetilde{\mathsf{Enc}}, \widetilde{\mathsf{Dec}})$ be the encryption scheme identical to Π except that a random function is used in place of F_k. That is, $\widetilde{\mathsf{Gen}}(1^n)$ chooses a uniform function $f \in \mathsf{Func}_n$, and $\widetilde{\mathsf{Enc}}$ encrypts just like Enc except that f is used instead of F_k. (Once again, neither $\widetilde{\mathsf{Gen}}$ nor $\widetilde{\mathsf{Enc}}$ is efficient but this does not matter for the purposes of defining an experiment involving $\widetilde{\Pi}$.)

As the first step of the proof, we claim that there is a negligible function negl such that

$$\left| \Pr\left[\mathsf{PrivK}^{\mathsf{LR\text{-}cpa}}_{\mathcal{A},\Pi}(n) = 1 \right] - \Pr\left[\mathsf{PrivK}^{\mathsf{LR\text{-}cpa}}_{\mathcal{A},\widetilde{\Pi}}(n) = 1 \right] \right| \leq \mathsf{negl}(n). \quad (3.13)$$

This is proved by reduction to the pseudorandomness of F in a way similar to the analogous step in the proof of Theorem 3.29, and so we omit the details.

We next claim that

$$\Pr\left[\mathsf{PrivK}^{\mathsf{LR\text{-}cpa}}_{\mathcal{A},\widetilde{\Pi}}(n) = 1 \right] \leq \frac{1}{2} + \frac{q(n)^2}{2^{3n/4+1}}. \quad (3.14)$$

Combined with Equation (3.13) this means that

$$\Pr\left[\mathsf{PrivK}^{\mathsf{LR\text{-}cpa}}_{\mathcal{A},\Pi}(n) = 1 \right] \leq \frac{1}{2} + \frac{q(n)^2}{2^{3n/4+1}} + \mathsf{negl}(n). \quad (3.15)$$

Since q is polynomial, $\frac{q(n)^2}{2^{3n/4+1}}$ is negligible and so this completes the proof.

To prove Equation (3.14), recall that a uniform IV is chosen for each of $\mathcal{A}$'s queries to its left-or-right oracle. Let IV_i be the IV used to answer the ith oracle query. There are two possibilities:

1. *Each IV is distinct, i.e., $IV_i \neq IV_j$ for all $i \neq j$:* The key observation is that in this case all the inputs to the random function f, across the entire experiment, are distinct. (If all IVs chosen are distinct, then the inputs to f when answering different oracle queries must be distinct; inputs to f when answering any particular oracle query are distinct from each other because of the counter.) Thus, the outputs of all the invocations of f are independent, uniform bit-strings. It follows that the ciphertexts returned by the left-or-right oracle are independent of the bit b determining which message is encrypted (by analogy with the one-time pad; see also the proof of Theorem 3.29). We conclude that the probability that $\mathcal{A}$ outputs $b' = b$ in this case is exactly $1/2$.

2. *Some IV is used more than once, i.e., $IV_i = IV_j$ for some $i \neq j$:* In this case, $\mathcal{A}$ can easily determine whether $b = 0$ or $b = 1$. However, this event occurs with only negligible probability. Specifically, since $\mathcal{A}$ makes at most $q(n)$ queries to its oracle and each IV is chosen uniformly from $\{0,1\}^{3n/4}$, the probability of this event is at most $\frac{q(n)^2}{2^{3n/4+1}}$ (using Lemma A.15).

Let repeat denote the event that some IV is used more than once. As just discussed, the probability that $\mathcal{A}$ succeeds in $\mathsf{PrivK}^{\mathsf{LR\text{-}cpa}}_{\mathcal{A},\widetilde{\Pi}}$ if repeat does not occur is exactly $1/2$, and $\Pr[\text{repeat}] \leq \frac{q(n)^2}{2^{3n/4+1}}$. Therefore:

$$\Pr[\text{PrivK}^{\text{LR-cpa}}_{\mathcal{A},\tilde{\Pi}}(n) = 1]$$

$$= \Pr[\text{PrivK}^{\text{LR-cpa}}_{\mathcal{A},\tilde{\Pi}}(n) = 1 \wedge \overline{\text{repeat}}] + \Pr[\text{PrivK}^{\text{LR-cpa}}_{\mathcal{A},\tilde{\Pi}}(n) = 1 \wedge \text{repeat}]$$

$$\leq \Pr[\text{PrivK}^{\text{LR-cpa}}_{\mathcal{A},\tilde{\Pi}}(n) = 1 \mid \overline{\text{repeat}}] + \Pr[\text{repeat}] \leq \frac{1}{2} + \frac{q(n)^2}{2^{3n/4+1}},$$

proving Equation (3.14). ∎

Practical Considerations

We conclude this section with a brief discussion of some issues that arise in practice when using block-cipher modes of operation.

Block length and concrete security. CBC, OFB, and CTR modes all use a uniform IV. This has the effect of randomizing the encryption process, and ensures that (with high probability) the underlying block cipher is always evaluated on *fresh* (i.e., new) inputs. This is important because, as we have noted in the proofs of Theorem 3.29 and Theorem 3.33, if an input to the block cipher is repeated an adversary may learn information about a message.

The block length of a block cipher thus has a significant impact on the concrete security of encryption schemes based on that cipher. Consider, e.g., CTR mode, whose concrete security when using a block cipher F with block length n is given by Equation (3.15). Since the IV is a uniform string of length $3n/4$, we expect an IV to repeat after encrypting $q(n) \approx 2^{3n/8}$ messages (cf. Lemma A.15). If n is too short, then the resulting concrete-security bound will be too weak for practical applications. Concretely, if $n = 64$ then after encrypting $q = 2^{24} \approx 17,000,000$ messages a repeated IV is expected to occur. Although this may seem like a lot, encrypting that many messages using a single key is commonplace nowadays.

The security bound may be weak even when n is large. For example, say $n = 128$ (which is the case for AES, a widely used block cipher we introduce in Chapter 7) and we want to use CTR mode while ensuring that an IV repeats with probability at most 2^{-32}. Solving $q^2/2^{3n/4+1} \leq 2^{-32}$ shows that we can safely encrypt only at most $q \approx 2^{32}$ messages.

We remark further that the proof of security for CTR mode given above assumes F is a pseudorandom *function*, but in practice F would be instantiated by a block cipher that is a pseudorandom *permutation*. Although every pseudorandom permutation F (with sufficiently large block length n) is also a pseudorandom function (cf. Proposition 3.26), using a pseudorandom permutation incurs a concrete-security loss of roughly $b^2/2^n$ where b denotes the number of invocations of F overall—e.g., in the case of CTR mode, b would be the total number of plaintext blocks encrypted. Thus, when b is large (even if q is small), the concrete security of CTR mode when using a block cipher may be unacceptably low.

IV misuse. In our description and discussion of the various (secure) modes, we have assumed a uniform IV of the appropriate length is chosen each time a message encrypted. What happens when this assumption fails, e.g., due to poor randomness generation or a mistaken implementation? The answer depends on the way the assumption fails, as well as the mode being used.

We first look at what happens if an IV repeats. For the "stream-cipher modes" (OFB and CTR), a repeated IV can be catastrophic: it implies that the entire pseudorandom stream (that is XORed with the plaintext) is repeated, which means that by XORing the two ciphertexts using the same IV the attacker learns the XOR of the underlying plaintexts (something we have seen previously is problematic). With CBC mode, however, one expects in practice that although some information is leaked when an IV repeats, the inputs to the block cipher in the two encryptions using the same IV will "diverge" after only a few plaintext blocks, and so the attacker will get no information about the plaintext blocks after that point.

Next, consider what happens if a scheme does not choose a *uniform IV* (even if we assume an IV never repeats); as an extreme case, imagine the IV is chosen in such a way that the attacker can predict it in advance—say, the IV is a monotonically increasing counter. CTR mode remains secure in this case, as the proof of security only requires that an IV never repeats. CBC mode, on the other hand, is no longer secure, as we have already discussed in the context of chained CBC mode.

One way to address potential IV misuse is to use *nonce-based encryption*, discussed in the following section.

Message tampering. In many texts, modes of operation are also compared based on how well they protect against adversarial modification of the ciphertext. We do *not* include such a comparison here because the issue of *message integrity* or *message authentication* must be dealt with separately from secrecy, and we do so in the next chapter. None of the above modes achieves message integrity in the sense we will define there.

With regard to the behavior of different modes in the presence of "benign" (i.e., non-adversarial) transmission errors, see Exercises 3.29 and 3.30. In general such errors can be addressed using standard non-cryptographic techniques (e.g., error correction or re-transmission).

3.6.4 *Nonce-Based Encryption

We have so far considered one particular syntax for private-key encryption—namely, Definition 3.7. Here we look at an alternate way of formalizing private-key encryption that is useful in some contexts. Specifically, we consider the notion of *nonce-based (private-key) encryption*, where the encryption and decryption algorithms additionally accept a *nonce* as input. (A "nonce" refers to a value that is supposed to be used once, and never repeated.) The syntax of nonce-based encryption does not specify where the nonce comes from; in practice, the nonce is provided by some higher-level application that must ensure that the same nonce is never used to encrypt more than once— e.g., the nonce may be a counter, or the current time.

DEFINITION 3.34 *A nonce-based (private-key) encryption scheme consists of probabilistic polynomial-time algorithms* (Gen, Enc, Dec) *such that:*

1. Gen *takes as input 1^n and outputs a key k with $|k| \geq n$.*

2. Enc *takes as input a key k, a nonce* nonce $\in \{0,1\}^*$, *and a message $m \in \{0,1\}^*$, and outputs a ciphertext c.*

3. Dec *takes as input a key k, a nonce* nonce $\in \{0,1\}^*$, *and a ciphertext c, and outputs a message $m \in \{0,1\}^*$ or $\perp$.*

We require that for every n, every k output by Gen(1^n), *every* nonce $\in \{0,1\}^*$, *and every $m \in \{0,1\}^*$, it holds that* Dec$_k($nonce, Enc$_k($nonce$, m)) = m$.

Some nonce-based encryption schemes only support nonces of a specific length; all the definitions we discuss can be adapted easily to that case.

Security for nonce-based encryption can be defined by suitably adapting any of the definitions we have seen before; for concreteness, we adapt the notion of CPA-security for multiple encryptions (Definition 3.22). The experiment we consider here is conceptually the same as the one considered in that earlier definition, and in particular we again provide the attacker with access to a "left-or-right" oracle that accepts two messages and encrypts either the "left" or "right" message. The difference here is that we also allow the attacker to *specify the nonce used during encryption*, subject to the constraint that the attacker may never repeat a nonce.

In the following experiment, the left-or-right oracle $\mathsf{LR}_{k,b}(\cdot, \cdot, \cdot)$ takes three inputs; $\mathsf{LR}_{k,b}($nonce$, m_0, m_1)$ computes $c \leftarrow$ Enc$_k($nonce$, m_b)$ and returns c. For any nonce-based encryption scheme Π, adversary $\mathcal{A}$, and security parameter n we define the following experiment:

The nonce-based LR-oracle experiment $\mathsf{PrivK}^{\mathsf{LR\text{-}ncpa}}_{\mathcal{A},\Pi}(n)$**:**

1. *A key k is generated by running* Gen(1^n).

2. *A uniform bit $b \in \{0,1\}$ is chosen.*

3. *The adversary $\mathcal{A}$ is given 1^n and oracle access to* $\mathsf{LR}_{k,b}(\cdot, \cdot, \cdot)$. *The adversary is not allowed to repeat the first input in any of its queries to the oracle.*

4. *The adversary $\mathcal{A}$ outputs a bit b'.*

5. *The output of the experiment is defined to be 1 if $b' = b$, in which case we say that $\mathcal{A}$* succeeds.

The definition of security is the same as usual, except that it now refers to the above experiment.

DEFINITION 3.35   *A nonce-based private-key encryption scheme* Π *is* CPA-secure for multiple encryptions *if for all probabilistic polynomial-time adversaries* $\mathcal{A}$ *there is a negligible function* negl *such that*

$$\Pr\left[\mathsf{PrivK}_{\mathcal{A},\Pi}^{\mathsf{LR\text{-}ncpa}}(n) = 1\right] \le \frac{1}{2} + \mathsf{negl}(n),$$

where the probability is taken over the randomness used by $\mathcal{A}$ *and the randomness used in the experiment.*

Because "CPA-security" and "CPA-security for multiple encryptions" are equivalent definitions (since an analogue of Theorem 3.23 can be shown for nonce-based encryption as well), we refer simply to CPA-security for brevity.

CPA-secure nonce-based encryption. It is easy to modify CTR mode to obtain a CPA-secure nonce-based encryption scheme: when encrypting, the *IV* is now set equal to the nonce that is provided as input, rather than being chosen uniformly. CPA-security can be shown exactly as in the proof of Theorem 3.33, using the fact that in this context repeat *cannot* occur (since the adversary is disallowed from repeating a nonce). Indeed, the concrete-security bound obtained here is better than what is obtained in the proof of Theorem 3.33 precisely because here repeat cannot occur. Of course, this is predicated on the assumption that the application using the encryption scheme ensures that nonces never repeat.

We see that a nonce-based encryption scheme can be CPA-secure even though it is *deterministic*. This does not contradict Theorem 3.20, since here we are considering an alternate syntax for encryption.

Advantages of nonce-based encryption. One may wonder what is gained by using nonce-based encryption, in particular since any nonce-based encryption scheme can be converted to a "standard" encryption scheme by simply choosing the nonce at random. There are several answers to this question.

First of all, CPA-secure nonce-based encryption is useful in settings where generating high-quality randomness is expensive or impossible. It may be much easier in such cases to use a counter as a nonce rather than to generate a nonce uniformly.

Somewhat similarly, there may be settings where using a *short* nonce is appropriate, e.g., when only very few messages will be encrypted. In such scenarios, choosing the nonce uniformly may result in a repeated nonce with probability that is unacceptably high.

Finally, we have already observed that tighter concrete-security bounds can sometimes be obtained by enforcing non-repeating nonces rather than by choosing a uniform nonce.

References and Additional Reading

The modern computational approach to cryptography was initiated in a groundbreaking paper by Goldwasser and Micali [87]. That paper introduced the notion of semantic security, and showed how that goal could be achieved in the setting of public-key encryption (see Chapters 11 and 12). The paper also proposed the notion of indistinguishability (cf. Definition 3.8), and showed that it implies semantic security. The converse was shown later [142]. Goldreich's book [83] contains further discussion of semantic security.

Blum and Micali [41] introduced the notion of pseudorandom generators and proved their existence based on a specific, number-theoretic assumption. In the same work, they also pointed out the connection between pseudorandom generators and private-key encryption as in Construction 3.17. The definition of pseudorandom generators given by Blum and Micali is different from the definition we use in this book (Definition 3.14); the latter definition originates in the work of Yao [205], who showed equivalence of the two formulations. Yao also showed constructions of pseudorandom generators based on general assumptions; we explore this topic in Chapter 8.

Formal definitions of security against chosen-plaintext attacks were given by Luby [131] and Bellare et al. [17]. See the work of Katz and Yung [112] for other notions of security for private-key encryption.

Pseudorandom functions were defined and constructed by Goldreich et al. [85], and their application to encryption was demonstrated in subsequent work by the same authors [84]. Pseudorandom permutations and strong pseudorandom permutations were studied by Luby and Rackoff [132]. These ideas are covered in Chapter 8. Stream ciphers and block ciphers had been used for many years before they began to be studied in the theoretical sense initiated by the above works. Practical constructions of stream ciphers and block ciphers are studied in Chapter 7.

The ECB, CBC, and OFB modes of operation (as well as CFB, a mode of operation not covered here) were standardized along with the DES block cipher [148]. CTR mode was standardized by NIST in 2001. CBC and CTR modes were proven CPA-secure by Bellare at al. [17]. The attack on chained CBC was first described by Rogaway (unpublished), and was used to attack SSL/TLS in the so-called "BEAST attack" by Duong and Rizzo. Nonce-based encryption was first explicitly highlighted by Rogaway [172].

Exercises

3.1 Prove Proposition 3.6.

3.2 Prove that Definition 3.8 cannot be satisfied if Π can encrypt arbitrary-length messages and the adversary is *not* restricted to outputting equal-length messages in experiment $\mathsf{PrivK}^{\mathsf{eav}}_{\mathcal{A},\Pi}$.

> **Hint:** Let $q(n)$ be a polynomial upper-bound on the length of the cipher-text when Π is used to encrypt a single bit. Then consider an adversary who outputs $m_0 \in \{0,1\}$ and a uniform $m_1 \in \{0,1\}^{q(n)+2}$.

3.3 Say $\Pi = (\mathsf{Gen}, \mathsf{Enc}, \mathsf{Dec})$ is such that for $k \in \{0,1\}^n$, algorithm Enc_k is only defined for messages of length at most $\ell(n)$ (for some polynomial ℓ). Construct a scheme satisfying Definition 3.8 even when the adversary is *not* restricted to outputting equal-length messages in $\mathsf{PrivK}^{\mathsf{eav}}_{\mathcal{A},\Pi}$.

3.4 Prove the equivalence of Definition 3.8 and Definition 3.9.

3.5 Define $G(s) \stackrel{\text{def}}{=} s \| s$ (where "$\|$" denotes concatenation). Describe and analyze an attack showing that G is not a pseudorandom generator.

3.6 Let G be a pseudorandom generator. In each of the following cases, say whether G' is necessarily a pseudorandom generator. If yes, give a proof; if not, show a counterexample.

   (a) Define $G'(s) \stackrel{\text{def}}{=} G(\bar{s})$, where $\bar{s}$ is the complement of s.

   (b) Define $G'(s) \stackrel{\text{def}}{=} \overline{G(s)}$.

   (c) Define $G'(s) \stackrel{\text{def}}{=} G\left(0^{|s|} \| s\right)$.

   (d) Define $G'(s) \stackrel{\text{def}}{=} G(s) \| G(s+1)$.

3.7 Let $|G(s)| = \ell(|s|)$ for some ℓ. Consider the following experiment:

> **The PRG indistinguishability experiment $\mathsf{PRG}_{\mathcal{A},G}(n)$:**
>
> *(a) A uniform bit $b \in \{0,1\}$ is chosen. If $b = 0$ then choose a uniform $r \in \{0,1\}^{\ell(n)}$; if $b = 1$ then choose a uniform $s \in \{0,1\}^n$ and set $r := G(s)$.*
>
> *(b) The adversary $\mathcal{A}$ is given r, and outputs a bit b'.*
>
> *(c) The output of the experiment is defined to be 1 if $b' = b$, and 0 otherwise.*

Provide a definition of a pseudorandom generator based on this experiment, and prove that your definition is equivalent to Definition 3.14. (That is, show that G satisfies your definition if and only if it satisfies Definition 3.14.)

3.8 Prove the converse of Theorem 3.16. Namely, show that if G is not a pseudorandom generator then Construction 3.17 does not have indistinguishable encryptions in the presence of an eavesdropper.

3.9 Consider a notion of indistinguishable encryption for multiple *distinct* messages, i.e., where a scheme need not hide whether the same message is encrypted twice.

(a) Modify Definition 3.18 to obtain a suitable definition of the above.

(b) Show that Construction 3.17 does not satisfy your definition.

(c) Give a construction of a *deterministic* (stateless) encryption scheme that satisfies your definition.

3.10 Prove *unconditionally* the existence of a pseudorandom function $F : \{0,1\}^* \times \{0,1\}^* \to \{0,1\}$ with $\ell_{key}(n) = n$ and $\ell_{in}(n) = \log n$.

Hint: Implement a *uniform* function with logarithmic input length.

3.11 Let F be a length preserving pseudorandom function. For the following constructions of a keyed function $F' : \{0,1\}^n \times \{0,1\}^{n-1} \to \{0,1\}^{2n}$, state whether F' is a pseudorandom function. If yes, prove it; if not, show an attack.

(a) $F'_k(x) \stackrel{\text{def}}{=} F_k(0\|x) \| F_k(0\|x)$.

(b) $F'_k(x) \stackrel{\text{def}}{=} F_k(0\|x) \| F_k(1\|x)$.

(c) $F'_k(x) \stackrel{\text{def}}{=} F_k(0\|x) \| F_k(x\|0)$.

(d) $F'_k(x) \stackrel{\text{def}}{=} F_k(0\|x) \| F_k(x\|1)$.

3.12 Assuming the existence of pseudorandom functions, prove that there is an encryption scheme that has indistinguishable multiple encryptions in the presence of an eavesdropper (i.e., satisfies Definition 3.18), but is not CPA-secure (i.e., does not satisfy Definition 3.21).

Hint: The scheme need not be "natural." You will need to use the fact that in a chosen-plaintext attack the adversary can choose its queries to the encryption oracle *adaptively*.

3.13 Let F be a keyed function and consider the following experiment:

The PRF indistinguishability experiment $\mathsf{PRF}_{\mathcal{A},F}(n)$:

(a) A uniform $b \in \{0,1\}$ is chosen. If $b = 0$, choose uniform $f \in \mathsf{Func}_n$; if $b = 1$, choose uniform $k \in \{0,1\}^n$.

(b) $\mathcal{A}$ is given 1^n as input. If $b = 0$ then $\mathcal{A}$ is given access to $f(\cdot)$. If $b = 1$ then $\mathcal{A}$ is given access to $F_k(\cdot)$.

(c) $\mathcal{A}$ outputs a bit b'.

(d) The output of the experiment is defined to be 1 if $b' = b$, and 0 otherwise.

Define pseudorandom functions using this experiment, and prove that your definition is equivalent to Definition 3.24.

3.14 Define the keyed function F as $F_k(x) \stackrel{\text{def}}{=} k \,\&\, x$, where "$\&$" denotes bitwise AND. Describe and analyze an attack showing that F is not a pseudorandom function.

3.15 Consider the following keyed function F: For security parameter n, the key is an $n \times n$ boolean matrix A and an n-bit boolean vector b. Define $F_{A,b} : \{0,1\}^n \to \{0,1\}^n$ by $F_{A,b}(x) \stackrel{\text{def}}{=} Ax + b$, where all operations are done modulo 2. Show that F is not a pseudorandom function.

3.16 Prove that if F is a length preserving pseudorandom function, then $G(s) \stackrel{\text{def}}{=} F_s(\langle 1 \rangle) \,\|\, F_s(\langle 2 \rangle) \,\|\, \cdots \,\|\, F_s(\langle \ell \rangle)$, where $\langle i \rangle$ is the n-bit encoding of i, is a pseudorandom generator with expansion factor $\ell \cdot n$.

3.17 Assume pseudorandom permutations exist. Show that there exists a keyed function F that is a pseudorandom permutation but is *not* a strong pseudorandom permutation.

> **Hint:** Construct F such that $F_k(k) = 0^{|k|}$.

3.18 Define a notion of perfect secrecy against chosen-plaintext attacks by adapting Definition 3.21. Show that the definition cannot be achieved.

3.19 Let F be a pseudorandom permutation, and define a fixed-length encryption scheme $(\mathsf{Enc}, \mathsf{Dec})$ as follows: On input a key $k \in \{0,1\}^n$ and message $m \in \{0,1\}^{n/2}$, algorithm Enc chooses a uniform string $r \in \{0,1\}^{n/2}$ and computes $c := F_k(r\|m)$.

Show how to decrypt, and prove that this scheme is CPA-secure for messages of length $n/2$.

3.20 Let F be a length preserving pseudorandom function and G be a pseudorandom generator with expansion factor $\ell(n) = n+1$. For each of the following encryption schemes, state whether the scheme is EAV-secure and whether it is CPA-secure. (In each case, the shared key is a uniform $k \in \{0,1\}^n$.) Explain your answer in each case.

   (a) To encrypt $m \in \{0,1\}^{n+1}$, choose uniform $r \in \{0,1\}^n$ and output the ciphertext $\langle r, G(r) \oplus m \rangle$.

   (b) To encrypt $m \in \{0,1\}^n$, output the ciphertext $m \oplus F_k(0^n)$.

   (c) To encrypt $m \in \{0,1\}^{2n}$, parse m as $m_1\|m_2$ with $|m_1| = |m_2|$, then choose uniform $r \in \{0,1\}^n$ and send $\langle r, m_1 \oplus F_k(r), m_2 \oplus F_k(r+1) \rangle$.

3.21 Let Π denote Construction 3.28 instantiated with the keyed function from Example 3.25. Describe and analyze an attack showing that Π is not CPA-secure.

3.22 Give a formal definition of CPA-security for stateful encryption, and prove that the synchronized stream-cipher mode of operation satisfies your definition if the underlying stream cipher is secure.

3.23 Prove that the unsynchronized stream-cipher mode of operation (Construction 3.31) is CPA-secure if the underlying stream cipher is secure.

3.24 Let F be a pseudorandom function, and consider the following construction of a stream cipher accepting an n-bit initialization vector:

- $\mathsf{Init}(s, IV)$ outputs $\mathsf{st} = (s, IV)$.
- $\mathsf{Next}(s, IV)$ outputs $y := F_s(IV)$ and $\mathsf{st}' = (s, IV + 1)$.

Show that this stream cipher is not secure.

3.25 Let F be a pseudorandom permutation. Consider the mode of operation in which a uniform value $IV \in \{0, 1\}^n$ is chosen, and the ith ciphertext block c_i is computed as $c_i := F_k(IV + i + m_i)$, where addition is modulo 2^n. Show that this scheme is not EAV-secure.

3.26 Say CBC-mode encryption is used with a block cipher having a 256-bit key and 128-bit block length to encrypt a 1024-bit message. What is the length of the resulting ciphertext?

3.27 Give the details of the proof by reduction of Equation (3.13).

3.28 For any function $g : \{0, 1\}^n \to \{0, 1\}^n$, define $g^\$(\cdot)$ to be a *probabilistic* oracle that, on input 1^n, chooses uniform $r \in \{0, 1\}^n$ and returns $\langle r, g(r) \rangle$. A keyed function F is a *weak pseudorandom function* if for all PPT algorithms D, there exists a negligible function negl such that:

$$\left| \Pr[D^{F_k^\$(\cdot)}(1^n) = 1] - \Pr[D^{f^\$(\cdot)}(1^n) = 1] \right| \leq \mathsf{negl}(n),$$

where $k \in \{0, 1\}^n$ and $f \in \mathsf{Func}_n$ are chosen uniformly.

(a) Prove that if F is pseudorandom then it is weakly pseudorandom.

(b) Let F' be a pseudorandom function, and define

$$F_k(x) \stackrel{\text{def}}{=} \begin{cases} F_k'(x) & \text{if } x \text{ is even} \\ F_k'(x + 1) & \text{if } x \text{ is odd.} \end{cases}$$

Prove that F is weakly pseudorandom, but *not* pseudorandom.

(c) Is CTR-mode encryption using a weak pseudorandom function necessarily CPA-secure? Prove your answer.

(d) Prove that Construction 3.28 is CPA-secure if F is a weak pseudorandom function.

3.29 What is the effect of a single bit flip in the ciphertext when using the CBC, OFB, and CTR modes of operation?

3.30 What is the effect of a dropped ciphertext block (e.g., if the transmitted ciphertext $c_1, c_2, c_3, \ldots$ is received as $c_1, c_3, \ldots$) when using the CBC, OFB, and CTR modes of operation?

3.31 Consider a variant of CTR mode where a uniform $IV \in \{0,1\}^n$ is chosen and the ith ciphertext block is computed as $c_i := m_i \oplus F_k(IV + i)$. Prove that this variant is CPA-secure. What concrete-security bound do you obtain?

3.32 Show that the scheme from Exercise 3.31 is *not* secure as a nonce-based encryption scheme if the nonce is used as the IV.

3.33 Show that CBC mode is not secure as a nonce-based encryption scheme if the nonce is used as the IV.

Chapter 4

Message Authentication Codes

4.1 Message Integrity

4.1.1 Secrecy vs. Integrity

A basic goal of cryptography is to enable parties to communicate securely. But what does "secure communication" entail? In Chapter 3 we showed how it is possible to achieve *secrecy*; that is, we showed how encryption can be used to prevent a passive eavesdropper from learning anything about messages sent over an open channel. However, not all security concerns are related to secrecy, and not all adversaries are limited to passive eavesdropping. In many cases, it is of equal or greater importance to guarantee *message integrity* (or *message authentication*) against an *active* adversary who can inject messages on the channel or modify messages in transit. We consider two motivating examples corresponding to the settings of Figures 1.1 and 1.2, respectively.

Imagine first a user communicating with her bank over the Internet. When the bank receives a request to transfer $1,000 from the user's account to the account of some other user X, the bank has to consider the following:

1. Is the request authentic? That is, did the user in question really issue this request, or was the request issued by an adversary (perhaps X itself) who is impersonating the legitimate user?

2. Assuming a transfer request *was* issued by the legitimate user, is the request received by the bank exactly the same as what was sent by that user? Or was, e.g., the transfer amount modified as the request was sent across the Internet?

Note that standard error-correction techniques do not suffice for the second concern. Error-correcting codes are only intended to detect and recover from "random" errors that affect a small portion of the transmission, but they do nothing to protect against a malicious adversary who can choose exactly where to introduce an arbitrary number of changes.

A very different scenario where the need for message integrity arises in practice is with regard to web cookies. The HTTP protocol used for web traffic is *stateless*, so when a client and server communicate in some session (e.g., when a user [client] shops at a merchant's [server's] website), any state generated as part of that session (e.g., the contents of the user's shopping cart) is often placed in a "cookie" that is stored by the user and included along with each message the user sends to the merchant. Assume the cookie stored by some user includes the items in the user's shopping cart along with a price for each item, as might be done if the merchant offers different prices to different users (reflecting discounts and promotions, or user-specific pricing). It would be undesirable for the user to be able to modify the cookie it stores so as to alter the prices of the items in its cart. The merchant thus needs a technique to ensure the integrity of the cookie that it stores at the user. Note that the contents of the cookie (namely, the items and their prices) are not secret and, in fact, must be known by the user. The problem here is purely one of integrity.

In general, one cannot assume the integrity of communication without taking specific measures to ensure it. Indeed, any unprotected online purchase order, online banking operation, email, or SMS message cannot, in general, be trusted to have originated from the claimed source and to have been unmodified in transit. Unfortunately, people are generally trusting and so information like the caller-ID or an email return address are taken to be "proofs of origin" in many cases, even though they are relatively easy to forge. This leaves the door open to potentially damaging attacks.

In this chapter we will show how to achieve message integrity by using cryptographic techniques to detect any spoofed messages or any tampering of messages sent over an unprotected communication channel. Note that we cannot hope to prevent message injection or message tampering altogether, as that can only be defended against at the physical level. Instead, what we will guarantee is that any such behavior will be detected by the honest parties.

4.1.2   Encryption vs. Message Authentication

Just as the goals of secrecy and message integrity are different, so are the techniques and tools for achieving them. Unfortunately, secrecy and integrity are often confused and unnecessarily intertwined, so let us be clear up front: encryption does *not* (in general) provide any integrity, and encryption should not be assumed to ensure message authentication unless it is specifically designed with that purpose in mind (something we will return to in Section 5.2).

One might mistakenly think that encryption solves the problem of message authentication. (In fact, this is a common error.) This is due to the fuzzy, and incorrect, reasoning that since a ciphertext completely hides the contents of the message, an adversary cannot possibly modify an encrypted message in any meaningful way. Despite its intuitive appeal, this reasoning is completely false. We illustrate this point by showing that all the encryption schemes we have seen thus far do not provide message integrity.

Encryption using stream ciphers. Consider encryption schemes in which the sender generates a pseudorandom pad based on a shared key (and possibly an *IV*) and then computes a ciphertext by XORing the resulting pad with a message, as in Constructions 3.17, 3.28, and 3.31 as well as OFB and CTR modes. Ciphertexts in this case are very easy to manipulate: flipping any bit in the ciphertext results in the same bit being flipped in the message that is recovered upon decryption. Thus, given a ciphertext c that encrypts a (possibly unknown) message m, it is possible for an adversary to generate a modified ciphertext c' such that $m' := \mathsf{Dec}_k(c')$ is the same as m but with a specific set of bits flipped. This simple attack can have severe consequences. As an example, consider the case of a user encrypting some dollar amount she wants to transfer from her bank account, where the amount is represented in binary. Flipping the least significant bit has the effect of changing this amount by \$1, and flipping the 11th least significant bit changes the amount by more than \$1,000! Interestingly, the adversary does not necessarily learn whether it is increasing or decreasing the initial amount, i.e., whether it is flipping a 0 to a 1 or vice versa. But if the adversary has some partial knowledge about the amount—say, that it is less than \$1,000 to begin with—then the modifications it introduces can have a predictable effect.

We stress that this attack does not contradict the *secrecy* of the encryption scheme. In fact, the exact same attack applies to the one-time pad encryption scheme, showing that even perfect secrecy is not sufficient to ensure the most basic level of message integrity.

Encryption using block ciphers. The attack described above exploits the fact that flipping a single bit in a ciphertext keeps the underlying plaintext unchanged except for the corresponding bit (which is also flipped). One might hope that encryption schemes using block ciphers in a more sophisticated way would prevent such attacks since, for example, if decryption involves inverting a (strong) pseudorandom permutation F on some portion x of the ciphertext then $F_k^{-1}(x)$ and $F_k^{-1}(x')$ will be completely uncorrelated if x and x' differ in even a single bit. Nevertheless, single-bit modifications of a ciphertext can still cause partially predictable changes in the plaintext. For example, when using ECB mode, flipping a bit in the ith block of a ciphertext affects *only* the ith block of the plaintext—all other blocks remain unchanged. (Of course, ECB mode does not even guarantee the most basic notion of secrecy, but that is irrelevant for the present discussion.) Although the effect on the ith block of the plaintext may be impossible to predict, changing that one block (while leaving everything else unchanged) may represent a harmful attack. Moreover, the order of plaintext blocks can be changed (without garbling any block) by simply changing the order of the corresponding ciphertext blocks, and the message can be truncated by dropping ciphertext blocks.

For CBC mode, flipping the jth bit of the *IV* changes only the jth bit of the first message block m_1 (since $m_1 := F_k^{-1}(c_1) \oplus IV'$, where IV' is the modified *IV*); all other plaintext blocks remain unchanged. Therefore, the first block of a CBC-encrypted message can be modified arbitrarily. We will see in Section 5.1.1 that this simple attack can have disastrous consequences.

Finally, observe that all the encryption schemes we have seen thus far have the property that *every* string of a certain length is a valid ciphertext, and so corresponds to *some* valid message. It is therefore trivial for an adversary to "spoof" a message on behalf of one of the communicating parties—by sending an arbitrary string of the correct length—even if the adversary has no idea what the underlying message will be. In the context of message integrity, even an attack of this sort should be ruled out.

4.2   Message Authentication Codes (MACs) – Definitions

We have seen that, in general, encryption does not solve the problem of message integrity. Rather, an additional mechanism is needed that will enable the communicating parties to know whether or not a message was tampered with. The right tool for this task is a *message authentication code* (MAC).

The aim of a message authentication code is to prevent an adversary from modifying a message sent by one party to another, or from injecting a new message, without the receiver detecting that the message did not originate from the intended party. As in the case of encryption, this is only possible if the communicating parties have some secret information that the adversary does not know (otherwise nothing can prevent an adversary from impersonating the party sending the message). Here, we continue to consider the private-key setting where the communicating parties share a secret key.

As in the case of private-key encryption, there are two canonical application scenarios for MACs (cf. Section 1.2): ensuring integrity for two parties communicating with each other (as in our earlier example of a user communicating with her bank), or for one user communicating "with himself" over time (as in our earlier example involving web cookies, or a user protecting the contents of his hard drive).

The Syntax of a Message Authentication Code

Before formally defining security of a message authentication code, we first define what a MAC is and how it is used. Two users who wish to communicate in an authenticated manner begin by generating and sharing a secret key k in advance of their communication. When one party wants to send a message m to the other, she computes a *tag* t based on the message and the shared key, and sends the message m along with t to the other party. The tag is computed using a *tag-generation algorithm* Mac; thus, rephrasing what we have just said, the sender of a message m computes $t \leftarrow \mathsf{Mac}_k(m)$ and transmits (m, t) to the receiver. Upon receiving (m, t), the second party *verifies* whether t is a valid tag on the message m (with respect to the shared key) or not. This is done by running a *verification algorithm* Vrfy that takes as input the shared key as well as a message m and a tag t, and indicates whether the given tag is valid. Formally:

DEFINITION 4.1 *A* message authentication code *(or* MAC*) consists of three probabilistic polynomial-time algorithms* (Gen, Mac, Vrfy) *such that:*

1. *The* key-generation algorithm Gen *takes as input the security parameter* 1^n *and outputs a key k with $|k| \geq n$.*

2. *The* tag-generation algorithm Mac *takes as input a key k and a message* $m \in \{0, 1\}^*$, *and outputs a tag t. Since this algorithm may be randomized, we write this as $t \leftarrow \mathsf{Mac}_k(m)$.*

3. *The deterministic* verification algorithm Vrfy *takes as input a key k, a message m, and a tag t. It outputs a bit b, with $b = 1$ meaning* valid *and $b = 0$ meaning* invalid. *We write this as $b := \mathsf{Vrfy}_k(m, t)$.*

It is required that for every n, every key k output by Gen(1^n)*, and every $m \in \{0, 1\}^*$, it holds that $\mathsf{Vrfy}_k(m, \mathsf{Mac}_k(m)) = 1$.*

If there is a function ℓ such that for every k output by Gen(1^n)*, algorithm* Mac_k *is only defined for messages $m \in \{0, 1\}^{\ell(n)}$, then we call the scheme a* fixed-length MAC for messages of length $\ell(n)$.

As with private-key encryption, Gen(1^n) almost always simply chooses a uniform key $k \in \{0, 1\}^n$, and we omit Gen in that case.

Canonical verification. For deterministic message authentication codes (i.e., where Mac is a deterministic algorithm), the canonical way to perform verification is simply to re-compute the tag and check for equality. In other words, $\mathsf{Vrfy}_k(m, t)$ first computes $\tilde{t} := \mathsf{Mac}_k(m)$ and then outputs 1 if and only if $\tilde{t} = t$. Even for deterministic MACs, though, it is useful to define a separate Vrfy algorithm to explicitly distinguish the semantics of *authenticating* a message to be sent vs. *verifying* authenticity of a message that was received.

Security of Message Authentication Codes

We now define the default notion of security for message authentication codes. The intuitive idea behind the definition is that no efficient adversary should be able to generate a valid tag on any "new" message that was not previously sent (and authenticated) by one of the communicating parties.

As with any security definition, to formalize this notion we need to define both the adversary's power as well as what should be considered a "break" of a scheme. As usual, we consider only probabilistic polynomial-time adversaries[1]

---

[1]See Section 4.6 for a discussion of information-theoretic message authentication, where no computational restrictions are placed on the adversary.

and so the real question is how we model the adversary's interaction with the communicating parties. In the setting of message authentication, an adversary observing the communication between the honest parties may be able to see all the messages sent by those parties along with their corresponding tags. The adversary may also be able to influence the *content* of those messages, whether directly or indirectly (if, e.g., external actions of the adversary affect the messages sent by the parties). This is true, for example, in the web cookie example from earlier, where the user's own actions influence the contents of the cookie being stored on his computer.

To model the above, we allow the adversary to request tags for *any* messages of its choice. Formally, we give the adversary access to a *MAC oracle* $\mathsf{Mac}_k(\cdot)$; the adversary can repeatedly submit any message m of its choice to this oracle, and is given in return a tag $t \leftarrow \mathsf{Mac}_k(m)$. (For a fixed-length MAC, only messages of the correct length can be submitted.)

An attacker "breaks" the scheme if it succeeds in outputting a forgery, i.e., if it outputs a message m along with a tag t such that (1) t is a valid tag on the message m (i.e., $\mathsf{Vrfy}_k(m, t) = 1$), and (2) the honest parties had not previously authenticated m (i.e., the adversary had not previously requested a tag on the message m from its oracle). These conditions imply that if the adversary were to send (m, t) to one of the honest parties, then that party would be mistakenly fooled into thinking that m originated from the other legitimate party (since $\mathsf{Vrfy}_k(m, t) = 1$) even though it did not.

A MAC that cannot be broken in the above sense is said to be *existentially unforgeable under an adaptive chosen-message attack*. "Existentially unforgeable" refers to the fact that the adversary is unable to forge a valid tag on *any* message; this should hold even if the attacker can carry out an "adaptive chosen-message attack" by which it is able to obtain tags on arbitrary messages chosen adaptively during its attack.

The above discussion leads us to consider the following experiment for a message authentication code $\Pi = (\mathsf{Gen}, \mathsf{Mac}, \mathsf{Vrfy})$, an adversary $\mathcal{A}$, and security parameter n:

The message authentication experiment $\mathsf{Mac\text{-}forge}_{\mathcal{A},\Pi}(n)$**:**

1. *A key k is generated by running* $\mathsf{Gen}(1^n)$.

2. *The adversary $\mathcal{A}$ is given input 1^n and oracle access to* $\mathsf{Mac}_k(\cdot)$. *The adversary eventually outputs (m, t). Let $\mathcal{Q}$ denote the set of all queries that $\mathcal{A}$ submitted to its oracle.*

3. $\mathcal{A}$ *succeeds if and only if (1)* $\mathsf{Vrfy}_k(m, t) = 1$ *and (2)* $m \notin \mathcal{Q}$. *In that case the output of the experiment is defined to be 1.*

A MAC is secure if no efficient adversary can succeed in the above experiment with non-negligible probability.

DEFINITION 4.2   *A message authentication code* $\Pi = (\mathsf{Gen}, \mathsf{Mac}, \mathsf{Vrfy})$ *is* existentially unforgeable under an adaptive chosen-message attack, *or just se-*

cure, *if for all probabilistic polynomial-time adversaries* $\mathcal{A}$, *there is a negligible function* negl *such that:*

$$\Pr[\mathsf{Mac\text{-}forge}_{\mathcal{A},\Pi}(n) = 1] \leq \mathsf{negl}(n).$$

Is the definition *too* strong? The above definition is rather strong in two respects. First, the adversary is allowed to repeatedly request tags for *any* messages of its choice. Second, the adversary is considered to have "broken" the scheme if it can output a valid tag on *any* previously unauthenticated message. One might object that both these components of the definition are unrealistic and overly strong, as in "real-world" usage of a MAC the honest parties would only authenticate "meaningful" messages (over which the adversary might have only limited control), and a forgery would only be damaging if it involved forging a valid tag on a "meaningful" message. Why not tailor the definition to capture this?

The crucial point is that what constitutes a meaningful message is entirely *application dependent*. While some applications of a MAC may only ever authenticate English-language messages, other applications may authenticate spreadsheet files, others database entries, and others raw data. Protocols may also be designed where *anything* will be authenticated—in fact, certain user-authentication protocols do exactly this. By making the definition of security for MACs as strong as possible, we ensure that secure MACs are broadly applicable for a wide range of purposes, without having to worry about compatibility of the MAC with the semantics of specific applications.

Replay attacks. The above definition, and message authentication codes by themselves, offer no protection against *replay attacks* in which an attacker simply re-sends a previously authenticated message along with its (valid) tag. The fact that replay attacks are not accounted for in the definition does *not* mean they are not a serious security concern! Consider again the scenario where a user (say, Alice) sends a request to her bank to transfer $1,000 from her account to some other user (say, Bob). In doing so, Alice can compute a tag and append it to her request so the bank knows the request is authentic. If the MAC is secure, Bob will be unable to intercept the request and change the amount to $10,000 because this would involve forging a valid tag on a previously unauthenticated message. However, nothing prevents Bob from *replaying* Alice's message (along with its tag) *ten times* to the bank. If the bank accepts each of those messages, the net effect is still that $10,000 will be transferred to Bob's account rather than the desired $1,000.

Despite the real threat that replay attacks represent, a MAC by itself *cannot* protect against such attacks since verification is *stateless* (and so every time a valid pair (m, t) is presented to the verification algorithm, it will always output 1). Instead, protection against replay attacks—if such protection is necessary in a given scenario—must be handled by some higher-level application. The reason the definition of a MAC is structured this way is, once again, because we are unwilling to assume any semantics for applications that use MACs; in particular, the decision as to whether or not a replayed message should be treated as "valid" may be application dependent.

Two common techniques for preventing replay attacks are to use *sequence numbers* (also known as *counters*) or *time-stamps*. The first approach requires the communicating users to maintain (synchronized) state, and can be problematic when users communicate over a lossy channel where messages are occasionally dropped (though this problem can be mitigated). In the second approach using time-stamps, the sender appends the current time T (say, to the nearest millisecond) to the message before authenticating, and sends T along with the message and the resulting tag t. When the receiver obtains T, m, t, it verifies that t is a valid tag on $m\|T$ and that T is within some acceptable clock skew of its own current time T'. This method has its own drawbacks, including the need for the sender and receiver to maintain closely synchronized clocks, and the possibility that a replay attack can still take place if it is done quickly enough (specifically, within the acceptable time window). We will discuss replay attacks further (in a more general context) in Section 5.4.

Strong unforgeability. As defined, a secure MAC ensures that an adversary cannot generate a valid tag on a message that was never previously authenticated. But it does not rule out the possibility that an attacker might be able to generate a new, valid tag on a previously authenticated message. In other words, a secure MAC guarantees that an attacker who learns tags $t_1, \ldots$ on messages $m_1, \ldots$ will be unable to forge a valid tag t on any message $m \notin \{m_1, \ldots\}$. However, it may be possible for that adversary to generate a *different* valid tag $t'_i \neq t_i$ on some previously authenticated message m_i. In standard applications of MACs, this type of adversarial behavior is not a concern. Nevertheless, in some settings it is useful to consider a stronger definition of security for MACs where such behavior is ruled out.

To model this formally, we consider a modified experiment Mac-sforge that is defined in exactly the same way as Mac-forge, except that now the set $\mathcal{Q}$ contains *pairs* of oracle queries and their associated responses. (That is, $(m, t) \in \mathcal{Q}$ if $\mathcal{A}$ queried $\mathsf{Mac}_k(m)$ and received in response the tag t.) The adversary $\mathcal{A}$ succeeds (and experiment Mac-sforge evaluates to 1) if and only if $\mathcal{A}$ outputs (m, t) such that $\mathsf{Vrfy}_k(m, t) = 1$ and $(m, t) \notin \mathcal{Q}$.

DEFINITION 4.3 *A message authentication code* $\Pi = (\mathsf{Gen}, \mathsf{Mac}, \mathsf{Vrfy})$ *is* strongly secure *if for all probabilistic polynomial-time adversaries* $\mathcal{A}$, *there is a negligible function* negl *such that:*

$$\Pr[\textsf{Mac-sforge}_{\mathcal{A}, \Pi}(n) = 1] \leq \mathsf{negl}(n).$$

It is not hard to see that if a secure MAC uses canonical verification then it is also strongly secure. This is important since many real-world MACs use canonical verification. We leave the proof of the following as an exercise.

PROPOSITION 4.4 *Let* $\Pi = (\mathsf{Gen}, \mathsf{Mac}, \mathsf{Vrfy})$ *be a secure (deterministic) MAC that uses canonical verification. Then* Π *is strongly secure.*

Verification queries. Definitions 4.2 and 4.3 consider an adversary given access to a MAC oracle, which corresponds to a real-world adversary who can influence an honest sender to generate a tag for some message m. One could also consider an adversary who interacts with an honest receiver, sending (m, t) to the receiver to learn whether $\mathsf{Vrfy}_k(m, t) = 1$. Such an adversary could be captured formally in the natural way by giving the adversary in the above definitions access to a verification oracle as well.

A definition that incorporates a verification oracle in this way is, perhaps, the "right" way to define security for message authentication codes. It turns out, however, that for MACs that use canonical verification it makes no difference: any such MAC that satisfies Definition 4.2 also satisfies the definitional variant in which verification queries are allowed. Moreover, any strongly secure MAC remains strongly secure even if verification queries are possible. In general, however, allowing verification queries can make a difference. Since most MACs covered in this book (as well as MACs used in practice) use canonical verification and/or are strongly secure, we use the traditional definitions that omit access to a verification oracle.

A potential timing attack. One issue not addressed by the above discussion of verification queries is the possibility of carrying out a *timing attack* on MAC verification. Here, we consider an adversary who can send message/tag pairs to the receiver—thus using the receiver as a verification oracle—and learn not only whether the receiver accepts or rejects, but also the *time* it takes for the receiver to make this decision. We show that if such an attack is possible then a natural implementation of MAC verification leads to an easily exploitable vulnerability. (In our usual cryptographic definitions of security, the attacker learns only the output of the oracles it has access to, but nothing else. The attack we describe here, which is an example of a *side-channel attack*, shows that certain real-world attacks are not captured by the usual definitions.)

Concretely, assume a MAC using canonical verification. To verify a tag t on a message m, the receiver computes $t' := \mathsf{Mac}_k(m)$ and then compares t' to t, outputting 1 if and only if t' and t are equal. Assume this comparison is implemented using a standard routine (like `strncmp` in C) that compares t and t' one byte at a time, and rejects as soon as the first unequal byte is encountered. The observation is that, when implemented in this way, the *time* to reject differs depending on the *position* of the first unequal byte.

It is possible to use this seemingly inconsequential information to forge a tag on any desired message m. Say the attacker knows the first i bytes of the (unique) valid tag for m. (At the outset, $i = 0$.) The attacker can learn the next byte of the valid tag by sending (m, t_0), ..., (m, t_{255}) to the receiver, where t_j is the string with the first i bytes set correctly, the $(i+1)$st byte equal to j (in hexadecimal), and the remaining bytes set to `0x00`. All these tags will likely be rejected (if not, then the attacker succeeds anyway); however, for exactly one of these tags the first $(i+1)$ bytes will be correct and rejection will take slightly longer than the rest. If t_j is the tag that caused rejection to take the longest, the attacker learns that the $(i+1)$st byte of the valid tag is j. In this way, the attacker learns each byte of the valid tag using at most 256 queries to the verification oracle. For a 16-byte tag, this attack requires at most $16 \cdot 256 = 4096$ verification queries to learn the entire tag.

One might wonder whether this attack is realistic, as it requires access to a verification oracle as well as the ability to measure the difference in time taken to compare i vs. $i+1$ bytes. In fact, such attacks have been carried out against real systems! As just one example, MACs were used to verify code updates in the Xbox 360, and the implementation of MAC verification took roughly 2.2 milliseconds to compare each byte. Attackers were able to exploit this and load pirated games onto the hardware.

Based on the above, we conclude that MAC verification should use *time-independent* string comparison that always compares *all* bytes.

4.3 Constructing Secure Message Authentication Codes

4.3.1 A Fixed-Length MAC

Pseudorandom functions are a natural tool for constructing secure message authentication codes. Intuitively, if the tag t is obtained by applying a pseudorandom function to the message m, then forging a tag on a previously unauthenticated message requires the adversary to correctly guess the value of the pseudorandom function at a "new" input point. The probability of guessing the value of a *random* function on a new point is 2^{-n} (if the output length of the function is n). The probability of guessing such a value for a *pseudorandom* function can be only negligibly greater.

The above idea, shown in Construction 4.5, gives a secure *fixed-length* MAC for *short* messages. In Section 4.3.2, we show how to extend this to handle messages of arbitrary length. We explore more efficient constructions of MACs for arbitrary-length messages in Sections 4.4, 4.5, and 6.3.2.

THEOREM 4.6 *If F is a pseudorandom function, then Construction 4.5 is a secure fixed-length MAC for messages of length n.*

PROOF As in the analysis of previous schemes based on pseudorandom functions, we first replace the pseudorandom function with a truly random function and show that this has limited impact on an adversary's success probability. We then analyze the scheme when using a truly random function.

Let $\mathcal{A}$ be a probabilistic polynomial-time adversary. Consider the message

---

CONSTRUCTION 4.5

Let F be a (length preserving) pseudorandom function. Define a fixed-length MAC for messages of length n as follows:

- Mac: on input a key $k \in \{0,1\}^n$ and a message $m \in \{0,1\}^n$, output the tag $t := F_k(m)$.

- Vrfy: on input a key $k \in \{0,1\}^n$, a message $m \in \{0,1\}^n$, and a tag $t \in \{0,1\}^n$, output 1 if and only if $t \stackrel{?}{=} F_k(m)$.

A fixed-length MAC from any pseudorandom function.

authentication code $\widetilde{\Pi} = (\widetilde{\mathsf{Gen}}, \widetilde{\mathsf{Mac}}, \widetilde{\mathsf{Vrfy}})$ which is the same as $\Pi = (\mathsf{Mac}, \mathsf{Vrfy})$ in Construction 4.5 except that a truly random function f is used instead of the pseudorandom function F_k. That is, $\widetilde{\mathsf{Gen}}(1^n)$ works by choosing a uniform function $f \in \mathsf{Func}_n$, and $\widetilde{\mathsf{Mac}}$ computes a tag just as Mac does except that f is used instead of F_k.

We show that there is a negligible function negl such that

$$\left| \Pr[\mathsf{Mac\text{-}forge}_{A,\Pi}(n) = 1] - \Pr[\mathsf{Mac\text{-}forge}_{A,\widetilde{\Pi}}(n) = 1] \right| \leq \mathsf{negl}(n). \qquad (4.1)$$

To prove this, we construct a polynomial-time distinguisher D that is given oracle access to some function $\mathcal{O}$, and whose goal is to determine whether $\mathcal{O}$ is pseudorandom (i.e., equal to F_k for uniform $k \in \{0,1\}^n$) or random (i.e., equal to f for uniform $f \in \mathsf{Func}_n$). To do this, D simulates the message authentication experiment for A and observes whether A succeeds in outputting a valid tag on a "new" message. If so, D guesses that its oracle is a pseudorandom function; otherwise, D guesses that its oracle is a random function. In detail:

Distinguisher D:
D is given input 1^n and access to an oracle $\mathcal{O} : \{0,1\}^n \rightarrow \{0,1\}^n$, and works as follows:

1. Run $A(1^n)$. Whenever A queries its MAC oracle on a message m (i.e., whenever A requests a tag on a message m), answer this query in the following way:

   Query $\mathcal{O}$ with m and obtain response t; return t to A.

2. When A outputs (m,t) at the end of its execution, do:

   (a) Query $\mathcal{O}$ with m and obtain response t'.

   (b) If (1) $t' = t$ and (2) A never queried its MAC oracle on m, then output 1; otherwise, output 0.

It is clear that D runs in polynomial time.

If D's oracle is F_k for a uniform k, then the view of A when run as a subroutine by D is distributed identically to the view of A in experiment

Mac-forge$_{\mathcal{A},\Pi}(n)$. Moreover, D outputs 1 exactly when Mac-forge$_{\mathcal{A},\Pi}(n) = 1$. Therefore

$$\Pr\left[D^{F_k(\cdot)}(1^n) = 1\right] = \Pr\left[\text{Mac-forge}_{\mathcal{A},\Pi}(n) = 1\right],$$

where $k \in \{0,1\}^n$ is chosen uniformly on the left-hand side above. If D's oracle is a random function, then the view of $\mathcal{A}$ when run as a subroutine by D is distributed identically to the view of $\mathcal{A}$ in experiment Mac-forge$_{\mathcal{A},\tilde{\Pi}}(n)$, and again D outputs 1 exactly when Mac-forge$_{\mathcal{A},\tilde{\Pi}}(n) = 1$. Thus,

$$\Pr\left[D^{f(\cdot)}(1^n) = 1\right] = \Pr\left[\text{Mac-forge}_{\mathcal{A},\tilde{\Pi}}(n) = 1\right],$$

where $f \in \mathsf{Func}_n$ is chosen uniformly. Since F is a pseudorandom function and D runs in polynomial time, there is a negligible function negl such that

$$\left|\Pr\left[D^{F_k(\cdot)}(1^n) = 1\right] - \Pr\left[D^{f(\cdot)}(1^n) = 1\right]\right| \leq \mathsf{negl}(n).$$

This implies Equation (4.1).

To complete the proof, we observe that

$$\Pr[\text{Mac-forge}_{\mathcal{A},\tilde{\Pi}}(n) = 1] \leq 2^{-n} \tag{4.2}$$

because for any message $m \notin \mathcal{Q}$ that $\mathcal{A}$ did not query to its MAC oracle, the tag $t' = f(m)$ is uniformly distributed in $\{0,1\}^n$ from $\mathcal{A}$'s point of view (since the values of f on all inputs are uniform and independent). Thus, the probability that $\mathcal{A}$ can correctly guess t' (for any $m \notin \mathcal{Q}$) is 2^{-n}.

Equations (4.1) and (4.2) together show that

$$\Pr[\text{Mac-forge}_{\mathcal{A},\Pi}(n) = 1] \leq 2^{-n} + \mathsf{negl}(n),$$

completing the proof of the theorem. ∎

4.3.2 Domain Extension for MACs

Construction 4.5 is important in that it shows a general paradigm for constructing secure message authentication codes from pseudorandom functions. Unfortunately, the construction is only capable of handling *fixed-length* messages that are furthermore rather short.[2] These limitations are unacceptable in most real-world applications. We show here how a MAC handling arbitrary-length messages can be constructed from any fixed-length MAC for messages

---

[2]Given a pseudorandom function taking arbitrary-length inputs, Construction 4.5 would yield a secure MAC for messages of arbitrary length. Likewise, a pseudorandom function with a larger domain would yield a secure MAC for longer messages. However, existing *practical* pseudorandom functions (i.e., block ciphers) take short, fixed-length inputs.

of length n. The construction we show is not very efficient and is unlikely to be used in practice; far more efficient constructions of secure MACs are known, as we will see later. We include the present construction for its simplicity and generality, and for pedagogical purposes.

Let $\Pi' = (\mathsf{Mac}', \mathsf{Vrfy}')$ be a secure fixed-length MAC for messages of length n. Before presenting the construction of a MAC for arbitrary-length messages based on Π', we rule out some simple ideas and describe some canonical attacks that must be prevented.

1. A natural first idea is to parse the message m as a sequence of n-bit blocks $m_1, \ldots, m_d$ and authenticate each block separately, i.e., compute $t_i := \mathsf{Mac}'_k(m_i)$ and output $\langle t_1, \ldots, t_d \rangle$ as the tag. This prevents an adversary from sending any previously unauthenticated block without being detected. However, it does not prevent a *block re-ordering attack* in which the attacker shuffles the order of blocks in an authenticated message. Specifically, if $\langle t_1, t_2 \rangle$ is a valid tag on the message m_1, m_2 (with $m_1 \neq m_2$), then an attacker can construct a valid tag $\langle t_2, t_1 \rangle$ on the (new) message m_2, m_1, something that is not allowed by Definition 4.2.

2. We can prevent the previous attack by authenticating a block index along with each block. That is, we now compute $t_i = \mathsf{Mac}'_k(i \| m_i)$ for all i, and output $\langle t_1, \ldots, t_d \rangle$ as the tag. (Note that now $|m_i| < n$.) This does not prevent a *truncation attack* whereby an attacker simply drops blocks from the end of the message (and drops the corresponding blocks of the tag as well).

3. A truncation attack can be thwarted by additionally authenticating the message length along with each block. (Authenticating the message length as a separate block does not work. Do you see why?) That is, compute $t_i = \mathsf{Mac}'_k(\ell \| i \| m_i)$ for all i, where ℓ denotes the length of the message in bits. (Once again, the block length $|m_i|$ will need to decrease.) This scheme is vulnerable to a *"mix-and-match" attack* where the adversary combines blocks from different messages. For example, if the adversary obtains tags $\langle t_1, \ldots, t_d \rangle$ and $\langle t'_1, \ldots, t'_d \rangle$ on messages $m = m_1, \ldots, m_d$ and $m' = m'_1, \ldots, m'_d$, respectively, it can output the valid tag $\langle t_1, t'_2, t_3, t'_4, \ldots \rangle$ on the message $m_1, m'_2, m_3, m'_4, \ldots$.

We can prevent this last attack by also including a random "message identifier" in each block that prevents the attacker from combining blocks from different messages. This leads us to Construction 4.7. (The scheme only handles messages of length less than $2^{n/4}$, but this is an exponential bound.)

THEOREM 4.8 *If Π' is a secure fixed-length MAC for messages of length n, then Construction 4.7 is a secure MAC (for arbitrary-length messages).*

CONSTRUCTION 4.7

Let $\Pi' = (\mathsf{Mac}', \mathsf{Vrfy}')$ be a fixed-length MAC for messages of length n. Define a MAC as follows:

- **Mac:** on input a key $k \in \{0,1\}^n$ and a message $m \in \{0,1\}^*$ of (nonzero) length $\ell < 2^{n/4}$, parse m as d blocks $m_1, \ldots, m_d$, each of length $n/4$. (The final block is padded with 0s if necessary.) Choose a uniform message identifier $r \in \{0,1\}^{n/4}$.

  For $i = 1, \ldots, d$, compute $t_i \leftarrow \mathsf{Mac}'_k(r\|\ell\|i\|m_i)$, where i, ℓ are encoded as strings of length $n/4$.[†] Output the tag $t := \langle r, t_1, \ldots, t_d \rangle$.

- **Vrfy:** on input a key $k \in \{0,1\}^n$, a message $m \in \{0,1\}^*$ of nonzero length $\ell < 2^{n/4}$, and a tag $t = \langle r, t_1, \ldots, t_{d'} \rangle$, parse m as d blocks $m_1, \ldots, m_d$, each of length $n/4$. (The final block is padded with 0s if necessary.) Output 1 if and only if $d' = d$ and $\mathsf{Vrfy}'_k(r\|\ell\|i\|m_i, t_i) = 1$ for $1 \leq i \leq d$.

[†] Note that i and ℓ can be encoded using $n/4$ bits because $i, \ell < 2^{n/4}$.

A MAC for arbitrary-length messages from any fixed-length MAC.

PROOF The intuition is that since Π' is secure, an adversary cannot introduce a new *block* with a valid tag (with respect to Π'). Furthermore, the extra information included in each block prevents the various attacks (dropping blocks, re-ordering blocks, etc.) sketched earlier. We prove security by showing that those attacks are the only ones possible.

Let Π be the MAC given by Construction 4.7, and let $\mathcal{A}$ be a probabilistic polynomial-time adversary. We show that $\Pr[\mathsf{Mac\text{-}forge}_{\mathcal{A},\Pi}(n) = 1]$ is negligible. We first introduce some notation that will be used in the proof. Let **repeat** denote the event that the same random identifier is used in two of the tags returned by the MAC oracle in experiment $\mathsf{Mac\text{-}forge}_{\mathcal{A},\Pi}(n)$. Denoting the final output of $\mathcal{A}$ by $(m, t = \langle r, t_1, \ldots \rangle)$, where m has length ℓ and is parsed as $m = m_1, \ldots$, we let **NewBlock** be the event that at least one of the blocks $r\|\ell\|i\|m_i$ was never previously authenticated by Mac' in the course of answering $\mathcal{A}$'s Mac queries. (Note that, by construction of Π, it is easy to tell exactly which blocks are authenticated by Mac'_k when computing $\mathsf{Mac}_k(m)$.) Informally, **NewBlock** is the event that $\mathcal{A}$ tries to forge a valid tag on a block that was never authenticated by the underlying fixed-length MAC Π'.

We have:

$$\Pr[\mathsf{Mac\text{-}forge}_{\mathcal{A},\Pi}(n) = 1] = \Pr[\mathsf{Mac\text{-}forge}_{\mathcal{A},\Pi}(n) = 1 \wedge \mathsf{repeat}]$$
$$+ \Pr[\mathsf{Mac\text{-}forge}_{\mathcal{A},\Pi}(n) = 1 \wedge \overline{\mathsf{repeat}} \wedge \mathsf{NewBlock}]$$
$$+ \Pr[\mathsf{Mac\text{-}forge}_{\mathcal{A},\Pi}(n) = 1 \wedge \overline{\mathsf{repeat}} \wedge \overline{\mathsf{NewBlock}}]$$
$$\leq \Pr[\mathsf{repeat}] \qquad\qquad (4.3)$$
$$+ \Pr[\mathsf{Mac\text{-}forge}_{\mathcal{A},\Pi}(n) = 1 \wedge \mathsf{NewBlock}]$$
$$+ \Pr[\mathsf{Mac\text{-}forge}_{\mathcal{A},\Pi}(n) = 1 \wedge \overline{\mathsf{repeat}} \wedge \overline{\mathsf{NewBlock}}].$$

We show that the first two terms of Equation (4.3) are negligible, and the final term is 0. This implies $\Pr[\text{Mac-forge}_{\mathcal{A},\Pi}(n) = 1]$ is negligible, as desired.

To see that $\Pr[\text{repeat}]$ is negligible, let $q = q(n)$ be the number of MAC oracle queries made by $\mathcal{A}$. To answer the ith oracle query of $\mathcal{A}$, the oracle chooses r_i uniformly from a set of size $2^{n/4}$. The probability of event repeat is exactly the probability that $r_i = r_j$ for some $i \neq j$. Applying Lemma A.15, we have $\Pr[\text{repeat}] \leq q^2/2^{n/4}$. Since q is polynomial (because $\mathcal{A}$ is a PPT adversary), this value is negligible.

We next consider the final term in Equation (4.3). We argue that if $\text{Mac-forge}_{\mathcal{A},\Pi}(n) = 1$, but repeat did not occur, then it must be the case that NewBlock occurred. In other words,

$$\Pr[\text{Mac-forge}_{\mathcal{A},\Pi}(n) = 1 \wedge \overline{\text{repeat}} \wedge \overline{\text{NewBlock}}] = 0.$$

This is, in some sense, the heart of the proof.

Again let $q = q(n)$ denote the number of MAC oracle queries made by $\mathcal{A}$, and let r_i denote the random identifier used to answer the ith oracle query of $\mathcal{A}$. If repeat does not occur then the values $r_1, \ldots, r_q$ are distinct. Recall that $(m, t = \langle r, t_1, \ldots \rangle)$ is the output of $\mathcal{A}$. If $r \notin \{r_1, \ldots, r_q\}$, then NewBlock clearly occurs. If not, then $r = r_j$ for some unique j (because repeat did not occur), and the blocks $r\|\ell\|1\|m_1, \ldots$ could then not possibly have been authenticated during the course of answering any Mac queries other than the jth such query. Let $m^{(j)}$ be the message that was used by $\mathcal{A}$ for its jth oracle query, and let ℓ_j be its length. There are two cases to consider:

Case 1: $\ell \neq \ell_j$. The blocks authenticated when answering the jth Mac query all have $\ell_j \neq \ell$ in the second position. So $r\|\ell\|1\|m_1$, in particular, was never authenticated in the course of answering the jth Mac query, and NewBlock occurs.

Case 2: $\ell = \ell_j$. If $\text{Mac-forge}_{\mathcal{A},\Pi}(n) = 1$, then we must have $m \neq m^{(j)}$. Let $m^{(j)} = m_1^{(j)}, \ldots$. Since m and $m^{(j)}$ have equal length, there must be at least one index i for which $m_i \neq m_i^{(j)}$. The block $r\|\ell\|i\|m_i$ was then never authenticated in the course of answering the jth Mac query. (Because i is included in the third position of the block, the block $r\|\ell\|i\|m_i$ could only possibly have been authenticated if $r\|\ell\|i\|m_i = r_j\|\ell_j\|i\|m_i^{(j)}$, but this is not true since $m_i \neq m_i^{(j)}$.)

To complete the proof of the theorem, we bound the second term on the right-hand side of Equation (4.3). Here we rely on the security of Π'. We construct a PPT adversary $\mathcal{A}'$ who attacks the fixed-length MAC Π' and succeeds in outputting a valid tag on a previously unauthenticated message with probability

$$\Pr[\text{Mac-forge}_{\mathcal{A}',\Pi'}(n) = 1] \geq \Pr[\text{Mac-forge}_{\mathcal{A},\Pi}(n) = 1 \wedge \text{NewBlock}]. \quad (4.4)$$

Security of Π' means that the left-hand side is negligible, implying that $\Pr[\mathsf{Mac\text{-}forge}_{\mathcal{A},\Pi}(n) = 1 \wedge \mathsf{NewBlock}]$ is negligible as well.

The construction of $\mathcal{A}'$ is the obvious one and so we describe it briefly. $\mathcal{A}'$ runs $\mathcal{A}$ as a subroutine, and answers the request by $\mathcal{A}$ for a tag on m by choosing $r \leftarrow \{0,1\}^{n/4}$ itself, parsing m appropriately, and making the necessary queries to its own MAC oracle $\mathsf{Mac}'_k(\cdot)$. When $\mathcal{A}$ outputs $(m, t = \langle r, t_1, \ldots \rangle)$, then $\mathcal{A}'$ checks whether $\mathsf{NewBlock}$ occurs. (This is easy to do since $\mathcal{A}'$ can keep track of all the queries it makes to its own oracle.) If so, then $\mathcal{A}'$ finds the first block $r\|\ell\|i\|m_i$ that was never previously authenticated by Mac' and outputs $(r\|\ell\|i\|m_i, t_i)$. (If not, $\mathcal{A}'$ outputs nothing.)

The view of $\mathcal{A}$ when run as a subroutine by $\mathcal{A}'$ is distributed identically to the view of $\mathcal{A}$ in experiment $\mathsf{Mac\text{-}forge}_{\mathcal{A},\Pi}(n)$, and so the probabilities of events $\mathsf{Mac\text{-}forge}_{\mathcal{A},\Pi}(n) = 1$ and $\mathsf{NewBlock}$ do not change. If $\mathsf{NewBlock}$ occurs then $\mathcal{A}'$ outputs a block $r\|\ell\|i\|m_i$ that was never previously authenticated by its own MAC oracle; if $\mathsf{Mac\text{-}forge}_{\mathcal{A},\Pi}(n) = 1$ then the tag on every block is valid (with respect to Π'), and so in particular this is true for the block output by $\mathcal{A}'$. This means that whenever $\mathsf{Mac\text{-}forge}_{\mathcal{A},\Pi}(n) = 1$ and $\mathsf{NewBlock}$ occur we have $\mathsf{Mac\text{-}forge}_{\mathcal{A}',\Pi'}(n) = 1$, proving Equation (4.4) and completing the proof of the theorem. ∎

4.4   CBC-MAC

Theorems 4.6 and 4.8 show that it is possible to construct a secure message authentication code for arbitrary-length messages from a pseudorandom function with block length n. This demonstrates, in principle, that secure MACs can be constructed from block ciphers. Unfortunately, the resulting construction is extremely inefficient: to compute a tag on a message of length dn, the block cipher is evaluated $4d$ times, and the tag is more than $4dn$ bits long. Fortunately, far more efficient constructions are available. We begin by exploring one such construction that relies solely on block ciphers.

4.4.1   The Basic Construction

CBC-MAC was one of the first message authentication codes to be standardized. A basic version of CBC-MAC, secure when authenticating messages of any *fixed* length, is given as Construction 4.9. (See also Figure 4.1.) We caution that this basic scheme is *not* secure in the general case when messages of different lengths may be authenticated; see further discussion below.

THEOREM 4.10   *Let ℓ be a polynomial. If F is a pseudorandom function, then Construction 4.9 is a secure MAC for messages of length $\ell(n) \cdot n$.*

CONSTRUCTION 4.9

Let F be a pseudorandom function, and fix a length function $\ell(n) > 0$. The basic CBC-MAC construction is as follows:

- Mac: on input a key $k \in \{0,1\}^n$ and a message m of length $\ell(n) \cdot n$, do the following (set $\ell = \ell(n)$ in what follows):

    1. Parse m as $m = m_1, \ldots, m_\ell$ where each m_i is of length n.

    2. Set $t_0 := 0^n$. Then, for $i = 1$ to ℓ, set $t_i := F_k(t_{i-1} \oplus m_i)$.

    Output t_ℓ as the tag.

- Vrfy: on input a key $k \in \{0,1\}^n$, a message m, and a tag t, do: If m is not of length $\ell(n) \cdot n$ then output 0. Otherwise, output 1 if and only if $t \stackrel{?}{=} \mathsf{Mac}_k(m)$.

Basic CBC-MAC (for fixed-length messages).

The proof of Theorem 4.10 is fairly complex. In the following section we will prove a more general result from which the above theorem follows.

Although Construction 4.9 can be extended in the obvious way to handle messages of different lengths, the construction is only secure when the length of the messages being authenticated is fixed and agreed upon in advance by the sender and receiver. (See Exercise 4.13.) The advantage of this construction over Construction 4.5, which also gives a fixed-length MAC, is that basic CBC-MAC can authenticate longer messages. Compared to Construction 4.7, basic CBC-MAC is much more efficient, requiring only d block-cipher evaluations for a message of length dn, and with a tag of length n.

CBC-MAC vs. CBC-mode encryption. Basic CBC-MAC is similar to the CBC mode of operation. There are, however, some important differences:

1. CBC-mode encryption uses a *random IV* and this is crucial for security. In contrast, CBC-MAC uses no *IV* (alternately, it can be viewed as

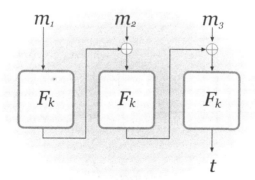

FIGURE 4.1: Basic CBC-MAC (for fixed-length messages).

using the fixed value $IV = 0^n$) and this is also crucial for security. Specifically, CBC-MAC using a random IV is not secure.

2. In CBC-mode encryption all intermediate values t_i (called c_i in the case of CBC-mode encryption) are output by the encryption algorithm as part of the ciphertext, whereas in CBC-MAC only the final block is output as the tag. If CBC-MAC is modified to output all the $\{t_i\}$ obtained during the course of the computation then it is no longer secure.

In Exercise 4.14 you are asked to verify that the modifications of CBC-MAC discussed above are insecure. These examples illustrate the fact that harmless-looking modifications to cryptographic constructions can render them insecure. One should always implement a cryptographic construction exactly as specified and not introduce any variations (unless the variations themselves can be proven secure). Furthermore, it is essential to understand the details of an implementation being used. In many cases cryptographic libraries provide the programmer with a "CBC function," but do not distinguish between the use of this function for encryption or message authentication.

Secure CBC-MAC for arbitrary-length messages. We briefly describe two ways Construction 4.9 can be modified, in a provably secure manner, to handle arbitrary-length messages. (Here for simplicity we assume that all messages being authenticated have length a multiple of n, and that Vrfy rejects any message whose length is not a multiple of n. In the following section we treat the more general case where messages can have arbitrary length.)

1. *Prepend* the message m with its length $|m|$ (encoded as an n-bit string), and then compute basic CBC-MAC on the result; see Figure 4.2. Security of this variant follows from the results proved in the next section.

   Note that appending $|m|$ to the *end* of the message and then computing basic CBC-MAC is *not* secure.

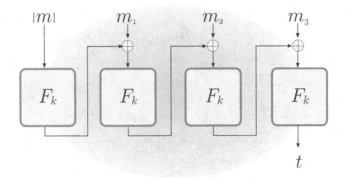

FIGURE 4.2:   A version of CBC-MAC secure for authenticating arbitrary-length messages.

2. Change the scheme so that key generation chooses two independent, uniform keys $k_1 \in \{0,1\}^n$ and $k_2 \in \{0,1\}^n$. Then to authenticate a message m, first compute the basic CBC-MAC of m using k_1 and let t be the result; output the tag $\hat{t} := F_{k_2}(t)$.

The second option has the advantage of not needing to know the message length in advance (i.e., when beginning to compute the tag). However, it has the drawback of using two keys for F. Note that, at the expense of two additional applications of F, it is possible to store a single key k and then derive the keys $k_1 := F_k(1)$ and $k_2 := F_k(2)$ at the beginning of the computation. Despite this, in practice, the operation of initializing a block cipher with a new key is considered relatively expensive, and so this option is not always desirable.

4.4.2 *Proof of Security

In this section we prove security of different variants of CBC-MAC. We begin by summarizing the results, and then give the details of the proof. The proof in this section is quite involved, and is intended for advanced readers.

Throughout this section, fix a keyed function F that, for security parameter n, maps n-bit keys and n-bit inputs to n-bit outputs. We define a keyed function CBC that, for security parameter n, maps n-bit keys and inputs in $(\{0,1\}^n)^+$ (i.e., nonempty strings whose length is a multiple of n) to n-bit outputs. This function is defined as

$$\mathsf{CBC}_k(x_1,\ldots,x_\ell) \stackrel{\text{def}}{=} F_k\left(F_k\left(\cdots F_k(F_k(x_1) \oplus x_2) \oplus \cdots\right) \oplus x_\ell\right),$$

where $|x_1| = \cdots = |x_\ell| = n$. (We leave CBC_k undefined on the empty string.) Note that CBC is computed in the same way as basic CBC-MAC, although here we explicitly allow inputs of different lengths.

A set of strings $P \subset (\{0,1\}^n)^*$ is *prefix-free* if it does not contain the empty string, and no string $X \in P$ is a prefix of any other string $X' \in P$. We show:

THEOREM 4.11 *If F is a pseudorandom function, then CBC is a pseudorandom function as long as the set of inputs on which it is queried is prefix-free. Formally, for any PPT distinguisher D that queries its oracle on a prefix-free set of inputs, there is a negligible function* negl *such that*

$$\left| \Pr[D^{\mathsf{CBC}_k(\cdot)}(1^n) = 1] - \Pr[D^{f(\cdot)}(1^n) = 1] \right| \leq \mathsf{negl}(n),$$

where k is chosen uniformly from $\{0,1\}^n$ and f is chosen uniformly from the set of functions mapping $(\{0,1\}^n)^*$ to $\{0,1\}^n$ (i.e., the value of f at each input is uniform and independent of the values of f at all other inputs).

Thus, we can convert a pseudorandom function F for fixed-length inputs into a pseudorandom function CBC for arbitrary-length inputs (subject to a constraint on which inputs can be queried). To use this for message authentication, we adapt the idea of Construction 4.5 as follows: to authenticate a message m, first apply some *encoding function* encode to obtain a string encode(m) $\in (\{0,1\}^n)^+$; then output the tag $\mathsf{CBC}_k(\mathsf{encode}(m))$. For this to be secure, the encoding needs to be *prefix-free*, namely, to have the property that for any distinct (allowed) messages m_1, m_2, the string encode(m_1) is not a prefix of encode(m_2). This implies that for any set of (allowed) messages $\{m_1, \ldots\}$, the set of encoded messages $\{\mathsf{encode}(m_1), \ldots\}$ is prefix-free.

We now examine two concrete applications of this idea:

- Fix ℓ, and let the set of allowed messages be $\{0,1\}^{\ell(n) \cdot n}$. Then we can take the trivial encoding encode(m) = m, which is prefix-free since a string cannot be a prefix of a different string of the same length. This is exactly basic CBC-MAC, and what we have said above implies that basic CBC-MAC is secure for messages of any fixed length (cf. Theorem 4.10).

- One way of handling arbitrary-length messages (technically, messages of length less than 2^n) is to encode a string $m \in \{0,1\}^*$ by prepending its length $|m|$ (encoded as an n-bit string), and then appending as many 0s as needed to make the length of the resulting string a multiple of n. (This is essentially what is shown in Figure 4.2.) This encoding is prefix-free, and we therefore obtain a secure MAC for arbitrary-length messages.

The rest of this section is devoted to a proof of Theorem 4.11. In proving the theorem, we analyze CBC when it is "keyed" with a *random function g* rather than a random key k for some underlying pseudorandom function F. That is, we consider the keyed function CBC_g defined as

$$\mathsf{CBC}_g(x_1, \ldots, x_\ell) \stackrel{\text{def}}{=} g\left(g\left(\cdots g(g(x_1) \oplus x_2) \oplus \cdots\right) \oplus x_\ell\right)$$

where, for security parameter n, the function g maps n-bit inputs to n-bit outputs, and $|x_1| = \cdots = |x_\ell| = n$. Note that CBC_g as defined here is not efficient (since the representation of g requires space exponential in n); nevertheless, it is still a well-defined, keyed function.

We show that if g is chosen uniformly from Func_n, then CBC_g is indistinguishable from a random function mapping $(\{0,1\}^n)^*$ to n-bit strings, as long as a prefix-free set of inputs is queried. More precisely:

THEOREM 4.12 *Fix any $n \geq 1$. For any distinguisher D that queries its oracle on a prefix-free set of q inputs, where the longest such input contains ℓ blocks, it holds that:*

$$\left| \Pr[D^{\mathsf{CBC}_g(\cdot)}(1^n) = 1] - \Pr[D^{f(\cdot)}(1^n) = 1] \right| \leq \frac{q^2 \ell^2}{2^n},$$

where g is chosen uniformly from Func_n, and f is chosen uniformly from the set of functions mapping $(\{0,1\}^n)^*$ to $\{0,1\}^n$.

(The theorem is unconditional, and does not impose any constraints on the running time of D. Thus we may take D to be deterministic.) The above implies Theorem 4.11 using standard techniques that we have already seen. In particular, for any D running in polynomial time $q(n)$ and $\ell(n)$ are polynomial and so $q(n)^2 \ell(n)^2 / 2^n$ is negligible.

PROOF (of Theorem 4.12) Fix some $n \geq 1$. The proof proceeds in two steps: We first define a notion of *smoothness* and prove that CBC is smooth; we then show that smoothness implies the claim.

Let $P = \{X_1, \ldots, X_q\}$ be a prefix-free set of q inputs, where each X_i is in $(\{0,1\}^n)^*$ and the longest string in P contains ℓ blocks (i.e., each $X_i \in P$ contains at most ℓ blocks of length n). Note that for any $t_1, \ldots, t_q \in \{0,1\}^n$ it holds that $\Pr[\forall i : f(X_i) = t_i] = 2^{-nq}$, where the probability is over uniform choice of the function f from the set of functions mapping $(\{0,1\}^n)^*$ to $\{0,1\}^n$. We say that CBC is (q, ℓ, δ)-*smooth* if for every prefix-free set $P = \{X_1, \ldots, X_q\}$ as above and every $t_1, \ldots, t_q \in \{0,1\}^n$, it holds that

$$\Pr[\forall i : \mathsf{CBC}_g(X_i) = t_i] \geq (1 - \delta) \cdot 2^{-nq},$$

where the probability is over uniform choice of $g \in \mathsf{Func}_n$.

In words, CBC is (q, ℓ, δ)-smooth if for every fixed set of input/output pairs $\{(X_i, t_i)\}$, where the $\{X_i\}$ form a prefix-free set and each contain at most ℓ blocks, the probability that $\mathsf{CBC}_g(X_i) = t_i$ for all i (where g is a random function from $\{0,1\}^n$ to $\{0,1\}^n$) is at least $1 - \delta$ times the probability that $f(X_i) = t_i$ for all i (where f is a random function from $(\{0,1\}^n)^*$ to $\{0,1\}^n$).

CLAIM 4.13 CBC *is* (q, ℓ, δ)-*smooth, for* $\delta = q^2 \ell^2 / 2^n$.

PROOF Fix P as above. For $X \in P$, with $X = x_1, \ldots$ and $x_i \in \{0,1\}^n$, let $\mathcal{C}_g(X)$ denote the ordered list of inputs on which g is evaluated during the computation of $\mathsf{CBC}_g(X)$; i.e., if $X \in (\{0,1\}^n)^m$ then

$$\mathcal{C}_g(X) \stackrel{\text{def}}{=} (x_1, \ \mathsf{CBC}_g(x_1) \oplus x_2, \ \ldots, \ \mathsf{CBC}_g(x_1, \ldots, x_{m-1}) \oplus x_m).$$

For $X \in (\{0,1\}^n)^m$ and $X' \in (\{0,1\}^n)^{m'}$, with $\mathcal{C}_g(X) = (I_1, \ldots, I_m)$ and $\mathcal{C}_g(X') = (I'_1, \ldots, I'_{m'})$, say there is a *non-trivial collision in* X if $I_i = I_j$ for some $i \neq j$, and say there is a *non-trivial collision between* X *and* X' if $I_i = I'_j$ but $(x_1, \ldots, x_i) \neq (x'_1, \ldots, x'_j)$. We say there is a *non-trivial collision in* P if there is a non-trivial collision in some $X \in P$ or between some pair of strings $X, X' \in P$. Let Coll be the event that there is a non-trivial collision in P.

We prove the claim in two steps. First, we show that conditioned on the event that Coll does not occur, the probability that $\mathsf{CBC}_g(X_i) = t_i$ for all i is exactly 2^{-nq}. Next, we show that $\Pr[\mathsf{Coll}] < \delta = q^2 \ell^2 / 2^n$.

Consider choosing a uniform g by choosing, one-by-one, uniform values for the outputs of g on different inputs. Determining whether there is a non-trivial collision between two strings $X, X' \in P$ can be done by first choosing the values of $g(I_1)$ and $g(I_1')$ (if $I_1' = I_1$, these values are the same), then choosing values for $g(I_2)$ and $g(I_2')$ (note that $I_2 = g(I_1) \oplus x_2$ and $I_2' = g(I_1') \oplus x_2'$ are defined once $g(I_1), g(I_1')$ have been fixed), and continuing in this way until we choose values for $g(I_{m-1})$ and $g(I_{m'-1}')$. Observe that the values of $g(I_m), g(I_{m'}')$ need not be chosen in order to determine whether there is a non-trivial collision between X and X'. Similarly, the value of $g(I_m)$ need not be chosen in order to determine whether there is a non-trivial collision in X. Thus, it is possible to determine whether Coll occurs by choosing the values of g on all but the final entries of each of $\mathcal{C}_g(X_1), \ldots, \mathcal{C}_g(X_q)$.

Assume Coll has not occurred after fixing the values of g on various inputs as described above. Consider the final entries in each of $\mathcal{C}_g(X_1), \ldots, \mathcal{C}_g(X_q)$. These entries are all distinct (this is immediate from the fact that Coll has not occurred), and we claim that the value of g on each of those points has not yet been fixed. Indeed, the only way the value of g could already be fixed on any of those points is if the final entry I_m of some $\mathcal{C}_g(X)$ is equal to a non-final entry I_j of some $\mathcal{C}_g(X')$. But since Coll has not occurred, this can only happen if $X \neq X'$ and $(x_1', \ldots, x_j') = (x_1, \ldots, x_m)$. But then X would be a prefix of X', contradicting the assumption that P is prefix-free.

Since g is a random function, the above means that $\mathsf{CBC}_g(X_1), \ldots, \mathsf{CBC}_g(X_q)$ are *uniform* and *independent* of each other as well as all the other values of g that have already been fixed. (This is because $\mathsf{CBC}_g(X_i)$ is the value of g when evaluated at the final entry of $\mathcal{C}_g(X_i)$, an input value which is different from all the other inputs at which g has already been fixed.) Thus, for any $t_1, \ldots, t_q \in \{0,1\}^n$ we have:

$$\Pr\left[\forall i : \mathsf{CBC}_g(X_i) = t_i \mid \overline{\mathsf{Coll}}\right] = 2^{-nq}. \tag{4.5}$$

We next show that $\overline{\mathsf{Coll}}$ occurs with high probability by upper-bounding $\Pr[\mathsf{Coll}]$. For distinct $X_i, X_j \in P$, let $\mathsf{Coll}_{i,j}$ be the event that there is a non-trivial collision in X_i or in X_j, or a non-trivial collision between X_i and X_j. We have $\mathsf{Coll} = \bigvee_{i,j} \mathsf{Coll}_{i,j}$ and so a union bound gives

$$\Pr[\mathsf{Coll}] \leq \sum_{i,j:\, i<j} \Pr[\mathsf{Coll}_{i,j}] < \frac{q^2}{2} \cdot \max_{i<j}\left\{\Pr[\mathsf{Coll}_{i,j}]\right\}. \tag{4.6}$$

Fixing distinct $X = X_i$ and $X' = X_j$ in P, we now bound $\max_{i<j}\{\Pr[\mathsf{Coll}_{i,j}]\}$. It is clear that the probability is maximized when X and X' are both as long as possible, and thus we assume they are each ℓ blocks long. Let $X = (x_1, \ldots, x_\ell)$ and $X' = (x_1', \ldots, x_\ell')$, and let t be the largest integer such that $(x_1, \ldots, x_t) = (x_1', \ldots, x_t')$. (Note that $t < \ell$ or else $X = X'$.) We assume $t > 0$, but the analysis below can be easily modified, giving the same result, if $t = 0$. We continue to let $I_1, I_2, \ldots$ (resp., $I_1', I_2', \ldots$) denote the inputs to g during the course of computing $\mathsf{CBC}_g(X)$ (resp., $\mathsf{CBC}_g(X')$); note that $(I'_1, \ldots, I'_t) = (I_1, \ldots, I_t)$. Consider choosing g by choosing uniform values for the outputs of g, one-by-one. We do this in $2\ell - t - 2$ steps as follows:

Steps 1 through $t - 1$ (if $t > 1$): In each step i, choose a uniform value for $g(I_i)$, thus defining I_{i+1} and I'_{i+1} (which are equal).

Step t: Choose a uniform value for $g(I_t)$, thus defining I_{t+1} and I'_{t+1}.

Steps $t + 1$ to $\ell - 1$ (if $t < \ell - 1$): Choose, in turn, uniform values for each of $g(I_{t+1})$, $g(I_{t+2})$, $\ldots$, $g(I_{\ell-1})$, thus defining I_{t+2}, I_{t+3}, $\ldots$, I_ℓ.

Steps ℓ to $2\ell - t - 2$ (if $t < \ell - 1$): Choose, in turn, uniform values for each of $g(I'_{t+1})$, $g(I'_{t+2})$, $\ldots$, $g(I'_{\ell-1})$, thus defining I'_{t+2}, I'_{t+3}, $\ldots$, I'_ℓ.

Let $\mathsf{Coll}(k)$ be the event that a non-trivial collision occurs by step k. Then

$$\Pr[\mathsf{Coll}_{i,j}] = \Pr\left[\bigvee_k \mathsf{Coll}(k)\right]$$

$$\leq \Pr[\mathsf{Coll}(1)] + \sum_{k=2}^{2\ell - t - 2} \Pr[\mathsf{Coll}(k) \mid \overline{\mathsf{Coll}(k-1)}], \qquad (4.7)$$

using Proposition A.9. For $k < t$, we claim $\Pr[\mathsf{Coll}(k) \mid \overline{\mathsf{Coll}(k-1)}] = k/2^n$; indeed, if no non-trivial collision has occurred by step $k - 1$, the value of $g(I_k)$ is chosen uniformly in step k; a non-trivial collision occurs only if it happens that $I_{k+1} = g(I_k) \oplus x_{k+1}$ is equal to one of $\{I_1, \ldots, I_k\}$ (which are all distinct, since $\mathsf{Coll}(k-1)$ has not occurred). By similar reasoning, we have $\Pr[\mathsf{Coll}(t) \mid \overline{\mathsf{Coll}(t-1)}] \leq 2t/2^n$ (here there are two values I_{t+1}, I'_{t+1} to consider; note that they cannot be equal to each other). Finally, arguing as before, for $k > t$ we have $\Pr[\mathsf{Coll}(k) \mid \overline{\mathsf{Coll}(k-1)}] = (k+1)/2^n$. Using Equation (4.7), we thus have

$$\Pr[\mathsf{Coll}_{i,j}] \leq 2^{-n} \cdot \left(\sum_{k=1}^{t-1} k + 2t + \sum_{k=t+1}^{2\ell - t - 2} (k+1)\right)$$

$$= 2^{-n} \cdot \sum_{k=2}^{2\ell - t - 1} k < 2\ell^2 \cdot 2^{-n}.$$

From Equation (4.6) we get $\Pr[\mathsf{Coll}] < q^2 \ell^2 \cdot 2^{-n} = \delta$. Finally, using Equation (4.5) we see that

$$\Pr\left[\forall i : \mathsf{CBC}_g(X_i) = t_i\right] \geq \Pr\left[\forall i : \mathsf{CBC}_g(X_i) = t_i \mid \overline{\mathsf{Coll}}\right] \cdot \Pr[\overline{\mathsf{Coll}}]$$

$$= 2^{-nq} \cdot \Pr[\overline{\mathsf{Coll}}] \geq (1 - \delta) \cdot 2^{-nq},$$

as claimed. ∎

We now show that smoothness implies the theorem. Assume without loss of generality that D always makes q (distinct) queries, each containing at most ℓ

blocks. D may choose its queries adaptively (i.e., depending on the answers to previous queries), but the set of D's queries must be prefix-free.

For distinct $X_1, \ldots, X_q \in (\{0,1\}^n)^*$ and arbitrary $t_1, \ldots, t_q \in \{0,1\}^n$, define $\alpha(X_1, \ldots, X_q; t_1, \ldots, t_q)$ to be 1 if and only if D outputs 1 when making queries $X_1, \ldots, X_q$ and getting responses $t_1, \ldots, t_q$. (If, say, D does not make query X_1 as its first query, then $\alpha(X_1, \ldots; \ldots) = 0$.) Letting $\bar{X} = (X_1, \ldots, X_q)$ and $\bar{t} = (t_1, \ldots, t_q)$, we then have

$$\Pr[D^{\mathsf{CBC}_g(\cdot)}(1^n) = 1] = \sum_{\bar{X} \text{ prefix-free};\ \bar{t}} \alpha(\bar{X}, \bar{t}) \cdot \Pr[\forall i : \mathsf{CBC}_g(X_i) = t_i]$$

$$\geq \sum_{\bar{X} \text{ prefix-free};\ \bar{t}} \alpha(\bar{X}, \bar{t}) \cdot (1 - \delta) \cdot \Pr[\forall i : f(X_i) = t_i]$$

$$= (1 - \delta) \cdot \Pr[D^{f(\cdot)}(1^n) = 1]$$

where, above, g is chosen uniformly from Func_n, and f is chosen uniformly from the set of functions mapping $(\{0,1\}^n)^*$ to $\{0,1\}^n$. This implies

$$\Pr[D^{f(\cdot)}(1^n) = 1] - \Pr[D^{\mathsf{CBC}_g(\cdot)}(1^n) = 1] \leq \delta \cdot \Pr[D^{f(\cdot)}(1^n) = 1] \leq \delta.$$

A symmetric argument for when D outputs 0 completes the proof. ∎

4.5   GMAC and Poly1305

One drawback of CBC-MAC is that it requires a number of cryptographic operations (specifically, block-cipher evaluations) *linear* in the length of the message being authenticated. We show here two (related) constructions of secure MACs that can be much more efficient. These MACs have been adopted by several internet standards.

We present a general paradigm for building secure MACs in Section 4.5.1, and then look at two concrete instantiations of that paradigm—GMAC and Poly1305—in Section 4.5.2.

4.5.1   MACs from Difference-Universal Functions

In this section we show a general approach for constructing MACs based on a combinatorial object called a *difference-universal function*. The paradigm we describe here is inspired by a construction of an information-theoretic MAC that we show in Section 4.6.2; nevertheless, our treatment is self contained and does not directly rely on any results from that section.

Let h be a keyed function that, for security parameter n, maps keys in $\mathcal{K}_n$ and inputs in $\mathcal{M}_n$ to outputs in $\mathcal{T}_n$. (We require also that h is efficiently computable, and that elements in $\{\mathcal{K}_n\}, \{\mathcal{M}_n\}$, and $\{\mathcal{T}_n\}$ can be sampled efficiently, but for simplicity we omit this from the definition.) As usual, we write $h_k(m)$ instead of $h(k, m)$. We assume that $\mathcal{T}_n$ is a *group* for each n. (The reader unfamiliar with the notion of a group can refer to Section 9.1; in this section, nothing beyond the definition of a group is needed.) We now define what it means for h to be *difference universal*.

DEFINITION 4.14 *A keyed function h as above is $\varepsilon(n)$-difference univer-sal if for all n, any distinct $m, m' \in \mathcal{M}_n$, and any $\Delta \in \mathcal{T}_n$ it holds that*

$$\Pr\left[h_k(m) - h_k(m') = \Delta \right] \leq \varepsilon(n),$$

where the probability is taken over uniform choice of $k \in \mathcal{K}_n$.

Note that we must have $\varepsilon(n) \geq 1/|\mathcal{T}_n|$. Difference-universal functions with $\varepsilon(n)$ negligible can be constructed without any assumptions. For now we simply assume their existence (though a simple example is given in Section 4.6.2), and defer further discussion to the next section.

In Construction 4.15 we show how to use a difference-universal function h in conjunction with a pseudorandom function F to construct a message authentication code. Roughly, the shared key consists of a key k_h for h as well as a key k_F for F; a tag on a message $m \in \mathcal{M}_n$ is computed by choosing a uniform value $r \in \{0, 1\}^n$ and masking $h_{k_h}(m)$ using $F_{k_F}(r)$. In the construction we assume for simplicity that for security parameter n the keyed function F maps n-bit keys and n-bit inputs to elements of $\mathcal{T}_n$.

Interestingly, this is the first (and only) example we will see of a *randomized* MAC. For this reason, we explicitly consider strong security.

THEOREM 4.16 *Let h be an $\varepsilon(n)$-difference-universal function for a neg-ligible function ε, and let F be a pseudorandom function. Then Construc-tion 4.15 is a strongly secure MAC for messages in $\{\mathcal{M}_n\}$.*

CONSTRUCTION 4.15

Let h, F be as in the text. Define a MAC for messages in $\{\mathcal{M}_n\}$ as follows:

- Gen: on input 1^n, choose uniform $k_h \in \mathcal{K}_n$ and $k_F \in \{0, 1\}^n$; output the key (k_h, k_F).

- Mac: on input a key (k_h, k_F) and a message $m \in \mathcal{M}_n$, choose a uniform $r \in \{0, 1\}^n$ and output the tag $t := \langle r, \; h_{k_h}(m) + F_{k_F}(r) \rangle$.

- Vrfy: on input a key (k_h, k_F), a message $m \in \mathcal{M}_n$, and a tag $t = \langle r, s \rangle$, output 1 if and only if $s \stackrel{?}{=} h_{k_h}(m) + F_{k_F}(r)$.

A MAC based on a difference-universal function.

PROOF   Let $\mathcal{A}$ be a PPT adversary, let $q = q(n)$ be a polynomial upper bound on the number of queries $\mathcal{A}$ makes to its Mac oracle, and let Π denote Construction 4.15. As usual, we first define a scheme $\widetilde{\Pi}$ that is the same as Π except that it uses a truly random function f with the appropriate domain and range in place of F_{k_F}. As in previous proofs involving pseudorandom functions, one can show that there is a negligible function negl such that

$$\left| \Pr\left[\text{Mac-sforge}_{\mathcal{A},\widetilde{\Pi}}(n) = 1\right] - \Pr\left[\text{Mac-sforge}_{\mathcal{A},\Pi}(n) = 1\right] \right| \leq \text{negl}(n).$$

In the remainder of the proof, we analyze $\widetilde{\Pi}$.

Let repeat denote the event that the same random value r is used to answer two different oracle queries in $\text{Mac-sforge}_{\mathcal{A},\widetilde{\Pi}}(n)$, and let new-r denote the event that $\mathcal{A}$ outputs $(m, \langle r, s \rangle)$ where r was not used to answer any oracle query. We have

$$\Pr\left[\text{Mac-sforge}_{\mathcal{A},\widetilde{\Pi}}(n) = 1\right] \leq \Pr[\text{repeat}] + \Pr\left[\text{Mac-sforge}_{\mathcal{A},\widetilde{\Pi}}(n) = 1 \wedge \text{new-r}\right]$$
$$+ \Pr\left[\text{Mac-sforge}_{\mathcal{A},\widetilde{\Pi}}(n) = 1 \wedge \overline{\text{repeat}} \wedge \overline{\text{new-r}}\right].$$

Bounding the first two terms of this sum is easy. Using Lemma A.15, we have $\Pr[\text{repeat}] \leq q^2/2^{n+1}$. Next, observe that if $\mathcal{A}$ outputs $(m, \langle r, s \rangle)$ where r was not used to answer any oracle query, then the value $f(r)$ is uniform in $\mathcal{T}_n$ and independent of $\mathcal{A}$'s view, and so the probability that $\langle r, s \rangle$ is a valid tag for m (i.e., the probability that $s = h_{k_h}(m) + f(r)$) is $1/|\mathcal{T}_n| \leq \varepsilon(n)$. It follows that $\Pr[\text{Mac-sforge}_{\mathcal{A},\widetilde{\Pi}}(n) = 1 \wedge \text{new-r}] \leq \varepsilon(n)$.

To complete the proof, we show that

$$\Pr\left[\text{Mac-sforge}_{\mathcal{A},\widetilde{\Pi}}(n) = 1 \wedge \overline{\text{repeat}} \wedge \overline{\text{new-r}}\right] \leq \varepsilon(n).$$

Here we rely on the fact that h is ε-difference universal. Consider an execution of experiment Mac-sforge in which neither repeat nor new-r occurs. Let $m_1, \ldots, m_q$ be the messages queried by $\mathcal{A}$ to its oracle, let $\langle r_1, s_1 \rangle, \ldots, \langle r_q, s_q \rangle$ be the responses, and let $(m, \langle r, s \rangle)$ be the final output of $\mathcal{A}$. Since repeat did not occur the $\{r_i\}$ are distinct; since new-r did not occur we therefore have $r = r_i$ for some unique i. Moreover, we may assume $m \neq m_i$ as otherwise there is no way $\mathcal{A}$'s output can be a valid forgery.

The crucial observations are:

1. Since the $\{r_i\}$ are all distinct, the values $\{f(r_i)\}$ are all uniform and independent. Those values thus serve to perfectly hide any information about k_h from $\mathcal{A}$ (by analogy with the one-time pad). Formally, this means that k_h is independent of $\mathcal{A}$'s view in experiment $\text{Mac-sforge}_{\mathcal{A},\widetilde{\Pi}}(n)$.

2. $\mathcal{A}$'s output is a valid forgery if and only if $h_{k_h}(m) - h_{k_h}(m_i) = s - s_i$.

Letting $\Delta = s - s_i$, the above imply that

$$\Pr\left[\mathsf{Mac\text{-}sforge}_{\mathcal{A},\widetilde{\Pi}}(n) = 1 \wedge \overline{\mathsf{repeat}} \wedge \overline{\mathsf{new\text{-}r}}\right] \leq \Pr_{k \leftarrow \mathcal{K}_n}[h_{k_h}(m) - h_{k_h}(m_i) = \Delta]$$

$$\leq \varepsilon(n).$$

Putting everything together, we conclude that

$$\Pr\left[\mathsf{Mac\text{-}sforge}_{\mathcal{A},\widetilde{\Pi}}(n) = 1\right] \leq 2 \cdot \varepsilon(n) + \frac{q^2}{2^{n+1}},$$

completing the proof. ∎

Nonce-based MACs. The only property of r used in the proof above is that r is unique across all tags (i.e., that repeat not occur). Thus, one can also prove security for Construction 4.15 in a nonce-based setting similar to what was formalized (for private-key encryption) in Section 3.6.4.

4.5.2   Instantiations

There is a clever and efficient way to instantiate the difference-universal function required by Construction 4.15 using polynomials over a finite field. (The reader unfamiliar with finite fields may consult Section A.5. The only results we require are that a finite field $\mathbb{F}_q$ containing q elements exists for any prime power q, and that a nonzero polynomial of degree ℓ over a finite field has at most ℓ roots.) Different realizations of that approach are, in turn, used by the standardized schemes GMAC and Poly1305.

For simplicity of exposition in this section (and because it matches current standards) we omit the security parameter and focus on a concrete setting.

An ε-difference-universal function. Fix a finite field $\mathbb{F}$. The idea is to let the key $k \in \mathcal{K}$ be a point in $\mathbb{F}$ and to view $m \in \mathcal{M}$ as a polynomial (of bounded degree) over $\mathbb{F}$; evaluating $h_k(m)$ then corresponds to evaluating m at the point k.

Formally, fix a constant ℓ and let $\mathcal{M} = \mathbb{F}^{<\ell}$, i.e., $\mathcal{M}$ consists of vectors over $\mathbb{F}$ containing fewer than ℓ entries. For any $m = (m_1, \ldots, m_{\ell'-1}) \in \mathcal{M}$, where $\ell' \leq \ell$, let $m_{\ell'} \in \mathbb{F}$ be an encoding of the length of m (i.e., $\ell' - 1$), and define the polynomial

$$m(X) \stackrel{\text{def}}{=} m_1 \cdot X^{\ell'} + m_2 \cdot X^{\ell'-1} + \cdots + m_{\ell'} \cdot X.$$

Finally, define the keyed function $h : \mathbb{F} \times \mathbb{F}^{<\ell} \to \mathbb{F}$ as

$$h_k(m) = m(k).$$

THEOREM 4.17 *The function h above is $\ell/|\mathbb{F}|$-difference universal.*

PROOF Fix distinct $m, m' \in \mathbb{F}^{<\ell}$ and $\Delta \in \mathbb{F}$. Define the polynomial

$$P(X) \stackrel{\text{def}}{=} m(X) - m'(X) - \Delta.$$

P is a nonzero polynomial of degree at most ℓ. (If the lengths of m and m' are equal then that fact that P is nonzero is immediate; otherwise, $m(X)$ and $m'(X)$ differ in their coefficients of the linear term.) So

$$\Pr_{k \in \mathbb{F}}[h_k(m) - h_k(m') = \Delta] = \Pr_{k \in \mathbb{F}}[P(k) = 0] \leq \ell/|\mathbb{F}|,$$

where the final inequality is because P has at most ℓ roots. ∎

Efficiency. The function h is extremely efficient in several respects. First, the key can be much shorter than the input. Second h can be evaluated quickly using Horner's rule. That is, to evaluate

$$m_1 \cdot k^{\ell'} + \cdots + m_{\ell'} \cdot k$$

set $y_0 := 0$ and then, for $i = 1$ to ℓ', set $y_i := (y_{i-1} + m_i) \cdot k$; output $y_{\ell'}$. This requires only $\ell' \leq \ell$ field multiplications and $O(1)$ memory, even if entries of m arrive in a streaming fashion and the length of m is not known in advance.

GMAC. The GMAC message authentication code is just[3] Construction 4.15 using a block cipher F with a 128-bit block length and the polynomial-based difference-universal function just described over the field $\mathbb{F} = \mathbb{F}_{2^{128}}$ with 2^{128} elements. Field elements are 128-bit strings; addition corresponds to bit-wise XOR, and multiplication can be done very efficiently using hardware-level instructions available in many modern processors.

Poly1305. The Poly1305 message authentication code is[4] defined similarly, but uses the field $\mathbb{F} = \mathbb{F}_p = \{0, \ldots, p-1\}$ where the prime $p = 2^{130} - 5$ was chosen for efficient implementation. Field operations here correspond to addition and multiplication modulo p. Observe that now there is a mismatch between the output of F (which is a 128-bit string) and the output of h (which lies in the range $\{0, \ldots, p-1\}$); to address this, the final tag is computed as

$$\langle r, [h_{k_h}(m) + F_{k_F}(r) \bmod 2^{128}] \rangle.$$

This small difference from Construction 4.15 can be accounted for in the security proof.

Comparison to CBC-MAC. Besides the fact that MACs based on Construction 4.15 can be more efficient than CBC-MAC, such MACs can also

---

[3]The GMAC standard does not correspond exactly to the construction described here; in particular, it supports messages whose length is not a multiple of 128.

[4]Again, we are omitting some details from the actual standard.

obtain a better concrete-security bound. Specifically, consider a setting in which q messages, each of length ℓ, are authenticated, and treat the block cipher F as a random function. The proof of security for CBC-MAC given in Section 4.4.2 guarantees that an attacker's probability of outputting a valid forgery is at most $q^2 \cdot \ell^2 / 2^n$, though this can be improved to $\mathcal{O}(q^2 \cdot \ell / 2^n)$ for small ℓ. In contrast, the security bounds obtained for the MACs described in this section show that an attacker's forgery probability is $\mathcal{O}((q^2 + \ell)/2^n)$, a significant improvement. Concretely, take $n = 128$, $q = 2^{40}$, and $\ell = 2^{20}$. CBC-MAC gives a security bound of approximately 2^{-8}, whereas GMAC and Poly1305 have security bounds of approximately 2^{-48}. The latter can be further improved in the nonce-based setting.

We remark further that in all cases the actual concrete-security bound includes a term that depends on an adversary's advantage in distinguishing the block cipher from a pseudorandom function. This term grows larger as the number of block-cipher evaluations increases. MACs based on Construction 4.15 have the advantage here as well in that they only evaluate the block cipher q times, as opposed to $q \cdot \ell$ times for CBC-MAC.

4.6   *Information-Theoretic MACs

Until now we have explored message authentication codes with *computational* security, i.e., where we assume bounds on the attacker's running time. But inspired by the results of Chapter 2, it is natural to ask whether message authentication in the presence of an *unbounded* adversary is possible. In this section, we show conditions under which *information-theoretic* (as opposed to computational) security is attainable.

A first observation is that it is impossible to achieve "perfect" security in this context. Namely, we cannot hope to have a message authentication code for which the probability that an adversary outputs a valid tag on a previously unauthenticated message is 0. The reason is that an adversary can simply guess a valid tag t on any message, and this guess will be correct with probability (at least) $1/|\mathcal{T}|$ (where $\mathcal{T}$ denotes the space of possible tags). Similarly, an attacker can always guess the key and generate a tag that is correct with probability $1/|\mathcal{K}|$ (where $\mathcal{K}$ denotes the space of possible keys).

The above examples tell us what we *can* hope to achieve: a MAC where the probability of forgery is at most $\max\{1/|\mathcal{T}|, 1/|\mathcal{K}|\}$, even for unbounded adversaries. We will see that this is achievable, but only under restrictions on how many messages are authenticated by the honest parties.

We first define information-theoretic security for message authentication codes. A starting point is to take experiment $\mathsf{Mac\text{-}forge}_{\mathcal{A},\Pi}(n)$ that is used to computationally secure MACs (cf. Definition 4.2), but drop the security parameter n and require simply that $\Pr[\text{Mac-forge}_{\mathcal{A},\Pi} = 1]$ be "small" for *all* adversaries $\mathcal{A}$ (and not just adversaries running in polynomial time). As mentioned above (and as will be proved formally in Section 4.6.3), however, such a definition is impossible to achieve unless we place some bound on the number of messages authenticated by the honest parties. We look here at the most basic setting, where the honest parties authenticate just a *single* message. We refer to this as *one-time message authentication*. The following experiment modifies $\text{Mac-forge}_{\mathcal{A},\Pi}(n)$ following the above discussion:

The one-time message authentication experiment $\text{Mac-forge}_{\mathcal{A},\Pi}^{\text{1-time}}$:

1. *A key k is generated by running* Gen.

2. *The adversary $\mathcal{A}$ outputs a message m', and is given in return a tag $t' \leftarrow \text{Mac}_k(m')$.*

3. *$\mathcal{A}$ outputs (m,t).*

4. *The output of the experiment is defined to be 1 if and only if (1) $\text{Vrfy}_k(m,t) = 1$ and (2) $m \neq m'$.*

DEFINITION 4.18 $\Pi = (\text{Gen}, \text{Mac}, \text{Vrfy})$ *is an ε-secure one-time MAC if for all (even unbounded) adversaries $\mathcal{A}$:*

$$\Pr\left[\text{Mac-forge}_{\mathcal{A},\Pi}^{\text{1-time}} = 1\right] \leq \varepsilon.$$

4.6.1 One-Time MACs from Strongly Universal Functions

In this section we show how to construct a one-time MAC based on any *strongly universal function*. We then show a simple construction of the latter.

Let $h : \mathcal{K} \times \mathcal{M} \to \mathcal{T}$ be a keyed function whose first input is a key $k \in \mathcal{K}$ and whose second input is taken from some domain $\mathcal{M}$; the output is in some set $\mathcal{T}$. As usual, we write $h_k(m)$ instead of $h(k,m)$. Then h is *strongly universal* (or *pairwise independent*) if for any two distinct inputs m, m' the values $h_k(m)$ and $h_k(m')$ are uniformly and independently distributed in $\mathcal{T}$ when k is a uniform key. This is equivalent to saying that the probability that $h_k(m), h_k(m')$ take on any particular values t, t' is exactly $1/|\mathcal{T}|^2$. That is:

DEFINITION 4.19 *A function $h : \mathcal{K} \times \mathcal{M} \to \mathcal{T}$ is strongly universal if for all distinct $m, m' \in \mathcal{M}$ and all (not necessarily distinct) $t, t' \in \mathcal{T}$ it holds that*

$$\Pr\left[h_k(m) = t \wedge h_k(m') = t'\right] = \frac{1}{|\mathcal{T}|^2},$$

where the probability is taken over uniform choice of $k \in \mathcal{K}$.

The above should motivate the construction of a one-time message authentication code from any strongly universal function h. The tag t on a message

m is obtained by computing $h_k(m)$, where the key k is uniform; see Construction 4.20. Intuitively, even after an adversary observes the tag $t' = h_k(m')$ for any message m', the correct tag $h_k(m)$ for any *other* message m is still uniformly distributed in $\mathcal{T}$ from the adversary's point of view. Thus, the adversary can do nothing more than blindly guess the tag, and this guess will be correct only with probability $1/|\mathcal{T}|$.

CONSTRUCTION 4.20

Let $h : \mathcal{K} \times \mathcal{M} \to \mathcal{T}$ be a strongly universal function. Define a MAC for messages in $\mathcal{M}$ as follows:

- Gen: choose uniform $k \in \mathcal{K}$ and output it as the key.
- Mac: on input a key $k \in \mathcal{K}$ and a message $m \in \mathcal{M}$, output the tag $t := h_k(m)$.
- Vrfy: on input a key $k \in \mathcal{K}$, a message $m \in \mathcal{M}$, and a tag $t \in \mathcal{T}$, output 1 if and only if $t \stackrel{?}{=} h_k(m)$. (If $m \notin \mathcal{M}$, then output 0.)

A one-time MAC from any strongly universal function.

The above construction can be viewed as analogous to Construction 4.5. This is because a strongly universal function h behaves like a random function as long as it is evaluated at most twice.

THEOREM 4.21 *Let $h : \mathcal{K} \times \mathcal{M} \to \mathcal{T}$ be a strongly universal function. Then Construction 4.20 is a $1/|\mathcal{T}|$-secure one-time MAC for messages in $\mathcal{M}$.*

PROOF Let $\mathcal{A}$ be an adversary and let Π denote Construction 4.20. Since $\mathcal{A}$ may be all-powerful, we may assume $\mathcal{A}$ is deterministic. So the message m' on which $\mathcal{A}$ requests a tag at the outset of the experiment is fixed. Furthermore, the pair (m, t) that $\mathcal{A}$ outputs at the end of the experiment is a deterministic function of the tag t' on m' that $\mathcal{A}$ receives. We thus have

$$
\Pr\left[\text{Mac-forge}_{\mathcal{A},\Pi}^{\text{1-time}} = 1\right] = \sum_{t' \in \mathcal{T}} \Pr\left[\text{Mac-forge}_{\mathcal{A},\Pi}^{\text{1-time}} = 1 \ \wedge \ h_k(m') = t'\right]
$$

$$
= \sum_{\substack{t' \in \mathcal{T} \\ (m,t) := \mathcal{A}(t')}} \Pr\left[h_k(m) = t \ \wedge \ h_k(m') = t'\right]
$$

$$
= \sum_{\substack{t' \in \mathcal{T} \\ (m,t) := \mathcal{A}(t')}} \frac{1}{|\mathcal{T}|^2} = \frac{1}{|\mathcal{T}|}.
$$

This proves the theorem. ∎

It remains to construct a strongly universal function. We assume some basic knowledge about arithmetic modulo a prime number; readers may refer to Sections 9.1.1 and 9.1.2 for necessary background. (Alternatively, everything we say generalizes to an arbitrary finite field, and the interested reader may consult Section A.5.) Fix a prime p, and let $\mathbb{Z}_p \stackrel{\text{def}}{=} \{0, \ldots, p-1\}$. We take as our message space $\mathcal{M} = \mathbb{Z}_p$; the space of possible tags will be $\mathcal{T} = \mathbb{Z}_p$. A key (a, b) consists of a pair of elements from $\mathbb{Z}_p$; thus, $\mathcal{K} = \mathbb{Z}_p \times \mathbb{Z}_p$. Define h as

$$h_{a,b}(m) \stackrel{\text{def}}{=} [a \cdot m + b \bmod p],$$

where the notation $[X \bmod p]$ refers to the result of reducing X modulo p.

THEOREM 4.22 *For any prime p, the function h defined above is strongly universal.*

PROOF Fix any distinct $m, m' \in \mathbb{Z}_p$ and any $t, t' \in \mathbb{Z}_p$. For which keys (a, b) does it hold that both $h_{a,b}(m) = t$ and $h_{a,b}(m') = t'$? This holds only if

$$a \cdot m + b = t \bmod p \quad \text{and} \quad a \cdot m' + b = t' \bmod p.$$

We thus have two linear equations in the two unknowns a, b. These two equations are both satisfied exactly when $a = [(t-t') \cdot (m-m')^{-1} \bmod p]$ and $b = [t - a \cdot m \bmod p]$; note that $[(m-m')^{-1} \bmod p]$ exists because $m \neq m'$ and so $m - m' \neq 0 \bmod p$. Restated, this means that for any m, m', t, t' as above there is a *unique* key (a, b) with $h_{a,b}(m) = t$ and $h_{a,b}(m') = t'$. We conclude that the probability (over uniform choice of the key) that $h_{a,b}(m) = t$ and $h_{a,b}(m') = t'$ is exactly $1/|\mathcal{K}| = 1/|\mathcal{T}|^2$ as required. ∎

Parameters of Construction 4.20. We briefly discuss the parameters of Construction 4.20 when instantiated with the strongly universal function described above. The construction is a $1/|\mathcal{T}|$-secure one-time MAC, so is optimal as far as the level of security achieved vs. the number of tags.

Let $\mathcal{M} = \mathbb{Z}_p$ be some message space for which we want to construct a one-time MAC. Construction 4.20 gives a $1/|\mathcal{M}|$-secure one-time MAC with keys that are (roughly) twice the message length. The reader may notice two problems here, at opposite ends of the spectrum: First, if $|\mathcal{M}|$ is small then a $1/|\mathcal{M}|$ probability of forgery may be unacceptably large. On the flip side, if $|\mathcal{M}|$ is large then a $1/|\mathcal{M}|$ probability of forgery may be overkill; one might be willing to accept a (somewhat) larger probability of forgery if that level of security can be achieved with shorter tags. The first problem (when $|\mathcal{M}|$ is small) is easy to deal with by simply embedding $\mathcal{M}$ into a larger message space $\mathcal{M}'$ by, e.g., padding messages with 0s. The second problem can be addressed as well by using Construction 4.20 and then truncating the tag. We omit details, and refer instead to the references at the end of this chapter.

4.6.2 One-Time MACs from Difference-Universal Functions

Here we explore a second construction of one-time MACs. In contrast to the construction given in the previous section, this approach can have shorter keys and better computational efficiency. Perhaps more importantly, it can be adapted to give a computationally secure scheme (for authenticating polynomially many messages), as shown in Section 4.5.

We begin by defining a *difference-universal function*. (In contrast to Definition 4.14, here we give a concrete version of the definition.) We assume familiarity with the notion of a group (cf. Section 9.1), but in this section nothing beyond the definition of a group is needed.

DEFINITION 4.23 *Let $\mathcal{T}$ be a group. A function $h : \mathcal{K} \times \mathcal{M} \to \mathcal{T}$ is ε-difference universal if for all distinct $m, m' \in \mathcal{M}$ and all $\Delta \in \mathcal{T}$ it holds that*

$$\Pr\left[h_k(m) - h_k(m') = \Delta\right] \le \varepsilon,$$

where the probability is taken over uniform choice of $k \in \mathcal{K}$.

Being difference universal is weaker than being strongly universal; in particular, any $h : \mathcal{K} \times \mathcal{M} \to \mathcal{T}$ that is strongly universal is also $1/|\mathcal{T}|$-difference universal, but the converse is not true. To see this, fix a prime p, let $\mathcal{K} = \mathcal{M} = \mathcal{T} = \mathbb{Z}_p$, and define h as

$$h_k(m) = [k \cdot m \bmod p].$$

It is easy to see that h is not strongly universal (since $h_k(0) = 0$ for all k), but for any distinct m, m' and any Δ we have

$$\begin{aligned} \Pr[h_k(m) - h_k(m') = \Delta] &= \Pr[k \cdot (m - m') = \Delta \bmod p] \\ &= \Pr[k = \Delta \cdot (m - m')^{-1} \bmod p] = 1/p, \end{aligned}$$

showing that h is $1/|\mathcal{T}|$-difference universal. (In Section 4.5 we show a construction of an ε-difference-universal function with $|\mathcal{K}| \ll |\mathcal{M}|$.)

Construction 4.24 shows how a difference-universal function $h : \mathcal{K} \times \mathcal{M} \to \mathcal{T}$ can be used to construct a one-time MAC. The shared key now consists of both a key $k \in \mathcal{K}$ for h as well as a uniform $r \in \mathcal{T}$ that will be used as a one-time pad. To authenticate a message $m \in \mathcal{M}$, the sender first computes $h_k(m)$ and then "masks" that value using r. (Note the similarity to Construction 4.15, which uses a block cipher to generate masks for polynomially many messages.) As intuition for the security of this scheme, note that even after observing the tag $t' = h_k(m') + r$ for a message m', an adversary learns nothing about $h_k(m')$. Moreover, if the attacker outputs a tag t on another message m, this is a successful forgery only if $t = h_k(m) + r$, i.e., if

$$t - t' = h_k(m) - h_k(m').$$

CONSTRUCTION 4.24

Let $h : \mathcal{K} \times \mathcal{M} \to \mathcal{T}$ be a difference-universal function. Define a MAC for messages in $\mathcal{M}$ as follows:

- **Gen**: choose uniform $k \in \mathcal{K}$ and $r \in \mathcal{T}$; output the key (k, r).
- **Mac**: on input a key (k, r) and a message $m \in \mathcal{M}$, output the tag $t := h_k(m) + r$. (Addition here is done in the group $\mathcal{T}$.)
- **Vrfy**: on input a key (k, r), a message $m \in \mathcal{M}$, and a tag $t \in \mathcal{T}$, output 1 if and only if $t \stackrel{?}{=} h_k(m) + r$.

A one-time MAC from any difference-universal function.

If h is ε-difference universal then the probability of the above (taken over choice of k) is at most ε.

THEOREM 4.25 *Let h be an ε-difference-universal function. Then Construction 4.24 is an ε-secure one-time MAC for messages in $\mathcal{M}$.*

PROOF Let Π denote Construction 4.24. The proof is similar to that of Theorem 4.21. As in that proof, fix an adversary $\mathcal{A}$ and let m' be the message whose tag is requested by $\mathcal{A}$ at the outset of the experiment. The message/tag pair output by $\mathcal{A}$ is then a deterministic function of the tag t' on m'. So

$$\Pr\left[\text{Mac-forge}_{\mathcal{A},\Pi}^{\text{1-time}} = 1\right] = \sum_{\substack{t' \in \mathcal{T} \\ (m,t) := \mathcal{A}(t')}} \Pr\left[h_k(m) + r = t \ \wedge \ h_k(m') + r = t'\right].$$

Now, for any $m \neq m'$ and t, t' we have $h_k(m) + r = t$ and $h_k(m') + r = t'$ if and only if $h_k(m) + r = t$ and $h_k(m) - h_k(m') = t - t' \stackrel{\text{def}}{=} \Delta$. Thus,

$$\Pr\left[h_k(m) + r = t \ \wedge \ h_k(m') + r = t'\right]$$
$$= \Pr\left[h_k(m) + r = t \ \wedge \ h_k(m) - h_k(m') = \Delta\right]$$
$$= \Pr\left[r = t - h_k(m) \mid h_k(m) - h_k(m') = \Delta\right] \cdot \Pr\left[h_k(m) - h_k(m') = \Delta\right]$$
$$= \left(\frac{1}{|\mathcal{T}|}\right) \cdot \varepsilon,$$

using the facts that h is ε-difference universal and that r is uniform and independent of k. The theorem follows. ∎

4.6.3 Limitations on Information-Theoretic MACs

Here we explore limitations on information-theoretic message authentication, showing that any ε-secure one-time MAC must have keys of length at least $1/\varepsilon^2$. An extension of the proof shows that any ε-secure ℓ-time MAC (where security is defined by modifying Definition 4.19 to allow the attacker to request tags on ℓ messages) requires keys of length at least $1/\varepsilon^{(\ell+1)}$. A corollary is that *no* MAC can provide information-theoretic security for authenticating an unbounded number of messages.

In the following, we assume the message space contains at least two messages; if not, there is no point in communicating, let alone authenticating.

THEOREM 4.26 *Let* $\Pi = (\mathsf{Gen}, \mathsf{Mac}, \mathsf{Vrfy})$ *be an* ε-secure one-time MAC *with key space* $\mathcal{K}$. *Then* $|K| \geq \varepsilon^{-2}$.

PROOF Fix distinct messages m_0, m_1. The intuition is that there must be at least ε^{-1} possibilities for the tag of m_0 (or else the adversary could guess it with probability better than ε); furthermore, even conditioned on the value of the tag for m_0, there must be ε^{-1} possibilities for the tag of m_1 (or else the adversary could forge a tag on m_1 with probability better than ε). Since each key defines tags for m_0 and m_1, this means there must be at least $\varepsilon^{-1} \times \varepsilon^{-1} = \varepsilon^{-2}$ keys. We make this formal below.

Let $\mathcal{K}$ denote the key space (i.e., the set of all possible keys that can be output by Gen). For any possible tag t_0, let $\mathcal{K}(t_0)$ denote the set of keys for which t_0 is a valid tag on m_0; i.e.,

$$\mathcal{K}(t_0) \stackrel{\text{def}}{=} \{k \mid \mathsf{Vrfy}_k(m_0, t_0) = 1\}.$$

For any t_0 we must have $|\mathcal{K}(t_0)| \leq \varepsilon \cdot |\mathcal{K}|$. Otherwise the adversary could simply output (m_0, t_0) as its forgery; this would be a valid forgery with probability at least $|\mathcal{K}(t_0)|/|\mathcal{K}| > \varepsilon$, contradicting the claimed security.

Consider now the adversary $\mathcal{A}$ who requests a tag on m_0, receives in return a tag t_0, chooses a uniform key $k \in \mathcal{K}(t_0)$, and outputs $(m_1, \mathsf{Mac}_k(m_1))$ as its forgery. The probability that $\mathcal{A}$ outputs a valid forgery is at least

$$\sum_{t_0} \Pr[\mathsf{Mac}_k(m_0) = t_0] \cdot \frac{1}{|\mathcal{K}(t_0)|} \geq \sum_{t_0} \Pr[\mathsf{Mac}_k(m_0) = t_0] \cdot \frac{1}{\varepsilon \cdot |\mathcal{K}|}$$

$$= \frac{1}{\varepsilon \cdot |\mathcal{K}|}.$$

By the claimed security of the scheme, the probability that the adversary can output a valid forgery is at most ε. Thus, we must have $|\mathcal{K}| \geq \varepsilon^{-2}$. ∎

As a corollary, a 2^{-n}-secure one-time MAC for which all keys have the same length must have keys of length at least $2n$.

References and Additional Reading

The definition of security for message authentication codes was adapted by Bellare et al. [20] from the definition of security for digital signatures [88] (see Chapter 13). Later work of Bellare et al. [19] highlighted the importance of the definitional variant where verification queries are allowed.

The paradigm of using pseudorandom functions for message authentication (as in Construction 4.5) was introduced by Goldreich et al. [84]. Construction 4.7 is due to Goldreich [83].

CBC-MAC was standardized in the early 1980s [102, 11]. Basic CBC-MAC was proven secure (for authenticating fixed-length messages) by Bellare et al. [20]. Bernstein [30] gives a more direct proof that we have adapted in Section 4.4.2. An improved bound on the security of basic CBC-MAC, which also directly takes into account reliance on a pseudorandom *permutation* rather than a pseudorandom *function*, was given by Bellare et al. [23].

As noted in this chapter, basic CBC-MAC is insecure when used to authenticate messages of different lengths. One way to fix this is to prepend the length to the message. Alternate approaches were explored by Petrank and Rackoff [158], Black and Rogaway [36], and Iwata and Kurosawa [103]; these led to a new proposed standard called CMAC [191].

GMAC was introduced as part of the GCM authenticated encryption scheme by McGrew and Viega [136], based on work of Kohno et al. [119]. Poly1305 is due to Bernstein [31].

Information-theoretic MACs were first studied by Gilbert et al. [80]. Carter and Wegman [48, 203] introduced the notion of strongly universal functions, and noted their application to one-time message authentication. They also showed how to reduce the key length for this task by using an *almost* strongly universal function. Construction 4.24 is based on an idea of Wegman and Carter [203], though difference-universal functions were not introduced until several years later [120, 121]. (Note that difference-universal functions are called *XOR-universal* or *almost* Δ-*universal* in the literature.) The reader interested in learning more about information-theoretic MACs is referred to the paper by Stinson [193], the survey by Simmons [187], or the first edition of Stinson's textbook [194, Chapter 10].

Exercises

4.1 Consider an extension of the definition of secure message authentication where the adversary is provided with both a Mac and a Vrfy oracle.

(a) Provide a formal definition of security for this case.

(b) Assume Π is a deterministic MAC using canonical verification that satisfies Definition 4.2. Prove that Π also satisfies your definition from part (a).

4.2 Assume secure MACs exist. Give a construction of a MAC that is secure with respect to Definition 4.2 but that is not secure when the adversary is additionally given access to a Vrfy oracle (cf. the previous exercise).

4.3 Prove Proposition 4.4.

4.4 Assume secure MACs exist. Prove that there exists a MAC that is secure (Definition 4.2) but is *not* strongly secure (Definition 4.3).

4.5 Consider the following MAC for messages of length $\ell(n) = 2n - 2$ using a pseudorandom function F: On input a message $m_0 \| m_1$ (with $|m_0| = |m_1| = n - 1$) and key $k \in \{0,1\}^n$, algorithm Mac outputs $t = F_k(0\|m_0) \, \| \, F_k(1\|m_1)$. Algorithm Vrfy is defined in the natural way. Is this MAC secure? Prove your answer.

4.6 Let F be a pseudorandom function. Show that each of the following MACs is insecure, even if used to authenticate fixed-length messages. (In each case Gen outputs a uniform $k \in \{0,1\}^n$; we let $\langle i \rangle$ denote an $n/2$-bit encoding of the integer i.)

(a) To authenticate a message $m = m_1, \ldots, m_\ell$, where $m_i \in \{0,1\}^n$, compute $t := F_k(m_1) \oplus \cdots \oplus F_k(m_\ell)$.

(b) To authenticate a message $m = m_1, \ldots, m_\ell$, where $m_i \in \{0,1\}^{n/2}$, compute $t := F_k(\langle 1 \rangle \| m_1) \oplus \cdots \oplus F_k(\langle \ell \rangle \| m_\ell)$.

(c) To authenticate a message $m = m_1, \ldots, m_\ell$, where $m_i \in \{0,1\}^{n/2}$, choose uniform $r \in \{0,1\}^n$, compute

$$t := F_k(r) \oplus F_k(\langle 1 \rangle \| m_1) \oplus \cdots \oplus F_k(\langle \ell \rangle \| m_\ell),$$

and let the tag be $\langle r, t \rangle$.

4.7 Let F be a pseudorandom function. Show that the following MAC for messages of length $2n$ is insecure: Gen outputs a uniform $k \in \{0,1\}^n$. To authenticate a message $m_1 \| m_2$ with $|m_1| = |m_2| = n$, compute the tag $F_k(m_1) \, \| \, F_k(F_k(m_2))$.

4.8 Given any *deterministic* MAC (Mac, Vrfy), we may view Mac as a keyed function. In both Constructions 4.5 and 4.9, Mac is a pseudorandom function. Give a construction of a secure, deterministic MAC in which Mac is *not* a pseudorandom function.

4.9 Is Construction 4.5 necessarily secure when instantiated using a weak pseudorandom function (cf. Exercise 3.28)? Explain.

4.10 Prove that Construction 4.7 is a secure MAC even when the adversary is additionally given access to a Vrfy oracle (cf. Exercise 4.1), assuming Π' is a secure MAC that uses canonical verification.

4.11 Prove that Construction 4.7 is strongly secure if Π' is strongly secure.

4.12 Prove that Construction 4.7 is secure if it is changed as follows: Set $t_i := F_k(r\|b\|i\|m_i)$ where b is a single bit such that $b = 0$ in all blocks but the last one, and $b = 1$ in the last block. (Assume for simplicity that the length of any message being authenticated is always an integer multiple of $n/2 - 1$.) What is the advantage of this modification?

4.13 We explore what happens when the basic CBC-MAC construction is used with messages of different lengths.

  (a) Say the sender and receiver do not agree on the message length in advance (and so $\mathsf{Vrfy}_k(m, t) = 1$ iff $t \stackrel{?}{=} \mathsf{Mac}_k(m)$, regardless of the length of m), but the sender is careful to only authenticate messages of length $2n$. Show that an adversary can forge a valid tag on a message of length $4n$.

  (b) Say the receiver only accepts 3-block messages (so $\mathsf{Vrfy}_k(m, t) = 1$ only if m has length $3n$ and $t \stackrel{?}{=} \mathsf{Mac}_k(m)$), but the sender authenticates messages of any length a multiple of n. Show that an adversary can forge a valid tag on a new message.

4.14 Prove that the following modifications of basic CBC-MAC do not yield a secure MAC (even for fixed-length messages):

  (a) Mac outputs all blocks $t_1, \ldots, t_\ell$, rather than just t_ℓ. (Verification only checks whether t_ℓ is correct.)

  (b) A random initial block is used each time a message is authenticated. That is, change Construction 4.9 by choosing uniform $t_0 \in \{0, 1\}^n$, computing t_ℓ as before, and then outputting the tag $\langle t_0, t_\ell \rangle$; verification is done in the natural way.

4.15 Show that appending the message length to the *end* of the message before applying basic CBC-MAC does not result in a secure MAC for arbitrary-length messages.

4.16 Define a version of CBC-MAC for messages of length at most $\ell \cdot 2^n$ as follows: given a message m, pad it with 0s so that it has length exactly $\ell \cdot 2^n$; apply basic CBC-MAC to the result. Is this secure?

4.17 Consider the following encoding that handles messages whose length is less than $n \cdot 2^n$: We encode a string $m \in \{0, 1\}^*$ by first appending as many 0s as needed to make the length of the resulting string $\hat{m}$ a nonzero multiple of n. Then we prepend the number of *blocks* in $\hat{m}$

(equivalently, prepend the integer $|\hat{m}|/n$), encoded as an n-bit string. Show that this encoding is *not* prefix-free.

4.18 Prove that the encoding for arbitrary-length messages described in Section 4.4.2 is prefix-free.

4.19 Prove that the following modification of basic CBC-MAC gives a secure MAC for arbitrary-length messages if F is a pseudorandom function. (Assume all messages have length a multiple of the block length.) $\mathsf{Mac}_k(m)$ first computes $k_\ell := F_k(\ell)$, where ℓ is the length of m. The tag is then computed using basic CBC-MAC with key k_ℓ.

4.20 Let F be a keyed function that is a secure (deterministic) MAC for messages of length n. (Note that F need not be a pseudorandom function.) Show that basic CBC-MAC is not necessarily a secure MAC (even for fixed-length messages) when instantiated with F.

4.21 Assume the same nonce r is used to authenticate two different messages in GMAC or Poly1305. Show how to construct a forgery in that case, with high probability.

> **Hint:** You may assume only single-block messages are authenticated.

4.22 Prove or disprove whether the following functions are $\ell/|\mathbb{F}|$-difference universal. In each case assume $\mathcal{K} = \mathbb{F}$ and $\mathcal{M} = \mathbb{F}^{<\ell}$, and for a message $m = (m_1, \ldots, m_{\ell'-1})$ let $m_{\ell'} \in \mathbb{F}$ be an encoding of $\ell' - 1$.

(a) $h'_k(m) = m'(k)$, where

$$m'(X) \stackrel{\text{def}}{=} m_1 \cdot X^\ell + m_2 \cdot X^{\ell-1} + \cdots + m_{\ell'} \cdot X^{\ell-\ell'+1}.$$

(b) $h''_k(m) = m''(k)$, where

$$m''(X) \stackrel{\text{def}}{=} m_1 \cdot X^{\ell'-1} + m_2 \cdot X^{\ell'-2} + \cdots + m_{\ell'}.$$

(c) $h'''_k(m) = m'''(k)$, where

$$m'''(X) \stackrel{\text{def}}{=} m_1 \cdot X + m_2 \cdot X^2 + \cdots + m_{\ell'} \cdot X^{\ell'}.$$

4.23 Show that the polynomial-based difference-universal function from Section 4.5.2 is not strongly universal.

4.24 Fix $\ell > 0$ and a prime p. Let $\mathcal{K} = \mathbb{Z}_p^{\ell+1}$, $\mathcal{M} = \mathbb{Z}_p^\ell$, and $\mathcal{T} = \mathbb{Z}_p$. Define $h : \mathcal{K} \times \mathcal{M} \to \mathcal{T}$ as

$$h_{k_0, k_1, \ldots, k_\ell}(m_1, \ldots, m_\ell) = \left[k_0 + \sum_i k_i m_i \bmod p \right].$$

Prove that h is strongly universal.

4.25 Fix $\ell, n > 0$. Let $\mathcal{K} = \{0,1\}^{\ell \times n} \times \{0,1\}^{\ell}$ (interpreted as a boolean $\ell \times n$ matrix and an ℓ-dimensional vector), let $\mathcal{M} = \{0,1\}^{n}$, and let $\mathcal{T} = \{0,1\}^{\ell}$. Define $h : \mathcal{K} \times \mathcal{M} \to \mathcal{T}$ as $h_{K,v}(m) = K \cdot m \oplus v$, where all operations are performed modulo 2. Prove that h is strongly universal.

4.26 A *Toeplitz matrix* K is a matrix in which $K_{i,j} = K_{i-1,j-1}$ when $i, j > 1$; i.e., the values along any diagonal are equal. So an $\ell \times n$ Toeplitz matrix has the form

$$
\begin{bmatrix}
K_n & K_{n-1} & K_{n-2} & \cdots & K_1 \\
K_{n+1} & K_n & K_{n-1} & \cdots & K_2 \\
K_{n+2} & K_{n+1} & K_n & \cdots & K_3 \\
\vdots & \vdots & \vdots & \vdots & \vdots \\
K_{n+\ell-1} & K_{n+\ell-2} & K_{n+\ell-3} & \cdots & K_\ell
\end{bmatrix}.
$$

Let $\mathcal{K} = T^{\ell \times n} \times \{0,1\}^{\ell}$ (where $T^{\ell \times n}$ denotes the set of $\ell \times n$ Toeplitz matrices), and let $\mathcal{M} = \{0,1\}^{n}$. Define $h : \mathcal{K} \times \mathcal{M} \to \{0,1\}^{\ell}$ as $h_{K,v}(m) = K \cdot m \oplus v$, where all operations are performed modulo 2. Prove that h is strongly universal. What is the advantage here as compared to the construction in the previous exercise?

4.27 Define an appropriate notion of a ε-secure *two-time* MAC, and give a construction that meets your definition.

Chapter 5

CCA-Security and Authenticated Encryption

In previous chapters we studied two different notions of security for parties communicating over an open communication channel. In Chapter 3 we focused on the goal of *secrecy* against a *passive* adversary who simply eavesdrops on the parties' communication, and showed CPA-secure encryption schemes realizing this goal. In Chapter 4 we explored *integrity* against an *active* adversary who can inject messages on the channel or otherwise tamper with the parties' communication, and described how message authentication codes can be used to achieve this notion. We consider the missing piece—*secrecy* in the presence of an *active* adversary—in Section 5.1, and introduce the notion of relevant notion of CCA-security there. Beginning in Section 5.2, we then consider the natural question of how to construct encryption schemes that achieve both secrecy and integrity *simultaneously*.

5.1 Chosen-Ciphertext Attacks and CCA-Security

We have so far considered encryption schemes secure only against passive (eavesdropping) adversaries. (Even though chosen-plaintext attacks allow an adversary to control what gets encrypted, the adversary in that setting is still limited to passively observing ciphertexts transmitted by the honest parties.) In the previous chapter, we discussed the importance of also defending against *active* attackers who may interfere with or modify the communication between the honest parties, focusing there on the case of message integrity. What might the effect of active attacks be when it comes to *secrecy*?

Consider a scenario in which a sender encrypts a message m and then transmits the resulting ciphertext c. An attacker who can tamper with the communication can modify c to generate another ciphertext c' that is received by the other party. This receiver will then decrypt c' to obtain a message m'. If $m' \neq m$ (and $m' \neq \perp$), this is a violation of integrity. What is of interest to us here, however, is the potential impact on secrecy. In particular, if the attacker learns partial information about m'—say, from subsequent behavior of the receiver—might that reveal information about the original message m?

This type of attack, in which an adversary causes a receiver to decrypt ciphertexts that the adversary generates, is called a *chosen-ciphertext attack*. Chosen-ciphertext attacks are possible, in principle, any time an attacker has the ability to inject traffic on the channel between the sender and receiver. There are many scenarios in which this can occur. (See also the discussion in Section 12.2.3 regarding chosen-ciphertext attacks in the public-key setting.) In the Midway example from Section 3.4.2, for example, US cryptanalysts could have sent encrypted messages containing the fragment AF to the Japanese; by monitoring their subsequent behavior (e.g., movement of troops and the like), the US could have learned information about what AF meant.

Alternatively, imagine a client sending encrypted messages to a server. If an adversary can impersonate the client and send ciphertexts to the server that appear to originate from the client, the server will decrypt those ciphertexts and the adversary may learn something about the result; for example, the attacker may be able to deduce when a ciphertext decrypts to an *ill-formed* plaintext (e.g., one that is not formatted correctly) based on the server's reaction (e.g., if the server sends an error message). In Section 5.1.1 we describe in detail an attack of exactly this sort where an attacker is able to leverage the information leaked from these decryptions to learn the entire contents of some other encrypted message! Such attacks have been carried out in practice on web servers to learn the contents of encrypted TLS sessions.

5.1.1 Padding-Oracle Attacks

We motivate the importance of security against chosen-ciphertext attacks by showing a real-world example where such attacks can be devastating. We consider a setting in which a client sends messages encrypted using CBC-mode encryption to a server. We assume the attacker can impersonate the client and send ciphertexts of its choice to the server, which the server will then decrypt. We assume further that the attacker can tell when the resulting decrypted messages are valid (in a sense we will define below) or not. Such information is frequently easy to obtain since, for example, the server might request retransmission or terminate a session if it receives a ciphertext that does not decrypt correctly, and either of those events would be detectable by the attacker. The attack has been shown to work in practice on various deployed protocols.

In our discussion of CBC-mode encryption in Section 3.6.3, we only dealt with the case where the message length was a multiple of the block length of the underlying block cipher F. If a message does not satisfy this property, it must be padded before CBC mode is applied; we refer to the result after padding as the *encoded data*. The padding must allow the receiver to unambiguously recover the original message from the encoded data. One popular padding scheme is defined by the PKCS #7 standard, and works as follows. Assume the original message has an integral number of bytes, and let L denote the block length (in bytes) of the block cipher F. Let $b > 0$ denote the number of bytes that need to be appended to the message in order to make the total length of the resulting encoded data a multiple of the block length. Then we append to the message the integer b (represented in one byte, i.e., two hexadecimal digits) repeated b times. That is, if one byte of padding is needed then the 1-byte string 0x01 (written in hexadecimal) is appended; if four bytes of padding are needed then 0x04040404 is appended; etc. (Note that b is an integer between 1 and L, inclusive—we cannot have $b = 0$ since this would lead to ambiguous padding. Thus, if the original message length is already a multiple of the block length, then $b = L$.) After padding, the encoded data is encrypted using regular CBC-mode encryption.

When decrypting, the server first uses CBC-mode decryption as usual to recover the encoded data, and then checks whether the encoded data is correctly padded. (This is easily done: simply read the value b of the final byte and then verify that the final b bytes of the result all have value b.) If so, the padding is stripped off and the original message returned. Otherwise, the standard procedure is to return a "bad padding" error. This means the server is serving as a "padding oracle" for the adversary: i.e., the adversary can send an arbitrary ciphertext to the server and learn (based on whether a "bad padding" error is returned) whether the underlying encoded data is padded correctly or not. Although this may seem like meaningless information, we show that it enables an adversary to completely recover the original message corresponding to any ciphertext of its choice.

We describe the attack on a 3-block ciphertext for simplicity. Let IV, c_1, c_2 be a ciphertext observed by the attacker, and let m_1, m_2 be the underlying encoded data (unknown to the attacker) that corresponds to a padded message, as discussed above. (Each block is L bytes long.) Note that

$$m_2 = F_k^{-1}(c_2) \oplus c_1, \tag{5.1}$$

where k is the key (which is, of course, not known to the attacker) being used by the honest parties. The second block m_2 ends in $\underbrace{\text{0x}b \cdots \text{0x}b}_{b \text{ times}}$, where we let 0x$b$ denote the 1-byte representation of some integer b. The key property used in the attack is that certain changes to the ciphertext yield predictable changes in the underlying encoded data after CBC-mode decryption. Specifically, let c_1' be identical to c_1 except for a modification in the final byte, and consider decryption of the modified ciphertext IV, c_1', c_2. This will result in encoded data m_1', m_2' where $m_2' = F_k^{-1}(c_2) \oplus c_1'$. Comparing to Equation (5.1) we see that m_2' will be identical to m_2 except for a modification in the final byte. (The value of m_1' is unpredictable, but this will not adversely affect the attack.) Similarly, if c_1' is the same as c_1 except for a change in its ith byte, then decryption of IV, c_1', c_2 will result in m_1', m_2' where m_2' is the same as m_2 except for a change in its ith byte. More generally, if $c_1' = c_1 \oplus \Delta$ for any string Δ, then decryption of IV, c_1', c_2 yields m_1', m_2' where $m_2' = m_2 \oplus \Delta$. The upshot is that the attacker can exercise significant control over the final block of the encoded data.

As a warmup, let us see how the adversary can exploit this to learn b, the amount of padding. (This reveals the length of the original message.) Recall that upon decryption, the server looks at the value b of the final byte of the encoded data, and then verifies that the final b bytes all have the same value. The attacker begins by modifying the first byte of c_1 and sending the resulting ciphertext IV, c_1', c_2 to the server. If decryption fails (i.e., the server returns an error) then it must be the case that the server is checking all L bytes of m_2', and therefore $b = L$! Otherwise, the attacker learns that $b < L$, and it can then repeat the process with the second byte, and so on. The left-most modified byte for which decryption fails reveals exactly the left-most byte being checked by the server, and so reveals exactly b.

With b known, the attacker can proceed to learn the bytes of the message one-by-one. We illustrate the idea for the final byte of the message, which we denote by M. The attacker knows that m_2 ends in $\text{0x}M\text{0x}b\cdots\text{0x}b$ (with $\text{0x}b$ repeated b times) and wishes to learn M. For $0 \le i < 2^8$ define

$$\Delta_i \stackrel{\text{def}}{=} \text{0x00} \cdots \text{0x00 } \text{0x}i \overbrace{\text{0x}(b+1) \cdots \text{0x}(b+1)}^{b \text{ times}}$$

$$\oplus \text{ 0x00} \cdots \text{0x00 0x00} \overbrace{\text{0x}b \quad \cdots \quad \text{0x}b}^{b \text{ times}};$$

i.e., the final $b+1$ bytes of Δ_i contain the integer i (in hexadecimal) followed by the value $(b+1) \oplus b$ (in hexadecimal) repeated b times. If the attacker submits the ciphertext $IV, c_1 \oplus \Delta_i, c_2$ to the server then, after CBC-mode decryption, the final $b+1$ bytes of the resulting encoded data will equal $\text{0x}(M \oplus i)\text{0x}(b+1)\cdots\text{0x}(b+1)$ (with $\text{0x}(b+1)$ repeated b times), and decryption will fail unless $\text{0x}(M \oplus i) = \text{0x}(b+1)$. The attacker tries at most 2^8 values $\Delta_0, \ldots, \Delta_{2^8-1}$ until decryption succeeds for some Δ_i, at which point it learns that $M = \text{0x}(b+1) \oplus \text{0x}i$. We leave it as an exercise to extend this attack so as to learn the next byte of m_2, as well as all of m_1.

A padding-oracle attack on CAPTCHAs. We have already mentioned that padding-oracle attacks have been carried out on encrypted web traffic. Here we give a second example.

A CAPTCHA is a distorted image of, say, an English word that is easy for humans to read, but hard for a computer to process. CAPTCHAs are used in order to ensure that a human user—and not some automated software—is interacting with a webpage.

CAPTCHAs can be provided as a separate service run on an independent server. To see how this works, we denote a web server by $\mathcal{S}_W$, a CAPTCHA server by $\mathcal{S}_C$, and a user by $\mathcal{U}$. When $\mathcal{U}$ loads a webpage served by $\mathcal{S}_W$, the following events occur: $\mathcal{S}_W$ encrypts a random English word w using a key k that was initially shared between $\mathcal{S}_W$ and $\mathcal{S}_C$, and sends the resulting ciphertext (along with the webpage) to the user. $\mathcal{U}$ forwards the ciphertext to $\mathcal{S}_C$, who decrypts it, obtains w, and renders a distorted image of w (i.e., the CAPTCHA) to $\mathcal{U}$. Finally, $\mathcal{U}$ sends w back to $\mathcal{S}_W$ for verification. Note that $\mathcal{S}_C$ decrypts any ciphertext it receives from $\mathcal{U}$ and will issue a "bad padding" error message if decryption fails, as described earlier. This presents $\mathcal{U}$ with an opportunity to carry out a padding-oracle attack, and thus to solve the CAPTCHA (i.e., to determine w) automatically *without any human involvement*, rendering the CAPTCHA ineffective.

5.1.2 Defining CCA-Security

What would it mean for an encryption scheme to be secure against chosen-ciphertext attacks? As usual, to define an appropriate notion of security we need to define two things: the assumed abilities of the attacker, and what constitutes a successful attack. For the latter, we will follow the approach we have taken in several previous definitions of security for encryption (e.g., in Definitions 3.8 and 3.21): namely, we give the attacker a challenge ciphertext c that is generated by encrypting one of two possible messages m_0, m_1 (each chosen with equal probability), and consider the scheme to be broken if the attacker can determine which message was encrypted with probability significantly better than $1/2$.

How should we model the attacker's capabilities in the present setting? Now, the adversary should have the ability not only to obtain the encryption of messages of its choice (as in a chosen-plaintext attack), but also to obtain the *decryption* of ciphertexts of its choice (with one exception discussed later). Formally, we give the adversary access to a *decryption oracle* $\mathsf{Dec}_k(\cdot)$ in addition to an encryption oracle $\mathsf{Enc}_k(\cdot)$. We present the formal definition and defer further discussion.

Consider the following experiment for any private-key encryption scheme $\Pi = (\mathsf{Gen}, \mathsf{Enc}, \mathsf{Dec})$, adversary $\mathcal{A}$, and value n for the security parameter.

The CCA indistinguishability experiment $\mathsf{PrivK}^{\mathsf{cca}}_{\mathcal{A},\Pi}(n)$:

1. *A key k is generated by running* $\mathsf{Gen}(1^n)$.

2. *$\mathcal{A}$ is given input 1^n and oracle access to* $\mathsf{Enc}_k(\cdot)$ *and* $\mathsf{Dec}_k(\cdot)$. *It outputs a pair of equal-length messages m_0, m_1.*

3. *A uniform bit $b \in \{0,1\}$ is chosen, and then a challenge ciphertext $c \leftarrow \mathsf{Enc}_k(m_b)$ is computed and given to $\mathcal{A}$.*

4. *The adversary $\mathcal{A}$ continues to have oracle access to* $\mathsf{Enc}_k(\cdot)$ *and* $\mathsf{Dec}_k(\cdot)$, *but is not allowed to query the latter on the challenge ciphertext itself. Eventually, $\mathcal{A}$ outputs a bit b'.*

5. *The output of the experiment is 1 if $b' = b$, and 0 otherwise. If the output of the experiment is 1, we say that $\mathcal{A}$ succeeds.*

DEFINITION 5.1 *A private-key encryption scheme Π has* indistinguishable encryptions under a chosen-ciphertext attack, *or is* CCA-secure, *if for all* *probabilistic polynomial-time adversaries $\mathcal{A}$ there is a negligible function* negl *such that:*

$$\Pr[\text{PrivK}_{\mathcal{A},\Pi}^{\text{cca}}(n) = 1] \leq \frac{1}{2} + \text{negl}(n),$$

where the probability is taken over all randomness used in the experiment.

For completeness, we remark that the natural analogue of Theorem 3.23 holds for CCA-security as well—namely, if a scheme has indistinguishable encryptions under a chosen-ciphertext attack then it has indistinguishable *multiple* encryptions under a chosen-ciphertext attack, defined appropriately.

Discussion. In the experiment considered above, the adversary is given access to a decryption oracle that returns the *entire result* of decrypting a ciphertext provided by the attacker. In general, this might be much more information than what is available to an attacker in the real world; for example, in the padding-oracle scenario described earlier, the attacker only learns whether decryption results in an error or not. As usual, however, we want to make cryptographic definitions as strong as possible so they are broadly applicable. Since we don't know what information an attacker might be able to learn when a ciphertext it sends is decrypted by a receiver, we make the worst-case assumption that the attacker learns *everything*.

There is one caveat. In the experiment, the adversary is allowed to submit any ciphertexts of its choice to the decryption oracle *except* that it may not request decryption of the challenge ciphertext itself. This restriction is clearly necessary or else there is no hope for any encryption scheme to satisfy the definition. Even with this restriction in place, the definition provides meaningful security. In particular, note that in the context of a padding-oracle attack the attacker does not learn anything by getting the receiver to decrypt the challenge ciphertext (since the attacker knows that it will not cause an error), and so a CCA-secure scheme would not be vulnerable to that attack.

Insecurity of the schemes we have studied. None of the encryption schemes we have seen thus far is CCA-secure. We demonstrate this for Construction 3.28, where encryption of a message m takes the form $\langle r, F_k(r) \oplus m \rangle$. Consider an adversary $\mathcal{A}$ running in the CCA indistinguishability experiment who chooses $m_0 = 0^n$ and $m_1 = 1^n$. Then, upon receiving a ciphertext $c = \langle r, s \rangle$, the adversary flips the first bit of s and asks for a decryption of the resulting ciphertext c'. Since $c' \neq c$, this query is allowed and the decryption oracle answers with either 10^{n-1} (in which case it is clear that $b = 0$) or 01^{n-1} (in which case $b = 1$). This example demonstrates that CCA-security is quite stringent. Any encryption scheme that allows ciphertexts to be "manipulated" in a controlled way cannot be CCA-secure. Thus, CCA-security implies a very important property called *non-malleability*. Loosely speaking, a non-malleable encryption scheme has the property that if the adversary modifies a given ciphertext, the result decrypts to a plaintext that bears no relation to the original one. This is a very useful property for encryption schemes used in complex cryptographic protocols.

5.2 Authenticated Encryption

CCA-security is extremely important, but is subsumed by an even stronger notion of security we introduce here. Until now, we have considered how to obtain secrecy (using encryption) and integrity (using message authentication codes) separately. The aim of *authenticated encryption*, defined below, is to achieve both goals *simultaneously*. It is best practice to *always ensure secrecy and integrity by default* in the private-key setting. Indeed, in many applications where secrecy is required it turns out that integrity is essential also. Moreover, a lack of integrity can sometimes lead to a breach of secrecy, as illustrated in the previous section.

5.2.1 Defining Authenticated Encryption

We begin, as usual, by defining precisely what we wish to achieve. One way to proceed is to define secrecy and integrity separately. Since we are explicitly concerned with an active adversary here, the natural notion of secrecy is CCA-security. The natural way to define integrity for encryption is via an analogue of the notion of existential unforgeability under an adaptive chosen-message attack that we considered for MACs. (We need a new definition because the syntax of an encryption scheme does not match the syntax of a MAC.) Consider the following experiment defined for a private-key encryption scheme $\Pi = (\mathsf{Gen}, \mathsf{Enc}, \mathsf{Dec})$, adversary $\mathcal{A}$, and value n for the security parameter:

The unforgeable encryption experiment $\mathsf{Enc\text{-}Forge}_{\mathcal{A},\Pi}(n)$:

1. *A key k is generated by running $\mathsf{Gen}(1^n)$.*

2. *The adversary $\mathcal{A}$ is given input 1^n and access to an encryption oracle $\mathsf{Enc}_k(\cdot)$. The adversary eventually outputs a ciphertext c. Let $m := \mathsf{Dec}_k(c)$ and let $\mathcal{Q}$ denote the set of all queries that $\mathcal{A}$ submitted to its oracle.*

3. *$\mathcal{A}$ succeeds if and only if (1) $m \neq\, \perp$ and (2) $m \notin \mathcal{Q}$. In that case the output of the experiment is defined to be 1.*

DEFINITION 5.2 *A private-key encryption scheme Π is* unforgeable *if for all probabilistic polynomial-time adversaries $\mathcal{A}$, there is a negligible function* negl *such that:*

$$\Pr[\mathsf{Enc\text{-}Forge}_{\mathcal{A},\Pi}(n) = 1] \leq \mathsf{negl}(n).$$

We may now define an *authenticated encryption* scheme.

DEFINITION 5.3   *A private-key encryption scheme is an* authenticated encryption (AE) scheme *if it is CCA-secure and unforgeable.*

It is also possible to capture both the above requirements in a definition involving a single experiment. The experiment is somewhat different from previous experiments we have considered, so we provide some motivation before giving the details. The idea is to consider two different scenarios, and require that they be indistinguishable to an attacker. In the first scenario, which can be viewed as corresponding to the real-world context in which the adversary operates, the attacker is given access to both an encryption oracle and a decryption oracle. In the second case, which can be viewed as corresponding to an "ideal" scenario, these two oracles are changed as follows:

- In place of an encryption oracle, the attacker is given access to an oracle that encrypts a 0-string of the correct length. Formally, the attacker is given access to an oracle $\mathsf{Enc}_k^0(\cdot)$ where $\mathsf{Enc}_k^0(m) = \mathsf{Enc}_k(0^{|m|})$. I.e., when requesting an encryption of m, the attacker is instead given an encryption of a 0-string of the same length as m.

- In place of a decryption oracle, the attacker is given access to an oracle $\mathsf{Dec}_\perp(\cdot)$ that always returns the error symbol $\perp$.

If an attacker cannot distinguish the first scenario from the second, then this means (1) any new ciphertexts the attacker generates in the real world will be invalid (i.e., will generate an error upon decryption). This not only implies a strong form of integrity, but also makes chosen-ciphertext attacks useless. Moreover, (2) the attacker cannot distinguish a real encryption oracle from an oracle that always encrypts 0s, which implies secrecy.

Formally, for a private-key encryption scheme Π, adversary $\mathcal{A}$, and value n for the security parameter, define the following experiment:

The authenticated-encryption experiment $\mathsf{PrivK}_{\mathcal{A},\Pi}^{\mathsf{ae}}(n)$:

1. *A key k is generated by running* $\mathsf{Gen}(1^n)$.

2. *A uniform bit $b \in \{0,1\}$ is chosen.*

3. *The adversary $\mathcal{A}$ is given input 1^n and access to two oracles:*

   (a) *If $b = 0$, then $\mathcal{A}$ is given access to* $\mathsf{Enc}_k(\cdot)$ *and* $\mathsf{Dec}_k(\cdot)$.
   (b) *If $b = 1$, then $\mathcal{A}$ is given access to* $\mathsf{Enc}_k^0(\cdot)$ *and* $\mathsf{Dec}_\perp(\cdot)$.

   $\mathcal{A}$ is not allowed to query a ciphertext c to its second oracle that it previously received as the response from its first oracle.

4. *The adversary outputs a bit b'.*

5. *The output of the experiment is defined to be 1 if* $b' = b$, *and* 0 *otherwise. In the former case, we say that* $\mathcal{A}$ succeeds.

In the experiment, the attacker is not allowed to submit ciphertexts to the decryption oracle that it received from its encryption oracle, since this would lead to a trivial way to distinguish the two scenarios. We remark that in the "real" case (i.e., when $b = 0$) the attacker already knows the decryption of those ciphertexts, so there is not much point in making such queries, anyway.

DEFINITION 5.4 *A private-key encryption scheme is an* authenticated encryption (AE) scheme *if for all probabilistic polynomial-time adversaries* $\mathcal{A}$ *there is a negligible function* negl *such that*

$$\Pr\left[\mathsf{PrivK}^{\mathsf{ae}}_{\mathcal{A},\Pi}(n) = 1\right] \leq \frac{1}{2} + \mathsf{negl}(n).$$

We have given two definitions of authenticated encryption. Fortunately, the definitions are equivalent:

THEOREM 5.5 *A private-key encryption scheme satisfies Definition 5.3 if and only if it satisfies Definition 5.4.*

Authenticated encryption with associated data. Often, a message m requires both secrecy and integrity but various *associated data* (e.g., header information) sent along with the message requires integrity only. While it is possible to simply concatenate the message and the associated data (in some way that allows for unambiguous parsing) and then use an AE scheme to encrypt them both, better efficiency can be achieved by providing the associated data with integrity protection only. We omit further details, but note that AE schemes with support for associated data are called *authenticated encryption with associated data (AEAD) schemes* in the literature.

5.2.2 CCA Security vs. Authenticated Encryption

It follows directly from Definition 5.3 that any authenticated encryption scheme is also CCA-secure. The converse, however, is not true, and there are private-key encryption schemes that are CCA-secure but that are *not* authenticated encryption schemes. You are asked to prove this in Exercise 5.9.

One can imagine applications where CCA-security is needed but authenticated encryption is not. One example might be when private-key encryption is used for *key transport*. As a concrete example, say a server gives a tamper-proof hardware token to a user, where the token stores a long-term key k. The server can share a fresh, short-term key k' with the token (that will remain unknown to the user) by giving the user $\mathsf{Enc}_k(k')$; the user is supposed to give this ciphertext to the token, which will decrypt it to obtain k'. CCA-security is necessary here because chosen-ciphertext attacks can be easily carried out by the user in this context. On the other hand, not much harm is done if the user can generate a valid ciphertext that causes the token to use some arbitrary key k'' that is uncorrelated with k'. (Of course, this depends on what the token does with this short-term key.)

Notwithstanding the above, most applications of private-key encryption in the presence of an active adversary do require integrity. Fortunately, most natural constructions of CCA-secure encryption schemes satisfy the stronger definition of authenticated encryption, anyway. Put differently, there is no reason to ever use a CCA-secure scheme that is *not* also an authenticated encryption scheme, simply because we don't have any constructions of the former that are more efficient than constructions of the latter.

From a *conceptual* point of view, however, the notions of CCA-security and authenticated encryption are distinct. With regard to CCA-security we are not interested in message integrity *per se*; rather, we wish to ensure *secrecy* even against an active adversary who can interfere with the communication from sender to receiver. In contrast, with regard to authenticated encryption we are explicitly interested in the twin goals of *secrecy* and *integrity*.

5.3 Authenticated Encryption Schemes

5.3.1 Generic Constructions

It is tempting to think that any reasonable combination of a CPA-secure encryption scheme and a secure message authentication code should result in an authenticated encryption scheme. In this section we show that this is not the case. This demonstrates that even secure cryptographic tools can be combined in such a way that the result is insecure, and highlights once again the importance of definitions and proofs of security. On the positive side, we show how encryption and message authentication *can* be combined properly to achieve joint secrecy and integrity.

Throughout, let $\Pi_E = (\mathsf{Enc}, \mathsf{Dec})$ be a CPA-secure encryption scheme and let $\Pi_M = (\mathsf{Mac}, \mathsf{Vrfy})$ denote a strongly secure MAC, where key generation in both schemes simply involves choosing a uniform n-bit key. There are three natural approaches to combining encryption and message authentication using independent[1] keys k_E and k_M for Π_E and Π_M, respectively:

1. *Encrypt-and-authenticate:* In this approach, encryption and message authentication are computed independently in parallel. That is, given

---

[1] *Independent* cryptographic keys should always be used when different schemes are combined. We return to this point at the end of this section.

a message m, the sender transmits the ciphertext $\langle c, t \rangle$ where:

$$c \leftarrow \mathsf{Enc}_{k_E}(m) \quad \text{and} \quad t \leftarrow \mathsf{Mac}_{k_M}(m).$$

The receiver decrypts c to recover m; assuming no error occurred, it then verifies the tag t. If $\mathsf{Vrfy}_{k_M}(m, t) = 1$, the receiver outputs m; otherwise, it outputs an error.

2. *Authenticate-then-encrypt:* Here a tag t is first computed, and then the message and tag are encrypted together. That is, given a message m, the sender transmits the ciphertext c computed as:

$$t \leftarrow \mathsf{Mac}_{k_M}(m) \quad \text{and} \quad c \leftarrow \mathsf{Enc}_{k_E}(m\|t).$$

The receiver decrypts c to obtain $m\|t$; assuming no error occurred, it then verifies the tag t. As before, if $\mathsf{Vrfy}_{k_M}(m, t) = 1$ the receiver outputs m; otherwise, it outputs an error.

3. *Encrypt-then-authenticate:* In this case, the message m is first encrypted and then a tag is computed over the result. That is, the ciphertext is the pair $\langle c, t \rangle$ where:

$$c \leftarrow \mathsf{Enc}_{k_E}(m) \quad \text{and} \quad t \leftarrow \mathsf{Mac}_{k_M}(c).$$

(See also Construction 5.6.) If $\mathsf{Vrfy}_{k_M}(c, t) = 1$, then the receiver decrypts c and outputs the result; otherwise, it outputs an error.

We analyze each of the above approaches when they are instantiated with "generic" secure components, i.e., when instantiated with an *arbitrary* CPA-secure encryption scheme and an *arbitrary* strongly secure MAC (cf. Definition 4.3). We are looking for an approach that provides joint secrecy and integrity when using *any* (secure) components, and so reject as "unsafe" any approach for which this is not the case. This reduces the likelihood of implementation flaws. Specifically, an approach might be implemented by making calls to an "encryption subroutine" and a "message authentication subroutine," and the implementation of those subroutines may be changed at some later point in time. (This commonly occurs when cryptographic libraries are updated, or when standards are modified.) An approach whose security depends on the details of how its underlying components are implemented—rather than on the security they provide—is therefore dangerous.

We stress that if an approach is rejected this does not mean that it is insecure for all possible instantiations of the components; it does, however, mean that any instantiation of the approach must be carefully analyzed and proven secure before it is used.

Encrypt-and-authenticate. Recall that in this approach encryption and message authentication are carried out independently. Given a message m, the ciphertext is $\langle c, t \rangle$ where $c \leftarrow \mathsf{Enc}_{k_E}(m)$ and $t \leftarrow \mathsf{Mac}_{k_M}(m)$. This approach is problematic since it may not achieve even the most basic level of secrecy. To see this, note that even a strongly secure MAC does not guarantee *any* secrecy and so it is possible for the tag t to leak information about m to an eavesdropper. (As a trivial example, consider a strongly secure MAC where the first bit of the tag is always equal to the first bit of the message.) So the encrypt-and-authenticate approach may yield a scheme that is not even EAV-secure.

The encrypt-and-authenticate approach is insecure against chosen-plaintext attacks even when instantiated with standard components (as opposed to the somewhat contrived example in the previous paragraph). In particular, if a *deterministic* MAC like CBC-MAC is used, then the tag computed on a message (for some fixed key k_M) is the same every time. This allows an eavesdropper to identify when the same message is sent twice, something that is not possible for a CPA-secure scheme. Many MACs used in practice are deterministic, so this represents a real concern.

Authenticate-then-encrypt. Here, a tag $t \leftarrow \mathsf{Mac}_{k_M}(m)$ is first computed; then $m\|t$ is encrypted and the ciphertext $c \leftarrow \mathsf{Enc}_{k_E}(m\|t)$ is transmitted. This combination also does not necessarily yield an authenticated encryption scheme. We have already encountered a CPA-secure encryption scheme for which this approach is insecure: the CBC-mode-with-padding scheme discussed in Section 5.1.1. (We assume in what follows that the reader is familiar with that section.) Recall that this scheme works by first padding the plaintext (which in our case will be $m\|t$) in a specific way so the result is a multiple of the block length, and then encrypting the result using CBC mode. During decryption, if an error in the padding is detected after performing the CBC-mode decryption, then a "bad padding" error is returned. With regard to the authenticate-then-encrypt approach, this means there are now *two* sources of potential decryption failure: the padding may be incorrect, or the tag may not verify. Schematically, the decryption algorithm Dec' in the combined scheme works as follows:

$\mathsf{Dec}'_{k_E, k_M}(c)$:

1. Compute $\tilde{m} := \mathsf{Dec}_{k_E}(c)$. If an error in the padding is detected (i.e., $\tilde{m} = \perp$), return "bad padding" and stop.

2. Otherwise, parse $\tilde{m}$ as $m\|t$. If $\mathsf{Vrfy}_{k_M}(m, t) = 1$ return m; else, output "authentication failure."

Assuming the attacker can distinguish between the two error messages, the attacker can apply the chosen-ciphertext attack described in Section 5.1.1 to recover the entire original plaintext from a given ciphertext. (This is due to the fact that the padding-oracle attack shown in Section 5.1.1 relies only on the ability to learn whether or not there was a padding error, something that is revealed by Dec'.) This type of attack has been carried out successfully in the real world in various settings, e.g., in configurations of TLS that used authenticate-then-encrypt.

One way to fix the above would be to ensure that only a *single* error message is returned, regardless of the source of decryption failure. This is an unsatisfying solution for several reasons: (1) there may be legitimate reasons (e.g., usability, debugging) to have multiple error messages; (2) forcing the error messages to be the same means that the combination is no longer truly generic, i.e., it requires the implementer of the authenticate-then-encrypt approach to be aware of what error messages are returned by the underlying CPA-secure encryption scheme; (3) most of all, it is extraordinarily hard to ensure that the different errors cannot be distinguished since, e.g., even a difference in the *time* to return each of these errors may allow an adversary to distinguish between them (cf. our earlier discussion of timing attacks at the end of Section 4.2). Some versions of TLS tried using only a single error message with the authenticate-then-encrypt approach, but a padding-oracle attack was still successfully carried out using small differences in timing.

Finally, we note that there are other counterexamples (that do not rely on distinguishing between different errors) showing that authenticate-then-encrypt does not necessarily provide authenticated encryption.

CONSTRUCTION 5.6

Let $\Pi_E = (\mathsf{Enc}, \mathsf{Dec})$ be a private-key encryption scheme and let $\Pi_M = (\mathsf{Mac}, \mathsf{Vrfy})$ be a message authentication code, where in each case key generation is done by simply choosing a uniform n-bit key. Define a private-key encryption scheme $(\mathsf{Gen}', \mathsf{Enc}', \mathsf{Dec}')$ as follows:

- Gen': on input 1^n, choose independent, uniform $k_E, k_M \in \{0,1\}^n$ and output the key (k_E, k_M).
- Enc': on input a key (k_E, k_M) and a plaintext message m, compute $c \leftarrow \mathsf{Enc}_{k_E}(m)$ and $t \leftarrow \mathsf{Mac}_{k_M}(c)$. Output the ciphertext $\langle c, t \rangle$.
- Dec': on input a key (k_E, k_M) and a ciphertext $\langle c, t \rangle$, first check if $\mathsf{Vrfy}_{k_M}(c, t) \overset{?}{=} 1$. If yes, output $\mathsf{Dec}_{k_E}(c)$; if no, output $\bot$.

The encrypt-then-authenticate approach.

Encrypt-then-authenticate. In this approach, the message is first encrypted and then a MAC is computed over the result. That is, the ciphertext is now the pair $\langle c, t \rangle$ where

$$c \leftarrow \mathsf{Enc}_{k_E}(m) \quad \text{and} \quad t \leftarrow \mathsf{Mac}_{k_M}(c).$$

Decryption of $\langle c, t \rangle$ outputs an error if $\mathsf{Vrfy}_{k_M}(c, t) \neq 1$, and otherwise outputs $\mathsf{Dec}_{k_E}(c)$. See Construction 5.6 for a formal description.

This approach *is* sound. As intuition for why, say a ciphertext $\langle c, t \rangle$ is *valid* if t is a valid tag on c. Strong security of the MAC ensures that an adversary will be unable to generate *any* valid ciphertext that it did not receive from its encryption oracle. This immediately implies that Construction 5.6 is unforgeable. Moreover, it effectively renders the decryption oracle useless: for every ciphertext $\langle c, t \rangle$ the adversary submits to its decryption oracle, the adversary either already knows the decryption (if it received $\langle c, t \rangle$ from its encryption oracle) or will receive an error. (Observe also that the tag is verified before decryption takes place; thus, errors during decryption cannot leak anything about the plaintext, in contrast to the padding oracle attack we saw against the authenticate-then-encrypt approach.) Therefore, CCA-security of the combined scheme reduces to CPA-security of Π_E.

THEOREM 5.7   *Let Π_E be a CPA-secure private-key encryption scheme, and let Π_M be a strongly secure message authentication code. Then Construction 5.6 is an authenticated encryption scheme.*

PROOF   We show that the scheme Π resulting from Construction 5.6 is unforgeable, and that it is CCA-secure. (See Definition 5.3.) Toward this end, we first show that strong security of Π_M implies that (except with negligible probability) any "new" ciphertexts an adversary submits to its decryption oracle will result in an error. This immediately implies unforgeability. (In fact, it is stronger than unforgeability.) It also renders the decryption oracle useless, and allows us to reduce CCA-security of Π to CPA-security of Π_E.

In more detail, let $\mathcal{A}$ be a PPT adversary attacking Construction 5.6 in a chosen-ciphertext attack (cf. Definition 5.1). Say a ciphertext that $\mathcal{A}$ submits to its decryption oracle is *new* if $\mathcal{A}$ did not receive it from its encryption oracle or as the challenge ciphertext. A ciphertext $\langle c, t \rangle$ is *valid* (with respect to the secret key (k_E, k_M) chosen as part of the experiment) if $\mathsf{Vrfy}_{k_M}(c, t) = 1$. Let ValidQuery be the event that $\mathcal{A}$ submits a new, valid ciphertext to its decryption oracle. We prove:

CLAIM 5.8 $\Pr[\mathsf{ValidQuery}]$ *is negligible.*

PROOF   Intuitively, this is because if ValidQuery occurs then the adversary has forged a new, valid pair (c, t) in the Mac-sforge experiment. Let $q(\cdot)$ be a polynomial upper bound on the number of decryption-oracle queries made by $\mathcal{A}$, and consider the following adversary $\mathcal{A}_M$ attacking the message authentication code Π_M:

Adversary $\mathcal{A}_M$:
$\mathcal{A}_M$ is given input 1^n and has access to a MAC oracle $\mathsf{Mac}_{k_M}(\cdot)$.

1. Choose uniform $k_E \in \{0, 1\}^n$ and $i \in \{1, \ldots, q(n)\}$.

2. Run $\mathcal{A}$ on input 1^n. When $\mathcal{A}$ makes an encryption-oracle query for the message m, answer it as follows:

   (i) Compute $c \leftarrow \mathsf{Enc}_{k_E}(m)$.

   (ii) Query c to the MAC oracle and receive t in response. Return $\langle c, t \rangle$ to $\mathcal{A}$.

   The challenge ciphertext is prepared in the exact same way (with a uniform bit $b \in \{0, 1\}$ chosen to select the message m_b that gets encrypted).

   When $\mathcal{A}$ makes a decryption-oracle query for the ciphertext $\langle c, t \rangle$, answer it as follows: If this is the ith decryption-oracle query, output (c, t) and halt. Otherwise:

   (i) If $\langle c, t \rangle$ was a response to a previous encryption-oracle query for a message m, return m.

   (ii) Otherwise, return $\bot$.

In essence, $\mathcal{A}_M$ is "guessing" that the ith decryption-oracle query of $\mathcal{A}$ is the first new, valid query $\mathcal{A}$ makes. In that case, $\mathcal{A}_M$ indeed outputs a new, valid message/tag pair (c, t).

Clearly $\mathcal{A}_M$ runs in polynomial time. We now analyze the probability that $\mathcal{A}_M$ outputs a new, valid message/tag pair. The key point is that the view of $\mathcal{A}$ when run as a subroutine by $\mathcal{A}_M$ is distributed identically to the view of $\mathcal{A}$ in experiment $\mathsf{PrivK}^{\mathsf{cca}}_{\mathcal{A},\Pi}(n)$ *until event* $\mathsf{ValidQuery}$ *occurs*. To see this, note that responses to the encryption-oracle queries of $\mathcal{A}$ (as well as the challenge ciphertext) are simulated perfectly by $\mathcal{A}_M$. As for the decryption-oracle queries of $\mathcal{A}$, until $\mathsf{ValidQuery}$ occurs these are all simulated properly. In case (i) this is obvious. As for case (ii), if the ciphertext $\langle c, t \rangle$ submitted to the decryption oracle is new, then as long as $\mathsf{ValidQuery}$ has not yet occurred the correct answer to that query is indeed $\bot$. (Recall also that $\mathcal{A}$ is disallowed from submitting the challenge ciphertext to the decryption oracle.)

Because the view of $\mathcal{A}$ when run as a subroutine by $\mathcal{A}_M$ is distributed identically to the view of $\mathcal{A}$ in experiment $\mathsf{PrivK}^{\mathsf{cca}}_{\mathcal{A},\Pi}(n)$ until event $\mathsf{ValidQuery}$ occurs, the probability of event $\mathsf{ValidQuery}$ in experiment $\mathsf{Mac\text{-}sforge}_{\mathcal{A}_M,\Pi_M}(n)$ is the same as the probability of that event in experiment $\mathsf{PrivK}^{\mathsf{cca}}_{\mathcal{A},\Pi}(n)$.

If $\mathcal{A}_M$ correctly guesses the first index i for which $\mathsf{ValidQuery}$ occurs, $\mathcal{A}_M$ outputs (c, t) for which $\mathsf{Vrfy}_{k_M}(c, t) = 1$ (since $\langle c, t \rangle$ is valid) and for which it was never given tag t in response to the query $\mathsf{Mac}_{k_M}(c)$ (since $\langle c, t \rangle$ is new). In this case, then, $\mathcal{A}_M$ succeeds in experiment $\mathsf{Mac\text{-}sforge}_{\mathcal{A}_M,\Pi_M}(n)$. The probability that $\mathcal{A}_M$ guesses i correctly is $1/q(n)$. Therefore

$$\Pr[\mathsf{Mac\text{-}sforge}_{\mathcal{A}_M,\Pi_M}(n) = 1] \geq \Pr[\mathsf{ValidQuery}]/q(n).$$

Since Π_M is a strongly secure MAC and q is polynomial, we conclude that $\Pr[\mathsf{ValidQuery}]$ is negligible. ∎

We use Claim 5.8 to prove security of Π. The easier step is to prove that Π is unforgeable. This follows immediately from the claim, and so we just provide informal reasoning. Observe first that the adversary in the unforgeable encryption experiment is a restricted version of the adversary in the chosen-ciphertext experiment. (In the former, the adversary only has access to an encryption oracle.) An attacker succeeds in the unforgeable encryption experiment only if it outputs a ciphertext $\langle c, t \rangle$ that is valid and new. But Claim 5.8 shows precisely that the probability of doing so is negligible.

It is slightly more involved to prove that Π is CCA-secure. Let $\mathcal{A}$ again be a probabilistic polynomial-time adversary attacking Π in a chosen-ciphertext attack. We have

$$\Pr[\mathsf{PrivK}^{\mathsf{cca}}_{\mathcal{A},\Pi}(n) = 1]$$
$$\leq \Pr[\mathsf{ValidQuery}] + \Pr[\mathsf{PrivK}^{\mathsf{cca}}_{\mathcal{A},\Pi}(n) = 1 \wedge \overline{\mathsf{ValidQuery}}]. \qquad (5.2)$$

We have already shown that $\Pr[\mathsf{ValidQuery}]$ is negligible. We show next that there is a negligible function negl such that

$$\Pr[\mathsf{PrivK}^{\mathsf{cca}}_{\mathcal{A},\Pi}(n) = 1 \wedge \overline{\mathsf{ValidQuery}}] \leq \frac{1}{2} + \mathsf{negl}(n).$$

To prove this, we rely on CPA-security of Π_E. Consider the following adversary $\mathcal{A}_E$ attacking Π_E in a chosen-plaintext attack:

Adversary $\mathcal{A}_E$:
$\mathcal{A}_E$ is given input 1^n and has access to $\mathsf{Enc}_{k_E}(\cdot)$.

1. Choose uniform $k_M \in \{0,1\}^n$.

2. Run $\mathcal{A}$ on input 1^n. When $\mathcal{A}$ makes an encryption-oracle query for the message m, answer it as follows:

   (i) Query m to $\mathsf{Enc}_{k_E}(\cdot)$ and receive c in response.

   (ii) Compute $t \leftarrow \mathsf{Mac}_{k_M}(c)$ and return $\langle c, t \rangle$ to $\mathcal{A}$.

   When $\mathcal{A}$ makes a decryption-oracle query for the ciphertext $\langle c, t \rangle$, answer it as follows: If $\langle c, t \rangle$ was a response to a previous encryption-oracle query for a message m, return m. Otherwise, return $\perp$.

3. When $\mathcal{A}$ outputs messages (m_0, m_1), output those same messages and receive a challenge ciphertext c in response. Compute $t \leftarrow \mathsf{Mac}_{k_M}(c)$, and return $\langle c, t \rangle$ to $\mathcal{A}$ as the challenge ciphertext. Continue answering $\mathcal{A}$'s oracle queries as above.

4. Output the same bit b' that is output by $\mathcal{A}$.

Notice that $\mathcal{A}_E$ does not need a decryption oracle because it simply assumes that any decryption query by $\mathcal{A}$ that was not the result of a previous encryption-oracle query is invalid.

Clearly, $\mathcal{A}_E$ runs in probabilistic polynomial time. Furthermore, the view of $\mathcal{A}$ when run as a subroutine by $\mathcal{A}_E$ is distributed identically to the view of $\mathcal{A}$ in experiment $\mathsf{PrivK}^{\mathsf{cca}}_{\mathcal{A},\Pi}(n)$ *as long as event* $\mathsf{ValidQuery}$ *never occurs.* Therefore, the probability that $\mathcal{A}_E$ succeeds is at least the probability that $\mathcal{A}$ succeeds and $\mathsf{ValidQuery}$ does not occur; i.e.,

$$\Pr[\mathsf{PrivK}^{\mathsf{cpa}}_{\mathcal{A}_E,\Pi_E}(n) = 1] \geq \Pr[\mathsf{PrivK}^{\mathsf{cpa}}_{\mathcal{A}_E,\Pi_E}(n) = 1 \wedge \overline{\mathsf{ValidQuery}}]$$
$$= \Pr[\mathsf{PrivK}^{\mathsf{cca}}_{\mathcal{A},\Pi}(n) = 1 \wedge \overline{\mathsf{ValidQuery}}].$$

Since Π_E is CPA-secure, there exists a negligible function negl such that $\Pr[\mathsf{PrivK}^{\mathsf{cpa}}_{\mathcal{A}_E,\Pi_E}(n) = 1] \leq \frac{1}{2} + \mathsf{negl}(n)$. Together with Equation (5.2), this proves that Π is CCA-secure. ∎

The need for independent keys. We conclude this section by stressing a basic principle of cryptography: *different instances of cryptographic primitives should always use independent keys.* To illustrate this, we consider what can happen to the encrypt-then-authenticate methodology if the same key k is used for both encryption and authentication. Let F be a strong pseudorandom permutation. It follows that F^{-1} is a strong pseudorandom permutation also. Define $\mathsf{Enc}_k(m) = F_k(m\|r)$ for $m \in \{0,1\}^{n/2}$ and a uniform $r \in \{0,1\}^{n/2}$; it can be shown that this encryption scheme is CPA-secure. (In fact, it is even CCA-secure; see Exercise 5.9.) Define $\mathsf{Mac}_k(c) = F_k^{-1}(c)$; this is just Construction 4.5, so is strongly secure. However, using these schemes with the same key k to encrypt-then-authenticate a message m yields:

$$\mathsf{Enc}_k(m), \mathsf{Mac}_k(\mathsf{Enc}_k(m)) = F_k(m\|r), F_k^{-1}(F_k(m\|r)) = F_k(m\|r), m\|r,$$

and so the message m is revealed in the clear! This does not in any way contradict Theorem 5.7, since Construction 5.6 expressly requires that k_M, k_E be chosen independently. We encourage the reader to determine where this independence is used in the proof of Theorem 5.7.

Authenticated encryption with associated data. As described at the end of Section 5.2.1, there are settings where a message m is encrypted along with associated data d that requires integrity but not secrecy. It is easy to modify the encrypt-then-authenticate approach to handle this: simply compute $c \leftarrow \mathsf{Enc}_{k_E}(m)$ followed by $t \leftarrow \mathsf{Mac}_{k_M}(d\|c)$.

5.3.2 Standardized Schemes

We close this chapter by briefly describing three AE schemes used in practice that are each inspired by one of the approaches discussed earlier. As usual, our aim here is not to provide an exact description of these schemes, but rather just a high-level understanding of the constructions.

GCM (Galois/counter mode). GCM can be viewed as following the *encrypt-then-authenticate* paradigm, with CTR mode (cf. Section 3.6.3) as the underlying encryption scheme and GMAC (cf. Section 4.5.2) as the underlying message authentication code. The main differences from the generic combination described in the previous section are that (1) the keys used for encryption and authentication are *not* independent and (2) the same IV is used both for CTR-mode encryption and as the nonce for GMAC. Both these changes can be proven secure for the particular way they are done by GCM.

One important property to be aware of when using GCM is that if the IV ever repeats, then not only does secrecy fail for the two messages encrypted using the same IV, but integrity of the scheme may be completely broken. This is due to a property of GMAC discussed in Exercise 4.21. For this reason, great care must be taken to ensure that IVs do not repeat when using GCM.

When GCM is instantiated with the AES block cipher (see Section 7.2.5) it is extremely fast on most modern processors due to dedicated hardware instructions for both AES and the field operations used in GMAC. The scheme is also highly parallelizable.

CCM (Counter with CBC-MAC). CCM follows the *authenticate-then-encrypt* approach, with CTR mode as the underlying encryption scheme and CBC-MAC (cf. Section 4.4.1) as the underlying message authentication code. Moreover, the same key k is used for both. Although—as discussed in the previous section—the authenticate-then-encrypt approach is not secure in general, and problems can occur when the keys used for encryption and authentication are not independent, CCM itself can be proven secure.

Because CCM relies only on a block cipher using a single key, it is easy to implement. However, CCM is relatively slow (it requires two block-cipher evaluations per plaintext block) and cannot be fully parallelized. In addition, it does not work in an on-line fashion, since it requires the message length to be known before encryption begins. (This is because the length is prepended to the message before CBC-MAC is computed, as discussed in Section 4.4.1.)

ChaCha20–Poly1305. This scheme relies on the *encrypt-then-authenticate* approach, where the underlying encryption is done using the stream cipher ChaCha20 (cf. Section 7.1.5) in unsynchronized mode (cf. Section 3.6.2) and the MAC used is Poly1305 (cf. Section 4.5.2) with ChaCha20 used here to instantiate the pseudorandom function. This scheme is extremely fast in software, and is becoming the method of choice on platforms where the dedicated hardware instructions used by GCM are not available.

5.4   Secure Communication Sessions

We briefly describe the application of authenticated encryption to the setting of two parties who wish to communicate "securely"—namely, with joint secrecy and integrity—over the course of a communication session. (For the purposes of this section, a *communication session* is simply a period of time during which the communicating parties maintain state.) In our treatment here we are deliberately informal; a formal definition is quite involved, and this topic arguably lies more in the area of network security than cryptography.

Let $\Pi = (\mathsf{Enc}, \mathsf{Dec})$ be an authenticated encryption scheme. Consider two parties A and B who share a key k and wish to use this key to secure their communication over the course of a session. The obvious thing to do here is to use Π: Whenever, say, A wants to transmit a message m to B, she computes $c \leftarrow \mathsf{Enc}_k(m)$ and sends c to B; in turn, B decrypts c to recover the result (ignoring the result if decryption returns $\perp$). Likewise, the same procedure is followed when B wants to send a message to A. This simple approach, however, is vulnerable to various potential attacks:

Re-ordering attack: An attacker can swap the order of messages. For example, if A transmits c_1 (an encryption of m_1) and subsequently transmits c_2 (an encryption of m_2), an attacker who has some control over the network can deliver c_2 before c_1 and thus cause B to output the messages in the wrong order.

Replay attack: An attacker can *replay* a (valid) ciphertext c sent previously by one of the parties. This would cause one party to output a message twice, even though the other party only sent it once.

Message-dropping attack: An attacker may drop some of the messages sent between A and B. Although nothing can prevent the attacker from doing this, we might at least hope that such behavior would be detected by the parties.

Reflection attack: An attacker can take a ciphertext c sent from A to B and send it back to A. This would cause A to output a message m, even though B never sent such a message.

The above list of attacks is not exhaustive, and is just an example of some of the challenges involved in achieving secure communication.

The above attacks can be addressed using *counters* to handle the first three and a *directionality bit* to prevent the fourth.[2] We describe these in tandem. Each party maintains two counters $\mathsf{ctr}_{A,B}$ and $\mathsf{ctr}_{B,A}$ keeping track of the number of messages sent from A to B (resp., B to A) during the session. These counters are initialized to 0 and incremented each time a party sends or receives a (valid) message. The parties also agree on a bit $b_{A,B}$, and define $b_{B,A}$ to be its complement. (One way to do this is to set $b_{A,B} = 0$ iff the identity of A is lexicographically smaller than the identity of B.)

---

[2]In practice, reflection attacks are often solved by simply having separate keys for each direction (i.e., the parties use a key k_A for messages sent from A to B, and an independent key k_B for messages sent from B to A).

When A wants to transmit a message m to B, she computes the ciphertext $c \leftarrow \mathsf{Enc}_k(b_{A,B}\|\mathsf{ctr}_{A,B}\|m)$ and sends c; she then increments $\mathsf{ctr}_{A,B}$. Upon receiving c, party B decrypts; if the result is $\bot$, he immediately rejects. Otherwise, he parses the decrypted message as $b\|\mathsf{ctr}\|m$. If $b = b_{A,B}$ and $\mathsf{ctr} = \mathsf{ctr}_{A,B}$, then B outputs m and increments $\mathsf{ctr}_{A,B}$; otherwise, B rejects. The above steps, *mutatis mutandis*, are applied when B sends a message to A.

References and Additional Reading

Chosen-ciphertext attacks (in the context of public-key encryption) were first formally defined by Naor and Yung [147] and Rackoff and Simon [168], and have received much subsequent attention as well [17, 68, 112]. The padding-oracle attack originated in the work of Vaudenay [199].

The importance of authenticated encryption was first explicitly highlighted by Katz and Yung [111] and Bellare and Namprempre [21]. Definition 5.4 is due to Shrimpton [184], who also proves Theorem 5.5. Bellare and Namprempre [21] analyze the three generic approaches discussed here, though the idea of using encrypt-then-authenticate for achieving CCA-security goes back at least to the work of Dolev et al. [68]. Krawczyk [122] examines other methods for achieving secrecy and authentication, and also analyzes specific instantiations of the authenticate-then-encrypt approach.

GCM is due to McGrew and Viega [136]. CCM was proposed by Whiting, Housley, and Ferguson [204] and proven secure by Jonsson [104]. ChaCha20–Poly1305 is specified in RFC 8439 [154].

Exercises

5.1 Show that the CBC, OFB, and CTR modes of operation do not give CCA-secure encryption schemes.

5.2 Write pseudocode for obtaining the entire plaintext for a 3-block ciphertext via a padding-oracle attack on CBC-mode encryption using PKCS #7 padding, as sketched in the text.

5.3 Describe a padding-oracle attack on CTR-mode encryption, assuming PKCS #7 padding is used to pad messages to a multiple of the block length before encrypting.

5.4 Show that Construction 5.6 is not necessarily CCA-secure if it is instantiated with a secure MAC that is not *strongly* secure.

5.5 Prove that Construction 5.6 is unforgeable when instantiated with any encryption scheme (even if not CPA-secure) and any secure MAC (even if not strongly secure).

5.6 Consider a strengthened version of unforgeability where $\mathcal{A}$ is additionally given access to a decryption oracle.

   (a) Write a formal definition for this version of unforgeability.

   (b) Prove that Construction 5.6 satisfies this stronger definition if Π_M is a strongly secure MAC.

   (c) Show by counterexample that Construction 5.6 need not satisfy this stronger definition if Π_M is a secure MAC that is not strongly secure. (Compare to the previous exercise.)

5.7 Prove that the authenticate-then-encrypt approach, instantiated with any CPA-secure encryption scheme and any secure MAC, yields a CPA-secure encryption scheme that is unforgeable.

5.8 Let F be a strong pseudorandom permutation, and define a fixed-length encryption scheme (Enc, Dec) as follows: On input $m \in \{0,1\}^{n/2}$ and key $k \in \{0,1\}^n$, algorithm Enc chooses a uniform string $r \in \{0,1\}^{n/2}$ of length $n/2$ and computes $c := F_k(r\|m)$.

   Show how to decrypt, and prove that this scheme is CCA-secure for messages of length $n/2$.

5.9 Show that the scheme in the previous exercise is not an authenticated encryption scheme.

5.10 Show a CPA-secure private-key encryption scheme that is unforgeable but is not CCA-secure.

Chapter 6

Hash Functions and Applications

In this chapter we look beyond the problem of secure communication that has occupied us until now, and consider a cryptographic primitive with many applications: *cryptographic hash functions*. At the most basic level, a hash function H provides a way to deterministically map a long input string to a shorter output string sometimes called a *digest*. The primary requirement is that it should be infeasible to find a *collision* in H: namely, two inputs that produce the same digest. As we will see, collision-resistant hash functions have numerous uses, including another approach—standardized as HMAC— for domain extension for message authentication codes.

Hash functions can be viewed as lying between the worlds of private- and public-key cryptography. On the one hand, as we will see in Chapter 7, they are (in practice) constructed using symmetric-key techniques. From a theoretical point of view, however, the existence of collision-resistant hash functions appears to be a qualitatively stronger assumption than the existence of other symmetric-key primitives, while at the same time being weaker than what is needed for public-key encryption. Hash functions have important applications in both the private- and public-key settings.

Hash functions have become ubiquitous in cryptography, and they are often used in scenarios that require properties much stronger than collision resistance. Indeed, it has become common to model cryptographic hash functions as being "completely unpredictable" (a.k.a., *random oracles*), and we discuss this model—and the controversy that surrounds it—in Section 6.5. In Section 6.6 we touch on a few applications of random oracles; we will encounter the random-oracle model again in the context of public-key cryptography.

6.1 Definitions

Hash functions are simply functions that take inputs of some length and *compress* them into short, fixed-length outputs. The classic use of (non-cryptographic) hash functions is in data structures, where they can be used to build hash tables that enable $\mathcal{O}(1)$ lookup time when storing a set of elements. Specifically, if the range of the hash function H is of size N, then element x

is stored in row $H(x)$ of a table of size N. To retrieve x, it suffices to compute $H(x)$ and probe that row of the table for the elements stored there. A "good" hash function for this purpose is one that yields few *collisions*, where a collision is a pair of distinct elements x and x' for which $H(x) = H(x')$; in this case we also say that x and x' *collide*. (When a collision occurs, two elements end up being stored in the same cell, increasing the lookup time.)

Collision-resistant hash functions are similar in spirit; again, the goal is to avoid collisions. However, there are fundamental differences. For one, the desire to minimize collisions in the setting of data structures becomes a requirement to *avoid* collisions in the setting of cryptography. Furthermore, in the context of data structures we assume that the set of elements being hashed is chosen independently of H and without any intention to cause collisions. In the context of cryptography, in contrast, we are faced with an adversary who may select elements with the explicit goal of causing collisions. This means that collision-resistant hash functions are much harder to design.

6.1.1 Collision Resistance

Informally, a function H is *collision resistant* if it is infeasible for any probabilistic polynomial-time algorithm to find a collision in H. We will only be interested in hash functions whose domain is larger than their range. In this case collisions must *exist*, but such collisions should be hard to find.

Formally, we consider *keyed* hash functions. That is, H is a two-input function that takes as input a key s and a string x, and outputs a string $H^s(x) \overset{\text{def}}{=} H(s, x)$. The requirement is that it must be hard to find a collision in H^s for a randomly generated key s. We highlight one major difference between keys in this context and the keys we have considered until now: In the present context, the key s is (generally) not kept secret, and collision resistance is required even when the adversary is given s. In order to emphasize that the key may not be secret, we superscript the key and write H^s rather than H_s.

DEFINITION 6.1   *A* hash function *(with output length $\ell(n)$) is a pair of probabilistic polynomial-time algorithms* (Gen, H) *satisfying the following:*

- Gen *is a probabilistic algorithm that takes as input a security parameter 1^n and outputs a key s. We assume that n is implicit in s.*

- H *is a deterministic algorithm that takes as input a key s and a string $x \in \{0,1\}^*$ and outputs a string $H^s(x) \in \{0,1\}^{\ell(n)}$ (where n is the value of the security parameter implicit in s).*

If H^s is defined only for inputs x of length $\ell'(n) > \ell(n)$, then we say that (Gen, H) *is a* fixed-length hash function for inputs of length $\ell'(n)$. *In this case, we also call H a* compression function.

In the fixed-length case we require that ℓ' be greater than ℓ. This ensures that H^s *compresses* its input. In the general case the function takes as input strings of arbitrary length; thus, it also compresses (albeit only inputs of length greater than $\ell(n)$). Note that without compression, collision resistance is trivial (since one can just take the identity function $H^s(x) = x$).

We now proceed to define security. As usual, we first define an experiment for a hash function $\mathcal{H} = (\mathsf{Gen}, H)$, an adversary $\mathcal{A}$, and a security parameter n:

The collision-finding experiment $\mathsf{Hash\text{-}coll}_{\mathcal{A},\mathcal{H}}(n)$:

1. *A key s is generated by running* $\mathsf{Gen}(1^n)$.

2. *The adversary $\mathcal{A}$ is given s, and outputs x, x'. (If $\mathcal{H}$ is a fixed-length hash function for inputs of length $\ell'(n)$, then we require $x, x' \in \{0,1\}^{\ell'(n)}$.)*

3. *The output of the experiment is defined to be 1 if and only if $x \neq x'$ and $H^s(x) = H^s(x')$. In such a case we say that $\mathcal{A}$ has found a collision.*

The definition of collision resistance states that no efficient adversary can find a collision in the above experiment except with negligible probability.

DEFINITION 6.2 *A hash function $\mathcal{H} = (\mathsf{Gen}, H)$ is* collision resistant *if for all probabilistic polynomial-time adversaries $\mathcal{A}$ there is a negligible function* negl *such that*

$$\Pr[\mathsf{Hash\text{-}coll}_{\mathcal{A},\mathcal{H}}(n) = 1] \leq \mathsf{negl}(n).$$

For simplicity, we sometimes refer to H or H^s as a "collision-resistant hash function," even though technically we should only say that $\mathcal{H} = (\mathsf{Gen}, H)$ is. This should not cause any confusion.

Cryptographic hash functions are designed with the explicit goal of being collision resistant (among other things). We will discuss some design principles for hash functions, along with some commonly used examples, in Chapter 7. In Section 9.4.2 we will see how it is possible to construct hash functions with proven collision resistance based on an assumption about the hardness of a certain number-theoretic problem.

Unkeyed hash functions. Cryptographic hash functions used in practice are generally *unkeyed* and have a fixed output length (by analogy with block ciphers), meaning that the hash function is just a fixed, deterministic function $H : \{0,1\}^* \to \{0,1\}^\ell$. This is problematic from a theoretical standpoint since for any such function there is always a constant-time algorithm that outputs a collision in H: the algorithm simply outputs a colliding pair (x, x') hardcoded into the algorithm itself. Using keyed hash functions solves this technical issue since it is impossible to hardcode a collision for every possible key using a reasonable amount of memory (and in an asymptotic setting, it would be impossible to hardcode a collision for every value of the security parameter).

Notwithstanding the above, the (unkeyed) cryptographic hash functions used in the real world are collision resistant for all practical purposes since colliding pairs are unknown (and computationally difficult to find) even though they must exist. Proofs of security for a scheme based on a collision-resistant hash function are still meaningful even when an unkeyed hash function H is used, as long as the proof shows that any efficient adversary "breaking" the primitive can be used to efficiently find a collision in H. (All the proofs in this book satisfy that condition.) In this case, the interpretation of the security proof is that if an adversary can break the scheme, then it can be used to find an explicit collision, something that is believed to be difficult.

In this chapter and throughout the rest of the book, we consider keyed hash functions when formally proving results that rely on collision resistance, but generally assume unkeyed hash functions otherwise.

6.1.2 Weaker Notions of Security

For some applications, security requirements weaker than collision resistance suffice. Security notions that are sometimes considered include:

- *Second-preimage resistance:* Informally, a hash function is said to be second-preimage resistant if given s and a uniform x it is infeasible for a PPT adversary to find $x' \neq x$ such that $H^s(x') = H^s(x)$.

- *Preimage resistance:* Informally, a hash function is preimage resistant if given s and $y = H^s(x)$ for a uniform x, it is infeasible for a PPT adversary to find a value x' (whether equal to x or not) with $H^s(x') = y$. (Looking ahead to Chapter 8, this basically means that H^s is *one-way*.)

It is immediate that any hash function that is collision resistant is also second-preimage resistant. It is also true that if a hash function is second-preimage resistant then it is preimage resistant. We do not formally define the above notions or prove these implications, since they are not used in the rest of the book. You are asked to formalize the above in Exercise 6.1.

6.2 The Merkle–Damgård Transform

Many applications require "full-fledged" collision-resistant hash functions that can handle very long inputs, or even inputs of arbitrary length. But it is much easier to construct fixed-length hash functions (i.e., compression functions) that only accept "short" inputs—something we will return to in Section 7.3. Fortunately, the *Merkle–Damgård transform* allows us to convert the latter to the former. This approach for domain extension of hash functions has been used frequently in practice, including for the hash function MD5 and the SHA hash family (cf. Section 7.3). The Merkle–Damgård transform is also interesting from a theoretical point of view since it implies that compressing by a single bit is as easy (or as hard) as compressing by an arbitrary amount.

For concreteness, assume the compression function (Gen, h) takes inputs of length $n + n' \geq 2n$, and generates outputs of length n. (The construction can be generalized for other input/output lengths, as long as h compresses.) Applying the Merkle–Damgård transform, defined in Construction 6.3 and depicted in Figure 6.1, yields a hash function (Gen, H) that maps inputs of *arbitrary* length to outputs of length n.

CONSTRUCTION 6.3

Let (Gen, h) be a compression function for inputs of length $n + n' \geq 2n$ with output length n. Fix $\ell \leq n'$ and $IV \in \{0, 1\}^n$. Construct hash function (Gen, H) as follows:

- Gen: remains unchanged.

- H: on input a key s and a string $x \in \{0, 1\}^*$ of length $L < 2^\ell$, do:

  1. Append a 1 to x, followed by enough zeros so that the length of the resulting string is ℓ less than a multiple of n'. Then append L, encoded as an ℓ-bit string. Parse the resulting string as the sequence of n'-bit blocks $x_1, \ldots, x_B$.

  2. Set $z_0 := IV$.

  3. For $i = 1, \ldots, B$, compute $z_i := h^s(z_{i-1} \| x_i)$.

  4. Output z_B.

The Merkle–Damgård transform.

THEOREM 6.4 *If* (Gen, h) *is collision resistant, then so is* (Gen, H).

PROOF We show that for any s, a collision in H^s yields a collision in h^s. Let x and x' be two different strings of length L and L', respectively, such that $H^s(x) = H^s(x')$. Let $x_1, \ldots, x_B$ be the B blocks of the padded x, and let $x'_1, \ldots, x'_{B'}$ be the B' blocks of the padded x'. Let $z_0, z_1, \ldots, z_B$ (resp., $z'_0, z'_1, \ldots, z'_{B'}$) be the intermediate results during computation of $H^s(x)$ (resp., $H^s(x')$). There are two cases to consider:

Case 1: $L \neq L'$. In this case, the last step of the computation of $H^s(x)$ is $z_B := h^s(z_{B-1} \| x_B)$, and the last step of the computation of $H^s(x')$ is $z'_{B'} := h^s(z'_{B'-1} \| x'_{B'})$. Since $H^s(x) = H^s(x')$ we have $h^s(z_{B-1} \| x_B) = h^s(z'_{B'-1} \| x'_{B'})$. However, $L \neq L'$ and so $x_B \neq x'_{B'}$. (Recall that the last ℓ bits of x_B encode L, and the last ℓ bits of $x'_{B'}$ encode L'.) Thus, $z_{B-1} \| x_B$ and $z'_{B'-1} \| x'_{B'}$ are a collision with respect to h^s.

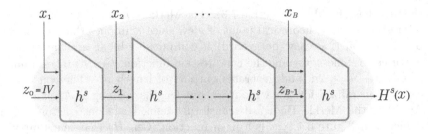

FIGURE 6.1:   The Merkle–Damgård transform.

Case 2: $L = L'$. This means that $B = B'$. Let $I_i \stackrel{\text{def}}{=} z_{i-1} \| x_i$ denote the ith input to h^s during computation of $H^s(x)$, and define $I_{B+1} \stackrel{\text{def}}{=} z_B$. Define $I'_1, \ldots, I'_{B+1}$ analogously with respect to x'. Let N be the largest index for which $I_N \neq I'_N$. Since $|x| = |x'|$ but $x \neq x'$, there is an i with $x_i \neq x'_i$ and so such an N certainly exists. Because

$$I_{B+1} = z_B = H^s(x) = H^s(x') = z'_B = I'_{B+1},$$

we have $N \leq B$. By maximality of N, we have $I_{N+1} = I'_{N+1}$ and in particular $z_N = z'_N$. But this means that I_N, I'_N collide under h^s.

We leave it as an exercise to turn the above into a proof by reduction. ∎

6.3   Message Authentication Using Hash Functions

We have already seen several constructions of message authentication codes for arbitrary-length messages. In this section we will see another approach that relies on collision-resistant hash functions. We then discuss a standardized and widely used scheme called HMAC that can be viewed as a specific instantiation of this approach.

6.3.1   Hash-and-MAC

Collision-resistant hash functions can naturally be used for domain extension of message authentication codes. Say we have a fixed-length MAC for $\ell(n)$-bit messages, and a collision-resistant hash function with $\ell(n)$-bit output length. Then we can authenticate an arbitrary-length message m by using the MAC to authenticate the *hash* of m. (See Construction 6.5.) Intuitively, this is secure because the MAC ensures that the attacker cannot authenticate any new hash value, while collision resistance ensures that the attacker will be unable to find any new message that hashes to a previously used hash value.

CONSTRUCTION 6.5

Let $\Pi = (\text{Mac}, \text{Vrfy})$ be a MAC for messages of length $\ell(n)$, and let $\mathcal{H} = (\text{Gen}_H, H)$ be a hash function with output length $\ell(n)$. Construct a MAC $\Pi' = (\text{Gen}', \text{Mac}', \text{Vrfy}')$ for arbitrary-length messages as follows:

- Gen': on input 1^n, choose uniform $k \in \{0,1\}^n$ and run $\text{Gen}_H(1^n)$ to obtain s; output the key (k, s).

- Mac': on input a key (k, s) and a message $m \in \{0,1\}^*$, output $t \leftarrow \text{Mac}_k(H^s(m))$.

- Vrfy': on input a key (k, s), a message $m \in \{0,1\}^*$, and a tag t, output 1 if and only if $\text{Vrfy}_k(H^s(m), t) \stackrel{?}{=} 1$.

The hash-and-MAC paradigm.

A bit more formally, say a sender uses Construction 6.5 to authenticate some set of messages $\mathcal{Q}$, and an attacker $\mathcal{A}$ is then able to forge a valid tag on a new message $m^* \notin \mathcal{Q}$. There are two possibilities:

Case 1: *there is a message $m \in \mathcal{Q}$ such that $H^s(m^*) = H^s(m)$.* Then $\mathcal{A}$ has found a collision in H^s, contradicting collision resistance of (Gen_H, H).

Case 2: *for every message $m \in \mathcal{Q}$ it holds that $H^s(m^*) \neq H^s(m)$.* Let $H^s(\mathcal{Q}) \stackrel{\text{def}}{=} \{H^s(m) \mid m \in \mathcal{Q}\}$. Then $H^s(m^*) \notin H^s(\mathcal{Q})$. In this case, $\mathcal{A}$ has forged a valid tag on the "new message" $h^* = H^s(m^*)$ with respect to the (fixed-length) message authentication code Π. This contradicts the assumption that Π is a secure MAC.

We now turn the above into a formal proof.

THEOREM 6.6 *If Π is a secure MAC for messages of length $\ell(n)$ and $\mathcal{H}$ is collision resistant, then Construction 6.5 is a secure MAC (for arbitrary-length messages).*

PROOF Let Π' denote Construction 6.5, and let $\mathcal{A}'$ be a PPT adversary attacking Π'. In an execution of experiment $\text{Mac-forge}_{\mathcal{A}', \Pi'}(n)$, let (k, s) denote the key (of Π'), let $\mathcal{Q}$ denote the set of messages whose tags were requested by $\mathcal{A}'$, and let (m^*, t) be the final output of $\mathcal{A}'$. We assume without loss of generality that $m^* \notin \mathcal{Q}$. Define coll to be the event that, in experiment $\text{Mac-forge}_{\mathcal{A}', \Pi'}(n)$, there is an $m \in \mathcal{Q}$ for which $H^s(m^*) = H^s(m)$. We have

$\Pr[\text{Mac-forge}_{\mathcal{A}', \Pi'}(n) = 1]$

$= \Pr[\text{Mac-forge}_{\mathcal{A}', \Pi'}(n) = 1 \wedge \text{coll}] + \Pr[\text{Mac-forge}_{\mathcal{A}', \Pi'}(n) = 1 \wedge \overline{\text{coll}}]$

$\leq \Pr[\text{coll}] + \Pr[\text{Mac-forge}_{\mathcal{A}', \Pi'}(n) = 1 \wedge \overline{\text{coll}}].$ \hfill (6.1)

We show that both terms in Equation (6.1) are negligible, thus completing the proof. Intuitively, the first term is negligible by collision resistance of $\mathcal{H}$, and the second term is negligible by security of Π.

Consider the following algorithm $\mathcal{C}$ for finding a collision in $\mathcal{H}$:

Algorithm $\mathcal{C}$:
The algorithm is given input s (with n implicit).

- Choose uniform $k \in \{0,1\}^n$.
- Run $\mathcal{A}'(1^n)$. When $\mathcal{A}'$ requests a tag on the ith message $m_i \in \{0,1\}^*$, compute $t_i \leftarrow \mathsf{Mac}_k(H^s(m_i))$ and give t_i to $\mathcal{A}'$.
- When $\mathcal{A}'$ outputs (m^*, t), then if there exists an i for which $H^s(m^*) = H^s(m_i)$, output (m^*, m_i).

It is clear that $\mathcal{C}$ runs in polynomial time. Let us analyze its behavior. When the input to $\mathcal{C}$ is generated by running $\mathsf{Gen}_H(1^n)$ to obtain s, the view of $\mathcal{A}'$ when run as a subroutine by $\mathcal{C}$ is distributed identically to the view of $\mathcal{A}'$ in experiment $\mathsf{Mac\text{-}forge}_{\mathcal{A}',\Pi'}(n)$. Thus, the probability that coll occurs is the same in both cases. Since $\mathcal{C}$ outputs a collision when coll occurs, we have

$$\Pr[\mathsf{Hash\text{-}coll}_{\mathcal{C},\mathcal{H}}(n) = 1] = \Pr[\mathsf{coll}].$$

Collision resistance of $\mathcal{H}$ thus implies that $\Pr[\mathsf{coll}]$ is negligible.

We now proceed to prove that the second term in Equation (6.1) is negligible. Consider the following adversary $\mathcal{A}$ attacking Π in $\mathsf{Mac\text{-}forge}_{\mathcal{A},\Pi}(n)$:

Adversary $\mathcal{A}$:
The adversary is given 1^n and access to an oracle $\mathsf{Mac}_k(\cdot)$.

- Compute $\mathsf{Gen}_H(1^n)$ to obtain s.
- Run $\mathcal{A}'(1^n)$. When $\mathcal{A}'$ requests a tag on the ith message $m_i \in \{0,1\}^*$, then: (1) compute $h_i := H^s(m_i)$; (2) obtain a tag t_i on h_i from the MAC oracle; and (3) give t_i to $\mathcal{A}'$.
- When $\mathcal{A}'$ outputs (m^*, t), set $h^* := H^s(m^*)$ and then output (h^*, t).

$\mathcal{A}$ runs in polynomial time. If $\mathcal{A}'$ outputs (m^*, t) with $\mathsf{Vrfy}_k(H^s(m^*), t) = 1$, and coll did not occur, then $\mathcal{A}$ outputs a valid forgery. (In that case t is a valid tag on $h^* = H^s(m^*)$ in scheme Π with respect to k. The fact that coll did not occur means that h^* was never asked by $\mathcal{A}$ to its own MAC oracle and so this is indeed a forgery.) Moreover, the view of $\mathcal{A}'$ when run as a subroutine by $\mathcal{A}$ in experiment $\mathsf{Mac\text{-}forge}_{\mathcal{A},\Pi}(n)$ is distributed identically to the view of $\mathcal{A}'$ in experiment $\mathsf{Mac\text{-}forge}_{\mathcal{A}',\Pi'}(n)$. We conclude that

$$\Pr[\mathsf{Mac\text{-}forge}_{\mathcal{A},\Pi}(n) = 1] = \Pr[\mathsf{Mac\text{-}forge}_{\mathcal{A}',\Pi'}(n) = 1 \wedge \overline{\mathsf{coll}}],$$

and security of Π implies that the former probability is negligible. This concludes the proof of the theorem. ∎

6.3.2 HMAC

In principle, the hash-and-MAC approach from the previous section could be instantiated by combining an arbitrary collision-resistant hash function with the fixed-length MAC of Construction 4.5. This way of realizing the hash-and-MAC approach has at least two drawbacks in practice. First, it requires implementing *two* cryptographic primitives: a hash function and a block cipher. (Recall that Construction 4.5 is based on a block cipher, and supports messages of length equal to the block length of the cipher.) This can be a problem, e.g., in constrained devices, where it is desirable to keep the size of the code implementing a cryptographic scheme as small as possible. A more fundamental difficulty is that there is often a mismatch between the output length of hash functions and the block length of block ciphers. (This is in part due to a difference between the parameters needed to achieve security for a block cipher vs. a hash function, as will be explored in the next section.) For example, the block cipher AES has a 128-bit block length, whereas modern hash functions have output lengths of at least 256 bits—and a 128-bit output length would be far too short to ensure meaningful collision resistance.

CONSTRUCTION 6.7

Let (Gen_H, H) be a hash function constructed by applying the Merkle–Damgård transform to a compression function (Gen_H, h) that takes inputs of length $n + n' > 2n + \log n + 2$ and generates output of length n. Fix distinct constants $\mathsf{opad}, \mathsf{ipad} \in \{0, 1\}^{n'}$. Define a MAC as follows:

- Gen: on input 1^n, run $\mathsf{Gen}_H(1^n)$ to obtain a key s. Also choose uniform $k \in \{0, 1\}^{n'}$. Output the key (s, k).

- Mac: on input a key (s, k) and a message $m \in \{0, 1\}^*$, output

$$t := H^s\Big((k \oplus \mathsf{opad}) \,\|\, H^s\big((k \oplus \mathsf{ipad}) \,\|\, m\big)\Big).$$

- Vrfy: on input a key (s, k), a message $m \in \{0, 1\}^*$, and a tag t, output 1 if and only if $t \overset{?}{=} H^s\big((k \oplus \mathsf{opad}) \,\|\, H^s\big((k \oplus \mathsf{ipad}) \,\|\, m\big)\big)$.

HMAC.

The above concerns motivated the design of HMAC, a message authentication code for arbitrary-length messages that can be based on any hash function (Gen_H, H) constructed using the Merkle–Damgård transform applied to a compression function (Gen_H, h). See Construction 6.7 for a high-level overview that abstracts out the underlying compression function, and Figure 6.2 for a graphical depiction that makes the compression function explicit.

Referring to Figure 6.2, we see that computation of HMAC on a message $m = m_1, m_2, \ldots$ using key k can be separated into an "inner" hash evaluation

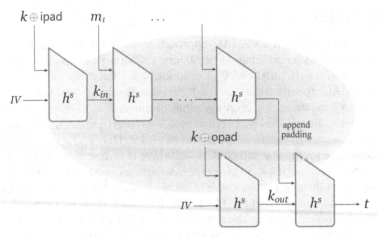

FIGURE 6.2: HMAC, pictorially.

and an "outer" hash evaluation. The inner hash evaluation involves computing $\hat{m} := H^s((k \oplus \mathsf{ipad}) \| m)$, where ipad is some fixed constant. As per the definition of the Merkle–Damgård transform, the input to H^s—which, in this case, is the string $(k \oplus \mathsf{ipad}) \| m$—is padded as part of the hash computation; this padding is left implicit in Figure 6.2. The outer hash evaluation involves computation of the tag $t := H^s((k \oplus \mathsf{opad}) \| \hat{m})$, where opad is another fixed constant; note that k, ipad, and opad are all exactly n' bits long. Once again, padding is applied to the $(n'+n)$-bit input string $(k \oplus \mathsf{opad}) \| \hat{m}$ as part of the hash computation; parameters are set such that the padded string is exactly two blocks long. That is, if we let $k_{out} \overset{\text{def}}{=} h^s(IV \| (k \oplus \mathsf{opad}))$ as in the figure, then $t = h^s(k_{out} \| \hat{m}^*)$, where $\hat{m}^*$ is the second block after padding.

Given this perspective, we see that HMAC can be viewed as an instantiation of the hash-and-MAC paradigm from the previous section, where the inner computation corresponds to hashing the message m to an n'-bit string $\hat{m}^*$ (including the padding), and the outer computation corresponds to computing a fixed-length message authentication code on $\hat{m}^*$. Formally, let $\widetilde{\Pi}^s = (\widetilde{\mathsf{Gen}}^s, \widetilde{\mathsf{Mac}}^s, \widetilde{\mathsf{Vrfy}}^s)$ be the message authentication code in which $\widetilde{\mathsf{Mac}}^s_{k_{out}}(\hat{m}^*) = h^s(k_{out} \| \hat{m}^*)$ (We view s here as a fixed, public value.) Intuitively, then, if (Gen_H, h) is collision resistant and $\widetilde{\Pi}^s$ is secure, HMAC is secure. As a technical matter, though, since the key k_{out} used by the MAC $\widetilde{\Pi}^s$ is derived from an underlying key k that is also used in the inner hash evaluation, we need one additional assumption regarding the "computational independence" of $k_{in} \overset{\text{def}}{=} h^s(IV \| (k \oplus \mathsf{ipad}))$ and k_{out}. Specifically, define

$$G^s(k) \overset{\text{def}}{=} h^s\left(IV \| (k \oplus \mathsf{ipad})\right) \| h^s\left(IV \| (k \oplus \mathsf{opad})\right) = k_{in} \| k_{out}.$$

Then it is possible to prove:

THEOREM 6.8 *Assume G^s is a pseudorandom generator, $\widetilde{\Pi}^s$ is a secure fixed-length MAC for messages of length n', and (Gen_H, h) is collision resistant. Then HMAC is a secure MAC (for arbitrary-length messages).*

(We require the first two assumptions in the theorem to hold for all s. Even if G^s is not expanding, it is still meaningful to speak of its output as being pseudorandom.) Because of the way the compression function h is typically designed (see Section 7.3.1), the first two assumptions are reasonable.

The roles of ipad and opad. One might wonder why it is necessary to incorporate k_{in} (or k itself) in the "inner" computation at all. In particular, for the hash-and-MAC approach all that is required is for the inner computation to be collision resistant, which does not require any secret key. The reason for including a secret key as part of the inner computation is that this allows security of HMAC to be based on the assumption that (Gen_H, H) is *weakly collision resistant*, where (informally) this refers to an experiment in which an attacker needs to find collisions in a secretly keyed hash function. This is a weaker condition than collision resistance, and hence is potentially easier to satisfy. The defensive design strategy of HMAC paid off when it was discovered that the hash function MD5 (see Section 7.3.2) used in HMAC–MD5 was *not* collision resistant. The attacks on MD5 did not violate *weak* collision resistance, and so HMAC–MD5 was not broken even though MD5 was. (Despite this, HMAC–MD5 should no longer be used now that weaknesses in MD5 are known.) This gave developers time to replace MD5 in HMAC implementations, without immediate fear of attack.

Ideally, *independent* keys k_{in}, k_{out} should have been used in the inner and outer computations. To reduce the key length of HMAC, a single key k is used to derive k_{in} and k_{out} using ipad and opad. (Moreover, in practice it is typical for the length of k to be much shorter than n' in which case k is simply padded with 0s before being XORed with ipad and opad.) If we assume that G^s (as defined above) is a pseudorandom generator for any s, then k_{in} and k_{out} can be treated as independent, uniform keys when k is uniform.

6.4 Generic Attacks on Hash Functions

In the context of the symmetric-key primitives we have studied so far (block ciphers, private-key encryption schemes, etc.), we noted that any scheme using an n-bit secret key is vulnerable to a *brute-force attack* in which an attacker enumerates all 2^n possible keys until it finds the right one. (Of course, this does not apply to information-theoretic schemes.) Put differently, if we want to achieve security against attackers running in time 2^n then we need to use secret keys that are at least n bits long.

What can we say about the security of hash functions against brute-force attacks? We show here that a *birthday attack* allows an attacker to find a collision in any hash function having an ℓ-bit output length in time $2^{\ell/2}$. Thus, if we want to ensure collision resistance against attackers running in time 2^n we need to use hash functions whose output is at least $2n$ bits long—twice the length of secret keys providing comparable security guarantees.

While on the topic of *generic attacks* (i.e., attacks that apply to arbitrary hash functions), we also consider attacks on preimage resistance, where the attacker's goal is to find an input x that hashes to a given value y. Here the question is complicated by the attacker's ability to use preprocessing and a large amount of storage to speed up the attack. This has important ramifications in practice when hashing users' passwords, something we touch on in Section 6.6.3.

6.4.1 Birthday Attacks for Finding Collisions

Let $H : \{0,1\}^* \rightarrow \{0,1\}^\ell$ be a hash function. For any such H, there is always a trivial collision-finding attack running in time $\mathcal{O}(2^\ell)$: simply evaluate H on $q = 2^\ell + 1$ distinct inputs; by the pigeonhole principle, two of the outputs must be equal. Is this the best possible attack?

Let us generalize the above algorithm by taking q as a parameter. Say we choose q uniform (distinct) inputs $x_1, \ldots, x_q$, compute $y_i := H(x_i)$ for all i, and check whether any of the $\{y_i\}$ are equal. As noted, if $q > 2^\ell$ then there is certainly a collision. When $q \leq 2^\ell$ we can no longer guarantee a collision, but there is clearly some nonzero probability that a collision occurs. It is somewhat difficult to analyze this probability when H is arbitrary, and so we instead consider the idealized case where H is treated as a random function. (It can be shown that this is the worst case, and collisions occur with higher probability if H deviates from random.) That is, for each i we assume that the value $y_i = H(x_i)$ is uniformly distributed in $\{0,1\}^\ell$ and independent of all the other values $\{y_j\}_{j \neq i}$ (recall all the $\{x_i\}$ are distinct). We have thus reduced our problem to the following: if we generate uniform $y_1, \ldots, y_q \in \{0,1\}^\ell$, what is the probability that there exist distinct i, j with $y_i = y_j$?

This question has been extensively studied, and is related to the so-called *birthday problem* discussed in detail in Appendix A.4; for this reason the collision-finding algorithm described above is one of a class of algorithms called *birthday attacks*. The birthday problem is this: if q people are in a room, what is the probability that some two of them share a birthday? (Assume birthdays are uniformly and independently distributed among the 365 days of a non-leap year.) This is analogous to our problem: if y_i is the birthday of person i, then we have uniform and independent $y_1, \ldots, y_q \in \{1, \ldots, 365\}$, and matching birthdays correspond to distinct i, j with $y_i = y_j$ (i.e., matching birthdays correspond to collisions).

In Appendix A.4 we show that when $y_1, \ldots, y_q$ are uniform in $\{1, \ldots, N\}$, then if $q = \Theta(N^{1/2})$ the probability of a collision is roughly $1/2$. (In the case of birthdays, once there are only 23 people the probability that some two of them have the same birthday is roughly 51%!) In our setting, this means that when the hash function H has output length ℓ (and so has range of size $N = 2^\ell$), evaluating H on $q = \Theta(2^{\ell/2})$ inputs yields a collision with probability roughly $1/2$. From a concrete-security perspective, this implies that for a hash function H to be collision resistant against attackers running in time 2^n it is required that H have output at least $2n$ bits long. Taking specific parameters: if we want finding collisions to be as difficult as an exhaustive search over 128-bit keys, then we need the output length of the hash function to be at least 256 bits. (We stress that having output this long is only a *necessary* condition, not a sufficient one.)

Finding meaningful collisions. The birthday attack just described gives a collision that is not necessarily very useful, since the colliding inputs are random. But the same idea can be used to find "meaningful" collisions as well. Assume Alice wishes to find two messages x and x' such that $H(x) = H(x')$, and furthermore x should be a letter from her employer explaining why she was fired from work, while x' should be a flattering letter of recommendation. (This might allow Alice to forge a tag on a letter of recommendation if the hash-and-MAC approach is being used by her employer to authenticate messages.) Note that the birthday attack only requires the hash inputs $x_1, \ldots, x_q$ to be distinct; they do not need to be random. Alice can carry out a birthday attack by generating $q = \Theta(2^{\ell/2})$ messages of the first type and q messages of the second type, and then looking for collisions between messages of the two types. A small change to the analysis from Appendix A.4 shows that this gives a collision between messages of different types with probability roughly $1/2$. A little thought shows that it is easy to write the same message in many different ways. For example, consider the following:

> It is *hard/difficult/challenging/impossible* to *imagine/believe* that we will *find/locate/hire* another *employee/person* having similar *abilities/skills/character* as Alice. She has done a *great/super* job.

Any combination of the italicized words is possible, and expresses the same idea. Thus, the sentence can be written in $4 \cdot 2 \cdot 3 \cdot 2 \cdot 3 \cdot 2 = 288$ different ways. This is just one sentence and so it is actually easy to generate a message that can be rewritten in 2^{64} different ways—all that is needed are 64 words with one synonym each. Alice can prepare $2^{\ell/2}$ letters explaining why she was fired and another $2^{\ell/2}$ letters of recommendation; with good probability, a collision between the two types of letters will be found.

6.4.2 Small-Space Birthday Attacks

The birthday attacks described above require a large amount of memory; specifically, they require the attacker to store all $\Theta(q) = \Theta(2^{\ell/2})$ values $\{y_i\}$, because the attacker does not know in advance which pair of values will yield a collision. This is a significant drawback because memory is, in general, a scarcer resource than time: one can always let a computation run as long as needed, whereas if a program requires more memory than is available then that program will simply halt. Furthermore, memory accesses are typically orders of magnitude slower than executing arithmetic instructions.

We show here a better birthday attack with drastically reduced memory requirements. In fact, it has similar time complexity and success probability as before, but uses only a *constant* amount of memory. The attack begins by choosing a uniform value x_0 and then computing $x_i := H(x_{i-1})$ and $x_{2i} := H(H(x_{2(i-1)}))$ for $i = 1, 2, \ldots$. (Note that $x_i = H^{(i)}(x_0)$ for all i, where $H^{(i)}$ refers to i-fold iteration of H.) In each step the values x_i and x_{2i} are compared; if they are equal then there is a collision somewhere in the sequence $x_0, x_1, \ldots, x_{2i-1}$. (The values x_{i-1} and x_{2i-1} might not be a collision because they may themselves be equal.) The algorithm then finds the least value of j for which $x_j = x_{j+i}$, and outputs x_{j-1}, x_{j+i-1} as a collision. This attack, described formally as Algorithm 6.9 and analyzed below, only requires storage of two hash values in each iteration.

ALGORITHM 6.9
A small-space birthday attack

Output: Distinct x, x' with $H(x) = H(x')$

$x_0 \leftarrow \{0,1\}^{\ell+1}$
$x' := x := x_0$
for $i = 1, 2, \ldots$ do:
    $x := H(x)$
    $x' := H(H(x'))$ // now $x = H^{(i)}(x_0)$ and $x' = H^{(2i)}(x_0)$
    if $x = x'$ **break**
$x' := x$, $x := x_0$
for $j = 1$ to i:
    if $H(x) = H(x')$ **return** x, x' and **halt**
    else $x := H(x)$, $x' := H(x')$
    // now $x = H^{(j)}(x_0)$ and $x' = H^{(j+i)}(x_0)$

How many iterations of the first loop do we expect before $x = x'$? Consider the sequence of values $x_1, x_2, \ldots$, where $x_i = H^{(i)}(x_0)$ as before. If we model H as a random function, then each x_i is uniform and independent of $x_1, \ldots, x_{i-1}$ as long as no repeat has yet occurred in this sequence. Thus, we expect a repeat to occur with probability $1/2$ in the first $q = \Theta(2^{\ell/2})$ elements of the sequence. When there *is* a repeat in the first q elements, the algorithm finds a repeat in at most q iterations of the first loop:

CLAIM 6.10   *Let $x_1, \ldots, x_q$ be a sequence of values with $x_m = H(x_{m-1})$. If $x_I = x_J$ with $1 \leq I < J \leq q$, then there is an $i < J$ such that $x_i = x_{2i}$.*

PROOF   The sequence $x_I, x_{I+1}, \ldots$ repeats with period $\Delta \overset{\text{def}}{=} J - I$. That is, for all $i \geq I$ and $k \geq 0$ it holds that $x_i = x_{i+k \cdot \Delta}$. Let i be the smallest multiple of Δ that is also greater than or equal to I. We have $i < J$ since the sequence of Δ values $I, I+1, \ldots I + (\Delta - 1) = J - 1$ contains a multiple of Δ. Since $i \geq I$ and $2i - i = i$ is a multiple of Δ, it follows that $x_i = x_{2i}$.   ■

Thus, if there is a repeated value in the sequence $x_1, \ldots, x_q$, there is some $i < q$ for which $x_i = x_{2i}$. But then in iteration i of Algorithm 6.9, we have $x = x'$ and the algorithm breaks out of the first loop. At that point in the algorithm, we know that $x_i = x_{i+i}$. The algorithm then sets $x' := x = x_i$ and $x := x_0$, and proceeds to find the *smallest* $j > 0$ for which $x_j = x_{j+i}$. (Note $x_0 \neq x_i$ because $|x_0| = \ell + 1$.) It outputs x_{j-1}, x_{j+i-1} as a collision.

Finding meaningful collisions. The algorithm just described may not seem amenable to finding meaningful collisions since it has no control over the $\{x_i\}$ values used. Nevertheless, we show that finding meaningful collisions is still possible. The trick is to find a collision in the right function!

Assume, as before, that Alice wants to find a collision between messages of two different "types," e.g., a letter explaining why she was fired and a flattering letter of recommendation. Alice writes each message so there are $\ell - 1$ interchangeable words in each; i.e., there are $2^{\ell-1}$ messages of each type. Define the function $g : \{0,1\}^\ell \rightarrow \{0,1\}^*$ such that the first bit of the input selects between messages of type 0 or type 1, and the remaining bits select between options for the interchangeable words in messages of the appropriate type. For example, if $\ell = 4$ we could consider the sentences:

type 0: Alice is a *good/great* and *honest/trustworthy* *worker/employee*.

type 1: Alice is a *bad/lousy* and *annoying/irritating* *worker/employee*.

The function g is then defined on 4-bit inputs, where the first bit determines the sentence type and the final three bits determine the words in the sentence. That is:

$g(0000)$ = Alice is a good and honest worker.

$g(1101)$ = Alice is a lousy and annoying employee.

Finally, define $f : \{0,1\}^\ell \rightarrow \{0,1\}^\ell$ by $f(x) \overset{\text{def}}{=} H(g(x))$. Alice can find a collision in f using a variant of the small-space birthday attack shown earlier. Note that any collision x, x' in f yields two messages $g(x), g(x')$ that collide under H. If x, x' is a random collision then we expect that with probability $1/2$ the colliding messages $g(x), g(x')$ will be of different types (since x and x' will differ in their first bit with probability $1/2$). If the colliding messages are not of different types, the process can be repeated.

6.4.3   *Time/Space Tradeoffs for Inverting Hash Functions

In this section we consider the question of preimage resistance, i.e., we are interested in algorithms for the problem of function inversion. Here, we have a hash function $H : \{0,1\}^* \to \{0,1\}^\ell$; an adversary is given $y = H(x)$ and its goal is to find any x' such that $H(x') = y$. (We call such an x' a *preimage* of y.) We begin by assuming that $x \in \{0,1\}^\ell$ for simplicity (and so view the domain of H as $\{0,1\}^\ell$), and consider the more general case at the end.

Finding a preimage of $y = H(x)$ can be done in time $\Theta(2^\ell)$ via exhaustive search over the domain of H, and this is optimal when H is modeled as a random function. However, it ignores the possibility of *preprocessing*. That is, it may be possible for an algorithm to perform a significant amount of work in an "off-line" preprocessing phase *before* y is known, and then to find a preimage x' in an "on-line" phase *after* being given y, using significantly less than $\Theta(2^\ell)$ computation. This can be a worthwhile tradeoff if work can be invested in advance, or if the algorithm will be used to find preimages of *multiple* values (since the same preprocessing can be used for all of them).

In fact, it is trivial to use preprocessing to improve the on-line time of function inversion. All we need to do is evaluate H on every point in $\{0,1\}^\ell$ during the preprocessing phase, and store all the pairs $\{(x, H(x))\}$ in a table, sorted by their second entry. Upon receiving a point y, a preimage of y can be found easily by using binary search to find a pair in the table with second entry y. The drawback here is that we need to allocate memory for storing 2^ℓ pairs, which can be prohibitive—if not impossible—for large ℓ.

Exhaustive search uses constant memory and $\Theta(2^\ell)$ on-line time, while the attack just described stores $\Theta(2^\ell)$ points in memory but enables inversion in essentially constant on-line time. We now show an approach that allows an attacker to *trade off* time and memory and interpolate between these extremes. Specifically, we show how to store $\mathcal{O}(2^{2\ell/3})$ values and find preimages in time $\mathcal{O}(2^{2\ell/3})$; other trade-offs are also possible.

A warmup. We begin by considering the simple case where the function H defines a cycle, meaning that $x, H(x), H(H(x)), \dots$ covers all of $\{0,1\}^\ell$ for any starting point x. (Note that most functions do not define a cycle, but we assume this in order to demonstrate the idea in a very simple case.) For clarity, let $N = 2^\ell$ denote the size of the domain and range.

In the preprocessing phase, the attacker simply exhausts the entire cycle, beginning at an arbitrary starting point x_0 and computing $x_1 := H(x_0)$, $x_2 := H(H(x_0))$, up to $x_N = H^{(N)}(x_0)$, where $H^{(i)}$ refers to i-fold evaluation of H. Let $x_i \overset{\text{def}}{=} H^{(i)}(x_0)$. We imagine partitioning the cycle into $\sqrt{N}$ segments of length $\sqrt{N}$ each, and having the attacker store the points at the beginning and end of each such segment. That is, the attacker stores in a table pairs of the form $(x_{i \cdot \sqrt{N}}, x_{(i+1) \cdot \sqrt{N}})$, for $i = 0$ to $\sqrt{N} - 1$, sorted by the second component of each pair. The resulting table contains $\mathcal{O}(\sqrt{N})$ points.

When the attacker is given a point y to invert in the on-line phase, it checks which of y, $H(y)$, $H^{(2)}(y)$, ... corresponds to the endpoint of a segment. (Each check just involves a table lookup on the second component of the stored pairs.) Since y lies in *some* segment, this is guaranteed to find an endpoint within $\sqrt{N}$ steps. Once an endpoint $x = x_{(i+1)\cdot\sqrt{N}}$ is identified, the attacker takes the starting point $x' = x_{i\cdot\sqrt{N}}$ of the corresponding segment and computes $H(x')$, $H^{(2)}(x')$, ... until y is reached; this immediately gives the desired preimage. This takes at most $\sqrt{N}$ additional evaluations of H.

In summary, this attack stores $\mathcal{O}(\sqrt{N}) = \mathcal{O}(2^{\ell/2})$ points and finds preimages with probability 1 using $\mathcal{O}(\sqrt{N}) = \mathcal{O}(2^{\ell/2})$ on-line hash computations.

Hellman's time/space tradeoff. Martin Hellman introduced a more general time/space tradeoff applicable to an arbitrary function H (though the analysis treats H as a random function). Hellman's attack still stores the starting point and endpoint of several segments, but in this case the segments are "independent" rather than being part of one large cycle. In more detail: let s, t be parameters we will set later. The attacker first chooses s uniform starting points $SP_1, \ldots, SP_s \in \{0,1\}^{\ell}$. For each such point SP_i, it computes a corresponding endpoint $EP_i := H^{(t)}(SP_i)$ using t-fold application of H. (See Figure 6.3.) The attacker then stores the values $\{(SP_i, EP_i)\}_{i=1}^{s}$ in a table, sorted by the second entry (i.e., the endpoint) of each pair.

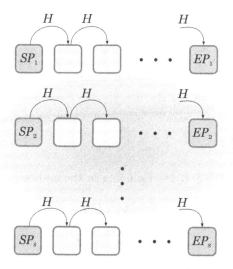

FIGURE 6.3: Table generation. Only the (SP_i, EP_i) pairs are stored.

Upon receiving a value y to invert, the attack proceeds as in the simple case discussed earlier. Specifically, it checks if any of y, $H(y)$, ..., $H^{(t-1)}(y)$ is equal to the endpoint of some segment (stopping as soon as the first such match is found). It is possible that none of these values is equal to an endpoint (as we discuss below). However, if $H^{(j)}(y) = EP_i = H^{(t)}(SP_i)$ for some i, j, then the attacker computes $H^{(t-j-1)}(SP_i)$ and checks whether this is a preimage of y. The entire process requires at most t evaluations of H.

This seems to work, but there are several subtleties we have ignored. First, it may happen that none of y, $H(y)$, ..., $H^{(t-1)}(y)$ is the endpoint of a segment. This can happen if y is not in the collection of at most $s \cdot t$ values (not counting the starting points) obtained during the initial process of generating the table. We can set $s \cdot t \geq N$ in an attempt to include every ℓ-bit string in the table, but this does not solve the problem since there can be collisions in the table itself—in fact, for $s \cdot t \geq N^{1/2}$ our previous analysis of the birthday problem tells us that collisions are likely—which will reduce the number of distinct points in the collection of values. A second problem, which arises even if y *is* in the table, is that even if we find a matching endpoint, and so $H^{(j)}(y) = EP_i = H^{(t)}(SP_i)$ for some i, j, this does not guarantee that $H^{(t-j-1)}(SP_i)$ is a preimage of y. The issue here is that the segment y, $H(y)$, ..., $H^{(t-1)}(y)$ might collide with the ith segment even though y itself is not in that segment; see Figure 6.4. (Even if y lies in some segment, the first matching endpoint may not be in that segment.) We call this a *false positive*. One might think this is unlikely to occur if H is collision resistant; again, however, we are dealing with a situation where more than $\sqrt{N}$ points are involved and so collisions actually become likely.

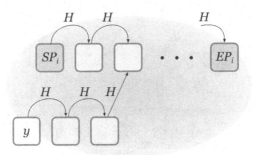

FIGURE 6.4:   Colliding in the on-line phase.

The problem of false positives can be addressed by modifying the algorithm so that it always computes the entire sequence y, $H(y)$, ..., $H^{(t-1)}(y)$, and checks whether $H^{(t-j-1)}(SP_i)$ is a preimage of y for *every* i, j such that $H^{(j)}(y) = EP_i$. This is guaranteed to find a preimage as long as y is in the collection of values (not including the starting points) generated during preprocessing. A concern now is that the running time of the algorithm might increase, since each false positive incurs an additional $\mathcal{O}(t)$ hash evaluations. One can show that the expected number of false positives is $\mathcal{O}(st^2/N)$. (There are at most t values in the sequence y, $H(y)$, ..., $H^{(t-1)}(y)$ and at most st distinct points in the table. Treating H as a random function, the probability that a given point in the sequence equals some point in the table is $1/N$.

The expected number of false positives is thus at most $t \cdot st \cdot 1/N = st^2/N$.)
So, as long as $st^2 \approx N$, which we will ensure for other reasons below, the
expected number of false positives is constant and dealing with false positives
is expected to require only $\mathcal{O}(t)$ additional hash computations in total.

Given the above modification, the probability of inverting $y = H(x)$ is
at least the probability that x is in the collection of points (not including
the endpoints) generated during preprocessing. We now lower bound this
probability, taken over the randomness of the preprocessing stage as well as
uniform choice of x, treating H as a random function in the analysis. We
first compute the expected number of distinct points in the table. Consider
what happens when the ith row of the table is generated. The starting point
SP_i is uniform and there are at most $(i-1) \cdot t$ distinct points (not including
the endpoints) in the table already, so the probability that SP_i is "new"
(i.e., not equal to any previous value) is at least $1 - (i-1) \cdot t/N$. What is
the probability that $H(SP_i)$ is new? If SP_i is not new, then almost surely
neither is $H(SP_i)$. On the other hand, if SP_i *is* new then $H(SP_i)$ is uniform
(because we treat H as a random function) and so is new with probability at
least $1 - ((i-1) \cdot t + 1)/N$. (We now have the additional point SP_i.) Thus,
the probability that $H(SP_i)$ is new is at least

$$\Pr\left[SP_i \text{ is new}\right] \cdot \Pr\left[H(SP_i) \text{ is new} \mid SP_i \text{ is new}\right]$$
$$\geq \left(1 - \frac{(i-1) \cdot t}{N}\right) \cdot \left(1 - \frac{(i-1) \cdot t + 1}{N}\right)$$
$$> \left(1 - \frac{(i-1) \cdot t + 1}{N}\right)^2.$$

Continuing in this way, the probability that $H^{(t-1)}(SP_i)$ is new is at least

$$\left(1 - \frac{i \cdot t}{N}\right)^t = \left[\left(1 - \frac{i \cdot t}{N}\right)^{\frac{N}{i \cdot t}}\right]^{\frac{i \cdot t^2}{N}} \approx e^{-it^2/N}.$$

The thing to notice here is that when $it^2 \leq N/2$, this probability is at least $1/2$;
on the other hand, once $it^2 > N$ the probability is relatively small. Considering the last row, when $i = s$, this means that we will not gain much additional
coverage if $st^2 > N$. A good setting of the parameters is thus $st^2 = N/2$.
Assuming this, the expected number of distinct points in the table is

$$\sum_{i=1}^{s} \sum_{j=0}^{t-1} \Pr\left[H^{(j)}(SP_i) \text{ is new}\right] \geq \sum_{i=1}^{s} \sum_{j=0}^{t-1} \frac{1}{2} = \frac{st}{2}.$$

The probability that x is "covered" is then at least $\frac{st}{2N} = \frac{1}{4t}$.

This gives a weak time/space tradeoff, in which we can use more space s
(and consequently less time t) while increasing the probability of inverting y.
But we can do even better by generating $T = 4t$ "independent" tables. (This

increases both the space and time by at most a factor of T.) As long as we can treat the probabilities of x being in each of these tables as independent, the probability that *at least one* of these tables contains x is

$$1 - \Pr[\text{no table contains } x] = 1 - \left(1 - \frac{1}{4t}\right)^{4t} \approx 1 - e^{-1} = 0.63.$$

The only remaining question is how to generate an independent table. (Note that generating a table exactly as before is the same as adding s additional rows to our original table, which we have already seen does not help.) We can do this for the ith such table by applying some function f_i after every evaluation of H, where $f_1, \ldots, f_T$ are all distinct. (A good choice might be to set $f_i(x) = x \oplus c_i$ for some fixed constant c_i that is different for each table.) Let $H_i \overset{\text{def}}{=} f_i \circ H$, i.e., $H_i(x) = f_i(H(x))$. Then for the ith table we again choose s random starting points, but for each such point we now compute $H_i(SP), H_i^{(2)}(SP)$, and so on. Upon receiving a value $y = H(x)$ to invert, the attacker first computes $y' = f_i(y)$ and then checks which of $y', H_i(y'), \ldots, H_i^{(t-1)}(y')$ corresponds to an endpoint in the ith table; this is repeated for $i = 1, \ldots, T$. (We omit further details.) While it is difficult to argue independence formally, this approach leads to good results in practice.

Choosing parameters. Summarizing the above, we see that as long as $st^2 = N/2$ we have an algorithm that stores $\mathcal{O}(s \cdot T) = \mathcal{O}(s \cdot t) = \mathcal{O}(N/t)$ points during a preprocessing phase, and can then invert y with constant probability in time $\mathcal{O}(t \cdot T) = \mathcal{O}(t^2)$. One setting of the parameters is $t = N^{1/3} = 2^{\ell/3}$, in which case we have an algorithm storing $\mathcal{O}(2^{2\ell/3})$ points that finds a preimage with constant probability using $\mathcal{O}(2^{2\ell/3})$ hash computations. If $\ell = 80$, this is feasible in practice.

Handling different domain and range. Consider the more general case where the original preimage x is chosen from a domain D that is different from the range $\{0,1\}^\ell$. This situation is quite common. One example is in the context of password cracking (see Section 6.6.3), where an attacker is given $H(pw)$ for a password pw composed of ASCII characters. (Not every bitstring corresponds to ASCII.) While it may be possible to artificially expand the domain, this will *not* be useful in general: In typical applications we would like to recover a preimage in D, but if the domain is artificially expanded then the algorithm above is likely to find a preimage that lies outside of D.

We can address this by applying a function f_i, as before, between each evaluation of H, though now we choose f_i mapping $\{0,1\}^\ell$ to D. This ensures that, when constructing the table, the values $f_i(H(SP)), (f_i \circ H)^{(2)}(SP), \ldots$ all lie in the desired domain D.

Application to key-recovery attacks. Time/space tradeoffs can lead to attacks on cryptographic primitives other than hash functions. A canonical example—in fact, the application originally considered by Hellman—is a key-recovery attack on an arbitrary block cipher F. Define $H(k) \stackrel{\text{def}}{=} F_k(m)$ where m is some arbitrary input that is used for building the table. If an attacker can subsequently obtain $F_k(m)$ for an unknown key k—either via a chosen-plaintext attack or by choosing m such that $F_k(m)$ is likely to be obtained in a known-plaintext attack—then by inverting H the attacker learns (a candidate value for) k. Note that it is possible for the key length of F to differ from its block length, but in this case we can use the technique just described for handling H with different domain and range.

6.5 The Random-Oracle Model

There are several examples of constructions based on cryptographic hash functions that cannot be proven secure based only on the assumption that the hash function is collision or preimage resistant. (We will see some in the following section.) In many cases, there appears to be *no* simple and reasonable assumption regarding the hash function that is sufficient for proving the construction secure.

Faced with this situation, there are several options. One is to look for schemes that *can* be proven secure based on some reasonable assumption about the underlying hash function. This is a good approach, but it leaves open the question of what to do until such schemes are found. Also, provably secure constructions may be significantly less efficient than other existing approaches that have not been proven secure. (This is a major issue we will encounter in the setting of public-key cryptography.)

Another possibility, of course, is to use an existing cryptosystem even if it has no justification for its security other than, perhaps, the fact that the designers tried to attack it and were unsuccessful. This flies in the face of everything we have said about the importance of the rigorous, modern approach to cryptography, and it should be clear that this is unacceptable.

An approach that has been hugely successful in practice, and which offers a "middle ground" between a fully rigorous proof of security on the one hand and no proof whatsoever on the other, is to introduce an *idealized model* in which to prove the security of cryptographic schemes. Although the idealization may not be an entirely accurate reflection of reality, we can at least derive some measure of confidence in the soundness of a scheme's design from a proof within the idealized model. As long as the model is reasonable, such proofs are certainly better than no proofs at all.

A popular example of this approach is the *random-oracle model*, which treats a cryptographic hash function H as a truly random function. (We have already seen an example of this in our discussion of birthday attacks, although there we were analyzing an attack rather than a construction.) More specifically, the random-oracle model posits the existence of a public, random function H that can be evaluated *only* by "querying" an oracle—which can be thought of as a "black box"—that returns $H(x)$ when given input x. (We will discuss how this is to be interpreted in the following section.) To differentiate things, the model we have been using until now (where no random oracle is present) is sometimes called the "standard model," although at this point the random-oracle model itself is considered quite standard in the literature.

No one claims that a random oracle exists, although there have been suggestions that a random oracle could be implemented in practice using a trusted party (i.e., some server on the Internet). Rather, the random-oracle model provides a formal *methodology* that can be used to design and validate cryptographic schemes using the following two-step approach:

1. First, a scheme is designed and proven secure in the random-oracle model. That is, we assume the world contains a random oracle, and construct and analyze a cryptographic scheme within this model. Standard cryptographic assumptions of the type we have seen until now may be utilized in the proof of security as well.

2. When we want to implement the scheme in the real world, a random oracle is not available. Instead, the random oracle is *instantiated* with an appropriately designed cryptographic hash function $\hat{H}$. (We return to this point at the end of this section.) That is, at each point where the scheme dictates that a party should query the oracle for the value $H(x)$, the party instead computes $\hat{H}(x)$ on its own.

The hope is that the cryptographic hash function used in the second step is "sufficiently good" at emulating a random oracle, so that the security proof given in the first step will carry over to the real-world instantiation of the scheme. The difficulty here is that there is no theoretical justification for this hope, and in fact there are (contrived) schemes that can be proven secure in the random-oracle model but are insecure *no matter how the random oracle is instantiated* in the second step. Furthermore, it is not clear (mathematically or heuristically) what it means for a hash function to be "sufficiently good" at emulating a random oracle, nor is it clear that this is an achievable goal. In particular, no concrete instantiation $\hat{H}$ can ever behave like a random function, since $\hat{H}$ is fixed and its code is known. For these reasons, a proof of security in the random-oracle model should be viewed as providing evidence that a scheme has no "inherent design flaws," but is *not* a rigorous proof that any real-world instantiation of the scheme is secure. Further discussion on how to interpret proofs in the random-oracle model is given in Section 6.5.2.

6.5.1 The Random-Oracle Model in Detail

Before continuing, let us pin down exactly what the random-oracle model entails. A good way to think about the random-oracle model is as follows: The oracle is simply a "black box" that takes a bit-string as input and returns a bit-string as output. The internal workings of the box are unknown and inscrutable. Everyone—honest parties as well as the adversary—can interact with the box, where such interaction consists of feeding in a binary string x as input and receiving a binary string y as output; we refer to this as *querying the oracle on x*, and call x a *query* made to the oracle. Queries to the oracle are assumed to be private so that if some party queries the oracle on input x then no one else learns x, or even learns that this party queried the oracle at all. This makes sense, because calls to the oracle correspond (in the real-world instantiation) to local evaluations of a cryptographic hash function.

An important property of this "box" is that it is *consistent*. That is, if the box ever outputs y for a particular input x, then it always outputs the same answer y when given the same input x again. This means that we can view the box as implementing a well-defined function H; i.e., we define the function H in terms of the input/output characteristics of the box. For convenience, we thus speak of "querying H" rather than querying the box. No one "knows" the entire function H (except the box itself); at best, all that is known are the values of H on the strings that have been explicitly queried thus far.

We have already discussed in Chapter 3 what it means to choose a random function H. We only reiterate here that there are two equivalent ways to think about the uniform selection of H: either view H as being chosen "in one shot" uniformly from the set of all functions on some specified domain and range, or imagine generating outputs for H "on-the-fly," as needed. Specifically, in the second case we can view the function as being defined by a table that is initially empty. When the oracle receives a query x it first checks whether $x = x_i$ for some pair (x_i, y_i) in the table; if so, the corresponding value y_i is returned. Otherwise, a *uniform* string $y \in \{0,1\}^\ell$ is chosen (for some specified ℓ), the answer y is returned, and the oracle stores (x, y) in its table. This second viewpoint is often conceptually easier to reason about, and is also technically easier to deal with if H is defined over an infinite domain (e.g., $\{0,1\}^*$).

When we defined pseudorandom functions in Section 3.5.1, we also considered algorithms having oracle access to a random function. Lest there be any confusion, we note that the usage of a random function there is very different from the usage of a random function here. There, a random function was used *as a way of defining* what it means for a (concrete) keyed function to be pseudorandom. In the random-oracle model, in contrast, the random function is used *as part of a construction itself* and must somehow be instantiated in the real world if we want a concrete realization of the construction. A pseudorandom function is not a random oracle because it is only pseudorandom if the key is *secret*. However, in the random-oracle model all parties need to be able to compute the function; thus there can be no secret key.

Definitions and Proofs in the Random-Oracle Model

Definitions in the random-oracle model are slightly different from their counterparts in the standard model because the probability spaces considered in each case are not the same. In the standard model a scheme Π is secure if for all PPT adversaries $\mathcal{A}$ the probability of some event is below some threshold, where *this probability is taken over the random choices of the parties running Π and those of the adversary $\mathcal{A}$*. Assuming the honest parties who use Π in the real world make random choices as directed by the scheme, satisfying a definition of this sort guarantees security for real-world usage of Π.

In the random-oracle model, in contrast, a scheme Π may rely on an oracle H. As before, Π is secure if for all PPT adversaries $\mathcal{A}$ the probability of some event is below some threshold, but now *this probability is taken over random choice of H* as well as the random choices of the parties running Π and those of the adversary $\mathcal{A}$. When using Π in the real world, some (instantiation of) H must be fixed. Unfortunately, security of Π is not guaranteed for any *particular* choice of H. This indicates one reason why it is difficult to argue that any concrete instantiation of the oracle H by some fixed function yields a secure scheme. (An additional, technical, difficulty is that once a concrete function H is fixed, the adversary $\mathcal{A}$ is no longer restricted to querying H as an oracle but can instead look at and use the *code* of H in its attack.)

Proofs in the random-oracle model can exploit the fact that H is chosen at random, and that the only way to evaluate $H(x)$ is to explicitly query x to H. Three properties of the random-oracle model are especially useful; we sketch them informally here, and show some simple applications of them in what follows, but caution that a full understanding will likely have to wait until we present formal proofs in the random-oracle model in later chapters.

A first useful property of the random-oracle model is:

> *If x has not been queried to H, then the value of $H(x)$ is* **uniform**.

This may seem superficially similar to the guarantee provided by a pseudorandom generator, but is actually much stronger. If G is a pseudorandom generator then $G(x)$ is pseudorandom to an observer *assuming x is chosen uniformly at random and is completely unknown to the observer*. If H is a random oracle, however, then $H(x)$ is truly uniform to an observer as long as the observer has not queried x. This is true even if x is known, or if x is not uniform but *is* hard to guess. (For example, if x is an n-bit string where the first half of x is known and the last half is random then $G(x)$ might be easy to distinguish from random but $H(x)$ will not be.)

The remaining two properties relate explicitly to *proofs by reduction* in the random-oracle model. (It may be helpful here to review Section 3.3.2.) As part of the reduction, the random oracle that the adversary $\mathcal{A}$ interacts with must be simulated. That is: $\mathcal{A}$ will submit queries to, and receive answers from, what it believes to be the oracle, but the reduction itself must now answer these queries. This turns out to give a lot of power. For starters:

> *If $\mathcal{A}$ queries x to H, the reduction can* **see this query** *and learn x*.

This is sometimes called "extractability." (This does not contradict the fact, mentioned earlier, that queries to the random oracle are "private." While that is true in the random-oracle model itself, here we are using $\mathcal{A}$ as a subroutine within a reduction that is simulating the random oracle for $\mathcal{A}$.) Finally:

> The reduction can **set** the value of $H(x)$ (i.e., the response to query x) to a value of its choice, as long as this value is correctly distributed, i.e., uniform.

This is called "programmability." There is no counterpart to extractability or programmability once H is instantiated with any concrete function.

Simple Illustrations of the Random-Oracle Model

At this point some examples may be helpful. The examples given here are relatively simple, and do not use the full power of the random-oracle model; they are intended merely to provide a gentle introduction. In what follows, we assume a random oracle mapping ℓ_{in}-bit inputs to ℓ_{out}-bit outputs, where $\ell_{in}, \ell_{out} > n$, the security parameter (so ℓ_{in}, ℓ_{out} are functions of n).

A random oracle as a pseudorandom generator. We first show that, for $\ell_{out} > \ell_{in}$, a random oracle can be used as a pseudorandom generator. (We do not say that a random oracle *is* a pseudorandom generator, since a random oracle is not a fixed function.) Formally, we claim that for any PPT adversary $\mathcal{A}$, there is a negligible function negl such that

$$\left| \Pr[\mathcal{A}^{H(\cdot)}(y) = 1] - \Pr[\mathcal{A}^{H(\cdot)}(H(x)) = 1] \right| \leq \mathsf{negl}(n),$$

where in the first case the probability is taken over uniform choice of H, uniform choice of $y \in \{0,1\}^{\ell_{out}(n)}$, and the randomness of $\mathcal{A}$, and in the second case the probability is taken over uniform choice of H, uniform choice of $x \in \{0,1\}^{\ell_{in}(n)}$, and the randomness of $\mathcal{A}$. We have explicitly indicated that $\mathcal{A}$ has oracle access to H in each case; once H has been chosen then $\mathcal{A}$ can freely make queries to it.

As a proof sketch, let S denote the set of points on which $\mathcal{A}$ queries H; of course, $|S|$ is polynomial in n. Observe that in the second case, the probability that $x \in S$ is negligible—this is because $\mathcal{A}$ starts with no information about x (note that $H(x)$ by itself reveals nothing about x because H is a random function), and S is exponentially smaller than $\{0,1\}^{\ell_{in}}$. Moreover, conditioned on $x \notin S$ in the second case, $\mathcal{A}$'s input in each case is a uniform string that is independent of the answers to $\mathcal{A}$'s queries.

A random oracle as a collision-resistant hash function. If $\ell_{out} < \ell_{in}$, a random oracle is collision resistant. That is, the success probability of any PPT adversary $\mathcal{A}$ in the following experiment is negligible:

1. A random function H is chosen.

2. $\mathcal{A}$ succeeds if it outputs distinct x, x' with $H(x) = H(x')$.

To see this, assume without loss of generality that $\mathcal{A}$ only outputs values x, x' that it had previously queried to the oracle, and that $\mathcal{A}$ never makes the same query to the oracle twice. Letting the oracle queries of $\mathcal{A}$ be $x_1, \ldots, x_q$, with $q = \mathsf{poly}(n)$, it is clear that the probability that $\mathcal{A}$ succeeds is upper-bounded by the probability that $H(x_i) = H(x_j)$ for some $i \neq j$. But this is exactly equal to the probability that if we pick q strings $y_1, \ldots, y_q \in \{0,1\}^{\ell_{out}}$ independently and uniformly at random, we have $y_i = y_j$ for some $i \neq j$. This is precisely the birthday problem, and so using the results of Appendix A.4 we see that $\mathcal{A}$ succeeds with negligible probability $\mathcal{O}(q^2/2^{\ell_{out}})$.

Constructing a pseudorandom function from a random oracle. It is also rather easy to construct a pseudorandom function in the random-oracle model. Suppose $\ell_{in}(n) = 2n$ and $\ell_{out}(n) = n$, and define

$$F_k(x) \stackrel{\text{def}}{=} H(k\|x),$$

where $|k| = |x| = n$. In Exercise 6.15 you are asked to show that this is a pseudorandom function, namely, for any polynomial-time $\mathcal{A}$ the success probability of $\mathcal{A}$ in the following experiment is $1/2 + \mathsf{negl}(n)$:

1. A function H and values $k \in \{0,1\}^n$ and $b \in \{0,1\}$ are chosen uniformly.

2. If $b = 0$, the adversary $\mathcal{A}$ is given access to an oracle for $F_k(\cdot) = H(k\|\cdot)$. If $b = 1$, then $\mathcal{A}$ is given access to a random function mapping n-bit inputs to n-bit outputs. (This random function is *independent* of H.)

3. $\mathcal{A}$ outputs a bit b', and succeeds if $b' = b$.

In step 2, $\mathcal{A}$ can access H in addition to the function oracle provided to it by the experiment. (A pseudorandom function in the random-oracle model must be indistinguishable from a random function that is independent of H.)

An interesting aspect of the above results is that they require no assumptions; they hold even for computationally unbounded adversaries as long as those adversaries are limited to making polynomially many queries to the oracle. This has no real-world counterpart, where computational assumptions are (currently) necessary to prove, e.g., the existence of pseudorandom generators.

6.5.2   Is the Random-Oracle Methodology Sound?

Schemes designed in the random-oracle model are implemented in the real world by instantiating H with some concrete function. With the mechanics of the random-oracle model behind us, we turn to a more fundamental question:

> *What do proofs of security in the random-oracle model guarantee as far as security of any real-world instantiation?*

This question does not have a definitive answer: there is currently debate within the cryptographic community about how to interpret proofs in the random-oracle model, and active research seeking to determine what, precisely, a proof of security in the random-oracle model implies vis-a-vis the real world. We can only hope to give a flavor of both sides of the debate.

Objections to the random-oracle model. The starting point for arguments against using random oracles is simple: as we have already noted, there is no formal justification for believing that a proof of security for some scheme Π in the random-oracle model says anything about the security of Π in the real world, once the random oracle H has been instantiated with any particular hash function $\hat{H}$. This is more than just theoretical uneasiness. A little thought shows that *no* hash function can ever act as a "true" random oracle. For example, in the random-oracle model the value $H(x)$ is "completely random" if x was not explicitly queried. The counterpart would be to require that $\hat{H}(x)$ is random (or pseudorandom) if $\hat{H}$ was not explicitly evaluated on x. How are we to interpret this in the real world? It is not even clear what it means to "explicitly evaluate" $\hat{H}$: what if an adversary knows a shortcut for computing $\hat{H}$ that does not involve running the actual code of $\hat{H}$? Moreover, $\hat{H}(x)$ cannot possibly be random (or even pseudorandom) since once the adversary learns the description of $\hat{H}$, the value of $\hat{H}$ on *all* inputs is immediately determined.

Limitations of the random-oracle model become clearer once we examine the proof techniques introduced earlier. Recall that one proof technique is to use the fact that a reduction can "see" the queries that an adversary $\mathcal{A}$ makes to the random oracle. If we replace the random oracle by a particular hash function $\hat{H}$, this means we must provide a description of $\hat{H}$ to the adversary at the beginning of the experiment. But then $\mathcal{A}$ can evaluate $\hat{H}$ on its own, without making any *explicit* queries, and so a reduction will no longer have the ability to "see" any queries made by $\mathcal{A}$. (In fact, as noted previously, the notion of $\mathcal{A}$ performing explicit evaluations of $\hat{H}$ may not be true and certainly cannot be formally defined.) Likewise, proofs of security in the random-oracle model allow the reduction to choose the outputs of H as it wishes, something that is clearly not possible when a concrete function is used.

Even if we are willing to overlook the above theoretical concerns, a practical problem is that we do not currently have a very good understanding of what it means for a concrete hash function to be "sufficiently good" at instantiating a random oracle. For concreteness, say we want to instantiate the random oracle using some appropriate modification of SHA-2. (SHA-2 is a cryptographic hash function discussed in Section 7.3.2.) While for some particular scheme Π it might be reasonable to assume that Π is secure when instantiated using SHA-2, it is much less reasonable to assume that SHA-2 can take the place of a random oracle in *every* scheme designed in the random-oracle model. Indeed, as we have said earlier, we *know* that SHA-2 is not a random oracle. And it is not hard to design a scheme that is secure in the random-oracle model, but is insecure when the random oracle is replaced by SHA-2.

We emphasize that an assumption of the form "SHA-2 acts like a random oracle" is qualitatively different from assumptions such as "SHA-2 is collision resistant" or "AES is a pseudorandom function." The problem lies partly with the fact that there is no satisfactory *definition* of what the first statement means, while we do have such definitions for the latter two statements.

Because of this, using the random-oracle model to prove security of a scheme is *qualitatively* different from, e.g., introducing a new cryptographic assumption in order to prove a scheme secure in the standard model. Therefore, proofs of security in the random-oracle model are less satisfying than proofs of security in the standard model.

Support for the random-oracle model. Given all the problems with the random-oracle model, why use it at all? More to the point: why has the random-oracle model been so influential in the development of modern cryptography (especially current practical usage of cryptography), and why does it continue to be so widely used? As we will see, the random-oracle model enables the design of substantially more-efficient schemes than those we know how to construct in the standard model. As such, there are few (if any) public-key cryptosystems used today having proofs of security in the standard model, while there are numerous deployed schemes having proofs of security in the random-oracle model. In addition, proofs in the random-oracle model are almost universally recognized as lending confidence to the security of schemes being considered for standardization.

The fundamental reason for this is the belief that:

> *A proof of security in the random-oracle model is significantly better than no proof at all.*

Although some disagree, we offer the following in support of this assertion:

- A proof of security for a scheme in the random-oracle model indicates that the scheme's design is "sound," in the sense that the only possible attacks on a real-world instantiation of the scheme are those that arise due to a weakness in the hash function used to instantiate the random oracle. Thus, if a "good enough" hash function is used to instantiate the random oracle, we should have confidence in the security of the scheme. Moreover, if a given instantiation of the scheme *is* successfully attacked, we can simply replace the hash function being used with a "better" one.

- Importantly, *there have been no successful real-world attacks on schemes proven secure in the random-oracle model,* when the random oracle was instantiated properly. (We remark that great care must be taken in instantiating the random oracle, as discussed next; see also Exercise 6.11.) This gives evidence of the usefulness of the random-oracle model in designing practical schemes.

Nevertheless, the above ultimately represent only intuitive speculation as to the usefulness of proofs in the random-oracle model and—all else being equal—proofs without random oracles are preferable.

Instantiating a Random Oracle

Properly instantiating a random oracle is subtle, and a full discussion is beyond the scope of this book. Here we only alert the reader that using an "off-the-shelf" cryptographic hash function without modification is, generally speaking, not a sound approach. For one thing, many cryptographic hash functions are constructed using the Merkle–Damgård transform (cf. Section 6.2), and can be distinguished easily from a random oracle when variable-length inputs are allowed. (See Exercise 6.11.) Also, in some constructions it is necessary for the output of the random oracle to lie in a certain range, which results in additional complications.

6.6 Additional Applications of Hash Functions

We conclude this chapter with a brief discussion of some additional applications of cryptographic hash functions in cryptography and computer security.

6.6.1 Fingerprinting and Deduplication

If H is a collision-resistant hash function, the hash (or *digest*) of a file serves as a unique identifier for that file. (If any other file is found to have the same digest, this implies a collision in H.) The hash $H(x)$ of a file x can thus serve as a "fingerprint" for x, and one can check whether two files are equal by comparing their digests. This simple idea has many applications.

- *Virus fingerprinting:* Virus scanners identify whether incoming files are potential viruses. Often, this is done not by analyzing the incoming file to determine whether it is malicious, but instead simply by checking whether the file is in a database of previously identified viruses. The observation here is that rather than comparing the file to each virus in the database, it suffices to compare the *hash* of the file to the hashes (i.e., fingerprints) of known viruses. This can lead to improved efficiency, as well as reduced communication if the database is stored remotely.

- *Deduplication:* Data deduplication is used to eliminate duplicate copies of data, especially in the context of cloud storage where multiple users rely on a single cloud service to store their data. The key insight is that if multiple users wish to store the same file (e.g., a popular video), then the file only needs to be uploaded and stored once and need not be uploaded and stored separately for each user. Deduplication can be achieved by first having a user upload a hash of the new file they want to store; if a file with this hash is already stored on the server, then the cloud-storage provider can simply add a pointer to the existing file to indicate that this specific user has also stored this file, thus saving both communication and storage. The soundness of this approach follows from collision resistance of the hash function.

- *Peer-to-peer (P2P) file sharing*: In P2P file-sharing systems, servers store different files and can advertise the files they hold by broadcasting the hashes of those files. Those hashes serves as unique identifiers for the files, and allow clients to easily find out which servers host a particular file (identified by its hash).

It may be surprising that a small digest can uniquely identify every file in the world. But this is the guarantee provided by collision-resistant hash functions, which makes them useful in the above settings.

6.6.2   Merkle Trees

Consider a client who uploads a file x to a server. When the client later retrieves x, it wants to make sure the server returns the original, unmodified file. The client could simply store x and check that the retrieved file is equal to x, but that defeats the purpose of using the server in the first place. We are looking for a solution in which the storage of the client is small.

A natural solution is to use the "fingerprinting" idea from the previous section. The client locally stores the short digest $h := H(x)$; when the server returns a candidate file x' the client need only check that $H(x') \stackrel{?}{=} h$.

What happens if we want to extend this solution to *multiple* files $x_1, \ldots, x_t$? There are two obvious ways of doing this. One is to simply hash each file individually; the client locally stores the digests $h_1, \ldots, h_t$, and verifies retrieved files as before. This has the disadvantage that the client's storage grows linearly in t. Another possibility is to hash all the files together. That is, the client computes $h := H(x_1, \ldots, x_t)$ and stores only a single digest h. (We assume the client concatenates the files in an unambiguous manner before hashing, so that from the input to h it is possible to determine the original files. This can be done using standard techniques.) The drawback now is that when the client wants to retrieve and verify the ith file x_i, it needs to retrieve *all* the files in order to recompute the digest and check the result.

Merkle trees, introduced by Ralph Merkle, give a tradeoff between these extremes. Assume t is a power of two for simplicity. (The idea can be easily extended when this is not the case.) A Merkle tree computed over input values $x_1, \ldots, x_t$ is simply a binary tree of depth $\log t$ in which hashes of the input values are placed at the leaves, and the value at each internal node is the hash of the values of its two children.

Referring to Figure 6.5 where $t = 8$, for example, each leaf i holds the value $h_i = H(x_i)$; the parent of leaves 3 and 4 holds the value $h_{3\ldots4} = H(h_3, h_4)$; and the parent of the right subtree holds the value

$$h_{5\ldots8} = H(h_{5\ldots6}, h_{7\ldots8}) = H(H(h_5, h_6), H(h_7, h_8)).$$

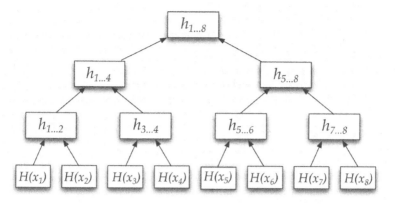

FIGURE 6.5: A Merkle tree.

Fixing some hash function H, we denote by $\mathcal{MT}_t$ the function that takes t input values $x_1, \ldots, x_t$, computes the resulting Merkle tree, and outputs the value of the root of the tree. (A keyed hash function yields a keyed function $\mathcal{MT}_t$ in the obvious way.) We have:

THEOREM 6.11 *If* (Gen_H, H) *is collision resistant, then* $(\mathsf{Gen}_H, \mathcal{MT}_t)$ *is collision resistant for any fixed* t.

Merkle trees thus provide an alternative to the Merkle–Damgård transform for domain extension of collision-resistant hash functions. (As described, however, Merkle trees are *not* collision resistant if the number of inputs t is allowed to vary. But they can be generalized fairly easily to handle that case.)

Merkle trees yield an efficient solution to our original problem. Specifically, the client will compute $h := \mathcal{MT}_t(x_1, \ldots, x_t)$, upload $x_1, \ldots, x_t$ to the server, and store h (along with the number of files t) locally. When the client wants to retrieve the ith file, the server sends x_i along with a "proof" π_i that this is the correct value. This proof consists of the values of the nodes in the Merkle tree adjacent to the path from the ith leaf to the root. From these values the client can recompute the value of the root and verify that it is equal to the stored value h. As an example, consider the Merkle tree in Figure 6.5. The client computes $h_{1\ldots8} := \mathcal{MT}_8(x_1, \ldots, x_8)$, uploads $x_1, \ldots, x_8$ to the server, and stores $h_{1\ldots8}$ locally. When the client retrieves x_3, the server sends x_3 along with h_4, $h_{1\ldots2}$, and $h_{5\ldots8}$. The client computes $h_3' := H(x_3)$, $h_{3\ldots4}' := H(h_3', h_4)$, $h_{1\ldots4}' := H(h_{1\ldots2}, h_{3\ldots4}')$, and $h_{1\ldots8}' := H(h_{1\ldots4}', h_{5\ldots8})$, and then verifies that $h_{1\ldots8}' \stackrel{?}{=} h_{1\ldots8}$. If H is collision resistant and the server tries to send an incorrect file $x_3' \neq x_3$, it will be infeasible for the server to send *any* proof that will cause verification to succeed. Using this approach, the client's local storage is *constant* (independent of t), and the communication overhead is logarithmic in t.

6.6.3   Password Hashing

One of the most common and important uses of hash functions in computer security is for password protection. Consider a user typing in a password before using their laptop. To authenticate the user, some form of the user's password must be stored somewhere on their laptop. If the user's password is stored in the clear, then an adversary who steals the laptop can read the user's password off the hard drive and then impersonate that user. (It may seem pointless to try to hide one's password from an attacker who can already read the contents of the hard drive. However, files on the hard drive may be encrypted with a key derived from the user's password, and would thus only be accessible after the password is entered. In addition, the user is likely to use the same password for other purposes.)

This risk can be mitigated by storing a *hash of the password* instead of the password itself. That is, the value $hpw = H(pw)$ is stored on the laptop in a password file; later, when the user enters its password pw, the operating system checks whether $H(pw) \overset{?}{=} hpw$ before granting access. The same basic approach is also used for password-based authentication over the web, with a login server holding the password file. Now, if an attacker steals the hard drive (or breaks into the login server), all it obtains is the hash of the password and not the password itself.

If the password is chosen from some relatively small space D of possibilities (e.g., D might be a dictionary of English words, in which case $|D| \approx 80,000$), an attacker can enumerate all possible passwords $pw_1, pw_2, \ldots \in D$ and, for each candidate pw_i, check whether $H(pw_i) = hpw$. We would like to claim that an attacker can do no better than this. (This would also ensure that the adversary could not learn the password of any user who chose a *strong* password from a large domain.) Unfortunately, preimage resistance (i.e., one-wayness) of H is not sufficient to imply what we want. For one thing, preimage resistance only says that $H(x)$ is hard to invert when x is chosen uniformly from a large domain. It says nothing about the hardness of inverting H when x is chosen from a small domain, or when x is chosen according to some other distribution. Moreover, preimage resistance says nothing about the *concrete* amount of time needed to find a preimage. For example, a hash function H for which recovering $x \in \{0,1\}^n$ from $H(x)$ requires time $2^{n/2}$ could still qualify as preimage resistant, yet this would mean that a 32-bit uniform password could be recovered in only 2^{16} time.

If we model H as a random oracle, though, we can formally prove the security we want: namely, recovering pw from hpw (assuming pw is chosen uniformly from D) requires $\mathcal{O}(|D|)$ evaluations of H, on average.

The above discussion assumes no preprocessing is done by the attacker. As we have seen in Section 6.4.3, though, preprocessing can be used to generate large tables that enable inversion (even of a random function!) faster than exhaustive search. The tables—called *rainbow tables*—only need to be generated once, and can be used to recover thousands of passwords in case of a server breach. This is a significant concern in practice: even if a user chooses their password as a random combination of 8 alphanumeric English characters— giving a password space of size $N = 62^8 \approx 2^{47.6}$—there is an attack using time and space $N^{2/3} \approx 2^{32}$ that will be highly effective at recovering the password. Such attacks are routinely carried out in practice.

Mitigation. We briefly describe two mechanisms used to mitigate the threat of password cracking. One technique is to use hash functions that are "moderately hard to compute," in the sense that they do not add significant overhead when evaluated once (as done by the server when authenticating a user) but are prohibitively expensive to evaluate tens of thousands of times (as would be done by a user in a brute-force attack).

A second mechanism is to introduce a *salt*. When a user registers their password, the laptop/server will generate a long random value s (a "salt") unique to that user, and store $(s, hpw = H(s, pw))$ instead of merely storing $H(pw)$ as before. Since s is unknown to the attacker in advance, preprocessing is ineffective and the best an attacker can do is to wait until it obtains the password file and then do a linear-time exhaustive search over the domain D. Note also that since a different salt is used for each user, a separate brute-force search is needed to recover each user's password.

6.6.4 Key Derivation

Symmetric-key cryptosystems require the secret key to be a *uniformly distributed* bit-string. Often, however, it is more convenient for two parties to rely on shared information such as a password or biometric data that is *not* uniformly distributed. (Jumping ahead, in Chapter 11 we will see how parties can interact over a public channel to generate a high-entropy shared secret that is also not necessarily uniformly distributed.) The parties could try to use their nonuniform shared information directly as a secret key, but in general this will not be secure. Moreover, the shared data may not even have the correct format to be used as a secret key (it may be too long, for example).

Truncating the shared secret, or mapping it in some other heuristic way to a string of the correct length, may lose a significant amount of entropy. (We define one notion of entropy more formally below, but for now one can think of entropy as the logarithm of the number of possible shared secrets.) For example, imagine two parties share a password composed of 28 random upper-case English letters, and want to use a cryptosystem with a 128-bit key. Since there are 26 possibilities for each character, there are $26^{28} > 2^{130}$ possible passwords. If the password is shared in ASCII format, each character is stored using 8 bits, and so the total length of the password is 224 bits. If the parties truncate their password to the first 128 bits, they will be using only the first 16 characters of their password. Even worse, this will not be a uniformly distributed 128-bit string! The ASCII representations of the letters A–Z lie between 0 x 41 and 0 x 5A; in particular, the first 3 bits of every byte are always 010. This means that *37.5% of the bits of the resulting key will be fixed*, and the 128-bit key the parties derive will have only about 75 bits of entropy (i.e., there are only 2^{75} or so possibilities for the key).

What we need is a generic solution for deriving a key of some desired length from a high-entropy (but not necessarily uniform) shared secret. Before continuing, we define the notion of entropy we consider here.

DEFINITION 6.12　*A probability distribution $\mathcal{X}$ has m bits of min-entropy if for every fixed value x it holds that $\Pr_{X \leftarrow \mathcal{X}}[X = x] \leq 2^{-m}$. In other words, even the most likely outcome occurs with probability at most 2^{-m}.*

The uniform distribution over a set of size S has min-entropy $\log S$. A distribution in which one element occurs with probability $1/10$ and 90 elements each occur with probability $1/100$ has min-entropy $\log 10 \approx 3.3$. The min-entropy of a distribution measures the probability with which an attacker can guess a value sampled from that distribution; the attacker's best strategy is to guess the most likely value, and so if the distribution has min-entropy m the attacker's guess is correct with probability at most 2^{-m}. This explains why min-entropy (rather than other notions of entropy) is useful in our context.

A *key-derivation function* provides a way to obtain a (close to) uniformly distributed string from any distribution with high min-entropy. It is not hard to see that if we model a hash function H as a random oracle, then H serves as a good key-derivation function. (As a technical point, we require the original distribution to be independent of H. This will normally be the case in practice.) Consider an attacker's uncertainty about $H(X)$, where X is sampled from a distribution with min-entropy m. Each of the attacker's queries to H can be viewed as a "guess" for the value of X; by assumption on the min-entropy of the distribution, an attacker making q queries to H will query $H(X)$ with probability at most $q \cdot 2^{-m}$. As long as the attacker does not query $H(X)$, the value $H(X)$ is uniform from the attacker's point of view.

It is also possible to design key-derivation functions without relying on the random-oracle model, by using keyed hash functions called *(strong) extractors*. The key for the extractor must be uniform, but need not be kept secret.

6.6.5　Commitment Schemes

A *commitment scheme* allows one party to "commit" to a value m by sending a *commitment* com, and then to reveal m (by "opening" the commitment) at a later point in time. We require the following properties to hold:

- *Hiding:* the commitment com reveals nothing about m.

- *Binding:* it is infeasible for the committer to output a commitment com that it can later "open" as two different messages m, m'. (In this sense, com truly "commits" the committer to at most one value.)

A commitment scheme can be viewed as a digital envelope: sealing a message m in an envelope and giving the envelope to another party hides m (until the envelope is opened) even though the value of m is fixed (since the contents of the envelope cannot be changed).

Formally, a (non-interactive) commitment scheme is defined by an algorithm Gen that outputs public parameters params, and a randomized algorithm Com that takes params and a message $m \in \{0,1\}^n$ and outputs a commitment com; when we make the randomness used by Com explicit, we denote it by r. A sender commits to m by choosing uniform r, computing com $:=$ Com(params, $m; r$), and sending it to a receiver. The sender can later open com and reveal m by sending m, r to the receiver; the receiver verifies that m is the committed value by checking that Com(params, $m; r$) $\stackrel{?}{=}$ com.

Hiding means that com reveals nothing about m. This is defined via the following experiment.

The commitment hiding experiment $\mathsf{Hiding}_{\mathcal{A},\mathsf{Com}}(n)$:

1. *Parameters* params $\leftarrow$ Gen(1^n) *are generated.*

2. *The adversary $\mathcal{A}$ is given input* params, *and outputs a pair of messages* $m_0, m_1 \in \{0,1\}^n$.

3. *A uniform* $b \in \{0,1\}$ *is chosen and* com $\leftarrow$ Com(params, m_b) *is computed.*

4. *The adversary $\mathcal{A}$ is given* com *and outputs a bit* b'.

5. *The output of the experiment is 1 if and only if* $b' = b$.

Binding means that it is impossible to output a commitment com that can be opened in two different ways.

The commitment binding experiment $\mathsf{Binding}_{\mathcal{A},\mathsf{Com}}(n)$:

1. *Parameters* params $\leftarrow$ Gen(1^n) *are generated.*

2. *$\mathcal{A}$ is given input* params *and outputs* (com, m, r, m', r').

3. *The output of the experiment is defined to be 1 if and only if* $m \neq m'$ *and* Com(params, $m; r$) $=$ com $=$ Com(params, $m'; r'$).

DEFINITION 6.13 *A commitment scheme* Com *is* secure *if for all* PPT *adversaries $\mathcal{A}$ there is a negligible function* negl *such that*

$$\Pr\left[\mathsf{Hiding}_{\mathcal{A},\mathsf{Com}}(n) = 1\right] \leq \frac{1}{2} + \mathsf{negl}(n)$$

and

$$\Pr\left[\mathsf{Binding}_{\mathcal{A},\mathsf{Com}}(n) = 1\right] \leq \mathsf{negl}(n).$$

It is easy to construct a secure commitment scheme from a random oracle H. To commit to a message m, the sender chooses uniform $r \in \{0,1\}^n$ and outputs com $:= H(m\|r)$. (In the random-oracle model, Gen and params are not needed since H, in effect, serves as the public parameters of the scheme.) Binding follows immediately from the fact that H is collision resistant. Intuitively, hiding follows from the fact that an adversary queries $H(\star\|r)$ with only negligible probability (since r is a uniform n-bit string); if it never makes a query of this form then com $= H(m\|r)$ reveals nothing about m.

Commitment schemes can be constructed without random oracles (in fact, from one-way functions), but the details are beyond the scope of this book.

References and Additional Reading

Collision-resistant hash functions were formally defined by Damgård [60]. As we have noted, other notions of security for hash functions can also be considered [137, 173]. The Merkle–Damgård transform was introduced independently by Merkle [140] and Damgård [61]

The hash-and-MAC paradigm is folklore. HMAC was introduced and analyzed by Bellare et al. [16], and subsequently standardized [149].

The small-space birthday attack described in Section 6.4.2 relies on a cycle-finding algorithm of Floyd. Related algorithms and results are described at http://en.wikipedia.org/wiki/Cycle_detection. The idea for finding meaningful collisions using the small-space attack is by Yuval [206]. The possibility of parallelizing collision-finding attacks, which can offer significant speedups in practice, is discussed in detail by van Oorschot and Wiener [198]. Time/space tradeoffs for function inversion were introduced by Hellman [95], with practical improvements—not discussed here—given by Rivest (unpublished) and Oechslin [155] (who coined the term "rainbow tables").

The first formal treatment of the random-oracle model was given by Bellare and Rogaway [24], although the idea of using a "random-looking" function in cryptographic applications had been suggested previously, most notably by Fiat and Shamir [72]. Proper instantiation of a random oracle from cryptographic hash functions is considered in several papers [24, 25, 26, 56]. The seminal negative result concerning the random-oracle model is that of Canetti et al. [47], who show (contrived) schemes that are secure in the random-oracle model but are insecure for *any* concrete instantiation of the random oracle.

Merkle trees go back at least to the 1980s [138]. Designing hash functions to make password cracking difficult is an active area of research; some popular examples of such hash functions include bcrypt and scrypt. A formal treatment of key derivation is given by Krawczyk [123]. Standardized key-derivation functions include HKDF and PBKDF2.

Exercises

6.1 Provide formal definitions for second-preimage resistance and preimage resistance. Then:

   (a) Prove that any hash function that is collision resistant is second-preimage resistant.

   (b) Prove that if a compression function mapping $2n$-bit inputs to n-bit outputs is second-preimage resistant then it is preimage resistant.

6.2 Let (Gen_1, H_1) and (Gen_2, H_2) be two hash functions. Define (Gen, H) so that Gen runs Gen_1 and Gen_2 to obtain keys s_1 and s_2, respectively. Then define $H^{s_1, s_2}(x) = H_1^{s_1}(x) \| H_2^{s_2}(x)$.

   (a) Prove that if at least one of (Gen_1, H_1) and (Gen_2, H_2) is collision resistant, then (Gen, H) is collision resistant.

   (b) Determine whether an analogous claim holds for second-preimage resistance and preimage resistance, respectively. Prove your answer in each case.

6.3 Let (Gen, H) be a collision-resistant hash function. Is $(\mathsf{Gen}, \hat{H})$ defined by $\hat{H}^s(x) \stackrel{\text{def}}{=} H^s(H^s(x))$ necessarily collision resistant?

6.4 Provide a formal proof of Theorem 6.4 (i.e., describe the reduction).

6.5 Generalize the Merkle–Damgård transform to the case where (Gen, h) takes inputs of length $n + 1$ and generates outputs of length n. (The hash function you construct should accept inputs of any length $L < 2^n$.) Prove that your transform yields a collision-resistant hash function for arbitrary-length inputs if (Gen, h) is collision resistant.

6.6 Consider the following modification of the Merkle–Damgård transform: append a 1 to the input x, followed by enough zeros so that the length of the resulting string is n more than a multiple of n'. Parse the resulting string as $z_0, x_1, \ldots, x_B$, where $|z_0| = n$ and $|x_i| = n'$. Then for $i = 1, \ldots, B$, compute $z_i := h^s(z_{i-1} \| x_i)$; output z_B.

   Show how to find a collision in the resulting hash function when this transform is applied to any compression function (Gen, h).

6.7 Consider the following modification of the Merkle–Damgård transform: append a 1 to the input x, followed by enough zeros so that the length of the resulting string is a multiple of n'. Parse the resulting string as the sequence of n'-bit blocks $x_1, \ldots, x_B$. Set $z_0 := 0^n$. Then for

$i = 1, \ldots, B$, compute $z_i := h^s(z_{i-1} \| x_i)$; output z_B. Assuming collision-resistant compression functions exist, show that there exists a collision-resistant compression function (Gen, h) such that this modified transform applied to (Gen, h) is not collision resistant.

Hint: Let h be such that $h^s(0^n, 0^{n'}) = 0^n$ for all s.

6.8 Assume collision-resistant hash functions exist. Show a construction of a fixed-length hash function (Gen, h) that is *not* collision resistant, but such that the hash function (Gen, H) obtained from applying the Merkle–Damgård transform to (Gen, h) *is* collision resistant.

6.9 Prove or disprove: if (Gen, h) is preimage resistant, then so is the hash function (Gen, H) obtained by applying the Merkle–Damgård transform to (Gen, h).

6.10 Prove or disprove: if (Gen, h) is second-preimage resistant, then so is the hash function (Gen, H) obtained by applying the Merkle–Damgård transform to (Gen, h).

6.11 Before HMAC, it was common to define a MAC for arbitrary-length messages by $\mathsf{Mac}_k(m) = H(k \| m)$ where H is a collision-resistant hash function.

    (a) Prove that this is a secure MAC if H is modeled as a random oracle.

    (b) Show that this is not a secure MAC when H is constructed via the Merkle–Damgård transform. (Assume $k \in \{0,1\}^n$.)

6.12 A student has 3,500 songs on her phone, and chooses songs to play at random. How many songs should the student expect to play before hearing some song twice (with probability at least 50%)?

6.13 Sample uniform $y_1, \ldots, y_q \in \{0,1\}^\ell$ and $y_1', \ldots, y_q' \in \{0,1\}^\ell$. What is the probability that there exist i, j such that $y_i = y_j'$?

6.14 Fix $H : \{0,1\}^n \to \{0,1\}^{2n}$, and define the keyed function $F : \{0,1\}^n \times \{0,1\}^n \to \{0,1\}^{2n}$ by $F_k(x) = H(k \oplus x)$. Show that an attacker given oracle access to $F_k(\cdot)$ can recover the n-bit key k with constant probability in time $\approx 2^{n/2}$ (which is better than a brute-force attack).

Hint: Use the previous exercise.

6.15 Prove that the keyed function F given in Section 6.5.1 is a pseudorandom function if H is modeled as a random oracle.

6.16 Prove Theorem 6.11.

6.17 Show how to find a collision in the Merkle tree construction if t is not fixed. Specifically, show how to find two sets of inputs $x_1, \ldots, x_t$ and $x_1', \ldots, x_{2t}'$ such that $\mathcal{MT}_t(x_1, \ldots, x_t) = \mathcal{MT}_{2t}(x_1', \ldots, x_{2t}')$.

6.18 Modify the construction of a Merkle tree so that it is collision resistant even when the number of inputs t may vary.

6.19 Consider the scenario introduced in Section 6.6.2 in which a client stores files on a server and wants to verify that files are returned unmodified.

    (a) Provide a formal definition of security for this setting.

    (b) Formalize the protocol based on Merkle trees as discussed in Section 6.6.2.

    (c) Prove that your construction is secure relative to your definition under the assumption that (Gen_H, H) is collision resistant.

6.20 Prove that the commitment scheme discussed in Section 6.6.5 is secure if H is modeled as a random oracle.

Chapter 7

Practical Constructions of Symmetric-Key Primitives

In previous chapters we have demonstrated how secure encryption schemes and message authentication codes can be constructed from cryptographic primitives such as pseudorandom generators (aka stream ciphers), pseudorandom permutations (aka block ciphers), and hash functions. One question we have not yet addressed, though, is how these cryptographic primitives are constructed in the first place, or even whether they exist at all! In the next chapter we will study this question from a theoretical point of view, and show constructions of pseudorandom generators and pseudorandom permutations based on quite weak assumptions. (It turns out that collision-resistant hash functions are more difficult to construct, and appear to require stronger assumptions. We will see a provably secure construction in Section 9.4.2.) In this chapter, our focus will be on comparatively heuristic—but far more efficient—constructions of these primitives that are widely used in practice.

The constructions we will explore in this chapter are heuristic in the sense that they cannot be proven secure based on any weaker assumption. Nevertheless, they are based on a number of sound design principles that can be justified by theoretical analysis. Perhaps more importantly, many of these constructions have withstood years of public scrutiny and attempted cryptanalysis; given this, it is quite reasonable to assume they are secure.

In some sense there is no fundamental difference between assuming, say, that factoring is hard and assuming that AES (a block cipher we will study later in this chapter) is a pseudorandom permutation. There is, however, a significant *qualitative* difference between these assumptions.[1] The primary difference is that the former assumption relates to a weaker requirement: the assumption that large integers are hard to factor is arguably simpler and more natural than the assumption that AES with a uniform key is indistinguishable from a random permutation. Other relevant differences are that factoring has been studied much longer than the problem of distinguishing AES from a random permutation, and that factoring was recognized as a hard problem by mathematicians independent of any cryptographic applications. The factoring problem has also been studied for a longer period of time.

---

[1]It should be clear that the discussion in this paragraph is informal, as we cannot formalize much given that we cannot prove factoring hard in the first place!

Aims of This Chapter

The main aims of this chapter are (1) to present some design principles used in the construction of modern cryptographic primitives, and (2) to introduce the reader to some popular schemes used in the real world. We caution that:

- It is *not* the aim of this chapter to teach readers how to design new cryptographic primitives. On the contrary, we believe that the design of new primitives requires significant expertise and effort, and is not something to be attempted lightly. Those who are interested in developing additional expertise in this area are advised to read the more advanced references included at the end of the chapter.

- It is *not* our intent to present all the low-level details of the various primitives we discuss here, and our descriptions should not be relied upon for implementation. In fact, our descriptions are sometimes purposefully inaccurate, as we omit certain details that are not relevant to the broader conceptual point we are trying to emphasize.

7.1 Stream Ciphers

Recall from Section 3.6.1 that a stream cipher is defined by two deterministic algorithms (Init, Next). The Init algorithm takes as input a key k (sometimes also called a seed) and optionally an initialization vector IV, and returns an initial state st. The Next algorithm can then be called repeatedly (updating the state after each invocation) to generate an unbounded stream of random-looking bits. A stream cipher that does not take an IV should behave like a pseudorandom generator: namely, when the key k is uniform then the sequence of generated bits should be indistinguishable from a sequence of uniform and independent bits. When a stream cipher takes an IV then it should act like a pseudorandom function; that is, for a uniform key k and distinct (known) initialization vectors $IV_1, IV_2, \ldots, IV_\ell$, the ℓ sequences of bits generated using k and each IV should be indistinguishable from ℓ sequences of independent, uniform bits. We refer to Section 3.6.1 for formal definitions.

In this section we consider three stream ciphers constructed in very different ways. *Trivium* is a standardized stream cipher that is very efficient in hardware. It is based on feedback shift registers, a topic of independent interest that we discuss in Sections 7.1.1 and 7.1.2. *RC4* is a software-optimized stream cipher developed in 1987 that was widely used for over twenty years. Although several weaknesses in RC4 have been discovered (and it should no longer be used), it is still interesting to study. We end with a discussion of *ChaCha20*, a modern stream cipher with good performance in software that has been adopted as a replacement for RC4 in several internet standards.

7.1.1 Linear-Feedback Shift Registers

We begin by discussing *linear-feedback shift registers* (LFSRs). These have been used historically for pseudorandom-number generation, as they are extremely efficient to implement in hardware, and generate output with good statistical properties. By themselves, however, they do *not* give cryptographically strong pseudorandom generators. Nevertheless, LFSRs (and their nonlinear generalizations that we discuss in the next section) can be used as a component of secure stream-cipher designs.

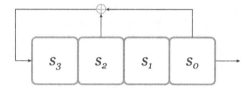

FIGURE 7.1: A linear-feedback shift register.

An LFSR consists of an array of n *registers* $s_{n-1}, \ldots, s_0$ along with a feedback loop specified by a set of n boolean *feedback coefficients* $c_{n-1}, \ldots, c_0$. (See Figure 7.1.) The size of the array is called the *degree* of the LFSR. Each register stores a single bit, and the state st of an LFSR at any point in time consists of the bits contained in its registers. The state of an LFSR is updated in each of a series of "clock ticks" by shifting the values in all the registers to the right, and setting the new value of the left-most register equal to the XOR of some subset of the current registers determined by the feedback coefficients. That is, if the state at some time t is $s_{n-1}^{(t)}, \ldots, s_0^{(t)}$, then the state after the next clock tick is $s_{n-1}^{(t+1)}, \ldots, s_0^{(t+1)}$ with

$$s_i^{(t+1)} := s_{i+1}^{(t)}, \qquad i = 0, \ldots, n-2$$

$$s_{n-1}^{(t+1)} := \bigoplus_{i=0}^{n-1} c_i\, s_i^{(t)}.$$

Figure 7.1 shows a degree-4 LFSR with $c_0 = c_2 = 1$ and $c_1 = c_3 = 0$.

At each clock tick, the LFSR outputs the value of the right-most register s_0. If the initial state of the LFSR is $s_{n-1}^{(0)}, \ldots, s_0^{(0)}$, the first n bits of the output stream are exactly $s_0^{(0)}, \ldots, s_{n-1}^{(0)}$. The next output bit is $s_{n-1}^{(1)} = \bigoplus_{i=0}^{n-1} c_i\, s_i^{(0)}$. In general, if we denote the output bits by $y_0, y_1, \ldots$, where $y_i = s_0^{(i)}$, then

$$y_i = s_i^{(0)} \qquad i = 0, \ldots, n-1$$

$$y_i = \bigoplus_{j=0}^{n-1} c_j\, y_{i-n+j} \qquad i > n-1.$$

As an example using the LFSR from Figure 7.1, if the initial state is $(s_3, s_2, s_1, s_0) = (0, 0, 1, 1)$ then the states for the first five time periods are

$$(0, 0, 1, 1)$$
$$(1, 0, 0, 1)$$
$$(1, 1, 0, 0)$$
$$(1, 1, 1, 0)$$
$$(1, 1, 1, 1)$$

and the output (which can be read off the right-most column of the above) is the stream of bits $1, 1, 0, 0, 1, \ldots$.

A degree-n LFSR can be used to define a stream cipher (Init, Next) in the natural way. Init takes as input an n-bit key k and sets the initial state of the LFSR to k. Next corresponds to one clock tick, outputting a single bit and updating the state of the LFSR accordingly.

A degree-n LFSR has 2^n possible states corresponding to the possible values of the bits in its registers. Define the *transition graph* of an LFSR to be a directed graph with a vertex corresponding to each state, and an edge from one vertex v to another vertex v' if updating the state corresponding to v in one clock tick results in the state corresponding to v'. (Thus, each vertex has a single outgoing edge.) We further label the edges of the graph with the bit that would be output by the LFSR when making the corresponding transition. For example, in the transition graph for the LFSR from Figure 7.1 the vertex $(1, 0, 0, 1)$ has an edge to the vertex $(1, 1, 0, 0)$ labeled with the bit '1.' Choosing a random initial state for the LFSR and then updating the LFSR in a series of clock ticks is thus equivalent to choosing a random initial vertex v and then following the path of directed edges (and outputting the corresponding bits on those edges) beginning at v.

A degree-n LFSR will eventually repeat some previous state; once it does, it will then repeatedly cycle among some set of states, and the bits it outputs will begin repeating as well. This corresponds to being in a cycle of the transition graph. The LFSR is *maximum length* if it cycles through all $2^n - 1$ nonzero states before repeating; i.e., its transition graph contains a cycle through all $2^n - 1$ nonzero states. (In the transition graph for any LFSR, the all-0 state has a self-loop. If the all-0 state is ever reached the LFSR remains in that state forever.) If an LFSR is maximum length then, when initialized in any nonzero state, it will cycle through all $2^n - 1$ nonzero states. Whether an LFSR is maximum length depends only on its feedback coefficients. It is well understood how to set the feedback coefficients so as to obtain a maximum-length LFSR, although the details are beyond the scope of this book.

Key-recovery attacks on LFSRs. The output of a maximum-length LFSR has good statistical properties; as just one example, the output stream contains roughly an equal number of 0s and 1s. Nevertheless, LFSRs are not secure stream ciphers. If we assume the feedback coefficients of the LFSR are known (as we should, following Kerckhoffs' principle), then the first n bits of output from a degree-n LFSR reveal the initial state (i.e., the key); once that is known, all future output bits can be computed. One might try to prevent this by using the key to also set the feedback coefficients; even in this case, however, the attacker can learn the entire key after observing at most $2n$ output bits. The first n output bits $y_0, \ldots, y_{n-1}$ of the LFSR reveal the entire initial state, as before. Given the next n output bits $y_n, \ldots, y_{2n-1}$, the attacker can set up a system of n linear equations in the n unknown feedback coefficients $c_{n-1}, \ldots, c_0$:

$$y_n = c_{n-1}\, y_{n-1} \oplus \cdots \oplus c_0\, y_0$$

$$\vdots$$

$$y_{2n-1} = c_{n-1}\, y_{2n-2} \oplus \cdots \oplus c_0\, y_{n-1}.$$

One can show that for a maximum-length LFSR the above equations are linearly independent (modulo 2), and so uniquely determine the feedback coefficients. The coefficients can thus be found efficiently using linear algebra. (If the LFSR is not maximum length, then variants of this attack still apply.) With the feedback coefficients and the initial state known, all subsequent output bits of the LFSR can again be easily determined.

7.1.2 Adding Nonlinearity

The linear relationships between the output bits of an LFSR enable an easy attack. To thwart such attacks, we must introduce some nonlinearity, i.e., using ANDs/ORs of secret values and not just their XOR. There are several different approaches to doing so, and we only explore some of them here. All the ideas we discuss can also be combined with each other in different ways.

Nonlinear feedback. One obvious way to introduce nonlinearity is to make the feedback loop nonlinear; we refer to the result simply as a feedback shift register (FSR). An FSR will again consist of an array of registers, each containing a single bit. As before, the state of the FSR is updated in each of a series of clock ticks by shifting the values in all the registers to the right; now, however, the new value of the left-most register will be a *nonlinear* function of the current registers. In other words, if the state at some time t is $s_{n-1}^{(t)}, \ldots, s_0^{(t)}$, then the state after the next clock tick is $s_{n-1}^{(t+1)}, \ldots, s_0^{(t+1)}$ with

$$s_i^{(t+1)} := s_{i+1}^{(t)}, \qquad i = 0, \ldots, n-2$$
$$s_{n-1}^{(t+1)} := g(s_{n-1}^{(t)}, \ldots, s_0^{(t)})$$

for some arbitrary (nonlinear) function g. As before, the FSR outputs the value of the right-most register s_0 at each clock tick. For security, g should be *balanced* in the sense that $\Pr[g(s_{n-1}, \ldots, s_0) = 1] \approx 1/2$, where the probability is over uniform choice of $s_{n-1}, \ldots, s_0$.

Nonlinear output. Another approach is to introduce nonlinearity in the output sequence. In the most basic case, we could have an LFSR as before (where the new value of the left-most register is again computed as a linear function of the current registers), but where the output at each clock tick is a nonlinear function g (called the *filter*) of the current registers, rather than just the right-most register. This construction is sometimes called a *filter generator*. As before, g should be balanced so that the output stream will not have any obvious bias.

Combination generators. Yet another possibility is to use more than one LFSR, and to generate the final output stream by combining the outputs of the individual LFSRs in some nonlinear way. This gives what is known as a (nonlinear) *combination generator*. The individual LFSRs need not have the same degree, and in fact the cycle length of the combination generator will be maximized if they do *not* have the same degree.

The way in which the output streams of the underlying LFSRs are combined must be done so as to ensure the final output is unbiased; simply computing the AND of the underlying output streams, for example, would result in output bits that are biased toward 0. Care must also be taken to ensure that the final output of the combination generator is not too highly correlated with any of the output streams of the underlying LFSRs, as high correlation can lead to attacks. For example, consider combining three LFSRs A, B, and C generating output streams $a_0, a_1, \ldots, b_0, b_1, \ldots$, and $c_0, c_1, \ldots$, respectively, by setting the ith output bit of the combination generator equal to $y_i := (a_i \wedge b_i) \oplus c_i$ (where $\wedge$ denotes binary AND). If the degrees of the individual LFSRs are n_a, n_b, and n_c, then the overall state has length $n_a + n_b + n_c$ and we might hope that the best attack distinguishing the output of the combination generator from uniform requires time $2^{n_a+n_b+n_c}$. But observe that if we treat each bit of each of the underlying output streams as uniform, then $a_i \wedge b_i$ is equal to 0 with probability $3/4$, and so $\Pr[c_i = y_i] = 3/4$. Thus, given a long output stream $y_0, y_1, \ldots$ of the combination generator, an attacker can enumerate all 2^{n_c} possible values of the initial state for LFSR C and compute the output sequence $c_0, c_1, \ldots$ for each one. The correct initial state for C will result in a sequence that agrees with the observed output stream roughly $3/4$ of the time; moreover, with high probability, no other candidate state will. The allows the attacker to obtain the initial state of C in time 2^{n_c}. Having done so, it can then recover the initial states of LFSRs A and B in time at most $2^{n_a+n_b}$. (See Exercise 7.4 for a better attack.)

7.1.3 Trivium

To illustrate the ideas from the previous section, we briefly describe the stream cipher Trivium. This stream cipher was selected as part of the portfolio of the eSTREAM project, a European effort completed in 2008 whose goal was to develop new stream ciphers. Trivium was designed to have a simple description and a compact hardware implementation.

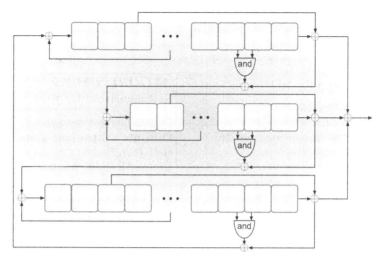

FIGURE 7.2:   A schematic illustration of Trivium with (from top to bottom) three coupled, nonlinear FSRs A, B, and C.

Trivium uses three coupled, nonlinear FSRs denoted by A, B, and C and having degrees 93, 84, and 111, respectively. (See Figure 7.2.) The state st of Trivium is simply the 288 bits comprising the values in all the registers of these FSRs. At each clock tick, the output of each FSR is the XOR of its right-most register and one additional register; the output of Trivium is the XOR of the output bits of the three FSRs. The FSRs are *coupled*: at each clock tick, the new value of the left-most register of each FSR is computed as a function of one of the registers in the same FSR and a subset of the registers from a second FSR. The feedback function in each case is nonlinear.

The Init algorithm of Trivium accepts an 80-bit key and an 80-bit IV. The key is loaded into the 80 left-most registers of A, and the IV is loaded into the 80 left-most registers of B. The remaining registers are set to 0, except for the three right-most registers of C, which are set to 1. The FSRs are then run for $4 \cdot 288$ clock ticks (with the output discarded), and the resulting state is taken as the initial state.

To date, no cryptanalytic attacks better than exhaustive search are known against Trivium.

7.1.4   RC4

LFSRs are efficient when implemented in hardware, but have poor performance in software. For this reason, alternate designs of stream ciphers have been explored. A prominent example is RC4, which was designed by Ron

Rivest in 1987. RC4 is remarkable for its speed and simplicity, and resisted serious attack for several years. While RC4 is still occasionally used, recent attacks have shown serious cryptographic weaknesses in RC4 and it is no longer recommended for cryptographic applications.

ALGORITHM 7.1
Init algorithm for RC4

Input: 16-byte key k
Output: Initial state (S, i, j)
(Note: All addition is modulo 256)

for $i = 0$ to 255:
  $S[i] := i$
  $k[i] := k[i \bmod 16]$
$j := 0$
for $i = 0$ to 255:
  $j := j + S[i] + k[i]$
  Swap $S[i]$ and $S[j]$
$i := 0, \ j := 0$
return initial state (S, i, j)

ALGORITHM 7.2
Next algorithm for RC4

Input: Current state (S, i, j)
Output: Output byte y; updated state (S, i, j)
(Note: All addition is modulo 256)

$i := i + 1$
$j := j + S[i]$
Swap $S[i]$ and $S[j]$
$t := S[i] + S[j]$
$y := S[t]$
return y and (S, i, j)

The state of RC4 consists of a 256-byte array S, which always contains a permutation of the elements $0, \ldots, 255$, along with two values $i, j \in \{0, \ldots, 255\}$. For simplicity we assume a 16-byte (128-bit) key k, although the algorithm can handle keys 1–256 bytes long. We index the bytes of S as $S[0], \ldots, S[255]$, and the bytes of the key as $k[0], \ldots, k[15]$.

The Init algorithm for RC4 is presented as Algorithm 7.1. During initialization, S is first set to the identity permutation (i.e., with $S[i] = i$ for all i) and k is expanded to 256 bytes by repeating it as many times as needed. Then each entry of S is swapped at least once with another entry of S at some "pseudorandom" location. The indices i, j are set to 0, and (S, i, j) is output as the initial state.

The initial state is used to generate a sequence of output bytes using the Next algorithm in Algorithm 7.2. Each time Next is called, the index i is simply incremented (modulo 256), and j is changed in some "pseudorandom" way. Entries $S[i]$ and $S[j]$ are swapped, and the value of S at position $S[i] + S[j]$ (again computed modulo 256) is output. Note that each entry of S is swapped with an entry of S (possibly itself) at least once every 256 iterations, ensuring good "mixing" of the permutation S.

RC4 was not designed to take an IV as input; however, in practice an IV is often incorporated by simply concatenating it with the actual key k' before initialization. That is, a random IV of the desired length is chosen, k is set equal to the concatenation of IV and k' (this can be done by either prepending or appending IV), and then Init is run as in Algorithm 7.1 to generate an initial state. Output bits are then produced using Algorithm 7.2 exactly as before. Assuming RC4 is being used in unsynchronized mode (see Section 3.6.2), the IV would then be sent in the clear to the receiver—who knows the actual key k'—thus enabling the sender and receiver to generate the same initial state and hence the same output stream. This method of incorporating an IV was used in the *Wired Equivalent Privacy* (WEP) encryption standard for protecting communications in 802.11 wireless networks.

One should be concerned by this unprincipled way of modifying RC4 to accept an IV. Even if RC4 were secure when used without an IV as originally intended, there is no reason to believe that it should be secure when modified to use an IV as just described. Indeed, contrary to the key, the IV is revealed to an attacker (since it is sent in the clear); furthermore, using different IVs with the same fixed key k'—as would be done when using RC4 in unsynchronized mode—means that *related* values k are being used to initialize the state of RC4. As we will see below, both of these issues lead to attacks when RC4 is used in this fashion.

Attacks on RC4. Various attacks on RC4 have been known for several years. Due to this, RC4 should no longer be used; instead, a more modern stream cipher or block cipher should be used in its place. We describe some basic attacks here to give a flavor for the techniques involved.

We begin by demonstrating a simple statistical attack on RC4 that does not rely on the honest parties' using an IV. Specifically, we show that the second output byte of RC4 is (slightly) biased toward 0. Let S_t denote the array S of the RC4 state after t iterations of Next, with S_0 denoting the initial array. Treating S_0 (heuristically) as a uniform permutation of $\{0, \ldots, 255\}$, with probability $1/256 \cdot (1 - 1/255) \approx 1/256$ it holds that $S_0[2] = 0$ and $X \stackrel{\text{def}}{=} S_0[1] \neq 2$. Assume for a moment that this is the case. Then in the first iteration of Next, the value of i is incremented to 1, and j is set equal to $S_0[i] = S_0[1] = X$. Then entries $S_0[1]$ and $S_0[X]$ are swapped, so that at the end of the iteration we have $S_1[X] = S_0[1] = X$. In the second iteration, i is incremented to 2 and j is assigned the value

$$j + S_1[i] = X + S_1[2] = X + S_0[2] = X,$$

since $S_0[2] = 0$. Then entries $S_1[2]$ and $S_1[X]$ are swapped, so that $S_2[X] = S_1[2] = S_0[2] = 0$ and $S_2[2] = S_1[X] = X$. Finally, the value of S_2 at position $S_2[i] + S_2[j] = S_2[2] + S_2[X] = X$ is output; this is exactly the value $S_2[X] = 0$.

When $S_0[2] \neq 0$ the second output byte is uniformly distributed. Overall, then, the probability that the second output byte is 0 is roughly

$$\Pr[S_0[2] = 0 \text{ and } S_0[1] \neq 2] + \frac{1}{256} \cdot \Pr[S_0[2] \neq 0] = \frac{1}{256} + \frac{1}{256} \cdot \left(1 - \frac{1}{256}\right)$$

$$\approx \frac{2}{256},$$

or roughly twice what would be expected for a uniform value.

By itself the above might not be viewed as a particularly serious attack, although it does indicate underlying structural problems with RC4. Moreover, statistical biases like the above have been found in other output bytes of RC4, and it has been shown that these biases are sufficiently large to allow for the recovery of plaintext when RC4 is used for encryption.

A more devastating attack against RC4 is possible when an IV is incorporated by prepending it to the key. This attack can be used to recover the key, regardless of its length, and is thus more serious than a distinguishing attack such as the one described above. Importantly, this attack can be used to completely break the WEP encryption standard mentioned earlier, and was influential in getting the standard replaced.

The core of the attack is a way to extend knowledge of the first n bytes of k to knowledge of the first $n+1$ bytes of k. Note that when an IV is prepended to the actual key k' (so $k = IV \| k'$), the first few bytes of k are given to the attacker for free! If the IV is n bytes long, then an adversary can use this attack to first recover the $(n+1)$st byte of k (which is the first byte of the real key k'), then the next byte of k, and so on, until it learns the entire key.

Assume the IV is 3 bytes long, as is the case for WEP. The attacker waits until the first two bytes of the IV have a specific form. The attack can be carried out with several possibilities for the first two bytes of the IV, but we look at the case where the IV takes the form $IV = (3, 255, X)$ for X an arbitrary byte. This means, of course, that $k[0] = 3, k[1] = 255$, and $k[2] = X$ in Algorithm 7.1. One can check that after the first four iterations of the second loop of Init, we have

$$S[0] = 3, \quad S[1] = 0, \quad S[3] = X + 6 + k[3].$$

In the next 252 iterations of the Init algorithm, i is always greater than 3. So the values of $S[0], S[1]$, and $S[3]$ are not subsequently modified as long as j never takes on the values 0, 1, or 3. If we (heuristically) treat j as taking on a uniform value in each iteration, this means that $S[0], S[1]$, and $S[3]$ are not subsequently modified with probability $(253/256)^{252} \approx 0.05$, or 5% of the time. Assuming this is the case, the first byte output by Next will be $S[3] = X + 6 + k[3]$; since X is known, this reveals $k[3]$.

So, the attacker knows that 5% of the time the first byte of the output is related to $k[3]$ as described above. (This is much better than random guessing, which is correct $1/256 = 0.4\%$ of the time.) By collecting sufficiently many samples of the first byte of the output—for several IVs of the correct form—the attacker obtains a high-confidence estimate for $k[3]$.

7.1.5   ChaCha20

ChaCha20, introduced in 2008, is a stream cipher intended to be extremely efficient in software. It is available as a replacement for RC4 in many systems and—as described in Section 5.3.2—is combined with the Poly1305 message authentication code to construct an authenticated encryption scheme widely used in the TLS protocol. We give a high-level description of ChaCha20 that gives the main ideas of the scheme, but refer elsewhere for the low-level details.

The core of ChaCha20 is a fixed permutation P that operates on 512-bit strings. This permutation is carefully constructed to be both highly efficient and "cryptographically strong." To improve efficiency, it was designed to rely primarily on only three assembly-level instructions operating on 32-bit words: Addition (modulo 2^{32}), bitwise (cyclic) Rotation, and XOR; P is thus an example of what is called an *ARX-based* design. From a cryptographic point of view, P is intended to be a suitable instantiation of a "random permutation," and constructions based on P can be analyzed in the so-called *random-permutation model*. By analogy with the random-oracle model (see Section 6.5), the random-permutation model assumes that all parties are given access to oracles for a uniform permutation P as well as its inverse P^{-1}. In this model, as in the random-oracle model, the *only* way to compute P (or P^{-1}) is to explicitly query those oracles. (We refer to Section 7.3.3 for an example of a proof of security in the random-permutation model.)

In ChaCha20, the permutation P is used to construct a pseudorandom function F taking a 256-bit key and mapping 128-bit inputs to 512-bit outputs. This keyed function F is defined as

$$F_k(x) \overset{\text{def}}{=} P(\mathsf{const}\|k\|x) \boxplus \mathsf{const}\|k\|x,$$

where const is a 128-bit constant. (Above, '$\boxplus$' denotes word-wise modular addition.) F can be shown to be a pseudorandom function if P is modeled as a random permutation.

The ChaCha20 stream cipher itself is then constructed from F as in Construction 3.30. Specifically, given a 256-bit seed s and an initialization vector $IV \in \{0,1\}^{64}$, the output of the stream cipher is $F_s(IV\|\langle 0\rangle), F_s(IV\|\langle 1\rangle), \ldots$, where the counter values $\langle 0\rangle, \langle 1\rangle$, etc., are encoded as 64-bit integers.

7.2 Block Ciphers

Recall from Section 3.5.1 that a block cipher is an efficient, keyed permutation $F : \{0,1\}^n \times \{0,1\}^\ell \to \{0,1\}^\ell$. This means the function F_k defined by $F_k(x) \overset{\text{def}}{=} F(k,x)$ is a bijection (i.e., a permutation), and moreover F_k and its inverse F_k^{-1} are efficiently computable given k. We refer to n as the *key length* and ℓ as the *block length* of F, and here we explicitly allow them to differ. The key length and block length are now fixed constants, whereas in Chapter 3 they were viewed as functions of a security parameter. This puts us in the setting of concrete security rather than asymptotic security.[2] The concrete-security requirements for block ciphers are quite stringent, and a block cipher is generally only considered "secure" if the best known attack (without pre-processing) has time complexity roughly equivalent to a brute-force search for the key. Thus, if a cipher with key length $n = 256$ can be broken in time 2^{128}, the cipher is (generally) considered insecure even though a 2^{128}-time attack is infeasible. (In contrast, in an asymptotic setting an attack of complexity $2^{n/2}$ is not considered efficient since it requires exponential time, and thus a cipher where such an attack is possible might still qualify as a pseudorandom permutation.) This is because in the concrete setting we care about the actual complexity of attacks, and not just their asymptotic behavior. Furthermore, there is a concern that existence of a better-than-brute-force attack may indicate some more fundamental weakness in the design of the cipher.

Block ciphers are designed to behave, at a minimum, as (strong) pseudorandom permutations; see Definition 3.27. (Often, block ciphers are designed and assumed to satisfy even stronger security properties, as we discuss in Section 7.3.1.) Modeling block ciphers as pseudorandom permutations allows proofs of security for constructions based on block ciphers, and also makes explicit the necessary requirements of a block cipher. A solid understanding of what block ciphers are supposed to achieve is instrumental in their design. The view that block ciphers should be modeled as pseudorandom permutations has, at least recently, served as a major influence in their design. As an example, the call for proposals for the Advanced Encryption Standard (AES) that we will encounter later in this chapter stated the following evaluation criterion:

> *The security provided by an algorithm is the most important factor. . . . Algorithms will be judged on the following factors: . . .*
>
> - *The extent to which the algorithm output is indistinguishable from a random permutation . . .*

Modern block ciphers are suitable for all the constructions using pseudorandom permutations (or pseudorandom functions) we have seen in this book.

Notwithstanding the fact that block ciphers are not, on their own, encryption schemes, the standard terminology for attacks on a block cipher F is:

- In a *known-plaintext attack*, the attacker is given pairs of inputs/outputs $\{(x_i, F_k(x_i))\}$ (for an unknown key k), with the $\{x_i\}$ outside the attacker's control.

- In a *chosen-plaintext attack*, the attacker is given $\{F_k(x_i)\}$ (again, for an unknown key k) for a series of inputs $\{x_i\}$ chosen by the attacker.

---

[2]Although a block cipher with fixed key length has no "security parameter" to speak of, we still view security as depending on the key length and thus denote that value by n. Viewing the key length as a parameter makes sense when comparing block ciphers with different key lengths, or when using a block cipher that supports keys of different lengths.

- In a *chosen-ciphertext attack*, the attacker is given $\{F_k(x_i)\}$ for $\{x_i\}$ chosen by the attacker, as well as $\{F_k^{-1}(y_i)\}$ for chosen $\{y_i\}$.

A cipher secure against chosen-plaintext attacks corresponds to a pseudorandom permutation, while one secure against chosen-ciphertext attacks corresponds to a strong pseudorandom permutation. In addition to attacks distinguishing F_k from a uniform permutation, we will also be interested in *key-recovery attacks* in which the attacker can recover the key k after interacting with F_k. (This is stronger than being able to distinguish F_k from uniform.)

7.2.1 Substitution-Permutation Networks

A secure block cipher (using a random key) must behave like a random permutation. There are $2^\ell!$ permutations on ℓ-bit strings, so representing an arbitrary permutation in this case requires $\log(2^\ell!) \approx \ell \cdot 2^\ell$ bits. This is impractical for $\ell > 20$ and infeasible for $\ell > 60$. (Looking ahead, modern block ciphers have block lengths $\ell \geq 128$.) The challenge when designing a block cipher is to construct permutations having a *concise* description (namely, a short key) that behave like *random* permutations. In particular, just as evaluating a random permutation at two inputs that differ in only a single bit should yield two (almost) independent outputs (they are not completely independent since they cannot be equal), so too changing one bit of the input to $F_k(\cdot)$, where k is uniform and unknown to an attacker, should yield an (almost) independent result. This implies that a one-bit change in the input should "affect" every bit of the output. (Note that this does not mean that all the output bits will be changed—that would be different behavior than one would expect for a random permutation. Rather, we just mean informally that each bit of the output is changed with probability roughly half.) This takes some work to achieve.

The confusion-diffusion paradigm. In addition to his work on perfect secrecy, Shannon also introduced a basic paradigm for constructing concise, random-looking permutations. The basic idea is to construct a random-looking permutation F with a large block length from many smaller random (or random-looking) permutations $\{f_i\}$ with small block length. Let us see how this works on the most basic level. Say we want F to have a block length of 128 bits. We can define F as follows: the key k for F will specify 16 permutations $f_1, \ldots, f_{16}$ that each have an 8-bit (1-byte) block length.[3] Given an input $x \in \{0,1\}^{128}$, we parse it as 16 bytes $x_1 \cdots x_{16}$ and then set

$$F_k(x) = f_1(x_1)\| \cdots \|f_{16}(x_{16}). \tag{7.1}$$

These *round functions* $\{f_i\}$ are said to introduce *confusion* into F.

---

[3] An arbitrary permutation on 8 bits can be represented using $\log(2^8!)$ bits, so the length of the key for F is about $16 \cdot \log(2^8!)$ bits, or about 3 kbytes. This is much smaller than the $\approx 128 \cdot 2^{128}$ bits that would be required to specify an arbitrary permutation on 128 bits.

It should be immediately clear, however, that F as defined above will *not* be pseudorandom. Specifically, if x and x' differ only in their first bit then $F_k(x)$ and $F_k(x')$ will differ only in their first byte (regardless of the key k). In contrast, for a truly random permutation changing the first bit of the input would be expected to affect all bytes of the output.

For this reason, a *diffusion* step is introduced whereby the bits of the output are permuted, or "mixed," using a *mixing permutation*. This has the effect of spreading a local change (e.g., a change in the first byte) throughout the entire block. In principle the mixing permutation could depend on the key, but in practice it is carefully designed and fixed.

The confusion/diffusion steps—together called a *round*—are repeated multiple times. This helps ensure that changing a single bit of the input will affect all the bits of the output. As an example, a two-round block cipher following this approach would operate as follows. First, confusion is introduced by computing the intermediate result $f_1(x_1)\|\cdots\|f_{16}(x_{16})$ as in Equation (7.1), where we stress again that the $\{f_i\}$ depend on the key. The bits of the result are then "shuffled," or re-ordered, using a mixing permutation to give $x' = x'_1 \cdots x'_{16}$. Then $f'_1(x'_1)\|\cdots\|f'_{16}(x'_{16})$ is computed, using possibly different functions $\{f'_i\}$ that again depend on the key, and the bits of the result are again permuted using a mixing permutation to give output x''.

Substitution-permutation networks. A substitution-permutation network (SPN) can be viewed as a direct implementation of the confusion-diffusion paradigm. The difference is that now the permutations (i.e., the $\{f_i\}, \{f'_i\}$) have a particular form rather than being chosen from the set of all possible permutations. Specifically, rather than having (a portion of) the key k specify an arbitrary permutation f, we instead fix a public "substitution function" (i.e., permutation) S called an *S-box*, and then let k define the function f given by $f(x) = S(k \oplus x)$. (If f takes 8-bit inputs as before, we have thus reduced the number of possibilities for f from $2^8!$ to 2^8.)

To see how this works concretely, consider an SPN with a 64-bit block length based on a collection of 8-bit (1-byte) S-boxes $S_1, \ldots, S_8$. (See Figure 7.3.) Evaluating the cipher proceeds in a series of rounds, where in each round we apply the following sequence of operations to the 64-bit input x of that round (the input to the first round is just the input to the cipher):

1. *Key mixing:* Set $x := x \oplus k$, where k is the current-round *sub-key*;

2. *Substitution:* Set $x := S_1(x_1)\|\cdots\|S_8(x_8)$, where x_i is the ith byte of x;

3. *Permutation:* Permute the bits of x to obtain the output of the round.

The output of each round is used as input to the next round. After the last round there is a final key-mixing step, and the result is the output of the cipher. (By Kerckhoffs' principle, we assume the S-boxes and the mixing permutation(s) are public and known to any attacker. Without the final key-mixing step, the substitution and permutation steps of the last round would

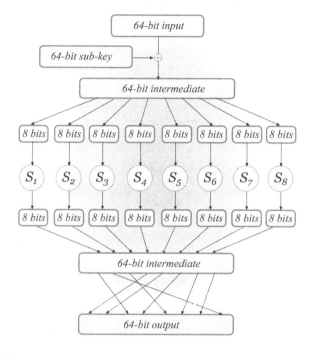

FIGURE 7.3: A single round of a substitution-permutation network.

offer no additional security since they do not depend on the key and can be inverted by an attacker.) Figure 7.4 shows three rounds of an SPN with a 16-bit block length and a different set of 4-bit S-boxes used in each round.

Different *sub-keys* (or *round keys*) are used in each round. The actual key of the block cipher is sometimes called the *master key*. The round keys are derived from the master key according to a *key schedule*. The key schedule is often simple and may just use different subsets of the bits of the master key as the various sub-keys, though more complex key schedules can also be defined. An r-round SPN has r rounds of key mixing, S-box substitution, and application of a mixing permutation, followed by a final key-mixing step. (This means that an r-round SPN uses $r+1$ sub-keys.)

Any SPN is invertible (given the key). To see this, it suffices to show that a single round can be inverted; this implies the entire SPN can be inverted by working from the final round back to the beginning. But inverting a single round is easy: the mixing permutation can easily be inverted since it is just a re-ordering of bits. Since the S-boxes are permutations (i.e., one-to-one), these too can be inverted. The result can then be XORed with the appropriate sub-key to obtain the original input. Summarizing:

PROPOSITION 7.3 *Let F be a keyed function defined by an SPN in which the S-boxes are all permutations. Then regardless of the key schedule and the number of rounds, F_k is a permutation for any k.*

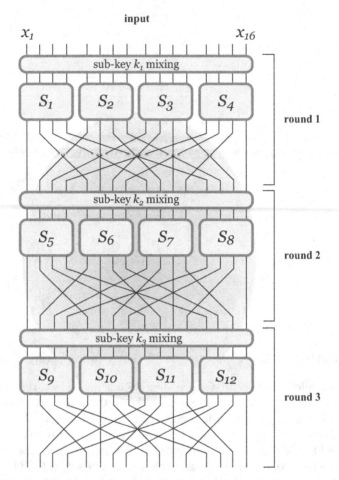

FIGURE 7.4:   Three rounds of a substitution-permutation network.

The number of rounds, along with the exact choices of the S-boxes, mixing permutations, and key schedule, are what ultimately determine whether a given block cipher is trivially breakable or highly secure. We now discuss a basic principle behind the design of the S-boxes and mixing permutations.

The avalanche effect. As noted repeatedly, an important property in any block cipher is that a small change in the input must "affect" every bit of the output. We refer to this as the *avalanche effect*. One way to induce the avalanche effect in a substitution-permutation network is to ensure that the following two properties hold (and sufficiently many rounds are used):

1. The S-boxes are designed so that changing a single bit of the input to an S-box changes at least *two bits* in the output of the S-box.

2. The mixing permutations are designed so that the bits output by any given S-box affect the input to *multiple* S-boxes in the next round. For

example, in Figure 7.4 the output from S_1 affects the input to S_5, S_6, S_7, and S_8.

To see how this yields the avalanche effect, at least heuristically, assume the S-boxes are all such that changing a single bit of the input to the S-box results in a change in exactly two bits in the output of the S-box, and that the mixing permutations are chosen as required above. For concreteness, assume the S-boxes have 8-bit input/output length, and that the block length of the cipher is 128 bits. Consider now what happens when the block cipher is applied to two inputs that differ in a single bit:

1. After the first round, the intermediate values differ in exactly two bits. This is because XORing the first-round sub-key maintains the 1-bit difference in the intermediate values, and so the inputs to all the S-boxes except one are identical. In the one S-box where the inputs differ, the output of the S-box causes a 2-bit difference. The mixing permutation applied to the results changes the positions of these differences, but maintains a 2-bit difference.

2. The mixing permutation applied at the end of the first round spreads the two bit-positions where the intermediate results differ into two *different* S-boxes in the second round. This remains true even after the second-round key mixing is done. So, in the second round there are now *two* S-boxes that receive inputs differing in a single bit. Thus, at the end of the second round the intermediate values differ in 4 bits.

3. Continuing the same argument, we expect 8 bits of the intermediate value to be affected after the 3rd round, 16 bits to be affected after the 4th round, and all 128 bits to be affected at the end of the 7th round.

The last point is not quite precise and it is certainly possible that there will be fewer differences than expected at the end of some round. (In fact, we want this to be the case because uncorrelated values should not differ in all their bits, either.) This can occur when the mixing permutation maps two bit-positions that differ in some intermediate result to the *same* S-box in the following round. For this reason, it is customary to use many more than the minimum number of rounds needed. But the above analysis gives a *lower bound*: if fewer than 7 rounds are used then there must be some set of output bits that are not affected by a single-bit change in the input, implying that it will be possible to distinguish the cipher from a random permutation.

One might expect that the "best" way to design S-boxes would be to choose them at random (subject to the restriction that they are permutations). Interestingly, this turns out not to be the case, at least if we want to satisfy the design criteria mentioned earlier. Consider the case of an S-box operating on 4-bit inputs and let x and x' be two distinct values. Let $y = S(x)$, and now consider choosing uniform $y' \neq y$ as the value of $S(x')$. There are 4 strings that differ from y in only 1 bit, and so with probability $4/15$ we will choose y'

that does *not* differ from y in two or more bits. The problem is compounded when we consider all pairs of inputs that differ in a single bit. We conclude based on this example that, as a general rule, the S-boxes must be designed carefully rather than being chosen at random. Random S-boxes are also not good for defending against attacks like the ones we will show in Section 7.2.6.

If a block cipher should also be *strongly* pseudorandom, then the avalanche effect must also apply to its *inverse*. That is, changing a single bit of the output should affect every bit of the input. For this it is useful if the S-boxes are designed so that changing a single bit of the output of an S-box changes at least two bits of the input to the S-box. Achieving the avalanche effect in both directions is another reason for further increasing the number of rounds.

Attacking Reduced-Round SPNs

Experience, along with many years of cryptanalytic effort, indicate that substitution-permutation networks are a good choice for constructing pseudorandom permutations as long as care is taken in the choice of the S-boxes, the mixing permutations, and the key schedule. The Advanced Encryption Standard, described in Section 7.2.5, is similar in structure to a substitution-permutation network as described above, and is widely believed to be a strong pseudorandom permutation.

The strength of a cipher F constructed as an SPN depends heavily on the number of rounds. In order to obtain more insight into substitution-permutation networks, we will demonstrate attacks on SPNs having very few rounds. These attacks are fairly simple, but are worth seeing as they demonstrate conclusively why a large number of rounds is needed.

A trivial case. We first consider a trivial case where F consists of one round and no final key-mixing step. We show that an adversary given only a *single* input/output pair (x, y) can easily learn the secret key k for which $y = F_k(x)$. The adversary begins with the output value y and then inverts the mixing permutation and the S-boxes. It can do this, as noted before, because the full specification of the mixing permutation and the S-boxes is public. The intermediate value that the adversary computes is exactly $x \oplus k$ (assuming, without loss of generality, that the master key is used as the sub-key in the only round of the network). Since the adversary also knows the input x, it can immediately derive the secret key k. This is therefore a complete break.

Although this is a trivial attack, it demonstrates that in any substitution-permutation network there is no security gained by performing S-box substitution or applying a mixing permutation after the last key-mixing step.

Attacking a one-round SPN. Now we have one round followed by a key-mixing step. For concreteness, we assume a 64-bit block length and S-boxes with 8-bit (1-byte) input/output length. We assume independent 64-bit sub-keys k_1, k_2 are used for the two key-mixing steps, and so the master key $k_1 \| k_2$ of the SPN is 128 bits long.

A first observation is that we can extend the attack from the trivial case above to give a key-recovery attack here using much less than 2^{128} work. The idea is as follows: Given a single input/output pair (x, y) as before, the attacker enumerates over all possible values for the first-round sub-key k_1. For each such value, the attacker can compute the first round of the SPN using k_1 to get a candidate intermediate value x'. The only second-round sub-key that is consistent with k_1 and output y is $k_2 = x' \oplus y$. Thus, for each possible choice of k_1 the attacker derives a unique corresponding k_2 for which $k_1 \| k_2$ might be the master key. In this way, the attacker obtains (in 2^{64} time) a list of 2^{64} possibilities for the master key. These can be narrowed down using additional input/output pairs in roughly 2^{64} additional time.

A better attack is possible by noting that individual bits of the output depend on only part of the sub-keys. Fix some given input/output pair (x, y) as before. Now, the adversary will enumerate over all possible values for the *first byte* of k_1. It can XOR each such value with the first byte of x to obtain a candidate value for the 1-byte input to the first S-box. Evaluating this S-box, the attacker learns a candidate value for the *output* of that S-box. Since the output of that S-box is XORed with 8 bits of k_2 to yield 8 bits of y (where the positions of those bits depend on the mixing permutation but are known to the attacker), this yields a candidate value for 8 bits of k_2.

To summarize: for each candidate value for the first byte of k_1, there is a *unique* possible corresponding value for some 8 bits of k_2. Put differently, this means that for some 16 bits of the master key, the attacker has reduced the number of possible values for those bits from 2^{16} to 2^8. The attacker can tabulate all those feasible values in 2^8 time. This can be repeated for each byte of k_1, giving 8 lists—each containing 2^8 16-bit values—that together characterize the possible values of the entire master key. In this way, the attacker has reduced the number of possible master keys to $(2^8)^8 = 2^{64}$, as in the earlier attack; the total time to do this, however, is now $8 \cdot 2^8 = 2^{11}$, a dramatic improvement.

The attacker can use additional input/output pairs to further reduce the space of possible keys. Importantly, this can be done for each list individually. Consider the list of 2^8 feasible values for some set of 16 bits of the master key. The attacker knows that the correct value from that list must be consistent with any additional input/output pairs the attacker learns, whereas any *incorrect* value in the list is expected to be consistent with another input/output pair (x', y') with probability no better than random. Since a 16-bit value from the list can be used to compute eight bits of the output given the input x', an incorrect value will be consistent with the actual output y' with probability roughly 2^{-8}. A small number of additional input/output pairs thus suffices to narrow down *all* the lists to just a single value each, at which point the entire master key is known.

This attack exploits the fact that the effects of different parts of the key can be isolated. Additional rounds are needed to ensure further *diffusion*, and to make sure that each bit of the key affects all of the bits of the output.

Attacking a two-round SPN. It is possible to extend the above ideas to give a better-than-brute-force attack on a two-round SPN using independent sub-keys in each round; we leave this as an exercise. Here we simply note that a two-round SPN will not be a good pseudorandom permutation, since the avalanche effect does not occur after only two rounds. (Of course, this depends on the block length of the cipher and the input/output length of the S-boxes, but with reasonable parameters this will be the case.) An attacker can distinguish a two-round SPN from a uniform permutation if it learns the result of evaluating the SPN on two inputs that differ in a single bit, since some predictable subset of the output bits will not change.

7.2.2 Feistel Networks

Feistel networks offer another approach for constructing block ciphers. An advantage of Feistel networks over substitution-permutation networks is that the underlying functions used in a Feistel network—in contrast to the S-boxes used in SPNs—need not be invertible. *A Feistel network thus provides a way to construct an invertible function from non-invertible components.* This is important because a good block cipher should have "unstructured" behavior (so it looks random), yet requiring all the components of a construction to be invertible inherently introduces structure. Requiring invertibility also introduces an additional constraint on S-boxes, making them harder to design.

A Feistel network operates in a series of rounds. In each round, a keyed *round function* is applied in the manner described below. Round functions need not be invertible. They will typically be constructed from components like S-boxes and mixing permutations, but a Feistel network can deal with *any* round functions irrespective of their design.

In a (balanced) Feistel network with ℓ-bit block length, the ith round function $\hat{f}_i$ takes as input a sub-key k_i and an $\ell/2$-bit string and generates an $\ell/2$-bit output. As in the case of SPNs, a master key k is used to derive sub-keys for each round. When some master key is chosen, thereby determining each sub-key k_i, we define $f_i : \{0,1\}^{\ell/2} \to \{0,1\}^{\ell/2}$ via $f_i(R) \overset{\text{def}}{=} \hat{f}_i(k_i, R)$. Note that the round functions $\hat{f}_i$ are fixed and publicly known, but the f_i depend on the master key and so are not known to the attacker.

The ith round of a Feistel network operates as follows. The ℓ-bit input to the round is divided into two halves denoted L_{i-1} and R_{i-1} (the "left" and "right" halves, respectively). The output (L_i, R_i) of the round is

$$L_i := R_{i-1} \quad \text{and} \quad R_i := L_{i-1} \oplus f_i(R_{i-1}). \tag{7.2}$$

In an r-round Feistel network, the ℓ-bit input to the network is parsed as (L_0, R_0), and the output is the ℓ-bit value (L_r, R_r) obtained after applying all r rounds. A three-round Feistel network is shown in Figure 7.5.

Inverting a Feistel network. A Feistel network is invertible *regardless of the $\{f_i\}$* (and thus regardless of the round functions $\{\hat{f}_i\}$). To show this we

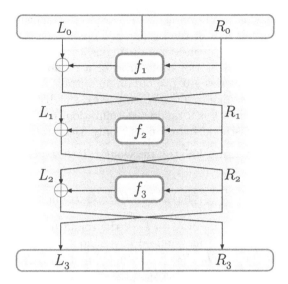

FIGURE 7.5:   A three-round Feistel network.

need only show that each round of the network can be inverted if the $\{f_i\}$ are known. Given the output (L_i, R_i) of the ith round, we can compute (L_{i-1}, R_{i-1}) as follows: first set $R_{i-1} := L_i$. Then compute

$$L_{i-1} := R_i \oplus f_i(R_{i-1}).$$

This gives the value (L_{i-1}, R_{i-1}) that was the input of this round (i.e., it computes the inverse of Equation (7.2)). Note that f_i is evaluated only in the forward direction, so it need not be invertible. We thus have:

PROPOSITION 7.4   *Let F be a keyed function defined by a Feistel network. Then regardless of the key schedule, the round functions $\{\hat{f}_i\}$, and the number of rounds, F_k is a permutation for any k.*

Attacking Reduced-Round Feistel Networks

As in the case of SPNs, attacks on Feistel networks are possible when the number of rounds is too low. Although it is not possible to show key-recovery attacks without knowing something about the round functions, we show here that one- and two-round Feistel networks can easily be distinguished from random functions. (In Section 8.6 we show that three- and four-round Feistel networks can be proven secure under certain conditions.)

Attacking a one-round Feistel network. If F is a one-round Feistel network then $F_k(L_0, R_0) = (R_0, f_1(R_0) \oplus L_0)$, where f_1 depends in some way on k. Although the attacker does not know f_1 (because it does not know k), it is clear that F_k (for a uniform key k) is easy to distinguish from a random function since the left half of the output of F_k is always equal to the right half of its input. Formally, consider a distinguisher given access to an oracle g that is either equal to F_k (for uniform k) or a random permutation. The distinguisher simply queries $g(0^\ell)$ to obtain an output y, and then outputs 1 iff the first half of y is equal to $0^{\ell/2}$. When g is F_k, the distinguisher outputs 1 with probability 1; when g is a random permutation, however, the value y is uniform and so the distinguisher outputs 1 only with probability $2^{-\ell/2}$.

Attacking a two-round Feistel network. If F is a two-round Feistel network then

$$F_k(L_0, R_0) = \left(f_1(R_0) \oplus L_0, \ R_0 \oplus f_2(f_1(R_0) \oplus L_0)\right),$$

where f_1, f_2 depend in some way on k. If the round functions $\hat{f}_1, \hat{f}_2$ are designed properly, then f_1, f_2 may indeed look random when k is unknown, in which case the output $F_k(L_0, R_0)$ for a single input may look random. Nevertheless, there are *correlations* between the outputs of F_k on related inputs that can be used to distinguish F_k from a random permutation. Specifically, consider evaluating F_k on the inputs $(0^{\ell/2}, 0^{\ell/2})$ and $(1^{\ell/2}, 0^{\ell/2})$. If we let

$$(L_2, R_2) \stackrel{\text{def}}{=} F_k(0^{\ell/2}, 0^{\ell/2}) \text{ and } (L'_2, R'_2) \stackrel{\text{def}}{=} F_k(1^{\ell/2}, 0^{\ell/2}),$$

then a little algebra gives

$$L_2 \oplus L'_2 = f_1(0^{\ell/2}) \oplus 0^{\ell/2} \oplus f_1(0^{\ell/2}) \oplus 1^{\ell/2} = 1^{\ell/2}.$$

This holds regardless of the key. On the other hand, for a random permutation f the probability that the XOR of the left halves of $f(0^{\ell/2}, 0^{\ell/2})$ and $f(1^{\ell/2}, 0^{\ell/2})$ is equal $1^{\ell/2}$ is roughly $2^{-\ell/2}$.

7.2.3 DES – The Data Encryption Standard

The Data Encryption Standard, or DES, was developed in the 1970s by IBM (with help from the National Security Agency) and adopted by the US in 1977 as a Federal Information Processing Standard. DES is of great historical significance. It has undergone intensive scrutiny within the cryptographic community, arguably more than any other cryptographic algorithm in history, and the consensus is that DES is an extremely well-designed cipher. Indeed, even after many years, the best attack on DES *in practice* is an exhaustive search over all 2^{56} possible keys. (There are important theoretical attacks on DES requiring less computation; however, those attacks assume certain conditions that seem difficult to realize in practice.) In its basic form, though, DES is no longer considered suitable since a 56-bit key is too short, i.e., brute-force attacks running in time 2^{56} are feasible today. The 64-bit block length of DES is also too small for modern applications. Nevertheless, DES remains in limited use in the strengthened form of triple-DES, described in Section 7.2.4.

In this section, we provide a high-level overview of the main components of DES. We do not provide a full specification, and we have simplified some parts of the design. The reader interested in the low-level details of DES can consult the references at the end of this chapter.

The Design of DES

The DES block cipher is a 16-round Feistel network with a block length of 64 bits and a key length of 56 bits. The same round function $\hat{f}$ is used in each of the 16 rounds. The round function takes a 48-bit sub-key and, as expected for a (balanced) Feistel network, a 32-bit input (namely, half a block). The *key schedule* of DES is used to derive a sequence of 48-bit sub-keys $k_1, \ldots, k_{16}$ from the 56-bit master key. The key schedule of DES is relatively simple, with each sub-key k_i being a permuted subset of 48 bits of the master key. For our purposes, it suffices to note that the 56 bits of the master key are divided into two halves—a "left half" and a "right half"—containing 28 bits each. (This division occurs after an initial permutation is applied to the key, but we ignore this in our description.) The left-most 24 bits of each round sub-key are taken as some subset of the 28 bits in the left half of the master key, and the right-most 24 bits of each round sub-key are taken as some subset of the 28 bits in the right half of the master key. The entire key schedule (including the manner in which the master key is divided into left and right halves, and which bits are used in forming each sub-key k_i) is fixed and public, and the only secret is the master key itself.

The DES round function. The DES round function $\hat{f}$—sometimes called the *DES mangler function*—is constructed using a paradigm we have previously analyzed: it is (basically) a substitution-permutation network! In more detail, computation of $\hat{f}(k_i, R)$ with $k_i \in \{0,1\}^{48}$ and $R \in \{0,1\}^{32}$ proceeds as follows: first, R is *expanded* to a 48-bit value R'. This is carried out by simply duplicating half the bits of R; we denote this by $R' := E(R)$ where E is called the *expansion function*. Following this, computation proceeds exactly as in our earlier discussion of SPNs: The expanded value R' is XORed with k_i, which is also 48 bits long, and the resulting value is divided into 8 blocks, each of which is 6 bits long. Each block is passed through a (different) S-box that takes a 6-bit input and yields a 4-bit output; concatenating the output from the 8 S-boxes gives a 32-bit result. A mixing permutation is then applied to the bits of this result to obtain the final output. See Figure 7.6.

One difference as compared to our original discussion of SPNs is that the S-boxes here are *not* invertible; indeed, they cannot be invertible since their inputs are longer than their outputs. Further discussion regarding the structural details of the S-boxes is given below.

We stress once again that everything in the above description (including the S-boxes themselves as well as the mixing permutation) is *publicly known*. The only secret is the master key which is used to derive all the sub-keys.

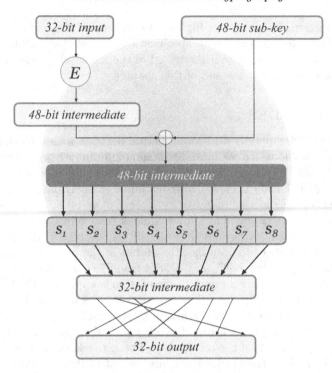

FIGURE 7.6:   The DES mangler function.

The *S*-boxes and the mixing permutation. The eight *S*-boxes that form the "core" of $\hat{f}$ are a crucial element of the DES construction and were very carefully designed. Studies of DES have shown that if the *S*-boxes were slightly modified, DES would have been much more vulnerable to attack. This should serve as a warning to anyone wishing to design a block cipher: seemingly arbitrary choices are not arbitrary at all, and if not made correctly may render the entire construction insecure.

Recall that each *S*-box maps a 6-bit input to a 4-bit output. Each *S*-box can be viewed as a table with 4 rows and 16 columns, where each cell of the table contains a 4-bit entry. A 6-bit input can be viewed as indexing one of the $2^6 = 64 = 4 \times 16$ cells of the table in the following way: The first and last input bits are used to choose the table row, and bits 2–5 are used to choose the table column. The 4-bit entry at some position of the table represents the output value for the input associated with that position.

The DES *S*-boxes have the following properties (among others):

1. Each *S*-box is a 4-to-1 function. (That is, exactly 4 inputs are mapped to each possible output.) This follows from the properties below.

2. Each row in the table contains each of the 16 possible 4-bit strings exactly once.

3. Changing *one bit* of any input to an S-box always changes at least *two bits* of the output.

The DES mixing permutation was also designed carefully. In particular it has the property that the four output bits from any S-box affect the input to six S-boxes in the next round. (This is possible because of the expansion function that is applied in the next round before the S-boxes are computed.)

The DES avalanche effect. The design of the mangler function ensures that DES exhibits a strong avalanche effect. In order to see this, we will trace the difference between the intermediate values in the DES computations of two inputs that differ by just a single bit. Let us denote the two inputs to the cipher by (L_0, R_0) and (L_0', R_0'), where we assume that $R_0 = R_0'$ and so the single-bit difference occurs in the left half of the inputs (it may help to refer to Equation (7.2) and Figure 7.6 in what follows). After the first round, the intermediate values (L_1, R_1) and (L_1', R_1') still differ by only a single bit, although now this difference is in the right half. In the second round of DES, the right half of each intermediate value is run through $\hat{f}$. Assuming that the bit where R_1 and R_1' differ is not duplicated in the expansion step, the intermediate values before applying the S-boxes still differ by only a single bit. By property 3 of the S-boxes, the intermediate values *after* the S-box computation differ in at least *two* bits. The result is that the intermediate values (L_2, R_2) and (L_2', R_2') differ in *three* bits: there is a 1-bit difference between L_2 and L_2' (carried over from the difference between R_1 and R_1') and a 2-bit difference between R_2 and R_2'.

The mixing permutation spreads the two-bit difference between R_2 and R_2' such that, in the following round, each of the two differing bits is used as input to a *different* S-box, resulting in a difference of at least 4 bits in the outputs from the S-boxes. (If either or both of the two bits in which R_2 and R_2' differ are duplicated by E, the difference may be even greater.) There is also now a 2-bit difference in the left halves.

As with a substitution-permutation network, the number of "affected" bits grows exponentially and so after 7 rounds we expect all 32 bits in the right half to be affected, and after 8 rounds we expect all 32 bits in the left half will be affected as well. DES has 16 rounds, and so the avalanche effect occurs very early in the computation. This ensures that the computation of DES on similar inputs yields independent-looking outputs.

Attacks on Reduced-Round DES

A useful exercise for understanding more about the DES construction and its security is to look at the behavior of DES with only a few rounds. We show attacks on one-, two-, and three-round variants of DES (recall that DES has 16 rounds). DES variants with three rounds or fewer cannot be pseudorandom functions because three rounds are not enough for the avalanche effect to occur. Thus, we will be interested in demonstrating more difficult (and more damaging) key-recovery attacks which compute the key k using only a relatively small number of input/output pairs computed using that key. Some of the attacks are similar to those we have seen in the context of substitution-permutation networks; here, however, we will see how they are applied to a concrete block cipher rather than to an abstract design.

The attacks below will be known-plaintext attacks in which the adversary knows some plaintext/ciphertext pairs $\{(x_i, y_i)\}$ computed using some secret key k. When we describe the attacks, we will focus on a particular plaintext/ciphertext pair (x, y) and describe the information about the key that the adversary can derive from this pair. Continuing to use the notation developed earlier, we denote the left and right halves of the input x as L_0 and R_0, respectively, and let L_i, R_i denote the left and right halves of the intermediate result after the ith round. Recall that E denotes the DES expansion function, k_i denotes the sub-key used in round i, and $f_i(R) = \hat{f}(k_i, R)$ denotes the actual function being applied in the Feistel network in the ith round.

One-round DES. Say we are given an input/output pair (x, y). In one-round DES, we have $y = (L_1, R_1)$, where $L_1 = R_0$ and $R_1 = L_0 \oplus f_1(R_0)$. We therefore know an input/output pair for f_1: specifically, we know that $f_1(R_0) = R_1 \oplus L_0$. By applying the inverse of the mixing permutation to the output $R_1 \oplus L_0$, we obtain the intermediate value consisting of the outputs from all the S-boxes, where the first 4 bits are the output from the first S-box, the next 4 bits are the output from the second S-box, and so on.

Consider the (known) 4-bit output of the first S-box. Since each S-box is a 4-to-1 function, this means there are exactly four possible inputs to this S-box that would result in the given output, and similarly for all the other S-boxes; each such input is 6 bits long. The input to the S-boxes is simply the XOR of $E(R_0)$ with the sub-key k_1. Since R_0, and hence $E(R_0)$, is known, we can compute a set of four possible values for each 6-bit portion of k_1. This means we have reduced the number of possible keys k_1 from 2^{48} to $4^{48/6} = 4^8 = 2^{16}$ (since there are four possibilities for each of the eight 6-bit portions of k_1). This is already a small number and so we can just try all the possibilities on a different input/output pair (x', y') to find the right key. We thus obtain the key using only two known plaintexts in time roughly 2^{16}.

Two-round DES. In two-round DES, the output y is equal to (L_2, R_2) where

$$L_1 = R_0$$
$$R_1 = L_0 \oplus f_1(R_0)$$
$$L_2 = R_1 = L_0 \oplus f_1(R_0)$$
$$R_2 = L_1 \oplus f_2(R_1).$$

L_0, R_0, L_2, and R_2 are known from the given input/output pair (x, y), and thus we also know $L_1 = R_0$ and $R_1 = L_2$. This means that we know the input/output of both f_1 and f_2, and so the same method used in the attack on one-round DES can be used here to determine both k_1 and k_2 in time roughly $2 \cdot 2^{16}$. This attack works even if k_1 and k_2 are completely independent keys, although in fact the key schedule of DES ensures that many of the bits of k_1 and k_2 are equal (which can be used to further speed up the attack).

Three-round DES. Referring to Figure 7.5, the output value y is now equal to (L_3, R_3). Since $L_1 = R_0$ and $R_2 = L_3$, the only unknown values in the figure are R_1 and L_2 (which are equal).

Now we no longer have the input/output to any round function f_i. For example, the output value of f_2 is equal to $L_1 \oplus R_2$, where both of these values are known. However, we do *not* know the value R_1 that is input to f_2. Similarly, we can determine the inputs to f_1 and f_3 but not the outputs of those functions. Thus, the attack we used to break one-round and two-round DES will not work here.

Instead of relying on full knowledge of the input and output of one of the round functions, we will use knowledge of a certain relation between the inputs and outputs of f_1 and f_3. Observe that the output of f_1 is equal to $L_0 \oplus R_1 = L_0 \oplus L_2$, and the output of f_3 is equal to $L_2 \oplus R_3$. Therefore,

$$f_1(R_0) \oplus f_3(R_2) = (L_0 \oplus L_2) \oplus (L_2 \oplus R_3) = L_0 \oplus R_3,$$

where both L_0 and R_3 are known. That is, *the XOR of the outputs of f_1 and f_3 is known.* Furthermore, the input to f_1 is R_0 and the input to f_3 is L_3, both of which are known. Summarizing: we can determine the inputs to f_1 and f_3, and the XOR of their outputs. We now describe an attack that finds the secret key based on this information.

Recall that the key schedule of DES has the property that the master key is divided into a "left half," which we denote by k_L, and a "right half" k_R, each containing 28 bits. Furthermore, the 24 left-most bits of the sub-key used in each round are taken only from k_L, and the 24 right-most bits of each sub-key are taken only from k_R. This means that k_L affects only the inputs to the first four S-boxes in any round, while k_R affects only the inputs to the last four S-boxes. Since the mixing permutation is known, we also know which bits of the output of each round function come from each S-box.

The idea behind the attack is to separately traverse the key space for each half of the master key, giving an attack with complexity roughly $2 \cdot 2^{28}$ rather than complexity 2^{56}. Such an attack will be possible if we can verify a guess of half the master key, and we now show how this can be done. Say we guess some value for k_L, the left half of the master key. We know the input R_0 of f_1, and so using our guess of k_L we can compute the input to the first four S-boxes. This means that we can compute half the output bits of f_1 (the mixing permutation spreads out the bits we know, but since the mixing permutation is known we know exactly which bits those are). Likewise, we can compute the same locations in the output of f_3 by using the known input L_3 to f_3 and the same guess for k_L. Finally, we can compute the XOR of these output values and check whether they match the appropriate bits in the known value of the XOR of the outputs of f_1 and f_3. If they are not equal, then our guess for k_L is incorrect. A correct guess for k_L will always pass this test, and so will not be eliminated, but an incorrect guess is expected to pass this test only with probability roughly 2^{-16} (since we check equality of 16 bits in two computed values). There are 2^{28} possible values for k_L, so if each incorrect value remains a viable candidate with probability 2^{-16} then we expect to be left with only $2^{28} \cdot 2^{-16} = 2^{12}$ possibilities for k_L after the above.

By performing the above for each half of the master key, we obtain in time $2 \cdot 2^{28}$ approximately 2^{12} candidates for the left half and 2^{12} candidates for the right half. Since each combination of the left and right halves is possible, we have 2^{24} candidate keys overall and can run a brute-force search over this set using an additional input/output pair (x', y'). (An alternative that is more efficient is to simply repeat the previous attack using the 2^{12} remaining candidates for each half of the key.) The time for the attack is roughly $2 \cdot 2^{28} + 2^{24} < 2^{30}$, much less than a 2^{56}-time brute-force attack.

Security of DES

After almost 30 years of intensive study, the best known practical attack on DES is still an exhaustive search through its key space. (We discuss some important theoretical attacks in Section 7.2.6. Those attacks require a large number of input/output pairs, which can be difficult to obtain in an attack on any real-world system using DES.) Unfortunately, the 56-bit key length of DES is short enough that an exhaustive search through all 2^{56} possible keys is now feasible. Already in the late 1970s there were strong objections to using such a short key for DES. Back then the objection was academic, as the computational power needed to search through 2^{56} keys was generally unavailable. (It has been estimated that in 1977 a computer that could crack DES in one day would cost \$20 million to build.) The practicality of a brute-force attack on DES, however, was demonstrated in 1997 when a DES challenge set up by RSA Security was solved by the DESCHALL project using thousands of computers coordinated across the Internet; the computation took 96 days. A second challenge was broken the following year in just 41 days by the `distributed.net` project. A significant breakthrough came in 1998 when a third challenge was solved in just *56 hours*. This impressive feat was achieved via a special-purpose DES-breaking machine called *Deep Crack* that was built by the Electronic Frontier Foundation at a cost of \$250,000. In 1999, a DES challenge was solved in just over 22 hours by a combined effort of Deep Crack and `distributed.net`. The current state-of-the-art is the DES cracking box by PICO Computing, which uses 48 FPGAs and can find a DES key in approximately 26 hours; see `https://crack.sh` for further details.

The time/space tradeoffs discussed in Section 6.4.3 show that exhaustive key-search attacks can be accelerated using pre-computation and additional memory. Due to the short key length of DES, time/space tradeoffs can be especially effective. Specifically, using pre-processing it is possible to generate a table a few terabytes large that enables recovery of a DES key with high probability from a single input/output pair using approximately 2^{38} DES evaluations (which can be computed in under a minute). The bottom line is that the key length of DES is far too short by modern standards, and DES cannot be considered secure for any serious application today.

A second cause for concern is the relatively short block length of DES. A short block length is problematic because the concrete security of many constructions based on block ciphers depends on the block length of the cipher— *even if the cipher is otherwise "perfect."* For example, the proof of CTR mode (cf. Theorem 3.33) shows that plaintext information can be leaked to an attacker if an *IV* repeats. If CTR mode is instantiated using DES, with a block length of only 64 bits, then security is compromised with high probability after encrypting only $\approx 2^{24}$ messages.

The insecurity of DES has nothing to do with its design *per se*, but rather is due to its short key length (and, to a lesser extent, its short block length). This is a great tribute to the designers of DES, who seem to have succeeded in constructing an almost "perfect" block cipher otherwise. Since DES itself seems not to have significant structural weaknesses, it makes sense to use DES as a building block for constructing block ciphers with longer keys. We discuss this further in Section 7.2.4.

The replacement for DES—the *Advanced Encryption Standard* (AES), covered later in this chapter—was explicitly designed to address concerns regarding the short key length and block length of DES. AES supports 128-, 192-, and 256-bit keys, and has a 128-bit block length.

7.2.4   3DES: Increasing the Key Length of a Block Cipher

The main weakness of DES is its short key. It thus makes sense to try to design a block cipher with a larger key length using DES as a building block. Some approaches to doing so are discussed in this section. Although we refer to DES frequently throughout the discussion, and DES is the most prominent block cipher to which these techniques have been applied, everything we say here applies generically to *any* block cipher.

Internal modifications vs. "black-box" constructions. There are two general approaches one could take to constructing another cipher based on DES. The first approach would be to somehow modify the *internal structure* of DES, while increasing the key length. For example, one could leave the round function untouched and simply use a 128-bit master key with a different key schedule (still choosing a 48-bit sub-key in each round). Or, one could change the *S*-boxes themselves and use a larger sub-key in each round. The disadvantage of such approaches is that by modifying DES—in even the smallest way—we lose the confidence we have gained in DES by virtue of the fact that it has remained resistant to attack for so many years. Cryptographic constructions are very sensitive; even mild, seemingly insignificant changes can render a construction completely insecure. (In fact, various results to this effect have been shown for DES; e.g., changing the S-boxes or the mixing permutation can make DES much more vulnerable to attack.) Tweaking the internal components of a block cipher is therefore not recommended.

An alternative approach that does not suffer from the above problem is to use DES as a "black box" and not touch its internal structure at all. In following this approach we treat DES as a "perfect" block cipher with a 56-bit key, and construct a new block cipher that only invokes the original, unmodified DES. Since DES itself is not tampered with, this is a much more prudent approach and is the one we will pursue here.

Double Encryption

Let F be a block cipher with an n-bit key length and ℓ-bit block length. Then a new block cipher F' with a key of length $2n$ can be defined by

$$F'_{k_1,k_2}(x) \stackrel{\text{def}}{=} F_{k_2}(F_{k_1}(x)),$$

where k_1 and k_2 are independent keys. If exhaustive key search were the best available attack, this would mean that the best attack would require time 2^{2n}. Unfortunately, we show an attack on F' that runs in time roughly 2^n. This means that F' is not any more secure against brute-force attacks than F, even though F' has a key that is twice as long.[4]

The attack is called a "meet-in-the-middle attack," for reasons that will soon become clear. Say the adversary is given a single input/output pair (x, y), where $y = F'_{k_1^*, k_2^*}(x) = F_{k_2^*}(F_{k_1^*}(x))$ for unknown k_1^*, k_2^*. The adversary can narrow down the set of possible keys in the following way:

1. For each $k_1 \in \{0,1\}^n$, compute $z := F_{k_1}(x)$ and store (z, k_1) in a list L.

2. For each $k_2 \in \{0,1\}^n$, compute $z := F_{k_2}^{-1}(y)$ and store (z, k_2) in a list L'.

3. Call entries $(z_1, k_1) \in L$ and $(z_2, k_2) \in L'$ a *match* if $z_1 = z_2$. For each such match, add (k_1, k_2) to a set S. (Matches can be found easily by first sorting the elements in L and L' by their first entry.)

See Figure 7.7 for a graphical depiction of the attack. The attack requires $2 \cdot 2^n$ evaluations of F, and uses $2 \cdot (n + \ell) \cdot 2^n$ bits of memory.

The set S output by this algorithm contains exactly those values (k_1, k_2) for which

$$F_{k_1}(x) = F_{k_2}^{-1}(y) \tag{7.3}$$

or, equivalently, for which $y = F'_{k_1,k_2}(x)$. In particular, $(k_1^*, k_2^*) \in S$. On the other hand, a pair $(k_1, k_2) \neq (k_1^*, k_2^*)$ is (heuristically) expected to satisfy

---

[4]This is not quite true since a brute-force attack on F can be carried out in time 2^n and constant memory, whereas the attack we show on F' requires 2^n time *and* 2^n memory. Nevertheless, the attack illustrates that F' does not achieve the desired level of security.

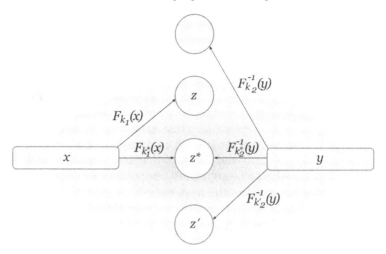

FIGURE 7.7: A meet-in-the-middle attack.

Equation (7.3) with probability $2^{-\ell}$ if we treat $F_{k_1}(x)$ and $F_{k_2}^{-1}(y)$ as uniform ℓ-bit strings, and so the expected size of S is $2^{2n} \cdot 2^{-\ell} = 2^{2n-\ell}$. Using another few input/output pairs, and taking the intersection of the sets that are obtained, the correct (k_1^*, k_2^*) can be identified with very high probability.

Triple Encryption

The obvious generalization of the preceding approach is to apply the block cipher *three* times in succession. Two variants of this approach are common:

Variant 1: three keys. The most natural thing to do is to choose three independent keys, i.e., to define $F''_{k_1,k_2,k_3}(x) \overset{\text{def}}{=} F_{k_3}(F_{k_2}^{-1}(F_{k_1}(x)))$.

Variant 2: two keys. As we explain below, another option is to choose two independent keys and define $F''_{k_1,k_2}(x) \overset{\text{def}}{=} F_{k_1}(F_{k_2}^{-1}(F_{k_1}(x)))$.

Note that the middle invocation of F is traditionally reversed. If F is a secure cipher this makes no difference as far as security is concerned (since if F is a strong pseudorandom permutation then F^{-1} is too). This is done for backward compatibility: by setting $k_3 = k_2 = k_1$, the resulting cipher is equivalent to a single invocation of F using the key k_1.

Security of the first variant. The key length of the first variant is $3n$, and so we might hope that the best attack requires time 2^{3n}. However, the cipher is susceptible to a meet-in-the-middle attack (just as in the case of double encryption) that here requires 2^{2n} time.

Security of the second variant. The key length of this variant is $2n$, and a meet-in-the-middle attack requires time 2^{2n}. Assuming $\ell \geq n$, this is the best known attack when the adversary is given only a few input/output pairs. (There is a known-plaintext attack using 2^t input/output pairs that runs in time $\approx 2^{n+\ell-t}$. See Exercise 7.16.)

Triple-DES (3DES). Triple-DES (or 3DES), standardized in 1999, is based on three invocations of DES using two or three keys, as described above. Two-key 3DES (which corresponds to the second variant) is no longer recommended, in part due to the known-plaintext attack mentioned above. Three-key 3DES is still used, though the current recommendation is to phase it out due to its small block length and the fact that it is relatively slow. These drawbacks have led 3DES to be supplanted in practice by the Advanced Encryption Standard, described in the next section.

7.2.5 AES – The Advanced Encryption Standard

In January 1997, the United States National Institute of Standards and Technology (NIST) announced that it would hold a competition to select a new block cipher—to be called the *Advanced Encryption Standard*, or AES— to replace DES. The competition began with an open call for teams to submit candidate block ciphers for evaluation. A total of 15 different algorithms were submitted from all over the world, including contributions from many of the best cryptographers and cryptanalysts. Each team's candidate cipher was intensively analyzed by members of NIST, the public, and (especially) the other teams. Two workshops were held, one in 1998 and one in 1999, to discuss and analyze the various submissions. Following the second workshop, NIST narrowed the field down to 5 "finalists" and the second round of the competition began. A third AES workshop was held in April 2000, inviting additional scrutiny on the five finalists. In October 2000, NIST announced that the winning algorithm was Rijndael (a block cipher designed by the Belgian cryptographers Vincent Rijmen and Joan Daemen), although NIST conceded that any of the 5 finalists would have made an excellent choice. In particular, no serious security vulnerabilities were found in any of the 5 finalists, and the selection of a "winner" was based in part on properties such as efficiency, performance in hardware, flexibility, etc.

The process of selecting AES was ingenious because any group that submitted an algorithm (and was therefore interested in having its algorithm adopted) had strong motivation to find attacks on the other submissions. This incentivized the world's best cryptanalysts to focus their attention on finding even the slightest weaknesses in the candidate ciphers submitted to the competition. After only a few years each candidate algorithm was already subjected to intensive study, thus increasing confidence in the security of the winner. Of course, the longer AES is used and studied without being broken, the more our confidence in it continues to grow. Today, AES is widely used and no significant security weaknesses have been discovered.

The AES construction. We present the high-level structure of AES. As with DES, we will not present a full specification and our description should not be used as a basis for implementation. Our aim is only to provide a general idea of how the algorithm works.

The AES block cipher has three variants called AES-128, AES-192, and AES-256 that use 128-, 192-, or 256-bit keys, respectively; they all have a 128-bit block length. The length of the key affects the key schedule (i.e., the way sub-keys are derived from the master key) as well as the number of rounds, but does not affect the high-level structure of each round.

In contrast to DES, which uses a Feistel structure, AES is essentially a substitution-permutation network. During computation of the AES algorithm, a 4-by-4 array of bytes called the *state* is modified in a series of rounds. The state is initially set equal to the input to the cipher (note that the input is 128 bits, which is exactly 16 bytes). In each round, the following operations are then applied to the state:

Stage 1 – AddRoundKey: A 128-bit sub-key is derived from the master key, and viewed as a 4-by-4 array of bytes. The state array is updated by XORing it with this sub-key.

Stage 2 – SubBytes: In this step, each byte of the state array is replaced by another byte according to a single, fixed lookup table S. This substitution table (or S-box) is a permutation on $\{0,1\}^8$.

Stage 3 – ShiftRows: Next, the bytes in each row of the state array are shuffled as follows: the first row of the array is untouched, each byte the second row is shifted one place to the left, the third row is shifted two places to the left, and the fourth row is shifted three places to the left. (All shifts are cyclic so that, e.g., in the second row the first byte becomes the fourth byte.)

Stage 4 – MixColumns: Finally, an invertible linear transformation is applied to the four bytes in each column. This transformation has the property that if two inputs differ in $b > 0$ bytes, then the resulting outputs differ in at least $5 - b$ bytes.

In the final round, MixColumns is replaced with AddRoundKey. This prevents an adversary from simply inverting the last three stages, which do not depend on the key.

By treating stages 3 and 4 as one step, we see that each round of AES has the structure of a substitution-permutation network: the round sub-key is first XORed with the input to the current round in a key-mixing step; next, an invertible S-box is applied to each byte of the resulting value; finally, the bits of the result are "permuted." The only difference is that, unlike our previous description of substitution-permutation networks, here the final step does not consist of simply shuffling the bits using a mixing permutation, but is instead carried out using a permutation plus an invertible linear transformation. Nevertheless, the net effect—namely, diffusion—is the same. Note that, as we have pointed out previously in our discussion of SPNs, a final key-mixing step is done after the last round.

The number of rounds depends on the key length. Ten rounds are used for a AES-128, 12 rounds for AES-192, and 14 rounds for a AES-256.

Security of AES. As we have mentioned, the AES cipher was subject to intense scrutiny during the selection process and has continued to be studied ever since. To date, there are no practical cryptanalytic attacks that are significantly better than an exhaustive search for the key.

We conclude that, as of today, AES constitutes an excellent choice for any cryptographic scheme that requires a (strong) pseudorandom permutation. It is free, standardized, efficient, and highly secure.

7.2.6   *Differential and Linear Cryptanalysis

Block ciphers are relatively complicated, and as such are difficult to analyze. Nevertheless, one should not be fooled into thinking that a complicated cipher is necessarily difficult to break. On the contrary, it is very hard to construct a secure block cipher, and surprisingly easy to find attacks on most constructions (no matter how complicated they appear). This should serve as a warning that non-experts should not try to construct new ciphers. Given the availability of AES, it is hard to justify using anything else.

In this section we describe two tools that are now a standard part of the cryptanalyst's toolbox. Our goal here is give a taste of some advanced cryptanalysis, as well as to reinforce the idea that designing a secure block cipher involves careful choice of its components.

Differential cryptanalysis. This technique, which can lead to a chosen-plaintext attack on a block cipher, was first presented in the late 1980s by Biham and Shamir, who used it to attack DES in 1993. The basic idea behind the attack is to tabulate *specific differences in the input* that lead to *specific differences in the output* with probability greater than would be expected for a random permutation. Specifically, say *the differential (Δ_x, Δ_y) occurs in some keyed permutation G with probability p* if for uniform inputs x_1 and x_2 satisfying $x_1 \oplus x_2 = \Delta_x$, and uniform choice of key k, the probability that $G_k(x_1) \oplus G_k(x_2) = \Delta_y$ is p. For any fixed (Δ_x, Δ_y) and x_1, x_2 satisfying $x_1 \oplus x_2 = \Delta_x$, if we choose a uniform function $f : \{0,1\}^\ell \to \{0,1\}^\ell$, we have $\Pr[f(x_1) \oplus f(x_2) = \Delta_y] = 2^{-\ell}$. In a weak block cipher, however, there may be differentials that occur with significantly higher probability. This can be leveraged to give a full key-recovery attack, as we now show for SPNs.

We describe the basic idea, and then work through a concrete example. Let F be an r-round SPN with an ℓ-bit block length, and let $G_k(x)$ denote the intermediate result in the computation of $F_k(x)$ after applying the key-mixing step of the last round. (That is, G excludes the S-box substitution and mixing permutation of the last round, as well as the final key-mixing step.) Assume there is a differential (Δ_x, Δ_y) in G that occurs with probability $p \gg 2^{-\ell}$. It is possible to exploit this high-probability differential to learn bits of the final sub-key k_{r+1}. The high-level idea is as follows: let $\{(x_1^i, x_2^i)\}_{i=1}^L$ be a collection of L pairs of random inputs with differential Δ_x, i.e., with $x_1^i \oplus x_2^i = \Delta_x$ for all i. Using a chosen-plaintext attack, obtain $y_1^i = F_k(x_1^i)$ and $y_2^i = F_k(x_2^i)$ for all i. Now, for each possible $k_{r+1}^* \in \{0,1\}^\ell$ do: for each pair y_1^i, y_2^i, invert the final key-mixing step using k_{r+1}^*, and also invert the mixing permutation and S-boxes of round r (which do not depend on the master key) to obtain $\tilde{y}_1^i, \tilde{y}_2^i$. Note that when $k_{r+1}^* = k_{r+1}$, we have $\tilde{y}_1^i = G_k(x_1^i)$ and $\tilde{y}_2^i = G_k(x_2^i)$, and in that case we expect that a p-fraction of the pairs will satisfy $\tilde{y}_1^i \oplus \tilde{y}_2^i = \Delta_y$. On the other hand, when $k^* \neq k_{r+1}$ we heuristically expect only a $2^{-\ell}$-fraction of the pairs to yield this differential. By setting L large enough, the correct value of the final sub-key k_{r+1} can be determined.

This works, but requires enumerating over 2^ℓ possible values for the final sub-key. We can do better by guessing portions of k_{r+1} at a time. More concretely, assume the S-boxes in F have 1-byte input/output length, and focus on the first byte of Δ_y. It is possible to verify if the differential holds in that byte by guessing only 8 bits of k_{r+1}, namely, the 8 bits that correspond (after the round-r mixing permutation) to the output of the first S-box. Thus, proceeding as above, we can learn these 8 bits by enumerating over all possible values for those bits, and seeing which value yields the desired differential in the first byte with the highest probability. Incorrect guesses for those 8 bits yield the expected differential in that byte with (heuristic) probability 2^{-8}, but the correct guess will give the expected differential with probability roughly $p + 2^{-8}$; this is because with probability p the differential holds on the entire block (so in particular for the first byte), and when this is not the case then we can treat the differential in the first byte as random. Note that different differentials may be needed to learn different portions of k_{r+1}.

In practice, various optimizations are performed to improve the effectiveness of the above test or, more specifically, to increase the gap between the probability that an incorrect guess for (bits of) k_{r+1} yields the differential vs. the probability that a correct guess does. One optimization is to use a *low-weight* differential in which Δ_y has many zero bytes. Any pairs $\tilde{y}_1, \tilde{y}_2$ satisfying such a differential have equal values entering many of the S-boxes in round r, and so will result in output values y_1, y_2 that are equal in the corresponding bit-positions (depending on the final mixing permutation). This means that the attacker can simply discard any pairs (y_1^i, y_2^i) that do not agree in those bit-positions (since the corresponding intermediate values $(\tilde{y}_1, \tilde{y}_2)$ cannot possibly satisfy the differential, for any choice of the final sub-key). This significantly improves the effectiveness of the attack.

Once k_{r+1} is known, the attacker can "peel off" the final key-mixing step, as well as the mixing permutation and S-box substitution steps of round r (since these do not depend on the master key), and then apply the same attack—using a different differential—to find the rth-round sub-key k_r, and so on, until it learns all sub-keys (or, equivalently, the master key). Relations between the sub-keys can be used to improve the efficiency of the attack.

A worked example. We work through a "toy" example, illustrating also how a good differential can be found. We use a four-round SPN with a block length of 16 bits, based on a single S-box with 4-bit input/output length. The S-box is defined as follows:

Input:	0000	0001	0010	0011	0100	0101	0110	0111
Output:	0000	1011	0101	0001	0110	1000	1101	0100

Input:	1000	1001	1010	1011	1100	1101	1110	1111
Output:	1111	0111	0010	1100	1001	0011	1110	1010

The mixing permutation, showing where each of the 16 bits in a block is moved, is:

In:	1	2	3	4	5	6	7	8	9	10	11	12	13	14	15	16
Out:	7	2	3	8	12	5	11	9	10	1	14	13	4	6	16	15

We first find a differential in the S-box. Let $S(x)$ denote the output of the S-box on input x. Consider the differential $\Delta_x = 1111$. Then, for example, we have $S(0000) \oplus S(1111) = 0000 \oplus 1010 = 1010$ and so in this case a difference of 1111 in the inputs leads to a difference of 1010 in the outputs. Let us see if this relation holds frequently. We have $S(0001) = 1011$ and $S(0001 \oplus 1111) = S(1110) = 1110$, and so here a difference of 1111 in the inputs does *not* lead to a difference of 1010 in the outputs. However, $S(0100) = 0110$ and $S(0100 \oplus 1111) = S(1011) = 1100$ and so in this case, a difference of 1111 in the inputs yields a difference of 1010 in the outputs. In Figure 7.8 we tabulate results for all possible inputs. We see that *half* the time a difference of 1111 in the inputs yields a difference of 1010 in the outputs. Thus, $(1111, 1010)$ is a differential in S that occurs with probability $1/2$.

x	S(x)	x ⊕ 1111	S(x ⊕ 1111)	S(x) ⊕ S(x ⊕ 1111)
0000	0000	1111	1010	1010
0001	1011	1110	1110	0101
0010	0101	1101	0011	0110
0011	0001	1100	1001	1000
0100	0110	1011	1100	1010
0101	1000	1010	0010	1010
0110	1101	1001	0111	1010
0111	0100	1000	1111	1011
1000	1111	0111	0100	1011
1001	0111	0110	1101	1010
1010	0010	0101	1000	1010
1011	1100	0100	0110	1010
1100	1001	0011	0001	1000
1101	0011	0010	0101	0110
1110	1110	0001	1011	0101
1111	1010	0000	0000	1010

FIGURE 7.8: The effect of the input difference $\Delta_x = 1111$ in our S-box.

Output Difference Δ_y

	0	1	2	3	4	5	6	7	8	9	A	B	C	D	E	F
0	16	0	0	0	0	0	0	0	0	0	0	0	0	0	0	0
1	0	0	0	0	4	0	0	0	2	2	2	2	0	0	4	0
2	0	0	0	0	0	2	0	2	0	2	2	4	2	2	0	0
3	0	2	2	4	0	4	0	0	0	0	0	0	0	2	2	0
4	0	0	0	2	2	2	6	0	2	0	0	0	2	0	0	0
5	0	2	2	0	0	0	0	0	4	0	0	0	4	2	2	0
6	0	2	0	2	0	0	0	0	0	2	0	2	0	4	0	4
7	0	2	0	0	2	4	2	2	0	2	0	0	0	2	0	0
8	0	0	0	2	0	0	0	2	0	0	0	2	2	2	2	4
9	0	2	0	2	2	2	0	4	0	2	2	0	0	0	0	0
A	0	0	4	0	2	0	2	4	2	0	2	0	0	0	0	0
B	0	0	2	0	0	0	2	0	0	2	0	0	4	2	4	0
C	0	0	0	0	0	0	0	0	4	4	0	4	0	0	0	4
D	0	4	2	2	0	0	2	2	0	0	0	0	0	0	0	4
E	0	2	4	2	4	0	0	0	0	0	0	0	2	0	2	0
F	0	0	0	0	0	2	2	0	2	0	8	2	0	0	0	0

Input Difference Δ_x

FIGURE 7.9: Differentials in our S-box.

This same process can be carried out for all 2^4 input differences Δ_x to calculate the probability of every differential. Namely, for each pair (Δ_x, Δ_y) we tabulate the number of inputs x for which $S(x) \oplus S(x \oplus \Delta_x) = \Delta_y$. We have done this for our example S-box in Figure 7.9. (For conciseness we use hexadecimal notation.) The table should be read as follows: entry (i, j) counts how many inputs with difference i map to outputs with difference j. Observe, for example, that there are 8 inputs with difference $\texttt{0xF} = 1111$ that map to output $\texttt{0xA} = 1010$, as we have shown above. This is the highest-probability differential (apart from the trivial differential $(\texttt{0x0}, \texttt{0x0})$). But there are also other differentials of interest: an input difference of $\Delta_x = \texttt{0x4} = 0100$ maps to an output difference of $\Delta_y = \texttt{0x6} = 0110$ with probability $6/16 = 3/8$, and there are several differentials with probability $4/16 = 1/4$.

We now extend this to find a good differential for the first three rounds of the SPN. Consider evaluating the SPN on two inputs that have a differential of 0000 1100 0000 0000, and tracing the differential between the intermediate values at each step of this evaluation. (Refer to Figure 7.10, which shows the first three rounds of the SPN.) The key-mixing step in the first round does not affect the differential, and so the inputs to the second S-box in the first round have differential 1100. We see from Figure 7.9 that a difference of $\texttt{0xC} = 1100$ in the inputs to the S-box yields a difference of $\texttt{0x8} = 1000$ in the outputs of the S-box with probability $1/4$. So with probability $1/4$ the differential in the output of the 2nd S-box after round 1 is a single bit which is moved by the mixing permutation from the 5th position to the 12th position. (The inputs to the other S-boxes are equal, so their outputs are equal and the differential of the outputs is 0000.) Assuming this to be the case, the input

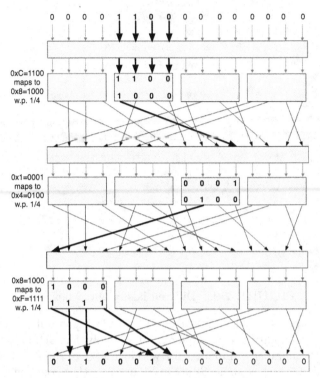

FIGURE 7.10:   Tracing differentials through the first three rounds of an SPN that uses the *S*-box and mixing permutation given in the text.

difference to the third *S*-box in the second round is 0x1 = 0001 (once again, the key-mixing step in the second round does not affect the differential); using Figure 7.9 we have that with probability 1/4 the output difference from that *S*-box is 0x4 = 0100. Thus, once again there is just a single output bit that is different, and it is moved from the 10th position to the first position by the mixing permutation. Finally, consulting Figure 7.9 yet again, we see that an input difference of 0x8 = 1000 to the *S*-box results in an output difference of 0xF = 1111 with probability 1/4. The bits in positions 1, 2, 3, and 4 are then moved by the mixing permutation to positions 7, 2, 3, and 8. Note that the key-mixing step in the fourth round does not affect the output differential.

Overall, then, we see that an input difference of $\Delta_x = 0000\ 1100\ 0000\ 0000$ yields the output difference $\Delta_y = 0110\ 0011\ 0000\ 0000$ after three rounds with probability at least $\frac{1}{4} \cdot \frac{1}{4} \cdot \frac{1}{4} = \frac{1}{64}$. (This is a lower bound on the probability of the differential, since there may be other differences in the intermediate values that result in the same difference in the outputs. We multiply the probabilities since we assume independence of the sub-keys used in each round.) For a random function, the probability that any given differential occurs is just $2^{-16} = 1/65536$. Thus, the differential we have found occurs with probability significantly higher than what would be expected for a random function. Observe also that we have found a low-weight differential.

We can use this differential to find 8 bits of the final sub-key k_5—namely, the bits at positions 2, 3, 5, 7, 8, 9, 11, and 12. (i.e., the positions that the outputs of the first two S-boxes from the 3rd round get mapped to by the mixing permutation.) As discussed earlier, we begin by letting $\{(x_1^i, x_2^i)\}_{i=1}^L$ be a set of L pairs of random inputs with differential Δ_x. Using a chosen-plaintext attack, we then obtain the values $y_1^i = F_k(x_1^i)$ and $y_2^i = F_k(x_2^i)$ for all i. Now, for all possible values of the specified 8 bits of k_5, we compute the initial 8 bits of $\tilde{y}_1^i, \tilde{y}_2^i$, the intermediate values after the key-mixing step of the 4th round. (We can do this because we only need to invert the two left-most S-boxes of the 4th round in order to derive those 8 bits.) When we guess the correct value for the specified 8 bits of k_5, we expect the 8-bit differential 0110 0011 to occur with probability at least $1/64$. Heuristically, an incorrect guess yields the expected differential only with probability $2^{-8} = 1/256$. By setting L large enough, we can (with high probability) identify the correct value.

Differential attacks in practice. Differential cryptanalysis is very powerful, and has been used to attack real ciphers. A prominent example is FEAL-8, which was proposed as an alternative to DES in 1987. A differential attack on FEAL-8 was found that requires just 1,000 chosen plaintexts. In 1991, it took less than 2 minutes using this attack to find the entire key. Today, any proposed cipher is tested for resistance to differential cryptanalysis.

A differential attack was also the first attack on DES to require less time than a simple brute-force search. While an interesting theoretical result, the attack is not very effective in practice since it requires 2^{47} chosen plaintexts, and it would be difficult for an attacker to obtain this many chosen plaintext/ciphertext pairs in most real-world applications. Interestingly, small modifications to the S-boxes of DES make the cipher much more vulnerable to differential attacks. Personal testimony of the DES designers (after differential attacks were discovered in the outside world) confirmed that the S-boxes of DES were designed specifically to thwart differential attacks.

Linear cryptanalysis. Linear cryptanalysis was developed by Matsui in the early 1990s. We will only describe the idea underlying the technique. The basic idea is to consider linear relationships between the input, output, and key that hold with high probability. In more detail, assume an n-bit key length and ℓ-bit block length, and let $I, O \subseteq \{1, \ldots, \ell\}$ and $K \subseteq \{1, \ldots, n\}$. For an ℓ-bit x, let x_I denote the XOR of the bits at the positions indicated by I; define k_K similarly for $k \in \{0,1\}^n$. We say that I, O, K have *linear bias* ε if, for uniform x and k, and $y \stackrel{\text{def}}{=} F_k(x)$, it holds that

$$\left| \Pr[x_I \oplus y_O \oplus k_K = 0] - \frac{1}{2} \right| = \varepsilon.$$

If such a bias can be identified, it will clearly be useful for determining bits of the key given a number of plaintext/ciphertext pairs. Besides giving another method for attacking ciphers, an important feature of this attack compared to differential cryptanalysis is that it uses *known* plaintexts rather than *chosen* plaintexts. This is very significant, since an encrypted file can provide a huge amount of known plaintext, whereas obtaining encryptions of chosen plaintexts is much more difficult. Matsui showed that DES can be broken using linear cryptanalysis with just 2^{43} plaintext/ciphertext pairs.

Impact on block-cipher design. Modern block ciphers are designed and evaluated based, in part, on their resistance to differential and linear cryptanalysis. When constructing a block cipher, designers choose S-boxes and other components so as to minimize differential probabilities and linear biases. It is not possible to eliminate *all* high-probability differentials in an S-box: any S-box will have *some* differential that occurs more frequently than others. Still, these deviations can be minimized. Moreover, increasing the number of rounds (and choosing the mixing permutation carefully) can both reduce the differential probabilities as well as make it more difficult for cryptanalysts to find any differentials to exploit.

7.3 Compression Functions and Hash Functions

Recall from Chapter 6 that the primary security requirement for a cryptographic hash function H is *collision resistance*: that is, it should be difficult to find a collision in H, i.e., distinct inputs x, x' such that $H(x) = H(x')$. (We drop mention of any key here, since real-world hash functions are generally unkeyed.) If the hash function has ℓ-bit output length, then the best we can hope for is that it should be infeasible to find a collision using substantially fewer than $2^{\ell/2}$ invocations of H. (See Section 6.4.1.)

We describe two approaches for constructing collision-resistant hash functions. In Section 7.3.1, we show how to build a *compression function* (i.e., a fixed-length hash function) from any block cipher. As we have seen in Section 6.2, any such compression function can be extended to a full-fledged hash function using the Merkle–Damgård transform. This approach has been used to design popular hash functions including MD5, SHA-1, and SHA-2.

In Section 7.3.3 we discuss a more recent approach for constructing hash functions based on the so-called *sponge construction*. This technique is used by the SHA-3 standard.

7.3.1 Compression Functions from Block Ciphers

Perhaps surprisingly, it is possible to build a collision-resistant compression function from a block cipher satisfying strong security properties. There are several ways to do this. One of the most common is via the *Davies–Meyer construction.* Let F be a block cipher with n-bit key length and ℓ-bit block length. The Davies–Meyer construction then defines the compression function $h : \{0,1\}^{n+\ell} \to \{0,1\}^\ell$ by $h(k,x) \stackrel{\text{def}}{=} F_k(x) \oplus x$. (See Figure 7.11.)

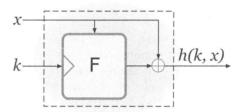

FIGURE 7.11: The Davies–Meyer construction.

We do not know how to prove collision resistance of h based only on the assumption that F is a strong pseudorandom permutation, and in fact there are reasons to believe such a proof is not possible. We *can*, however, prove collision resistance if we are willing to model F as an *ideal cipher*. The ideal-cipher model is a strengthening of the random-oracle model (see Section 6.5), in which we posit that all parties have access to an oracle for a random keyed permutation $F : \{0,1\}^n \times \{0,1\}^\ell \to \{0,1\}^\ell$ as well as its inverse F^{-1} (i.e., $F^{-1}(k, F(k,x)) = x$ for all k, x). Another way to think of this is that each key $k \in \{0,1\}^n$ specifies an independent, uniform permutation $F(k,\cdot)$ on ℓ-bit strings. As in the random-oracle model, the *only* way to compute F (or F^{-1}) is to explicitly query the oracle with (k,x) and receive back $F(k,x)$ (or $F^{-1}(k,x)$). The ideal-cipher model is stronger than the random-permutation model that we encountered briefly in Section 7.1.5.

Analyzing constructions in the ideal-cipher model comes with all the advantages and disadvantages of working in the random-oracle model, as discussed at length in Section 6.5. We only add here that the ideal-cipher model implies the absence of *related-key attacks* on F, in the sense that the permutations $F(k,\cdot)$ and $F(k',\cdot)$ must behave independently even if, for example, k and k' differ in only a single bit. In addition, there can be no "weak keys" k (say, the all-0 key) for which $F(k,\cdot)$ is easily distinguishable from random. It also means that $F(k,\cdot)$ should "behave randomly" *even when k is known.* These requirements are not part of the definition of a (strong) pseudorandom permutation. Moreover, these properties do *not* necessarily hold for real-world block ciphers, and the reader may note that we have not discussed these properties in any of our analysis of block-cipher constructions. (In fact, DES and triple-DES do not satisfy these properties.) Any block cipher being considered for instantiating an ideal cipher must be evaluated with respect to these more stringent requirements.

The following shows that, when F is modeled as an ideal cipher, the Davies–

Meyer construction is collision resistant as long as ℓ is sufficiently large.

THEOREM 7.5 *If F is modeled as an ideal cipher, then any attacker making q queries to F or F^{-1} can find a collision in the Davies–Meyer construction with probability at most $q^2/2^\ell$.*

PROOF To be clear, we consider here the probabilistic experiment in which a uniform F is sampled (more precisely, for each $k \in \{0,1\}^n$ the function $F(k, \cdot) : \{0,1\}^\ell \to \{0,1\}^\ell$ is chosen uniformly from the set Perm_ℓ of permutations on ℓ-bit strings) and then the attacker is given oracle access to F and F^{-1}. The attacker then tries to find a colliding pair (k, x), (k', x'), i.e., for which $F(k, x) \oplus x = F(k', x') \oplus x'$. No computational bounds are placed on the attacker other than bounding the number of oracle queries it makes. We assume the attacker never makes the same query more than once, and never queries $F^{-1}(k, y)$ once it has learned that $y = F(k, x)$ (or vice versa). We assume that if the attacker outputs a candidate collision (k, x), (k', x') then it has previously made the oracle queries necessary to compute the values $h(k, x)$ and $h(k', x')$. All these assumptions are without much loss of generality.

Consider the ith query the attacker makes to one of its oracles. A query (k_i, x_i) to F reveals only the hash value $h_i \overset{\text{def}}{=} h(k_i, x_i) = F(k_i, x_i) \oplus x_i$; similarly, a query (k_i, y_i) to F^{-1} giving the result $x_i = F^{-1}(k_i, y_i)$ yields only the hash value $h_i \overset{\text{def}}{=} h(k_i, x_i) = y_i \oplus F^{-1}(k_i, y_i)$. The key observation is that no matter which kind of query the attacker makes, the hash value h_i it learns is almost uniformly distributed (since the result of the oracle query to F or F^{-1} is almost uniformly distributed—with the only deviation from uniform being that $F(k, x)$ cannot be equal to $F(k, x')$ for any $x \neq x'$). This makes finding a collision hard since the attacker does not obtain a collision unless $h_i = h_j$ for some $i \neq j$.

In detail: Fix i, j with $i > j$ and consider the probability that $h_i = h_j$. At the time of the ith query, the value of h_j is fixed. A collision between h_i and h_j is obtained on the ith query only if the attacker queries (k_i, x_i) to F and obtains the result $F(k_i, x_i) = h_j \oplus x_i$, or queries (k_i, y_i) to F^{-1} and obtains the result $F^{-1}(k_i, y_i) = h_j \oplus y_i$. Either event occurs with probability at most $1/(2^\ell - (i-1))$ since, for example, $F(k_i, x_i)$ is uniform over $\{0,1\}^\ell$ except that it cannot be equal to any value $F(k_i, x)$ already defined by the attacker's (at most) $i-1$ previous oracle queries using key k_i. Assuming $i \leq q < 2^{\ell/2}$ (if not, the theorem is trivially true), the probability that $h_i = h_j$ is at most $2/2^\ell$.

Taking a union bound over all $\binom{q}{2} < q^2/2$ distinct pairs i, j gives the result stated in the theorem. ∎

Davies–Meyer and DES. As we have mentioned above, one must take care when instantiating the Davies–Meyer construction with any concrete block cipher, since the cipher must satisfy additional properties (beyond being a strong pseudorandom permutation) in order for the resulting construction to be secure. In Exercise 7.24 we explore what goes wrong when DES is used in the Davies–Meyer construction.

This should serve as a warning that the proof of security for the Davies–Meyer construction in the ideal-cipher model does not necessarily translate into real-world security when instantiated with a specific cipher. Nevertheless, as we will describe below, this paradigm has been used to construct practical hash functions that have resisted attack (although in those cases the block cipher used was designed specifically for this purpose).

In conclusion, the Davies–Meyer construction is a useful paradigm for constructing collision-resistant compression functions. However, it should *not* be applied to block ciphers not designed to behave like an ideal cipher.

7.3.2 MD5, SHA-1, and SHA-2

Several prominent and widely used hash functions have been constructed by applying the Davies–Meyer construction to some underlying block cipher to obtain a compression function, and then applying the Merkle–Damgård transform. Examples include the hash functions MD5, SHA-1, and SHA-2, which we discuss next.

MD5. MD5 is a hash function with a 128-bit output length. It was designed in 1991 and for some time was believed to be collision resistant. Over a period of several years, various weaknesses began to be found in MD5 but these did not appear to lead to any easy way to find collisions. Shockingly, in 2004 a team of Chinese cryptanalysts presented a new method for finding collisions in MD5 and demonstrated an explicit collision. Since then, the attack has been improved and today collisions in MD5 can be found in under a minute on a desktop PC. In addition, the attacks have been extended so that even "controlled collisions" (e.g., two pdf files) can be found. Due to these attacks, MD5 should not be used anywhere cryptographic security is needed. We mention MD5 only because it is still found in legacy code.

SHA-1. The *Secure Hash Algorithms* (SHA) refer to a set of cryptographic hash functions standardized by NIST. The hash function SHA-1, standardized in 1995, has a 160-bit output length and was considered secure for many years. Beginning in 2005, theoretical analysis indicated that collisions in SHA-1 could be found using roughly 2^{69} hash-function evaluations, which is much lower than the 2^{80} hash-function evaluations that would be needed for a birthday attack. This prompted researchers to recommend migrating away from SHA-1; nevertheless, since even 2^{69} operations is still significant, an explicit collision in SHA-1 remained out of reach. It was not until 2017 that an improvement in the collision-finding attack, along with tremendous computational resources devoted by Google, enabled researchers to find an explicit collision. The attack required the equivalent of 2^{63} hash-function evaluations, and took 6,500 CPU years (along with 100 GPU years) to execute on a distributed cluster of machines. As of the time of this writing, more-devastating attacks have been found, and SHA-1 is no longer recommended for use.

SHA-2. The SHA-2 hash family, introduced in 2001, consists of the two related hash functions SHA-256 and SHA-512 with 256- and 512-bit output lengths, respectively. (The outputs can be truncated if smaller hash values are desired.) These hash functions do not currently appear to have the same weaknesses that led to attacks on SHA-1; moreover, because of their long output lengths, it will remain difficult to find collisions even if small weaknesses are discovered. SHA-2, or the more recent standard SHA-3 (see below), are currently recommended when collision-resistant hashing is needed.

7.3.3 The Sponge Construction and SHA-3 (Keccak)

In the aftermath of the collision attack on MD5 and the theoretical weaknesses found in SHA-1, NIST announced in 2007 a public competition to design a new cryptographic hash function. As in the case of the AES competition from roughly a decade earlier, the competition was completely open and transparent; anyone could submit an algorithm for consideration, and the public was invited to give their opinions on any of the candidates. The 51 first-round candidates were narrowed down to 14 in December 2008, and these were further reduced to five finalists in 2010. These remaining candidates were subject to intense scrutiny by the cryptographic community over the next two years. In October 2012, NIST announced the selection of *Keccak* as the winner of the competition. The resulting standard SHA-3, released in 2015, supports 224-, 256-, 384-, and 512-bit output lengths.

The structure of Keccak is very different from the structure of SHA-1 and SHA-2, and in particular it does not use the Merkle–Damgård transform. (Interestingly, this may have been one of the reasons it was chosen.) The core primitive of Keccak is an unkeyed permutation P with a large block length of 1600 bits. P is used to build a hash function directly (i.e., without first building a compression function in an intermediate step) via what is known as the *sponge construction*. The resulting hash function can be proven to be collision resistant if P is modeled as a *random permutation*. (We have already seen the random-permutation model in Section 7.1.5.) By analogy with the random-oracle and ideal-cipher models, the random-permutation model assumes that all parties are given access to oracles for a uniform permutation P as well as its inverse P^{-1}; the *only* way to compute P or P^{-1} is to explicitly query those oracles. Note that the random-permutation model is weaker than the ideal-cipher model; indeed, we can easily obtain a random permutation P from an ideal cipher F by defining $P(x) \stackrel{\text{def}}{=} F(0^n, x)$, i.e., by simply fixing the key for F to any constant value.

We now describe the construction. Fix a permutation $P : \{0,1\}^\ell \to \{0,1\}^\ell$, and let $r, c, v \geq 1$ be such that $r + c = \ell$ and $v \leq \ell$. The sponge construction accepts as input a sequence of r-bit blocks $m_1, \ldots, m_t$. (See Figure 7.12.)

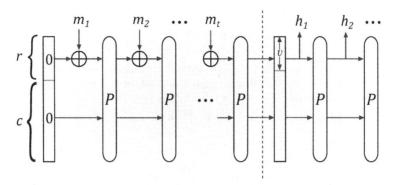

FIGURE 7.12: The sponge construction. The *absorbing phase* is to the left of the dashed line, and the *squeezing phase* is to the right.

During its computation, the construction maintains an ℓ-bit state, initialized to zero. This state is modified in an input-dependent way during an "absorbing phase," and the final state is then used to generate output in a "squeezing phase." (Hence the name "sponge.") When processing the ith block during the absorbing phase, the state is updated from y_{i-1} to y_i by XORing m_i with the first r bits of y_{i-1} to obtain an intermediate value x_i, and then setting $y_i := P(x_i)$. The final state y_t is then used to generate output in the squeezing phase by repeatedly outputting the initial v bits of the state followed by an application of P.

The sponge construction can be used for many purposes. In Construction 7.6 we provide a formal description of how to use it to build a hash function. That construction includes a parameter $\lambda \geq 1$ that affects how many times the squeezing step is run, and thus determines the output length of the hash. (Namely, the output is a string of length $\lambda \cdot v$.) The construction also incorporates an initial padding step so that the resulting hash function can accept inputs of arbitrary length.

Hash functions following the sponge construction can be shown to satisfy several security properties when P is modeled as a random permutation. Here we prove collision resistance as long as r and c are sufficiently large, assuming for simplicity that $\lambda = 1$ (which is the case for the SHA-3 standard).

THEOREM 7.7 *Let H denote Construction 7.6 with $\lambda = 1$. If P is modeled as a random permutation, then any attacker making q queries to P or P^{-1} can find a collision in H with probability at most $\frac{q^2}{2^v} + \frac{q \cdot (q+1)}{2^c}$.*

PROOF Consider an attacker that is given oracle access to a random permutation P and its inverse P^{-1}, and then outputs a pair of distinct messages; let $m_1, \ldots, m_t$ and $m'_1, \ldots, m'_{t'}$ denote the results after padding. (Note that, because of the way padding is done, these padded messages are also distinct.)

CONSTRUCTION 7.6

Fix $P : \{0,1\}^\ell \to \{0,1\}^\ell$ and constants r, c, v as in the text and $\lambda \geq 1$. Hash function H, on input $\hat{m} \in \{0,1\}^*$, does:

(Padding) Append a 1 to $\hat{m}$, followed by enough zeros so that the length of the resulting string is a multiple of r. Parse the resulting string as the sequence of r-bit blocks $m_1, \ldots, m_t$.

(Absorbing phase) Set $y_0 := 0^\ell$. Then for $i = 1, \ldots, t$ do:

- $x_i := y_{i-1} \oplus (m_i \| 0^c)$.
- $y_i := P(x_i)$.

(Squeezing phase) Set $y_1^* := y_t$, and let h_1 be the first v bits of y_1^*. Then for $i = 2, \ldots, \lambda$ do

- $y_i^* := P(y_{i-1}^*)$.
- Let h_i be the first v bits of z_i^*.

(Output) Output $h_1 \| \cdots \| h_\lambda$.

A hash function based on the sponge construction.

We assume the attacker never makes the same query to P or P^{-1} more than once, and never queries $P^{-1}(y)$ once it has learned that $y = P(x)$ (and vice versa). We further assume that by the end of its execution the attacker has made the oracle queries necessary to evaluate H on the messages it outputs.

Define the following three events:

E1: The attacker makes two distinct queries to P whose results agree on their first v bits.

E2: The attacker makes a query to P or P^{-1} whose result has its last c bits equal to 0^c.

E3: The attacker makes two distinct queries (to either P or P^{-1}) whose results agree on their last c bits.

We show that if the attacker outputs a collision then one of the above events occurs; we complete the proof by bounding the probabilities of these events.

CLAIM 7.8 *If the attacker outputs a collision then* **E1, E2,** *or* **E3** *occurs.*

PROOF Consider the execution of Construction 7.6 on the padded message $m_1, \ldots, m_t$. Let $y_0, x_1, y_1, \ldots, x_t, y_t$ be the values of the variables during the course of the execution, so that $y_0 = 0^\ell$ and, for $i \geq 1$, the last c bits of y_{i-1} and x_i are equal and $y_i = P(x_i)$. Define $y_0', x_1', y_1', \ldots, x_{t'}', y_{t'}'$ analogously with respect to the padded message $m_1', \ldots, m_{t'}'$. If, for some i, the attacker queried $P^{-1}(y_i)$ to obtain x_i or queried $P^{-1}(y_i')$ to obtain x_i' then we say *an inverse query occurred*. We consider two cases:

Case 1: An inverse query occurred. Assume without loss of generality an inverse query occurred for the first padded message. Let i be minimal such that the attacker queried $P^{-1}(y_i)$ to obtain x_i. If $i = 1$ then the last c bits of x_1 are 0^c and **E2** occurred. Otherwise, the last c bits of $y_{i-1} = P(x_{i-1})$ and $x_i = P^{-1}(y_i)$ are equal and so **E3** occurred.

Case 2: No inverse query occurred. If $y_t \neq y'_{t'}$, then the first v bits of y_t and $y'_{t'}$ are equal (since the attacker output a collision) even though $x_t \neq x'_{t'}$. Since no inverse query occurred, the attacker must have queried $P(x_t)$ and $P(x'_{t'})$ and so **E1** occurred.

If $y_t = y'_{t'}$, assume without loss of generality that $t' \geq t$. Let $C(z)$ denote the last c bits of an ℓ-bit string z. Take $0 \leq i \leq t$ maximal such that $(C(y_{t-i}), \ldots, C(y_t)) = (C(y'_{t'-i}), \ldots, C(y'_{t'}))$. If $i < t$ then $C(y_{t-i-1}) \neq C(y'_{t'-i-1})$ and hence $x_{t-i} \neq x'_{t'-i}$, but

$$C(P(x_{t-i})) = C(y_{t-i}) = C(y'_{t'-i}) = C(P(x'_{t'-i}))$$

and so **E3** occurred. If $i = t$ and $t' > t$ then $C(P(x'_{t'-i})) = C(y'_{t'-i}) = C(y_0) = 0^c$; thus, **E2** occurred. If $i = t$ and $t' = t$ then we have $(C(y_0), \ldots, C(y_t)) = (C(y'_0), \ldots, C(y'_t))$. Let j be minimal such that $m_j \neq m'_j$ (such a j must exist since the padded messages are distinct). Then $y_{j-1} = y'_{j-1}$ but $x_j \neq x'_j$, and yet

$$C(P(x_j)) = C(y_j) = C(y'_j) = C(P(x'_j))$$

and so **E3** occurred. ∎

CLAIM 7.9 $\Pr[\mathbf{E1} \vee \mathbf{E2} \vee \mathbf{E3}] \leq \frac{q^2}{2^v} + \frac{q \cdot (q+1)}{2^c}$.

PROOF We bound the probability of each event; a union bound yields the claim. It is easy to see that $\Pr[\mathbf{E2}] \leq q/2^c$. To bound $\Pr[\mathbf{E1}]$ we use an analysis similar to the one used to prove the birthday bound (cf. Appendix A.4). Let $\mathsf{Coll}_{i,j}$ be the event that the results of the ith and jth queries of the attacker agree on their first v bits. We have $\Pr[\mathsf{Coll}_{i,j}] \leq 2^{\ell-v}/(2^\ell - 1) \leq 2 \cdot 2^{-v}$. (Taking into account that P is a random *permutation*.) So

$$\Pr[\mathbf{E1}] = \Pr\left[\bigvee_{i<j} \mathsf{Coll}_{i,j}\right] \leq \sum_{i<j} \Pr[\mathsf{Coll}_{i,j}] \leq \binom{q}{2} \cdot 2 \cdot 2^{-v} \leq q^2/2^v.$$

A similar argument gives $\Pr[\mathbf{E3}] \leq q^2/2^c$. ∎

This concludes the proof of the theorem. ∎

References and Additional Reading

Lidl and Niederreiter [130] give the standard treatment of LFSRs. Additional information on LFSRs in the context of cryptography can be found in the *Handbook of Applied Cryptography* [137] or the text by Paar and Pelzl [156]. Further details about eSTREAM, as well as a document describing the design of Trivium, can be found at https://www.ecrypt.eu.org/stream.

The work of Alfardan et al. [9] surveys recent attacks on RC4. ChaCha20 is due to Bernstein [29], and is described in RFC 8439 [154]. It can be analyzed (in the random-permutation model) as an Even-Mansour cipher [70].

The confusion-diffusion paradigm and substitution-permutation networks were introduced by Shannon [177] and Feistel [71]. See the thesis of Heys [98] for further information regarding SPN design. A theoretical analysis of block ciphers based on SPNs has recently been given by Cogliati et al. [52]. We remark that SPNs are useful not only for building ciphers, but also for increasing the block length of an existing cipher.

Feistel networks were first described in 1973 [71]. A theoretical analysis of Feistel networks was given by Luby and Rackoff [132]; see Chapter 8.

More details on DES, AES, and block-cipher constructions in general can be found in the text by Knudsen and Robshaw [117]. The meet-in-the-middle attack on double encryption is due to Diffie and Hellman [66]. The attack on two-key triple encryption mentioned in the text (and explored in Exercise 7.16) is by Merkle and Hellman [141] and has been developed further [197, 144]. Positive results about double/triple encryption are also known [6, 27].

Work of Bhargavan and Leurent [33] demonstrates real-world security implications of using ciphers (like DES or 3DES) with small block length. See https://sweet32.info for further information.

Differential cryptanalysis was introduced by Biham and Shamir [34], and its application to DES is described in a book by those authors [35]. Coppersmith [53] describes design principles of the DES S-boxes in light of the public discovery of differential cryptanalysis. Linear cryptanalysis was discovered by Matsui [134]. For more information on these advanced cryptanalytic techniques, we refer the reader to the tutorial on differential and linear cryptanalysis by Heys [99] or the book by Knudsen and Robshaw [117].

Menezes et al. [137] give further information about MD5 and SHA-1; note, though, that their treatment pre-dates the recent attacks on those hash functions. Various other constructions of compression functions from block ciphers are known [164, 37]. The sponge construction is described and analyzed by Bertoni et al. [32]. For additional details about the SHA-3 competition, see https://csrc.nist.gov/projects/hash-functions/sha-3-project.

The first explicit collision in SHA-1 was found in 2017 by Stevens et al. [192]; improved attacks, which have serious practical security implications, have been shown even more recently [127, 128].

Exercises

7.1 Consider a degree-6 LFSR where only c_5 and c_0 are set to 1.

    (a) What are the first 10 bits output by this LFSR if it starts in initial state $(s_5, s_4, s_3, s_2, s_1, s_0) = (1, 1, 1, 1, 1, 1)$?

    (b) Is this LFSR maximum length?

7.2 Consider a degree-7 LFSR where only c_6, c_1, and c_0 are set to 1.

    (a) What are the first 10 bits output by this LFSR if it starts in the initial state $(s_6, s_5, s_4, s_3, s_2, s_1, s_0) = (0, 0, 0, 0, 0, 0, 1)$?

    (b) Show that this LFSR is not maximum length.

        Hint: Find a nonzero state with a self-loop in the transition graph.

7.3 Consider a stream cipher constructed from a degree-n LFSR where the output at each clock tick is not s_0, but instead $g(s_{n-1}, \ldots, s_0)$ for some nonlinear function g. The n-bit key of the stream cipher is used as the initial state of the LFSR. Show that this does not result in a secure stream cipher for the following choices of g:

    (a) $g(s_{n-1}, \ldots, s_0) = s_0 \wedge s_1$.

    (b) $g(s_{n-1}, \ldots, s_0) = s_2 \oplus (s_1 \wedge s_0)$.

7.4 Consider a stream cipher constructed from two LFSRs A and B of degrees n_a and n_b, respectively, where the output at each clock tick is computed by taking the AND of the outputs of the two LFSRs. The key $k \in \{0, 1\}^{n_a + n_b}$ is used to set the initial states of the two LFSRs.

    (a) Show that this is never a secure stream cipher.

    (b) Show that given a long enough output from this stream cipher, it is possible to recover the key in time $\approx 2^{n_a} + 2^{n_b}$.

7.5 Fix a public, invertible permutation P, and define the keyed function $F_k(x) \stackrel{\text{def}}{=} P(\text{const} \| k \| x)$. Show that F is not a pseudorandom function.

7.6 Let F be a block cipher with n-bit key length and block length. Say there is a key-recovery attack on F that succeeds with probability 1 using n chosen plaintexts and minimal computational effort. Prove formally that F cannot be a pseudorandom permutation.

7.7 In our attack on a one-round SPN, we considered a block length of 64 bits and 8 S-boxes that each take a 8-bit input. Repeat the analysis for the case of 16 S-boxes, each taking an 4-bit input. What is the complexity of the attack now? Repeat the analysis again with a 128-bit block length and 16 S-boxes that each take an 8-bit input.

7.8 Consider a modified SPN that first applies r rounds of key mixing (using independent sub-keys), then carries out r rounds of substitution (using different S-boxes in each round), and finally applies r (different) mixing permutations. Show an attack on this construction.

7.9 In this question we assume a two-round SPN with 64-bit block length.

   (a) Assume independent 64-bit sub-keys are used in each round, so the master key is 192 bits long. Show a key-recovery attack using much less than 2^{192} time.

   (b) Assume the first and third sub-keys are equal, and the second sub-key is independent, so the master key is 128 bits long. Show a key-recovery attack using much less than 2^{128} time.

7.10 What is the output of an r-round Feistel network when the input is (L_0, R_0) in each of the following two cases:

   (a) Each round function outputs all 0s, regardless of the input.

   (b) Each round function is the identity function.

7.11 Let $\mathsf{Feistel}_{f_1, f_2}(\cdot)$ denote a two-round Feistel network using functions f_1 and f_2 (in that order). Define $\mathsf{swap}(L, R) = (R, L)$.

   (a) Show that if

$$(L_2, R_2) = \mathsf{swap}(\mathsf{Feistel}_{f_1, f_2}(L_0, R_0))$$

   then $(L_0, R_0) = \mathsf{swap}(\mathsf{Feistel}_{f_2, f_1}(L_2, R_2))$.

   (b) Show that if

$$(L_{16}, R_{16}) = \mathsf{swap}\left(\mathsf{Feistel}_{f_{15}, f_{16}}(\cdots(\mathsf{Feistel}_{f_1, f_2}(L_0, R_0))\cdots)\right)$$

   then

$$(L_0, R_0) = \mathsf{swap}\left(\mathsf{Feistel}_{f_2, f_1}(\cdots\mathsf{Feistel}_{f_{16}, f_{15}}(L_{16}, R_{16})\cdots)\right).$$

7.12 For this exercise, rely on the description of DES given in this chapter. However, use the fact that in the actual construction of DES the two halves of the output of the final round of the Feistel network are swapped. That is, if the output of the final round of the Feistel network is (L_{16}, R_{16}), then the output of DES is (R_{16}, L_{16}).

   (a) Show that the only difference between computation of DES_k and DES_k^{-1} is the order in which sub-keys are used. (Rely on the previous exercise.)

   (b) Show that when $k = 0^{56}$ then $DES_k(DES_k(x)) = x$ for all x.
       Hint: Consider the sub-keys generated from this key.

(c) Find three other DES keys with the same property. These keys are known as *weak keys* for DES. (Note: the keys you find will differ from the actual weak keys of DES because of differences in our description of the DES key schedule.)

(d) Do these 4 weak keys represent a serious vulnerability in the use of triple-DES as a pseudorandom permutation? Explain.

7.13 (This exercise relies on Exercise 7.12.) Our goal is to show that for any weak key k of DES, it is easy to find an input x such that $DES_k(x) = x$.

(a) Assume we evaluate DES_k on input (L_0, R_0), and the intermediate result after 8 rounds of the Feistel network is (L_8, R_8) with $L_8 = R_8$. Show that $(L_0, R_0) = DES_k(L_0, R_0)$. (Recall from Exercise 7.12 that DES swaps the two halves of the output of the 16th round of the Feistel network.)

(b) Show how to find an input (L_0, R_0) with the property in part (a).

7.14 Show that DES has the property that $DES_k(x) = \overline{DES_{\bar{k}}(\bar{x})}$ for every key k and input x (where $\bar{z}$ denotes the bitwise complement of z). (This is called the *complementarity property* of DES.) Does this represent a serious vulnerability in the use of triple-DES as a pseudorandom permutation? Explain.

7.15 Describe attacks on the following modifications of DES:

(a) Each sub-key is 32 bits long, and the round function simply XORs the sub-key with the input to the round (i.e., $\hat{f}(k, R) = k_i \oplus R$). For this question, the key schedule is unimportant and you can treat the sub-keys k_i as independent keys.

(b) Instead of using different sub-keys in every round, the same 48-bit sub-key is used in every round. Show how to distinguish the cipher from a random permutation using two chosen plaintexts and negligible work.

Hint: Exercises 7.11 and 7.12 may help...

7.16 This question develops an attack on two-key triple encryption. Let F be a block cipher with ℓ-bit block length and n-bit key length (where $\ell \geq n$), and set $F'_{k_1, k_2}(x) \overset{\text{def}}{=} F_{k_1}(F_{k_2}(F_{k_1}(x)))$. Assume an attacker is given $N \ll 2^\ell$ input/output pairs $\{(x_i, y_i)\}_{i=1}^N$ where the $\{x_i\}$ are uniform and $y_i = F'_{k_1, k_2}(x_i)$ for unknown keys k_1, k_2.

(a) Assume the attacker knows $z \in \{0, 1\}^\ell$ such that $F_{k_1}(x_i) = z$ for some i. (The attacker does not know i.) Show how the attacker can find k_1, k_2 using $2^{n+1} + \mathcal{O}(N) \approx 2^{n+1}$ evaluations of F/F^{-1}.

Hint: Start by computing $\{F_k(z)\}$ for all possible keys.

(b) In general, the attacker does not know z as required for part (a). Show how the attacker can nevertheless learn k_1, k_2 using roughly $2^{n+\ell+1}/N$ evaluations of F/F^{-1}.

   Hint: What happens if the attacker chooses a random z?

7.17 Say the key schedule of DES is modified as follows: the left half of the master key is used to derive all the sub-keys in rounds 1–8, while the right half of the master key is used to derive all the sub-keys in rounds 9–16. Show an attack on this modified scheme that recovers the entire key in time roughly 2^{28}.

7.18 Fix arbitrary $G_1, G_2 : \{0,1\}^n \to \{0,1\}^{4n}$, and define

$$G(s_1\|s_2) = G_1(s_1) \oplus G_2(s_2).$$

Show how to distinguish the output of G from random in time $\approx 2^{n+1}$.

   Hint: Adapt the meet-in-the-middle attack.

7.19 Let $f : \{0,1\}^m \times \{0,1\}^\ell \to \{0,1\}^\ell$ and $g : \{0,1\}^n \times \{0,1\}^\ell \to \{0,1\}^\ell$ be secure block ciphers with $m > n$, and define $F_{k_1,k_2}(x) = f_{k_1}(g_{k_2}(x))$. Show a key-recovery attack on F using time $\mathcal{O}(2^m)$ and space $\mathcal{O}(\ell \cdot 2^n)$.

7.20 Define $DESY_{k,k'}(x) = DES_k(x \oplus k')$. The key length of $DESY$ is 120 bits. Show a key-recovery attack on $DESY$ taking $\approx 2^{56}$ time and $\mathcal{O}(1)$ memory.

7.21 Choose random S-boxes and mixing permutations for SPNs of different sizes, and develop differential attacks against them. We recommend trying five-round SPNs with 16-bit and 24-bit block lengths, using S-boxes with 4-bit input/output. Write code to compute the differential tables, and to carry out the attack.

7.22 Implement the time/space tradeoff from Section 6.4.3 for a key-recovery attack on 40-bit DES (e.g., fix the first 16 bits of the key to 0). Calculate the time and memory needed, and empirically estimate the probability of success. Experimentally verify the increase in success probability as the number of tables is increased. (Warning: this is a big project!)

7.23 For each of the following constructions of a compression function h from a block cipher $F : \{0,1\}^n \times \{0,1\}^n \to \{0,1\}^n$, either show an attack or prove collision resistance in the ideal-cipher model:

(a) $h(k, x) = F_k(x)$.

(b) $h(k, x) = F_k(x) \oplus k \oplus x$.

(c) $h(k, x) = F_k(x) \oplus k$.

7.24 Let F be a block cipher for which it is easy to find fixed points for some key: namely, there is a key k for which it is easy to find inputs x for which $F_k(x) = x$. Find a collision in the Davies–Meyer construction when applied to F. (Consider this in light of Exercise 7.13.)

7.25 Consider using DES to construct a compression function in the following way: Define $h : \{0,1\}^{112} \to \{0,1\}^{64}$ as $h(x_1, x_2) \overset{\text{def}}{=} DES_{x_1}(DES_{x_2}(0^{64}))$ where $|x_1| = |x_2| = 56$.

    (a) Write down an explicit collision in h.

        Hint: Use Exercise 7.12(a–b).

    (b) Show how to find a preimage of an arbitrary value y (that is, x_1, x_2 such that $h(x_1 \| x_2) = y$) in roughly 2^{56} time.

    (c) Show a more clever preimage attack that runs in roughly 2^{32} time and succeeds with high probability.

        Hint: Rely on the results of Appendix A.4.

7.26 Say $S_1, \ldots, S_8 : \{0,1\}^n \to \{0,1\}^n$ are modeled as random permutations, and say $P : \{0,1\}^{8n} \to \{0,1\}^{8n}$ is constructed by defining

$$P(x_1 \| \cdots \| x_8) = S_1(x_1) \| \cdots \| S_8(x_n).$$

Show that it is easy to find a collision in Construction 7.6 for $\lambda = 1$ and $r = c = v = 4n$ when using this P.

Chapter 8

*Theoretical Constructions of Symmetric-Key Primitives

In Chapter 3 we introduced the notion of pseudorandomness and defined some basic cryptographic primitives including pseudorandom generators, functions, and permutations. We further showed in Chapters 3–5 that these primitives can serve as building blocks for all of private-key cryptography. As such, it is of great importance to understand these primitives from a theoretical point of view. In this chapter we formally introduce the concept of *one-way functions*—functions that are, informally, easy to compute but hard to invert—and show how pseudorandom generators, functions, and permutations can be constructed under the sole assumption that one-way functions exist.[1] Moreover, we will see that one-way functions are necessary for "non-trivial" private-key cryptography. That is: *the existence of one-way functions is equivalent to the existence of all (non-trivial) private-key cryptography.* This is one of the major contributions of modern cryptography.

The constructions we show in this chapter should be viewed as complementary to the constructions of stream ciphers and block ciphers discussed in the previous chapter. The focus of the previous chapter was on how various cryptographic primitives are currently realized in practice, and the intent of that chapter was to introduce some basic approaches and design principles that are used. Somewhat disappointing, though, was the fact that none of the constructions we showed could be *proven* secure based on any weaker (i.e., more reasonable) assumptions. In contrast, in this chapter we will show constructions that can be proven secure starting from the very mild assumption that one-way functions exist. That assumption is more appealing than assuming, say, that AES is a pseudorandom permutation, both because it is a qualitatively weaker assumption and also because we have a number of candidate, number-theoretic one-way functions that have been studied for many years, even before the advent of cryptography. (See the very beginning of Chapter 7 for further discussion of this point.) The downside, however, is that the constructions we show here are all far less efficient than those of Chapter 7, and thus (currently) have little practical significance. It remains an important challenge for cryptographers to "bridge this gap" and develop provably secure

---

[1] Although we will for the most part rely on the stronger assumption of one-way *permutations* in this chapter, it is known that one-way functions suffice.

constructions of pseudorandom generators and permutations whose efficiency is comparable to the best available stream ciphers and block ciphers.

Collision-resistant hash functions. Unlike the previous chapter, here we do not consider collision-resistant hash functions. The reason is that constructions of such hash functions from one-way functions are unknown and, in fact, there is evidence suggesting that such constructions are impossible. We will see a provably secure construction of a collision-resistant hash function—based on a specific, number-theoretic assumption—in Section 9.4.2.

A note regarding this chapter. The material in this chapter is somewhat more advanced than the material in the rest of this book. This material is not used explicitly anywhere else in the book, and so can be skipped if desired. Having said this, we have tried to present the material in such a way that it is understandable (with effort) to an advanced undergraduate or beginning graduate student. We encourage all readers to peruse Sections 8.1 and 8.2, which introduce one-way functions and provide an overview of the rest of this chapter. We believe that familiarity with at least some of the topics covered here is important enough to warrant the effort.

8.1 One-Way Functions

In this section we formally define one-way functions, and then briefly discuss some candidates that are believed to satisfy this definition. (We will see more examples of conjectured one-way functions in Chapter 9.) We next introduce the notion of *hard-core predicates*, which can be viewed as encapsulating the hardness of inverting a one-way function and will be used extensively in the constructions that follow in subsequent sections.

8.1.1 Definitions

A one-way function $f : \{0,1\}^* \to \{0,1\}^*$ is easy to compute, yet hard to invert. The first condition is easy to formalize: we will simply require that f be computable in polynomial time. Since we are ultimately interested in building cryptographic schemes that are hard for a probabilistic polynomial-time adversary to break except with negligible probability, we will formalize the second condition by requiring that it be infeasible for any probabilistic polynomial-time algorithm to invert f—that is, to find a preimage of a given value y—except with negligible probability. A technical point is that this probability is taken over an experiment in which y is generated by choosing a uniform element x in the domain of f and then setting $y := f(x)$ (rather than choosing y uniformly from the range of f). The reason for this should become clear from the constructions we will see in the remainder of the chapter.

Let $f : \{0,1\}^* \to \{0,1\}^*$ be a function. Consider the following experiment defined for any algorithm $\mathcal{A}$ and any value n for the security parameter:

The inverting experiment $\mathsf{Invert}_{\mathcal{A},f}(n)$

1. *Choose uniform $x \in \{0,1\}^n$, and compute $y := f(x)$.*

2. *$\mathcal{A}$ is given 1^n and y as input, and outputs x'.*

3. *The output of the experiment is defined to be 1 if $f(x') = y$, and 0 otherwise.*

We stress that $\mathcal{A}$ need not find the original preimage x; it suffices for $\mathcal{A}$ to find any value x' for which $f(x') = y = f(x)$. The security parameter 1^n is given to $\mathcal{A}$ in the second step to stress that $\mathcal{A}$ may run in time polynomial in the security parameter n, regardless of the length of y.

We can now define what it means for a function f to be one-way.

DEFINITION 8.1   *A function $f : \{0,1\}^* \to \{0,1\}^*$ is* one-way *if the following two conditions hold:*

1. **(Easy to compute:)** *There exists a polynomial-time algorithm M_f computing f; that is, $M_f(x) = f(x)$ for all x.*

2. **(Hard to invert:)** *For every probabilistic polynomial-time algorithm $\mathcal{A}$, there is a negligible function* negl *such that*

$$\Pr[\mathsf{Invert}_{\mathcal{A},f}(n) = 1] \le \mathsf{negl}(n).$$

Notation. In this chapter we will often make the probability space more explicit by subscripting (part of) it in the probability notation. For example, we can succinctly express the second requirement in the definition above as follows: For every probabilistic polynomial-time algorithm $\mathcal{A}$, there exists a negligible function negl such that

$$\Pr_{x \leftarrow \{0,1\}^n} \left[\mathcal{A}(1^n, f(x)) \in f^{-1}(f(x)) \right] \le \mathsf{negl}(n).$$

(Recall that $x \leftarrow \{0,1\}^n$ means that x is chosen uniformly from $\{0,1\}^n$.) The probability above is also taken over the randomness used by $\mathcal{A}$, which here is left implicit.

Successful inversion of one-way functions. A function that is *not* one-way is not necessarily easy to invert all the time (or even "often"). Rather, the converse of the second condition of Definition 8.1 is that there exists a probabilistic polynomial-time algorithm $\mathcal{A}$ and a non-negligible function γ such that $\mathcal{A}$ inverts $f(x)$ with probability at least $\gamma(n)$ (where the probability is taken over uniform choice of $x \in \{0,1\}^n$ and the randomness of $\mathcal{A}$). This means, in turn, that there exists a positive polynomial $p(\cdot)$ such that for *infinitely many values of n*, algorithm $\mathcal{A}$ inverts f with probability at least $1/p(n)$. Thus, if there exists an $\mathcal{A}$ that inverts f with probability n^{-10} for all even values of n (but always fails to invert f when n is odd), then f is not one-way—even though $\mathcal{A}$ only succeeds on half the values of n, and only succeeds with probability n^{-10} (for values of n where it succeeds at all).

Exponential-time inversion. Any one-way function can be inverted at any point y in exponential time, by simply trying all values $x \in \{0,1\}^n$ until a value x is found such that $f(x) = y$. Thus, the existence of one-way functions is inherently an assumption about *computational complexity* and *computational hardness*. That is, it concerns a problem that can be solved in principle but is assumed to be hard to solve efficiently.

One-way permutations. We will often be interested in one-way functions with additional structural properties. We say a function f is *length-preserving* if $|f(x)| = |x|$ for all x. A one-way function that is length-preserving and one-to-one is called a *one-way permutation*. If f is a one-way permutation, then any value y has a unique preimage $x = f^{-1}(y)$. Nevertheless, it is still hard to find x in polynomial time.

One-way function/permutation families. The above definitions of one-way functions and permutations are convenient in that they consider a single function over an infinite domain and range. However, most candidate one-way functions and permutations do not fit neatly into this framework. Instead, there is an algorithm generating some set of parameters I that define a function f_I; one-wayness here means essentially that f_I should be one-way with all but negligible probability over choice of I. Because each value of I defines a different function, we now refer to *families* of one-way functions (resp., permutations). We give the definition now, and refer the reader to the next section for a concrete example. (See also Section 9.4.1.)

DEFINITION 8.2 *A tuple* $\Pi = (\mathsf{Gen}, \mathsf{Samp}, f)$ *of probabilistic polynomial-time algorithms is a* function family *if the following hold:*

1. *The* parameter-generation algorithm Gen, *on input* 1^n, *outputs parameters* I *with* $|I| \geq n$. *Each value of* I *output by* Gen *defines sets* $\mathcal{D}_I$ *and* $\mathcal{R}_I$ *that constitute the domain and range, respectively, of a function* f_I.

2. *The* sampling algorithm Samp, *on input* I, *outputs a uniformly distributed element of* $\mathcal{D}_I$.

3. *The deterministic* evaluation algorithm f, *on input* I *and* $x \in \mathcal{D}_I$, *outputs an element* $y \in \mathcal{R}_I$. *We write this as* $y := f_I(x)$.

Π *is a* permutation family *if for each value of* I *output by* $\mathsf{Gen}(1^n)$, *it holds that* $\mathcal{D}_I = \mathcal{R}_I$ *and the function* $f_I : \mathcal{D}_I \to \mathcal{D}_I$ *is a bijection.*

Let Π be a function family. What follows is the natural analogue of the experiment introduced previously.

The inverting experiment $\mathsf{Invert}_{\mathcal{A},\Pi}(n)$**:**

1. $\mathsf{Gen}(1^n)$ *is run to obtain* I, *and then* $\mathsf{Samp}(I)$ *is run to obtain a uniform* $x \in \mathcal{D}_I$. *Finally,* $y := f_I(x)$ *is computed.*

2. $\mathcal{A}$ *is given* I *and* y *as input, and outputs* x'.

3. *The output of the experiment is 1 if* $f_I(x') = y$.

DEFINITION 8.3   *A function/permutation family* $\Pi = (\mathsf{Gen}, \mathsf{Samp}, f)$ *is* one-way *if for all probabilistic polynomial-time algorithms* $\mathcal{A}$ *there exists a negligible function* negl *such that*

$$\Pr[\mathsf{Invert}_{\mathcal{A},\Pi}(n) = 1] \leq \mathsf{negl}(n).$$

Throughout this chapter we work with one-way functions/permutations over an infinite domain (as in Definition 8.1), rather than working with families of one-way functions/permutations. This is primarily for convenience, and does not significantly affect any of the results. (See Exercise 8.7.)

8.1.2   Candidate One-Way Functions

One-way functions are of interest only if they exist. We do not know how to prove they exist unconditionally (this would be a major breakthrough in complexity theory), so we must conjecture or assume their existence. Such a conjecture is based on the fact that several natural computational problems have received much attention, yet still are not known to be solvable by any polynomial-time algorithm. Perhaps the most famous such problem is *integer factorization*, i.e., finding the prime factors of a large integer. It is easy to multiply two numbers and obtain their product, but difficult to take a number and find its factors. This leads us to define the function $f_{\mathsf{mult}}(x, y) = x \cdot y$. If we do not restrict the lengths of x and y, however, f_{mult} is easy to invert: with high probability $x \cdot y$ will be *even*, in which case $(2, xy/2)$ is an inverse. This issue can be addressed by restricting the domain of f_{mult} to equal-length *primes* x and y. We return to this idea in Section 9.2.

Another candidate one-way function, not relying directly on number theory, is based on the *subset-sum problem* and is defined by

$$f_{\mathsf{ss}}(x_1, \ldots, x_n, J) = \left(x_1, \ldots, x_n, \left[\sum_{j \in J} x_j \bmod 2^n\right]\right),$$

where each x_i is an n-bit string interpreted as an integer, and J is an n-bit string interpreted as specifying a subset of $\{1, \ldots, n\}$. Inverting f_{ss} on an output $(x_1, \ldots, x_n, y)$ requires finding a subset $J' \subseteq \{1, \ldots, n\}$ such that

$\sum_{j \in J'} x_j = y \bmod 2^n$. Readers who have studied $\mathcal{NP}$-completeness may recall that this problem is $\mathcal{NP}$-complete. But even $\mathcal{P} \neq \mathcal{NP}$ would not imply that f_{ss} is one-way: $\mathcal{P} \neq \mathcal{NP}$ would mean that every polynomial-time algorithm fails to solve the subset-sum problem on *at least one* input, whereas for f_{ss} to be one-way it is required that every polynomial-time algorithm fails to solve the subset-sum problem (at least for certain parameters) *almost always*. Thus, our belief that the function above is one-way is based on the lack of known algorithms to solve this problem even with "small" probability on random inputs, and not merely the fact that the problem is $\mathcal{NP}$-complete.

We conclude by showing a family of *permutations* that is believed to be one-way. Let Gen be a probabilistic polynomial-time algorithm that, on input 1^n, outputs an n-bit prime p along with a special element $g \in \{2, \dots, p-1\}$. (The element g should be a *generator* of $\mathbb{Z}_p^*$; see Section 9.3.3.) Let Samp be an algorithm that, given p and g, outputs a uniform integer $x \in \{1, \dots, p-1\}$. Finally, define

$$f_{p,g}(x) = [g^x \bmod p].$$

(The fact that $f_{p,g}$ can be computed efficiently follows from the results in Appendix B.2.3.) It can be shown that this function is one-to-one, and thus a permutation. The presumed difficulty of inverting this function is based on the conjectured hardness of the *discrete-logarithm problem*; we will have much more to say about this in Section 9.3.

Finally, we remark that very efficient one-way functions can be obtained from practical cryptographic constructions such as SHA-2 or AES under the assumption that they are collision resistant or a pseudorandom permutation, respectively; see Exercises 8.4 and 8.5. (Technically speaking, they cannot satisfy the definition of one-wayness since they have fixed-length input/output and so their asymptotic behavior is undefined. Nevertheless, it is plausible to conjecture that they are one-way in a concrete sense.)

8.1.3 Hard-Core Predicates

By definition, a one-way function is hard to invert. Stated differently: given $y = f(x)$, the value x cannot be computed *in its entirety* by any polynomial-time algorithm (except with negligible probability; we ignore this here). One might get the impression that nothing about x can be determined from $f(x)$ in polynomial time. This is *not* necessarily the case. Indeed, it is possible for $f(x)$ to "leak" a lot of information about x even if f is one-way. For a trivial example, let g be a one-way function and define $f(x_1, x_2) \stackrel{\text{def}}{=} (x_1, g(x_2))$, where $|x_1| = |x_2|$. It is easy to show that f is also a one-way function (this is left as an exercise), even though it reveals half its input.

For our applications, we will need to identify a specific piece of information about x that is "hidden" by $f(x)$. This motivates the notion of a *hard-core predicate*. A hard-core predicate $\mathsf{hc} : \{0,1\}^* \to \{0,1\}$ of a function f has the property that $\mathsf{hc}(x)$ is hard to compute with probability significantly better than $1/2$ given $f(x)$. (Since hc is a boolean function, it is always possible to compute $\mathsf{hc}(x)$ with probability $1/2$ by random guessing.) Formally:

DEFINITION 8.4 *A function* $\mathsf{hc} : \{0,1\}^* \rightarrow \{0,1\}$ *is a* hard-core *predicate of a function* f *if* hc *can be computed in polynomial time, and for every probabilistic polynomial-time algorithm* $\mathcal{A}$ *there is a negligible function* negl *such that*

$$\Pr_{x \leftarrow \{0,1\}^n} [\mathcal{A}(1^n, f(x)) = \mathsf{hc}(x)] \leq \frac{1}{2} + \mathsf{negl}(n),$$

where the probability is taken over the uniform choice of x *in* $\{0,1\}^n$ *and the randomness of* $\mathcal{A}$.

We stress that $\mathsf{hc}(x)$ is efficiently computable given x (since the function hc can be computed in polynomial time); the definition requires that $\mathsf{hc}(x)$ is hard to compute given $f(x)$. The above definition does not require f to be one-way; if f is a permutation, however, then it cannot have a hard-core predicate unless it is one-way. (See Exercise 8.13.)

Simple ideas don't work. Consider the predicate $\mathsf{hc}(x) \stackrel{\text{def}}{=} \bigoplus_{i=1}^{n} x_i$ where $x_1, \ldots, x_n$ denote the bits of x. One might hope that this is a hard-core predicate of any one-way function f: if f cannot be inverted, then $f(x)$ must hide at least one of the bits x_i of its preimage x, which would seem to imply that the XOR of all of the bits of x is hard to compute. Despite its appeal, this argument is incorrect. To see this, let g be a one-way function and define $f(x) \stackrel{\text{def}}{=} (g(x), \bigoplus_{i=1}^{n} x_i)$. It is not hard to show that f is one-way. However, it is clear that $f(x)$ does not hide the value of $\mathsf{hc}(x) = \bigoplus_{i=1}^{n} x_i$ because this is part of its output; therefore, $\mathsf{hc}(x)$ is not a hard-core predicate of f. Extending this, one can show that for any fixed predicate hc, there is a one-way function f for which hc is not a hard-core predicate of f.

Trivial hard-core predicates. Some functions have "trivial" hard-core predicates. For example, let f be the function that drops the last bit of its input (i.e., $f(x_1 \cdots x_n) = x_1 \cdots x_{n-1}$). It is hard to determine x_n given $f(x)$ since x_n is independent of the output; thus, $\mathsf{hc}(x) = x_n$ is a hard-core predicate of f. However, f is not one-way. When we use hard-core predicates in our constructions, it will become clear why trivial hard-core predicates of this sort are of no use.

8.2 From One-Way Functions to Pseudorandomness

In this chapter we show how to construct pseudorandom generators, functions, and permutations from any one-way function/permutation. In this section, we give an overview of these constructions. Details are given in the sections that follow.

A hard-core predicate from any one-way function. The first step is to show that a hard-core predicate exists for any one-way function. Actually, it remains open whether this is true; we show something weaker that suffices for our purposes: Namely, we show that given a one-way function f we can construct *another* one-way function g along with a hard-core predicate of g.

THEOREM 8.5 (Goldreich–Levin theorem)   *Assume one-way functions (resp., permutations) exist. Then there exists a one-way function (resp., permutation) g and a hard-core predicate* gl *of g.*

Let f be a one-way function. Functions g and gl are constructed as follows: set $g(x, r) \stackrel{\text{def}}{=} (f(x), r)$, for $|x| = |r|$, and define

$$\text{gl}(x, r) \stackrel{\text{def}}{=} \bigoplus_{i=1}^{n} x_i \cdot r_i,$$

where x_i (resp., r_i) denotes the ith bit of x (resp., r). Notice that if r is uniform, then $\text{gl}(x, r)$ outputs the XOR of a *random subset* of the bits of x. (When $r_i = 1$ the bit x_i is included in the XOR, and otherwise it is not.) The Goldreich–Levin theorem thus states that if f is a one-way function then $f(x)$ hides the XOR of a *random subset* of the bits of x.

Pseudorandom generators from one-way permutations. The next step is to show how a hard-core predicate of a one-way *permutation* can be used to construct a pseudorandom generator. (It is known that a hard-core predicate of a one-way *function* suffices, but the proof is extremely complicated and well beyond the scope of this book.) Specifically, we show:

THEOREM 8.6   *Let f be a one-way permutation and let* hc *be a hard-core predicate of f. Then, G defined by $G(s) \stackrel{\text{def}}{=} f(s) \,\|\, \text{hc}(s)$ is a pseudorandom generator with expansion factor $\ell(n) = n + 1$.*

As intuition for why G is a pseudorandom generator, note first that the initial n bits of $G(s)$ (i.e., the bits of $f(s)$) are uniformly distributed when s is uniformly distributed, since f is a permutation. Next, the fact that hc is a hard-core predicate of f means that $\text{hc}(s)$ "looks random"—i.e., is *pseudorandom*—even given $f(s)$ (assuming again that s is uniform). Putting these observations together, we see that the entire output of G is pseudorandom.

Pseudorandom generators with arbitrary expansion. The existence of a pseudorandom generator that stretches its seed by even a single bit (as we have just seen) is already highly non-trivial. But for applications (e.g., for efficient encryption of large messages as in Section 3.3), we need a pseudorandom generator with much larger expansion. Fortunately, we can obtain any polynomial expansion factor we like:

THEOREM 8.7 *If there exists a pseudorandom generator with expansion factor $\ell(n) = n + 1$, then for any polynomial* poly *there exists a pseudorandom generator with expansion factor* poly(n).

We conclude that pseudorandom generators with arbitrary (polynomial) expansion can be constructed from any one-way permutation.

Pseudorandom permutations from pseudorandom generators. Pseudorandom generators suffice for constructing EAV-secure private-key encryption schemes. For CPA-secure private-key encryption (not to mention message authentication codes), however, we relied on pseudorandom functions. The following result shows that the latter can be constructed from the former:

THEOREM 8.8 *If there exists a pseudorandom generator with expansion factor $\ell(n) = 2n$, then there exists a pseudorandom function.*

In fact, we can do even more:

THEOREM 8.9 *If there exists a pseudorandom function, then there exists a strong pseudorandom permutation.*

Combining the above theorems and the results of Chapters 3–5 we have:

COROLLARY 8.10 *Assuming the existence of one-way permutations:*

- *There exist pseudorandom generators with any expansion factor, pseudorandom functions, and strong pseudorandom permutations.*

- *There exist authenticated encryption schemes and secure message authentication codes.*

As noted earlier, even one-way *functions* suffice.

8.3 Hard-Core Predicates from One-Way Functions

In this section, we prove Theorem 8.5 by showing the following:

THEOREM 8.11 *Let f be a one-way function and define $g(x, r) \stackrel{\text{def}}{=} (f(x), r)$, where $|x| = |r|$, and $\mathsf{gl}(x, r) \stackrel{\text{def}}{=} \bigoplus_{i=1}^{n} x_i \cdot r_i$. Then gl is a hard-core predicate of g.*

Due to the complexity of the proof, we prove three successively stronger results culminating in what is claimed in the theorem.

8.3.1　A Simple Case

We first show that if there exists a polynomial-time adversary $\mathcal{A}$ that *always* correctly computes $\mathsf{gl}(x, r)$ given $g(x, r) = (f(x), r)$, then it is possible to invert f in polynomial time. (Note that such an $\mathcal{A}$ can only possibly exist if f is one-to-one.) Given the assumption that f is a one-way function, it follows that no such adversary $\mathcal{A}$ exists.

PROPOSITION 8.12　*Let f and gl be as in Theorem 8.11. If there exists a polynomial-time algorithm $\mathcal{A}$ such that $\mathcal{A}(f(x), r) = \mathsf{gl}(x, r)$ for all n and all $x, r \in \{0, 1\}^n$, then there exists a polynomial-time algorithm $\mathcal{A}'$ such that $\mathcal{A}'(1^n, f(x)) = x$ for all n and all $x \in \{0, 1\}^n$.*

PROOF　We construct $\mathcal{A}'$ as follows. $\mathcal{A}'(1^n, y)$ computes $x_i := \mathcal{A}(y, e^i)$ for $i = 1, \ldots, n$, where e^i denotes the n-bit string with 1 in the ith position and 0 everywhere else. Then $\mathcal{A}'$ outputs $x = x_1 \cdots x_n$. Clearly $\mathcal{A}'$ runs in polynomial time.

In the execution of $\mathcal{A}'(1^n, f(\hat{x}))$, the value x_i computed by $\mathcal{A}'$ satisfies

$$x_i = \mathcal{A}(f(\hat{x}), e^i) = \mathsf{gl}(\hat{x}, e^i) = \bigoplus_{j=1}^{n} \hat{x}_j \cdot e_j^i = \hat{x}_i.$$

Thus, $x_i = \hat{x}_i$ for all i and so $\mathcal{A}'$ outputs the correct inverse $x = \hat{x}$.　∎

If f is one-way, it is impossible for any probabilistic polynomial-time algorithm to invert f with non-negligible probability. Thus, we conclude that there is no polynomial-time algorithm that always correctly computes $\mathsf{gl}(x, r)$ from $(f(x), r)$. This is a rather weak result that is very far from our ultimate goal of showing that $\mathsf{gl}(x, r)$ cannot be computed with probability significantly better than $1/2$ given $(f(x), r)$.

8.3.2　A More Involved Case

We now show that it is hard for any probabilistic polynomial-time algorithm $\mathcal{A}$ to compute $\mathsf{gl}(x, r)$ from $(f(x), r)$ with probability significantly better than $3/4$. We will again show that any such $\mathcal{A}$ would imply the existence of a polynomial-time algorithm $\mathcal{A}'$ that inverts f with non-negligible probability. Notice that the strategy in the proof of Proposition 8.12 fails here because it may be that $\mathcal{A}$ *never* succeeds when $r = e^i$ (although it may succeed, say, on all other values of r). Furthermore, in the present case $\mathcal{A}'$ does not know if the result $\mathcal{A}(f(x), r)$ is equal to $\mathsf{gl}(x, r)$ or not; the only thing $\mathcal{A}'$ knows is that with high probability, algorithm $\mathcal{A}$ is correct. This further complicates the proof.

PROPOSITION 8.13   *Let f and gl be as in Theorem 8.11. If there exists a probabilistic polynomial-time algorithm $\mathcal{A}$ and a polynomial $p(\cdot)$ such that*

$$\Pr_{x, r \leftarrow \{0,1\}^n} \left[\mathcal{A}(f(x), r) = \mathsf{gl}(x, r) \right] \geq \frac{3}{4} + \frac{1}{p(n)}$$

for infinitely many values of n, then there exists a probabilistic polynomial-time algorithm $\mathcal{A}'$ such that

$$\Pr_{x \leftarrow \{0,1\}^n} \left[\mathcal{A}'(1^n, f(x)) \in f^{-1}(f(x)) \right] \geq \frac{1}{4 \cdot p(n)}$$

for infinitely many values of n.

PROOF   The main observation underlying the proof of this proposition is that for every $r \in \{0, 1\}^n$, the values $\mathsf{gl}(x, r \oplus e^i)$ and $\mathsf{gl}(x, r)$ together can be used to compute the ith bit of x. (Recall that e^i denotes the n-bit string with 0s everywhere except the ith position.) This is true because

$$\mathsf{gl}(x, r) \oplus \mathsf{gl}(x, r \oplus e^i)$$

$$= \left(\bigoplus_{j=1}^n x_j \cdot r_j \right) \oplus \left(\bigoplus_{j=1}^n x_j \cdot (r_j \oplus e_j^i) \right) = x_i \cdot r_i \oplus \left(x_i \cdot \bar{r}_i \right) = x_i \,,$$

where $\bar{r}_i$ is the complement of r_i, and the second equality is due to the fact that for $j \neq i$, the value $x_j \cdot r_j$ appears in both sums and so is canceled out.

The above demonstrates that if $\mathcal{A}$ answers correctly on both $(f(x), r)$ and $(f(x), r \oplus e^i)$, then $\mathcal{A}'$ can correctly compute x_i. Unfortunately, $\mathcal{A}'$ does not know when $\mathcal{A}$ answers correctly and when it does not; $\mathcal{A}'$ knows only that $\mathcal{A}$ answers correctly with "high" probability. For this reason, $\mathcal{A}'$ will use multiple random values of r, using each one to obtain an estimate of x_i, and then take the estimate occurring a majority of the time as its final guess for x_i.

As a preliminary step, we show that for many x's the probability that $\mathcal{A}$ answers correctly for both $(f(x), r)$ and $(f(x), r \oplus e^i)$, when r is uniform, is sufficiently high. This allows us to fix x and then focus solely on uniform choice of r, which makes the analysis easier.

CLAIM 8.14   *Let n be such that*

$$\Pr_{x, r \leftarrow \{0,1\}^n} \left[\mathcal{A}(f(x), r) = \mathsf{gl}(x, r) \right] \geq \frac{3}{4} + \frac{1}{p(n)} \,.$$

Then there exists a set $S_n \subseteq \{0,1\}^n$ of size at least $\frac{1}{2p(n)} \cdot 2^n$ such that for every $x \in S_n$ it holds that

$$\Pr_{r \leftarrow \{0,1\}^n} [\mathcal{A}(f(x), r) = \mathsf{gl}(x,r)] \geq \frac{3}{4} + \frac{1}{2p(n)}.$$

PROOF   Let $\varepsilon(n) = 1/p(n)$, and define $S_n \subseteq \{0,1\}^n$ to be the set of all x's for which

$$\Pr_{r \leftarrow \{0,1\}^n} [\mathcal{A}(f(x), r) = \mathsf{gl}(x,r)] \geq \frac{3}{4} + \frac{\varepsilon(n)}{2}.$$

We have:

$$\Pr_{x,r \leftarrow \{0,1\}^n} \left[\mathcal{A}(f(x), r) = \mathsf{gl}(x,r) \right] = \frac{1}{2^n} \sum_{x \in \{0,1\}^n} \Pr_{r \leftarrow \{0,1\}^n} \left[\mathcal{A}(f(x), r) = \mathsf{gl}(x,r) \right]$$

$$= \frac{1}{2^n} \sum_{x \in S_n} \Pr_{r \leftarrow \{0,1\}^n} \left[\mathcal{A}(f(x), r) = \mathsf{gl}(x,r) \right]$$

$$+ \frac{1}{2^n} \sum_{x \notin S_n} \Pr_{r \leftarrow \{0,1\}^n} \left[\mathcal{A}(f(x), r) = \mathsf{gl}(x,r) \right]$$

$$\leq \frac{|S_n|}{2^n} + \frac{1}{2^n} \cdot \sum_{x \notin S_n} \left(\frac{3}{4} + \frac{\varepsilon(n)}{2} \right)$$

$$\leq \frac{|S_n|}{2^n} + \left(\frac{3}{4} + \frac{\varepsilon(n)}{2} \right).$$

Since $\frac{3}{4} + \varepsilon(n) \leq \Pr_{x,r \leftarrow \{0,1\}^n} \left[\mathcal{A}(f(x), r) = \mathsf{gl}(x,r) \right]$, straightforward algebra gives $|S_n| \geq \frac{\varepsilon(n)}{2} \cdot 2^n$. ∎

The following is an easy consequence.

CLAIM 8.15   *Let n be such that*

$$\Pr_{x,r \leftarrow \{0,1\}^n} \left[\mathcal{A}(f(x), r) = \mathsf{gl}(x,r) \right] \geq \frac{3}{4} + \frac{1}{p(n)}.$$

Then there exists a set $S_n \subseteq \{0,1\}^n$ of size at least $\frac{1}{2p(n)} \cdot 2^n$ such that for every $x \in S_n$ and every i it holds that

$$\Pr_{r \leftarrow \{0,1\}^n} \left[\mathcal{A}(f(x), r) = \mathsf{gl}(x,r) \bigwedge \mathcal{A}(f(x), r \oplus e^i) = \mathsf{gl}(x, r \oplus e^i) \right] \geq \frac{1}{2} + \frac{1}{p(n)}.$$

PROOF   Let $\varepsilon(n) = 1/p(n)$, and take S_n to be the set guaranteed by the previous claim. Fix any $x \in S_n$. We have:

$$\Pr_{r \leftarrow \{0,1\}^n}[\mathcal{A}(f(x), r) \neq \mathsf{gl}(x, r)] \leq \frac{1}{4} - \frac{\varepsilon(n)}{2} \,.$$

Fix any $i \in \{1, \ldots, n\}$. If r is uniform then so is $r \oplus e^i$; thus

$$\Pr_{r \leftarrow \{0,1\}^n}[\mathcal{A}(f(x), r \oplus e^i) \neq \mathsf{gl}(x, r \oplus e^i)] \leq \frac{1}{4} - \frac{\varepsilon(n)}{2} \,.$$

We are interested in lower-bounding the probability that $\mathcal{A}$ outputs the correct answer for *both* $\mathsf{gl}(x, r)$ and $\mathsf{gl}(x, r \oplus e^i)$; equivalently, we want to upper-bound the probability that $\mathcal{A}$ fails to output the correct answer in *either* of these cases. Note that r and $r \oplus e^i$ are not independent, so we cannot just multiply the probabilities of failure. However, we can apply the union bound (see Proposition A.7) and sum the probabilities of failure. That is, the probability that $\mathcal{A}$ is *incorrect* on either $\mathsf{gl}(x, r)$ or $\mathsf{gl}(x, r \oplus e^i)$ is at most

$$\left(\frac{1}{4} - \frac{\varepsilon(n)}{2} \right) + \left(\frac{1}{4} - \frac{\varepsilon(n)}{2} \right) = \frac{1}{2} - \varepsilon(n),$$

and so $\mathcal{A}$ is correct on *both* $\mathsf{gl}(x, r)$ and $\mathsf{gl}(x, r \oplus e^i)$ with probability *at least* $1/2 + \varepsilon(n)$. This proves the claim. $\blacksquare$

For the rest of the proof we set $\varepsilon(n) = 1/p(n)$ and consider only those values of n for which

$$\Pr_{x, r \leftarrow \{0,1\}^n}\left[\mathcal{A}(f(x), r) = \mathsf{gl}(x, r)\right] \geq \frac{3}{4} + \varepsilon(n) \,. \tag{8.1}$$

The previous claim states that for an $\varepsilon(n)/2$ fraction of inputs x, and any i, algorithm $\mathcal{A}$ answers correctly on both $(f(x), r)$ and $(f(x), r \oplus e^i)$ with probability at least $1/2 + \varepsilon(n)$ over uniform choice of r, and from now on we focus only on such values of x. We construct a probabilistic polynomial-time algorithm $\mathcal{A}'$ that inverts $f(x)$ with probability at least $1/2$ when $x \in S_n$. This suffices to prove Proposition 8.13 since then, for infinitely many values of n,

$$\Pr_{x \leftarrow \{0,1\}^n}[\mathcal{A}'(1^n, f(x)) \in f^{-1}(f(x))]$$

$$\geq \Pr_{x \leftarrow \{0,1\}^n}[\mathcal{A}'(1^n, f(x)) \in f^{-1}(f(x)) \mid x \in S_n] \cdot \Pr_{x \leftarrow \{0,1\}^n}[x \in S_n]$$

$$\geq \frac{1}{2} \cdot \frac{\varepsilon(n)}{2} = \frac{1}{4p(n)} \,.$$

Algorithm $\mathcal{A}'$, given as input 1^n and y, works as follows:

1. For $i = 1, \ldots, n$ do:

- Repeatedly choose a uniform $r \in \{0,1\}^n$ and compute $\mathcal{A}(y,r) \oplus \mathcal{A}(y, r \oplus e^i)$ as an "estimate" for the ith bit of the preimage of y. After doing this sufficiently many times (see below), let x_i be the "estimate" that occurs a majority of the time.

2. Output $x = x_1 \cdots x_n$.

We sketch an analysis of the probability that $\mathcal{A}'$ correctly inverts its given input y. (We allow ourselves to be a bit laconic, since a full proof for a more difficult case is given in the following section.) Say $y = f(\hat{x})$ and recall that we assume here that n is such that Equation (8.1) holds and $\hat{\imath} \in S_n$. Fix some i. The previous claim implies that the estimate $\mathcal{A}(y,r) \oplus \mathcal{A}(y, r \oplus e^i)$ is equal to $\mathsf{gl}(\hat{x}, e^i)$ with probability at least $\frac{1}{2} + \varepsilon(n)$ over choice of r. By obtaining sufficiently many estimates and letting x_i be the majority value, $\mathcal{A}'$ can ensure that x_i is equal to $\mathsf{gl}(\hat{x}, e^i)$ with probability at least $1 - \frac{1}{2n}$. Since $\varepsilon(n) = 1/p(n)$ for some polynomial p, and an independent value of r is used for obtaining each estimate, the *Chernoff bound* (cf. Proposition A.14) shows that polynomially many estimates suffice.

Summarizing, we have that for each i the value x_i computed by $\mathcal{A}'$ is incorrect with probability at most $\frac{1}{2n}$. A union bound thus shows that $\mathcal{A}'$ is incorrect for *some* i with probability at most $n \cdot \frac{1}{2n} = \frac{1}{2}$. That is, $\mathcal{A}'$ is correct for all i—and thus correctly inverts y—with probability at least $1 - \frac{1}{2} = \frac{1}{2}$. This completes the proof of Proposition 8.13. ∎

A corollary of Proposition 8.13 is that if f is a one-way function, then for any polynomial-time algorithm $\mathcal{A}$ the probability that $\mathcal{A}$ correctly guesses $\mathsf{gl}(x,r)$ when given $(f(x),r)$ is at most negligibly more than $3/4$.

8.3.3 The Full Proof

We assume familiarity with the simplified proofs in the previous sections, and build on the ideas developed there. We rely on some terminology and standard results from probability theory discussed in Appendix A.3.

PROPOSITION 8.16 *Let f and gl be as in Theorem 8.11. If there exists a probabilistic polynomial-time algorithm $\mathcal{A}$ and a polynomial $p(\cdot)$ such that*

$$\Pr_{x,r \leftarrow \{0,1\}^n} \left[\mathcal{A}(f(x),r) = \mathsf{gl}(x,r) \right] \geq \frac{1}{2} + \frac{1}{p(n)}$$

for infinitely many values of n, then there exists a probabilistic polynomial-time algorithm $\mathcal{A}'$ and a polynomial $p'(\cdot)$ such that

$$\Pr_{x \leftarrow \{0,1\}^n} \left[\mathcal{A}'(1^n, f(x)) \in f^{-1}(f(x)) \right] \geq \frac{1}{p'(n)}$$

for infinitely many values of n.

PROOF   Once again we set $\varepsilon(n) = 1/p(n)$ and consider only those values of n for which $\Pr_{x,r \leftarrow \{0,1\}^n}\left[\mathcal{A}(f(x),r) = \mathsf{gl}(x,r)\right] \geq \frac{1}{2} + \frac{1}{p(n)}$. The following is analogous to Claim 8.14 and is proved in the same way.

CLAIM 8.17   *Let n be such that*

$$\Pr_{x,r \leftarrow \{0,1\}^n}\left[\mathcal{A}(f(x),r) = \mathsf{gl}(x,r)\right] \geq \frac{1}{2} + \varepsilon(n).$$

Then there exists a set $S_n \subseteq \{0,1\}^n$ of size at least $\frac{\varepsilon(n)}{2} \cdot 2^n$ such that for every $x \in S_n$ it holds that

$$\Pr_{r \leftarrow \{0,1\}^n}[\mathcal{A}(f(x),r) = \mathsf{gl}(x,r)] \geq \frac{1}{2} + \frac{\varepsilon(n)}{2}. \tag{8.2}$$

If we start by trying to prove an analogue of Claim 8.15, the best we can claim here is that when $x \in S_n$ we have

$$\Pr_{r \leftarrow \{0,1\}^n}\left[\mathcal{A}(f(x),r) = \mathsf{gl}(x,r) \bigwedge \mathcal{A}(f(x),r \oplus e^i) = \mathsf{gl}(x,r \oplus e^i)\right] \geq \varepsilon(n)$$

for any i. Thus, if we try to use $\mathcal{A}(f(x),r) \oplus \mathcal{A}(f(x),r \oplus e^i)$ as an estimate for x_i, all we can claim is that this estimate will be correct with probability at least $\varepsilon(n)$, which may not be any better than taking a random guess! We cannot claim that flipping the result gives a good estimate, either.

Instead, we design $\mathcal{A}'$ so that it computes $\mathsf{gl}(x,r)$ and $\mathsf{gl}(x,r \oplus e^i)$ by invoking $\mathcal{A}$ only once. We do this by having $\mathcal{A}'$ run $\mathcal{A}(f(x),r \oplus e^i)$, and then simply "guessing" the value $\mathsf{gl}(x,r)$ itself. The naive way to do this would be to choose the r's independently, as before, and to have $\mathcal{A}'$ make an independent guess of $\mathsf{gl}(x,r)$ for each value of r. But then the probability that all such guesses are correct—which, as we will see, is necessary if $\mathcal{A}'$ is to output the correct inverse—would be negligible because polynomially many r's are used.

The crucial observation of the present proof is that $\mathcal{A}'$ can generate the r's in a *pairwise-independent* manner and make its guesses in a particular way so that with non-negligible probability all its guesses are correct. Specifically, in order to generate m values of r, we have $\mathcal{A}'$ select $\ell = \lceil \log(m+1) \rceil$ independent and uniformly distributed strings $s^1, \ldots, s^\ell \in \{0,1\}^n$. Then, for every nonempty subset $I \subseteq \{1, \ldots, \ell\}$, we set $r^I := \oplus_{i \in I} s^i$. Since there are $2^\ell - 1$ nonempty subsets, this defines a collection of $2^{\lceil \log(m+1) \rceil} - 1 \geq m$ strings. Each such string is uniformly distributed. The strings are not independent, but they *are* pairwise independent. To see this, notice that for every two subsets $I \neq J$ there is an index $j \in I \cup J$ such that $j \notin I \cap J$. Without loss of generality, assume $j \notin I$. Then the value of s^j is uniform and independent of the value of r^I. Since s^j is included in the XOR that defines r^J, this implies that r^J is uniform and independent of r^I as well.

We now have the following two important observations:

1. Given $\mathsf{gl}(x, s^1), \ldots, \mathsf{gl}(x, s^\ell)$, it is possible to compute $\mathsf{gl}(x, r^I)$ for every subset $I \subseteq \{1, \ldots, \ell\}$. This is because

$$\mathsf{gl}(x, r^I) = \mathsf{gl}(x, \oplus_{i \in I} s^i) = \oplus_{i \in I} \mathsf{gl}(x, s^i).$$

2. If $\mathcal{A}'$ simply guesses the values of $\mathsf{gl}(x, s^1), \ldots, \mathsf{gl}(x, s^\ell)$ by choosing a uniform bit for each, then *all* these guesses will be correct with probability $1/2^\ell$. If m is polynomial in the security parameter n, then $1/2^\ell$ is not negligible, and so with *non-negligible probability* $\mathcal{A}'$ correctly guesses all the values $\mathsf{gl}(x, s^1), \ldots, \mathsf{gl}(x, s^\ell)$.

Combining the above yields a way of obtaining $m = \mathsf{poly}(n)$ uniform and pairwise-independent strings $\{r^I\}$ along with *correct* values for $\{\mathsf{gl}(x, r^I)\}$ with non-negligible probability. These values can then be used to compute x_i in the same way as in the proof of Proposition 8.13. Details follow.

The inversion algorithm $\mathcal{A}'$. We now provide a full description of an algorithm $\mathcal{A}'$ that receives inputs $1^n, y$ and tries to compute an inverse of y. The algorithm proceeds as follows:

1. Set $\ell := \lceil \log(2n/\varepsilon(n)^2 + 1) \rceil$.

2. Choose uniform, independent $s^1, \ldots, s^\ell \in \{0,1\}^n$ and $\sigma^1, \ldots, \sigma^\ell \in \{0,1\}$.

3. For every nonempty subset $I \subseteq \{1, \ldots, \ell\}$, compute $r^I := \oplus_{i \in I} s^i$ and $\sigma^I := \oplus_{i \in I} \sigma^i$.

4. For $i = 1, \ldots, n$ do:

    (a) For every nonempty subset $I \subseteq \{1, \ldots, \ell\}$, set

    $$x_i^I := \sigma^I \oplus \mathcal{A}(y, r^I \oplus e^i).$$

    (b) Set $x_i := \mathsf{majority}_I \{x_i^I\}$ (i.e., take the bit that appeared a majority of the time in the previous step).

5. Output $x = x_1 \cdots x_n$.

It remains to compute the probability that $\mathcal{A}'$ outputs $x \in f^{-1}(y)$. As in the proof of Proposition 8.13, we focus only on n as in Claim 8.17 and assume $y = f(\hat{x})$ for some $\hat{x} \in S_n$. Each σ^i represents a "guess" for the value of $\mathsf{gl}(\hat{x}, s^i)$. As noted earlier, with non-negligible probability all these guesses are correct; we show that conditioned on this event, $\mathcal{A}'$ outputs $x = \hat{x}$ with probability at least $1/2$.

Assume $\sigma^i = \mathsf{gl}(\hat{x}, s^i)$ for all i. Then $\sigma^I = \mathsf{gl}(\hat{x}, r^I)$ for all I. Fix an index $i \in \{1, \ldots, n\}$ and consider the probability that $\mathcal{A}'$ obtains the correct value $x_i = \hat{x}_i$. For any nonempty I we have $\mathcal{A}(y, r^I \oplus e^i) = \mathsf{gl}(\hat{x}, r^I \oplus e^i)$ with probability at least $\frac{1}{2} + \varepsilon(n)/2$ over choice of r; this follows because $\hat{x} \in S_n$

and $r^I \oplus e^i$ is uniformly distributed. Thus, for any nonempty subset I we have $\Pr[x_i^I = \hat{x}_i] \geq \frac{1}{2} + \varepsilon(n)/2$. Moreover, the $\{x_i^I\}_{I \subseteq \{1,\ldots,\ell\}}$ are pairwise independent because the $\{r^I\}_{I \subseteq \{1,\ldots,\ell\}}$ (and hence the $\{r^I \oplus e^i\}_{I \subseteq \{1,\ldots,\ell\}}$) are pairwise independent. Since x_i is defined to be the value that occurs a majority of the time among the $\{x_i^I\}_{I \subseteq \{1,\ldots,\ell\}}$, we can apply Proposition A.13 to obtain

$$\Pr[x_i \neq \hat{x}_i] \leq \frac{1}{4 \cdot (\varepsilon(n)/2)^2 \cdot (2^\ell - 1)}$$
$$\leq \frac{1}{4 \cdot (\varepsilon(n)/2)^2 \cdot (2n/\varepsilon(n)^2)}$$
$$= \frac{1}{2n}.$$

The above holds for all i, so by applying a union bound we see that the probability that $x_i \neq \hat{x}_i$ for *some* i is at most $1/2$. That is, $x_i = \hat{x}_i$ for *all* i (and hence $x = \hat{x}$) with probability at least $1/2$.

Putting everything together: Let n be as in Claim 8.17 and $y = f(\hat{x})$. With probability at least $\varepsilon(n)/2$ we have $\hat{x} \in S_n$. All the guesses σ^i are correct with probability at least

$$\frac{1}{2^\ell} \geq \frac{1}{2 \cdot (2n/\varepsilon(n)^2 + 1)} > \frac{\varepsilon(n)^2}{5n}$$

for n sufficiently large. Conditioned on both the above, $\mathcal{A}'$ outputs $x = \hat{x}$ with probability at least $1/2$. The overall probability with which $\mathcal{A}'$ inverts its input is thus at least $\varepsilon(n)^3/20n = 1/(20\,np(n)^3)$ for infinitely many n. Since $20\,np(n)^3$ is polynomial in n, this proves Proposition 8.16. ∎

8.4 Constructing Pseudorandom Generators

We first show how to construct pseudorandom generators that stretch their input by a single bit, under the assumption that one-way *permutations* exist. We then show how to extend this to obtain any polynomial expansion factor.

8.4.1 Pseudorandom Generators with Minimal Expansion

Let f be a one-way permutation with hard-core predicate hc. This means that hc(s) "looks random" given $f(s)$, when s is uniform. Furthermore, since f is a permutation, $f(s)$ itself is uniformly distributed. (Applying a permutation to a uniformly distributed value yields a uniformly distributed value.) So if s is a uniform n-bit string, the $(n+1)$-bit string $f(s)\|\mathsf{hc}(s)$ consists of a uniform n-bit string plus an additional bit that looks uniform even conditioned on the initial n bits; in other words, this $(n+1)$-bit string is *pseudorandom*. Thus, the algorithm G defined by $G(s) = f(s)\|hc(s)$ is a pseudorandom generator.

THEOREM 8.18 *Let f be a one-way permutation with hard-core predicate hc. Then algorithm G defined by $G(s) = f(s)\|hc(s)$ is a pseudorandom generator with expansion factor $\ell(n) = n + 1$.*

PROOF Let D be a probabilistic polynomial-time algorithm. We prove that there is a negligible function negl such that

$$\Pr_{r \leftarrow \{0,1\}^{n+1}}[D(r) = 1] - \Pr_{s \leftarrow \{0,1\}^n}[D(G(s)) = 1] \leq \mathsf{negl}(n). \qquad (8.3)$$

A similar argument shows that there is a negligible function negl' for which

$$\Pr_{s \leftarrow \{0,1\}^n}[D(G(s)) = 1] - \Pr_{r \leftarrow \{0,1\}^{n+1}}[D(r) = 1] \leq \mathsf{negl'}(n),$$

which completes the proof.

Observe first that

$$\Pr_{r \leftarrow \{0,1\}^{n+1}}[D(r) = 1] = \Pr_{r \leftarrow \{0,1\}^n, r' \leftarrow \{0,1\}}[D(r\|r') = 1]$$

$$= \Pr_{s \leftarrow \{0,1\}^n, r' \leftarrow \{0,1\}}[D(f(s)\|r') = 1]$$

$$= \frac{1}{2} \cdot \Pr_{s \leftarrow \{0,1\}^n}[D(f(s) \| hc(s)) = 1]$$

$$+ \frac{1}{2} \cdot \Pr_{s \leftarrow \{0,1\}^n}[D(f(s) \| \overline{hc}(s)) = 1],$$

using the fact that f is a permutation for the second equality, and that a uniform bit r' is equal to $hc(s)$ with probability exactly $1/2$ for the third equality. Since

$$\Pr_{s \leftarrow \{0,1\}^n}[D(G(s)) = 1] = \Pr_{s \leftarrow \{0,1\}^n}[D(f(s)\|hc(s)) = 1]$$

(by definition of G), this means that Equation (8.3) is equivalent to

$$\frac{1}{2} \cdot \left(\Pr_{s \leftarrow \{0,1\}^n}[D(f(s)\|\overline{hc}(s)) = 1] - \Pr_{s \leftarrow \{0,1\}^n}[D(f(s)\|hc(s)) = 1] \right) \leq \mathsf{negl}(n).$$

Consider the following algorithm $\mathcal{A}$ that is given as input a value $y = f(s)$ and tries to predict the value of $hc(s)$:

1. Choose uniform $r' \in \{0, 1\}$.

2. Run $D(y\|r')$. If D outputs 0, output r'; otherwise output $\bar{r}'$.

Clearly $\mathcal{A}$ runs in polynomial time. By definition of $\mathcal{A}$, we have

$$\Pr_{s\leftarrow\{0,1\}^n}[\mathcal{A}(f(s)) = \mathsf{hc}(s)]$$

$$= \frac{1}{2} \cdot \Pr_{s\leftarrow\{0,1\}^n}[\mathcal{A}(f(s)) = \mathsf{hc}(s) \mid r' = \mathsf{hc}(s)]$$

$$+ \frac{1}{2} \cdot \Pr_{s\leftarrow\{0,1\}^n}[\mathcal{A}(f(s)) = \mathsf{hc}(s) \mid r' \neq \mathsf{hc}(s)]$$

$$= \frac{1}{2} \cdot \left(\Pr_{s\leftarrow\{0,1\}^n}[D(f(s)\|\mathsf{hc}(s)) = 0] + \Pr_{s\leftarrow\{0,1\}^n}[D(f(s)\|\overline{\mathsf{hc}}(s)) = 1] \right)$$

$$= \frac{1}{2} \cdot \left(\left(1 - \Pr_{s\leftarrow\{0,1\}^n}[D(f(s)\|\mathsf{hc}(s)) = 1] \right) + \Pr_{s\leftarrow\{0,1\}^n}[D(f(s)\|\overline{\mathsf{hc}}(s)) = 1] \right)$$

$$= \frac{1}{2} + \frac{1}{2} \cdot \left(\Pr_{s\leftarrow\{0,1\}^n}[D(f(s)\|\overline{\mathsf{hc}}(s)) = 1] - \Pr_{s\leftarrow\{0,1\}^n}[D(f(s)\|\mathsf{hc}(s)) = 1] \right).$$

Since hc is a hard-core predicate of f, it follows that there exists a negligible function negl for which

$$\frac{1}{2} \cdot \left(\Pr_{s\leftarrow\{0,1\}^n}[D(f(s)\|\overline{\mathsf{hc}}(s)) = 1] - \Pr_{s\leftarrow\{0,1\}^n}[D(f(s)\|\mathsf{hc}(s)) = 1] \right) \leq \mathsf{negl}(n),$$

as desired. ∎

8.4.2 Increasing the Expansion Factor

We now show that the expansion factor of a pseudorandom generator can be increased by any desired (polynomial) amount. This means that the previous construction, with expansion factor $\ell(n) = n + 1$, suffices for constructing a pseudorandom generator with arbitrary (polynomial) expansion factor.

THEOREM 8.19 *If there exists a pseudorandom generator G with expansion factor $n + 1$, then for any polynomial* poly *there exists a pseudorandom generator $\hat{G}$ with expansion factor* poly(n).

PROOF We first consider constructing a pseudorandom generator $\hat{G}$ that outputs $n + 2$ bits. $\hat{G}$ works as follows: Given an initial seed $s \in \{0,1\}^n$, it computes $t_1 := G(s)$ to obtain $n + 1$ pseudorandom bits. The initial n bits of t_1 are then used again as a seed for G; the resulting $n+1$ bits, concatenated with the final bit of t_1, yield the $(n + 2)$-bit output. (See Figure 8.1.) The second application of G uses a pseudorandom seed rather than a random one. The proof of security we give next shows that this does not impact the pseudorandomness of the output.

We now prove that $\hat{G}$ is a pseudorandom generator. Define three sequences of distributions $\{H_n^0\}_{n=1,\ldots}$, $\{H_n^1\}_{n=1,\ldots}$, and $\{H_n^2\}_{n=1,\ldots}$, where each of H_n^0, H_n^1, and H_n^2 is a distribution on strings of length $n + 2$. In distribution H_n^0, a uniform string $t_0 \in \{0,1\}^n$ is chosen and the output is $t_2 := \hat{G}(t_0)$. In distribution H_n^1, a uniform string $t_1 \in \{0,1\}^{n+1}$ is chosen and parsed as $s_1 \| \sigma_1$ (where s_1 is the initial n bits of t_1 and σ_1 is the final bit). The output is $t_2 := G(s_1) \| \sigma_1$. In distribution H_n^2, the output is a uniform string $t_2 \in \{0,1\}^{n+2}$. We denote by $t_2 \leftarrow H_n^i$ the process of generating an $(n+2)$-bit string t_2 according to distribution H_n^i.

Fix an arbitrary probabilistic polynomial-time distinguisher D. We first claim that there is a negligible function negl' such that

$$\left| \Pr_{t_2 \leftarrow H_n^0}[D(t_2) = 1] - \Pr_{t_2 \leftarrow H_n^1}[D(t_2) = 1] \right| \leq \mathsf{negl}'(n). \qquad (8.4)$$

To see this, consider the polynomial-time distinguisher D' that, on input $t_1 \in \{0,1\}^{n+1}$, parses t_1 as $s_1 \| \sigma_1$ with $|s_1| = n$, computes $t_2 := G(s_1) \| \sigma_1$, and outputs $D(t_2)$. Clearly D' runs in polynomial time. Observe that:

1. If t_1 is uniform, the distribution on t_2 generated by D' is exactly that of distribution H_n^1. Thus,

$$\Pr_{t_1 \leftarrow \{0,1\}^{n+1}}[D'(t_1) = 1] = \Pr_{t_2 \leftarrow H_n^1}[D(t_2) = 1].$$

2. If $t_1 = G(s)$ for uniform $s \in \{0,1\}^n$, the distribution on t_2 generated by D' is exactly that of distribution H_n^0. That is,

$$\Pr_{s \leftarrow \{0,1\}^n}[D'(G(s)) = 1] = \Pr_{t_2 \leftarrow H_n^0}[D(t_2) = 1].$$

Pseudorandomness of G implies that there is a negligible function negl' with

$$\left| \Pr_{s \leftarrow \{0,1\}^n}[D'(G(s)) = 1] - \Pr_{t_1 \leftarrow \{0,1\}^{n+1}}[D'(t_1) = 1] \right| \leq \mathsf{negl}'(n).$$

Equation (8.4) follows.

We next claim that there is a negligible function negl'' such that

$$\left| \Pr_{t_2 \leftarrow H_n^1}[D(t_2) = 1] - \Pr_{t_2 \leftarrow H_n^2}[D(t_2) = 1] \right| \leq \mathsf{negl}''(n). \qquad (8.5)$$

To see this, consider the polynomial-time distinguisher D'' that, on input $w \in \{0,1\}^{n+1}$, chooses uniform $\sigma_1 \in \{0,1\}$, sets $t_2 := w \| \sigma_1$, and outputs $D(t_2)$. If w is uniform then so is t_2; thus,

$$\Pr_{w \leftarrow \{0,1\}^{n+1}}[D''(w) = 1] = \Pr_{t_2 \leftarrow H_n^2}[D(t_2) = 1].$$

On the other hand, if $w = G(s)$ for uniform $s \in \{0,1\}^n$, then t_2 is distributed exactly according to H_n^1 and so

$$\Pr_{s \leftarrow \{0,1\}^n}[D''(G(s)) = 1] = \Pr_{t_2 \leftarrow H_n^1}[D(t_2) = 1].$$

As before, pseudorandomness of G implies Equation (8.5).

Putting everything together, we have

$$\left| \Pr_{s \leftarrow \{0,1\}^n}[D(\hat{G}(s)) = 1] - \Pr_{r \leftarrow \{0,1\}^{n+2}}[D(r) = 1] \right| \tag{8.6}$$

$$= \left| \Pr_{t_2 \leftarrow H_n^0}[D(t_2) = 1] - \Pr_{t_2 \leftarrow H_n^2}[D(t_2) = 1] \right|$$

$$\leq \left| \Pr_{t_2 \leftarrow H_n^0}[D(t_2) = 1] - \Pr_{t_2 \leftarrow H_n^1}[D(t_2) = 1] \right|$$

$$+ \left| \Pr_{t_2 \leftarrow H_n^1}[D(t_2) = 1] - \Pr_{t_2 \leftarrow H_n^2}[D(t_2) = 1] \right|$$

$$\leq \mathsf{negl}'(n) + \mathsf{negl}''(n),$$

using Equations (8.4) and (8.5). Since D was an arbitrary polynomial-time distinguisher, this proves that $\hat{G}$ is a pseudorandom generator.

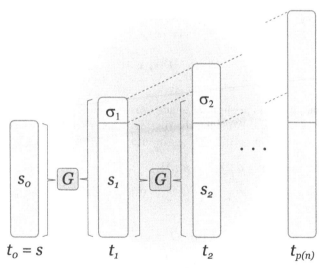

FIGURE 8.1:   Increasing the expansion of a pseudorandom generator.

The general case. The same idea as above can be iteratively applied to generate as many pseudorandom bits as desired. Formally, say we wish to construct a pseudorandom generator $\hat{G}$ with expansion factor $n + p(n)$, for some polynomial p. On input $s \in \{0,1\}^n$, algorithm $\hat{G}$ does (cf. Figure 8.1):

1. Set $t_0 := s$. For $i = 1, \ldots, p(n)$ do:

   (a) Let s_{i-1} be the first n bits of t_{i-1}, and let σ_{i-1} denote the remaining $i - 1$ bits. (When $i = 1$, $s_0 = t_0$ and σ_0 is the empty string.)

   (b) Set $t_i := G(s_{i-1}) \| \sigma_{i-1}$.

2. Output $t_{p(n)}$.

We show that $\hat{G}$ is a pseudorandom generator. The proof uses a common technique known as a *hybrid argument*. (Actually, even the case of $p(n) = 2$, above, used a simple hybrid argument.) The main difference with respect to the previous proof is a technical one. Previously, we could define and explicitly work with three sequences of distributions $\{H_n^0\}$, $\{H_n^1\}$, and $\{H_n^2\}$. Here that is not possible since the number of distributions to consider grows with n.

For any n and $0 \le j \le p(n)$, let H_n^j be the distribution on strings of length $n + p(n)$ defined as follows: choose uniform $t_j \in \{0,1\}^{n+j}$, then run $\hat{G}$ starting from iteration $j + 1$ and output $t_{p(n)}$. (When $j = p(n)$ this means we simply choose uniform $t_{p(n)} \in \{0,1\}^{n+p(n)}$ and output it.) The crucial observation is that H_n^0 corresponds to outputting $\hat{G}(s)$ for uniform $s \in \{0,1\}^n$, while $H_n^{p(n)}$ corresponds to outputting a uniform $(n + p(n))$-bit string. Fixing any polynomial-time distinguisher D, this means that

$$
\left| \Pr_{s \leftarrow \{0,1\}^n}[D(\hat{G}(s)) = 1] - \Pr_{r \leftarrow \{0,1\}^{n+p(n)}}[D(r) = 1] \right|
$$

$$
= \left| \Pr_{t \leftarrow H_n^0}[D(t) = 1] - \Pr_{t \leftarrow H_n^{p(n)}}[D(t) = 1] \right|. \tag{8.7}
$$

We prove the above is negligible, hence $\hat{G}$ is a pseudorandom generator.

Fix D as above, and consider the distinguisher D' that does the following when given a string $w \in \{0,1\}^{n+1}$ as input:

1. Choose uniform $j \in \{1, \ldots, p(n)\}$.

2. Choose uniform $\sigma'_j \in \{0,1\}^{j-1}$. (When $j = 1$ then σ'_j is the empty string.)

3. Set $t_j := w \| \sigma'_j$. Then run $\hat{G}$ starting from iteration $j + 1$ to compute $t_{p(n)} \in \{0,1\}^{n+p(n)}$. Output $D(t_{p(n)})$.

Clearly D' runs in polynomial time. Analyzing the behavior of D' is more complicated than before, although the underlying ideas are the same. Fix n and say D' chooses $j = j^*$. If w is uniform, then t_{j^*} is uniform and so the distribution on $t \stackrel{\text{def}}{=} t_{p(n)}$ is exactly that of distribution $H_n^{j^*}$. That is,

$$
\Pr_{w \leftarrow \{0,1\}^{n+1}}[D'(w) = 1 \mid j = j^*] = \Pr_{t \leftarrow H_n^{j^*}}[D(t) = 1].
$$

Since each value for j is chosen with equal probability,

$$\Pr_{w \leftarrow \{0,1\}^{n+1}}[D'(w) = 1] = \frac{1}{p(n)} \cdot \sum_{j^*=1}^{p(n)} \Pr_{w \leftarrow \{0,1\}^{n+1}}[D'(w) = 1 \mid j = j^*]$$

$$= \frac{1}{p(n)} \cdot \sum_{j^*=1}^{p(n)} \Pr_{t \leftarrow H_n^{j^*}}[D(t) = 1]. \qquad (8.8)$$

On the other hand, say D' chooses $j = j^*$ and $w = G(s)$ for uniform $s \in \{0,1\}^n$. Defining $t_{j^*-1} = s \| \sigma'_{j^*}$, we see that t_{j^*-1} is uniform and so the experiment involving D' is equivalent to running $\hat{G}$ from iteration j^* to compute $t_{p(n)}$. That is, the distribution on $t \stackrel{\text{def}}{=} t_{p(n)}$ is now exactly that of distribution $H_n^{j^*-1}$, and so

$$\Pr_{s \leftarrow \{0,1\}^n}[D'(G(s)) = 1 \mid j = j^*] = \Pr_{t \leftarrow H_n^{j^*-1}}[D(t) = 1].$$

Therefore,

$$\Pr_{s \leftarrow \{0,1\}^n}[D'(\hat{G}(s)) = 1] = \frac{1}{p(n)} \cdot \sum_{j^*=1}^{p(n)} \Pr_{s \leftarrow \{0,1\}^n}[D'(G(s)) = 1 \mid j = j^*]$$

$$= \frac{1}{p(n)} \cdot \sum_{j^*=1}^{p(n)} \Pr_{t \leftarrow H_n^{j^*-1}}[D(t) = 1]$$

$$= \frac{1}{p(n)} \cdot \sum_{j^*=0}^{p(n)-1} \Pr_{t \leftarrow H_n^{j^*}}[D(t) = 1]. \qquad (8.9)$$

We can now analyze how well D' distinguishes outputs of G from random:

$$\left| \Pr_{s \leftarrow \{0,1\}^n}[D'(G(s)) = 1] - \Pr_{w \leftarrow \{0,1\}^{n+1}}[D'(w) = 1] \right| \qquad (8.10)$$

$$= \frac{1}{p(n)} \cdot \left| \sum_{j^*=0}^{p(n)-1} \Pr_{t \leftarrow H_n^{j^*}}[D(t) = 1] - \sum_{j^*=1}^{p(n)} \Pr_{t \leftarrow H_n^{j^*}}[D(t) = 1] \right|$$

$$= \frac{1}{p(n)} \cdot \left| \Pr_{t \leftarrow H_n^0}[D(t) = 1] - \Pr_{t \leftarrow H_n^{p(n)}}[D(t) = 1] \right|,$$

relying on Equations (8.8) and (8.9) for the first equality. (The second equality holds because the same terms are included in each sum, except for the first term of the left sum and the last term of the right sum.) Since G is a pseudorandom generator, the term on the left-hand side of Equation (8.10) is negligible; because p is polynomial, this implies that Equation (8.7) is negligible, completing the proof that $\hat{G}$ is a pseudorandom generator. ∎

Putting it all together. Let f be a one-way permutation. Taking the pseudorandom generator with expansion factor $n + 1$ from Theorem 8.18, and increasing the expansion factor to $n + \ell$ using the approach from the proof of Theorem 8.19, we obtain the following pseudorandom generator $\hat{G}$:

$$\hat{G}(s) = f^{(\ell)}(s) \, \| \, \mathsf{hc}(f^{(\ell-1)}(s)) \, \| \, \cdots \, \| \, \mathsf{hc}(s),$$

where $f^{(i)}$ refers to i-fold iteration of f. Note that $\hat{G}$ uses ℓ evaluations of f, and generates one pseudorandom bit per evaluation using the hard-core predicate hc.

Connection to stream ciphers. Recall from Section 3.6.1 that a stream cipher (without an IV) is defined by algorithms (Init, Next), where Init takes a seed $s \in \{0,1\}^n$ and returns initial state st, and Next takes as input the current state st and outputs a bit σ and updated state st′. The construction $\hat{G}$ from the preceding proof fits nicely into this paradigm: take Init to be the trivial algorithm that outputs st $= s$, and define Next(st) to compute $G(\mathsf{st})$, parse the result as st′$\|\sigma$ with $|\mathsf{st}'| = n$, and output the bit σ and updated state st′. (If we use this stream cipher to generate $p(n)$ output bits starting from seed s, then we get exactly the final $p(n)$ bits of $\hat{G}(s)$ in reverse order.) The preceding proof shows that this yields a pseudorandom generator.

Hybrid arguments. A *hybrid argument* is a basic tool for proving indistinguishability when a primitive is (or several different primitives are) applied multiple times. Somewhat informally, the technique works by defining a series of intermediate "hybrid distributions" that bridge between two "extreme distributions" that we wish to prove indistinguishable. (In the proof above, these extreme distributions correspond to the output of $\hat{G}$ and a random string.) To apply the proof technique, three conditions should hold: First, the extreme distributions should match the original cases of interest. (In the proof above, H_n^0 was equal to the distribution induced by $\hat{G}$, while $H_n^{p(n)}$ was the uniform distribution.) Second, it must be possible to translate the capability of distinguishing consecutive hybrid distributions into breaking some underlying assumption. (Intuitively, we showed that distinguishing H_n^i from H_n^{i+1} was equivalent to distinguishing the output of G from random.) Finally, the number of hybrid distributions should be polynomial. See also Theorem 8.31.

8.5 Constructing Pseudorandom Functions

We now show how to construct a pseudorandom function from any (length-doubling) pseudorandom generator. Recall that a pseudorandom function is an efficiently computable, keyed function F that is indistinguishable from a truly random function in the sense described in Section 3.5.1. For simplicity, we restrict our attention here to the case where F is length preserving, meaning that for $k \in \{0,1\}^n$ the function F_k maps n-bit inputs to n-bit outputs.

A length-preserving pseudorandom function can be viewed, informally, as a pseudorandom generator with expansion factor $n \cdot 2^n$; given such a pseudorandom generator G we could define $F_k(i)$ (for $0 \le i < 2^n$) to be the ith n-bit block of $G(k)$. One reason this does not work is that F must be efficiently computable; there are exponentially many blocks, and we need a way to compute the ith block without having to compute all other blocks. We show how to do this by computing "blocks" of the output by walking down a binary tree. We exemplify the idea by first showing a pseudorandom function taking 2-bit inputs.

Let G be a pseudorandom generator with expansion factor $2n$. If we use G as in the proof of Theorem 8.19 we can obtain a pseudorandom generator $\hat{G}$ with expansion factor $4n$ that uses three invocations of G. (We produce n additional pseudorandom bits each time G is applied.) If we define $F_k'(i)$ (where $0 \le i < 4$ and i is encoded as a 2-bit binary string) to be the ith block of $\hat{G}(k)$, then computation of $F_k'(0)$ would require computing $\hat{G}$ in its entirety and hence three invocations of G. We show how to construct a pseudorandom function F using only two invocations of G on any input.

Let G_0 and G_1 be functions denoting the first and second halves of the output of G; i.e., $G(k) = G_0(k) \| G_1(k)$ where $|G_0(k)| = |G_1(k)| = |k|$. Define F as follows:

$$F_k(00) = G_0(G_0(k)) \qquad F_k(10) = G_0(G_1(k))$$
$$F_k(01) = G_1(G_0(k)) \qquad F_k(11) = G_1(G_1(k)).$$

We claim that the four strings above are indistinguishable from four uniform, independent n-bit strings. (This suffices to prove that F is pseudorandom.) Intuitively, this is because $G_0(k)\|G_1(k) = G(k)$ is pseudorandom and hence indistinguishable from a uniform $2n$-bit string $k_0\|k_1$. But then

$$G_0(G_0(k)) \| G_1(G_0(k)) \| G_0(G_1(k)) \| G_1(G_1(k))$$

is indistinguishable from

$$G_0(k_0) \| G_1(k_0) \| G_0(k_1) \| G_1(k_1) = G(k_0) \| G(k_1).$$

Since G is a pseudorandom generator, the above is indistinguishable from a uniform $4n$-bit string. (A formal proof uses a hybrid argument.)

Generalizing this idea, we can obtain a pseudorandom function on n-bit inputs by defining

$$F_k(x) = G_{x_n}(\cdots G_{x_1}(k) \cdots),$$

where $x = x_1 \cdots x_n$; see Construction 8.20. The intuition for why this function is pseudorandom is the same as before, but the formal proof is complicated by the fact that there are now exponentially many inputs to consider.

CONSTRUCTION 8.20

Let G be a pseudorandom generator with expansion factor $\ell(n) = 2n$, and define G_0, G_1 as in the text. For $k \in \{0,1\}^n$, define the function $F_k : \{0,1\}^n \to \{0,1\}^n$ as:

$$F_k(x_1 x_2 \cdots x_n) = G_{x_n}\left(\cdots \left(G_{x_2}(G_{x_1}(k))\right) \cdots\right).$$

A pseudorandom function from a pseudorandom generator.

It is useful to view this construction as defining, for each key $k \in \{0,1\}^n$, a complete binary tree of depth n in which each node contains an n-bit value. (See Figure 8.2, where $n = 3$.) The root has value k, and every non-leaf node with value v has left child with value $G_0(v)$ and right child with value $G_1(v)$. The result $F_k(x)$ for $x = x_1 \cdots x_n$ is defined to be the value on the leaf node reached by traversing the tree according to the bits of x, where $x_i = 0$ means "go left" and $x_i = 1$ means "go right." (The function is only defined for inputs of length n, and thus only values at the leaves are ever output.) The size of the tree is exponential in n. Nevertheless, to compute $F_k(x)$ the entire tree need not be constructed or stored; only n evaluations of G are needed.

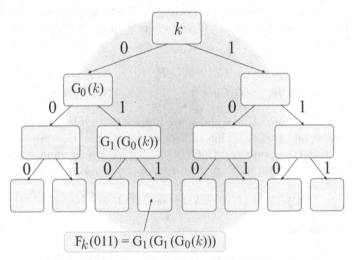

FIGURE 8.2: Constructing a pseudorandom function.

THEOREM 8.21 *If G is a pseudorandom generator with expansion factor $\ell(n) = 2n$, then Construction 8.20 is a pseudorandom function.*

PROOF We first show that for any polynomial t it is infeasible to distinguish $t(n)$ uniform $2n$-bit strings from $t(n)$ pseudorandom strings; i.e., for any polynomial t and any PPT algorithm A, the following is negligible:

$$\left| \Pr\left[A\left(r_1 \| \cdots \| r_{t(n)} \right) = 1 \right] - \Pr\left[A\left(G(s_1) \| \cdots \| G(s_{t(n)}) \right) = 1 \right] \right|,$$

where the first probability is over uniform choice of $r_1, \ldots, r_{t(n)} \in \{0,1\}^{2n}$, and the second probability is over uniform choice of $s_1, \ldots, s_{t(n)} \in \{0,1\}^n$.

The proof is by a hybrid argument. Fix a polynomial t and a PPT algorithm A, and consider the following algorithm A':

Distinguisher A':
A' is given as input a string $w \in \{0,1\}^{2n}$.

1. Choose uniform $j \in \{1, \ldots, t(n)\}$.

2. Choose uniform, independent values $r_1, \ldots, r_{j-1} \in \{0,1\}^{2n}$ and $s_{j+1}, \ldots, s_{t(n)} \in \{0,1\}^n$.

3. Output $A\left(r_1 \| \cdots \| r_{j-1} \| w \| G(s_{j+1}) \| \cdots \| G(s_{t(n)})\right)$.

For any n and $0 \leq i \leq t(n)$, let G_n^i denote the distribution on strings of length $2n \cdot t(n)$ in which the first i "blocks" of length $2n$ are uniform and the remaining $t(n) - i$ blocks are pseudorandom. Note that $G_n^{t(n)}$ corresponds to the distribution in which all $t(n)$ blocks are uniform, while G_n^0 corresponds to the distribution in which all $t(n)$ blocks are pseudorandom. That is,

$$\left| \Pr_{y \leftarrow G_n^{t(n)}}[A(y) = 1] - \Pr_{y \leftarrow G_n^0}[A(y) = 1] \right| \tag{8.11}$$
$$= \left| \Pr\left[A\left(r_1 \| \cdots \| r_{t(n)}\right) = 1\right] - \Pr\left[A\left(G(s_1) \| \cdots \| G(s_{t(n)})\right) = 1\right] \right|$$

Say A' chooses $j = j^*$. If its input w is a uniform $2n$-bit string, then A is run on an input distributed according to $G_n^{j^*}$. If, on the other hand, $w = G(s)$ for uniform s, then A is run on an input distributed according to $G_n^{j^*-1}$. This means that

$$\Pr_{r \leftarrow \{0,1\}^{2n}}[A'(r) = 1] = \frac{1}{t(n)} \cdot \sum_{j=1}^{t(n)} \Pr_{y \leftarrow G_n^j}[A(y) = 1]$$

and

$$\Pr_{s \leftarrow \{0,1\}^n}[A'(G(s)) = 1] = \frac{1}{t(n)} \cdot \sum_{j=0}^{t(n)-1} \Pr_{y \leftarrow G_n^j}[A(y) = 1].$$

Therefore,

$$\left| \Pr_{r \leftarrow \{0,1\}^{2n}}[A'(r) = 1] - \Pr_{s \leftarrow \{0,1\}^n}[A'(G(s)) = 1] \right| \tag{8.12}$$
$$= \frac{1}{t(n)} \cdot \left| \Pr_{y \leftarrow G_n^{t(n)}}[A(y) = 1] - \Pr_{y \leftarrow G_n^0}[A(y) = 1] \right|.$$

Since G is a pseudorandom generator and A' runs in polynomial time, we know that the left-hand side of Equation (8.12) must be negligible; because

$t(n)$ is polynomial, this implies that the left-hand side of Equation (8.11) is negligible as well.

Turning to the crux of the proof, we now show that F as in Construction 8.20 is a pseudorandom function. Let D be an arbitrary PPT distinguisher that is given 1^n as input. We show that D cannot distinguish between the case when it is given oracle access to a function that is equal to F_k for a uniform k, or a function chosen uniformly from Func_n. (See Section 3.5.1.) To do so, we use another hybrid argument. Here, we define distributions over n-bit values at the leaves of a complete binary tree of depth n. By associating each leaf of these binary trees with an n-bit input as in Construction 8.20, we can equivalently view these as distributions over functions mapping n-bit inputs to n-bit outputs. For any n and $0 \leq i \leq n$, let H_n^i be the following distribution over the values at the leaves of a binary tree of depth n: first choose values for the nodes at level i independently and uniformly from $\{0,1\}^n$. Then for every node at level i or below with value k, its left child is given value $G_0(k)$ and its right child is given value $G_1(k)$. Note that H_n^n corresponds to the distribution in which all values at the leaves are chosen uniformly and independently, and thus corresponds to choosing a uniform function from Func_n, whereas H_n^0 corresponds to choosing a uniform key k in Construction 8.20 since in that case only the value at the root (at level 0) is chosen uniformly. That is,

$$\left| \Pr_{k \leftarrow \{0,1\}^n}[D^{F_k(\cdot)}(1^n) = 1] - \Pr_{f \leftarrow \mathsf{Func}_n}[D^{f(\cdot)}(1^n) = 1] \right|$$

$$= \left| \Pr_{f \leftarrow H_n^0}[D^{f(\cdot)}(1^n) = 1] - \Pr_{f \leftarrow H_n^n}[D^{f(\cdot)}(1^n) = 1] \right|. \qquad (8.13)$$

We show that Equation (8.13) is negligible, completing the proof.

Let $t = t(n)$ be a polynomial upper bound on the number of queries D makes to its oracle on input 1^n. Define a distinguisher A that tries to distinguish $t(n)$ uniform $2n$-bit strings from $t(n)$ pseudorandom strings, as follows:

Distinguisher A:
A is given as input a $2n \cdot t(n)$-bit string $w_1 \| \cdots \| w_{t(n)}$.

1. Choose uniform $j \in \{0, \ldots, n-1\}$. In what follows, A (implicitly) maintains a binary tree of depth n with n-bit values at (a subset of) the internal nodes at depth $j+1$ and below.

2. Run $D(1^n)$. When D makes oracle query $x = x_1 \cdots x_n$, look at the prefix $x_1 \cdots x_j$. There are two cases:

   - If D has never made a query with this prefix before, then use $x_1 \cdots x_j$ to reach a node v on the jth level of the tree. Take the next unused $2n$-bit string w and set the value of the left child of node v to the first half of w, and the value of the right child of v to the second half of w.

- If D has made a query with prefix $x_1 \cdots x_j$ before, then node $x_1 \cdots x_{j+1}$ has already been assigned a value.

  Using the value at node $x_1 \cdots x_{j+1}$, compute the value at the leaf corresponding to $x_1 \cdots x_n$ as in Construction 8.20, and return this value to D.

3. When execution of D is done, output the bit returned by D.

A runs in polynomial time. It is important here that A does not need to store the entire binary tree of exponential size. Instead, it "fills in" the values of at most $2t(n)$ nodes in the tree.

Say A chooses $j = j^*$. Observe that:

1. If A's input is a uniform $2n \cdot t(n)$-bit string, then the answers it gives to D are distributed exactly as if D were interacting with a function chosen from distribution $H_n^{j^*+1}$. This holds because the values of the nodes at level $j^* + 1$ of the tree are uniform and independent.

2. If A's input consists of $t(n)$ pseudorandom strings—i.e., $w_i = G(s_i)$ for uniform seed s_i—then the answers it gives to D are distributed exactly as if D were interacting with a function chosen from distribution $H_n^{j^*}$. This holds because the values of the nodes at level j^* of the tree (namely, the $\{s_i\}$) are uniform and independent. (The $\{s_i\}$ are unknown to A, but that makes no difference.)

Proceeding as before, one can show that

$$\left| \Pr\left[A\left(r_1\| \cdots \|r_{t(n)}\right) = 1\right] - \Pr\left[A\left(G(s_1)\| \cdots \|G(s_{t(n)})\right) = 1\right]\right| \quad (8.14)$$

$$= \frac{1}{n} \cdot \left| \Pr_{f \leftarrow H_n^0}[D^{f(\cdot)}(1^n) = 1] - \Pr_{f \leftarrow H_n^n}[D^{f(\cdot)}(1^n) = 1]\right|.$$

We have shown earlier that Equation (8.14) must be negligible. The above thus implies that Equation (8.13) must be negligible as well. ∎

8.6 Constructing (Strong) Pseudorandom Permutations

We next show how pseudorandom permutations and strong pseudorandom permutations can be constructed from any pseudorandom function. Recall from Section 3.5.1 that a pseudorandom permutation is a pseudorandom function that is also efficiently invertible, while a *strong* pseudorandom permutation is additionally hard to distinguish from a random permutation even by an adversary given oracle access to both the permutation *and its inverse*.

Feistel networks revisited. A Feistel network, introduced in Section 7.2.2, provides a way of constructing an efficiently invertible permutation from an arbitrary set of functions. A Feistel network operates in a series of rounds. The input to the ith round is a string of length $2n$, divided into two n-bit halves L_{i-1} and R_{i-1} (the "left half" and the "right half," respectively). The output of the ith round is the $2n$-bit string (L_i, R_i), where

$$L_i := R_{i-1} \quad \text{and} \quad R_i := L_{i-1} \oplus f_i(R_{i-1})$$

for some efficiently computable (but not necessarily invertible) function f_i mapping n-bit inputs to n-bit outputs. We denote by $\mathsf{Feistel}_{f_1,\ldots,f_r}$ the r-round Feistel network using functions $f_1,\ldots,f_r$. (That is, $\mathsf{Feistel}_{f_1,\ldots,f_r}(L_0, R_0)$ outputs the $2n$-bit string (L_r, R_r).) We saw in Section 7.2.2 that $\mathsf{Feistel}_{f_1,\ldots,f_r}$ is an efficiently invertible permutation regardless of the $\{f_i\}$.

We can define a keyed permutation by using a Feistel network in which the $\{f_i\}$ depend on a key. For example, let $F : \{0,1\}^n \times \{0,1\}^n \to \{0,1\}^n$ be a pseudorandom function, and define the keyed *permutation* $F^{(1)}$ as

$$F_k^{(1)}(x) \overset{\text{def}}{=} \mathsf{Feistel}_{F_k}(x).$$

(Note that $F_k^{(1)}$ has an n-bit key and maps $2n$-bit inputs to $2n$-bit outputs.) Is $F^{(1)}$ pseudorandom? A little thought shows that it is decidedly *not*. For any key $k \in \{0,1\}^n$, the first n bits of the output of $F_k^{(1)}$ (that is, L_1) are equal to the last n bits of the input (i.e., R_0), something that occurs with only negligible probability for a random permutation.

Trying again, define $F^{(2)} : \{0,1\}^{2n} \times \{0,1\}^{2n} \to \{0,1\}^{2n}$ as follows:

$$F_{k_1,k_2}^{(2)}(x) \overset{\text{def}}{=} \mathsf{Feistel}_{F_{k_1},F_{k_2}}(x). \tag{8.15}$$

(Note that k_1 and k_2 are independent keys.) Unfortunately, $F^{(2)}$ is not pseudorandom either, as you are asked to show in Exercise 8.16.

Given this, it may be somewhat surprising that a *three*-round Feistel network *is* pseudorandom. Define the keyed permutation $F^{(3)}$, taking a key of length $3n$ and mapping $2n$-bit inputs to $2n$-bit outputs, as follows:

$$F_{k_1,k_2,k_3}^{(3)}(x) \overset{\text{def}}{=} \mathsf{Feistel}_{F_{k_1},F_{k_2},F_{k_3}}(x) \tag{8.16}$$

where, once again, $k_1, k_2,$ and k_3 are independent. We have:

THEOREM 8.22 *If F is a pseudorandom function, then $F^{(3)}$ is a pseudorandom permutation.*

PROOF In the standard way, we can replace the pseudorandom functions used in the construction of $F^{(3)}$ with functions chosen uniformly at random

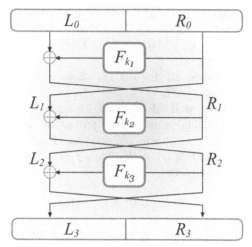

FIGURE 8.3:   A three-round Feistel network, as used to construct a pseudorandom permutation from a pseudorandom function.

instead. Pseudorandomness of F implies that this has only a negligible effect on the output of any probabilistic polynomial-time distinguisher interacting with $F^{(3)}$ as an oracle. We leave the details as an exercise.

Let D be a probabilistic polynomial-time distinguisher. In the remainder of the proof, we show the following is negligible:

$$\left| \Pr[D^{\mathsf{Feistel}_{f_1,f_2,f_3}(\cdot)}(1^n) = 1] - \Pr[D^{\pi(\cdot)}(1^n) = 1] \right|,$$

where the first probability is taken over uniform and independent choice of f_1, f_2, f_3 from Func_n, and the second probability is taken over uniform choice of π from Perm_{2n}. Fix some value for the security parameter n, and let $q = q(n)$ denote a polynomial upper bound on the number of oracle queries made by D. We assume without loss of generality that D never makes the same oracle query twice. Focusing on D's interaction with $\mathsf{Feistel}_{f_1,f_2,f_3}(\cdot)$, let (L_0^i, R_0^i) denote the ith query D makes to its oracle, and let (L_1^i, R_1^i), (L_2^i, R_2^i), and (L_3^i, R_3^i) denote the intermediate values after rounds 1, 2, and 3, respectively, that result from that query. (See Figure 8.3.) Note that D chooses (L_0^i, R_0^i) and sees the result (L_3^i, R_3^i), but does not directly observe (L_1^i, R_1^i) or (L_2^i, R_2^i).

We say there is *a collision at* R_1 if $R_1^i = R_1^j$ for some distinct i, j. We first prove that a collision at R_1 occurs with only negligible probability. Consider any fixed, distinct i, j. If $R_0^i = R_0^j$ then $L_0^i \neq L_0^j$, but then

$$R_1^i = L_0^i \oplus f_1(R_0^i) \neq L_0^j \oplus f_1(R_0^j) = R_1^j.$$

If $R_0^i \neq R_0^j$ then $f_1(R_0^i)$ and $f_1(R_0^j)$ are uniform and independent, so

$$\Pr[R_1^i = R_1^j] = \Pr\left[f_1(R_0^j) = L_0^i \oplus f_1(R_0^i) \oplus L_0^j\right] = 2^{-n}.$$

Taking a union bound over all distinct i, j shows that the probability of a collision at R_1 is at most $q^2/2^n$.

Say there is *a collision at* R_2 if $R_2^i = R_2^j$ for some distinct i, j. We prove that conditioned on no collision at R_1, the probability of a collision at R_2 is negligible. The analysis is as above: consider any fixed i, j, and note that if there is no collision at R_1 then $R_1^i \neq R_1^j$. Thus $f_2(R_1^i)$ and $f_2(R_1^j)$ are uniform and independent, and therefore

$$\Pr\left[L_1^i \oplus f_2(R_1^i) = L_1^j \oplus f_2(R_1^j) \mid \text{no collision at } R_1\right] = 2^{-n}.$$

(Note that f_2 is independent of f_1, making the above calculation easy.) Taking a union bound over all distinct i, j gives

$$\Pr[\text{collision at } R_2 \mid \text{no collision at } R_1] \leq q^2/2^n.$$

Note that $L_3^i = R_2^i = L_1^i \oplus f_2(R_1^i)$; so, conditioned on there being no collision at R_1, the values $L_3^1, \ldots, L_3^q$ are all independent and uniformly distributed in $\{0,1\}^n$. If we additionally condition on the event that there is no collision at R_2, then the values $L_3^1, \ldots, L_3^q$ are uniformly distributed among all sequences of q *distinct* values in $\{0,1\}^n$. Similarly, $R_3^i = L_2^i \oplus f_3(R_2^i)$; thus, conditioned on there being no collision at R_2, the values $R_3^1, \ldots, R_3^q$ are all uniformly distributed in $\{0,1\}^n$, independent of each other as well as $L_3^1, \ldots, L_3^q$.

To summarize: when querying $F^{(3)}$ (with uniform round functions) on a series of q distinct inputs, except with negligible probability the output values $(L_3^1, R_3^1), \ldots, (L_3^q, R_3^q)$ are distributed such that the $\{L_3^i\}$ are uniform and independent, but distinct, n-bit values, and the $\{R_3^i\}$ are uniform and independent n-bit values. In contrast, when querying a random permutation on a series of q distinct inputs, the output values $(L_3^1, R_3^1), \ldots, (L_3^q, R_3^q)$ are uniform and independent, but distinct, $2n$-bit values. It can be shown that the best distinguishing attack for D, then, is to guess that it is interacting with a random permutation if $L_3^i = L_3^j$ for some distinct i, j. But that event occurs with negligible probability even in that case. ∎

$F^{(3)}$ is not a *strong* pseudorandom permutation, as you are asked to demonstrate in Exercise 8.17. Fortunately, adding a fourth round *does* yield a strong pseudorandom permutation. The details are given as Construction 8.23.

THEOREM 8.24 *If F is a pseudorandom function, then Construction 8.23 is a strong pseudorandom permutation that maps $2n$-bit inputs to $2n$-bit outputs (and uses a $4n$-bit key).*

CONSTRUCTION 8.23

Let F be a keyed, length-preserving function. Define the keyed permutation $F^{(4)}$ as follows:

- **Inputs:** A key $k = (k_1, k_2, k_3, k_4)$ with $|k_i| = n$, and an input $x \in \{0,1\}^{2n}$ parsed as (L_0, R_0) with $|L_0| = |R_0| = n$.
- **Computation:**
    1. Compute $L_1 := R_0$ and $R_1 := L_0 \oplus F_{k_1}(R_0)$.
    2. Compute $L_2 := R_1$ and $R_2 := L_1 \oplus F_{k_2}(R_1)$.
    3. Compute $L_3 := R_2$ and $R_3 := L_2 \oplus F_{k_3}(R_2)$.
    4. Compute $L_4 := R_3$ and $R_4 := L_3 \oplus F_{k_4}(R_3)$.
    5. Output (L_4, R_4).

A strong pseudorandom permutation from any pseudorandom function.

8.7 Assumptions for Private-Key Cryptography

We have shown that (1) if there exist one-way permutations, then there exist pseudorandom generators; (2) if there exist pseudorandom generators, then there exist pseudorandom functions; and (3) if there exist pseudorandom functions, then there exist (strong) pseudorandom permutations. Although we did not prove it here, it is possible to construct pseudorandom generators from one-way *functions*. We thus have the following fundamental theorem:

THEOREM 8.25 *If one-way functions exist, then so do pseudorandom generators, pseudorandom functions, and strong pseudorandom permutations.*

All the private-key schemes we have studied in Chapters 3–5 can be constructed from pseudorandom generators/functions. We therefore have:

THEOREM 8.26 *If one-way functions exist, then so do authenticated encryption schemes and secure message authentication codes.*

That is, *one-way functions are sufficient for all private-key cryptography.* Here, we show that one-way functions are also *necessary.*

Pseudorandomness implies one-way functions. We begin by showing that pseudorandom generators imply the existence of one-way functions:

PROPOSITION 8.27 *If a pseudorandom generator exists, then so do one-way functions.*

PROOF   Let G be a pseudorandom generator with expansion factor $\ell(n) = 2n$. (By Theorem 8.19, we know that the existence of a pseudorandom generator implies the existence of one with this expansion factor.) We show that G itself is one-way. Efficient computability is straightforward (since G can be computed in polynomial time). We show that the ability to invert G can be translated into the ability to distinguish the output of G from uniform. Intuitively, this holds because the ability to invert G implies the ability to find the seed used by the generator.

Let $\mathcal{A}$ be an arbitrary probabilistic polynomial-time algorithm. We show that $\Pr[\mathsf{Invert}_{\mathcal{A},G}(n) = 1]$ is negligible (cf. Definition 8.1). To see this, consider the following PPT distinguisher D: on input a string $w \in \{0,1\}^{2n}$, run $\mathcal{A}(w)$ to obtain output s. If $G(s) = w$ then output 1; otherwise, output 0.

We now analyze the behavior of D. First consider the probability that D outputs 1 when its input string w is uniform. Since there are at most 2^n values in the range of G (namely, the values $\{G(s)\}_{s\in\{0,1\}^n}$), the probability that w is in the range of G is at most $2^n/2^{2n} = 2^{-n}$. When w is not in the range of G, it is impossible for $\mathcal{A}$ to compute an inverse of w and thus impossible for D to output 1. We conclude that $\Pr_{w\leftarrow\{0,1\}^{2n}}[D(w) = 1] \leq 2^{-n}$.

On the other hand, if $w = G(s)$ for a seed $s \in \{0,1\}^n$ chosen uniformly at random then, by definition, $\mathcal{A}$ computes a correct inverse (and so D outputs 1) with probability exactly equal to $\Pr[\mathsf{Invert}_{\mathcal{A},G}(n) = 1]$. Thus,

$$\left| \Pr_{w\leftarrow\{0,1\}^{2n}}[D(w) = 1] - \Pr_{s\leftarrow\{0,1\}^n}[D(G(s)) = 1] \right| \geq \Pr[\mathsf{Invert}_{\mathcal{A},G}(n) = 1] - 2^{-n}.$$

Since G is a pseudorandom generator, the above must be negligible. Since 2^{-n} is negligible, this implies that $\Pr[\mathsf{Invert}_{\mathcal{A},G}(n) = 1]$ is negligible as well and so G is one-way.   ∎

Non-trivial private-key encryption implies one-way functions. Proposition 8.27 does not imply that one-way functions are needed for constructing secure private-key encryption schemes, since it may be possible to construct the latter without relying on a pseudorandom generator. Furthermore, it *is* possible to construct perfectly secret encryption schemes (see Chapter 2), as long as the plaintext is no longer than the key. Thus, a proof that secure private-key encryption implies one-way functions requires more care.

PROPOSITION 8.28   *If there exists an EAV-secure private-key encryption scheme that encrypts messages twice as long as its key, then a one-way function exists.*

PROOF   Let $\Pi = (\mathsf{Enc}, \mathsf{Dec})$ be a private-key encryption scheme that has indistinguishable encryptions in the presence of an eavesdropper and encrypts messages of length $2n$ when the key has length n. (We assume for simplicity that the key is chosen uniformly.) Let $\ell(n)$ be a bound on the number of random bits used by Enc. Denote the encryption of a message m using key k and randomness r by $\mathsf{Enc}_k(m; r)$.

Define the following function f:

$$f(k, m, r) \overset{\text{def}}{=} \mathsf{Enc}_k(m; r) \parallel m,$$

where $|k| = n$, $|m| = 2n$, and $|r| = \ell(n)$. We claim that f is a one-way function. Clearly it can be efficiently computed; we show that it is hard to invert. Letting $\mathcal{A}$ be an arbitrary PPT algorithm, we show that $\Pr[\mathsf{Invert}_{\mathcal{A}, f}(n) = 1]$ is negligible (cf. Definition 8.1).

Consider the following probabilistic polynomial-time adversary $\mathcal{A}'$ attacking private-key encryption scheme Π (i.e., in experiment $\mathsf{PrivK}^{\mathsf{eav}}_{\mathcal{A}', \Pi}(n)$):

Adversary $\mathcal{A}'(1^n)$

1. Choose uniform $m_0, m_1 \leftarrow \{0, 1\}^{2n}$ and output them. Receive in return a challenge ciphertext c.

2. Run $\mathcal{A}(c \parallel m_0)$ to obtain (k', m', r'). If $f(k', m', r') = c \parallel m_0$, output 0; else, output 1.

We now analyze the behavior of $\mathcal{A}'$. When c is an encryption of m_0, then $c \parallel m_0$ is distributed exactly as $f(k, m_0, r)$ for uniform k, m_0, and r. Therefore, $\mathcal{A}$ outputs a valid inverse of $c \parallel m_0$ (and hence $\mathcal{A}'$ outputs 0) with probability exactly equal to $\Pr[\mathsf{Invert}_{\mathcal{A}, f}(n) = 1]$.

On the other hand, when c is an encryption of m_1 then c is independent of m_0. For any fixed value of the challenge ciphertext c, there are at most 2^n possible messages (one for each possible key) to which c can correspond. Since m_0 is a uniform $2n$-bit string, the probability that there *exists* some key k for which $\mathsf{Dec}_k(c) = m_0$ is at most $2^n / 2^{2n} = 2^{-n}$. This gives an upper bound on the probability with which $\mathcal{A}$ can possibly output a valid inverse of $c \parallel m_0$ under f, and hence an upper bound on the probability with which $\mathcal{A}'$ outputs 0 in that case.

Putting the above together, we have:

$$\Pr\left[\mathsf{PrivK}^{\mathsf{eav}}_{\mathcal{A}', \Pi}(n) = 1\right]$$
$$= \frac{1}{2} \cdot \Pr\left[\mathcal{A}' \text{ outputs } 0 \mid b = 0\right] + \frac{1}{2} \cdot \Pr\left[\mathcal{A}' \text{ outputs } 1 \mid b = 1\right]$$
$$\geq \frac{1}{2} \cdot \Pr[\mathsf{Invert}_{\mathcal{A}, f}(n) = 1] + \frac{1}{2} \cdot \left(1 - 2^{-n}\right)$$
$$= \frac{1}{2} + \frac{1}{2} \cdot \left(\Pr[\mathsf{Invert}_{\mathcal{A}, f}(n) = 1] - 2^{-n}\right).$$

Security of Π means that $\Pr\left[\mathsf{PrivK}^{\mathsf{eav}}_{\mathcal{A}', \Pi}(n) = 1\right] \leq \frac{1}{2} + \mathsf{negl}(n)$ for some negligible function negl. This, in turn, implies that $\Pr[\mathsf{Invert}_{\mathcal{A}, f}(n) = 1]$ is negligible, completing the proof that f is one-way. ∎

Message authentication codes imply one-way functions. It is also true that message authentication codes satisfying Definition 4.2 imply the existence of one-way functions. As in the case of private-key encryption, a proof of this fact is somewhat subtle because unconditional message authentication codes *do* exist when there is a bound on the number of messages that will be authenticated. (See Section 4.6.) Thus, a proof relies on the fact that Definition 4.2 requires security even when the adversary sees tags for an *arbitrary* (polynomial) number of messages. The proof is somewhat involved, so we do not give it here.

Discussion. We conclude that the existence of one-way functions is necessary and sufficient for all (non-trivial) private-key cryptography. In other words, one-way functions are a *minimal* assumption as far as private-key cryptography is concerned. Interestingly, this appears not to be the case for hash functions and public-key encryption, where one-way functions are known to be necessary but are not known (or believed) to be sufficient.

8.8   Computational Indistinguishability

The notion of computational indistinguishability is central to the theory of cryptography, and underlies much of what we have seen in Chapter 3 and this chapter. Informally, two probability distributions are *computationally indistinguishable* if no efficient algorithm can tell them apart (or *distinguish* them). In more detail, consider two distributions X and Y over strings of some length ℓ; that is, X and Y each assigns some probability to every string in $\{0,1\}^\ell$. When we say that some algorithm D cannot distinguish these two distributions, we mean that D cannot tell whether it is given a string sampled according to distribution X or whether it is given a string sampled according to distribution Y. Put differently, if we imagine D outputting "0" when it believes its input was sampled according to X and outputting "1" if it thinks its input was sampled according to Y, then the probability that D outputs "1" should be roughly the same regardless of whether D is provided with a sample from X or from Y. In other words, we want

$$\left| \Pr_{s \leftarrow X}[D(s) = 1] - \Pr_{s \leftarrow Y}[D(s) = 1] \right|$$

to be small.

This should be reminiscent of the way we defined pseudorandom generators and, indeed, we will soon formally redefine the notion of a pseudorandom generator using this terminology.

The formal definition of computational indistinguishability refers to *probability ensembles*, which are infinite sequences of probability distributions.

(This formalism is necessary for a meaningful asymptotic approach.) Although the notion can be generalized, for our purposes we consider probability ensembles in which the underlying distributions are indexed by natural numbers. If for every natural number n we have a distribution X_n, then $\mathcal{X} = \{X_n\}_{n \in \mathbb{N}}$ is a probability ensemble. It is often the case that $X_n = Y_{t(n)}$ for some function t, in which case we write $\{Y_{t(n)}\}_{n \in \mathbb{N}}$ in place of $\{X_n\}_{n \in \mathbb{N}}$.

We will only be interested in *efficiently sampleable* probability ensembles. An ensemble $\mathcal{X} = \{X_n\}_{n \in \mathbb{N}}$ is efficiently sampleable if there is a probabilistic polynomial-time algorithm S such that the random variables $S(1^n)$ and X_n are identically distributed. That is, algorithm S is an efficient way of sampling $\mathcal{X}$.

We can now formally define what it means for two ensembles to be computationally indistinguishable.

DEFINITION 8.29 *Two probability ensembles* $\mathcal{X} = \{X_n\}_{n \in \mathbb{N}}$ *and* $\mathcal{Y} = \{Y_n\}_{n \in \mathbb{N}}$ *are* computationally indistinguishable, *denoted* $\mathcal{X} \stackrel{c}{\equiv} \mathcal{Y}$, *if for every probabilistic polynomial-time distinguisher D there exists a negligible function* negl *such that:*

$$\left| \Pr_{x \leftarrow X_n}[D(1^n, x) = 1] - \Pr_{y \leftarrow Y_n}[D(1^n, y) = 1] \right| \leq \mathsf{negl}(n).$$

In the definition, D is given the unary input 1^n so it can run in time polynomial in n. This is important when the outputs of X_n and Y_n may have length less than n. As shorthand in probability expressions, we will sometimes write X as a placeholder for a random sample from distribution X. That is, we would write $\Pr[D(1^n, X_n) = 1]$ in place of $\Pr_{x \leftarrow X_n}[D(1^n, x) = 1]$.

Pseudorandomness and pseudorandom generators. Pseudorandomness is just a special case of computational indistinguishability. For any integer ℓ, let U_ℓ denote the uniform distribution over $\{0,1\}^\ell$. We can define a pseudorandom generator as follows:

DEFINITION 8.30 *Let $\ell(\cdot)$ be a polynomial and let G be a (deterministic) polynomial-time algorithm where for all s it holds that $|G(s)| = \ell(|s|)$. We say that G is a* pseudorandom generator *if the following two conditions hold:*

1. **(Expansion.)** *For every n it holds that $\ell(n) > n$.*

2. **(Pseudorandomness.)** *The ensemble $\{G(U_n)\}_{n \in \mathbb{N}}$ is computationally indistinguishable from the ensemble $\{U_{\ell(n)}\}_{n \in \mathbb{N}}$.*

Many of the other definitions and assumptions in this book can also be cast as special cases or variants of computational indistinguishability.

Multiple samples. An important theorem regarding computational indistinguishability is that polynomially many samples of (efficiently sampleable) computationally indistinguishable ensembles are also computationally indistinguishable.

THEOREM 8.31 *Let $\mathcal{X}$ and $\mathcal{Y}$ be efficiently sampleable probability ensembles that are computationally indistinguishable. Then, for every polynomial t, the ensemble $\overline{\mathcal{X}} = \{(X_n^{(1)}, \ldots, X_n^{(t(n))})\}_{n \in \mathbb{N}}$ is computationally indistinguishable from the ensemble $\overline{\mathcal{Y}} = \{(Y_n^{(1)}, \ldots, Y_n^{(t(n))})\}_{n \in \mathbb{N}}$.*

For example, let G be a pseudorandom generator with expansion factor $2n$, in which case the ensembles $\{G(U_n)\}_{n \in \mathbb{N}}$ and $\{U_{2n}\}_{n \in \mathbb{N}}$ are computationally indistinguishable. In the proof of Theorem 8.21 we showed that for any polynomial t the ensembles

$$\{(\underbrace{G(U_n), \ldots, G(U_n)}_{t(n)})\}_{n \in \mathbb{N}} \quad \text{and} \quad \{(\underbrace{U_{2n}, \ldots, U_{2n}}_{t(n)})\}_{n \in \mathbb{N}}$$

are also computationally indistinguishable. Theorem 8.31 is proved by a hybrid argument in exactly the same way.

References and Additional Reading

The notion of a one-way function was first proposed by Diffie and Hellman [65] and later formalized by Yao [205]. Hard-core predicates were introduced by Blum and Micali [41], and the fact that there exists a hard-core predicate for every one-way function was proved by Goldreich and Levin [86].

The first construction of pseudorandom generators (under a specific number-theoretic hardness assumption) was given by Blum and Micali [41]. The construction of a pseudorandom generator from any one-way permutation was given by Yao [205], and the result that pseudorandom generators can be constructed from any one-way function was shown by Håstad et al. [93]. Pseudorandom functions were defined and constructed by Goldreich, Goldwasser and Micali [85] and their extension to (strong) pseudorandom permutations was shown by Luby and Rackoff [132]. The fact that one-way functions are a necessary assumption for most of private-key cryptography was shown in [101]. The proof of Proposition 8.28 is from [79].

Our presentation is heavily influenced by Goldreich's book [82], which is highly recommended for those interested in exploring the topics of this chapter in greater detail.

Exercises

8.1 Prove that if there exists a one-way function, then there exists a one-way function f such that $f(0^n) = 0^n$ for every n. Note that for infinitely many values y, it is easy to compute $f^{-1}(y)$. Why does this not contradict one-wayness?

8.2 Prove that if f is a one-way function, then the function g defined by $g(x_1, x_2) \stackrel{\text{def}}{=} (f(x_1), x_2)$, where $|x_1| = |x_2|$, is also a one-way function. Observe that g reveals half of its input, but is nevertheless one-way.

8.3 Prove that if there exists a one-way function, then there exists a length-preserving one-way function.

> **Hint:** Let f be a one-way function and let $p(\cdot)$ be a polynomial such that $|f(x)| \leq p(|x|)$. (Justify the existence of such a p.) Define $f'(x) \stackrel{\text{def}}{=} f(x) \| 10^{p(|x|) - |f(x)|}$. Further modify f' to get a length-preserving function that remains one-way.

8.4 Let (Gen, H) be a collision-resistant hash function, where H maps strings of length $2n$ to strings of length n. Prove that the function family $(\mathsf{Gen}, \mathsf{Samp}, H)$ is one-way (cf. Definition 8.3), where Samp is the trivial algorithm that samples a uniform string of length $2n$.

> **Hint:** Choosing uniform $x \in \{0,1\}^{2n}$ and finding an inverse of $y = H^s(x)$ does not guarantee a collision. But it does yield a collision most of the time...

8.5 Let F be a (length-preserving) pseudorandom permutation.

   (a) Show that the function $f(x, y) = F_x(y)$ is not one-way.

   (b) Show that the function $f(y) = F_{0^n}(y)$ (where $n = |y|$) is not one-way.

   (c) Prove that the function $f(x) = F_x(0^n)$ (where $n = |x|$) is one-way.

8.6 Let f be a length-preserving one-way *function*, and let hc be a hard-core predicate of f. Define G as $G(x) = f(x) \| \mathsf{hc}(x)$. Is G necessarily a pseudorandom generator? Prove your answer.

8.7 Prove that there exist one-way functions if and only if there exist one-way function families. Discuss why your proof does not carry over to the case of one-way permutations.

8.8 Let f be a length-preserving one-way function. Is $g(x) \stackrel{\text{def}}{=} f(f(x))$ necessarily one-way? What about $g'(x) \stackrel{\text{def}}{=} f(x) \| f(f(x))$?

8.9 Let $\Pi = (\text{Gen}, \text{Samp}, f)$ be a function family. A function $\text{hc} : \{0,1\}^* \rightarrow \{0,1\}$ is a *hard-core predicate of* Π if it is efficiently computable and if for every PPT algorithm $\mathcal{A}$ there is a negligible function negl such that

$$\Pr_{I \leftarrow \text{Gen}(1^n),\, x \leftarrow \text{Samp}(I)} [\mathcal{A}(I, f_I(x)) = \text{hc}(I, x)] \leq \frac{1}{2} + \text{negl}(n).$$

Prove a version of the Goldreich–Levin theorem for this setting, namely, if a one-way function (resp., permutation) family Π exists, then there exists a one-way function (resp., permutation) family Π' and a hard-core predicate hc of Π'.

8.10 Show a construction of a pseudorandom generator from any one-way permutation family. You may use the result of the previous exercise.

8.11 This exercise is for students who have taken a course in complexity theory or are otherwise familiar with $\mathcal{NP}$-completeness.

 (a) Show that the existence of one-way functions implies $\mathcal{P} \neq \mathcal{NP}$.

 (b) Assume that $\mathcal{P} \neq \mathcal{NP}$. Show that there exists a function f that is: (1) computable in polynomial time, (2) hard to invert *in the worst case* (i.e., for all probabilistic polynomial-time $\mathcal{A}$, $\Pr_{x \leftarrow \{0,1\}^n}[f(\mathcal{A}(f(x))) = f(x)] \neq 1$), but (3) is *not* one-way.

8.12 For $x \in \{0,1\}^n$ let $x = x_1 \cdots x_n$. Prove that if there exists a one-way function, then there exists a one-way function f such that for every i there is an algorithm A_i such that

$$\Pr_{x \leftarrow \{0,1\}^n} [A_i(f(x)) = x_i] \geq \frac{1}{2} + \frac{1}{2n}.$$

(This exercise demonstrates that it is not possible to claim that every one-way function hides at least one *specific* bit of the input.)

8.13 Show that if an efficiently computable one-to-one function f has a hard-core predicate, then f is one-way.

8.14 Show that if Construction 8.20 is modified in the natural way so that $F_k(x)$ is defined for every nonempty string x of length at most n, then the construction is no longer a pseudorandom function.

8.15 Prove that if there exists a pseudorandom function that, using a key of length n, maps n-bit inputs to single-bit outputs, then there exists a pseudorandom function that maps n-bit inputs to n-bit outputs.

>   **Hint:** Use a key of length n^2, and prove your construction secure using a hybrid argument.

8.16 Prove that a two-round Feistel network using pseudorandom round functions (as in Equation (8.15)) is not a pseudorandom permutation.

8.17 Prove that a three-round Feistel network using pseudorandom round functions (as in Equation (8.16)) is not a *strong* pseudorandom permutation.

> **Hint:** This is significantly more difficult than the previous exercise. Use a distinguisher that makes two queries to the permutation and one query to its inverse.

8.18 Consider the keyed permutation F^* defined by

$$F_k^*(x) \stackrel{\text{def}}{=} \mathsf{Feistel}_{F_k, F_k, F_k}(x).$$

(Note that the same key is used in each round.) Show that F^* is not a pseudorandom permutation.

8.19 Let $\mathcal{X}, \mathcal{Y}, \mathcal{Z}$ be probability ensembles. Prove that if $\mathcal{X} \stackrel{c}{\equiv} \mathcal{Y}$ and $\mathcal{Y} \stackrel{c}{\equiv} \mathcal{Z}$, then $\mathcal{X} \stackrel{c}{\equiv} \mathcal{Z}$.

8.20 Prove Theorem 8.31.

8.21 Let $\mathcal{X} = \{X_n\}_{n \in \mathbb{N}}$ and $\mathcal{Y} = \{Y_n\}_{n \in \mathbb{N}}$ be computationally indistinguishable probability ensembles. Prove that for any probabilistic polynomial-time algorithm $\mathcal{A}$, the ensembles $\{\mathcal{A}(X_n)\}_{n \in \mathbb{N}}$ and $\{\mathcal{A}(Y_n)\}_{n \in \mathbb{N}}$ are computationally indistinguishable.

Part III

Public-Key (Asymmetric) Cryptography

Chapter 9

Number Theory and Cryptographic Hardness Assumptions

Modern cryptosystems are invariably based on an assumption that *some* problem is hard. In Chapters 3–5, for example, we saw that private-key cryptography—both encryption schemes and message authentication codes—can be based on the assumption that pseudorandom permutations (a.k.a. block ciphers) exist. On the face of it, the assumption that pseudorandom permutations exist seems quite strong and unnatural, and it is reasonable to ask whether this assumption is true or whether there is any evidence to support it. In Chapter 7 we explored how block ciphers are constructed in practice. The fact that these constructions have resisted attack serves as an indication that the existence of pseudorandom permutations is plausible. Still, it may be difficult to believe that there are *no* efficient distinguishing attacks on existing block ciphers. Moreover, the current state of our theory is such that we do not know how to prove the pseudorandomness of any of the existing practical constructions relative to any "simpler" or "more reasonable" assumption. All in all, this is not an entirely satisfying state of affairs.

In contrast, as mentioned in Chapter 3 (and investigated in detail in Chapter 8) it is possible to *prove* that pseudorandom permutations exist based on the much milder assumption that one-way functions exist. (Informally, a function is *one-way* if it is easy to compute but hard to invert; see Section 9.4.1.) Apart from a brief discussion in Section 8.1.2, however, we have not seen any concrete examples of functions believed to be one-way.

One goal of this chapter is to introduce various problems believed to be "hard," and to present conjectured one-way functions based on those problems.[1] As such, this chapter can be viewed as a culmination of a "top down" approach to private-key cryptography. (See Figure 9.1.) That is, in Chapters 3–5 we have shown that private-key cryptography can be based on pseudorandom functions and permutations. We have then seen that the latter can be instantiated in practice using block ciphers, as explored in Chapter 7, or can be provably constructed from any one-way function, as shown in Chapter 8. Here, we take this one step further and show how one-way functions can be based on certain hard mathematical problems.

---

[1]Recall we currently do not know how to *prove* that one-way functions exist, so the best we can do is base one-way functions on assumptions regarding the hardness of certain problems.

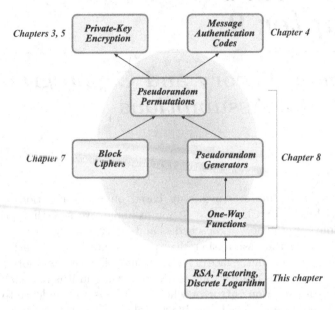

FIGURE 9.1: Private-key cryptography: a top-down approach.

The examples we explore are *number theoretic* in nature, and we therefore begin with a short introduction to number theory. Because we are also interested in problems that can be solved efficiently (even a one-way function must be easy to compute in one direction, and cryptographic schemes must admit efficient algorithms for the honest parties), we also initiate a study of *algorithmic* number theory. Even the reader who is familiar with number theory is encouraged to read this chapter, since algorithmic aspects are typically ignored in a purely mathematical treatment of these topics.

A second goal of this chapter is to develop the material needed for *public-key cryptography*, whose study we will begin in Chapter 11. Strikingly, although in the private-key setting there exist efficient constructions of the necessary primitives (both block ciphers and hash functions) without invoking any number theory, in the public-key setting *all known constructions rely on hard number-theoretic problems*. The material in this chapter thus serves not only as a culmination of our study of private-key cryptography, but also as the foundation for our treatment of public-key cryptography.

9.1   Preliminaries and Basic Group Theory

We begin with a review of prime numbers and basic modular arithmetic. Even the reader who has seen these topics before should skim the next two sections since some of the material may be new and we include proofs for most of the stated results.

9.1.1   Primes and Divisibility

The set of integers is denoted by $\mathbb{Z}$. For $a, b \in \mathbb{Z}$, we say that a *divides* b, written $a \mid b$, if there exists an integer c such that $ac = b$. If a does not divide b, we write $a \nmid b$. (We are primarily interested in the case where a, b, and c are all positive, although the definition makes sense even when one or more of them is negative or zero.) A simple observation is that if $a \mid b$ and $a \mid c$ then $a \mid (Xb + Yc)$ for any $X, Y \in \mathbb{Z}$.

If $a \mid b$ and a is positive, we call a a *divisor* of b. If in addition $a \notin \{1, b\}$ then a is called a *nontrivial* divisor, or a *factor*, of b. A positive integer $p > 1$ is *prime* if it has no factors; i.e., it has only two divisors: 1 and itself. A positive integer greater than 1 that is not prime is called *composite*. By convention, the number 1 is neither prime nor composite.

A fundamental theorem of arithmetic is that every integer greater than 1 can be expressed *uniquely* (up to ordering) as a product of primes. That is, any positive integer $N > 1$ can be written as $N = \prod_i p_i^{e_i}$, where the $\{p_i\}$ are distinct primes and $e_i \geq 1$ for all i; furthermore, the $\{p_i\}$ (and $\{e_i\}$) are uniquely determined up to ordering.

We are familiar with the process of *division with remainder* from elementary school. The following proposition formalizes this notion.

PROPOSITION 9.1   *Let a be an integer and let b be a positive integer. Then there exist unique integers q, r for which $a = qb + r$ and $0 \leq r < b$.*

Furthermore, given integers a and b as in the proposition it is possible to compute q and r in polynomial time; see Appendix B.1. (An algorithm's running time is measured as a function of the length(s) of its input(s). An important point in the context of algorithmic number theory is that integer inputs are always assumed to be represented in binary. The running time of an algorithm taking as input an integer N is therefore measured in terms of $\|N\|$, the *length of the binary representation of N*. Note that $\|N\| = \lfloor \log N \rfloor + 1$.)

The *greatest common divisor* of two integers a, b, written $\gcd(a, b)$, is the largest integer c such that $c \mid a$ and $c \mid b$. (We leave $\gcd(0, 0)$ undefined.) The notion of greatest common divisor makes sense when either or both of a, b are negative but we will typically have $a, b \geq 1$; anyway, $\gcd(a, b) = \gcd(|a|, |b|)$. Note that $\gcd(b, 0) = \gcd(0, b) = b$; also, if p is prime then $\gcd(a, p)$ is either equal to 1 or p. If $\gcd(a, b) = 1$ we say that a and b are *relatively prime*.

The following is a useful result:

PROPOSITION 9.2   *Let a, b be positive integers. Then there exist integers X, Y such that $Xa + Yb = \gcd(a, b)$. Furthermore, $\gcd(a, b)$ is the smallest positive integer that can be expressed in this way.*

PROOF   Consider the set $I \overset{\text{def}}{=} \{\hat{X}a + \hat{Y}b \mid \hat{X}, \hat{Y} \in \mathbb{Z}\}$. Note that $a, b \in I$, and so I certainly contains some positive integers. Let d be the smallest positive integer in I. We show that $d = \gcd(a, b)$; since d can be written as $d = Xa + Yb$ for some $X, Y \in \mathbb{Z}$ (because $d \in I$), this proves the theorem.

To show that $d = \gcd(a, b)$, we must prove that $d \mid a$ and $d \mid b$, and that d is the largest integer with this property. In fact, we can show that d divides every element in I. To see this, take an arbitrary $c \in I$ and write $c = X'a + Y'b$ with $X', Y' \in \mathbb{Z}$. Using division with remainder (Proposition 9.1) we have that $c = qd + r$ with q, r integers and $0 \leq r < d$. Then

$$r = c - qd = X'a + Y'b - q(Xa + Yb) = (X' - qX)a + (Y' - qY)b \in I.$$

If $r \neq 0$, this contradicts our choice of d as the *smallest* positive integer in I (because $r < d$). So, $r = 0$ and hence $d \mid c$. This shows that d divides every element of I.

Since $a \in I$ and $b \in I$, the above shows that $d \mid a$ and $d \mid b$ and so d is a common divisor of a and b. It remains to show that it is the greatest common divisor. Assume there is an integer $d' > d$ such that $d' \mid a$ and $d' \mid b$. Then by the observation made earlier, $d' \mid Xa + Yb$. Since the latter is equal to d, this means $d' \mid d$. But this is impossible if d' is larger than d. We conclude that d is the largest integer dividing both a and b, and hence $d = \gcd(a, b)$.   ∎

Given a and b, the *Euclidean algorithm* can be used to compute $\gcd(a, b)$ in polynomial time. The *extended Euclidean algorithm* can be used to compute X, Y (as in the above proposition) in polynomial time as well. See Appendix B.1.2 for details.

The preceding proposition is very useful in proving additional results about divisibility. We show two examples now.

PROPOSITION 9.3   *If $c \mid ab$ and $\gcd(a, c) = 1$, then $c \mid b$. Thus, if p is prime and $p \mid ab$ then either $p \mid a$ or $p \mid b$.*

PROOF   Since $c \mid ab$ we have $\gamma c = ab$ for some integer γ. If $\gcd(a, c) = 1$ then, by the previous proposition, we know there exist integers X, Y such that $1 = Xa + Yc$. Multiplying both sides by b, we obtain

$$b = Xab + Ycb = X\gamma c + Ycb = c \cdot (X\gamma + Yb).$$

Since $(X\gamma + Yb)$ is an integer, it follows that $c \mid b$.

The second part of the proposition follows from the fact that if $p \nmid a$ and p is prime then $\gcd(a, p) = 1$.   ∎

PROPOSITION 9.4   *If $a \mid N$, $b \mid N$, and $\gcd(a, b) = 1$, then $ab \mid N$.*

PROOF   Write $ac = N$, $bd = N$, and (using Proposition 9.2) $1 = Xa + Yb$, where c, d, X, Y are all integers. Multiplying both sides of the last equation by N we obtain

$$N = XaN + YbN = Xabd + Ybac = ab(Xd + Yc),$$

showing that $ab \mid N$.   ∎

9.1.2   Modular Arithmetic

Let $a, b, N \in \mathbb{Z}$ with $N > 1$. We use the notation $[a \bmod N]$ to denote the remainder of a upon division by N. In more detail: by Proposition 9.1 there exist unique q, r with $a = qN + r$ and $0 \le r < N$, and we define $[a \bmod N]$ to be equal to this r. Note therefore that $0 \le [a \bmod N] < N$. We refer to the process of mapping a to $[a \bmod N]$ as *reduction modulo N*.

We say that a and b are *congruent modulo N*, written $a = b \bmod N$, if $[a \bmod N] = [b \bmod N]$, i.e., if the remainder when a is divided by N is the same as the remainder when b is divided by N. Note that $a = b \bmod N$ if and only if $N \mid (a - b)$. By way of notation, in an expression such as

$$a = b = c = \cdots = z \bmod N,$$

the understanding is that *every* equal sign in this sequence (and not just the last) refers to congruence modulo N.

Note that $a = [b \bmod N]$ implies $a = b \bmod N$, but not vice versa. For example, $36 = 21 \bmod 15$ but $36 \neq [21 \bmod 15] = 6$. On the other hand, $[a \bmod N] = [b \bmod N]$ if and only if $a = b \bmod N$.

Congruence modulo N is an equivalence relation, i.e., it is reflexive ($a = a \bmod N$ for all a), symmetric ($a = b \bmod N$ implies $b = a \bmod N$), and transitive (if $a = b \bmod N$ and $b = c \bmod N$, then $a = c \bmod N$). Congruence modulo N also obeys the standard rules of arithmetic with respect to addition, subtraction, and multiplication; so, for example, if $a = a' \bmod N$ and $b = b' \bmod N$ then $(a + b) = (a' + b') \bmod N$ and $ab = a'b' \bmod N$. A consequence is that we can "reduce and then add/multiply" instead of having to "add/multiply and then reduce," which can often simplify calculations.

Example 9.5
Let us compute $[1093028 \cdot 190301 \bmod 100]$. Since $1093028 = 28 \bmod 100$ and $190301 = 1 \bmod 100$, we have

$$1093028 \cdot 190301 = [1093028 \bmod 100] \cdot [190301 \bmod 100] \bmod 100$$
$$= 28 \cdot 1 = 28 \bmod 100.$$

The alternate way of calculating the answer (i.e., computing the product $1093028 \cdot 190301$ and then reducing the result modulo 100) is less efficient. ◇

Congruence modulo N does *not* (in general) respect division. That is, if $a = a' \bmod N$ and $b = b' \bmod N$ then it is not necessarily true that $a/b = a'/b' \bmod N$; in fact, the expression "$a/b \bmod N$" is not necessarily well-defined. As a specific example that often causes confusion, $ab = cb \bmod N$ does *not* necessarily imply that $a = c \bmod N$.

Example 9.6
Take $N = 24$. Then $3 \cdot 2 = 6 = 15 \cdot 2 \bmod 24$, but $3 \neq 15 \bmod 24$. ◇

In certain cases, however, we can define a meaningful notion of division. If for a given integer b there exists an integer c such that $bc = 1 \bmod N$, we say that b is *invertible* modulo N and call c a (multiplicative) *inverse* of b modulo N. Clearly, 0 is never invertible. It is also not difficult to show that if c is a multiplicative inverse of b modulo N then so is $[c \bmod N]$. Furthermore, if c' is another multiplicative inverse of b then $[c \bmod N] = [c' \bmod N]$. When b is invertible we can therefore simply let b^{-1} denote the *unique* multiplicative inverse of b that lies in the range $\{1, \ldots, N - 1\}$.

When b is invertible modulo N, we define division by b modulo N as multiplication by b^{-1} (i.e., we define $[a/b \bmod N] \stackrel{\text{def}}{=} [ab^{-1} \bmod N]$). We stress that division by b is *only defined* when b is invertible. If $ab = cb \bmod N$ and b is invertible, then we may divide each side of the equation by b (or, really, multiply each side by b^{-1}) to obtain

$$(ab) \cdot b^{-1} = (cb) \cdot b^{-1} \bmod N \quad \Rightarrow \quad a = c \bmod N.$$

We see that in this case, division works as expected. Thus, invertible integers modulo N are "nicer" to work with, in some sense.

The natural question is: which integers are invertible modulo a given modulus N? We can fully answer this question using Proposition 9.2:

PROPOSITION 9.7   *Let b, N be integers, with $b \geq 1$ and $N > 1$. Then b is invertible modulo N if and only if $\gcd(b, N) = 1$.*

PROOF   Assume b is invertible modulo N, and let c denote its inverse. Since $bc = 1 \bmod N$, this implies that $bc - 1 = \gamma N$ for some $\gamma \in \mathbb{Z}$. Equivalently, $bc - \gamma N = 1$. Since, by Proposition 9.2, $\gcd(b, N)$ is the smallest positive integer that can be expressed in this way, and there is no positive integer smaller than 1, this implies that $\gcd(b, N) = 1$.

Conversely, if $\gcd(b, N) = 1$ then by Proposition 9.2 there exist integers X, Y such that $Xb + YN = 1$. Reducing each side of this equation modulo N gives $Xb = 1 \bmod N$, and we see that X is a multiplicative inverse of b. (In fact, this gives an efficient algorithm to compute inverses.) ■

Example 9.8
Let $b = 11$ and $N = 17$. Then $(-3) \cdot 11 + 2 \cdot 17 = 1$, and so $14 = [-3 \bmod 17]$ is the inverse of 11. One can verify that $14 \cdot 11 = 1 \bmod 17$.   ◇

Addition, subtraction, multiplication, and computation of inverses (when they exist) modulo N can all be carried out in polynomial time; see Appendix B.2. Exponentiation (i.e., computing $[a^b \bmod N]$ for $b > 0$ an integer) can also be computed in polynomial time; see Appendix B.2.3.

9.1.3 Groups

Let $\mathbb{G}$ be a set. A *binary operation* $\circ$ on $\mathbb{G}$ is simply a function $\circ(\cdot, \cdot)$ that maps two elements of $\mathbb{G}$ to another element of $\mathbb{G}$. If $g, h \in \mathbb{G}$ then instead of using the cumbersome notation $\circ(g, h)$, we write $g \circ h$.

We now introduce the important notion of a *group*.

DEFINITION 9.9   *A* group *is a set $\mathbb{G}$ along with a binary operation $\circ$ for which the following conditions hold:*

- **(Closure:)** *For all $g, h \in \mathbb{G}$, $g \circ h \in \mathbb{G}$.*

- **(Existence of an identity:)** *There exists an* identity *$e \in \mathbb{G}$ such that for all $g \in \mathbb{G}$, $e \circ g = g = g \circ e$.*

- **(Existence of inverses:)** *For all $g \in \mathbb{G}$ there exists an element $h \in \mathbb{G}$ such that $g \circ h = e = h \circ g$. Such an h is called an* inverse *of g.*

- **(Associativity:)** *For all $g_1, g_2, g_3 \in \mathbb{G}$, $(g_1 \circ g_2) \circ g_3 = g_1 \circ (g_2 \circ g_3)$.*

When $\mathbb{G}$ has a finite number of elements, we say $\mathbb{G}$ is finite *and let $|\mathbb{G}|$ denote the* order *of the group (that is, the number of elements in $\mathbb{G}$).*

A group $\mathbb{G}$ with operation $\circ$ is abelian *if the following holds:*

- **(Commutativity:)** *For all $g, h \in \mathbb{G}$, $g \circ h = h \circ g$.*

When the binary operation is understood, we simply call the set $\mathbb{G}$ a group.

We will always deal with finite, abelian groups. We will be careful to specify, however, when a result requires these assumptions.

Associativity implies that we do not need to include parentheses when writing long expressions; that is, the notation $g_1 \circ g_2 \circ \cdots \circ g_n$ is unambiguous since it does not matter in what order we evaluate the operation $\circ$.

One can show that the identity element in a group $\mathbb{G}$ is *unique*, and so we can therefore refer to *the* identity of a group. One can also show that each element g of a group has a *unique* inverse. See Exercise 9.1.

If $\mathbb{G}$ is a group, a set $\mathbb{H} \subseteq \mathbb{G}$ is a *subgroup of* $\mathbb{G}$ if $\mathbb{H}$ itself forms a group under the same operation associated with $\mathbb{G}$. To check that $\mathbb{H}$ is a subgroup, we need to verify closure, existence of identity and inverses, and associativity as per Definition 9.9. (In fact, associativity—as well as commutativity if $\mathbb{G}$ is abelian—is inherited automatically from $\mathbb{G}$.) Every group $\mathbb{G}$ always has the trivial subgroups $\mathbb{G}$ and $\{1\}$. We call $\mathbb{H}$ a *strict* subgroup of $\mathbb{G}$ if $\mathbb{H} \neq \mathbb{G}$.

In general, we will not use the notation $\circ$ to denote the group operation. Instead, we will use either *additive* notation or *multiplicative* notation depending on the group under discussion. *This does not imply that the group operation corresponds to integer addition or multiplication; it is merely useful notation.* When using additive notation, the group operation applied to two elements g, h is denoted $g + h$; the identity is denoted by 0; the inverse of an element g is denoted by $-g$; and we write $h - g$ in place of $h + (-g)$. When using multiplicative notation, the group operation applied to g, h is denoted by $g \cdot h$ or simply gh; the identity is denoted by 1; the inverse of an element g is denoted by g^{-1}; and we sometimes write h/g in place of hg^{-1}.

At this point, it may be helpful to see some examples.

Example 9.10
A set may be a group under one operation, but not another. For example, the set of integers $\mathbb{Z}$ is an abelian group under addition: the identity is the element 0, and every integer g has inverse $-g$. On the other hand, it is not a group under multiplication since, for example, the integer 2 does not have a multiplicative inverse in the integers.   ◇

Example 9.11
The set of real numbers $\mathbb{R}$ is not a group under multiplication, since 0 does not have a multiplicative inverse. The set of *nonzero* real numbers, however, is an abelian group under multiplication with identity 1.   ◇

The following example introduces the group $\mathbb{Z}_N$ that we will use frequently.

Example 9.12
Let $N > 1$ be an integer. The set $\{0, \ldots, N - 1\}$ with respect to addition modulo N (i.e., where $a + b \stackrel{\text{def}}{=} [a + b \bmod N]$) is an abelian group of order N. Closure is obvious; associativity and commutativity follow from the fact that the integers satisfy these properties; the identity is 0; and, since $a + (N - a) = 0 \bmod N$, it follows that the inverse of any element a is $[(N - a) \bmod N]$. We denote this group by $\mathbb{Z}_N$. (We will also sometimes use $\mathbb{Z}_N$ to denote the set $\{0, \ldots, N - 1\}$ without regard to any particular group operation.)   ◇

We end this section with an easy lemma that formalizes a "cancelation law" for groups.

LEMMA 9.13 *Let $\mathbb{G}$ be a group and $a, b, c \in \mathbb{G}$. If $ac = bc$, then $a = b$. In particular, if $ac = c$ then a is the identity in $\mathbb{G}$.*

PROOF We know $ac = bc$. Multiplying both sides by the unique inverse c^{-1} of c, we obtain $a = b$. In detail:

$$ac = bc \;\Rightarrow\; (ac)c^{-1} = (bc) \cdot c^{-1} \;\Rightarrow\; a(cc^{-1}) = b(cc^{-1}) \;\Rightarrow\; a \cdot 1 = b \cdot 1,$$

i.e., $a = b$. ∎

Compare the above proof to the discussion (preceding Proposition 9.7) regarding a cancelation law for division modulo N. As indicated by the similarity, the *invertible* elements modulo N form a group under multiplication modulo N. We will return to this example in more detail shortly.

Group Exponentiation

It is often useful to be able to describe the group operation applied m times to a fixed element g, where m is a positive integer. When using additive notation, we express this as $m \cdot g$ or mg; that is,

$$mg = m \cdot g \overset{\text{def}}{=} \underbrace{g + \cdots + g}_{m \text{ times}}.$$

Note that m is an *integer*, while g is a *group element*. So mg does *not* represent the group operation applied to m and g (indeed, we are working in a group where the group operation is written additively). Thankfully, however, the notation "behaves as it should"; so, for example, if $g \in \mathbb{G}$ and m, m' are integers then $(mg) + (m'g) = (m + m')g$, $m(m'g) = (mm')g$, and $1 \cdot g = g$. In an abelian group $\mathbb{G}$ with $g, h \in \mathbb{G}$, $(mg) + (mh) = m(g + h)$.

When using multiplicative notation, we express application of the group operation m times to an element g by g^m. That is,

$$g^m \overset{\text{def}}{=} \underbrace{g \cdots g}_{m \text{ times}}.$$

The familiar rules of exponentiation hold: $g^m \cdot g^{m'} = g^{m+m'}$, $(g^m)^{m'} = g^{mm'}$, and $g^1 = g$. Also, if $\mathbb{G}$ is an abelian group and $g, h \in \mathbb{G}$ then $g^m \cdot h^m = (gh)^m$. All these are simply "translations" of the results from the previous paragraph to the setting of groups written multiplicatively rather than additively.

The above notation is extended in the natural way to the case when m is zero or a negative integer. When using additive notation we define $0 \cdot g \overset{\text{def}}{=} 0$ (note that the 0 on the left-hand side is the integer 0 while the 0 on the right-hand side is the identity element of the group) and define $(-m) \cdot g \overset{\text{def}}{=} m \cdot (-g)$ for m a positive integer. Observe that $-g$ is the inverse of g and, as one would expect, $(-m) \cdot g = -(mg)$. When using multiplicative notation, $g^0 \overset{\text{def}}{=} 1$ and $g^{-m} \overset{\text{def}}{=} (g^{-1})^m$. Again, g^{-1} is the inverse of g, and we have $g^{-m} = (g^m)^{-1}$.

Let $g \in \mathbb{G}$ and $b \geq 0$ be an integer. Then the exponentiation g^b can be computed using polynomially many group operations in $\mathbb{G}$. Thus, if the group operation can be computed in polynomial time then so can exponentiation. This is discussed in Appendix B.2.3.

We now know enough to prove the following remarkable result:

THEOREM 9.14 *Let $\mathbb{G}$ be a finite group with $m = |\mathbb{G}|$, the order of the group. Then for any element $g \in \mathbb{G}$, it holds that $g^m = 1$.*

PROOF We prove the theorem only when $\mathbb{G}$ is abelian (although it holds for any finite group). Fix arbitrary $g \in \mathbb{G}$, and let $g_1, \ldots, g_m$ be the elements of $\mathbb{G}$. We claim that

$$g_1 \cdot g_2 \cdots g_m = (gg_1) \cdot (gg_2) \cdots (gg_m).$$

To see this, note that $gg_i = gg_j$ implies $g_i = g_j$ by Lemma 9.13. So each of the m elements in parentheses on the right-hand side is distinct. Because there are exactly m elements in $\mathbb{G}$, the m elements being multiplied together on the right-hand side are simply all elements of $\mathbb{G}$ in some permuted order. Since $\mathbb{G}$ is abelian, the order in which elements are multiplied does not matter, and so the right-hand side is equal to the left-hand side.

Again using the fact that $\mathbb{G}$ is abelian, we can "pull out" all occurrences of g and obtain

$$g_1 \cdot g_2 \cdots g_m = (gg_1) \cdot (gg_2) \cdots (gg_m) = g^m \cdot (g_1 \cdot g_2 \cdots g_m).$$

Appealing once again to Lemma 9.13, this implies $g^m = 1$. ∎

An important corollary of the above is that we can work "modulo the group order" in the exponent:

COROLLARY 9.15 *Let $\mathbb{G}$ be a finite group with $m = |\mathbb{G}| > 1$. Then for any $g \in \mathbb{G}$ and any integer x, we have $g^x = g^{[x \bmod m]}$.*

PROOF Say $x = qm + r$, where q, r are integers and $r = [x \bmod m]$. Then

$$g^x = g^{qm+r} = g^{qm} \cdot g^r = (g^m)^q \cdot g^r = 1^q \cdot g^r = g^r$$

(using Theorem 9.14), as claimed. ∎

Example 9.16
Written additively, the above corollary says that if g is an element in a group of order m, then $x \cdot g = [x \bmod m] \cdot g$. As an example, consider the group $\mathbb{Z}_{15}$

of order $m = 15$, and take $g = 11$. The corollary says that

$$152 \cdot 11 = [152 \bmod 15] \cdot 11 = 2 \cdot 11 = 11 + 11 = 22 = 7 \bmod 15.$$

The above agrees with the fact (cf. Example 9.5) that we can "reduce and then multiply" rather than having to "multiply and then reduce." ◇

Another corollary that will be extremely useful for cryptographic applications is the following:

COROLLARY 9.17 *Let $\mathbb{G}$ be a finite group with $m = |\mathbb{G}| > 1$. Let $e > 0$ be an integer, and define the function $f_e : \mathbb{G} \to \mathbb{G}$ by $f_e(g) = g^e$. If $\gcd(e, m) = 1$, then f_e is a permutation (i.e., a bijection). Moreover, if $d = e^{-1} \bmod m$ then f_d is the inverse of f_e. (Note by Proposition 9.7, $\gcd(e, m) = 1$ implies e is invertible modulo m.)*

PROOF Since $\mathbb{G}$ is finite, the second part of the claim implies the first; thus, we need only show that f_d is the inverse of f_e. This is true because for any $g \in \mathbb{G}$, we have

$$f_d\left(f_e(g)\right) = f_d(g^e) = (g^e)^d = g^{ed} = g^{[ed \bmod m]} = g^1 = g,$$

where the fourth equality follows from Corollary 9.15. ■

9.1.4 The Group $\mathbb{Z}_N^*$

As discussed in Example 9.12, the set $\mathbb{Z}_N = \{0, \dots, N-1\}$ is a group under addition modulo N. Can we define a group with respect to *multiplication* modulo N? In doing so, we will have to eliminate those elements in $\mathbb{Z}_N$ that are not invertible; e.g., we will have to eliminate 0 since it has no multiplicative inverse. Nonzero elements may also fail to be invertible (cf. Proposition 9.7).

Which elements $b \in \{1, \dots, N-1\}$ are invertible modulo N? Proposition 9.7 says that these are exactly the elements b for which $\gcd(b, N) = 1$. We have also seen in Section 9.1.2 that whenever b is invertible, it has an inverse lying in the range $\{1, \dots, N-1\}$. This leads us to define, for any $N > 1$, the set

$$\mathbb{Z}_N^* \overset{\text{def}}{=} \{b \in \{1, \dots, N-1\} \mid \gcd(b, N) = 1\};$$

i.e., $\mathbb{Z}_N^*$ consists of integers in the set $\{1, \dots, N-1\}$ that are relatively prime to N. The group operation is multiplication modulo N; i.e., $ab \overset{\text{def}}{=} [ab \bmod N]$.

We claim that $\mathbb{Z}_N^*$ is an abelian group with respect to this operation. Since 1 is always in $\mathbb{Z}_N^*$, the set clearly contains an identity element. The discussion above shows that each element in $\mathbb{Z}_N^*$ has a multiplicative inverse in the same set. Commutativity and associativity follow from the fact that these properties hold over the integers. To show that closure holds, let $a, b \in \mathbb{Z}_N^*$; then $[ab \bmod N]$ has inverse $[b^{-1}a^{-1} \bmod N]$, which means that $\gcd([ab \bmod N], N) = 1$ and so $ab \in \mathbb{Z}_N^*$. Summarizing:

PROPOSITION 9.18   *Let $N > 1$ be an integer. Then $\mathbb{Z}_N^*$ is an abelian group under multiplication modulo N.*

Define $\phi(N) \stackrel{\text{def}}{=} |\mathbb{Z}_N^*|$, the order of the group $\mathbb{Z}_N^*$. (ϕ is called the *Euler phi function*.) What is the value of $\phi(N)$? First consider the case when $N = p$ is prime. Then *all* elements in $\{1, \ldots, p-1\}$ are relatively prime to p, and so $\phi(p) = |\mathbb{Z}_p^*| = p - 1$. Next consider the case that $N = pq$, where p, q are distinct primes. If an integer $a \in \{1, \ldots, N-1\}$ is not relatively prime to N, then either $p \mid a$ or $q \mid a$ (a cannot be divisible by both p and q since this would imply $pq \mid a$ but $a < N = pq$). The elements in $\{1, \ldots, N-1\}$ divisible by p are exactly the $(q-1)$ elements $p, 2p, 3p, \ldots, (q-1)p$, and the elements divisible by q are exactly the $(p-1)$ elements $q, 2q, \ldots, (p-1)q$. The number of elements remaining (i.e., those that are neither divisible by p nor q) is therefore given by

$$(N-1) - (q-1) - (p-1) = pq - p - q + 1 = (p-1)(q-1).$$

We have thus proved that $\phi(N) = (p-1)(q-1)$ when N is the product of two distinct primes p and q.

You are asked to prove the following general result (used only rarely in the rest of the book) in Exercise 9.4:

THEOREM 9.19   *Let $N = \prod_i p_i^{e_i}$, where the $\{p_i\}$ are distinct primes and $e_i \geq 1$. Then $\phi(N) = \prod_i p_i^{e_i-1}(p_i - 1)$.*

Example 9.20
Take $N = 15 = 5 \cdot 3$. Then $\mathbb{Z}_{15}^* = \{1, 2, 4, 7, 8, 11, 13, 14\}$ and $|\mathbb{Z}_{15}^*| = 8 = 4 \cdot 2 = \phi(15)$. The inverse of 8 in $\mathbb{Z}_{15}^*$ is 2, since $8 \cdot 2 = 16 = 1 \bmod 15$.   ◇

We have shown that $\mathbb{Z}_N^*$ is a group of order $\phi(N)$. The following are now easy corollaries of Theorem 9.14 and Corollary 9.17:

COROLLARY 9.21   *Take arbitrary integer $N > 1$ and $a \in \mathbb{Z}_N^*$. Then*

$$a^{\phi(N)} = 1 \bmod N.$$

For the specific case that $N = p$ is prime and $a \in \{1, \ldots, p-1\}$, we have

$$a^{p-1} = 1 \bmod p.$$

COROLLARY 9.22 *Fix $N > 1$. For integer $e > 0$ define $f_e : \mathbb{Z}_N^* \to \mathbb{Z}_N^*$ by $f_e(x) = [x^e \bmod N]$. If e is relatively prime to $\phi(N)$ then f_e is a permutation. Moreover, if $d = e^{-1} \bmod \phi(N)$ then f_d is the inverse of f_e.*

9.1.5 *Isomorphisms and the Chinese Remainder Theorem

Two groups are *isomorphic* if they have the same underlying structure. From a mathematical point of view, an isomorphism of a group $\mathbb{G}$ provides an alternate, but equivalent, way of thinking about $\mathbb{G}$. From a computational perspective, an isomorphism provides a different way to *represent* elements in $\mathbb{G}$, which can often have a significant impact on algorithmic efficiency.

DEFINITION 9.23 *Let $\mathbb{G}, \mathbb{H}$ be groups with respect to the operations $\circ_\mathbb{G}, \circ_\mathbb{H}$, respectively. A function $f : \mathbb{G} \to \mathbb{H}$ is an* isomorphism *from $\mathbb{G}$ to $\mathbb{H}$ if:*

1. f is a bijection, and

2. For all $g_1, g_2 \in \mathbb{G}$ we have $f(g_1 \circ_\mathbb{G} g_2) = f(g_1) \circ_\mathbb{H} f(g_2)$.

If there exists an isomorphism from $\mathbb{G}$ to $\mathbb{H}$ then we say that these groups are isomorphic *and write $\mathbb{G} \simeq \mathbb{H}$.*

In essence, an isomorphism from $\mathbb{G}$ to $\mathbb{H}$ is just a *renaming* of elements of $\mathbb{G}$ as elements of $\mathbb{H}$. Note that if $\mathbb{G}$ is finite and $\mathbb{G} \simeq \mathbb{H}$, then $\mathbb{H}$ must be finite and of the same size as $\mathbb{G}$. Also, if there exists an isomorphism f from $\mathbb{G}$ to $\mathbb{H}$ then f^{-1} is an isomorphism from $\mathbb{H}$ to $\mathbb{G}$. It is possible, however, that f is efficiently computable while f^{-1} is not (or vice versa).

The aim of this section is to use the language of isomorphisms to better understand the group structure of $\mathbb{Z}_N$ and $\mathbb{Z}_N^*$ when $N = pq$ is a product of two distinct primes. We first need to introduce the notion of a *direct product* of groups. Given groups $\mathbb{G}, \mathbb{H}$ with group operations $\circ_\mathbb{G}, \circ_\mathbb{H}$, respectively, we define a new group $\mathbb{G} \times \mathbb{H}$ (the *direct product of $\mathbb{G}$ and $\mathbb{H}$*) as follows. The elements of $\mathbb{G} \times \mathbb{H}$ are ordered pairs (g, h) with $g \in \mathbb{G}$ and $h \in \mathbb{H}$; thus, if $\mathbb{G}$ has n elements and $\mathbb{H}$ has n' elements, $\mathbb{G} \times \mathbb{H}$ has $n \cdot n'$ elements. The group operation $\circ$ on $\mathbb{G} \times \mathbb{H}$ is applied component-wise; that is:

$$(g, h) \circ (g', h') \stackrel{\text{def}}{=} (g \circ_\mathbb{G} g', \ h \circ_\mathbb{H} h').$$

We leave it to Exercise 9.8 to verify that $\mathbb{G} \times \mathbb{H}$ is indeed a group. The above notation can be extended to direct products of more than two groups in the natural way, although we will not need this for what follows.

We may now state and prove the *Chinese remainder theorem*.

THEOREM 9.24 (Chinese remainder theorem)   *Let $N = pq$ where $p, q > 1$ are relatively prime. Then*

$$\mathbb{Z}_N \simeq \mathbb{Z}_p \times \mathbb{Z}_q \quad and \quad \mathbb{Z}_N^* \simeq \mathbb{Z}_p^* \times \mathbb{Z}_q^*.$$

Moreover, let f be the function mapping elements $x \in \{0, \ldots, N-1\}$ to pairs (x_p, x_q) with $x_p \in \{0, \ldots, p-1\}$ and $x_q \in \{0, \ldots, q-1\}$ defined by

$$f(x) \stackrel{\text{def}}{=} ([x \bmod p], [x \bmod q]).$$

Then f is an isomorphism from $\mathbb{Z}_N$ to $\mathbb{Z}_p \times \mathbb{Z}_q$, and the restriction of f to $\mathbb{Z}_N^*$ is an isomorphism from $\mathbb{Z}_N^*$ to $\mathbb{Z}_p^* \times \mathbb{Z}_q^*$.

PROOF   For any $x \in \mathbb{Z}_N$ the output $f(x)$ is a pair of elements (x_p, x_q) with $x_p \in \mathbb{Z}_p$ and $x_q \in \mathbb{Z}_q$. We claim that if $x \in \mathbb{Z}_N^*$, then $(x_p, x_q) \in \mathbb{Z}_p^* \times \mathbb{Z}_q^*$. Indeed, if $x_p \notin \mathbb{Z}_p^*$ then this means that $\gcd([x \bmod p], p) \neq 1$. But then $\gcd(x, p) \neq 1$. This implies $\gcd(x, N) \neq 1$, contradicting the assumption that $x \in \mathbb{Z}_N^*$. (An analogous argument holds if $x_q \notin \mathbb{Z}_q^*$.)

We now show that f is an isomorphism from $\mathbb{Z}_N$ to $\mathbb{Z}_p \times \mathbb{Z}_q$. (The proof that it is an isomorphism from $\mathbb{Z}_N^*$ to $\mathbb{Z}_p^* \times \mathbb{Z}_q^*$ is similar.) Let us start by proving that f is one-to-one. Say $f(x) = (x_p, x_q) = f(x')$. Then $x = x_p = x' \bmod p$ and $x = x_q = x' \bmod q$. This in turn implies that $(x - x')$ is divisible by both p and q. Since $\gcd(p, q) = 1$, Proposition 9.4 says that $pq = N$ divides $(x - x')$. But then $x = x' \bmod N$. For $x, x' \in \mathbb{Z}_N$, this means that $x = x'$ and so f is indeed one-to-one. Since $|\mathbb{Z}_N| = N = p \cdot q = |\mathbb{Z}_p| \cdot |\mathbb{Z}_q|$, the sizes of $\mathbb{Z}_N$ and $\mathbb{Z}_p \times \mathbb{Z}_q$ are the same. This in combination with the fact that f is one-to-one implies that f is bijective.

In the following paragraph, let $+_N$ denote addition modulo N, and let $\boxplus$ denote the group operation in $\mathbb{Z}_p \times \mathbb{Z}_q$ (i.e., addition modulo p in the first component and addition modulo q in the second component). To conclude the proof that f is an isomorphism from $\mathbb{Z}_N$ to $\mathbb{Z}_p \times \mathbb{Z}_q$, we need to show that for all $a, b \in \mathbb{Z}_N$ it holds that $f(a +_N b) = f(a) \boxplus f(b)$.

To see that this is true, note that

$$f(a +_N b) = \Big([(a +_N b) \bmod p], [(a +_N b) \bmod q] \Big)$$

$$= \Big([(a + b) \bmod p], [(a + b) \bmod q] \Big)$$

$$= \Big([a \bmod p], [a \bmod q] \Big) \boxplus \Big([b \bmod p], [b \bmod q] \Big) = f(a) \boxplus f(b).$$

(For the second equality, above, we use the fact that $[[X \bmod N] \bmod p] = [[X \bmod p] \bmod p]$ when $p \mid N$; see Exercise 9.9.) ∎

An extension of the Chinese remainder theorem says that if $p_1, p_2, \ldots, p_\ell$ are pairwise relatively prime (i.e., $\gcd(p_i, p_j) = 1$ for all $i \neq j$) and $N \stackrel{\text{def}}{=} \prod_{i=1}^{\ell} p_i$, then

$$\mathbb{Z}_N \simeq \mathbb{Z}_{p_1} \times \cdots \times \mathbb{Z}_{p_\ell} \quad \text{and} \quad \mathbb{Z}_N^* \simeq \mathbb{Z}_{p_1}^* \times \cdots \times \mathbb{Z}_{p_\ell}^*.$$

An isomorphism in each case is obtained by a natural extension of the one used in the theorem above.

By way of notation, with N understood and $x \in \{0, 1, \dots, N-1\}$ we write $x \leftrightarrow (x_p, x_q)$ for $x_p = [x \bmod p]$ and $x_q = [x \bmod q]$. That is, $x \leftrightarrow (x_p, x_q)$ if and only if $f(x) = (x_p, x_q)$, where f is as in the theorem above. One way to think about this notation is that it means "x (in $\mathbb{Z}_N$) *corresponds to* (x_p, x_q) (in $\mathbb{Z}_p \times \mathbb{Z}_q$)." The same notation is used when dealing with $x \in \mathbb{Z}_N^*$.

Example 9.25

Take $15 = 5 \cdot 3$, and consider $\mathbb{Z}_{15}^* = \{1, 2, 4, 7, 8, 11, 13, 14\}$. The Chinese remainder theorem says this group is isomorphic to $\mathbb{Z}_5^* \times \mathbb{Z}_3^*$. We can compute

$$
\begin{array}{llll}
1 \leftrightarrow (1,1) & 2 \leftrightarrow (2,2) & 4 \leftrightarrow (4,1) & 7 \leftrightarrow (2,1) \\
8 \leftrightarrow (3,2) & 11 \leftrightarrow (1,2) & 13 \leftrightarrow (3,1) & 14 \leftrightarrow (4,2)
\end{array},
$$

where each pair (a, b) with $a \in \mathbb{Z}_5^*$ and $b \in \mathbb{Z}_3^*$ appears exactly once. $\diamondsuit$

Using the Chinese Remainder Theorem

If two groups are isomorphic, then they both serve as representations of the same underlying "algebraic structure." Nevertheless, the choice of which representation to use can affect the *computational efficiency* of group operations. We discuss this abstractly, and then in the specific context of $\mathbb{Z}_N$ and $\mathbb{Z}_N^*$.

Let $\mathbb{G}, \mathbb{H}$ be groups with operations $\circ_{\mathbb{G}}, \circ_{\mathbb{H}}$, respectively, and say f is an isomorphism from $\mathbb{G}$ to $\mathbb{H}$ where both f and f^{-1} can be computed efficiently. Then for $g_1, g_2 \in \mathbb{G}$ we can compute $g = g_1 \circ_{\mathbb{G}} g_2$ in two ways: either by directly computing the group operation in $\mathbb{G}$, or via the following steps:

1. Compute $h_1 = f(g_1)$ and $h_2 = f(g_2)$;

2. Compute $h = h_1 \circ_{\mathbb{H}} h_2$ using the group operation in $\mathbb{H}$;

3. Compute $g = f^{-1}(h)$.

The above extends in the natural way when we want to compute multiple group operations in $\mathbb{G}$ (e.g., to compute g^x for some integer x). Which method is better depends on the relative efficiency of computing the group operation in each group, as well as the efficiency of computing f and f^{-1}.

We now turn to the specific case of computations modulo N, when $N = pq$ is a product of distinct primes. The Chinese remainder theorem shows that addition, multiplication, or exponentiation (which is just repeated multiplication) modulo N can be "transformed" to analogous operations modulo p and q. Building on Example 9.25, we show some simple examples with $N = 15$.

Example 9.26
Say we want to compute the product $14 \cdot 13$ modulo 15 (i.e., in $\mathbb{Z}_{15}^*$). Example 9.25 gives $14 \leftrightarrow (4, 2)$ and $13 \leftrightarrow (3, 1)$. In $\mathbb{Z}_5^* \times \mathbb{Z}_3^*$, we have

$$(4, 2) \cdot (3, 1) = ([4 \cdot 3 \bmod 5], [2 \cdot 1 \bmod 3]) = (2, 2).$$

Note $(2, 2) \leftrightarrow 2$, which is the correct answer since $14 \cdot 13 = 2 \bmod 15$.   ◇

Example 9.27
Say we want to compute $11^{53} \bmod 15$. Example 9.25 gives $11 \leftrightarrow (1, 2)$. Notice that $2 = -1 \bmod 3$ and so

$$(1, 2)^{53} = ([1^{53} \bmod 5], [(-1)^{53} \bmod 3]) = (1, [-1 \bmod 3]) = (1, 2).$$

Thus, $11^{53} \bmod 15 = 11$.   ◇

Example 9.28
Say we want to compute $[29^{100} \bmod 35]$. We first compute the correspondence $29 \leftrightarrow ([29 \bmod 5], \quad [29 \bmod 7]) = ([-1 \bmod 5], 1)$. Using the Chinese remainder theorem, we have

$$([-1 \bmod 5], 1)^{100} = ([(-1)^{100} \bmod 5], [1^{100} \bmod 7]) = (1, 1),$$

and it is immediate that $(1, 1) \leftrightarrow 1$. We conclude that $[29^{100} \bmod 35] = 1$.   ◇

Example 9.29
Say we want to compute $[18^{25} \bmod 35]$. We have $18 \leftrightarrow (3, 4)$ and so

$$18^{25} \bmod 35 \leftrightarrow (3, 4)^{25} = ([3^{25} \bmod 5], [4^{25} \bmod 7]).$$

Since $\mathbb{Z}_5^*$ is a group of order 4, we can "work modulo 4 in the exponent" (cf. Corollary 9.15) and see that

$$3^{25} = 3^{[25 \bmod 4]} = 3^1 = 3 \bmod 5.$$

Similarly,
$$4^{25} = 4^{[25 \bmod 6]} = 4^1 = 4 \bmod 7.$$

Thus, $([3^{25} \bmod 5], [4^{25} \bmod 7]) = (3, 4) \leftrightarrow 18$ and so $[18^{25} \bmod 35] = 18$.   ◇

One thing we have not yet discussed is how to convert back and forth between the representation of an element modulo N and its representation modulo p and q. The conversion can be carried out efficiently provided the factorization of N is known. Assuming p and q are known, it is easy to map an element x modulo N to its corresponding representation modulo p and q:

the element x corresponds to $([x \bmod p], [x \bmod q])$, and both the modular reductions can be carried out efficiently (cf. Appendix B.2).

For the other direction, we make use of the following observation: an element with representation (x_p, x_q) can be written as

$$(x_p, x_q) = x_p \cdot (1, 0) + x_q \cdot (0, 1).$$

So, if we can find elements $1_p, 1_q \in \{0, \ldots, N-1\}$ such that $1_p \leftrightarrow (1, 0)$ and $1_q \leftrightarrow (0, 1)$, then (appealing to the Chinese remainder theorem) we know that

$$(x_p, x_q) \leftrightarrow [(x_p \cdot 1_p + x_q \cdot 1_q) \bmod N].$$

Since p, q are distinct primes, $\gcd(p, q) = 1$. We can use the extended Euclidean algorithm (cf. Appendix B.1.2) to find integers X, Y such that

$$Xp + Yq = 1.$$

Note that $Yq = 0 \bmod q$ and $Yq = 1 - Xp = 1 \bmod p$. This means that $[Yq \bmod N] \leftrightarrow (1, 0)$; i.e., $[Yq \bmod N] = 1_p$. Similarly, $[Xp \bmod N] = 1_q$.

In summary, we can convert an element represented as (x_p, x_q) to its representation modulo N in the following way (assuming p and q are known):

1. Compute X, Y such that $Xp + Yq = 1$.

2. Set $1_p := [Yq \bmod N]$ and $1_q := [Xp \bmod N]$.

3. Compute $x := [(x_p \cdot 1_p + x_q \cdot 1_q) \bmod N]$.

If many such conversions will be performed, then $1_p, 1_q$ can be computed once-and-for-all in a preprocessing phase.

Example 9.30
Take $p = 5$, $q = 7$, and $N = 5 \cdot 7 = 35$. Say we are given the representation $(4, 3)$ and want to convert this to the corresponding element of $\mathbb{Z}_{35}$. Using the extended Euclidean algorithm, we compute

$$3 \cdot 5 - 2 \cdot 7 = 1.$$

Thus, $1_p = [-2 \cdot 7 \bmod 35] = 21$ and $1_q = [3 \cdot 5 \bmod 35] = 15$. (We can check that these are correct: e.g., for $1_p = 21$ we can verify that $[21 \bmod 5] = 1$ and $[21 \bmod 7] = 0$.) Using these values, we can then compute

$$
\begin{aligned}
(4, 3) &= 4 \cdot (1, 0) + 3 \cdot (0, 1) \\
&\leftrightarrow [4 \cdot 1_p + 3 \cdot 1_q \bmod 35] \\
&= [4 \cdot 21 + 3 \cdot 15 \bmod 35] = 24.
\end{aligned}
$$

Since $24 = 4 \bmod 5$ and $24 = 3 \bmod 7$, this is indeed the correct result.   ◇

9.2   Primes, Factoring, and RSA

In this section, we show the first examples of number-theoretic problems that are conjectured to be "hard." We begin with a discussion of one of the oldest problems: *integer factorization* or just *factoring*.

Given a composite integer N, the factoring problem is to find integers $p, q > 1$ such that $pq = N$. Factoring is a classic example of a hard problem, both because it is so simple to describe and since it has been recognized as a hard computational problem for a long time (even before its use in cryptography). The problem can be solved in *exponential* time $\mathcal{O}(\sqrt{N} \cdot \mathsf{polylog}(N))$ using *trial division*: that is, by exhaustively checking whether p divides N for $p = 2, \ldots, \lfloor \sqrt{N} \rfloor$. (This method requires $\sqrt{N}$ divisions, each one taking $\mathsf{polylog}(N) = \|N\|^c$ time for some constant c.) This always succeeds because although the *largest* prime factor of N may be as large as $N/2$, the *smallest* prime factor of N can be at most $\lfloor \sqrt{N} \rfloor$. Although algorithms with better running time are known (see Chapter 10), no *polynomial-time* algorithm for factoring has been demonstrated despite many years of effort.

Consider the following experiment for a given algorithm $\mathcal{A}$ and parameter n:

The weak factoring experiment w-Factor$_{\mathcal{A}}(n)$**:**

1. *Choose two uniform n-bit integers x_1, x_2.*
2. *Compute $N := x_1 \cdot x_2$.*
3. *$\mathcal{A}$ is given N, and outputs $x_1', x_2' > 1$.*
4. *The output of the experiment is defined to be 1 if $x_1' \cdot x_2' = N$, and 0 otherwise.*

We have just said that the factoring problem is believed to be hard. Does this mean that

$$\Pr[\text{w-Factor}_{\mathcal{A}}(n) = 1] \leq \mathsf{negl}(n)$$

is negligible for every PPT algorithm $\mathcal{A}$? Not at all. For starters, the number N in the above experiment is *even* with probability $3/4$ (this occurs when either x_1 or x_2 is even); it is, of course, easy for $\mathcal{A}$ to factor N in this case. While we can make $\mathcal{A}$'s job more difficult by requiring $\mathcal{A}$ to output integers x_1', x_2' of length n, it remains the case that x_1 or x_2 (and hence N) might have small prime factors that can still be easily found. For cryptographic applications, we will need to prevent this.

As this discussion indicates, the "hardest" numbers to factor are those having only large prime factors. This suggests redefining the above experiment so that x_1, x_2 are random n-bit *primes* rather than random n-bit *integers*, and in fact such an experiment will be used when we formally define the factoring assumption in Section 9.2.3. For this experiment to be useful in a cryptographic setting, however, it is necessary to be able to generate random n-bit primes *efficiently*. This is the topic of the next two sections.

9.2.1 Generating Random Primes

A natural approach to generating a random n-bit prime is to repeatedly choose random n-bit integers until we find one that is prime; we repeat this at most t times or until we are successful. See Algorithm 9.31 for a high-level description of the process.

ALGORITHM 9.31

Generating a random prime – high-level outline

Input: Length n; parameter t
Output: A uniform n-bit prime

for $i = 1$ to t:
    $p' \leftarrow \{0,1\}^{n-1}$
    $p := 1 \| p'$
    if p is prime **return** p
return fail

Note that the algorithm forces the output to be an integer of length *exactly* n (rather than length *at most* n) by fixing the high-order bit of p to "1." Our convention throughout this book is that an "integer of length n" means an integer whose binary representation *with most significant bit equal to* 1 is exactly n bits long.

Given a way to determine whether or not a given integer p is prime, the above algorithm outputs a *uniform* n-bit prime conditioned on the event that it does not output fail. The probability that the algorithm outputs fail depends on t, and for our purposes we will want to set t so as to obtain a failure probability that is negligible in n. To show that Algorithm 9.31 leads to an *efficient* (i.e., polynomial-time in n) algorithm for generating primes, we need a better understanding of two issues: (1) the probability that a uniform n-bit integer is prime and (2) how to efficiently test whether a given integer p is prime. We discuss these issues briefly now, and defer a more in-depth exploration of the second topic to the following section.

The distribution of primes. The *prime number theorem*, an important result in mathematics, gives fairly precise bounds on the fraction of integers of a given length that are prime. We state a corollary (without proof) that suffices for our purposes:

THEOREM 9.32 *For any $n > 1$, the fraction of n-bit integers that are prime is at least $1/3n$.*

Returning to the approach for generating primes described above, this implies that if we set $t = 3n^2$ then the probability that a prime is *not* chosen in all t

iterations of the algorithm is at most

$$\left(1 - \frac{1}{3n}\right)^t = \left(\left(1 - \frac{1}{3n}\right)^{3n}\right)^n \leq \left(e^{-1}\right)^n = e^{-n}$$

(using Inequality A.2), which is negligible in n. Thus, using $\mathsf{poly}(n)$ iterations we obtain an algorithm for which the probability of outputting fail is negligible in n. (Tighter results than Theorem 9.32 are known, and so in practice even fewer iterations are needed.)

Testing primality. The problem of efficiently determining whether a given number is prime has a long history. In the 1970s the first efficient algorithms for testing primality were developed. These algorithms were *probabilistic* and had the following guarantee: if the input p were a prime number, the algorithm would always output "prime." On the other hand, if p were composite, then the algorithm would almost always output "composite," but might output the wrong answer ("prime") with probability negligible in the length of p. Put differently, if the algorithm outputs "composite" then p is definitely composite, but if the output is "prime" then it is very likely that p is prime but it is also possible that a mistake has occurred (and p is really composite).

When using a randomized primality test of this sort in Algorithm 9.31 (the prime-generation algorithm shown earlier), the output of the algorithm is a uniform prime of the desired length so long as the algorithm does not output fail *and* the randomized primality test did not err during the execution of the algorithm. This means that an additional source of error (besides the possibility of outputting fail) is introduced, and the algorithm may now output a composite number by mistake. Since we can ensure that this happens with only negligible probability, this remote possibility is of no practical concern and we can safely ignore it.

A *deterministic* polynomial-time algorithm for testing primality was demonstrated in a breakthrough result in 2002. That algorithm, although running in polynomial time, is slower than the probabilistic tests mentioned above. For this reason, probabilistic primality tests are still used exclusively in practice for generating large prime numbers.

In Section 9.2.2 we describe and analyze one of the most commonly used probabilistic primality tests: the *Miller–Rabin* algorithm. This algorithm takes two inputs: an integer p and a parameter t (in unary) that determines the error probability. The Miller–Rabin algorithm runs in time polynomial in $\|p\|$ and t, and satisfies:

THEOREM 9.33   *If p is prime, then the Miller–Rabin test always outputs "prime." If p is composite, the algorithm outputs "composite" except with probability at most 2^{-t}.*

Putting it all together. Given the preceding discussion, we can now describe a polynomial-time prime-generation algorithm that, on input n, outputs an n-bit prime except with probability negligible in n; moreover, conditioned on the output p being prime, p is a uniformly distributed n-bit prime. The full procedure is described in Algorithm 9.34.

ALGORITHM 9.34
Generating a random prime

Input: Length n
Output: A uniform n-bit prime

for $i = 1$ to $3n^2$:
    $p' \leftarrow \{0, 1\}^{n-1}$
    $p := 1\|p'$
    run the Miller–Rabin test on input p and parameter 1^n
    if the output is "prime," **return** p
return fail

Generating primes of a particular form. It is sometimes desirable to generate a random n-bit prime p of a particular form, for example, satisfying $p = 3 \bmod 4$ or such that $p = 2q + 1$ where q is also prime (p of the latter type are called *strong primes*). In this case, appropriate modifications of the prime-generation algorithm shown above can be used. (For example, in order to obtain a prime of the form $p = 2q + 1$, modify the algorithm to generate a random prime q, compute $p := 2q + 1$, and then output p if it too is prime.) While these modified algorithms work well in practice, rigorous proofs that they run in polynomial time and fail with only negligible probability are more complex (and, in some cases, rely on unproven number-theoretic conjectures regarding the density of primes of a particular form). A detailed exploration of these issues is beyond the scope of this book, and we will simply assume the existence of appropriate prime-generation algorithms when needed.

9.2.2 *Primality Testing

We now describe the Miller–Rabin primality test and prove Theorem 9.33. (We rely on the material presented in Section 9.1.5.) This material is not used directly in the rest of the book.

The key to the Miller–Rabin algorithm is to find a property that distinguishes primes and composites. Let N denote the input number to be tested. We start with the following observation: if N is prime then $|\mathbb{Z}_N^*| = N - 1$, and so for any $a \in \{1, \ldots, N-1\}$ we have $a^{N-1} = 1 \bmod N$ by Theorem 9.14. This suggests testing whether N is prime by choosing a uniform element a and checking whether $a^{N-1} \stackrel{?}{=} 1 \bmod N$. If $a^{N-1} \neq 1 \bmod N$, then N cannot be prime. Conversely, we might hope that if N is not prime then there is a reasonable chance that we will pick a with $a^{N-1} \neq 1 \bmod N$, and so by repeating this test many times we can determine whether N is prime or not with high confidence. The above approach is shown as Algorithm 9.35. (Recall that exponentiation modulo N and computation of greatest common divisors can be carried out in polynomial time. Choosing a uniform element of $\{1, \ldots, N-1\}$ can also be done in polynomial time. See Appendix B.2.)

ALGORITIIM 9.35
Primality testing – first attempt

Input: Integer N and parameter 1^t
Output: A decision as to whether N is prime or composite

for $i = 1$ to t:
    $a \leftarrow \{1, \ldots, N-1\}$
    if $a^{N-1} \neq 1 \bmod N$ **return** "composite"
return "prime"

If N is prime the algorithm always outputs "prime." If N is composite, the algorithm outputs "composite" if in any iteration it finds an $a \in \{1, \ldots, N-1\}$ such that $a^{N-1} \neq 1 \bmod N$. Observe that if $a \notin \mathbb{Z}_N^*$ then $a^{N-1} \neq 1 \bmod N$. (If $\gcd(a, N) \neq 1$ then $\gcd(a^{N-1}, N) \neq 1$ and so $[a^{N-1} \bmod N]$ cannot equal 1.) For now, we therefore restrict our attention to $a \in \mathbb{Z}_N^*$. We refer to any such a with $a^{N-1} \neq 1 \bmod N$ as a *witness that N is composite*, or simply a *witness*. We might hope that when N is composite there are *many* witnesses, and thus the algorithm finds such a witness with "high" probability. This intuition is correct *provided there is at least one witness*. Before proving this, we need two group-theoretic lemmas.

PROPOSITION 9.36   *Let $\mathbb{G}$ be a finite group, and $\mathbb{H} \subseteq \mathbb{G}$. Assume $\mathbb{H}$ is nonempty, and for all $a, b \in \mathbb{H}$ we have $ab \in \mathbb{H}$. Then $\mathbb{H}$ is a subgroup of $\mathbb{G}$.*

PROOF   We need to verify that $\mathbb{H}$ satisfies all the conditions of Definition 9.9. By assumption, $\mathbb{H}$ is closed under the group operation. Associativity in $\mathbb{H}$ is inherited automatically from $\mathbb{G}$. Let $m = |\mathbb{G}|$ (here is where we use the fact that $\mathbb{G}$ is finite), and consider an arbitrary element $a \in \mathbb{H}$. Closure of $\mathbb{H}$ means that $\mathbb{H}$ contains $a^{m-1} = a^{-1}$ as well as $a^m = 1$. Thus, $\mathbb{H}$ contains the inverse of each of its elements, as well as the identity.   ■

LEMMA 9.37   *Let $\mathbb{H}$ be a strict subgroup of a finite group $\mathbb{G}$ (i.e., $\mathbb{H} \neq \mathbb{G}$). Then $|\mathbb{H}| \leq |\mathbb{G}|/2$.*

PROOF   Let $\bar{h}$ be an element of $\mathbb{G}$ that is *not* in $\mathbb{H}$; since $\mathbb{H} \neq \mathbb{G}$, we know such an $\bar{h}$ exists. Consider the set $\bar{\mathbb{H}} \stackrel{\text{def}}{=} \{\bar{h}h \mid h \in \mathbb{H}\}$. We show that (1) $|\bar{\mathbb{H}}| = |\mathbb{H}|$, and (2) every element of $\bar{\mathbb{H}}$ lies outside of $\mathbb{H}$; i.e., the intersection of $\mathbb{H}$ and $\bar{\mathbb{H}}$ is empty. Since both $\mathbb{H}$ and $\bar{\mathbb{H}}$ are subsets of $\mathbb{G}$, these imply $|\mathbb{G}| \geq |\mathbb{H}| + |\bar{\mathbb{H}}| = 2|\mathbb{H}|$, proving the lemma.

For any $h_1, h_2 \in \mathbb{H}$, if $\bar{h}h_1 = \bar{h}h_2$ then, multiplying by $\bar{h}^{-1}$ on each side, we have $h_1 = h_2$. This shows that every distinct element $h \in \mathbb{H}$ corresponds to a distinct element $\bar{h}h \in \bar{\mathbb{H}}$, proving (1).

Assume toward a contradiction that $\bar{h}h \in \mathbb{H}$ for some h. This means $\bar{h}h = h'$ for some $h' \in \mathbb{H}$, and so $\bar{h} = h'h^{-1}$. Now, $h'h^{-1} \in \mathbb{H}$ since $\mathbb{H}$ is a subgroup and $h', h^{-1} \in \mathbb{H}$. But this means that $\bar{h} \in \mathbb{H}$, in contradiction to the way $\bar{h}$ was chosen. This proves (2) and completes the proof of the lemma. ∎

The following theorem will enable us to analyze the algorithm given earlier.

THEOREM 9.38 *Fix N. Say there exists a witness that N is composite. Then at least half the elements of $\mathbb{Z}_N^*$ are witnesses that N is composite.*

PROOF Let Bad be the set of elements in $\mathbb{Z}_N^*$ that are *not* witnesses; that is, $a \in$ Bad means $a^{N-1} = 1 \bmod N$. Clearly, $1 \in$ Bad. If $a, b \in$ Bad, then $(ab)^{N-1} = a^{N-1} \cdot b^{N-1} = 1 \cdot 1 = 1 \bmod N$ and hence $ab \in$ Bad. By Lemma 9.36, we conclude that Bad is a subgroup of $\mathbb{Z}_N^*$. Since (by assumption) there is at least one witness, Bad is a *strict* subgroup of $\mathbb{Z}_N^*$. Lemma 9.37 then shows that $|\text{Bad}| \leq |\mathbb{Z}_N^*|/2$, showing that at least half the elements of $\mathbb{Z}_N^*$ are *not* in Bad (and hence are witnesses). ∎

Let N be composite. If there exists a witness that N is composite, then there are at least $|\mathbb{Z}_N^*|/2$ witnesses. The probability that we find either a witness or an element not in $\mathbb{Z}_N^*$ in any given iteration of the algorithm is thus at least $1/2$, and so the probability that the algorithm does not find a witness in any of the t iterations (and hence the probability that the algorithm mistakenly outputs "prime") is at most 2^{-t}.

The above, unfortunately, does not give a complete solution since there are infinitely many composite numbers N that do not have *any* witnesses that they are composite! Such values N are known as *Carmichael numbers*; a detailed discussion is beyond the scope of this book.

Happily, a refinement of the above test can be shown to work for all N. Let $N - 1 = 2^r u$, where u is odd and $r \geq 1$. (It is easy to compute r and u given N. Also, restricting to $r \geq 1$ means that N is odd, but testing primality is easy when N is even!) The algorithm shown previously tests only whether $a^{N-1} = a^{2^r u} = 1 \bmod N$. A more refined algorithm looks at the *sequence* of $r + 1$ values $a^u, a^{2u}, \ldots, a^{2^r u}$ (all modulo N). Each term in this sequence is the square of the preceding term; thus, if some value is equal to ± 1 then all subsequent values will be equal to 1.

Say that $a \in \mathbb{Z}_N^*$ is a *strong witness that N is composite* (or simply a *strong witness*) if (1) $a^u \neq \pm 1 \bmod N$ and (2) $a^{2^i u} \neq -1 \bmod N$ for all $i \in \{1, \ldots, r-1\}$. Note that when an element a is *not* a strong witness then the sequence $(a^u, a^{2u}, \ldots, a^{2^r u})$ (all taken modulo N) takes one of the following forms:

$$(\pm 1, 1, \ldots, 1) \quad \text{or} \quad (\star, \ldots, \star, -1, 1, \ldots, 1),$$

where $\star$ is an arbitrary term. If a is *not* a strong witness then we have $a^{2^{r-1}u} = \pm 1 \bmod N$ and

$$a^{N-1} = a^{2^r u} = \left(a^{2^{r-1}u} \right)^2 = 1 \bmod N,$$

and so a is not a witness that N is composite, either. Put differently, if a is a witness then it is also a strong witness and so there can only possibly be *more* strong witnesses than witnesses.

We first show that if N is prime then there does not exist a strong witness that N is composite. In doing so, we rely on the following easy lemma (which is a special case of Proposition 15.16 proved subsequently in Chapter 15):

LEMMA 9.39   *Say $x \in \mathbb{Z}_N^*$ is a square root of 1 modulo N if $x^2 = 1 \bmod N$. If N is an odd prime then the only square roots of 1 modulo N are $[\pm 1 \bmod N]$.*

PROOF   Say $x^2 = 1 \bmod N$ with $x \in \{1, \ldots, N-1\}$. Then $0 = x^2 - 1 = (x+1)(x-1) \bmod N$, implying that $N \mid (x+1)$ or $N \mid (x-1)$ by Proposition 9.3. This can only possibly occur if $x = [\pm 1 \bmod N]$.   ∎

Let N be an odd prime and fix arbitrary $a \in \mathbb{Z}_N^*$. Let $i \geq 0$ be the minimum value for which $a^{2^i u} = 1 \bmod N$; since $a^{2^r u} = a^{N-1} = 1 \bmod N$ we know that some such $i \leq r$ exists. If $i = 0$ then $a^u = 1 \bmod N$ and a is not a strong witness. Otherwise,

$$\left(a^{2^{i-1}u} \right)^2 = a^{2^i u} = 1 \bmod N$$

and $a^{2^{i-1}u}$ is a square root of 1. If N is an odd prime, the only square roots of 1 are ± 1; by choice of i, however, $a^{2^{i-1}u} \neq 1 \bmod N$. So $a^{2^{i-1}u} = -1 \bmod N$, and a is not a strong witness. We conclude that when N is an odd prime there is no strong witness that N is composite.

A composite integer N is a *prime power* if $N = p^r$ for some prime p and integer $r \geq 1$. We now show that every odd, composite N that is not a prime power has many strong witnesses.

THEOREM 9.40   *Let N be an odd number that is not a prime power. Then at least half the elements of $\mathbb{Z}_N^*$ are strong witnesses that N is composite.*

PROOF Let Bad $\subseteq \mathbb{Z}_N^*$ denote the set of elements that are not strong witnesses. We define a set Bad$'$ and show that: (1) Bad is a subset of Bad$'$, and (2) Bad$'$ is a strict subgroup of $\mathbb{Z}_N^*$. This suffices because by combining (2) and Lemma 9.37 we have that $|\mathsf{Bad}'| \leq |\mathbb{Z}_N^*|/2$. Furthermore, by (1) it holds that Bad $\subseteq$ Bad$'$, and so $|\mathsf{Bad}| \leq |\mathsf{Bad}'| \leq |\mathbb{Z}_N^*|/2$ as in Theorem 9.38. Thus, at least half the elements of $\mathbb{Z}_N^*$ are strong witnesses. (We stress that we do not claim that Bad is a subgroup of $\mathbb{Z}_N^*$.)

Note first that $-1 \in$ Bad since $(-1)^u = -1 \bmod N$ (recall u is odd). Let $i \in \{0, \ldots, r-1\}$ be the largest integer for which there exists an $a \in$ Bad with $a^{2^i u} = -1 \bmod N$; alternatively, i is the largest integer for which there exists an $a \in$ Bad with

$$(a^u, a^{2u}, \ldots, a^{2^r u}) = (\underbrace{\star, \ldots, \star, -1}_{i+1 \text{ terms}}, 1, \ldots, 1).$$

Since $-1 \in$ Bad and $(-1)^{2^0 u} = -1 \bmod N$, some such i exists.

Fix i as above, and define

$$\mathsf{Bad}' \stackrel{\text{def}}{=} \{a \mid a^{2^i u} = \pm 1 \bmod N\}.$$

We now prove what we claimed above.

CLAIM 9.41 Bad $\subseteq$ Bad$'$.

Let $a \in$ Bad. Then either $a^u = 1 \bmod N$ or $a^{2^j u} = -1 \bmod N$ for some $j \in \{0, \ldots, r-1\}$. In the first case, $a^{2^i u} = (a^u)^{2^i} = 1 \bmod N$ and so $a \in$ Bad$'$. In the second case, we have $j \leq i$ by choice of i. If $j = i$ then clearly $a \in$ Bad$'$. If $j < i$ then $a^{2^i u} = (a^{2^j u})^{2^{i-j}} = 1 \bmod N$ and $a \in$ Bad$'$. Since a was arbitrary, this shows Bad $\subseteq$ Bad$'$.

CLAIM 9.42 Bad$'$ *is a subgroup of* $\mathbb{Z}_N^*$.

Clearly $1 \in$ Bad$'$. Furthermore, if $a, b \in$ Bad$'$ then

$$(ab)^{2^i u} = a^{2^i u} b^{2^i u} = (\pm 1)(\pm 1) = \pm 1 \bmod N$$

and so $ab \in$ Bad$'$. By Lemma 9.36, Bad$'$ is a subgroup.

CLAIM 9.43 Bad$'$ *is a strict subgroup of* $\mathbb{Z}_N^*$.

If N is an odd, composite integer that is not a prime power, then N can be written as $N = N_1 N_2$ with $N_1, N_2 > 1$ odd and $\gcd(N_1, N_2) = 1$. Appealing to the Chinese remainder theorem, let $a \leftrightarrow (a_1, a_2)$ denote the representation of $a \in \mathbb{Z}_N^*$ as an element of $\mathbb{Z}_{N_1}^* \times \mathbb{Z}_{N_2}^*$; that is, $a_1 = [a \bmod N_1]$ and $a_2 =$

$[a \bmod N_2]$. Take $a \in \mathsf{Bad}'$ such that $a^{2^i u} = -1 \bmod N$ (such an a must exist by the way we defined i), and say $a \leftrightarrow (a_1, a_2)$. Since $-1 \leftrightarrow (-1, -1)$ we have

$$(a_1, a_2)^{2^i u} = (a_1^{2^i u}, a_2^{2^i u}) = (-1, -1),$$

and so

$$a_1^{2^i u} = -1 \bmod N_1 \quad \text{and} \quad a_2^{2^i u} = -1 \bmod N_2.$$

Consider the element $b \in \mathbb{Z}_N^*$ with $b \leftrightarrow (a_1, 1)$. Then

$$b^{2^i u} \leftrightarrow (a_1, 1)^{2^i u} = ([a_1^{2^i u} \bmod N_1], 1) = (-1, 1) \not\leftrightarrow \pm 1.$$

That is, $b^{2^i u} \neq \pm 1 \bmod N$ and so we have found an element $b \notin \mathsf{Bad}'$. This proves that Bad' is a *strict* subgroup of $\mathbb{Z}_N^*$ and so, by Lemma 9.37, the size of Bad' (and thus the size of Bad) is at most half the size of $\mathbb{Z}_N^*$. ∎

An integer N is a *perfect power* if $N = \hat{N}^e$ for integers $\hat{N}$ and $e \geq 2$ (here it is not required for $\hat{N}$ to be prime, although of course any prime power is also a perfect power). Algorithm 9.44 gives the Miller–Rabin primality test. Exercises 9.16 and 9.17 ask you to show that testing whether N is a perfect power, and testing whether a particular a is a strong witness, can be done in polynomial time. Given these results, the algorithm clearly runs in time polynomial in $\|N\|$ and t. We can now complete the proof of Theorem 9.33:

ALGORITHM 9.44

The Miller–Rabin primality test

Input: Integer $N > 2$ and parameter 1^t
Output: A decision as to whether N is prime or composite

if N is even, **return** "composite"
if N is a perfect power, **return** "composite"
compute $r \geq 1$ and u odd such that $N - 1 = 2^r u$
for $j = 1$ to t:
$\quad a \leftarrow \{1, \ldots, N - 1\}$
$\quad$**if** $a^u \neq \pm 1 \bmod N$ and $a^{2^i u} \neq -1 \bmod N$ for $i \in \{1, \ldots, r - 1\}$
$\quad\quad$**return** "composite"
return "prime"

PROOF If N is an odd prime, there are no strong witnesses and so the Miller–Rabin algorithm always outputs "prime." If N is even or a prime power, the algorithm always outputs "composite." The interesting case is when N is an odd, composite integer that is not a prime power. Consider any iteration of the inner loop. Note first that if $a \notin \mathbb{Z}_N^*$ then $a^u \neq \pm 1 \bmod N$ and

$a^{2^i u} \neq -1 \bmod N$ for $i \in \{1, \ldots, r-1\}$. The probability of finding either a strong witness or an element not in $\mathbb{Z}_N^*$ is at least $1/2$ (invoking Theorem 9.40). Thus, the probability that the algorithm never outputs "composite" in any of the t iterations is at most 2^{-t}.   ■

9.2.3   The Factoring Assumption

Let GenModulus be a polynomial-time algorithm that, on input 1^n, outputs (N, p, q) where $N = pq$, and p and q are n-bit primes except with probability negligible in n. (The natural way to do this is to generate two uniform n-bit primes, as discussed previously, and then multiply them to obtain N.) Then consider the following experiment for a given algorithm $\mathcal{A}$ and parameter n:

The factoring experiment Factor$_{\mathcal{A},\mathsf{GenModulus}}(n)$**:**

1. *Run* GenModulus(1^n) *to obtain* (N, p, q).

2. *$\mathcal{A}$ is given N, and outputs $p', q' > 1$.*

3. *The output of the experiment is defined to be 1 if $p' \cdot q' = N$, and 0 otherwise.*

Note that if the output of the experiment is 1 then $\{p', q'\} = \{p, q\}$, unless p or q are composite (which happens with only negligible probability).

We now formally define the factoring assumption:

DEFINITION 9.45   Factoring is hard relative to GenModulus *if for all probabilistic polynomial-time algorithms $\mathcal{A}$ there exists a negligible function* negl *such that*
$$\Pr[\mathsf{Factor}_{\mathcal{A},\mathsf{GenModulus}}(n) = 1] \leq \mathsf{negl}(n).$$

The *factoring assumption* is the assumption that there exists a GenModulus relative to which factoring is hard.

9.2.4   The RSA Assumption

The factoring problem has been studied for hundreds of years without an efficient algorithm being found. Although the factoring assumption does give a one-way function (see Section 9.4.1), it unfortunately does not *directly* yield practical cryptosystems. (In Section 15.5.2, however, we show how to construct efficient cryptosystems based on a problem whose hardness is *equivalent* to that of factoring.) This has motivated a search for other problems whose difficulty is related to the hardness of factoring. The best known of these is a problem introduced in 1978 by Rivest, Shamir, and Adleman and now called the *RSA problem* in their honor.

Given a modulus N and an integer $e > 2$ relatively prime to $\phi(N)$, Corollary 9.22 shows that exponentiation to the eth power modulo N is a *permutation*. We can therefore define $[y^{1/e} \bmod N]$ (for any $y \in \mathbb{Z}_N^*$) as the unique element of $\mathbb{Z}_N^*$ that yields y when raised to the eth power modulo N; that is, $x = y^{1/e} \bmod N$ if and only if $x^e = y \bmod N$. The RSA problem, informally, is to compute $[y^{1/e} \bmod N]$ for a modulus N of unknown factorization.

Formally, let GenRSA be a probabilistic polynomial-time algorithm that, on input 1^n, outputs a modulus N that is the product of two n-bit primes, as well as integers $e, d > 0$ with $\gcd(e, \phi(N)) = 1$ and $ed = 1 \bmod \phi(N)$. (Such a d exists since e is invertible modulo $\phi(N)$. The purpose of d will become clear later.) The algorithm may fail with probability negligible in n. Consider the following experiment for a given algorithm $\mathcal{A}$ and security parameter n:

The RSA experiment RSA-inv$_{\mathcal{A}, \mathsf{GenRSA}}(n)$:

1. *Run* GenRSA(1^n) *to obtain* (N, e, d).

2. *Choose a uniform* $y \in \mathbb{Z}_N^*$.

3. $\mathcal{A}$ *is given* N, e, y, *and outputs* $x \in \mathbb{Z}_N^*$.

4. *The output of the experiment is defined to be* 1 *if* $x^e = y \bmod N$, *and* 0 *otherwise.*

DEFINITION 9.46 The RSA problem is hard relative to GenRSA *if for all probabilistic polynomial-time algorithms* $\mathcal{A}$ *there exists a negligible function* negl *such that* $\Pr[\text{RSA-inv}_{\mathcal{A}, \mathsf{GenRSA}}(n) = 1] \leq \text{negl}(n)$.

The *RSA assumption* is that there exists a GenRSA algorithm relative to which the RSA problem is hard. A suitable GenRSA algorithm can be constructed from any algorithm GenModulus that generates a composite modulus along with its factorization. A high-level outline is provided as Algorithm 9.47, where the only thing left unspecified is how exactly e is chosen. In fact, the RSA problem is believed to be hard for *any* e that is relatively prime to $\phi(N)$. We discuss some typical choices of e below.

ALGORITHM 9.47
GenRSA – high-level outline

Input: Security parameter 1^n
Output: N, e, d as described in the text

$(N, p, q) \leftarrow$ GenModulus(1^n)
$\phi(N) := (p - 1)(q - 1)$
choose $e > 1$ such that $\gcd(e, \phi(N)) = 1$
compute $d := [e^{-1} \bmod \phi(N)]$
return N, e, d

Example 9.48

Say GenModulus outputs $(N, p, q) = (143, 11, 13)$. Then $\phi(N) = 120$. Next, we need to choose an e that is relatively prime to $\phi(N)$; say we take $e = 7$. The next step is to compute d such that $d = [e^{-1} \bmod \phi(N)]$. This can be done as shown in Appendix B.2.2 to obtain $d = 103$. (One can check that $7 \cdot 103 = 721 = 1 \bmod 120$.) Our GenRSA algorithm in this case thus outputs $(N, e, d) = (143, 7, 103)$.

As an example of the RSA problem relative to these parameters, take $y = 64$ and so the problem is to compute the 7th root of 64 modulo 143 *without* knowledge of d or the factorization of N.   ◇

Computing eth roots modulo N becomes easy if d, $\phi(N)$, or the factorization of N is known. (As we show in the next section, any of these can be used to efficiently compute the others.) This follows from Corollary 9.22, which shows that $[y^d \bmod N]$ is the eth root of y modulo N. This asymmetry—namely, that the RSA problem appears to be hard when d or the factorization of N is unknown, but becomes easy when d *is* known—serves as the basis for applications of the RSA problem to public-key cryptography.

Example 9.49

Continuing the previous example, we can compute the 7th root of 64 modulo 143 using the value $d = 103$; the answer is $25 = 64^d = 64^{103} \bmod 143$. We can verify that this is the correct solution since $25^e = 25^7 = 64 \bmod 143$.   ◇

On the choice of e. There does not appear to be any difference in the hardness of the RSA problem for different exponents e and, as such, different methods have been suggested for selecting it. One popular choice is to set $e = 3$, since then computing eth powers modulo N requires only two multiplications (see Appendix B.2.3). If e is to be set equal to 3, then p and q must be chosen with $p, q \neq 1 \bmod 3$ so that $\gcd(e, \phi(N)) = 1$. For similar reasons, another popular choice is $e = 2^{16} + 1 = 65537$, a prime number with low Hamming weight (in Appendix B.2.3, we explain why such exponents are preferable). As compared to choosing $e = 3$, this makes exponentiation slightly more expensive but reduces the constraints on p and q, and avoids some "low-exponent attacks" (described at the end of Section 12.5.1) that can result from poorly implemented cryptosystems based on the RSA problem.

Note that choosing d small (that is, changing GenRSA to choose small d and then compute $e := [d^{-1} \bmod \phi(N)]$) is a bad idea. If d lies in a very small range then a brute-force search for d can be carried out (and, as noted, once d is known the RSA problem can be solved easily). Even if d is chosen so that $d \approx N^{1/4}$, and so brute-force attacks are ruled out, there are known algorithms that can be used to recover d from N and e in this case. For similar reasons, choosing d with low Hamming weight is also not recommended.

9.2.5 *Relating the Factoring and RSA Assumptions

Say GenRSA is constructed as in Algorithm 9.47. If N can be factored, then we can compute $\phi(N)$ and use this to compute $d := [e^{-1} \bmod \phi(N)]$ for any given e (using Algorithm B.11). So for the RSA problem to be hard relative to GenRSA, the factoring problem must be hard relative to GenModulus. Put differently, the RSA problem cannot be *more* difficult than factoring; hardness of factoring (relative to GenModulus) can only potentially be a *weaker* assumption than hardness of the RSA problem (relative to GenRSA).

What about the other direction? That is, is hardness of the RSA problem implied by hardness of factoring? That remains an open question. The best we can show is that computing an RSA private key from an RSA public key (i.e., computing d from N and e) is as hard as factoring. We start by proving a slightly more powerful result.

THEOREM 9.50 *Fix N, and assume there is a subroutine that, given $x \in \mathbb{Z}_N^*$, outputs an integer $k > 0$ with $x^k = 1 \bmod N$. Then there is an algorithm that finds a factor of N in time $\mathsf{poly}(\|N\|)$ (counting each call to the subroutine as one step), except with probability negligible in $\|N\|$.*

PROOF For simplicity (and because it is most relevant to cryptography) we focus on factoring N that are a product of two distinct, odd primes p and q. We use the Chinese remainder theorem (Section 9.1.5), and rely on Proposition 9.36 and Lemma 9.37 as well as the following facts (which follow from more-general results proved in Sections 15.4.2 and 15.5.2):

- For N of the above form, 1 has exactly four square roots modulo N. Two of these are the "trivial" square roots $[\pm 1 \bmod N]$, and two of these are "nontrivial" square roots. In the Chinese remaindering representation, the nontrivial square roots are $(1, -1)$ and $(-1, 1)$.

- Any nontrivial square root of 1 can be used to (efficiently) compute a factor of N. This is by virtue of the fact that $y^2 = 1 \bmod N$ implies

$$0 = y^2 - 1 = (y - 1)(y + 1) \bmod N,$$

  and so $N \mid (y - 1)(y + 1)$. However, $N \nmid (y - 1)$ and $N \nmid (y + 1)$ because $y \neq \pm 1 \bmod N$. So it must be the case that $\gcd(y - 1, N)$ is equal to one of the prime factors of N.

We use the following strategy to factor N: repeatedly choose a uniform $x \in \mathbb{Z}_N^*$, compute $k > 0$ with $x^k = 1 \bmod N$ (using the assumed subroutine for doing so), write $k = 2^s \cdot v$ for v an odd integer, and compute the sequence

$$x^v, x^{2v}, \ldots, x^{2^s v}$$

modulo N. Each term in this sequence is the square of the preceding term, and the final term is 1. Let j be largest with $y \stackrel{\text{def}}{=} [x^{2^j v} \bmod N] \neq 1$. (If there is no such j, then start again by choosing another x.) By choice of j, we have $y^2 = 1 \bmod N$. If $y \neq -1 \bmod N$ we have found a nontrivial square root of N, and can then factor N as discussed earlier. All the above can be done in polynomial time, and so it only remains to determine the probability, over choice of x, that y exists and is a nontrivial square root of N.

We first observe that the probability that the sequence constructed above contains a nontrivial square root of 1 indeed depends only on x, and not on k. To see this, fix x and let λ be the smallest positive integer for which $x^\lambda = 1 \bmod N$. Write $\lambda = 2^\alpha \cdot \beta$ with β odd, and assume there is a $j \geq 0$ for which $[x^{2^j \beta} \bmod N]$ is a nontrivial square root of 1. Without loss of generality, assume $x^{2^j \beta} \leftrightarrow (-1, 1)$. Now take any $k > 0$ for which $x^k = 1 \bmod N$, and write $k = 2^s \cdot v$ as before. Since k must be a multiple of λ, we have $v = \beta \cdot \gamma$ for some odd γ. But then $x^{2^j v} = x^{2^j \beta \gamma} \leftrightarrow (-1, 1)^\gamma = (-1, 1)$, and so $[x^{2^j v} \bmod N]$ is a nontrivial square root of N. A similar argument shows that the implication goes in the other direction as well.

Let $\phi(N) = 2^r \cdot u$ with u odd. We know that $x^{\phi(N)} = x^{2^r u} = 1 \bmod N$ for all $x \in \mathbb{Z}_N^*$. Let $i \in \{0, \ldots, r-1\}$ be the largest integer for which there exists an $x \in \mathbb{Z}_N^*$ such that $x^{2^i u} \neq 1 \bmod N$. (Since u is odd $(-1)^u = -1 \neq 1 \bmod N$, and so the definition is not vacuous.) Then for all $x \in \mathbb{Z}_N^*$, we have $x^{2^{i+1} u} = 1 \bmod N$ and so $[x^{2^i u} \bmod N]$ is a square root of 1. Define

$$\mathsf{Bad} \overset{\text{def}}{=} \{x \mid x^{2^i u} = \pm 1 \bmod N\}.$$

By the argument above, we know that if our algorithm chooses $x \notin \mathsf{Bad}$ then it finds a nontrivial square root of 1. We show that Bad is a *strict* subgroup of $\mathbb{Z}_N^*$; by Lemma 9.37, this implies $|\mathsf{Bad}| \leq |\mathbb{Z}_N^*|/2$. This means that $x \notin \mathsf{Bad}$ (and the algorithm factors N) with probability at least $1/2$ in each iteration. Using sufficiently many iterations gives the result of the theorem.

We now prove that Bad is a strict subgroup of $\mathbb{Z}_N^*$. If $x, x' \in \mathsf{Bad}$ then

$$(xx')^{2^i u} = x^{2^i u}(x')^{2^i u} = (\pm 1) \cdot (\pm 1) = \pm 1 \bmod N,$$

and so $xx' \in \mathsf{Bad}$ and Bad is a subgroup. To see that Bad is a *strict* subgroup, let $x \in \mathbb{Z}_N^*$ be such that $x^{2^i u} \neq 1 \bmod N$ (such an x must exist by our definition of i). If $x^{2^i u} \neq -1 \bmod N$, then $x \notin \mathsf{Bad}$ and we are done. Otherwise, let $x \leftrightarrow (x_p, x_q)$ be the Chinese remaindering representation of x. Since $x^{2^i u} = -1 \bmod N$, we know that

$$(x_p, x_q)^{2^i u} = (x_p^{2^i u}, x_q^{2^i u}) = (-1, -1) \leftrightarrow -1.$$

But then the element corresponding to $(x_p, 1)$ is not in Bad since

$$(x_p, 1)^{2^i u} = (x_p^{2^i u}, 1) = (-1, 1) \not\leftrightarrow \pm 1.$$

This completes the proof. ■

COROLLARY 9.51   *There is a probabilistic polynomial-time algorithm that, given as input an integer N and integers e, d with $ed = 1 \bmod \phi(N)$, factors N except with probability negligible in $\|N\|$.*

PROOF   Let $k = ed - 1 > 0$ and note that $\phi(N) \mid k$. Since $x^k = 1 \bmod N$ for all $x \in \mathbb{Z}_N^*$ (cf. Corollary 9.21), we can trivially implement the subroutine needed by the previous theorem by always outputting k.   ∎

Assuming factoring is hard, the above result rules out the possibility of efficiently solving the RSA problem by first computing d from N and e. However, it does not rule out the possibility that there might be some completely different way of attacking the RSA problem that does not involve (or imply) factoring N. Thus, based on our current knowledge, the RSA assumption is stronger than the factoring assumption—that is, it may be that the RSA problem can be solved in polynomial time even though factoring cannot. Nevertheless, when GenRSA is constructed based on GenModulus as in Algorithm 9.47, the prevailing conjecture is that the RSA problem is hard relative to GenRSA whenever factoring is hard relative to GenModulus.

9.3   Cryptographic Assumptions in Cyclic Groups

In this section we introduce a class of cryptographic hardness assumptions in *cyclic groups*. We begin with a general discussion of cyclic groups, followed by abstract definitions of the relevant assumptions. We then look at two concrete and widely used examples of cyclic groups in which these assumptions are believed to hold.

9.3.1   Cyclic Groups and Generators

Let $\mathbb{G}$ be a finite group of order m. For arbitrary $g \in \mathbb{G}$, consider the set

$$\langle g \rangle \stackrel{\text{def}}{=} \{g^0, g^1, \ldots\}.$$

(We warn the reader that if $\mathbb{G}$ is an infinite group, $\langle g \rangle$ is defined differently.) By Theorem 9.14, we have $g^m = 1$. Let $i \leq m$ be the smallest positive integer for which $g^i = 1$. Then the above sequence repeats after i terms (i.e., $g^i = g^0$, $g^{i+1} = g^1$, etc.), and so

$$\langle g \rangle = \{g^0, \ldots, g^{i-1}\}.$$

We see that $\langle g \rangle$ contains at most i elements. In fact, it contains exactly i elements since if $g^j = g^k$ with $0 \leq j < k < i$ then $g^{k-j} = 1$ and $0 < k - j < i$, contradicting our choice of i as the smallest positive integer for which $g^i = 1$.

It is not hard to verify that $\langle g \rangle$ is a subgroup of $\mathbb{G}$ for any g (see Exercise 9.3); we call $\langle g \rangle$ the *subgroup generated by* g. If the order of the subgroup $\langle g \rangle$ is i, then i is called the *order of* g; that is:

DEFINITION 9.52 *Let* $\mathbb{G}$ *be a finite group and* $g \in \mathbb{G}$. *The* order of g *is the smallest positive integer* i *with* $g^i = 1$.

The following is a useful analogue of Corollary 9.15 (the proof is identical):

PROPOSITION 9.53 *Let* $\mathbb{G}$ *be a finite group, and* $g \in \mathbb{G}$ *an element of order* i. *Then for any integer* x, *we have* $g^x = g^{[x \bmod i]}$.

We can prove something stronger:

PROPOSITION 9.54 *Let* $\mathbb{G}$ *be a finite group, and* $g \in \mathbb{G}$ *an element of order* i. *Then* $g^x = g^y$ *if and only if* $x = y \bmod i$.

PROOF If $x = y \bmod i$ then $[x \bmod i] = [y \bmod i]$ and the previous proposition says that
$$g^x = g^{[x \bmod i]} = g^{[y \bmod i]} = g^y.$$
For the more interesting direction, say $g^x = g^y$. Then $1 = g^{x-y} = g^{[x-y \bmod i]}$ (using the previous proposition). Since $[x - y \bmod i] < i$, but i is the smallest positive integer with $g^i = 1$, we must have $[x - y \bmod i] = 0$. ∎

The identity element of any group $\mathbb{G}$ is the only element of order 1, and generates the group $\langle 1 \rangle = \{1\}$. At the other extreme, if there is an element $g \in \mathbb{G}$ that has order m (where m is the order of $\mathbb{G}$), then $\langle g \rangle = \mathbb{G}$. In this case, we call $\mathbb{G}$ a *cyclic group* and say that g is a *generator* of $\mathbb{G}$. (A cyclic group may have multiple generators, and so we cannot speak of *the* generator.) If g is a generator of $\mathbb{G}$ then, by definition, every element $h \in \mathbb{G}$ is equal to g^x for some $x \in \{0, \ldots, m-1\}$, a point we will return to in the next section.

Different elements of the same group $\mathbb{G}$ may have different orders. We can, however, place some restrictions on what these possible orders might be.

PROPOSITION 9.55 *Let* $\mathbb{G}$ *be a finite group of order* m, *and say* $g \in \mathbb{G}$ *has order* i. *Then* $i \mid m$.

PROOF By Theorem 9.14 we know that $g^m = 1 = g^0$. Proposition 9.54 implies that $m = 0 \bmod i$. ∎

The next corollary illustrates the power of this result:

COROLLARY 9.56   *If $\mathbb{G}$ is a group of prime order p, then $\mathbb{G}$ is cyclic. Furthermore, all elements of $\mathbb{G}$ except the identity are generators of $\mathbb{G}$.*

PROOF   By Proposition 9.55, the only possible orders of elements in $\mathbb{G}$ are 1 and p. Only the identity has order 1, and so all other elements have order p and generate $\mathbb{G}$.   ∎

Groups of prime order form one class of cyclic groups. The additive group $\mathbb{Z}_N$, for $N > 1$, gives another example of a cyclic group (the element 1 is always a generator). The next theorem—a special case of Theorem A.21—gives an important additional class of cyclic groups; a proof is outside the scope of this book, but can be found in any standard abstract algebra text.

THEOREM 9.57   *If p is prime then $\mathbb{Z}_p^*$ is a cyclic group of order $p - 1$.*

For $p > 3$ prime, $\mathbb{Z}_p^*$ does not have prime order and so the above does not follow from the preceding corollary.

Example 9.58
Consider the (additive) group $\mathbb{Z}_{15}$. As we have noted, $\mathbb{Z}_{15}$ is cyclic and the element 1 is a generator since $15 \cdot 1 = 0 \bmod 15$ and $i \cdot 1 = i \neq 0 \bmod 15$ for any $0 < i < 15$ (recall that in this group the identity is 0).

$\mathbb{Z}_{15}$ has other generators. For example, $\langle 2 \rangle = \{0, 2, 4, \ldots, 14, 1, 3, \ldots, 13\}$ and so 2 is also a generator.

Not every element generates $\mathbb{Z}_{15}$. For example, the element 3 has order 5 since $5 \cdot 3 = 0 \bmod 15$, and so 3 does not generate $\mathbb{Z}_{15}$. The subgroup $\langle 3 \rangle$ consists of the 5 elements $\{0, 3, 6, 9, 12\}$, and this is indeed a subgroup under addition modulo 15. The element 10 has order 3 since $3 \cdot 10 = 0 \bmod 15$, and the subgroup $\langle 10 \rangle$ consists of the 3 elements $\{0, 5, 10\}$. The orders of the subgroups (i.e., 5 and 3) divide $|\mathbb{Z}_{15}| = 15$ as required by Proposition 9.55. ◇

Example 9.59
Consider the (multiplicative) group $\mathbb{Z}_{15}^*$ of order $(5 - 1)(3 - 1) = 8$. We have $\langle 2 \rangle = \{1, 2, 4, 8\}$, and so the order of 2 is 4. As required by Proposition 9.55, 4 divides 8.   ◇

Example 9.60
Consider the (additive) group $\mathbb{Z}_p$ of prime order p. We know this group is cyclic, but Corollary 9.56 tells us more: namely, *every* element except 0 is a generator. Indeed, for any $h \in \{1, \ldots, p - 1\}$ and integer $i > 0$ we have $ih = 0 \bmod p$ if and only if $p \,|\, ih$. But then Proposition 9.3 says that either $p \,|\, h$ or $p \,|\, i$. The former cannot occur (since $h < p$), and the smallest positive

integer for which the latter can occur is $i = p$. We have thus shown that every nonzero element h has order p (and so generates $\mathbb{Z}_p$), in accordance with Corollary 9.56. $\diamond$

Example 9.61

Consider the (multiplicative) group $\mathbb{Z}_7^*$, which is cyclic by Theorem 9.57. We have $\langle 2 \rangle = \{1, 2, 4\}$, and so 2 is *not* a generator. However,

$$\langle 3 \rangle = \{1, 3, 2, 6, 4, 5\} = \mathbb{Z}_7^*,$$

and so 3 is a generator of $\mathbb{Z}_7^*$. $\diamond$

The following example relies on the material of Section 9.1.5.

Example 9.62

Let $\mathbb{G}$ be a cyclic group of order n, and let g be a generator of $\mathbb{G}$. Then the mapping $f : \mathbb{Z}_n \to \mathbb{G}$ given by $f(a) = g^a$ is an isomorphism between $\mathbb{Z}_n$ and $\mathbb{G}$. Indeed, for $a, a' \in \mathbb{Z}_n$ we have

$$f(a + a') = g^{[a+a' \bmod n]} = g^{a+a'} = g^a \cdot g^{a'} = f(a) \cdot f(a').$$

Bijectivity of f can be proved using the fact that n is the order of g. $\diamond$

The previous example shows that all cyclic groups of the same order are isomorphic and thus the same from an *algebraic* point of view. We stress that this is not true in a *computational* sense, and in particular an isomorphism $f^{-1} : \mathbb{G} \to \mathbb{Z}_n$ (which we know must exist) need not be efficiently computable. This point should become clearer from the discussion in the sections below as well as Chapter 10.

9.3.2 The Discrete-Logarithm/Diffie–Hellman Assumptions

We now introduce several computational problems that can be defined for any class of cyclic groups. We will keep the discussion in this section abstract, and consider specific examples of groups in which these problems are believed to be hard in Sections 9.3.3 and 9.3.4.

We let $\mathcal{G}$ denote a generic, polynomial-time, *group-generation algorithm*. This is an algorithm that, on input 1^n, outputs a description of a cyclic group $\mathbb{G}$, its order q (with $\|q\| = n$), and a generator $g \in \mathbb{G}$. The description of a cyclic group specifies how elements of the group are represented as bit-strings; we assume that each group element is represented by a unique bit-string. We require that there are efficient algorithms (namely, algorithms running in time polynomial in n) for testing whether a given bit-string represents an element of $\mathbb{G}$, as well as for computing the group operation. This implies efficient algorithms for exponentiation in $\mathbb{G}$ (see Appendix B.2.3), computing inverses (the inverse of g is g^{q-1}) and for sampling a uniform element $h \in \mathbb{G}$ (simply choose uniform $x \in \mathbb{Z}_q$ and set $h := g^x$). As discussed at the end of the previous section, although all cyclic groups of a given order are isomorphic, the *representation* of the group determines the computational complexity of mathematical operations in that group.

If $\mathbb{G}$ is a cyclic group of order q with generator g, then $\{g^0, g^1, \ldots, g^{q-1}\}$ is all of $\mathbb{G}$. Equivalently, for every $h \in \mathbb{G}$ there is a *unique* $x \in \mathbb{Z}_q$ such that $g^x = h$. When the underlying group $\mathbb{G}$ is understood from the context, we call this x the *discrete logarithm of h with respect to g* and write $x = \log_g h$. (Logarithms in this case are called "discrete" since they take on integer values, as opposed to "standard" logarithms from calculus whose values range over the real numbers.) Note that if $g^{x'} = h$ for some arbitrary integer x', then $[x' \bmod q] = \log_g h$.

Discrete logarithms obey many of the same rules as "standard" logarithms. For example, $\log_g 1 = 0$ (where 1 is the identity of $\mathbb{G}$); for any integer r, we have $\log_g h^r = [r \cdot \log_g h \bmod q]$; and $\log_g(h_1 h_2) = [(\log_g h_1 + \log_g h_2) \bmod q]$.

The *discrete-logarithm problem* in a cyclic group $\mathbb{G}$ with generator g is to compute $\log_g h$ for a uniform element $h \in \mathbb{G}$. Consider the following experiment for a group-generation algorithm $\mathcal{G}$, algorithm $\mathcal{A}$, and parameter n:

The discrete-logarithm experiment $\mathsf{DLog}_{\mathcal{A},\mathcal{G}}(n)$:

1. *Run $\mathcal{G}(1^n)$ to obtain $(\mathbb{G}, q, g)$, where $\mathbb{G}$ is a cyclic group of order q (with $\|q\| = n$), and g is a generator of $\mathbb{G}$.*
2. *Choose a uniform $h \in \mathbb{G}$.*
3. *$\mathcal{A}$ is given $\mathbb{G}, q, g, h$, and outputs $x \in \mathbb{Z}_q$.*
4. *The output of the experiment is defined to be 1 if $g^x = h$, and 0 otherwise.*

DEFINITION 9.63 *We say the discrete-logarithm problem is hard relative to $\mathcal{G}$ if for all probabilistic polynomial-time algorithms $\mathcal{A}$ there exists a negligible function negl such that $\Pr[\mathsf{DLog}_{\mathcal{A},\mathcal{G}}(n) = 1] \leq \mathsf{negl}(n)$.*

The discrete-logarithm assumption is simply the assumption that there exists a $\mathcal{G}$ for which the discrete-logarithm problem is hard. The following two sections discuss some candidate group-generation algorithms $\mathcal{G}$ for which this is believed to be the case.

The Diffie–Hellman problems. The so-called *Diffie–Hellman* problems are related, but not known to be equivalent, to the problem of computing discrete logarithms. There are two important variants: the *computational* Diffie–Hellman (CDH) problem and the *decisional* Diffie–Hellman (DDH) problem.

Fix a cyclic group $\mathbb{G}$ and a generator $g \in \mathbb{G}$. Given elements $h_1, h_2 \in \mathbb{G}$, define $\mathsf{DH}_g(h_1, h_2) \stackrel{\text{def}}{=} g^{\log_g h_1 \cdot \log_g h_2}$. That is, if $h_1 = g^{x_1}$ and $h_2 = g^{x_2}$ then

$$\mathsf{DH}_g(h_1, h_2) = g^{x_1 \cdot x_2} = h_1^{x_2} = h_2^{x_1}.$$

The *CDH problem* is to compute $\mathsf{DH}_g(h_1, h_2)$ for uniform h_1 and h_2. Hardness of this problem can be formalized by the natural experiment; we leave the details as an exercise.

If the discrete-logarithm problem relative to some $\mathcal{G}$ is easy, then the CDH problem is, too: given h_1 and h_2, first compute $x_1 := \log_g h_1$ and then output the answer $h_2^{x_1}$. In contrast, it is not clear (in general) whether hardness of the discrete-logarithm problem implies that the CDH problem is hard as well.

The *DDH problem*, roughly speaking, is to distinguish $\mathsf{DH}_g(h_1, h_2)$ from a uniform group element when h_1, h_2 are uniform. That is, given uniform h_1, h_2 and a third group element h', the problem is to decide whether $h' = \mathsf{DH}_g(h_1, h_2)$ or whether h' was chosen uniformly from $\mathbb{G}$. Formally:

DEFINITION 9.64 *We say the DDH problem is hard relative to $\mathcal{G}$ if for all probabilistic polynomial-time algorithms $\mathcal{A}$ there is a negligible function* negl *such that*

$$\Big| \Pr[\mathcal{A}(\mathbb{G}, q, g, g^x, g^y, g^z) = 1] - \Pr[\mathcal{A}(\mathbb{G}, q, g, g^x, g^y, g^{xy}) = 1] \Big| \leq \mathsf{negl}(n),$$

where in each case the probabilities are taken over the experiment in which $\mathcal{G}(1^n)$ outputs $(\mathbb{G}, q, g)$, and then uniform $x, y, z \in \mathbb{Z}_q$ are chosen. (Note that when z is uniform in $\mathbb{Z}_q$, then g^z is uniformly distributed in $\mathbb{G}$.)

We have already seen that if the discrete-logarithm problem is easy relative to some $\mathcal{G}$, then the CDH problem is too. Similarly, if the CDH problem is easy relative to $\mathcal{G}$ then so is the DDH problem; you are asked to show this in Exercise 9.19. The converse, however, does not appear to be true, and there are examples of groups in which the discrete-logarithm and CDH problems are believed to be hard even though the DDH problem is easy; see Exercise 15.16.

Using Prime-Order Groups

There are various (classes of) cyclic groups in which the discrete-logarithm and Diffie–Hellman problems are believed to be hard. There is a preference, however, for cyclic groups of *prime order*, for reasons we now explain.

One reason for preferring groups of prime order is because, in a certain sense, the discrete-logarithm problem is hardest in such groups. This is a consequence of the *Pohlig–Hellman algorithm*, described in Chapter 10, which shows that the discrete-logarithm problem in a group of order q becomes easier if q has (small) prime factors. This does not necessarily mean that the discrete-logarithm problem is *easy* in groups of nonprime order; it merely means that the problem becomes *easier*.

Related to the above is the fact that the DDH problem is easy if the group order q has small prime factors. We refer to Exercise 15.16 for one example of this phenomenon.

A second motivation for using prime-order groups is because finding a generator in such groups is trivial. This follows from Corollary 9.56, which says that *every* element of a prime-order group (except the identity) is a generator. In contrast, efficiently finding a generator of an arbitrary cyclic group requires the factorization of the group order to be known (see Appendix B.3).

Proofs of security for some cryptographic constructions require computing multiplicative inverses of certain exponents. When the group order is prime, any nonzero exponent will be invertible, making this computation possible. In particular, we have the following useful result:

LEMMA 9.65   *Fix a group $\mathbb{G}$ of prime order q, and elements $g, h \in \mathbb{G}$ with $g \neq 1$. Given distinct pairs $(x, y), (x', y') \in \mathbb{Z}_q \times \mathbb{Z}_q$ with $g^x h^y = g^{x'} h^{y'}$, it is possible to efficiently compute $\log_g h$.*

PROOF   Note that g is a generator of $\mathbb{G}$. Simple algebra gives

$$g^{x-x'} = h^{y'-y}. \tag{9.1}$$

Note that $y' - y \neq 0 \bmod q$; otherwise, we would have $x - x' = 0 \bmod q$ and then the pairs (x, y) and (x', y') would not be distinct. Since q is prime, the inverse $\Delta \stackrel{\text{def}}{=} [(y' - y)^{-1} \bmod q]$ exists. Raising each side of Equation (9.1) to this power gives:

$$g^{(x-x') \cdot \Delta} = \left(h^{y'-y} \right)^{\Delta} = h^1 = h.$$

So $\log_g h = [(x - x') \cdot (y' - y)^{-1} \bmod q]$, which is easy to compute.   ∎

A final reason for working with prime-order groups is relevant in situations when the *decisional* Diffie–Hellman problem should be hard. Fixing a group $\mathbb{G}$ with generator g, the DDH problem boils down to distinguishing between tuples of the form $(h_1, h_2, \mathsf{DH}_g(h_1, h_2))$ for uniform h_1, h_2, and tuples of the form (h_1, h_2, y), for uniform h_1, h_2, y. A necessary condition for the DDH problem to be hard is that $\mathsf{DH}_g(h_1, h_2)$ by itself (i.e., even without h_1, h_2) should be indistinguishable from a uniform group element. One can show that $\mathsf{DH}_g(h_1, h_2)$ is "close" to uniform (in a sense we do not define here) when the group order q is prime, something that is not necessarily true otherwise.

9.3.3   Working in (Subgroups of) $\mathbb{Z}_p^*$

Groups of the form $\mathbb{Z}_p^*$, for p prime, give one class of cyclic groups in which the discrete-logarithm problem is believed to be hard. Concretely, let $\mathcal{G}$ be an algorithm that, on input 1^n, chooses a uniform n-bit prime p, and outputs p and the group order $q = p - 1$ along with a generator g of $\mathbb{Z}_p^*$. (Section 9.2.1 discusses efficient algorithms for choosing a random prime, and Appendix B.3

shows how to efficiently find a generator of $\mathbb{Z}_p^*$ given the factorization of $p-1$.) The representation of $\mathbb{Z}_p^*$ here is the trivial one where elements are represented as integers between 1 and $p-1$. It is conjectured that the discrete-logarithm problem is hard relative to $\mathcal{G}$ of this sort.

The cyclic group $\mathbb{Z}_p^*$ (for $p > 3$ prime), however, does *not* have prime order. (The preference for groups of prime order was discussed in the previous section.) More problematic, the decisional Diffie–Hellman problem is, in general, *not hard* in such groups (see Exercise 15.16), and they are therefore unacceptable for the cryptographic applications based on the DDH assumption that we will explore in later chapters.

These issues can be addressed by using a prime-order *subgroup* of $\mathbb{Z}_p^*$. Let $p = rq + 1$ where both p and q are prime; r is called the *cofactor*. We prove that $\mathbb{Z}_p^*$ has a subgroup $\mathbb{G}$ of order q given by the set of rth *residues modulo p*, i.e., the set of elements $\{[h^r \bmod p] \mid h \in \mathbb{Z}_p^*\}$ that are equal to the rth power of some $h \in \mathbb{Z}_p^*$.

THEOREM 9.66   *Let $p = rq + 1$ with p, q prime. Then*

$$\mathbb{G} \stackrel{\text{def}}{=} \left\{ [h^r \bmod p] \mid h \in \mathbb{Z}_p^* \right\}$$

is a subgroup of $\mathbb{Z}_p^*$ of order q.

PROOF   The proof that $\mathbb{G}$ is a subgroup is straightforward and is omitted. We prove that $\mathbb{G}$ has order q by showing that the function $f_r : \mathbb{Z}_p^* \to \mathbb{G}$ defined by $f_r(g) = [g^r \bmod p]$ is an r-to-1 function. (Since $|\mathbb{Z}_p^*| = p - 1$, this shows that $|\mathbb{G}| = (p-1)/r = q$.) To see this, let g be a generator of $\mathbb{Z}_p^*$ so that $g^0, \ldots, g^{p-2}$ are all the elements of $\mathbb{Z}_p^*$. By Proposition 9.54 we have $\left(g^i\right)^r = \left(g^j\right)^r$ if and only if $ir = jr \bmod (p-1)$ or, equivalently, $p-1 \mid (i-j)r$. Since $p-1 = rq$, this is equivalent to $q \mid (i-j)$. For any fixed $j \in \{0, \ldots, p-2\}$, this means that the set of values $i \in \{0, \ldots, p-2\}$ for which $\left(g^i\right)^r = \left(g^j\right)^r$ is exactly the set of r distinct values

$$\{j,\ j+q,\ j+2q,\ \ldots,\ j+(r-1)q\},$$

all reduced modulo $p - 1$. (Note that $j + rq = j \bmod (p-1)$.) This proves that f_r is an r-to-1 function. ∎

Besides showing existence of an appropriate subgroup, the proof of the theorem also implies that it is easy to generate a uniform element of $\mathbb{G}$ and to test whether a given element of $\mathbb{Z}_p^*$ lies in $\mathbb{G}$. Specifically, choosing a uniform element of $\mathbb{G}$ can be done by choosing a uniform $h \in \mathbb{Z}_p^*$ and computing $[h^r \bmod p]$. Determining whether a given $h \in \mathbb{Z}_p^*$ is also in the subgroup $\mathbb{G}$ can be done by checking whether $h^q \stackrel{?}{=} 1 \bmod p$. To see that this works, let

$h = g^i$ for g a generator of $\mathbb{Z}_p^*$ and $i \in \{0, \ldots, p-2\}$. Then

$$h^q = 1 \bmod p \Longleftrightarrow g^{iq} = 1 \bmod p$$
$$\Longleftrightarrow iq = 0 \bmod (p-1) \Longleftrightarrow rq \,|\, iq \Longleftrightarrow r \,|\, i,$$

using Proposition 9.54. So $h = g^i = g^{cr} = (g^c)^r$ for some c, and $h \in \mathbb{G}$.

Algorithm 9.67 encapsulates the above discussion. In the algorithm, we let n denote the length of q (the order of the group), and let ℓ denote the length of p (the modulus being used). The relationship between these parameters is discussed below.

ALGORITHM 9.67

A group-generation algorithm $\mathcal{G}$

Input: Security parameter 1^n, parameter $\ell = \ell(n)$
Output: Cyclic group $\mathbb{G}$, its (prime) order q, and a generator g

choose ℓ-bit prime p and n-bit prime q such that $q \,|\, (p-1)$
    // we omit the details of how this is done
until $g \neq 1$ **do:**
    choose uniform $h \in \mathbb{Z}_p^*$
    set $g := [h^{(p-1)/q} \bmod p]$
return p, q, g // $\mathbb{G}$ is the order-q subgroup of $\mathbb{Z}_p^*$ generated by g

---

Choosing ℓ. Let $n = \|q\|$ and $\ell = \|p\|$. Two types of algorithms are known for computing discrete logarithms in order-q subgroups of $\mathbb{Z}_p^*$ (see Section 10.2): those that run in time $\mathcal{O}(\sqrt{q}) = \mathcal{O}(2^{n/2})$ and those that run in time $2^{\mathcal{O}((\log p)^{1/3} \cdot (\log \log p)^{2/3})} = 2^{\mathcal{O}(\ell^{1/3} \cdot (\log \ell)^{2/3})}$. Fixing some desired security parameter n, the value of ℓ should be chosen so as to balance these times. (If ℓ is any smaller, security is reduced; if ℓ is any larger, operations in $\mathbb{G}$ will be less efficient without any gain in security.) See also Section 10.4.

In practice, standardized values (e.g., recommended by NIST) for p, q, and a generator g are used, and there is no need to generate parameters of one's own.

Example 9.68
Consider the group $\mathbb{Z}_{11}^*$ of order 10. Let us try to find a generator of this group. Consider trying 2:

Powers of 2:	2^0	2^1	2^2	2^3	2^4	2^5	2^6	2^7	2^8	2^9
Values:	1	2	4	8	5	10	9	7	3	6

(All values above are computed modulo 11.) We got lucky the first time—2 is a generator! Let's try 3:

Powers of 3:	3^0	3^1	3^2	3^3	3^4	3^5	3^6	3^7	3^8	3^9
Values:	1	3	9	5	4	1	3	9	5	4

We see that 3 is not a generator of the entire group. Rather, it generates a *subgroup* $\mathbb{G} = \{1, 3, 4, 5, 9\}$ of order 5. Now, let's see what happens with 10:

Powers of 10:	10^0	10^1	10^2	10^3	10^4	10^5	10^6	10^7	10^8	10^9
Values:	1	10	1	10	1	10	1	10	1	10

In this case we generate a subgroup of order 2.

For cryptographic purposes we want to work in a prime-order group. Since $11 = 2 \cdot 5 + 1$ we can apply Theorem 9.66 with $q = 5$ and $r = 2$, or with $q = 2$ and $r = 5$. In the first case, the theorem tells us that the *squares* of all the elements of $\mathbb{Z}_{11}^*$ should give a subgroup of order 5. This can be easily verified:

Element:	1	2	3	4	5	6	7	8	9	10
Square:	1	4	9	5	3	3	5	9	4	1

We have seen above that 3 is a generator of this subgroup. (In fact, since the subgroup has prime order, every element of the subgroup besides 1 is a generator of the subgroup.) Taking $q = 2$ and $r = 5$, Theorem 9.66 tells us that taking 5th powers will give a subgroup of order 2. One can check that this gives the order-2 subgroup generated by 10. ◇

Subgroups of finite fields. The discrete-logarithm problem is also believed to be hard in the multiplicative group of a finite field of large characteristic when the polynomial representation is used. (Appendix A.5 provides a brief background on finite fields.) Recall that for any prime p and integer $k \geq 1$ there is a (unique) field $\mathbb{F}_{p^k}$ of order p^k; the multiplicative group $\mathbb{F}_{p^k}^*$ of that field is a cyclic group of order $p^k - 1$ (cf. Theorem A.21). If q is a large prime factor of $p^k - 1$, then Theorem 9.66 shows that $\mathbb{F}_{p^k}^*$ has a cyclic subgroup of order q. (The only property of $\mathbb{Z}_p^*$ we used in the proof of that theorem was that $\mathbb{Z}_p^*$ is cyclic.) This offers another choice of prime-order groups in which the discrete-logarithm and Diffie–Hellman problems are believed to be hard. Our treatment of $\mathbb{Z}_p^*$ in this section corresponds to the special case $k = 1$. (Appropriate choice of parameters for cryptographic applications when $k > 1$ is outside the scope of this book.)

9.3.4 Elliptic Curves

The groups we have concentrated on thus far have all been based directly on modular arithmetic. Another class of groups important for cryptography is given by groups consisting of *points on elliptic curves*. Such groups are especially interesting from a cryptographic perspective since, in contrast to $\mathbb{Z}_p^*$ or the multiplicative group of a finite field, there are currently no known sub-exponential time algorithms for solving the discrete-logarithm problem in appropriately chosen elliptic-curve groups. (See Section 10.4 for further discussion.) For cryptosystems based on the discrete-logarithm or Diffie–Hellman assumptions, this means that implementations based on elliptic-curve groups can be much more efficient—in terms of both computation and, especially, communication—than implementations based on prime-order subgroups of $\mathbb{Z}_p^*$ at any given level of security. In this section we provide a brief introduction to elliptic-curve cryptography. A deeper understanding of the issues discussed here requires more sophisticated mathematics than we are willing to assume on the part of the reader. Those interested in further exploring this topic are advised to consult the references at the end of this chapter.

Throughout this section, let $p \geq 5$ be a prime.[2] For our purposes, an *elliptic curve* is defined by a cubic equation (modulo p) in two variables x and y; the *points* on the curve are the solutions to the equation. For example, consider an equation E in the variables x and y of the form

$$y^2 = x^3 + Ax + B \bmod p, \tag{9.2}$$

where $A, B \in \mathbb{Z}_p$ satisfy $4A^3 + 27B^2 \neq 0 \bmod p$. (This condition ensures that the equation $x^3 + Ax + B = 0 \bmod p$ has no repeated roots.) Equation (9.2) is called the *Weierstrass representation* of an elliptic curve, and any elliptic curve can be written in this form by applying an invertible affine transformation to the variables x and y. Let $E(\mathbb{Z}_p)$ denote the set of pairs $(x, y) \in \mathbb{Z}_p \times \mathbb{Z}_p$ satisfying the above equation along with a special value $\mathcal{O}$ whose purpose we will discuss shortly; that is,

$$E(\mathbb{Z}_p) \stackrel{\text{def}}{=} \{(x, y) \mid x, y \in \mathbb{Z}_p \text{ and } y^2 = x^3 + Ax + B \bmod p\} \cup \{\mathcal{O}\}.$$

The elements $E(\mathbb{Z}_p)$ are called the *points* on the *elliptic curve* E defined by Equation (9.2), and $\mathcal{O}$ is called the *point at infinity*.

Example 9.69
An element $y \in \mathbb{Z}_p^*$ is a *quadratic residue modulo* p if there is an $x \in \mathbb{Z}_p^*$ such that $x^2 = y \bmod p$; in that case, we say x is a *square root of* y. If y is not a quadratic residue then we say it is a *quadratic non-residue*. For $p > 2$ prime, exactly half the elements in $\mathbb{Z}_p^*$ are quadratic residues, and every quadratic residue has exactly two square roots. (See Section 15.4.1.)

Let $f(x) \stackrel{\text{def}}{=} x^3 + 3x + 3$ and consider the curve $E : y^2 = f(x) \bmod 7$. Each value of x for which $f(x)$ is a quadratic residue modulo 7 yields two points on the curve; values x where $f(x)$ is not a quadratic residue have no corresponding point on the curve; values of x for which $f(x) = 0 \bmod 7$ give one point on the curve. This allows us to determine the points on the curve:

- $f(0) = 3 \bmod 7$, a quadratic non-residue modulo 7.

---

[2]The theory can be adapted to deal with $p \in \{2, 3\}$ but this introduces additional complications. Elliptic curves can, in fact, be defined over arbitrary *fields* (cf. Section A.5), and our discussion largely carries over to fields of characteristic not equal to 2 or 3.

- $f(1) = 0 \bmod 7$, so we obtain the point $(1,0) \in E(\mathbb{Z}_7)$.

- $f(2) = 3 \bmod 7$, a quadratic non-residue modulo 7.

- $f(3) = 4 \bmod 7$, a quadratic residue modulo 7 with square roots 2 and 5. This yields the points $(3,2), (3,5) \in E(\mathbb{Z}_7)$.

- $f(4) = 2 \bmod 7$, a quadratic residue modulo 7 with square roots 3 and 4. This yields the points $(4,3), (4,4) \in E(\mathbb{Z}_7)$.

- $f(5) = 3 \bmod 7$, a quadratic non-residue modulo 7.

- $f(6) = 6 \bmod 7$, a quadratic non-residue modulo 7.

Including the point at infinity $\mathcal{O}$, there are 6 points in $E(\mathbb{Z}_7)$. ◇

A useful way to think about $E(\mathbb{Z}_p)$ is to look at the graph of Equation (9.2) over the reals (i.e., the equation $y^2 = x^3 + Ax + B$ without reduction modulo p) as in Figure 9.2. This figure does not correspond exactly to $E(\mathbb{Z}_p)$ because, for example, $E(\mathbb{Z}_p)$ has a finite number of points ($\mathbb{Z}_p$ is, after all, a finite set) while there are an infinite number of solutions to the same equation if we allow x and y to range over all real numbers. Nevertheless, the picture provides useful intuition. In such a figure, one can think of the "point at infinity" $\mathcal{O}$ as sitting at the top of the y-axis and lying on every vertical line.

It can be shown that every line intersecting $E(\mathbb{Z}_p)$ at two points must also intersect it at a third point, where (1) a point P is counted twice if the line is tangent to the curve at P, and (2) the point at infinity is also counted when the line is vertical. This fact is used to define a binary operation, called "addition" and denoted by $+$, on points of $E(\mathbb{Z}_p)$ in the following way:

- The point $\mathcal{O}$ is defined to be an (additive) identity; that is, for all $P \in E(\mathbb{Z}_p)$ we define $P + \mathcal{O} = \mathcal{O} + P = P$.

- For two points $P_1, P_2 \neq \mathcal{O}$ on E, we evaluate their sum $P_1 + P_2$ by drawing the line through P_1, P_2 (if $P_1 = P_2$ then draw the line tangent to the curve at P_1) and finding the third point of intersection P_3 of this line with $E(\mathbb{Z}_p)$; the third point of intersection may be $P_3 = \mathcal{O}$ if the line is vertical. If $P_3 = (x, y) \neq \mathcal{O}$ then we define $P_1 + P_2 \stackrel{\text{def}}{=} (x, -y)$. (Graphically, this corresponds to reflecting P_3 in the x-axis.) If $P_3 = \mathcal{O}$ then $P_1 + P_2 \stackrel{\text{def}}{=} \mathcal{O}$.

If $P = (x, y) \neq \mathcal{O}$ is a point of $E(\mathbb{Z}_p)$, then $-P \stackrel{\text{def}}{=} (x, -y)$ (which is clearly also a point of $E(\mathbb{Z}_p)$) is the unique inverse of P. Indeed, the line through (x, y) and $(x, -y)$ is vertical, and so the addition rule implies that $P + (-P) = \mathcal{O}$. (If $y = 0$ then $P = (x, y) = (x, -y) = -P$ but then the tangent line at P will be vertical and so $P + (-P) = \mathcal{O}$ here as well.) Of course, $-\mathcal{O} = \mathcal{O}$.

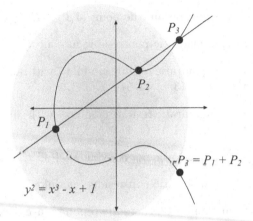

P_3

P_2

P_1

$P_3 = P_1 + P_2$

$y^2 = x^3 - x + 1$

FIGURE 9.2: An elliptic curve over the reals.

It is straightforward, but tedious, to work out the addition law concretely for an elliptic curve in Weierstrass form. Let $P_1 = (x_1, y_1)$ and $P_2 = (x_2, y_2)$ be two points in $E(\mathbb{Z}_p)$, with $P_1, P_2 \neq \mathcal{O}$ and E as in Equation (9.2). To keep matters simple, suppose $x_1 \neq x_2$ (dealing with the case $x_1 = x_2$ is even more tedious). The slope of the line through these points is

$$s \overset{\text{def}}{=} \left[\frac{y_2 - y_1}{x_2 - x_1} \bmod p \right] ;$$

our assumption that $x_1 \neq x_2$ means that the inverse of $(x_2 - x_1)$ modulo p exists. The line passing through P_1 and P_2 has the equation

$$y = s \cdot (x - x_1) + y_1 \bmod p. \tag{9.3}$$

To find the third point of intersection of this line with E, substitute the above into the equation for E to obtain

$$\left(s \cdot (x - x_1) + y_1 \right)^2 = x^3 + Ax + B \bmod p.$$

The values of x that satisfy this equation are x_1, x_2, and

$$x_3 \overset{\text{def}}{=} [s^2 - x_1 - x_2 \bmod p].$$

The first two solutions correspond to the original points P_1 and P_2, while the third is the x-coordinate of the third point of intersection P_3. Plugging x_3 into Equation (9.3) we find that the y-coordinate corresponding to x_3 is $y_3 = [s \cdot (x_3 - x_1) + y_1 \bmod p]$. To obtain the desired answer $P_1 + P_2$, we flip the sign of the y-coordinate to obtain:

$$(x_1, y_1) + (x_2, y_2) = \left([s^2 - x_1 - x_2 \bmod p], \ [s \cdot (x_1 - x_3) - y_1 \bmod p] \right).$$

We summarize and extend this in the following proposition.

PROPOSITION 9.70 *Let $p \geq 5$ be prime and let E be the elliptic curve given by $y^2 = x^3 + Ax + B \bmod p$ where $4A^3 + 27B^2 \neq 0 \bmod p$. Let $P_1, P_2 \neq \mathcal{O}$ be points on E, with $P_1 = (x_1, y_1)$ and $P_2 = (x_2, y_2)$.*

1. *If $x_1 \neq x_2$, then $P_1 + P_2 = (x_3, y_3)$ with*

$$x_3 = [s^2 - x_1 - x_2 \bmod p] \quad and \quad y_3 = [s \cdot (x_1 - x_3) - y_1 \bmod p],$$

where $s = \left[\frac{y_2 - y_1}{x_2 - x_1} \bmod p \right]$.

2. *If $x_1 = x_2$ but $y_1 \neq y_2$ then $P_1 = -P_2$ and so $P_1 + P_2 = \mathcal{O}$.*

3. *If $P_1 = P_2$ and $y_1 \neq 0$ then $P_1 + P_2 = 2P_1 = (x_3, y_3)$ with*

$$x_3 = [s^2 - x_1 - x_2 \bmod p] \quad and \quad y_3 = [s \cdot (x_1 - x_3) - y_1 \bmod p],$$

where $s = \left[\frac{3x_1^2 + A}{2y_1} \bmod p \right]$.

4. *If $P_1 = P_2$ and $y_1 = 0$ then $P_1 + P_2 = 2P_1 = \mathcal{O}$.*

Somewhat amazingly, the set of points $E(\mathbb{Z}_p)$ under the addition rule defined above forms an abelian group, called the *elliptic-curve group of $E(\mathbb{Z}_p)$*. Commutativity follows from the way addition is defined, $\mathcal{O}$ acts as the identity, and we have already seen that each point in $E(\mathbb{Z}_p)$ has an inverse in $E(\mathbb{Z}_p)$. The difficult property to verify is associativity, which the disbelieving reader can check through tedious calculation. (A more illuminating proof that does not involve explicit calculation relies on algebraic geometry.)

Example 9.71
Consider the curve from Example 9.69. We show associativity for three specific points. Let $P_1 = (1, 0)$, $P_2 = P_3 = (4, 3)$. When computing $P_1 + P_2$ we get $s = [(3 - 0) \cdot (4 - 1)^{-1} \bmod 7] = 1$ and $[1^2 - 1 - 4 \bmod 7] = 3$. Thus,

$$Q \overset{\text{def}}{=} P_1 + P_2 = (3, \ [1 \cdot (1 - 3) - 0 \bmod 7]) = (3, 5);$$

note that this is indeed a point on $E(\mathbb{Z}_7)$. If we then compute $Q + P_3$ we get $s = [(3 - 5) \cdot (4 - 3)^{-1} \bmod 7] = 5$ and $[5^2 - 3 - 4 \bmod 7] = 4$. Thus,

$$(P_1 + P_2) + P_3 = Q + P_3 = (4, \ [5 \cdot (3 - 4) - 5 \bmod 7]) = (4, 4).$$

If we compute $P_2 + P_3 = 2P_2$ we obtain $s = [(3 \cdot 4^2 + 3) \cdot (2 \cdot 3)^{-1} \bmod 7] = 5$ and $[5^2 - 2 \cdot 4 \bmod 7] = 3$. Thus,

$$Q' \overset{\text{def}}{=} P_2 + P_3 = (3, \ [5 \cdot (4 - 3) - 3 \bmod 7]) = (3, 2).$$

If we then compute the value $P_1 + Q'$ we find $s = [2 \cdot (3-1)^{-1} \bmod 7] = 1$ and $[1^2 - 1 - 3 \bmod 7] = 4$. So

$$P_1 + (P_2 + P_3) = P_1 + Q' = (4, \ [1 \cdot (1-4) - 0 \bmod 7]) = (4, 4),$$

and $P_1 + (P_2 + P_3) = (P_1 + P_2) + P_3$. ◇

Recall that when a group is written additively, "exponentiation" corresponds to repeated addition. Thus, if we fix some point P in an elliptic-curve group, the discrete-logarithm problem becomes (informally) the problem of computing the integer x from xP, while the decisional Diffie–Hellman problem becomes (informally) the problem of distinguishing tuples of the form (aP, bP, abP) from those of the form (aP, bP, cP). These problems are believed to be hard in elliptic-curve groups (or subgroups thereof) of large prime order, subject to a few technical conditions we will mention below.

Montgomery representation. The Weierstrass representation is not the only way to define an elliptic curve, and other representations are often used for reasons of efficiency and/or implementation-level security (e.g., better resistance to side-channel attacks). The *Montgomery representation* involves equations of the form

$$By^2 = x^3 + Ax^2 + x \bmod p,$$

where $B \neq 0 \bmod p$ and $A \neq \pm 2 \bmod p$. Once again, given an equation E of the above form we let $E(\mathbb{Z}_p)$ denote the set of points (with coordinates in $\mathbb{Z}_p$) satisfying the equation plus the point at infinity $\mathcal{O}$; it is possible to define addition of these points in a way analogous to before. (Note that the addition law will *not* take the same form as in Proposition 9.70. Instead, addition is defined geometrically as before, and then the corresponding equations must be derived.) In contrast to the Weierstrass representation, not every curve can be expressed in Montgomery representation; in particular, the order of any elliptic-curve group written in Montgomery form is a multiple of 4.

(Twisted) Edwards representation. The *twisted Edwards representation* of an elliptic curve involves an equation E of the form

$$ax^2 + y^2 = 1 + dx^2 y^2 \bmod p,$$

with $a, d \neq 0 \bmod p$ and $a \neq d \bmod p$; the special case where $a = 1$ is called the *Edwards representation*. The twisted Edwards representation can express the same set of elliptic curves as the Montgomery representation.

$E(\mathbb{Z}_p)$ again denotes the elliptic-curve group containing the points satisfying equation E; interestingly, here there is no need for a "special" point at infinity since one can show that the point $(0, 1)$ on the curve is the identity. A nice feature of the twisted Edwards representation is that when a is a quadratic residue modulo p, but d is a quadratic non-residue, the addition law is simple: the sum of $P_1 = (x_1, y_1)$ and $P_2 = (x_2, y_2)$ is

$$(x_3, y_3) = \left(\frac{x_1 y_2 + x_2 y_1}{1 + d x_1 x_2 y_1 y_2}, \frac{y_1 y_2 - a x_1 x_2}{1 - d x_1 x_2 y_1 y_2} \right).$$

Here addition is computed using a *single* equation, rather than having to consider various subcases as in Proposition 9.70. This can greatly simplify the process of writing code for elliptic-curve operations.

Choosing an elliptic-curve group. For cryptographic purposes, we need an elliptic-curve group of large order. Thus, the first question we must address is: how large are elliptic-curve groups? Consider the Weierstrass representation. (Recall that any elliptic curve can be expressed in that way.) As noted in Example 9.69, the equation $y^2 = f(x) \bmod p$ has two solutions whenever $f(x)$ is a quadratic residue, and one solution when $f(x) = 0$. Since half the elements in $\mathbb{Z}_p^*$ are quadratic residues, we thus heuristically expect to find $2 \cdot (p-1)/2 + 1 + 1 = p + 1$ points (including the point at infinity) on the curve. The *Hasse bound* says that this heuristic estimate is accurate, in the sense that every elliptic-curve group has "almost" this many points.

THEOREM 9.72 (Hasse bound) *Let p be prime, and let E be an elliptic curve over $\mathbb{Z}_p$. Then $p + 1 - 2\sqrt{p} \leq |E(\mathbb{Z}_p)| \leq p + 1 + 2\sqrt{p}$.*

In other words, we have $|E(\mathbb{Z}_p)| = p + 1 - t$ for $|t| \leq 2\sqrt{p}$. The value $t = p + 1 - |E(\mathbb{Z}_p)|$ is called the *trace* of the elliptic curve E. Efficient algorithms for computing the order (or, equivalently, the trace) of a given elliptic-curve group $E(\mathbb{Z}_p)$ are known, but are beyond the scope of this book.

The Hasse bound implies that it is always easy to *find* a point on a given elliptic curve $y^2 = f(x) \bmod p$: simply choose uniform $x \in \mathbb{Z}_p$, check whether $f(x)$ is 0 or a quadratic residue, and—if so—let y be a square root of $f(x)$. (Algorithms for deciding quadratic residuosity and computing square roots modulo a prime are discussed in Chapter 15.) Since points on the elliptic curve are plentiful, we will not have to try very many values of x before finding a point.

For cryptographic purposes, we want to work in an elliptic-curve (sub)group of prime order. If $|E(\mathbb{Z}_p)|$ is prime, we can simply work in the group $E(\mathbb{Z}_p)$. Otherwise, if $|E(\mathbb{Z}_p)|$ has a large prime factor then we can work in an appropriate subgroup of $E(\mathbb{Z}_p)$. Concretely, say $|E(\mathbb{Z}_p)| = rq$ with q prime and $r < q$ (so, in particular, $\gcd(r, q) = 1$); r is called the *cofactor*. Then it is possible to show that

$$\mathbb{G} \stackrel{\text{def}}{=} \{ rP \mid P \in E(\mathbb{Z}_p) \} \subset E(\mathbb{Z}_p)$$

is a subgroup of $E(\mathbb{Z}_p)$ of order q. (Note the parallel with Theorem 9.66, although here the larger group $E(\mathbb{Z}_p)$ may not be cyclic.)

Finally, we also want an elliptic-curve group in which the discrete-logarithm problem is as hard as possible, namely, for which the best-known algorithm for computing discrete logarithms in that group is an exponential-time "generic" algorithm. (See Section 10.2 for further discussion.) Several classes of elliptic curves are cryptographically weak and should be avoided. These include curves over $\mathbb{Z}_p$ whose order is equal to p (as discrete logarithms can be computed in polynomial time in that case), as well as curves whose order divides $p^k - 1$ for "small" k (since in that case the discrete-logarithm problem in $E(\mathbb{Z}_p)$ can be reduced to a discrete-logarithm problem in the field $\mathbb{F}_{p^k}$, which can in turn be solved by non-generic algorithms in sub-exponential time).

In practice, standardized curves recommended by NIST or other international standards organizations are used (see below); generating a curve of one's own for cryptographic purposes is not recommended.

Practical Considerations

We conclude this section with a brief discussion of some efficiency optimizations when using elliptic curves, and other practical aspects.

Point compression. A useful observation is that the number of bits needed to represent a point on an elliptic curve can be reduced almost by half. We illustrate the idea for curves using the Weierstrass representation. For any $x \in \mathbb{Z}_p$ there are at most two points on the curve with x as their x-coordinate: namely, $(x, \pm y)$ for some y. (It is possible that $y = 0$ in which case these are the same point.) Thus, we can specify any point $P = (x, y)$ on the curve by its x-coordinate and a bit b that distinguishes between the (at most) two possibilities for the value of its y-coordinate. One convenient way to do this is to set $b = 0$ if y is even and $b = 1$ if y is odd. Given x and b we can recover P by computing the two square roots y_1, y_2 of the equation $y^2 = f(x) \bmod p$; since $y_1 = -y_2 \bmod p$ and p is odd, either $y_1 = y_2 = 0$ or exactly one of y_1, y_2 will be even and the other will be odd.

Projective coordinates. Representing elliptic-curve points as we have been doing until now—in which a point P on an elliptic curve is described by a pair of elements (x, y)—is called using *affine coordinates*. There are alternate ways to represent points using *projective coordinates* that can offer efficiency improvements. While these alternate representations can be motivated mathematically, we treat them simply as useful computational aids. We continue to assume the Weierstrass representation for the elliptic curve.

Points in projective coordinates are represented using *three* elements of $\mathbb{Z}_p$. Specifically, a point $P = (x, y) \neq \mathcal{O}$ in affine coordinates is represented using (standard) projective coordinates by any tuple $(X, Y, Z) \in \mathbb{Z}_p^3$ for which $X/Z = x \bmod p$ and $Y/Z = y \bmod p$. (An interesting feature of using projective coordinates is that each point now has multiple representations.) The point at infinity $\mathcal{O}$ is represented by any tuple $(0, Y, 0)$ with $Y \neq 0$, and these are the only points (X, Y, Z) with $Z = 0$. We can easily translate between coordinate systems: $(x, y) \neq \mathcal{O}$ in affine coordinates becomes $(x, y, 1)$ in projective coordinates, and (X, Y, Z) (with $Z \neq 0$) in projective coordinates is mapped to $([X/Z \bmod p], [Y/Z \bmod p])$ in affine coordinates.

The advantage of using projective coordinates is that we can add points without computing inverses modulo p. (Adding points in affine coordinates requires computing inverses; see Proposition 9.70. Although computing inverses modulo p can be done in polynomial time, it is more expensive than addition or multiplication modulo p.) This is done by exploiting the fact that points have multiple representations. To see this, let us work out the addition law for two points $P_1 = (X_1, Y_1, Z_1)$ and $P_2 = (X_2, Y_2, Z_2)$ with $P_1, P_2 \neq \mathcal{O}$ (so $Z_1, Z_2 \neq 0$) and $P_1 \neq \pm P_2$ (so $X_1/Z_1 \neq X_2/Z_2 \bmod p$). (If either P_1 or P_2 is equal to $\mathcal{O}$, addition is trivial. The case of $P_1 = \pm P_2$ can be handled as well, but we omit details here.) We can express P_1 and P_2 as $(X_1/Z_1, Y_1/Z_1)$ and $(X_2/Z_2, Y_2/Z_2)$, respectively, in affine coordinates, so (using Proposition 9.70)

$$P_3 \overset{\text{def}}{=} P_1 + P_2 = \left(s^2 - X_1/Z_1 - X_2/Z_2, \right.$$
$$\left. s \cdot (X_1/Z_1 - s^2 + X_1/Z_1 + X_2/Z_2) - Y_1/Z_1, \ 1 \right),$$

where

$$s = (Y_2/Z_2 - Y_1/Z_1)(X_2/Z_2 - X_1/Z_1)^{-1} = (Y_2 Z_1 - Y_1 Z_2)(X_2 Z_1 - X_1 Z_2)^{-1}$$

and all computations are done modulo p. Note we use projective coordinates (X_3, Y_3, Z_3) to represent P_3, setting $Z_3 = 1$ above. But using projective coordinates means we are not limited to $Z_3 = 1$. Multiplying each coordinate by $Z_1 Z_2 (X_2 Z_1 - X_1 Z_2)^3 \neq 0 \bmod p$, we find that P_3 can also be represented as

$$P_3 = \left(vw, \ u(v^2 X_1 Z_2 - w) - v^3 Y_1 Z_2, \ Z_1 Z_2 v^3 \right) \tag{9.4}$$

where

$$u = Y_2 Z_1 - Y_1 Z_2, \qquad v = X_2 Z_1 - X_1 Z_2,$$
$$w = u^2 Z_1 Z_2 - v^3 - 2v^2 X_1 Z_2. \tag{9.5}$$

The key observation is that the computations in Equations (9.4) and (9.5) can be carried out without having to perform any modular inversions.

Several other coordinate systems have also been developed, with the goal of minimizing the cost of elliptic-curve operations. Further details are beyond the scope of the book.

When points have multiple representations, some subtleties can arise. (Note that, until now, we have explicitly assumed that group elements have unique representations as bit-strings. That is no longer true when working in projective coordinates.) Specifically, a point expressed in projective coordinates may reveal information about how that point was computed, which may in turn leak some secret information. To address this, affine coordinates should be used for transmitting and storing points, with projective coordinates used only as an intermediate representation during the course of a computation.

Popular elliptic curves. As noted earlier, in practice people typically do not generate their own elliptic curves, but instead use standardized curves that have been carefully selected to ensure both good security and efficient implementation. Some popular choices include:

- The *P-256 curve* (also known as secp256r1) is an elliptic curve over $\mathbb{Z}_p$ for the 256-bit prime $p = 2^{256} - 2^{224} + 2^{192} + 2^{96} - 1$. The prime was chosen to have this form because it allows for efficient implementation of arithmetic modulo p. The curve has the equation $y^2 = x^3 - 3x + B \bmod p$ where B is a specified constant; $A = -3$ was chosen to enable optimization of elliptic-curve operations. This curve has prime order (so cannot be represented using Montgomery or twisted Edwards form) that, by the Hasse bound, is of the same magnitude as p.

  P-384 (secp384r1) and P-521 (secp521r1) are analogous curves defined modulo 384- and 521-bit primes, respectively.

- *Curve25519* is an elliptic curve that can be represented in Montgomery form; it can also be represented in twisted Edwards form, where it is known as *Ed25519*. This curve is defined over $\mathbb{Z}_p$ for the 255-bit prime $p = 2^{255} - 19$, where again the prime was chosen to have this form because it allows for efficient implementation of arithmetic modulo p. This elliptic-curve group does not have prime order, but cryptographic operations can be carried out in a subgroup of large prime order.

- The *secp256k1* curve is a prime-order curve defined over $\mathbb{Z}_p$ where $p = 2^{256} - 2^{32} - 2^9 - 2^8 - 2^7 - 2^6 - 2^4 - 1$. This is a *Koblitz curve* with equation $y^2 = x^3 + 7 \bmod p$; a Koblitz curve has certain algebraic properties that allow for efficient implementation. This curve is most notable for being used by Bitcoin.

9.4 *Cryptographic Applications

We have spent a fair bit of time discussing number theory and group theory, and introducing computational hardness assumptions that are widely believed to hold. Applications of these assumptions will occupy us for the rest of the book, but we provide some brief examples here.

9.4.1 One-Way Functions and Permutations

One-way functions are the minimal cryptographic primitive, and they are both necessary and sufficient for private-key encryption and message authentication codes. A more complete discussion of the role of one-way functions in cryptography appears in Chapter 8; here we only provide a definition of one-way functions and demonstrate that their existence follows from the number-theoretic hardness assumptions we have seen in this chapter.

Informally, a function f is *one-way* if it is easy to compute but hard to invert. The following experiment and definition, a restatement of Definition 8.1, formalizes this.

> **The inverting experiment** $\mathsf{Invert}_{\mathcal{A},f}(n)$:
>
> 1. *Choose uniform $x \in \{0,1\}^n$ and compute $y := f(x)$.*
> 2. *$\mathcal{A}$ is given 1^n and y as input, and outputs x'.*
> 3. *The output of the experiment is 1 if and only if $f(x') = y$.*

DEFINITION 9.73   *A function $f : \{0,1\}^* \to \{0,1\}^*$ is one-way if the following two conditions hold:*

1. **(Easy to compute:)** *There is a polynomial-time algorithm that on input x outputs $f(x)$.*

2. **(Hard to invert:)** *For all* PPT *algorithms $\mathcal{A}$ there is a negligible function* negl *such that* $\Pr[\mathsf{Invert}_{\mathcal{A},f}(n) = 1] \leq \mathsf{negl}(n)$.

We now show formally that the factoring assumption implies the existence of a one-way function. Let Gen be a polynomial-time algorithm that, on input 1^n, outputs (N, p, q) where $N = pq$ and p and q are n-bit primes except with probability negligible in n. (We use Gen rather than GenModulus here purely for notational convenience.) Since Gen runs in polynomial time, there is a polynomial upper bound on the number of random bits the algorithm uses. For simplicity, and in order to get the main ideas across, we assume Gen always uses at most n random bits on input 1^n. In Algorithm 9.74 we define a function f_{Gen} that uses its input as the random bits for running Gen. Thus, f_{Gen} is a *deterministic* function (as required).

If the factoring problem is hard relative to Gen then f_{Gen} is a one-way function. Certainly f_{Gen} is easy to compute. As for the hardness of inverting this function, note that the following distributions are identical:

1. The modulus N output by $f_{\mathsf{Gen}}(x)$, when $x \in \{0,1\}^n$ is chosen uniformly.

2. The modulus N output by (the randomized algorithm) $\mathsf{Gen}(1^n)$.

If moduli N generated according to the second distribution are hard to factor, then the same holds for moduli N generated according to the first distribution.

ALGORITHM 9.74
Algorithm computing f_{Gen}

Input: String x of length n
Output: Integer N

compute $(N, p, q) := Gen(1^n; x)$
// i.e., run $Gen(1^n)$ using x as the random tape
return N

Moreover, given any preimage x' of N with respect to f_{Gen} (i.e., an x' for which $f_{Gen}(x') = N$; note that we do not require $x' = x$), it is easy to recover a factor of N by running $Gen(1^n; x')$ to obtain (N, p, q) and outputting the factors p and q. Thus, finding a preimage of N with respect to f_{Gen} is as hard as factoring N. One can easily turn this into a formal proof.

One-Way Permutations

We can also use number-theoretic assumptions to construct a family of one-way *permutations*. We begin with a restatement of Definitions 8.2 and 8.3, specialized to the case of permutations:

DEFINITION 9.75 *A triple* $\Pi = (Gen, Samp, f)$ *of probabilistic polynomial-time algorithms is a* family of permutations *if the following hold:*

1. *The* parameter-generation algorithm Gen, *on input* 1^n, *outputs parameters* I *with* $|I| \geq n$. *Each value of* I *defines a set* $\mathcal{D}_I$ *that constitutes the domain and range of a permutation (i.e., bijection)* $f_I : \mathcal{D}_I \to \mathcal{D}_I$.

2. *The* sampling algorithm Samp, *on input* I, *outputs a uniformly distributed element of* $\mathcal{D}_I$.

3. *The deterministic* evaluation algorithm f, *on input* I *and* $x \in \mathcal{D}_I$, *outputs an element* $y \in \mathcal{D}_I$. *We write this as* $y := f_I(x)$.

Given a family of functions Π, consider the following experiment for any algorithm $\mathcal{A}$ and parameter n:

The inverting experiment $\mathsf{Invert}_{\mathcal{A},\Pi}(n)$:

1. $Gen(1^n)$ *is run to obtain* I, *and then* $Samp(I)$ *is run to choose a uniform* $x \in \mathcal{D}_I$. *Finally,* $y := f_I(x)$ *is computed.*

2. $\mathcal{A}$ *is given* I *and* y *as input, and outputs* x'.

3. *The output of the experiment is* 1 *if and only if* $f_I(x') = y$.

DEFINITION 9.76 *The family of permutations* $\Pi = (Gen, Samp, f)$ *is* one-way *if for all probabilistic polynomial-time algorithms* $\mathcal{A}$ *there exists a*

CONSTRUCTION 9.77

Let GenRSA be as before. Define a family of permutations as follows:

- **Gen:** on input 1^n, run GenRSA(1^n) to obtain (N, e, d) and output $I = \langle N, e \rangle$. Set $\mathcal{D}_I = \mathbb{Z}_N^*$.
- **Samp:** on input $I = \langle N, e \rangle$, choose a uniform element of $\mathbb{Z}_N^*$.
- f: on input $I = \langle N, e \rangle$ and $x \in \mathbb{Z}_N^*$, output $[x^e \bmod N]$.

A family of permutations based on the RSA problem.

negligible function negl *such that*

$$\Pr[\mathsf{Invert}_{A,\Pi}(n) = 1] \leq \mathsf{negl}(n).$$

Given GenRSA as in Section 9.2.4, Construction 9.77 defines a family of permutations. It is immediate that if the RSA problem is hard relative to GenRSA then this family is one-way. It can similarly be shown that hardness of the discrete-logarithm problem in $\mathbb{Z}_p^*$, with p prime, implies the existence of a one-way family of permutations; see Section 8.1.2.

9.4.2 Collision-Resistant Hash Functions

Collision-resistant hash functions were introduced in Section 6.1. Although we have discussed constructions of collision-resistant hash functions used in practice in Section 7.3, we have not yet seen constructions that can be rigorously based on simpler assumptions. We show here a construction based on the discrete-logarithm assumption in prime-order groups. (A construction based on the RSA problem is described in Exercise 9.27.) Although these constructions are less efficient than the hash functions used in practice, they are important since they illustrate the *feasibility* of achieving collision resistance based on standard and well-studied number-theoretic assumptions.

Let $\mathcal{G}$ be a polynomial-time algorithm that, on input 1^n, outputs a (description of a) cyclic group $\mathbb{G}$, its order q (with $\|q\| = n$), and a generator g. Here we also require that q is *prime* except possibly with negligible probability. We define a fixed-length hash function (Gen, H) by choosing a uniform $h \in \mathbb{G}$ as part of the key s, and defining $H^s(x_1, x_2) = g^{x_1} h^{x_2}$; see Construction 9.78.

Note that Gen and H can be computed in polynomial time. Before continuing with an analysis of the construction, we make some technical remarks:

- For a given $s = \langle \mathbb{G}, q, g, h \rangle$ with $n = \|q\|$, the function H^s is described as taking elements of $\mathbb{Z}_q \times \mathbb{Z}_q$ as input. However, H^s can be viewed as taking bit-strings of length $2 \cdot (n-1)$ as input if we parse an input $x \in \{0,1\}^{2(n-1)}$ as two strings x_1, x_2, each of length $n-1$, and then view x_1, x_2 as elements of $\mathbb{Z}_q$ in the natural way.

- The output of H^s is similarly specified as being an element of $\mathbb{G}$, but we can view this as a bit-string if we fix some representation of $\mathbb{G}$. To

---

CONSTRUCTION 9.78

Let $\mathcal{G}$ be as described in the text. Define a fixed-length hash function (Gen, H) as follows:

- **Gen:** on input 1^n, run $\mathcal{G}(1^n)$ to obtain $(\mathbb{G}, q, g)$ and then select a uniform $h \in \mathbb{G}$. Output $s := \langle \mathbb{G}, q, g, h \rangle$ as the key.
- **H:** given a key $s = \langle \mathbb{G}, q, g, h \rangle$ and input $(x_1, x_2) \in \mathbb{Z}_q \times \mathbb{Z}_q$, output $H^s(x_1, x_2) := g^{x_1} h^{x_2} \in \mathbb{G}$.

A fixed-length hash function.

satisfy the requirements of Definition 6.2 (which requires the output length to be fixed as a function of n) we can pad the output as needed.

- Given the above, the construction only compresses its input when elements of $\mathbb{G}$ can be represented using fewer than $2n - 2$ bits. A generalization of Construction 9.78 can be used to obtain compression from *any* $\mathcal{G}$ for which the discrete-logarithm problem is hard, regardless of the number of bits required to represent group elements; see Exercise 9.28.

THEOREM 9.79   *Say $\mathcal{G}$ outputs prime-order groups, and the discrete-logarithm problem is hard relative to $\mathcal{G}$. Then Construction 9.78 is a fixed-length collision-resistant hash function (subject to the discussion regarding compression, above).*

PROOF   Let $\Pi = (\mathsf{Gen}, H)$ be as in Construction 9.78, and let $\mathcal{A}$ be a probabilistic polynomial-time algorithm with

$$\varepsilon(n) \overset{\text{def}}{=} \Pr[\mathsf{Hash\text{-}coll}_{\mathcal{A},\Pi}(n) = 1]$$

(cf. Definition 6.2). We show how $\mathcal{A}$ can be used by an algorithm $\mathcal{A}'$ to solve the discrete-logarithm problem with success probability $\varepsilon(n)$:

Algorithm $\mathcal{A}'$:
The algorithm is given $\mathbb{G}, q, g, h$ as input.

1. Let $s := \langle \mathbb{G}, q, g, h \rangle$. Run $\mathcal{A}(s)$ and obtain output x and x'.
2. If $x \neq x'$ and $H^s(x) = H^s(x')$ then parse x as (x_1, x_2) and parse x' as (x_1', x_2'), where $x_1, x_2, x_1', x_2' \in \mathbb{Z}_q$. Use Lemma 9.65 to compute $\log_g h$.

Clearly, $\mathcal{A}'$ runs in polynomial time. Furthermore, the input s given to $\mathcal{A}$ when run as a subroutine by $\mathcal{A}'$ is distributed exactly as in experiment $\mathsf{Hash\text{-}coll}_{\mathcal{A},\Pi}$ for the same value of the security parameter n. (The input to $\mathcal{A}'$ is generated by running $\mathcal{G}(1^n)$ to obtain $\mathbb{G}, q, g$ and then choosing uniform $h \in \mathbb{G}$. This is exactly how s is generated by $\mathsf{Gen}(1^n)$.) So, with probability exactly $\varepsilon(n)$

there is a *collision*; i.e., $x \neq x'$ and $H^s(x) = H^s(x')$. Lemma 9.65 implies that whenever there is a collision, $\mathcal{A}'$ returns the correct answer $\log_g h$.

In summary, $\mathcal{A}'$ correctly solves the discrete-logarithm problem with probability exactly $\varepsilon(n)$. Since, by assumption, the discrete-logarithm problem is hard relative to $\mathcal{G}$, we conclude that $\varepsilon(n)$ is negligible. ■

References and Additional Reading

The book by Childs [51] has excellent coverage of the group theory discussed in this chapter (and more), in greater depth but at a similar level of exposition. Shoup [183] gives a more advanced, yet still accessible, treatment of much of this material also, with special focus on algorithmic aspects. Relatively gentle introductions to abstract algebra and group theory that go well beyond what we have space for here are available in the books by Fraleigh [74] and Herstein [97]; the interested reader will have no trouble finding more-advanced algebra texts if they are so inclined.

The first efficient primality test was by Solovay and Strassen [190]. The Miller–Rabin test is due to Miller [143] and Rabin [167]. A deterministic primality test was discovered by Agrawal et al. [5]. See Dietzfelbinger [64] for a comprehensive survey of this area.

The RSA problem was publicly introduced by Rivest, Shamir, and Adleman [171], although it was revealed in 1997 that Ellis, Cocks, and Williamson, three members of the British intelligence agency GCHQ, had explored similar ideas—without fully recognizing their importance— several years earlier, in a classified setting.

The discrete-logarithm and Diffie–Hellman problems were first considered, at least implicitly, by Diffie and Hellman [65] in the group $\mathbb{Z}_p^*$. Current practical guidance for that setting can be found in various standards [15, 150, 151]. Most treatments of elliptic curves require advanced mathematical background; the book by Silverman and Tate [185] is perhaps an exception. As with many books on the subject written for mathematicians, however, that book has little coverage of elliptic curves over *finite* fields, which is the case most relevant to cryptography. The text by Washington [202], although a bit more advanced, deals heavily (but not exclusively) with the finite-field case. Implementation issues related to elliptic-curve cryptography are covered by Hankerson et al. [91]. Recommended elliptic curves are given by NIST [50].

The collision-resistant hash function based on the discrete-logarithm problem is due to Chaum et al. [49], and an earlier construction based on the hardness of factoring is given by Goldwasser et al. [88] (see also Exercise 9.27).

Exercises

9.1 Let $\mathbb{G}$ be an abelian group. Prove that there is a *unique* identity in $\mathbb{G}$, and that every element $g \in \mathbb{G}$ has a *unique* inverse.

9.2 Show that Proposition 9.36 does not necessarily hold when $\mathbb{G}$ is infinite.

> **Hint:** Consider the set $\{1\} \cup \{2, 4, 6, 8, \ldots\} \subset \mathbb{R}$ under multiplication.

9.3 Let $\mathbb{G}$ be a finite group, and $g \in \mathbb{G}$. Show that $\langle g \rangle$ is a subgroup of $\mathbb{G}$. Is the set $\{g^0, g^1, \ldots\}$ necessarily a subgroup of $\mathbb{G}$ when $\mathbb{G}$ is infinite?

9.4 This question concerns the Euler phi function.

(a) Let p be prime and $e \geq 1$ an integer. Show that $\phi(p^e) = p^{e-1}(p-1)$.

(b) Let p, q be relatively prime. Show that $\phi(pq) = \phi(p) \cdot \phi(q)$. (You may use the Chinese remainder theorem.)

(c) Prove Theorem 9.19.

9.5 Compute the final two (decimal) digits of 3^{1000} (by hand).

> **Hint:** The answer is $[3^{1000} \bmod 100]$.

9.6 Compute $[101^{4,800,000,002} \bmod 35]$ (by hand).

9.7 Compute $[46^{51} \bmod 55]$ (by hand) using the Chinese remainder theorem.

9.8 Prove that if $\mathbb{G}, \mathbb{H}$ are groups, then $\mathbb{G} \times \mathbb{H}$ is a group.

9.9 Let p, N be integers with $p \mid N$. Prove that for any integer X,

$$[[X \bmod N] \bmod p] = [X \bmod p].$$

Show that, in contrast, $[[X \bmod p] \bmod N]$ need not equal $[X \bmod N]$.

9.10 This question concerns the group $\mathbb{Z}_{24}$.

(a) List the elements of this group.

(b) Is this group cyclic?

(c) Is 18 a generator of this group? What about 5?

9.11 This question concerns the group $\mathbb{Z}_{21}^*$.

(a) How many elements are in this group? List the elements.

(b) What is $\phi(21)$?

(c) Compute $[11^{-1} \bmod 21]$.

(d) Compute $[2^{2403} \bmod 21]$ (by hand).

9.12 This question concerns the group $\mathbb{Z}_{23}^*$.

    (a) What is the order of this group?

    (b) Compute $\left[3^{46} \bmod 23\right]$ (by hand).

    (c) Is this group cyclic? Is 2 a generator? What about 5?

9.13 This question concerns the group $\mathbb{Z}_{55}^*$.

    (a) Compute $\phi(55)$.

    (b) Is exponentiating to the 3rd power a permutation of $\mathbb{Z}_{55}^*$?

    (c) Compute $\left[2^{1/3} \bmod 55\right]$ (i.e., the 3rd root of 2 modulo 55).

    (d) Is exponentiating to the 5th power a permutation of $\mathbb{Z}_{55}^*$?

9.14 Corollary 9.21 shows that if $N = pq$ for distinct primes p and q, and $ed = 1 \bmod \phi(N)$, then for all $x \in \mathbb{Z}_N^*$ we have $(x^e)^d = x \bmod N$. Show that this holds for all $x \in \{0, \ldots, N-1\}$.

    Hint: Use the Chinese remainder theorem.

9.15 Complete the details of the proof of the Chinese remainder theorem, showing that $\mathbb{Z}_N^*$ is isomorphic to $\mathbb{Z}_p^* \times \mathbb{Z}_q^*$.

9.16 This exercise develops an efficient algorithm for testing whether an integer is a perfect power.

    (a) Show that if $N = \hat{N}^e$ for some integers $\hat{N}, e > 1$ then $e \leq \|N\|$.

    (b) Given N and e with $2 \leq e \leq \|N\| + 1$, show how to determine in $\mathsf{poly}(\|N\|)$ time whether there exists an integer $\hat{N}$ with $\hat{N}^e = N$.

        Hint: Use binary search.

    (c) Given N, show how to test in $\mathsf{poly}(\|N\|)$ time whether N is a perfect power.

9.17 Given N and $a \in \mathbb{Z}_N^*$, show how to test in polynomial time whether a is a strong witness that N is composite.

9.18 Fix N, e with $\gcd(e, \phi(N)) = 1$, and assume there is an adversary $\mathcal{A}$ running in time t for which

$$\Pr\left[\mathcal{A}\left([x^e \bmod N]\right) = x\right] = 0.01,$$

where the probability is taken over uniform choice of $x \in \mathbb{Z}_N^*$. Show that it is possible to construct an adversary $\mathcal{A}'$ for which

$$\Pr\left[\mathcal{A}'\left([x^e \bmod N]\right) = x\right] = 0.99$$

for *all* x. The running time t' of $\mathcal{A}'$ should be polynomial in t and $\|N\|$.

    Hint: Use the fact that $y^{1/e} \cdot r = (y \cdot r^e)^{1/e} \bmod N$.

9.19 Formally define the CDH assumption. Prove that hardness of the CDH problem relative to $\mathcal{G}$ implies hardness of the discrete-logarithm problem relative to $\mathcal{G}$, and that hardness of the DDH problem relative to $\mathcal{G}$ implies hardness of the CDH problem relative to $\mathcal{G}$.

9.20 This question concerns the cyclic group $\mathbb{Z}_{47}^*$, in which $g = 5$ is a generator. You may use a calculator.

   (a) Let $h_1 = g^4$. What is the value of h_1?
   (b) Let $h_2 = g^{32}$. What is the value of h_2?
   (c) What is the value of $\mathsf{DH}_g(h_1, h_2)$?

9.21 Can the following problem be solved in polynomial time? Given a prime p, an integer $e \in \mathbb{Z}_{p-1}^*$, and $y := [g^e \bmod p]$ (where g is a uniform value in $\mathbb{Z}_p^*$), find g, i.e., compute $y^{1/e} \bmod p$. If your answer is "yes," give a polynomial-time algorithm. If your answer is "no," show a reduction to one of the assumptions introduced in this chapter.

9.22 Determine the points on the elliptic curve $E : y^2 = x^3 + 2x + 1$ over $\mathbb{Z}_{11}$. How many points are on this curve?

9.23 Prove the third statement in Proposition 9.70.

9.24 When using the twisted Edwards representation, show that the inverse of a point (x, y) is the point $(-x, y)$.

9.25 Consider the elliptic-curve group from Example 9.69. (See also Example 9.71.) Compute $(1, 0) + (4, 3) + (4, 3)$ in this group by first converting to projective coordinates and then using Equations (9.4) and (9.5).

9.26 Fix N, an element $y \in \mathbb{Z}_N^*$, and e with $\gcd(e, \phi(N)) = 1$. Show that given $w \in \mathbb{Z}_N^*$ and an integer k with $\gcd(k, e) = 1$ and $w^e = y^k \bmod N$, it is possible to efficiently compute x such that $x^e = y \bmod N$.

   Hint: Apply Proposition 9.2 to k, e, and express y^1 as a power of e.

9.27 Let GenRSA be as in Section 9.2.4. Prove that if the RSA problem is hard relative to GenRSA then Construction 9.80 is a fixed-length collision-resistant hash function.

CONSTRUCTION 9.80

Define (Gen, H) as follows:

- **Gen:** on input 1^n, run $\mathsf{GenRSA}(1^n)$ to obtain N, e, d, and select $y \leftarrow \mathbb{Z}_N^*$. The key is $s := \langle N, e, y \rangle$.

- H: if $s = \langle N, e, y \rangle$, then H^s maps inputs in $\{0, 1\}^{3n}$ to outputs in $\mathbb{Z}_N^*$. Let $f_0^s(x) \overset{\text{def}}{=} [x^e \bmod N]$ and $f_1^s(x) \overset{\text{def}}{=} [y \cdot x^e \bmod N]$. For a 3n-bit long string $x = x_1 \cdots x_{3n}$, define

$$H^s(x) \overset{\text{def}}{=} f_{x_1}^s\left(f_{x_2}^s\left(\cdots\left(1\right)\cdots\right)\right).$$

9.28 Consider the following generalization of Construction 9.78:

CONSTRUCTION 9.81

Define a fixed-length hash function (Gen, H) as follows:

(a) Gen: on input 1^n, run $\mathcal{G}(1^n)$ to obtain $(\mathbb{G}, q, h_1)$ and then select $h_2, \ldots, h_t \leftarrow \mathbb{G}$. Output $s := \langle \mathbb{G}, q, (h_1, \ldots, h_t) \rangle$ as the key.

(b) H: given a key $s = \langle \mathbb{G}, q, (h_1, \ldots, h_t) \rangle$ and input $(x_1, \ldots, x_t)$ with $x_i \in \mathbb{Z}_q$, output $H^s(x_1, \ldots, x_t) := \prod_i h_i^{x_i}$.

(a) Prove that if the discrete-logarithm problem is hard relative to $\mathcal{G}$ and q is prime, then for any $t = \mathsf{poly}(n)$ this construction is a fixed-length collision-resistant hash function.

(b) Discuss how this construction can be used to obtain *compression* regardless of the number of bits needed to represent elements of $\mathbb{G}$ (as long as it is polynomial in n).

Chapter 10

*Algorithms for Factoring and Computing Discrete Logarithms

In the last chapter, we introduced several number-theoretic problems—most prominently, *factoring* the product of two large primes and *computing discrete logarithms* in certain groups—that are widely believed to be hard. As defined there, this means there are presumed to be no polynomial-time algorithms for these problems. This *asymptotic* notion of hardness, however, tells us little about how to set the security parameter—sometimes called the *key length*, although the terms are not interchangeable—to achieve some desired, *concrete* level of security in practice. A proper understanding of this issue is extremely important for the real-world deployment of cryptosystems based on these problems. Setting the security parameter too low means a cryptosystem may be vulnerable to attacks more efficient than anticipated; being overly conservative and setting the security parameter too high will give good security, but at the expense of efficiency for the honest users. The relative difficulty of different number-theoretic problems can also play a role in determining which problems to use as the basis for building cryptosystems in the first place.

The fundamental issue, of course, is that a brute-force search may not be the best algorithm for solving a given problem; thus, using key length n does not, in general, give security against attackers running for 2^n time. This is in contrast to the private-key setting where the best attacks on existing block ciphers have roughly the complexity of brute-force search. As a consequence, the key lengths used in the public-key setting tend to be significantly larger than those used in the private-key setting.

To gain a better appreciation of this point, we explore in this chapter several algorithms for factoring and computing discrete logarithms that do not run in polynomial time, but nevertheless perform far better than brute-force search. The goal is merely to give a taste of existing algorithms for these problems, as well as to provide some basic guidance for setting parameters in practice. Our focus is on the high-level ideas, and we consciously do not address many important implementation-level details that would be critical to deal with if these algorithms were to be used in practice. We also concentrate exclusively on *classical* algorithms here, deferring a discussion about the effect of *quantum* algorithms to Chapter 14.

The reader may also notice that we only describe algorithms for factoring and computing discrete logarithms, and not algorithms for, say, solving the

RSA or decisional Diffie–Hellman problems. Our choice is justified by the facts that the best known algorithms for solving RSA require factoring the modulus, and (in the groups discussed in Sections 9.3.3 and 9.3.4) the best known approaches for solving the decisional Diffie–Hellman problem require computing discrete logarithms.

10.1 Algorithms for Factoring

Throughout this chapter, we assume that $N = pq$ is a product of two distinct primes with $p < q$. We will be most interested in the case when p and q each has the same (known) length n, and so $n = \Theta(\log N)$.

We will frequently use the Chinese remainder theorem along with the notation developed in Section 9.1.5. The Chinese remainder theorem states that

$$\mathbb{Z}_N \simeq \mathbb{Z}_p \times \mathbb{Z}_q \quad \text{and} \quad \mathbb{Z}_N^* \simeq \mathbb{Z}_p^* \times \mathbb{Z}_q^*,$$

with isomorphism given by $f(x) \stackrel{\text{def}}{=} ([x \bmod p], [x \bmod q])$. The fact that f is an isomorphism means, in particular, that it gives a bijection between elements $x \in \mathbb{Z}_N$ and pairs $(x_p, x_q) \in \mathbb{Z}_p \times \mathbb{Z}_q$. We write $x \leftrightarrow (x_p, x_q)$ to denote this bijection, with $x_p = [x \bmod p]$ and $x_q = [x \bmod q]$.

Recall from Section 9.2 that *trial division*—a trivial, brute-force factoring method—finds a factor of a given number N in time $\mathcal{O}(N^{1/2} \cdot \text{polylog}(N))$. (This is an exponential-time algorithm, since the size of the input is $\|N\|$, the length of the binary representation of N, and $\|N\| = \mathcal{O}(\log N)$.[1]) We show here three factoring algorithms with better performance:

- *Pollard's $p-1$ method* is effective if $p-1$ has only "small" prime factors.

- *Pollard's rho method* applies to arbitrary N. (As such, it is called a *general-purpose* factoring algorithm.) Its running time for N of the form discussed at the beginning of this section is $\mathcal{O}(N^{1/4} \cdot \text{polylog}(N))$. Note this is still *exponential* in n, the length of N.

- The *quadratic sieve algorithm* is a general-purpose factoring algorithm that runs in time *sub-exponential* in the length of N. We give a high-level overview of how this algorithm works, but the details are somewhat complex and beyond the scope of this book.

The fastest known general-purpose factoring algorithm is the *general number field sieve*. Heuristically, this algorithm factors its input N in expected time $2^{\mathcal{O}((\log N)^{1/3} \cdot (\log \log N)^{2/3})}$, which is sub-exponential in the length of N.

---

[1] Thus, a running time of $N^{\mathcal{O}(1)} = 2^{\mathcal{O}(\|N\|)}$ is exponential, a running time of $2^{o(\log N)} = 2^{o(\|N\|)}$ is sub-exponential, and a running time of $(\log N)^{\mathcal{O}(1)} = \|N\|^{\mathcal{O}(1)}$ is polynomial.

10.1.1 Pollard's $p-1$ Algorithm

If $N = pq$ and $p-1$ has *only* "small" prime factors, Pollard's $p-1$ algorithm can be used to efficiently factor N. The basic idea is simple. Let B be an integer for which $(p-1) \mid B$ and $(q-1) \nmid B$; we defer to below the details of how such a B is computed. Say $B = \gamma \cdot (p-1)$ for some integer γ. Choose a uniform $x \in \mathbb{Z}_N^*$ and compute $y := [x^B - 1 \bmod N]$. (Note that y can be computed using the efficient exponentiation algorithm from Appendix B.2.3.) Since $1 \leftrightarrow (1,1)$, we have

$$
\begin{aligned}
y = [x^B - 1 \bmod N] &\leftrightarrow (x_p, x_q)^B - (1,1) \\
&= (x_p^B - 1 \bmod p, \ x_q^B - 1 \bmod q) \\
&= ((x_p^{p-1})^\gamma - 1 \bmod p, \ x_q^B - 1 \bmod q) \\
&= (0, [x_q^B - 1 \bmod q]),
\end{aligned}
$$

using Theorem 9.14 and the fact that the order of $\mathbb{Z}_p^*$ is $p-1$. We show below that, with high probability, $x_q^B \neq 1 \bmod q$. Assuming this is the case, we have obtained an integer $y \in \mathbb{Z}_N^*$ for which

$$
y = 0 \bmod p \quad \text{but} \quad y \neq 0 \bmod q;
$$

that is, $p \mid y$ but $q \nmid y$. This, in turn, implies that $\gcd(y, N) = p$. Thus, a simple gcd computation (which can be done efficiently as described in Appendix B.1.2) yields a prime factor of N.

ALGORITHM 10.1
Pollard's $p-1$ algorithm for factoring

Input: Integer N
Output: A nontrivial factor of N

$x \leftarrow \mathbb{Z}_N^*$
$y := [x^B - 1 \bmod N]$
$\quad // \ B$ is as in the text
$p := \gcd(y, N)$
if $p \notin \{1, N\}$ **return** p

We now argue that the algorithm works with high probability. Because $(q-1) \nmid B$, as long as $x_q \stackrel{\text{def}}{=} [x \bmod q]$ is a generator of $\mathbb{Z}_q^*$ we must have $x_q^B \neq 1 \bmod q$. (This follows from Proposition 9.53.) It remains to analyze the probability that x_q is a generator. Here we rely on some results proved in Appendix B.3.1. Since q is prime, $\mathbb{Z}_q^*$ is a cyclic group of order $q-1$ that has exactly $\phi(q-1)$ generators (cf. Theorem B.16). If x is chosen uniformly from $\mathbb{Z}_N^*$, then x_q is uniformly distributed in $\mathbb{Z}_q^*$. (This is a consequence of the fact that the Chinese remainder theorem gives a bijection between $\mathbb{Z}_N^*$ and $\mathbb{Z}_p^* \times \mathbb{Z}_q^*$.) Thus, the probability that x_q is a generator is $\frac{\phi(q-1)}{q-1} =$

$\Omega(1/\log q) = \Omega(1/n)$ (cf. Theorem B.15). Multiple values of x can be chosen to boost the probability of success.

We are left with the problem of finding B such that $(p-1) \mid B$ but $(q-1) \nmid B$. One possibility is to choose $B = \prod_{i=1}^{k} p_i^{\lfloor n/\log p_i \rfloor}$ for some k, where p_i denotes the ith prime (i.e., $p_1 = 2, p_2 = 3, p_3 = 5, \ldots$) and n is the length of p. (Note that $p_i^{\lfloor n/\log p_i \rfloor}$ is the largest power of p_i that can possibly divide $p-1$.) If $p-1$ can be written as $\prod_{i=1}^{k} p_i^{e_i}$ with $e_i \geq 0$ (that is, if the largest prime factor of $p-1$ is less than p_k), then it will hold that $(p-1) \mid B$. In contrast, if $q-1$ has *any* prime factor larger than p_k, then $(q-1) \nmid B$.

Choosing a larger value for k increases B and so increases the running time of the algorithm (which performs a modular exponentiation to the power B). A larger value of k also makes it more likely that $(p-1) \mid B$, but at the same time makes it less likely that $(q-1) \nmid B$. It is, of course, possible to run the algorithm repeatedly using multiple choices for k.

Pollard's $p-1$ algorithm is thwarted if both $p-1$ and $q-1$ have any large prime factors. (More precisely, the algorithm still works but only for B so large that the algorithm becomes impractical.) For this reason, when generating a modulus $N = pq$ for cryptographic applications, p and q are sometimes chosen to be *strong* primes, namely, with $(p-1)/2$ and $(q-1)/2$ themselves prime. This ensures that both $p-1$ and $q-1$ have a large prime factor, and so the resulting modulus will not be vulnerable to Algorithm 10.1. Selecting p and q in this way is markedly less efficient than choosing p and q as *arbitrary* primes. Moreover, if p and q are uniform n-bit primes, it is unlikely that either $p-1$ or $q-1$ will have only small prime factors and so unlikely that Algorithm 10.1 will apply. Finally, better factoring algorithms are available anyway (as we will see below). For these reasons, the current consensus is that the added computational cost of generating p and q as strong primes does not yield any appreciable security gains.

10.1.2   Pollard's Rho Algorithm

In contrast to Algorithm 10.1, which is only effective for certain moduli, Pollard's *rho algorithm* can be used to factor an arbitrary integer $N = pq$; in that sense, it is a *general-purpose* factoring algorithm. Heuristically, the algorithm factors N with constant probability in $\mathcal{O}\left(N^{1/4} \cdot \text{polylog}(N)\right)$ time; this is still exponential, but a vast improvement over trial division.

The core idea of the approach is to find distinct values $x, x' \in \mathbb{Z}_N^*$ that are equivalent modulo p (i.e., for which $x = x' \bmod p$); call such a pair *good*. Note that for a good pair x, x' it holds that $\gcd(x - x', N) = p$ (since $x \neq x' \bmod N$), so computing the gcd gives a nontrivial factor of N.

How can we find a good pair? Say we choose values $x^{(1)}, \ldots, x^{(k)}$ uniformly from $\mathbb{Z}_N^*$, where $k = 2^{n/2} = \mathcal{O}(\sqrt{p})$. Viewing these in their Chinese-remaindering representation as $(x_p^{(1)}, x_q^{(1)}), \ldots, (x_p^{(k)}, x_q^{(k)})$, we have that each $x_p^{(i)} \stackrel{\text{def}}{=} [x^{(i)} \bmod p]$ is uniform in $\mathbb{Z}_p^*$. (This follows from bijection between

$\mathbb{Z}_N^*$ and $\mathbb{Z}_p^* \times \mathbb{Z}_q^*$.) Thus, using the birthday bound of Lemma A.15, we see that with high probability there exist distinct i, j with $x_p^{(i)} = x_p^{(j)}$ or, equivalently, $x^{(i)} = x^{(j)} \bmod p$. Moreover, Lemma A.15 shows that $x^{(i)} \neq x^{(j)}$ except with negligible probability. Thus, with high probability we obtain a good pair $x^{(i)}, x^{(j)}$ that can be used to find a nontrivial factor of N, as discussed earlier.

ALGORITHM 10.2
Pollard's rho algorithm for factoring

Input: Integer N, a product of two n-bit primes
Output: A nontrivial factor of N

$x \leftarrow \mathbb{Z}_N^*, \quad x' := x$
for $i = 1$ to $2^{n/2}$:
$\quad x := F(x)$
$\quad x' := F(F(x'))$
$\quad p := \gcd(x - x', N)$
$\quad$ if $p \notin \{1, N\}$ **return** p and **stop**

We can generate $k = \mathcal{O}(\sqrt{p})$ uniform elements of $\mathbb{Z}_N^*$ in $\mathcal{O}(\sqrt{p}) = \mathcal{O}(N^{1/4})$ time. *Testing* all pairs of elements in order to identify a good pair, however, would require $\binom{k}{2} = \mathcal{O}(k^2) = \mathcal{O}(p) = \mathcal{O}\left(N^{1/2}\right)$ time! (Note that since p is unknown we cannot simply compute $x_p^{(1)}, \ldots, x_p^{(k)}$ explicitly and then sort the $x_p^{(i)}$ to find a good pair. Instead, for all distinct pairs i, j we must compute $\gcd(x^{(i)} - x^{(j)}, N)$ to see whether this gives a nontrivial factor of N.) Without further optimizations, this will be no better than trial division.

Pollard's idea was to use a technique we have seen in Section 6.4.2 in the context of small-space birthday attacks. Specifically, we compute the sequence $x^{(1)}, x^{(2)}, \ldots$ by letting each value be a function of the one before it, i.e., we fix some function $F : \mathbb{Z}_N^* \to \mathbb{Z}_N^*$, choose a uniform $x^{(0)} = x \in \mathbb{Z}_N^*$, and then set $x^{(i)} := F(x^{(i-1)})$ for $i = 1, \ldots, k$. We require F to have the property that if $x = x' \bmod p$, then $F(x) = F(x') \bmod p$; this ensures that once equivalence modulo p occurs, it persists. (A standard choice is $F(x) = [x^2 + 1 \bmod N]$, but any polynomial modulo N will have this property.) If we heuristically model F as a random function, then with high probability there is a good pair in the first k elements of this sequence. Proceeding roughly as in Algorithm 6.9 from Section 6.4.2, we can detect a good pair (if there is one) using only $\mathcal{O}(k)$ gcd computations; see Algorithm 10.2.

10.1.3 The Quadratic Sieve Algorithm

Pollard's rho algorithm is better than trial division, but still runs in exponential time. The *quadratic sieve* algorithm runs in sub-exponential time. It was the fastest known factoring algorithm until the early 1990s and remains the factoring algorithm of choice for numbers up to about 300 bits long. We describe the general principles of the algorithm but caution the reader that several important details are omitted.

An element $z \in \mathbb{Z}_N^*$ is a *quadratic residue modulo N* if there is an $x \in \mathbb{Z}_N^*$ such that $x^2 = z \bmod N$; in this case, we say that x is a *square root of z*. The following observations serve as our starting point:

- If N is a product of two distinct, odd primes, then every quadratic residue modulo N has exactly four square roots. (See Section 15.4.2.)

- Given x, y with $x^2 = y^2 \bmod N$ and $x \neq \pm y \bmod N$, it is possible to compute a nontrivial factor of N in polynomial time. This is by virtue of the fact that $x^2 = y^2 \bmod N$ implies

$$0 = x^2 - y^2 = (x - y)(x + y) \bmod N,$$

and so $N \mid (x - y)(x + y)$. However, $N \nmid (x - y)$ and $N \nmid (x + y)$ because $x \neq \pm y \bmod N$. So it must be the case that $\gcd(x - y, N)$ is equal to one of the prime factors of N. (See also Lemma 15.35.)

The quadratic sieve algorithm tries to generate x, y with $x^2 = y^2 \bmod N$ and $x \neq \pm y \bmod N$. A naive way of doing this—which forms the basis of an older factoring algorithm due to Fermat—is to choose an $x \in \mathbb{Z}_N^*$, compute $q := [x^2 \bmod N]$, and then check whether q is a square *over the integers* (i.e., without reduction modulo N). If so, then $q = y^2$ for some integer y and so $x^2 = y^2 \bmod N$. Unfortunately, the probability that $[x^2 \bmod N]$ is a square is so low that this process must be repeated exponentially many times.

A significant improvement is obtained by generating a sequence of values $q_1 := [x_1^2 \bmod N], \ldots$ and identifying a subset of those values whose *product* is a square over the integers. In the quadratic sieve algorithm this is accomplished using the following two steps:

Step 1. Fix some bound B. Say an integer is *B-smooth* if all its prime factors are less than or equal to B. In the first phase of the algorithm, we search for integers of the form $q_i = [x_i^2 \bmod N]$ that are B-smooth and factor them. (Although factoring is hard, finding and factoring B-smooth numbers is feasible when B is small enough.) These $\{x_i\}$ are chosen by successively trying $x = \sqrt{N}+1, \sqrt{N}+2, \ldots$; this ensures a nontrivial reduction modulo N (since $x > \sqrt{N}$) and has the advantage that $q \stackrel{\text{def}}{=} [x^2 \bmod N] = x^2 - N$ is "small" so that q is more likely to be B-smooth.

Let $\{p_1, \ldots, p_k\}$ be the set of prime numbers less than or equal to B. Once we have found and factored the B-smooth $\{q_i\}$ as described above, we have a set of equations of the form:

$$q_1 = [x_1^2 \bmod N] = \prod_{i=1}^{k} p_i^{e_{1,i}}$$

$$\vdots \qquad (10.1)$$

$$q_\ell = [x_\ell^2 \bmod N] = \prod_{i=1}^{k} p_i^{e_{\ell,i}}.$$

(Note that the above equations are over the integers.)

Step 2. We next want to find some subset of the $\{q_i\}$ whose product is a square. If we multiply some subset S of the $\{q_i\}$, we see that the result

$$z = \prod_{j \in S} q_j = \prod_{i=1}^{k} p_i^{\sum_{j \in S} e_{j,i}}$$

is a square if and only if the exponent of each prime p_i is even. This suggests that we care about the exponents $\{e_{j,i}\}$ in Equation (10.1) only modulo 2; moreover, we can use linear algebra to find a subset of the $\{q_i\}$ whose "exponent vectors" sum to the 0-vector modulo 2.

In more detail: if we reduce the exponents in Equation (10.1) modulo 2, we obtain the 0/1-matrix Γ given by

$$\begin{pmatrix} \gamma_{1,1} & \gamma_{1,2} & \cdots & \gamma_{1,k} \\ \vdots & \vdots & \ddots & \vdots \\ \gamma_{\ell,1} & \gamma_{\ell,2} & \cdots & \gamma_{\ell,k} \end{pmatrix} \stackrel{\text{def}}{=} \begin{pmatrix} [e_{1,1} \bmod 2] & [e_{1,2} \bmod 2] & \cdots & [e_{1,k} \bmod 2] \\ \vdots & \vdots & \ddots & \vdots \\ [e_{\ell,1} \bmod 2] & [e_{\ell,2} \bmod 2] & \cdots & [e_{\ell,k} \bmod 2] \end{pmatrix}.$$

If $\ell = k + 1$, then Γ has more rows than columns and there must be some nonempty subset S of the rows that sum to the 0-vector modulo 2. Such a subset can be found efficiently using linear algebra. Then:

$$z \stackrel{\text{def}}{=} \prod_{j \in S} q_j = \prod_{i=1}^{k} p_i^{\sum_{j \in S} e_{j,i}} = \left(\prod_{i=1}^{k} p_i^{\left(\sum_{j \in S} e_{j,i}\right)/2} \right)^2,$$

using the fact that all the $\left\{ \sum_{j \in S} e_{j,i} \right\}$ are even. Since

$$z = \prod_{j \in S} q_j = \prod_{j \in S} x_j^2 = \left(\prod_{j \in S} x_j \right)^2 \bmod N,$$

we have obtained two square roots (modulo N) of z. Although there is no guarantee that these square roots will enable factorization of N (for reasons discussed at the beginning of this section), heuristically they do with constant probability. By taking $\ell > k + 1$ we can obtain multiple subsets S with the desired property and try to factor N using each possibility.

Example 10.3
Take $N = 377753$. We have $6647 = [620^2 \bmod N]$, and we can factor 6647 (over the integers, without any modular reduction) as

$$[620^2 \bmod N] = 6647 = 17^2 \cdot 23.$$

Similarly,

$$[621^2 \bmod N] = 2^4 \cdot 17 \cdot 29$$
$$[645^2 \bmod N] = 2^7 \cdot 13 \cdot 23$$
$$[655^2 \bmod N] = 2^3 \cdot 13 \cdot 17 \cdot 29.$$

Letting our subset S include all four of the above equations, we see that

$$620^2 \cdot 621^2 \cdot 645^2 \cdot 655^2 = 2^{14} \cdot 13^2 \cdot 17^4 \cdot 23^2 \cdot 29^2 \bmod N$$
$$\Rightarrow [620 \cdot 621 \cdot 645 \cdot 655 \bmod N]^2 = [2^7 \cdot 13 \cdot 17^2 \cdot 23 \cdot 29 \bmod N]^2 \bmod N$$
$$\Rightarrow 127194^2 = 45335^2 \bmod N,$$

with $127194 \neq \pm 45335 \bmod N$. Computing $\gcd(127194 - 45335, 377753) = 751$ yields a nontrivial factor of N. $\diamondsuit$

Running time. Choosing a larger value of B makes it more likely that a uniform value $q = [x^2 \bmod N]$ is B-smooth; on the other hand, it means we will have to work harder to identify and factor B-smooth numbers, and we will have to find more of them (since we require $\ell > k$, where k is the number of primes less than or equal to B). It also means that the matrix Γ will be larger, and so the linear-algebraic step will be slower. Choosing the optimal value of B gives an algorithm that (heuristically, at least) factors N in time $2^{O(\sqrt{\log N \log \log N})}$. (In fact, the constant term in the exponent can be determined quite precisely.) The important point for our purposes is that this is sub-exponential in the length of N.

10.2 Algorithms for Computing Discrete Logarithms

Let $\mathbb{G}$ be a cyclic group of known order q. An instance of the discrete-logarithm problem in $\mathbb{G}$ specifies a generator $g \in \mathbb{G}$ and an element $h \in \mathbb{G}$; the goal is to find $x \in \mathbb{Z}_q$ such that $g^x = h$. (See Section 9.3.2.) The solution

x is called the *discrete logarithm of h with respect to g*. A trivial brute-force search for x can be done in time $\mathcal{O}(q)$, and so we are interested in algorithms whose running time is better than this.

Algorithms for solving the discrete-logarithm problem fall into two categories: those that are *generic* and apply to any group $\mathbb{G}$, and those that are tailored to work for some *specific* class of groups. We begin in this section by discussing three generic algorithms:

- When the group order q is not prime and a (partial or full) factorization of q is known, the *Pohlig–Hellman algorithm* reduces the problem of finding discrete logarithms in $\mathbb{G}$ to that of finding discrete logarithms in *subgroups* of $\mathbb{G}$. When the complete factorization of q is known, the effect is to reduce the complexity of computing discrete logarithms in a group of order q to the complexity of computing discrete logarithms in a group of order q', where q' is the largest prime dividing q. This explains the preference for using prime-order groups (cf. Section 9.3.2).

- The *baby-step/giant-step* method, due to Shanks, computes the discrete logarithm in a group of order q using $\mathcal{O}(\sqrt{q})$ group operations. It also requires $\mathcal{O}(\sqrt{q})$ memory.

- *Pollard's rho algorithm* also computes discrete logarithms with $\mathcal{O}(\sqrt{q})$ group operations, but using *constant* memory. It can be viewed as exploiting the connection between the discrete-logarithm problem and collision-resistant hashing that we have seen in Section 9.4.2.

It can be shown that the time complexity of the latter two algorithms is *optimal* as far as generic algorithms are concerned. Thus, to have any hope of doing better we must look at algorithms for specific groups that exploit the *binary representation* of elements in those groups, i.e., the way group elements are encoded as bit-strings. This point bears some discussion. From a mathematical point of view, any two cyclic groups of the same order are isomorphic, meaning that the groups are identical up to a "renaming" of the group elements. From a computational/algorithmic point of view, however, this "renaming" can have a significant impact. For example, consider the cyclic group $\mathbb{Z}_q$ of integers $\{0, \ldots, q-1\}$ under *addition* modulo q. Computing discrete logarithms in this group is trivial: Say we are given $g, h \in \mathbb{Z}_q$ with g a generator, and we want to find x such that $x \cdot g = h \bmod q$. We must have $\gcd(g, q) = 1$ (cf. Theorem B.16) and so g has a multiplicative inverse g^{-1} modulo q. Moreover, g^{-1} can be computed efficiently, as described in Appendix B.2.2. But then $x = h \cdot g^{-1} \bmod q$ is the desired solution. Note that, formally, x here denotes an integer and not a group element—after all, the group operation is addition, not multiplication. Nevertheless, in solving the discrete-logarithm problem in $\mathbb{Z}_q$ we can make use of the fact that another operation (namely, multiplication) can be defined on the elements of that group. The main takeaway point is that *the group representation matters*.

Turning to groups with cryptographic significance, in Section 10.3 we focus our attention on (subgroups of) $\mathbb{Z}_p^*$ for p prime. (See Section 9.3.3.) As a nontrivial example of an algorithm that is not generic, we give a high-level overview of the *index calculus algorithm* for solving the discrete-logarithm problem in such groups in sub-exponential time. Currently, the best known algorithm for this class of groups is the *general number field sieve*,[2] which heuristically runs in time $2^{\mathcal{O}((\log p)^{1/3} \cdot (\log \log p)^{2/3})}$. Sub-exponential algorithms for computing discrete logarithms in multiplicative subgroups of arbitrary finite fields are also known, but these are beyond our scope.

Importantly, no sub-exponential algorithms are known for computing discrete logarithms in general elliptic-curve groups. This explains why smaller parameters can be used (at the same level of security) when working in elliptic-curve groups than when working in $\mathbb{Z}_p^*$, resulting in more-efficient cryptosystems in the former case.

10.2.1 The Pohlig–Hellman Algorithm

The Pohlig–Hellman algorithm can be used to speed up the computation of discrete logarithms in a group $\mathbb{G}$ when any nontrivial factors of the group order q are known. Recall that the order of an element g, which we denote here by $\mathsf{ord}(g)$, is the smallest positive integer i for which $g^i = 1$. We will need the following lemma:

LEMMA 10.4 Let $\mathsf{ord}(g) = q$, and say $p \mid q$. Then $\mathsf{ord}(g^p) = q/p$.

PROOF Since $(g^p)^{q/p} = g^q = 1$, the order of g^p is at most q/p. Let $i > 0$ be such that $(g^p)^i = 1$. Then $g^{pi} = 1$ and, since q is the order of g, we must have $pi \geq q$ or equivalently $i \geq q/p$. The order of g^p is thus exactly q/p. ∎

We will also use a generalization of the Chinese remainder theorem: if $q = \prod_{i=1}^{k} q_i$ and $\gcd(q_i, q_j) = 1$ for all $i \neq j$ then

$$\mathbb{Z}_q \simeq \mathbb{Z}_{q_1} \times \cdots \times \mathbb{Z}_{q_k} \quad \text{and} \quad \mathbb{Z}_q^* \simeq \mathbb{Z}_{q_1}^* \times \cdots \times \mathbb{Z}_{q_k}^*.$$

(This can be proved by induction on k, using the basic Chinese remainder theorem for $k = 2$.) Moreover, by an extension of the algorithm in Section 9.1.5 it is possible to convert efficiently between the representation of an element as an element of $\mathbb{Z}_q$ and its representation as an element of $\mathbb{Z}_{q_1} \times \cdots \times \mathbb{Z}_{q_k}$ when the factorization $q = \prod_{i=1}^{k} q_i$ is known.

We now describe the Pohlig–Hellman algorithm. We are given a generator g and an element h and wish to find x such that $g^x = h$. Say a factorization

---

[2]The algorithm is related to the general number field sieve for factoring.

$q = \prod_{i=1}^{k} q_i$ is known with the $\{q_i\}$ pairwise relatively prime. (This need not be the complete prime factorization of q.) We know that

$$\left(g^{q/q_i} \right)^x = (g^x)^{q/q_i} = h^{q/q_i} \quad \text{for } i = 1, \ldots, k. \tag{10.2}$$

Letting $g_i \stackrel{\text{def}}{=} g^{q/q_i}$ and $h_i \stackrel{\text{def}}{=} h^{q/q_i}$, we thus have k instances of a discrete-logarithm problem in k *smaller* groups. Specifically, each problem $g_i^x = h_i$ is in a subgroup of size $\text{ord}(g_i) = q_i$ (by Lemma 10.4). We can solve each of the k resulting instances using any algorithm for solving the discrete-logarithm problem. Solving these instances gives a set of answers $\{x_i\}_{i=1}^{k}$, with $x_i \in \mathbb{Z}_{q_i}$, for which $g_i^{x_i} = h_i = g_i^x$. Proposition 9.54 implies that $x = x_i \bmod q_i$ for all i. By the generalized Chinese remainder theorem discussed earlier, the constraints

$$x = x_1 \bmod q_1$$

$$\vdots$$

$$x = x_k \bmod q_k$$

uniquely determine x modulo q, and so the desired solution x can be efficiently reconstructed from the $\{x_i\}$.

Example 10.5

Consider the problem of computing discrete logarithms in $\mathbb{Z}_{31}^*$, a group of order $q = 30 = 5 \cdot 3 \cdot 2$. Say $g = 3$ and $h = 26 = g^x$ with x unknown. We have:

$$\begin{aligned}
(g^{30/5})^x = h^{30/5} &\to (3^6)^x = 26^6 &\Rightarrow& \quad 16^x = 1 \\
(g^{30/3})^x = h^{30/3} &\Rightarrow (3^{10})^x = 26^{10} &\Rightarrow& \quad 25^x = 5 \\
(g^{30/2})^x = h^{30/2} &\Rightarrow (3^{15})^x = 26^{15} &\Rightarrow& \quad 30^x = 30.
\end{aligned}$$

(All the above equations are modulo 31.) We have $\text{ord}(16) = 5$, $\text{ord}(25) = 3$, and $\text{ord}(30) = 2$. Solving each equation, we obtain

$$x = 0 \bmod 5, \quad x = 2 \bmod 3, \quad \text{and} \quad x = 1 \bmod 2,$$

and so $x = 5 \bmod 30$. Indeed, $3^5 = 26 \bmod 31$. $\diamondsuit$

If q has (known) prime factorization $q = \prod_{i=1}^{k} p_i^{e_i}$ then, by using the Pohlig–Hellman algorithm, the time to compute discrete logarithms in a group of order q is dominated by the computation of a discrete logarithm in a subgroup of size $\max_i\{p_i^{e_i}\}$. This can be further reduced to computation of a discrete logarithm in a subgroup of size $\max_i\{p_i\}$; see Exercise 10.5.

10.2.2 The Baby-Step/Giant-Step Algorithm

The baby-step/giant-step algorithm computes discrete logarithms in a group of order q using $\mathcal{O}(\sqrt{q})$ group operations. The idea is simple. Given a generator $g \in \mathbb{G}$, we can imagine the powers of g as forming a cycle

$$1 = g^0, g^1, g^2, \ldots, g^{q-2}, g^{q-1}, g^q = 1.$$

We know that h must lie somewhere in this cycle. Computing all the points in this cycle to find h would take $\Omega(q)$ time. Instead, we "mark off" the cycle at intervals of size $t \stackrel{\text{def}}{=} \lfloor \sqrt{q} \rfloor$; more precisely, we compute and store the $\lfloor q/t \rfloor + 1 = \mathcal{O}(\sqrt{q})$ elements

$$g^0, g^t, g^{2t}, \ldots, g^{\lfloor q/t \rfloor \cdot t}.$$

(These are the "giant steps.") Note that the gap between any consecutive "marks" (wrapping around at the end) is at most t. Furthermore, we know that $h = g^x$ lies in one of these gaps. Thus, if we take "baby steps" and compute the t elements

$$h \cdot g^1, \ldots, h \cdot g^t,$$

each of which corresponds to a "shift" of h, we know that one of these values will be equal to one of the marked points. Say we find $h \cdot g^i = g^{k \cdot t}$. We can then easily compute $\log_g h := [(kt - i) \bmod q]$. Pseudocode for this algorithm follows.

ALGORITHM 10.6
The baby-step/giant-step algorithm

Input: Elements $g, h \in \mathbb{G}$; the order q of $\mathbb{G}$
Output: $\log_g h$

$t := \lfloor \sqrt{q} \rfloor$
for $i = 0$ to $\lfloor q/t \rfloor$:
    compute $g_i := g^{i \cdot t}$
sort the pairs (i, g_i) by their second component
for $i = 1$ to t:
    compute $h_i := h \cdot g^i$
    if $h_i = g_k$ for some k, **return** $[(kt - i) \bmod q]$

The algorithm requires $\mathcal{O}(\sqrt{q})$ exponentiations/multiplications in $\mathbb{G}$. (In fact, after computing $g_1 = g^t$, each subsequent value g_i can be computed using a single multiplication as $g_i := g_{i-1} \cdot g_1$. Similarly, each h_i can be computed as $h_i := h_{i-1} \cdot g$.) Sorting the $\mathcal{O}(\sqrt{q})$ pairs $\{(i, g_i)\}$ takes time $\mathcal{O}(\sqrt{q} \cdot \log q)$, and we can then use binary search to check if each h_i is equal to some g_k in time $\mathcal{O}(\log q)$. The overall algorithm thus runs in time $\mathcal{O}(\sqrt{q} \cdot \text{polylog}(q))$.

Example 10.7
We show an application of the algorithm in the cyclic group $\mathbb{Z}_{29}^*$ of order $q = 29 - 1 = 28$. Take $g = 2$ and $h = 17$. We set $t = 5$ and compute:

$$2^0 = 1, \quad 2^5 = 3, \quad 2^{10} = 9, \quad 2^{15} = 27, \quad 2^{20} = 23, \quad 2^{25} = 11.$$

(It should be understood that all operations are in $\mathbb{Z}_{29}^*$.) Then compute:

$$17 \cdot 2^1 = 5, \quad 17 \cdot 2^2 = 10, \quad 17 \cdot 2^3 = 20, \quad 17 \cdot 2^4 = 11,$$

and notice that $17 \cdot 2^4 = 11 = 2^{25}$. We thus have $\log_2 17 = 25 - 4 = 21$. $\Diamond$

10.2.3 Discrete Logarithms from Collisions

A drawback of the baby-step/giant-step algorithm is that it uses a large amount of memory, as it requires storage of $\mathcal{O}(\sqrt{q})$ points. We can obtain an algorithm that uses constant memory—and has the same asymptotic running time—by exploiting the connection between the discrete-logarithm problem and collision-resistant hashing shown in Section 9.4.2, and recalling the small-space birthday attack for finding collisions from Section 6.4.2.

We describe the high-level idea. Fix a generator $g \in \mathbb{G}$ and an element h. If we define the hash function $H_{g,h} : \mathbb{Z}_q \times \mathbb{Z}_q \to \mathbb{G}$ by $H_{g,h}(x_1, x_2) = g^{x_1} h^{x_2}$, then finding a collision in $H_{g,h}$ implies the ability to compute $\log_g h$ (cf. Lemma 9.65 and Theorem 9.79). We have thus reduced the problem of computing $\log_g h$ to that of finding a collision in a hash function, something we know how to do in time $\mathcal{O}(\sqrt{|\mathbb{G}|}) = \mathcal{O}(\sqrt{q})$ using a birthday attack! Moreover, a small-space birthday attack will give a collision in the same time and constant space.

It only remains to address a few technical details. One is that the small-space birthday attack described in Section 6.4.2 assumes that the range of the hash function is a subset of its domain; that is not the case here, and in fact (depending on the representation being used for elements of $\mathbb{G}$) it could even be that $H_{g,h}$ is not compressing. A second issue is that the analysis in Section 6.4.2 treated the hash function as a random function, whereas $H_{g,h}$ has a significant amount of algebraic structure.

Pollard's rho algorithm provides one way to deal with these issues. We describe a different algorithm that can be viewed as a more direct implementation of the above ideas. (In practice, Pollard's algorithm would be more efficient, although both algorithms use only $\mathcal{O}(\sqrt{q})$ group operations.) Let $F : \mathbb{G} \to \mathbb{Z}_q \times \mathbb{Z}_q$ denote a cryptographic hash function obtained by, e.g., a suitable modification of SHA-2. Define $H : \mathbb{G} \to \mathbb{G}$ by $H(k) \stackrel{\text{def}}{=} H_{g,h}(F(k))$. We can use Algorithm 6.9, with natural modifications, to find a collision in H using an expected $\mathcal{O}(\sqrt{|\mathbb{G}|}) = \mathcal{O}(\sqrt{q})$ evaluations of H (and constant memory). With overwhelming probability, this yields a collision in $H_{g,h}$. You are asked to flesh out the details in Exercise 10.7.

It is interesting to observe here a certain duality: the proof that hardness of the discrete-logarithm implies a collision-resistant hash function leads to a better algorithm for solving the discrete-logarithm problem! A little reflection should convince us that this is not surprising: a proof by reduction demonstrates that an attack on some construction (in this case, finding collisions in the hash function) directly yields an attack on the underlying assumption (here, the hardness of the discrete-logarithm problem), which is exactly the property exploited by the above algorithm.

10.3 Index Calculus

We conclude with a brief look at the (non-generic) *index calculus algorithm* for computing discrete logarithms in the cyclic group $\mathbb{Z}_p^*$ (for p prime). In contrast to the preceding (generic) algorithms, this approach has running time *sub-exponential* in the size of the group. The algorithm bears some resemblance to the quadratic sieve algorithm introduced in Section 10.1.3, and we assume readers are familiar with the discussion there. As in that case, we discuss the main ideas of the index calculus method but leave a detailed analysis outside the scope of our treatment. Also, some simplifications are introduced to clarify the presentation.

As in the quadratic sieve algorithm, the index calculus method uses a two-step process. Importantly, the first step requires knowledge only of the modulus p and the base g and so it can be run as a preprocessing step before h—the value whose discrete logarithm we wish to compute—is known. For the same reason, it suffices to run the first step only once in order to solve multiple instances of the discrete-logarithm problem (as long as all those instances share the same p and g).

Step 1. Fix some bound B, and let $\{p_1, \ldots, p_k\}$ be the set of prime numbers less than or equal to B. In this step, we find $\ell \geq k$ distinct values $x_1, \ldots, x_\ell \in \mathbb{Z}_{p-1}$ for which $g_i \stackrel{\text{def}}{=} [g^{x_i} \bmod p]$ is B-smooth. This is done by simply choosing uniform $\{x_i\}$ until suitable values are found.

Factoring the resulting B-smooth numbers, we have the ℓ equations:

$$g^{x_1} = \prod_{i=1}^{k} p_i^{e_{1,i}} \bmod p$$

$$\vdots$$

$$g^{x_\ell} = \prod_{i=1}^{k} p_i^{e_{\ell,i}} \bmod p.$$

Taking discrete logarithms, we can transform these into the linear equations

$$x_1 = \sum_{i=1}^{k} e_{1,i} \cdot \log_g p_i \mod (p-1)$$

$$\vdots \tag{10.3}$$

$$x_\ell = \sum_{i=1}^{k} e_{\ell,i} \cdot \log_g p_i \mod (p-1).$$

Note that the $\{x_i\}$ and the $\{e_{i,j}\}$ are known, while the $\{\log_g p_i\}$ are unknown.

Step 2. Now we are given an element h and want to compute $\log_g h$. Here, we find a value $x \in \mathbb{Z}_{p-1}$ for which $[g^x \cdot h \mod p]$ is B-smooth. (Once again, this is done simply by choosing x uniformly.) Say

$$g^x \cdot h = \prod_{i=1}^{k} p_i^{e_i} \mod p$$

$$\Rightarrow x + \log_g h = \sum_{i=1}^{k} e_i \cdot \log_g p_i \mod (p-1),$$

where x and the $\{e_i\}$ are known. Combined with Equation (10.3), we have $\ell + 1 \geq k + 1$ linear equations in the $k+1$ unknowns $\{\log_g p_i\}_{i=1}^{k}$ and $\log_g h$. Using linear-algebraic[3] methods (and assuming the system of equations is not under-defined), we can solve for each of the unknowns and in particular obtain the desired solution $\log_g h$.

Example 10.8
Let $p = 101$, $g = 3$, and $h = 87$. We have $[3^{10} \mod 101] = 65 = 5 \cdot 13$. Similarly, $[3^{12} \mod 101] = 80 = 2^4 \cdot 5$ and $[3^{14} \mod 101] = 13$. We thus have the linear equations

$$10 = \log_3 5 + \log_3 13 \mod 100$$
$$12 = 4 \cdot \log_3 2 + \log_3 5 \mod 100$$
$$14 = \log_3 13 \mod 100.$$

We also have $3^5 \cdot 87 = 32 = 2^5 \mod 101$, or

$$5 + \log_3 87 = 5 \cdot \log_3 2 \mod 100. \tag{10.4}$$

Adding the second and third equations and subtracting the first, we derive $4 \cdot \log_3 2 = 16 \mod 100$. This doesn't determine $\log_3 2$ uniquely (since 4 is not

---

[3]Technically, things are slightly more complicated here since the linear equations are all modulo $p-1$, which is not prime. Nevertheless, there exist techniques for dealing with this.

invertible modulo 100), but it does tell us that $\log_3 2 = 4, 29, 54$, or 79 (cf. Exercise 10.3). Trying all possibilities gives $\log_3 2 = 29$. Plugging this into Equation (10.4) gives $\log_3 87 = 40$. ◇

Running time. Choosing a larger value of B makes it more likely that a uniform value in $\mathbb{Z}_p^*$ is B-smooth; however, it means we will have to work harder to identify and factor B-smooth numbers, and we will have to find more of them. Because the system of equations will be larger, solving the system will take longer. Choosing the optimal value of B gives an algorithm that (heuristically, at least) computes discrete logarithms in $\mathbb{Z}_p^*$ in time $2^{\mathcal{O}(\sqrt{\log p \log \log p})}$. The important point for our purposes is that this is sub-exponential in the length of p.

10.4 Recommended Key Lengths

Understanding the best available algorithms for solving various cryptographic problems is essential for determining the appropriate key length for achieving a desired level of security. Figure 10.1 summarizes the key lengths currently recommended by the US National Institute of Standards and Technology[4] (NIST) [14]. The "effective key length" is a value n such that the best known algorithm for solving a problem takes time roughly 2^n, i.e., the computational difficulty of solving a problem is approximately equivalent to that of performing a brute-force search against a symmetric-key scheme with an n-bit key, or the time to find collisions in a hash function with a $2n$-bit output length. NIST deems a 112-bit effective key length acceptable for security until the year 2030, but recommends 128-bit or higher key lengths for applications where security is required beyond then.

Given what we have learned in this chapter, it is instructive to look more closely at some of the numbers in the table. One thing to notice is that elliptic-curve groups can be used to realize any given level of security with smaller parameters than for RSA or subgroups of $\mathbb{Z}_p^*$. This is simply because no sub-exponential algorithms are known for solving the discrete-logarithm problem in elliptic-curve groups (when chosen appropriately). Achieving n-bit security, however, requires an elliptic-curve group whose order q is $2n$-bits long. This is a consequence of the generic algorithms we have seen in this chapter, which solve the discrete-logarithm problem (in any group) in time $\mathcal{O}(\sqrt{q})$.

Turning to the case of $\mathbb{Z}_p^*$ we see that here, too, a $2n$-bit value of q is needed for n-bit security (for the same reason). The length of p, however, must be

---

[4]Other groups have made their own recommendations; see http://keylength.com.

	RSA	Discrete Logarithm	
Effective Key Length	Modulus N	Order-q Subgroup of $\mathbb{Z}_p^*$	Elliptic-Curve Group Order q
112	2048	p: 2048, q: 224	224
128	3072	p: 3072, q: 256	256
192	7680	p: 7680, q: 384	384
256	15360	p: 15360, q: 512	512

FIGURE 10.1: All values are in bits, e.g., for a 112-bit effective key length in the RSA setting, a 2048-bit modulus N should be used.

significantly larger, because non-generic algorithms like the index calculus method or the number field sieve can be used to compute discrete logarithms in $\mathbb{Z}_p^*$ in time sub-exponential in the length of p. That is, p and q are chosen such that the running time of the number field sieve, which depends on the length of p, and the running time of a generic algorithm, which depends on the length of q, are approximately equal and both around 2^n. The practical ramifications of this are that, for any desired security level, elliptic-curve cryptosystems can use significantly smaller parameters (and thus give better efficiency for honest users) than cryptosystems based on subgroups of $\mathbb{Z}_p^*$.

References and Additional Reading

Pollard's $p-1$ algorithm was published in 1974 [160], and his rho method for factoring was described the following year [161]. The quadratic sieve algorithm is due to Pomerance [163], based on earlier ideas of Dixon [67].

The Pohlig–Hellman algorithm was published in 1978 [159]. The baby-step/giant-step algorithm is due to Shanks [176]. Pollard's paper introducing the rho algorithm for computing discrete logarithms [162] also includes his famous "kangaroo" algorithm for the same problem. A nice feature of the kangaroo method is that it is more flexible; in particular, it can be used to compute discrete logarithms known to lie in a given interval $[a, b]$ using $\mathcal{O}(\sqrt{b-a})$ steps. (Although the baby-step/giant-step algorithm can also be adapted for that case—see Exercise 10.6—the kangaroo algorithm stores only a constant number of group elements.) Lower bounds on the running time of generic algorithms for computing discrete logarithms, which asymptotically match the running times of the algorithms described in this chapter, were given by Nechaev [152] and Shoup [179].

The index calculus algorithm as we have described it is by Adleman [4]. The texts by Wagstaff [201], Shoup [183], Crandall and Pomerance [59], Joux [105], and Galbraith [76] provide further information on algorithms for factoring and computing discrete logarithms in finite fields, including descriptions of the (general) number field sieve. The current state-of-the-art for factoring and computing discrete logarithms in $\mathbb{Z}_p^*$ for large p is surveyed in a recent article by Boudot et al. [45].

Recently, improved algorithms for solving the discrete-logarithm problem in finite fields of small characteristic [12] or even any fixed characteristic [116] have been announced. It seems prudent to avoid using such groups for cryptographic applications.

Lenstra and Verheul [126] provide a comprehensive discussion, somewhat dated but still relevant, of how known algorithms for factoring and computing discrete logarithms affect the choice of cryptographic parameters in practice.

Exercises

10.1 In order to speed up the key-generation algorithm for RSA, it has been suggested to generate a large prime number by generating many small random primes, multiplying them together, and adding one (of course, then checking that the result is prime). What do you think of the security implications of this method?

10.2 In an execution of Algorithm 10.2, define $x^{(i)} \stackrel{\text{def}}{=} F^{(i)}(x)$. Show that if, in a given execution, there exist $i, j \leq 2^{n/2}$ such that $x^{(i)} \neq x^{(j)}$ but $x^{(i)} = x^{(j)} \bmod p$, then that execution of the algorithm outputs p with overwhelming probability. (The analysis is a little different from the analysis of Algorithm 6.9, since the algorithms—and their goals—are slightly different.)

10.3 (a) Show that if $ab = c \bmod N$ and $\gcd(b, N) = d$, then:

   i. $d \mid c$;
   ii. $a \cdot (b/d) = (c/d) \bmod (N/d)$; and
   iii. $\gcd(b/d, N/d) = 1$.

   (b) Describe how to use the above to compute $\log_g h$ in $\mathbb{Z}_N$ even when g is not a generator of $\mathbb{Z}_N$ (but $h \in \langle g \rangle$).

10.4 Here we consider how to solve the discrete-logarithm problem in a cyclic group $\mathbb{G}$ of order $q = p^e$ using $\mathcal{O}(e\sqrt{p})$ group operations. We are given as input a generator g and an element h, and want to compute $x = \log_g h$.

   (a) Show how to compute $[x \bmod p]$ using $\mathcal{O}(\sqrt{p})$ group operations.

Hint: Solve the equation

$$\left(g^{p^{e-1}}\right)^{x_0} = h^{p^{e-1}}$$

and use the same ideas as in the Pohlig–Hellman algorithm.

(b) Say $x = x_0 + x_1 \cdot p + \cdots + x_{e-1} \cdot p^{e-1}$ with $0 \le x_i < p$. (i.e., write x in base p.) In the previous step we determined x_0. Show how to compute a value h_1 such that $(g^p)^{x_1 + x_2 \cdot p + \cdots + x_{e-1} \cdot p^{e-2}} = h_1$.

(c) Show a recursive algorithm computing the discrete logarithm x in the claimed running time.

10.5 Let q have prime factorization $q = \prod_{i=1}^{k} p_i^{e_i}$. Using the result from the previous problem, show a modification of the Pohlig–Hellman algorithm that solves the discrete-logarithm problem in a group of order q using $\mathcal{O}\left(\sum_{i=1}^{k} e_i \sqrt{p_i}\right)$ group operations.

10.6 Let $\mathbb{G}$ be a cyclic group of order q, with generator g. Let $h \in \mathbb{G}$ be given, where it is known that $h = g^x$ for $x \in [a, b]$ (and a, b are known). Show how to modify the baby-step/giant-step algorithm to compute $\log_g h$ using $\mathcal{O}(\sqrt{b-a})$ group operations.

10.7 Based on the ideas described in Section 10.2.3, give pseudocode for a generic algorithm that computes discrete logarithms in a group of order q using $\mathcal{O}(\sqrt{q})$ group operations and $O(1)$ memory. Also give a heuristic analysis of the probability with which your algorithm succeeds.

Chapter 11

Key Management and the Public-Key Revolution

11.1 Key Distribution and Key Management

In previous chapters we have seen how private-key cryptography can be used to ensure secrecy and integrity for two parties communicating over an insecure channel, if we are willing to assume those two parties hold a shared, secret key. The question we have deferred since Chapter 1, however, is:

How can the parties share a secret key in the first place?

Clearly, the key cannot simply be sent over the insecure communication channel because an eavesdropping adversary would then be able to observe the key and it would no longer be secret. Some other mechanism must be used.

In some situations, the parties may have access to a secure channel that they can use to reliably share a secret key. One common example is when the two parties are physically co-located at some point in time, during which they can share a key. Alternatively, the parties might be able to use a trusted courier service as a secure channel. We stress that even if the parties have access to a secure channel at some point, this does *not* make private-key cryptography useless: in the first example, the parties have a secure channel at one point in time but not later; in the second example, utilizing the secure channel might be slower and more costly than communicating over an insecure channel.

The above approaches have been used to share keys in government, diplomatic, and military settings. As an example, the "red phone" connecting Moscow and Washington in the 1960s was encrypted using a one-time pad, with keys shared by couriers who flew from one country to the other carrying briefcases full of print-outs. Such approaches can also be used in corporations, e.g., to set up a shared key between a central database and a new employee on his/her first day of work. (We return to this example in the next section.)

Relying on a secure channel to distribute keys, however, does not work well in many other situations. For example, consider a large, multinational corporation in which *every pair* of employees might need the ability to communicate securely, with their communication protected from other employees as well. It will be inconvenient, to say the least, for each pair of employees to meet so they can securely share a key; for employees working in different cities, this may even be impossible. Even if the current set of employees could somehow share keys with each other, it would be impractical for them to share keys with new employees who join after this initial sharing is done.

Even assuming these N employees are somehow able to securely share keys with each other, another significant drawback is that each employee would have to manage and store $N - 1$ secret keys (one for each other employee in the company). In fact, this may significantly under-count the number of keys stored by each user, because employees may also need keys to communicate securely with remote resources such as databases, servers, printers, and so on. The proliferation of so many secret keys is a significant logistical problem. Moreover, all these keys must be stored *securely*. The more keys there are, the harder it is to protect them, and the higher the chance of some keys being stolen by an attacker. Computer systems are often infected by viruses, worms, and other forms of malicious software that can steal secret keys and send them quietly over the network to an attacker. Thus, storing keys on employees' personal computers is not always a safe solution.

To be clear, potential compromise of secret keys is always a concern, irrespective of the number of keys each party holds. When only a few keys need to be stored, however, there are good solutions available for dealing with this threat. A typical solution today is to store keys on *secure hardware* such as a smartcard. A smartcard can carry out cryptographic computations using the stored secret keys, ensuring that these keys never make their way onto users' personal computers. If designed properly, the smartcard can be much more resilient to attack than a personal computer—for example, it typically cannot be infected by malware—and so offers a good means of protecting users' secret keys. Unfortunately, smartcards are typically quite limited in memory, and so cannot store hundreds (or thousands) of keys; they may also be somewhat expensive and difficult to replace if lost.

The concerns outlined above can all be addressed—in principle, even if not in practice—in "closed" organizations consisting of a well-defined population of users, all of whom are willing to follow the same policies for distributing and storing keys. They break down, however, in "open systems" where users have transient interactions, cannot arrange a physical meeting, and may not even be aware of each other's existence until the time they first want to communicate. This is, in fact, a more common situation than one might initially realize: consider sending credit-card information to an Internet merchant from whom you have never previously purchased anything, or sending email to someone whom you have never met in person. In such cases, private-key cryptography alone simply does not provide a solution, and we must look further for adequate solutions.

To summarize, there are at least three distinct problems related to the use of private-key cryptography. The first is that of *key distribution*, the second is that of *storing and managing large numbers of secret keys*, and the third is the *inapplicability of private-key cryptography to open systems*.

11.2 A Partial Solution: Key-Distribution Centers

One way to address some of the concerns from the previous section is to use a *key-distribution center (KDC)* to establish shared keys. Consider again the case of a large corporation where all pairs of employees must be able to communicate securely. In such a setting, we can leverage the fact that all employees may *trust* some entity—say, the system administrator—at least with respect to the security of work-related information. This trusted entity can then act as a KDC and help all the employees share pairwise keys.

When a new employee joins, the KDC can share a key with that employee (in person, in a secure location) as part of that employee's first day of work. At the same time, the KDC could also distribute shared keys between that employee and all existing employees. That is, when the ith employee joins, the KDC could (in addition to sharing a key between itself and this new employee) generate $i - 1$ keys $k_1, \ldots, k_{i-1}$, give these keys to the new employee, and then send key k_j to the jth existing employee by encrypting it using the key that employee already shares with the KDC. Following this, the new employee shares a key with every other employee (as well as with the KDC).

A better approach, which avoids requiring employees to store and manage multiple keys, is to utilize the KDC in an *online* fashion to generate keys "on demand" whenever two employees wish to communicate securely. As before, the KDC will share a (different) key with each employee, something that can be done securely on each employee's first day of work. Say the KDC shares key k_A with employee Alice, and k_B with employee Bob. At some later time, when Alice wishes to communicate securely with Bob, she can simply send the message ``I, Alice, want to talk to Bob'' to the KDC. (If desired, this message can be authenticated using the key shared by Alice and the KDC.) The KDC then chooses a new, random key—called a *session key*—and sends this key k to Alice encrypted using k_A, and to Bob encrypted using k_B. (This protocol is too simplistic to be used in practice; see further discussion below.) Once Alice and Bob both recover this session key, they can use it to communicate securely. When they are done with their conversation, they can (and should) erase the session key because they can always contact the KDC again if they wish to communicate at some later time.

Consider the advantages of this approach:

1. Each employee needs to store only *one* long-term secret key (namely, the one they share with the KDC). Employees still need to manage and store session keys, but these are short-term keys that are erased once a communication session concludes.

   The KDC needs to store many long-term keys. However, the KDC can be kept in a secure location and be given the highest possible protection against network attacks.

2. When an employee joins the organization, all that must be done is to set up a key between this employee and the KDC. No other employees need to update the set of keys they hold.

Thus, KDCs can alleviate two of the problems we have seen with regard to private-key cryptography: they can simplify key distribution (since only one new key must be shared when an employee joins, and it is reasonable to assume a secure channel between the KDC and that employee on their first day of work), and can reduce the complexity of key storage (since each employee only needs to store a single key). KDCs go a long way toward making private-key cryptography practical in large organizations where there is a single entity who is trusted by everyone.

There are, however, some drawbacks to relying on KDCs:

1. A successful attack on the KDC will result in a complete break of the system: an attacker can compromise all keys and subsequently eavesdrop on all network traffic. This makes the KDC a high-value target. Note that even if the KDC is well-protected against external attacks, there is always the possibility of an insider attack by an employee who has access to the KDC (for example, the IT manager).

2. The KDC is a single point of failure: if the KDC is down, secure communication is temporarily impossible. If employees are constantly contacting the KDC and asking for session keys to be established, the load on the KDC can be very high, thereby increasing the chances that it may fail or be slow to respond.

A simple solution to the second problem is to replicate the KDC. This works (and is done in practice), but also means that there are now more points of attack on the system. Adding more KDCs also makes it more difficult to add new employees, since updates must be securely propagated to every KDC.

Protocols for key distribution using a KDC. There are a number of protocols in the literature for secure key distribution using a KDC. We mention in particular the *Needham–Schroeder protocol*, which forms the core of *Kerberos*, an important and widely used service for performing authentication and supporting secure communication. (Kerberos is used in many universities and corporations, and is the default mechanism for supporting secure networked authentication and communication in Windows and many UNIX systems.) We only highlight one feature of this protocol. When Alice contacts the KDC and asks to communicate with Bob, the KDC does not send the encrypted session key to both Alice and Bob as we have described earlier. Instead, the KDC sends to Alice the session key encrypted under Alice's key *in addition to* the session key encrypted under Bob's key. Alice then forwards the second ciphertext to Bob as in Figure 11.1. The second ciphertext is sometimes called a *ticket*, and can be viewed as a credential that allows Alice to talk to Bob (and allows Bob to be assured that he is talking to Alice). Indeed,

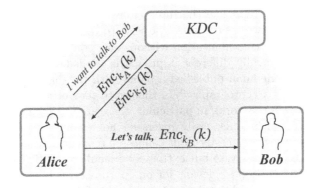

FIGURE 11.1: A general template for key-distribution protocols.

although we have not stressed this point in our discussion, a KDC-based approach can provide a useful means of performing authentication as well. Note also that Alice and Bob need not both be users; Alice might be a user and Bob a resource such as a remote server, a database, or a printer.

The protocol was designed in this way to reduce the load on the KDC. In the protocol as described, the KDC does not need to initiate a second connection to Bob, and need not worry whether Bob is on-line when Alice initiates the protocol. Moreover, if Alice retains the ticket (and her copy of the session key), then she can re-initiate secure communication with Bob by simply re-sending the ticket to Bob, without the involvement of the KDC at all. (In practice, tickets expire and eventually need to be renewed. But a session could be re-established within some acceptable time period.)

We conclude by noting that in practice the key that Alice shares with the KDC might be a short, easy-to-memorize password. In this case, many additional security problems arise that must be dealt with. We have also been implicitly assuming an attacker who only passively eavesdrops, rather than one who might actively try to interfere with the protocol. We refer the interested reader to the references at the end of this chapter for more information about how such issues can be addressed.

11.3 Key Exchange and the Diffie–Hellman Protocol

KDCs and protocols like Kerberos are used in practice. But these approaches to the key-distribution problem still require, at some point, a *private* and *authenticated* channel that can be used to share keys. (In particular, we assumed the existence of such a channel between the KDC and an employee on his or her first day.) Thus, they still cannot solve the problem of key distribution in open systems like the Internet, where there may be no private channel available between two users who wish to communicate.

To achieve private communication without ever communicating over a private channel, a radically different approach is needed. In 1976, Whitfield Diffie and Martin Hellman published a paper with the innocent-looking title "New Directions in Cryptography." In that work they observed that there is often *asymmetry* in the world; in particular, there are certain actions that can be easily performed but not easily reversed. For example, padlocks can be locked without a key (i.e., easily), but cannot be reopened. More strikingly, it is easy to shatter a glass vase but extremely difficult to put it back together. Algorithmically (and more relevant for our purposes), it is easy to multiply two large primes but difficult to recover those primes from their product. (This is exactly the factoring problem discussed in previous chapters.) Diffie and Hellman realized that such phenomena could be used to derive interactive protocols for *secure key exchange* that allow two parties to share a secret key, via communication over a public channel, by having the parties perform operations that an eavesdropper cannot reverse.

The existence of secure key-exchange protocols is quite amazing. It means that you and a friend could agree on a secret by simply shouting across a room (and performing some local computation); the secret would be unknown to anyone else, even if they had listened to everything that was said. Indeed, until 1976 it was generally believed that secure communication could not be done without first sharing some secret information using a private channel.

The influence of Diffie and Hellman's paper was enormous. In addition to introducing a fundamentally new way of looking at cryptography, it was one of the first steps toward moving cryptography out of the private domain and into the public one. We quote the first two paragraphs of their paper:

> *We stand today on the brink of a revolution in cryptography. The development of cheap digital hardware has freed it from the design limitations of mechanical computing and brought the cost of high grade cryptographic devices down to where they can be used in such commercial applications as remote cash dispensers and computer terminals.*
>
> *In turn, such applications create a need for new types of cryptographic systems which minimize the necessity of secure key distribution channels. ... At the same time, theoretical developments in information theory and computer science show promise of providing provably secure cryptosystems, changing this ancient art into a science.*

Diffie and Hellman were not exaggerating, and the revolution they spoke of was due in great part to their work.

In this section we present the Diffie–Hellman key-exchange protocol. We prove its security against eavesdropping adversaries or, equivalently, under the assumption that the parties communicate over a public but *authenticated* channel (so an attacker cannot interfere with their communication). Security against an eavesdropping adversary is a relatively weak guarantee, and in practice key-exchange protocols must satisfy stronger notions of security that are beyond our present scope. (Moreover, we are interested here in the setting where the communicating parties have *no* prior shared information, in which case there is nothing that can be done to prevent an adversary from impersonating one of the parties. We return to this point later.)

The setting and definition of security. We consider a setting with two parties—traditionally called Alice and Bob—who run a probabilistic protocol Π in order to generate a shared, secret key; Π can be viewed as the set of instructions for Alice and Bob in the protocol. Alice and Bob begin by holding the security parameter 1^n; they then run Π using (independent) random bits. At the end of the protocol, Alice and Bob output keys $k_A, k_B \in \{0,1\}^n$, respectively. The basic correctness requirement is that $k_A = k_B$. Since we will only deal with protocols that satisfy this requirement, we will speak simply of *the* key $k = k_A = k_B$ generated in some honest execution of Π. (Since Π is randomized the key will, in general, be different every time Π is run.)

We now turn to defining security. Intuitively, a key-exchange protocol is secure if the key output by Alice and Bob is completely hidden from an eavesdropping adversary. This is formally defined by requiring that an adversary who has eavesdropped on an execution of the protocol should be unable to distinguish the key k generated by that execution (and now shared by Alice and Bob) from a *uniform* key of length n. This is much stronger than simply requiring that the adversary be unable to *guess* k exactly, and this stronger notion is necessary if the parties will subsequently use k for some cryptographic application (e.g., as a key for a private-key encryption scheme).

Formalizing the above, let Π be a key-exchange protocol, $\mathcal{A}$ an adversary, and n the security parameter. We have the following experiment:

The key-exchange experiment $\mathsf{KE}^{\mathsf{eav}}_{\mathcal{A},\Pi}(n)$:

1. *Two parties holding 1^n execute protocol Π. This results in a transcript trans containing all the messages sent by the parties, and a key k output by each of the parties.*

2. *A uniform bit $b \in \{0,1\}$ is chosen. If $b = 0$ set $\hat{k} := k$, and if $b = 1$ then choose uniform $\hat{k} \in \{0,1\}^n$.*

3. *$\mathcal{A}$ is given trans and $\hat{k}$, and outputs a bit b'.*

4. *The output of the experiment is defined to be 1 if $b' = b$, and 0 otherwise. (In case $\mathsf{KE}^{\mathsf{eav}}_{\mathcal{A},\Pi}(n) = 1$, we say that $\mathcal{A}$ succeeds.)*

$\mathcal{A}$ is given trans to capture the fact that $\mathcal{A}$ eavesdrops on the entire execution of the protocol and thus sees all messages exchanged by the parties. In the real world, $\mathcal{A}$ would not be given any key; in the experiment the adversary is given

$\hat{k}$ only as a means of defining what it means for $\mathcal{A}$ to "break" the security of Π. That is, the adversary succeeds in "breaking" Π if it can correctly determine whether the key $\hat{k}$ is the real key corresponding to the given execution of the protocol, or whether $\hat{k}$ is a uniform key that is independent of the transcript.

As expected, we say Π is secure if the adversary succeeds with probability that is at most negligibly greater than $1/2$. That is:

DEFINITION 11.1   *A key-exchange protocol Π is* secure in the presence of an eavesdropper *if for all probabilistic polynomial-time adversaries $\mathcal{A}$ there is a negligible function* negl *such that*

$$\Pr\left[\mathsf{KE}^{\mathsf{eav}}_{\mathcal{A},\Pi}(n) = 1\right] \leq \frac{1}{2} + \mathsf{negl}(n).$$

The aim of a key-exchange protocol is almost always to generate a shared key k that will be used by the parties for some further cryptographic purpose, e.g., to encrypt and authenticate their subsequent communication using, say, an authenticated encryption scheme. Intuitively, using a shared key generated by a secure key-exchange protocol should be "as good as" using a key shared over a private channel. It is possible to prove this formally; see Exercise 11.1.

The Diffie–Hellman key-exchange protocol. We now describe the key-exchange protocol that appeared in the original paper by Diffie and Hellman (although they were less formal than we will be here). Let $\mathcal{G}$ be a probabilistic polynomial-time algorithm that, on input 1^n, outputs a description of a cyclic group $\mathbb{G}$, its order q (with $\|q\| = n$), and a generator $g \in \mathbb{G}$. (See Section 9.3.2.) The Diffie–Hellman key-exchange protocol is described formally as Construction 11.2 and illustrated in Figure 11.2.

CONSTRUCTION 11.2

- **Common input:** The security parameter 1^n
- **The protocol:**
  1. Alice runs $\mathcal{G}(1^n)$ to obtain $(\mathbb{G}, q, g)$.
  2. Alice chooses a uniform $x \in \mathbb{Z}_q$, and computes $h_A := g^x$.
  3. Alice sends $(\mathbb{G}, q, g, h_A)$ to Bob.
  4. Bob receives $(\mathbb{G}, q, g, h_A)$. He chooses a uniform $y \in \mathbb{Z}_q$, and computes $h_B := g^y$. Bob sends h_B to Alice and outputs the key $k_B := h_A^y$.
  5. Alice receives h_B and outputs the key $k_A := h_B^x$.

The Diffie–Hellman key-exchange protocol.

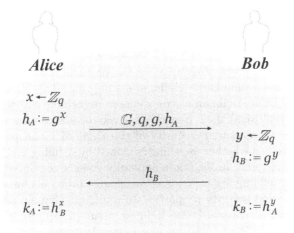

FIGURE 11.2: The Diffie–Hellman key-exchange protocol.

In our description, we have assumed that Alice generates $(\mathbb{G}, q, g)$ and sends these parameters to Bob as part of her first message. In practice, these parameters are *standardized* and known to both parties before the protocol begins. In that case Alice need only send h_A, and Bob need not wait to receive Alice's message before computing and sending h_B.

It is not hard to see that the protocol is correct: Bob computes the key

$$k_B = h_A^y = (g^x)^y = g^{xy}$$

and Alice computes the key

$$k_A = h_B^x = (g^y)^x = g^{xy},$$

and so $k_A = k_B$. (The observant reader will note that the shared key is a group element, not a bit-string. We will return to this point later.)

Diffie and Hellman did not prove security of their protocol; indeed, the appropriate notions (both the definitional framework as well as the idea of formulating precise assumptions) were not yet in place. Let us see what sort of assumption will be needed in order for the protocol to be secure. A first observation, made by Diffie and Hellman, is that a *minimal* requirement for security here is that the discrete-logarithm problem be hard relative to $\mathcal{G}$. If not, then an adversary given the transcript (which, in particular, includes h_A) can compute the secret value of one of the parties (i.e., x) and then easily compute the shared key using that value. So, hardness of the discrete-logarithm problem is necessary for the protocol to be secure. It is not, however, sufficient, as it is possible that there are other ways of computing the key $k_A = k_B$ without explicitly computing x or y. The *computational Diffie–Hellman assumption*—which would only guarantee that the key g^{xy} is hard to compute in its *entirety* from the transcript—does not suffice either.

What is required by Definition 11.1 is that the shared key g^{xy} should be *indistinguishable from uniform* for any adversary given g, g^x, and g^y. This is exactly the *decisional* Diffie–Hellman assumption introduced in Section 9.3.2.

As we will see, a proof of security for the protocol follows almost immediately from the decisional Diffie–Hellman assumption. This should not be surprising, as the Diffie–Hellman assumptions were introduced—well after Diffie and Hellman published their paper—as a way of abstracting the properties underlying the (conjectured) security of the Diffie–Hellman protocol. Given this, it is fair to ask whether anything is gained by defining and proving security here. By this point in the book, hopefully you are convinced the answer is *yes*. Precisely defining security for key-exchange protocols forces us to think about exactly what security properties we want; specifying a precise assumption (namely, the decisional Diffie–Hellman assumption) means we can study that assumption independently of any particular application and—once we are convinced of its plausibility—construct other protocols based on it; finally, proving security shows that the assumption does, indeed, suffice for the protocol to meet our desired notion of security.

In our proof of security, we use a modified version of Definition 11.1 in which it is required that the shared key be indistinguishable from a *uniform element of* $\mathbb{G}$ rather than from a *uniform n-bit string*. This discrepancy will need to be addressed before the protocol can be used in practice—after all, group elements are not typically useful as cryptographic keys, and the representation of a uniform group element will not, in general, be a uniform bit-string—and we briefly discuss one standard way to do so following the proof. For now, we let $\widehat{\mathsf{KE}}^{\mathrm{eav}}_{\mathcal{A},\Pi}(n)$ denote a modified experiment where if $b = 1$ then $\hat{k}$ is chosen uniformly from $\mathbb{G}$ rather than uniformly from $\{0,1\}^n$.

THEOREM 11.3   *If the decisional Diffie–Hellman problem is hard relative to $\mathcal{G}$, then the Diffie–Hellman key-exchange protocol Π is secure in the presence of an eavesdropper (with respect to the modified experiment $\widehat{\mathsf{KE}}^{\mathrm{eav}}_{\mathcal{A},\Pi}$).*

PROOF   Let $\mathcal{A}$ be a PPT adversary. Since $\Pr[b = 0] = \Pr[b = 1] = 1/2$, we have

$$\Pr\left[\widehat{\mathsf{KE}}^{\mathrm{eav}}_{\mathcal{A},\Pi}(n) = 1\right]$$

$$= \frac{1}{2} \cdot \Pr\left[\widehat{\mathsf{KE}}^{\mathrm{eav}}_{\mathcal{A},\Pi}(n) = 1 \mid b = 0\right] + \frac{1}{2} \cdot \Pr\left[\widehat{\mathsf{KE}}^{\mathrm{eav}}_{\mathcal{A},\Pi}(n) = 1 \mid b = 1\right].$$

In experiment $\widehat{\mathsf{KE}}^{\mathrm{eav}}_{\mathcal{A},\Pi}(n)$ the adversary $\mathcal{A}$ receives $(\mathbb{G}, q, g, h_A, h_B, \hat{k})$, where $(\mathbb{G}, q, g, h_A, h_B)$ represents the transcript of the protocol execution, and $\hat{k}$ is either the actual key computed by the parties (if $b = 0$) or a uniform group element (if $b = 1$). Distinguishing between these two cases is exactly equivalent to solving the decisional Diffie–Hellman problem. That is

$$\Pr\left[\widehat{\mathsf{KE}}^{\mathsf{eav}}_{\mathcal{A},\Pi}(n) = 1\right]$$

$$= \frac{1}{2} \cdot \Pr\left[\widehat{\mathsf{KE}}^{\mathsf{eav}}_{\mathcal{A},\Pi}(n) = 1 \mid b = 0\right] + \frac{1}{2} \cdot \Pr\left[\widehat{\mathsf{KE}}^{\mathsf{eav}}_{\mathcal{A},\Pi}(n) = 1 \mid b = 1\right]$$

$$= \frac{1}{2} \cdot \Pr[\mathcal{A}(\mathbb{G}, q, g, g^x, g^y, g^{xy}) = 0] + \frac{1}{2} \cdot \Pr[\mathcal{A}(\mathbb{G}, q, g, g^x, g^y, g^z) = 1]$$

$$= \frac{1}{2} \cdot \left(1 - \Pr[\mathcal{A}(\mathbb{G}, q, g, g^x, g^y, g^{xy}) = 1]\right) + \frac{1}{2} \cdot \Pr[\mathcal{A}(\mathbb{G}, q, g, g^x, g^y, g^z) = 1]$$

$$= \frac{1}{2} + \frac{1}{2} \cdot \left(\Pr[\mathcal{A}(\mathbb{G}, q, g, g^x, g^y, g^z) = 1] - \Pr[\mathcal{A}(\mathbb{G}, q, g, g^x, g^y, g^{xy}) = 1]\right)$$

$$\leq \frac{1}{2} + \frac{1}{2} \cdot \left| \Pr[\mathcal{A}(\mathbb{G}, q, g, g^x, g^y, g^z) = 1] - \Pr[\mathcal{A}(\mathbb{G}, q, g, g^x, g^y, g^{xy}) = 1]\right|,$$

where the probabilities are all taken over $(\mathbb{G}, q, g)$ output by $\mathcal{G}(1^n)$, and uniform choice of $x, y, z \in \mathbb{Z}_q$. (Note that since g is a generator, g^z is a uniform element of $\mathbb{G}$ when z is uniformly distributed in $\mathbb{Z}_q$.) If the decisional Diffie–Hellman assumption is hard relative to $\mathcal{G}$, that exactly means that there is a negligible function negl for which

$$\left| \Pr[\mathcal{A}(\mathbb{G}, q, g, g^x, g^y, g^z) = 1] - \Pr[\mathcal{A}(\mathbb{G}, q, g, g^x, g^y, g^{xy}) = 1]\right| \leq \mathsf{negl}(n).$$

We conclude that

$$\Pr\left[\widehat{\mathsf{KE}}^{\mathsf{eav}}_{\mathcal{A},\Pi}(n) = 1\right] \leq \frac{1}{2} + \frac{1}{2} \cdot \mathsf{negl}(n),$$

completing the proof. ∎

Uniform group elements vs. uniform bit-strings. The previous theorem shows that the key output by Alice and Bob in the Diffie–Hellman protocol is indistinguishable (for a polynomial-time eavesdropper) from a uniform group element. In order to use the key for subsequent cryptographic applications—as well as to meet Definition 11.1—the key output by the parties should instead be indistinguishable from a uniform bit-string of the appropriate length. The Diffie–Hellman protocol can be modified to achieve this by having the parties apply an appropriate *key-derivation function* (cf. Section 6.6.4) to the shared group element g^{xy} they each compute.

Active adversaries. So far we have considered only an eavesdropping adversary. Although eavesdropping attacks are by far the most common (as they are the easiest to carry out), they are by no means the only possible attack. *Active* attacks, in which the adversary sends messages of its own to one or both of the parties, are also a concern, and any protocol used in practice must be resilient to such attacks as well. When considering active attacks, it is useful to distinguish, informally, between *impersonation* attacks where the adversary impersonates one party while interacting with the other party, and *man-in-the-middle* attacks where both honest parties are executing the protocol and the adversary is intercepting and modifying messages being sent from one party to the other. We will not formally define security against either class of attacks, as such definitions are rather involved and cannot be achieved without the parties sharing *some* information in advance. Nevertheless, it is worth remarking that the Diffie–Hellman protocol is *completely insecure* against a man-in-the-middle attack. In fact, a man-in-the-middle adversary can act in such a way that Alice and Bob terminate the protocol with different keys k_A and k_B that are both known to the adversary, yet neither Alice nor Bob can detect that any attack was carried out. We leave the details of this attack as an exercise.

Diffie–Hellman key exchange in practice. The Diffie–Hellman protocol in its basic form is typically not used in practice due to its insecurity against man-in-the-middle attacks, as discussed above. This does not detract in any way from its importance. The Diffie–Hellman protocol served as the first demonstration that asymmetric techniques (and number-theoretic problems) could be used to alleviate the problems of key distribution in cryptography. Furthermore, the Diffie–Hellman protocol is at the core of standardized key-exchange protocols that are resilient to man-in-the-middle attacks and are in wide use today. One notable example is TLS; see Section 13.7.

11.4   The Public-Key Revolution

In addition to key exchange, Diffie and Hellman also introduced in their ground-breaking work the notion of *public-key* (or *asymmetric*) cryptography. In the public-key setting (in contrast to the private-key setting we have studied until now), a party who wishes to communicate securely generates a *pair* of keys: a *public key* that is widely disseminated, and a *private key* that it keeps secret. (The fact that there are now two different keys is what makes the scheme asymmetric.) Having generated these keys, a party can use them to ensure secrecy for messages it receives using a *public-key encryption scheme*, or integrity for messages it sends using a *digital signature scheme*. (See Figure 11.3.) We provide a brief taste of these primitives here, and discuss them in extensive detail in Chapters 12 and 13, respectively.

In a public-key encryption scheme, the public key generated by some party serves as an *encryption key*; anyone who knows that public key can use it to encrypt messages and generate corresponding ciphertexts. The private key serves as a *decryption key* and is used by the party who knows it to recover the original message from any ciphertext generated using the matching public key. Furthermore—and it is amazing that something like this exists!—the secrecy of encrypted messages is preserved *even against an adversary who knows the public key* (but not the private key). In other words, the (public) encryption

	Private-Key Setting	**Public-Key Setting**
Secrecy	Private-key encryption	Public-key encryption
Integrity	Message authentication codes	Digital signature schemes

FIGURE 11.3: Cryptographic primitives in the private-key and the public-key settings.

key is of no use for an attacker trying to decrypt ciphertexts encrypted using that key. To allow for secret communication, then, a receiver can simply send her public key to a potential sender (without having to worry about an eavesdropper who observes it), or publicize her public key on her webpage or in some central database. A public-key encryption scheme thus enables private communication without relying on a private channel for key distribution.[1]

A digital signature scheme is a public-key analogue of a message authentication code (MAC). Here, the private key serves as an "authentication key" (called a *signing key*) that enables the party who knows this key to generate "authentication tags" (aka *signatures*) for messages it sends. The public key acts as a *verification key*, allowing anyone who knows it to verify signatures issued by the sender. As with MACs, a digital signature scheme can be used to prevent undetected tampering of a message; here, however, security holds even against an adversary who knows the public key. The fact that verification is public (i.e., can be done by anyone who knows the public key of the sender) has far-reaching ramifications, as it makes it possible to take a document signed by Alice and present it to a third party (say, a judge) for verification. This property is called *non-repudiation* and has extensive applications in e-commerce (e.g., for signing legal documents). Digital signatures are also used for the secure distribution of public keys as part of a *public-key infrastructure*, as discussed in more detail in Section 13.6.

In their paper, Diffie and Hellman set forth the notion of public-key cryptography but did not give any candidate constructions, A year later, Ron Rivest, Adi Shamir, and Len Adleman proposed the *RSA problem* and presented the first public-key encryption and digital signature schemes based on the hardness of that problem. Variants of their schemes are now among the most widely used cryptosystems today. In 1985, Taher El Gamal presented an encryption scheme that is essentially a slight twist on the Diffie–Hellman key-exchange protocol, variants of which are now also widely used. Thus, although Diffie and Hellman did not succeed in constructing a (non-interactive) public-key encryption scheme, they came very close.

We conclude by summarizing how public-key cryptography addresses the limitations of the private-key setting discussed in Section 11.1:

---

[1] For now, however, we do assume an *authenticated* channel that allows the sender to obtain a legitimate copy of the receiver's public key. In Section 13.6 we show how public-key cryptography can be used to solve that problem as well.

1. Public-key cryptography allows key distribution to be done over public (but authenticated) channels. This can simplify the distribution and updating of shared, secret keys.

2. Public-key cryptography reduces the need for users to store many secret keys. Consider again the setting of a large corporation where each pair of employees needs the ability to communicate securely. Using public-key cryptography, it suffices for each employee to store just a *single* private key (their own) and the public keys of all other employees. Importantly, these latter keys do *not* need to be kept secret; they could even be stored in some central (public) repository.

3. Finally, public-key cryptography is (more) suitable for open environments where parties who have never previously interacted want the ability to communicate securely. As one commonplace example, a company can post its public key on-line; a user making a purchase can obtain the company's public key, as needed, when they need to encrypt their credit-card information to send to that company.

The invention of public-key encryption was a revolution in cryptography. It is no coincidence that until the late 1970s and early 1980s, encryption and cryptography in general belonged to the domain of intelligence and military organizations, and only with the advent of public-key techniques did the use of cryptography become widespread.

Why study private-key cryptography? It should be apparent that public-key cryptography is strictly stronger than private-key cryptography; in particular, any public-key encryption scheme could be used as a private-key encryption scheme. (The communicating users can simply share both the public key and the private key. If secrecy for encrypted messages holds even when the eavesdropper knows the public key, then it clearly holds when the public key is kept secret!) So why did we bother studying private-key cryptography at all? The answer is simple: *private-key cryptography is much more efficient than public-key cryptography, and should be used in settings where it is appropriate.* That is, in cases where it *is* possible for communicating parties to share a key, private-key cryptography should be used. This includes small-scale, closed systems of users as well as applications like disk encryption. Moroever, as we will see in Sections 12.3 and 13.7, private-key encryption is used *in the public-key setting* to obtain better efficiency.

References and Additional Reading

We have only briefly discussed the problems of key distribution and key management. For more information, we recommend looking at textbooks on network security, such as the one by Kaufman et al. [113].

We have not made any attempt to capture the full history of the development of public-key cryptography. Others besides Diffie and Hellman were working on similar ideas in the 1970s. One researcher in particular doing similar and independent work was Ralph Merkle, considered by many to be a co-inventor of public-key cryptography (although he published after Diffie and Hellman). We also mention Michael Rabin, who developed constructions of signature schemes and public-key encryption schemes based on the hardness of factoring about one year after the work of Rivest, Shamir, and Adleman [171]. We highly recommend reading the original paper by Diffie and Hellman [65], and refer the reader to the book by Levy [129] for more on the political and historical aspects of the public-key revolution.

Interestingly, aspects of public-key cryptography were discovered in the intelligence community before being published in the open scientific literature. In the early 1970s, James Ellis, Clifford Cocks, and Malcolm Williamson of the British intelligence agency GCHQ invented the notion of public-key cryptography, a variant of RSA encryption, and a variant of the Diffie–Hellman key-exchange protocol. Their work was not declassified until 1997. Although the underlying mathematics of public-key cryptography may have been discovered before 1976, it is fair to say that the widespread ramifications of this new technology were not appreciated until Diffie and Hellman came along.

Exercises

11.1 Let Π be a key-exchange protocol, and $(\mathsf{Enc}, \mathsf{Dec})$ be a private-key encryption scheme. Consider the following *interactive* protocol Π' for encrypting a message: first, the sender and receiver run Π to generate a shared key k. Next, the sender computes $c \leftarrow \mathsf{Enc}_k(m)$ and sends c to the other party, who decrypts and recovers m using k.

   (a) Formulate a definition of indistinguishable encryptions in the presence of an eavesdropper (cf. Definition 3.8) appropriate for this interactive setting.

   (b) Prove that if Π is secure in the presence of an eavesdropper and $(\mathsf{Enc}, \mathsf{Dec})$ has indistinguishable encryptions in the presence of an eavesdropper, then Π' satisfies your definition.

11.2 Show that, for either of the groups considered in Sections 9.3.3 or 9.3.4, a uniform group element (expressed using the natural representation) is easily distinguishable from a uniform bit-string of the same length.

11.3 Describe a man-in-the-middle attack on the Diffie–Hellman protocol where the adversary shares a key k_A with Alice and a (different) key k_B with Bob, and Alice and Bob cannot detect that anything is wrong.

11.4 Consider the following key-exchange protocol:

(a) Alice chooses uniform $k, r \in \{0,1\}^n$, and sends $s := k \oplus r$ to Bob.

(b) Bob chooses uniform $t \in \{0,1\}^n$, and sends $u := s \oplus t$ to Alice.

(c) Alice computes $w := u \oplus r$ and sends w to Bob.

(d) Alice outputs k and Bob outputs $w \oplus t$.

Show that Alice and Bob output the same key. Analyze the security of this protocol against a passive eavesdropper.

Chapter 12

Public-Key Encryption

12.1 Public-Key Encryption – An Overview

The introduction of public-key encryption marked a revolution in cryptography. Until that time, cryptographers had relied exclusively on shared, *secret* keys to achieve private communication. Public-key techniques, in contrast, enable parties to communicate privately without having agreed on *any* secret information in advance. As we have already noted, it is quite amazing and counterintuitive that this is possible: it means that two people on opposite sides of a room who can only communicate by shouting to each other, and have no initial shared secret, can talk in such a way that no one else in the room learns anything about what they are saying!

In the setting of private-key encryption, two parties agree on a secret key that can be used, by either party, for both encryption and decryption. Public-key encryption is *asymmetric* in both these respects. One party (the *receiver*) generates a *pair* of keys (pk, sk), called the *public key* and the *private key*, respectively. The public key is used by a *sender* to encrypt a message; the receiver uses the private key to decrypt the resulting ciphertext.

Since the goal is to avoid the need for two parties to meet in advance to agree on any information, how does the sender learn pk? At an abstract level, there are two ways this can occur. Say Alice is the receiver, and Bob is the sender. In the first approach, when Alice learns that Bob wants to communicate with her, she can at that point generate (pk, sk) (assuming she hasn't done so already) and then send pk to Bob *in the clear*; Bob can then use pk to encrypt his message. We emphasize that the channel between Alice and Bob may be public, but is assumed to be authenticated, meaning that the adversary cannot modify the public key sent by Alice to Bob (and, in particular, cannot replace pk with its own public key). In Section 13.6 we discuss how public keys can be distributed over unauthenticated channels.

An alternative approach is for Alice to generate her keys (pk, sk) in advance, *independently* of any particular sender. (In fact, at the time of key generation Alice need not be aware that Bob wants to talk to her, or even that Bob exists.) Alice can widely disseminate her public key pk by, say, publishing it on her webpage, putting it on her business cards, or placing it in a public directory. Now, *anyone* who wishes to communicate privately with Alice can look up her public key and proceed as above. Multiple senders can communicate multiple times with Alice using the same public key pk for encrypting all their communication.

Note that pk is inherently public—and can thus be learned easily by an attacker—in either of the above scenarios. In the first case, an adversary eavesdropping on the communication between Alice and Bob obtains pk directly; in the second case, an adversary could just as well look up Alice's public key on its own. We see that the security of public-key encryption cannot rely on secrecy of pk, but must instead rely on secrecy of sk. It is therefore crucial that Alice not reveal her private key to anyone, including the sender Bob.

Comparison to Private-Key Encryption

Perhaps the most obvious difference between private- and public-key encryption is that the former assumes *complete* secrecy of all cryptographic keys, whereas the latter requires secrecy only for the private key sk. Although this may seem like a minor distinction, the ramifications are huge: in the private-key setting the communicating parties must somehow be able to share the secret key without allowing any third party to learn it, whereas in the public-key setting the public key can be sent from one party to the other over a public channel without compromising security. For parties shouting across a room or, more realistically, communicating over a public WiFi network or the Internet, public-key encryption is the only option.

Another important distinction is that private-key encryption schemes use the same key for both encryption and decryption, whereas public-key encryption schemes use different keys for each operation. That is, public-key encryption is inherently *asymmetric*. This asymmetry in the public-key setting means that the roles of sender and receiver are *not* interchangeable as they are in the private-key setting: a single key-pair allows communication in one direction only. (Bidirectional communication can be achieved in a number of ways; the point is that a single invocation of a public-key encryption scheme forces a distinction between one user who acts as a receiver and other users who act as senders.) In addition, a single instance of a public-key encryption scheme enables multiple senders to communicate privately with a single receiver, in contrast to the private-key case where a secret key shared between two parties enables private communication between those two parties only.

Summarizing and elaborating the preceding discussion, we see that public-key encryption has the following advantages relative to private-key encryption:

- Public-key encryption addresses (to some extent) the *key-distribution problem*, since communicating parties do not need to secretly share a key in advance of their communication. Two parties can communicate secretly even if *all* communication between them is monitored.

- When a single receiver is communicating with N senders (e.g., an on-line merchant processing credit-card orders from multiple purchasers), it is much more convenient for the receiver to store a *single* private key sk rather than to share, store, and manage N different secret keys (i.e., one for each sender).

- When using public-key encryption the number and identities of potential senders need not be known at the time of key generation. This allows enormous flexibility in "open systems."

The fact that public-key encryption schemes allow anyone to act as a sender can be a drawback when a receiver only wants to receive messages from one specific individual. In that case, an authenticated (private-key) encryption scheme would be a better choice than public-key encryption.

The main disadvantage of public-key encryption is that it is roughly *2–3 orders of magnitude slower* than private-key encryption. (It is difficult to give an exact comparison since the relative efficiency depends on the exact schemes under consideration as well as various implementation details.) It can be a challenge to implement public-key encryption in severely resource-constrained devices like smartcards or radio-frequency identification (RFID) tags. Even when a desktop computer is performing cryptographic operations, carrying out thousands of such operations per second (as in the case of a website processing credit-card transactions) may be prohibitive. Thus, when private-key encryption is an option (i.e., if two parties *can* securely share a key in advance), it should be used.

As we will see in Section 12.3, private-key encryption is used *in the public-key setting* to improve the efficiency of (public-key) encryption for long messages. A thorough understanding of private-key encryption is therefore crucial for appreciating how public-key encryption is implemented in practice.

Secure Distribution of Public Keys

In our entire discussion thus far, we have implicitly assumed that the adversary is *passive*; that is, the adversary only eavesdrops on communication between the sender and receiver but does not *actively* interfere with the communication. Equivalently, we assume the communication channel between the sender and receiver is *authenticated*, at least for the initial sharing of the public key. If the adversary has the ability to tamper with *all* communication between the honest parties, and the honest parties share no keys in advance, then privacy simply cannot be achieved. For example, if a receiver Alice sends her public key pk to Bob but the adversary replaces it with a key pk' of its own (for which it knows the matching private key sk'), then even though Bob encrypts his message using pk' the adversary will easily be able to recover the message (using sk'). A similar attack works if an adversary is able to change the value of Alice's public key that is stored in some public directory, or if the adversary can tamper with the public key as it is transmitted from the public directory to Bob. If Alice and Bob do not share any information in advance, and are not willing to rely on some mutually trusted third party, there is nothing Alice or Bob can do to prevent active attacks of this sort, or even to tell that such an attack is taking place.[1]

Importantly, our treatment of public-key encryption in this chapter *assumes that senders are able to obtain a legitimate copy of the receiver's public key.* (This will be implicit in the security definitions we provide.) That is, we assume *secure key distribution*. This assumption is made not because active attacks of the type discussed above are of no concern—in fact, they represent a serious threat that must be dealt with in any real-world system that uses public-key encryption. Rather, this assumption is made because there exist other mechanisms for preventing active attacks (see, for example, Section 13.6), and it is therefore convenient (and useful) to decouple the study of secure public-key encryption from the study of secure public-key distribution.

12.2 Definitions

We begin by defining the syntax of public-key encryption. The definition is very similar to Definition 3.7, with the exception that instead of working with just one key, we now have distinct encryption and decryption keys.

DEFINITION 12.1 *A public-key encryption scheme is a triple of probabilistic polynomial-time algorithms* (Gen, Enc, Dec) *such that:*

1. *The* key-generation algorithm Gen *takes as input the security parameter* 1^n *and outputs a pair of keys* (pk, sk). *We refer to the first of these as the* public key *and the second as the* private key. *We assume for convenience that* pk *and* sk *each has length at least* n, *and that* n *can be determined from* pk, sk.

   The public key pk *defines a message space* $\mathcal{M}_{pk}$.

2. *The* encryption algorithm Enc *takes as input a public key* pk *and message* $m \in \mathcal{M}_{pk}$, *and outputs a ciphertext* c; *we denote this by* $c \leftarrow \mathsf{Enc}_{pk}(m)$. *(Looking ahead,* Enc *will need to be probabilistic in order to achieve meaningful security.)*

3. *The deterministic* decryption algorithm Dec *takes as input a private key* sk *and a ciphertext* c, *and outputs a message* m *or a special symbol* $\perp$ *denoting failure. We write this as* $m := \mathsf{Dec}_{sk}(c)$.

It is required that, except with negligible probability over the randomness of Gen *and* Enc, *we have* $\mathsf{Dec}_{sk}(\mathsf{Enc}_{pk}(m)) = m$ *for any message* $m \in \mathcal{M}_{pk}$.

---

[1] In our "shouting-across-a-room" scenario, Alice and Bob can detect when an adversary interferes with the communication. But this is only because: (1) the adversary cannot prevent Alice's messages from reaching Bob, and (2) Alice and Bob "share" in advance information (e.g., the sound of their voices) that allows them to "authenticate" their communication.

The important difference from the private-key setting is that the key-generation algorithm Gen now outputs *two* keys instead of one. The public key *pk* is used for encryption, while the private key *sk* is used for decryption. Reiterating our earlier discussion, *pk* is assumed to be widely distributed so that anyone can encrypt messages for the party who generated this key, but *sk* must be kept private by the receiver in order for security to possibly hold.

We allow for a negligible probability of decryption error and, indeed, some of the schemes we present will have a negligible error probability (e.g., if a prime needs to be chosen, but with negligible probability a composite is obtained instead). Despite this, we will generally ignore the issue from here on.

For practical usage of public-key encryption, we will want the message space to be bit-strings of some length (and, in particular, to be independent of the public key). When we describe encryption schemes with some message space $\mathcal{M}_{pk}$, we will in such cases also specify how to encode bit-strings as elements of $\mathcal{M}$ (unless it is obvious). This encoding must be both efficiently computable and efficiently reversible, so the receiver can recover the bit-string that was encrypted.

12.2.1 Security against Chosen-Plaintext Attacks

We initiate our treatment of security by introducing the "natural" counterpart of Definition 3.8 in the public-key setting. Since extensive motivation for this definition (as well as the others we will see) has already been given in Chapter 3, the discussion here will be relatively brief and will focus primarily on the differences between the private-key and the public-key settings.

Given a public-key encryption scheme $\Pi = (\mathsf{Gen}, \mathsf{Enc}, \mathsf{Dec})$ and an adversary $\mathcal{A}$, consider the following experiment:

The eavesdropping indistinguishability experiment $\mathsf{PubK}^{\mathsf{eav}}_{\mathcal{A},\Pi}(n)$:

1. $\mathsf{Gen}(1^n)$ *is run to obtain keys* (pk, sk).

2. *Adversary* $\mathcal{A}$ *is given* pk, *and outputs a pair of equal-length messages* $m_0, m_1 \in \mathcal{M}_{pk}$.

3. *A uniform bit* $b \in \{0, 1\}$ *is chosen, and then a ciphertext* $c \leftarrow \mathsf{Enc}_{pk}(m_b)$ *is computed and given to* $\mathcal{A}$. *We call* c *the* challenge ciphertext.

4. *$\mathcal{A}$ outputs a bit* b'. *The output of the experiment is 1 if* $b' = b$, *and 0 otherwise. If* $b' = b$ *we say that* $\mathcal{A}$ succeeds.

DEFINITION 12.2 *A public-key encryption scheme* $\Pi = (\mathsf{Gen}, \mathsf{Enc}, \mathsf{Dec})$ *has* indistinguishable encryptions in the presence of an eavesdropper *if for all probabilistic polynomial-time adversaries* $\mathcal{A}$ *there is a negligible function* negl *such that*

$$\Pr[\mathsf{PubK}^{\mathsf{eav}}_{\mathcal{A},\Pi}(n) = 1] \leq \frac{1}{2} + \mathsf{negl}(n).$$

The main difference between the above definition and Definition 3.8 is that here $\mathcal{A}$ *is given the public key pk.* Furthermore, we allow $\mathcal{A}$ to choose its messages m_0 and m_1 based on this public key. This is essential when defining security of public-key encryption since, as discussed previously, it makes sense to assume that the adversary knows the public key of the recipient.

The seemingly "minor" modification of giving the adversary $\mathcal{A}$ the public key pk has a tremendous impact: it effectively gives $\mathcal{A}$ access to an encryption oracle *for free.* (The concept of an encryption oracle is explained in Section 3.4.2.) This is true because the adversary, given pk, can encrypt any message m on its own by simply computing $\mathsf{Enc}_{pk}(m)$. (As always, $\mathcal{A}$ is assumed to know the algorithm Enc.) The upshot is that Definition 12.2 is *equivalent* to CPA-security (i.e., security against chosen-plaintext attacks), where this is defined in a manner analogous to Definition 3.21 with the only difference being that the attacker is given the public key in the corresponding experiment. We thus have:

PROPOSITION 12.3   *If a public-key encryption scheme has indistinguishable encryptions in the presence of an eavesdropper, it is CPA-secure.*

This is in contrast to the private-key setting, where there exist schemes that have indistinguishable encryptions in the presence of an eavesdropper but are insecure under a chosen-plaintext attack (see Proposition 3.19). Further differences from the private-key setting that follow almost immediately as consequences of the above are discussed next.

Impossibility of perfectly secret public-key encryption. *Perfectly secret* public-key encryption could be defined analogously to Definition 2.3 by conditioning on the entire view of an eavesdropper (i.e., including the public key). Equivalently, it could be defined by extending Definition 12.2 to require that for *all* adversaries $\mathcal{A}$ (not only efficient ones), we have:

$$\Pr[\mathsf{PubK}^{\mathsf{eav}}_{\mathcal{A},\Pi}(n) = 1] = \frac{1}{2}.$$

In contrast to the private-key setting, however, perfectly secret public-key encryption is *impossible*, regardless of how long the keys are or how small the message space is. In fact, an unbounded adversary given pk and a ciphertext c computed via $c \leftarrow \mathsf{Enc}_{pk}(m)$ can determine m with probability 1 (assuming errorless encryption). A proof of this is left as Exercise 12.1.

Insecurity of deterministic public-key encryption. As noted in the context of private-key encryption, no deterministic encryption scheme can be CPA-secure. The same is true here:

THEOREM 12.4   *No deterministic public-key encryption scheme is CPA-secure.*

Because Theorem 12.4 is so important, it merits more discussion. The theorem is *not* an "artefact" of our security definition, or an indication that our definition is too strong. Deterministic public-key encryption schemes are vulnerable to *practical* attacks in *realistic* scenarios and should not be used. The reason is that a deterministic scheme not only allows the adversary to determine when the same message is sent twice (as in the private-key setting), but also allows the adversary to recover the message if the set of possible messages is small. For example, consider a professor encrypting students' grades. Here, an eavesdropper knows that each student's grade is one of $\{A, B, C, D, F\}$. If the professor uses a deterministic public-key encryption scheme, an eavesdropper can determine any student's grade by encrypting all possible grades and comparing the results to the corresponding ciphertext.

Although Theorem 12.4 seems deceptively simple, for a long time *many real-world systems were designed using deterministic public-key encryption*. When public-key encryption was introduced, it is fair to say that the importance of probabilistic encryption was not yet fully realized. The seminal work of Goldwasser and Micali, in which (something equivalent to) Definition 12.2 was proposed and Theorem 12.4 was stated, marked a turning point in the field of cryptography. The importance of pinning down one's intuition in a formal definition and looking at things the right way for the first time—even if seemingly simple in retrospect—should not be underestimated.

12.2.2 Multiple Encryptions

As in Chapter 3, it is important to understand the effect of using the same key (in this case, the same public key) for encrypting multiple messages. We could formulate security in such a setting by having an adversary output two *lists* of plaintexts, as in Definition 3.18. For the reasons discussed in Section 3.4.3, however, we choose instead to use a definition in which the attacker is given access to a "left-or-right" oracle $\mathsf{LR}_{pk,b}$ that, on input a pair of equal-length messages m_0, m_1, computes the ciphertext $c \leftarrow \mathsf{Enc}_{pk}(m_b)$ and returns c. The attacker is allowed to query this oracle as many times as it likes, and the definition therefore models security when multiple (unknown) messages are encrypted using the same public key.

Formally, consider the following experiment defined for an adversary $\mathcal{A}$ and a public-key encryption scheme $\Pi = (\mathsf{Gen}, \mathsf{Enc}, \mathsf{Dec})$:

The LR-oracle experiment $\mathsf{PubK}_{\mathcal{A},\Pi}^{\mathsf{LR\text{-}cpa}}(n)$:

1. $\mathsf{Gen}(1^n)$ *is run to obtain keys* (pk, sk).
2. *A uniform bit* $b \in \{0, 1\}$ *is chosen.*
3. *The adversary* $\mathcal{A}$ *is given input* pk *and oracle access to* $\mathsf{LR}_{pk,b}(\cdot, \cdot)$.
4. *The adversary* $\mathcal{A}$ *outputs a bit* b'.
5. *The output of the experiment is defined to be* 1 *if* $b' = b$, *and* 0 *otherwise. If* $\mathsf{PubK}_{\mathcal{A},\Pi}^{\mathsf{LR\text{-}cpa}}(n) = 1$, *we say that* $\mathcal{A}$ succeeds.

DEFINITION 12.5   *A public-key encryption scheme* $\Pi = (\text{Gen}, \text{Enc}, \text{Dec})$ *has* indistinguishable multiple encryptions *if for all probabilistic polynomial-time adversaries* $\mathcal{A}$ *there exists a negligible function* negl *such that:*

$$\Pr[\text{PubK}_{\mathcal{A},\Pi}^{\text{LR-cpa}}(n) = 1] \leq \frac{1}{2} + \text{negl}(n).$$

We will show that any CPA-secure scheme *automatically* has indistinguishable multiple encryptions; that is, in the public-key setting, security for encryption of a single message implies security for encryption of multiple messages. This means if we can prove security of some scheme with respect to Definition 12.2, which is simpler and easier to work with, we may then conclude that the scheme satisfies Definition 12.5, a seemingly stronger definition that more accurately models real-world usage of public-key encryption. A proof of the following theorem is given below.

THEOREM 12.6   *If public-key encryption scheme* Π *is CPA-secure, then it also has indistinguishable* multiple *encryptions.*

An analogous result in the private-key setting was stated, but not proved, as Theorem 3.23.

Encrypting arbitrary-length messages.   An immediate consequence of Theorem 12.6 is that a CPA-secure public-key encryption scheme for *fixed-length* messages implies a public-key encryption scheme for *arbitrary-length* messages satisfying the same notion of security. We illustrate this in the extreme case when the original scheme encrypts only 1-bit messages. Say $\Pi = (\text{Gen}, \text{Enc}, \text{Dec})$ is an encryption scheme for single-bit messages. We can construct a new scheme $\Pi' = (\text{Gen}, \text{Enc}', \text{Dec}')$ that has message space $\{0,1\}^*$ by defining Enc' as follows:

$$\text{Enc}'_{pk}(m) = \text{Enc}_{pk}(m_1), \ldots, \text{Enc}_{pk}(m_\ell), \tag{12.1}$$

where $m = m_1 \cdots m_\ell$. (The decryption algorithm Dec' is constructed in the obvious way.) We have:

CLAIM 12.7   *Let* Π *and* Π' *be as above. If* Π *is CPA-secure, then so is* Π'.

The claim follows since we can view encryption of the message m using Π' as encryption of ℓ messages $(m_1, \ldots, m_\ell)$ using scheme Π.

A note on terminology.   We have introduced three definitions of security for public-key encryption schemes—indistinguishable encryptions in the presence of an eavesdropper, CPA-security, and indistinguishable *multiple* encryptions—that are all equivalent. Following the usual convention in the cryptographic literature, we will simply use the term "CPA-security" to refer to schemes meeting these notions of security.

*Proof of Theorem 12.6

The proof of Theorem 12.6 is rather involved. We therefore provide some intuition before turning to the details. For this intuitive discussion we assume for simplicity that $\mathcal{A}$ makes only *two* calls to the LR oracle in experiment $\mathsf{PubK}_{\mathcal{A},\Pi}^{\mathsf{LR\text{-}cpa}}(n)$. (In the full proof, the number of calls can be arbitrary.)

Fix an arbitrary PPT adversary $\mathcal{A}$ and a CPA-secure public-key encryption scheme Π, and consider an experiment $\mathsf{PubK}_{\mathcal{A},\Pi}^{\mathsf{LR\text{-}cpa2}}(n)$ where $\mathcal{A}$ can make only *two* queries to the LR oracle. Denote the queries made by $\mathcal{A}$ to the oracle by $(m_{1,0}, m_{1,1})$ and $(m_{2,0}, m_{2,1})$; note that the second pair of messages may depend on the first ciphertext obtained by $\mathcal{A}$ from the oracle. In the experiment, $\mathcal{A}$ receives either a pair of ciphertexts $(\mathsf{Enc}_{pk}(m_{1,0}), \mathsf{Enc}_{pk}(m_{2,0}))$ (if $b = 0$), or a pair of ciphertexts $(\mathsf{Enc}_{pk}(m_{1,1}), \mathsf{Enc}_{pk}(m_{2,1}))$ (if $b = 1$). We write $\mathcal{A}(pk, \mathsf{Enc}_{pk}(m_{1,0}), \mathsf{Enc}_{pk}(m_{2,0}))$ to denote the output of $\mathcal{A}$ in the first case, and analogously for the second.

Let $\vec{C}_0$ denote the distribution of ciphertext pairs in the first case, and $\vec{C}_1$ the distribution of ciphertext pairs in the second case. To show that Definition 12.5 holds (for $\mathsf{PubK}^{\mathsf{LR\text{-}cpa2}}$), we need to prove that $\mathcal{A}$ cannot distinguish between being given a pair of ciphertexts distributed according to $\vec{C}_0$, or a pair of ciphertexts distributed according to $\vec{C}_1$. That is, we need to prove that there is a negligible function negl such that

$$\big| \Pr[\mathcal{A}(pk, \mathsf{Enc}_{pk}(m_{1,0}), \mathsf{Enc}_{pk}(m_{2,0})) = 1]$$
$$- \Pr[\mathcal{A}(pk, \mathsf{Enc}_{pk}(m_{1,1}), \mathsf{Enc}_{pk}(m_{2,1})) = 1]\big| \leq \mathsf{negl}(n). \quad (12.2)$$

(This is equivalent to Definition 12.5 for the same reason that Definition 3.9 is equivalent to Definition 3.8.) To prove this, we will show that

1. CPA-security of Π implies that $\mathcal{A}$ cannot distinguish between the case when it is given a pair of ciphertexts distributed according to $\vec{C}_0$, or a pair of ciphertexts $(\mathsf{Enc}_{pk}(m_{1,0}), \mathsf{Enc}_{pk}(m_{2,1}))$, which corresponds to encrypting the *first* message in $\mathcal{A}$'s first oracle query and the *second* message in $\mathcal{A}$'s second oracle query. (Although this cannot occur in $\mathsf{PubK}_{\mathcal{A},\Pi}^{\mathsf{LR\text{-}cpa2}}(n)$, we can still ask what $\mathcal{A}$'s behavior would be if given such a ciphertext pair.) Let $\vec{C}_{01}$ denote the distribution of ciphertext pairs in this latter case.

2. Similarly, CPA-security of Π implies that $\mathcal{A}$ cannot distinguish between the case when it is given a pair of ciphertexts distributed according to $\vec{C}_{01}$, or a pair of ciphertexts distributed according to $\vec{C}_1$.

The above says that $\mathcal{A}$ cannot distinguish between distributions $\vec{C}_0$ and $\vec{C}_{01}$, nor between distributions $\vec{C}_{01}$ and $\vec{C}_1$. We conclude (using simple algebra) that $\mathcal{A}$ cannot distinguish between distributions $\vec{C}_0$ and $\vec{C}_1$.

The crux of the proof, then, is showing that $\mathcal{A}$ cannot distinguish between being given a pair of ciphertexts distributed according to $\vec{C}_0$, or a pair of ciphertexts distributed according to $\vec{C}_{01}$. (The other case follows similarly.) That is, we want to show that there is a negligible function negl for which

$$\big| \Pr[\mathcal{A} \, (pk, \mathsf{Enc}_{pk}(m_{1,0}), \mathsf{Enc}_{pk}(m_{2,0})) = 1]$$
$$- \Pr[\mathcal{A} \, (pk, \mathsf{Enc}_{pk}(m_{1,0}), \mathsf{Enc}_{pk}(m_{2,1})) = 1] \big| \leq \mathsf{negl}(n). \qquad (12.3)$$

Note that the only difference between the input of the adversary $\mathcal{A}$ in each case is in the second element. Intuitively, indistinguishability follows from the single-message case since $\mathcal{A}$ can generate $\mathsf{Enc}_{pk}(m_{1,0})$ by itself. Formally, consider the following PPT adversary $\mathcal{A}'$ running in experiment $\mathsf{PubK}^{\mathsf{eav}}_{\mathcal{A}',\Pi}(n)$:

Adversary $\mathcal{A}'$:

1. On input pk, adversary $\mathcal{A}'$ runs $\mathcal{A}(pk)$ as a subroutine.

2. When $\mathcal{A}$ makes its first query $(m_{1,0}, m_{1,1})$ to the LR oracle, $\mathcal{A}'$ computes $c_1 \leftarrow \mathsf{Enc}_{pk}(m_{1,0})$ and returns c_1 to $\mathcal{A}$ as the response from the oracle.

3. When $\mathcal{A}$ makes its second query $(m_{2,0}, m_{2,1})$ to the LR oracle, $\mathcal{A}'$ outputs $(m_{2,0}, m_{2,1})$ and receives back a challenge ciphertext c_2. This is returned to $\mathcal{A}$ as the response from the LR oracle.

4. $\mathcal{A}'$ outputs the bit b' output by $\mathcal{A}$.

Looking at experiment $\mathsf{PubK}^{\mathsf{eav}}_{\mathcal{A}',\Pi}(n)$, we see that when $b = 0$ then the challenge ciphertext c_2 is computed as $\mathsf{Enc}_{pk}(m_{2,0})$. Thus,

$$\Pr[\mathcal{A}' \, (\mathsf{Enc}_{pk}(m_{2,0})) = 1] = \Pr[\mathcal{A} \, (\mathsf{Enc}_{pk}(m_{1,0}), \mathsf{Enc}_{pk}(m_{2,0})) = 1]. \quad (12.4)$$

(We suppress explicit mention of pk to save space.) In contrast, when $b = 1$ in experiment $\mathsf{PubK}^{\mathsf{eav}}_{\mathcal{A}',\Pi}(n)$, then c_2 is computed as $\mathsf{Enc}_{pk}(m_{2,1})$ and so

$$\Pr[\mathcal{A}' \, (\mathsf{Enc}_{pk}(m_{2,1})) = 1] = \Pr[\mathcal{A} \, (\mathsf{Enc}_{pk}(m_{1,0}), \mathsf{Enc}_{pk}(m_{2,1})) = 1]. \quad (12.5)$$

CPA-security of Π implies that there is a negligible function negl such that

$$|\Pr[\mathcal{A}'(\mathsf{Enc}_{pk}(m_{2,0})) = 1] - \Pr[\mathcal{A}'(\mathsf{Enc}_{pk}(m_{2,1})) = 1]| \leq \mathsf{negl}(n).$$

This, together with Equations (12.4) and (12.5), yields Equation (12.3). In almost exactly the same way, we can prove that:

$$\big| \Pr[\mathcal{A} \, (pk, \mathsf{Enc}_{pk}(m_{1,0}), \mathsf{Enc}_{pk}(m_{2,1})) = 1]$$
$$- \Pr[\mathcal{A} \, (pk, \mathsf{Enc}_{pk}(m_{1,1}), \mathsf{Enc}_{pk}(m_{2,1})) = 1] \big| \leq \mathsf{negl}(n). \qquad (12.6)$$

Equation (12.2) follows by combining Equations (12.3) and (12.6).

The main complication that arises in the general case is that the number of queries to the LR oracle is no longer fixed but may instead be an arbitrary polynomial of n. In the formal proof this is handled using a *hybrid argument*. (Hybrid arguments were used also in Chapter 8.)

PROOF (of Theorem 12.6)   Let Π be a CPA-secure public-key encryption scheme and $\mathcal{A}$ an arbitrary PPT adversary in experiment $\mathsf{PubK}^{\mathsf{LR\text{-}cpa}}_{\mathcal{A},\Pi}(n)$. Let $t = t(n)$ be a polynomial upper bound on the number of queries made by $\mathcal{A}$ to the LR oracle, and assume without loss of generality that $\mathcal{A}$ always queries the oracle *exactly* this many times. For a given public key pk and $0 \le i \le t$, let LR^i_{pk} denote the oracle that on input (m_0, m_1) returns $\mathsf{Enc}_{pk}(m_0)$ for the first i queries it receives, and returns $\mathsf{Enc}_{pk}(m_1)$ for the next $t - i$ queries it receives. (That is, for the first i queries the first message in the input pair is encrypted, and for the remaining queries the second message in the input pair is encrypted.) We stress that each encryption is computed using uniform, independent randomness. Using this notation, we have

$$\Pr\left[\mathsf{PubK}^{\mathsf{LR\text{-}cpa}}_{\mathcal{A},\Pi}(n) = 1\right] = \frac{1}{2} \cdot \Pr[\mathcal{A}^{\mathsf{LR}^t_{pk}}(pk) = 0] + \frac{1}{2} \cdot \Pr[\mathcal{A}^{\mathsf{LR}^0_{pk}}(pk) = 1]$$

because oracle LR^t_{pk} is equivalent to $\mathsf{LR}_{pk,0}$, and oracle LR^0_{pk} is equivalent to $\mathsf{LR}_{pk,1}$. To prove that Π satisfies Definition 12.5, we will show that for any PPT $\mathcal{A}$ there is a negligible function negl' such that

$$\left| \Pr[\mathcal{A}^{\mathsf{LR}^t_{pk}}(pk) = 1] - \Pr[\mathcal{A}^{\mathsf{LR}^0_{pk}}(pk) = 1] \right| \le \mathsf{negl}'(n). \tag{12.7}$$

(As before, this is equivalent to Definition 12.5 for the same reason that Definition 3.9 is equivalent to Definition 3.8.)

Consider the following PPT adversary $\mathcal{A}'$ that eavesdrops on the encryption of a *single* message:

Adversary $\mathcal{A}'$:

1. $\mathcal{A}'$, given pk, chooses a uniform index $i \leftarrow \{1,\ldots,t\}$.

2. $\mathcal{A}'$ runs $\mathcal{A}(pk)$, answering its jth oracle query $(m_{j,0}, m_{j,1})$ as follows:

    (a) For $j < i$, adversary $\mathcal{A}'$ computes $c_j \leftarrow \mathsf{Enc}_{pk}(m_{j,0})$ and returns c_j to $\mathcal{A}$ as the response from its oracle.

    (b) For $j = i$, adversary $\mathcal{A}'$ outputs $(m_{j,0}, m_{j,1})$ and receives back a challenge ciphertext c_j. This is returned to $\mathcal{A}$ as the response from its oracle.

    (c) For $j > i$, adversary $\mathcal{A}'$ computes $c_j \leftarrow \mathsf{Enc}_{pk}(m_{j,1})$ and returns c_j to $\mathcal{A}$ as the response from its oracle.

3. $\mathcal{A}'$ outputs the bit b' that is output by $\mathcal{A}$.

Consider experiment $\mathsf{PubK}^{\mathsf{eav}}_{\mathcal{A}',\Pi}(n)$. Fixing some choice of $i = i^*$, note that if c_{i^*} is an encryption of $m_{i^*,0}$ then the interaction of $\mathcal{A}$ with its oracle is identical to an interaction with oracle $\mathsf{LR}_{pk}^{i^*}$. Thus,

$$\Pr[\mathcal{A}' \text{ outputs } 1 \mid b = 0] = \sum_{i^*=1}^{t} \Pr[i = i^*] \cdot \Pr[\mathcal{A}' \text{ outputs } 1 \mid b = 0 \wedge i = i^*]$$

$$= \sum_{i^*=1}^{t} \frac{1}{t} \cdot \Pr\left[\mathcal{A}^{\mathsf{LR}_{pk}^{i^*}}(pk) = 1\right].$$

On the other hand, if c_{i^*} is an encryption of $m_{i^*,1}$ then the interaction of $\mathcal{A}$ with its oracle is identical to an interaction with oracle $\mathsf{LR}_{pk}^{i^*-1}$, and so

$$\Pr[\mathcal{A}' \text{ outputs } 1 \mid b = 1] = \sum_{i^*=1}^{t} \Pr[i = i^*] \cdot \Pr[\mathcal{A}' \text{ outputs } 1 \mid b = 1 \wedge i = i^*]$$

$$= \sum_{i^*=1}^{t} \frac{1}{t} \cdot \Pr\left[\mathcal{A}^{\mathsf{LR}_{pk}^{i^*-1}}(pk) = 1\right]$$

$$= \sum_{i^*=0}^{t-1} \frac{1}{t} \cdot \Pr\left[\mathcal{A}^{\mathsf{LR}_{pk}^{i^*}}(pk) = 1\right].$$

Since $\mathcal{A}'$ runs in polynomial time, the assumption that Π is CPA-secure means that there exists a negligible function negl such that

$$\left|\Pr[\mathcal{A}' \text{ outputs } 1 \mid b = 0] - \Pr[\mathcal{A}' \text{ outputs } 1 \mid b = 1]\right| \leq \mathsf{negl}(n).$$

But this means that

$$\mathsf{negl}(n) \geq \left| \sum_{i^*=1}^{t} \frac{1}{t} \cdot \Pr\left[\mathcal{A}^{\mathsf{LR}_{pk}^{i^*}}(pk) = 1\right] - \sum_{i^*=0}^{t-1} \frac{1}{t} \cdot \Pr\left[\mathcal{A}^{\mathsf{LR}_{pk}^{i^*}}(pk) = 1\right] \right|$$

$$= \frac{1}{t} \cdot \left|\Pr\left[\mathcal{A}^{\mathsf{LR}_{pk}^{t}}(pk) = 1\right] - \Pr\left[\mathcal{A}^{\mathsf{LR}_{pk}^{0}}(pk) = 1\right]\right|,$$

since all but one of the terms in each summation cancel. We conclude that

$$\left|\Pr\left[\mathcal{A}^{\mathsf{LR}_{pk}^{t}}(pk) = 1\right] - \Pr\left[\mathcal{A}^{\mathsf{LR}_{pk}^{0}}(pk) = 1\right]\right| \leq t(n) \cdot \mathsf{negl}(n).$$

Because t is polynomial, the function $t \cdot \mathsf{negl}(n)$ is negligible. Since $\mathcal{A}$ was an arbitrary PPT adversary, this shows that Equation (12.7) holds and so completes the proof that Π has indistinguishable multiple encryptions. ∎

12.2.3 Security against Chosen-Ciphertext Attacks

Chosen-ciphertext attacks, in which an adversary is able to obtain the decryption of arbitrary ciphertexts of its choice (with one technical restriction described below), are a concern in the public-key setting just as they are in the private-key setting. In fact, they are arguably *more* of a concern in the public-key setting since in that context a receiver expects to receive ciphertexts from multiple senders who are possibly unknown in advance, whereas a receiver in the private-key setting intends to communicate only with a single, known sender using any particular secret key.

Assume an eavesdropper $\mathcal{A}$ observes a ciphertext c sent by a sender $\mathcal{S}$ to a receiver $\mathcal{R}$. Broadly speaking, in the public-key setting there are two ways in which $\mathcal{A}$ might carry out a chosen-ciphertext attack:

- $\mathcal{A}$ might send a modified ciphertext c' to $\mathcal{R}$ *on behalf of* $\mathcal{S}$. (For example, in the context of encrypted e-mail, $\mathcal{A}$ might construct an encrypted e-mail c' and forge the "`From`" field so that it appears the e-mail originated from $\mathcal{S}$.) In this case, although it is unlikely that $\mathcal{A}$ would be able to obtain the entire decryption m' of c', it might be possible for $\mathcal{A}$ to infer some information about m' based on the subsequent behavior of $\mathcal{R}$. Based on this information, $\mathcal{A}$ might be able to learn something about the original message m.

- $\mathcal{A}$ might send a modified ciphertext c' to $\mathcal{R}$ *in its own name*. In this case, $\mathcal{A}$ might obtain the entire decryption m' of c' if $\mathcal{R}$ responds directly to $\mathcal{A}$. Even if $\mathcal{A}$ learns nothing about m', this modified message may have a known *relation* to the original message m that can be exploited by $\mathcal{A}$; see the third scenario below for an example.

The second class of attacks is specific to the setting of public-key encryption, and has no analogue in the private-key case.

It is not hard to identify a number of realistic scenarios illustrating the above types of attacks:

Scenario 1. Say a user $\mathcal{S}$ logs in to her bank account by sending to her bank an encryption of her password pw concatenated with a timestamp. Assume further that there are two types of error messages the bank sends: it returns "password incorrect" if the encrypted password does not match the stored password of $\mathcal{S}$, and "timestamp incorrect" if the password is correct but the timestamp is not.

If an adversary obtains a ciphertext c sent by $\mathcal{S}$ to the bank, the adversary can now mount a chosen-ciphertext attack by sending ciphertexts c' to the bank on behalf of $\mathcal{S}$ and observing the error messages that are sent in response. (This is similar to the padding-oracle attack that we saw in Section 5.1.1.) In some cases, this information may be enough to allow the adversary to determine the user's entire password.

Scenario 2. Say $\mathcal{S}$ sends an encrypted e-mail c to $\mathcal{R}$, and this e-mail is observed by $\mathcal{A}$. If $\mathcal{A}$ sends, in its own name, an encrypted e-mail c' to $\mathcal{R}$, then $\mathcal{R}$ might reply to this e-mail *and quote the decrypted text m' corresponding* *to c'.* In this case, $\mathcal{R}$ is essentially acting as a decryption oracle for $\mathcal{A}$ and might potentially decrypt any ciphertext that $\mathcal{A}$ sends it.

Scenario 3. An issue that is closely related to that of chosen-ciphertext security is potential *malleability* of ciphertexts. We do not provice a formal definition but instead only give the intuitive idea. An encryption scheme is *malleable* if it has the following property: given an encryption c of some unknown message m, it is possible to come up with a ciphertext c' that is an encryption of a message m' *that is related in some known way to m.* For example, perhaps given an encryption of m, it is possible to construct an encryption of $m + 1$. (Later we will see natural examples of CPA-secure schemes that are malleable; see also Section 15.2.3.)

Now imagine that $\mathcal{R}$ is running an auction, where two parties $\mathcal{S}$ and $\mathcal{A}$ submit their bids by encrypting them using the public key of $\mathcal{R}$. If a malleable encryption scheme is used, it may be possible for an adversary $\mathcal{A}$ to always place the higher bid (without bidding the maximum) by carrying out the following attack: wait until $\mathcal{S}$ sends a ciphertext c corresponding to its bid m (that is unknown to $\mathcal{A}$); then send a ciphertext c' corresponding to the bid $m' = m + 1$. Note that m (and m', for that matter) remain unknown to $\mathcal{A}$ until $\mathcal{R}$ announces the results, and so the possibility of such an attack does not contradict the fact that the encryption scheme is CPA-secure. Schemes secure against chosen-ciphertext attacks, on the other hand, can be shown to be non-malleable and so are not vulnerable to such attacks.

The definition. Security against chosen-ciphertext attacks is defined by suitable modification of the analogous definition from the private-key setting (Definition 5.1). Given a public-key encryption scheme Π and an adversary $\mathcal{A}$, consider the following experiment:

The CCA indistinguishability experiment $\mathsf{PubK}^{\mathsf{cca}}_{\mathcal{A},\Pi}(n)$:

1. $\mathsf{Gen}(1^n)$ *is run to obtain keys (pk, sk).*

2. *The adversary $\mathcal{A}$ is given pk and access to a decryption oracle $\mathsf{Dec}_{sk}(\cdot)$. It outputs a pair of messages $m_0, m_1 \in \mathcal{M}_{pk}$ of the same length.*

3. *A uniform bit $b \in \{0,1\}$ is chosen, and then a ciphertext $c \leftarrow \mathsf{Enc}_{pk}(m_b)$ is computed and given to $\mathcal{A}$.*

4. *$\mathcal{A}$ continues to interact with the decryption oracle, but may not request a decryption of c itself. Finally, $\mathcal{A}$ outputs a bit b'.*

5. *The output of the experiment is defined to be 1 if $b' = b$, and 0 otherwise.*

DEFINITION 12.8 *A public-key encryption scheme $\Pi = (\mathsf{Gen}, \mathsf{Enc}, \mathsf{Dec})$ has* indistinguishable encryptions under a chosen-ciphertext attack *(or is CCA-secure) if for all probabilistic polynomial-time adversaries $\mathcal{A}$ there exists a* *negligible function* negl *such that*

$$\Pr[\mathsf{PubK}^{\mathsf{cca}}_{\mathcal{A},\Pi}(n) = 1] \leq \frac{1}{2} + \mathsf{negl}(n).$$

The natural analogue of Theorem 12.6 holds for CCA-security as well. That is, if a scheme has indistinguishable encryptions under a chosen-ciphertext attack then it has indistinguishable *multiple* encryptions under a chosen-ciphertext attack (defined appropriately). Interestingly, however, the analogue of Claim 12.7 does *not* hold for CCA-security.

As in Definition 5.1, we must prevent the attacker from submitting the challenge ciphertext c to the decryption oracle in order for the definition to be achievable. But this restriction does not make the definition meaningless and, in particular, for each of the three motivating scenarios given earlier one can argue that setting $c' = c$ is of no benefit to the attacker:

- In the first scenario involving password-based login, the attacker learns nothing about $\mathcal{S}$'s password by replaying c since in this case it already knows that the error message "timestamp incorrect" will be returned.

- In the second scenario involving encrypted email, sending $c' = c$ to the receiver would likely make the receiver suspicious and so it would refuse to respond at all.

- In the final scenario involving an auction, $\mathcal{R}$ could easily detect cheating if the adversary's encrypted bid is identical to the other party's encrypted bid. Even if $\mathcal{R}$ ignores such cheating, all the attacker achieves by replaying c is to submit the *same* bid as the honest party.

An analogue of authenticated encryption? In the setting of private-key encryption, we introduced the notion of *authenticated encryption* (cf. Section 5.2) and noted that it was even stronger than CCA-security. This notion cannot be translated directly to the context of public-key encryption, where a single public key is used by many senders to communicate to one receiver (in contrast to the private-key case where a given key is used by only two parties to communicate). Nevertheless, an analogue of authenticated encryption can be considered in the public-key setting; see Section 13.8.

12.3 Hybrid Encryption and the KEM/DEM Paradigm

Claim 12.7 shows that any CPA-secure public-key encryption scheme for ℓ'-bit messages can be used to obtain a CPA-secure public-key encryption scheme for messages of arbitrary length. Encrypting an ℓ-bit message using this approach requires $\gamma \overset{\text{def}}{=} \lceil \ell/\ell' \rceil$ invocations of the original encryption scheme, meaning that both the computation and the ciphertext length are increased by a multiplicative factor of γ relative to the underlying scheme.

It is possible to do better by using private-key encryption *in tandem with* public-key encryption. This improves efficiency because private-key encryption is significantly faster than public-key encryption, and improves bandwidth because private-key schemes have lower ciphertext expansion. The resulting combination is called *hybrid encryption* and is used extensively in practice. The basic idea is to use public-key encryption to obtain a shared key k, and then encrypt the message m using a private-key encryption scheme and key k. The receiver uses its long-term (asymmetric) private key to derive k, and then uses private-key decryption (with key k) to recover the original message. We stress that although private-key encryption is used as a component, this is a full-fledged public-key encryption scheme by virtue of the fact that the sender and receiver do not share any secret key *in advance*.

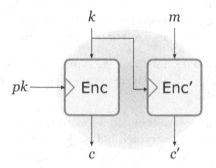

FIGURE 12.1:   Hybrid encryption. Enc denotes a public-key encryption scheme, while Enc′ is a private-key encryption scheme.

In a direct implementation of this idea (see Figure 12.1), the sender would share k by (1) choosing a uniform value k and then (2) encrypting k using a public-key encryption scheme. A more direct approach is to use a public-key primitive called a *key-encapsulation mechanism* (KEM) to accomplish both of these "in one shot." This is advantageous both from a conceptual point of view and in terms of efficiency, as we will see later.

A KEM has three algorithms similar in spirit to those of a public-key encryption scheme. As before, the key-generation algorithm Gen is used to generate a pair of public and private keys. In place of encryption, we now have an *encapsulation* algorithm Encaps that takes only a public key as input (and no message), and outputs a ciphertext c along with a key k. A corresponding *decapsulation* algorithm Decaps is run by the receiver to recover k from the ciphertext c using the private key. Formally:

DEFINITION 12.9   *A* key-encapsulation mechanism (KEM) *is a tuple of* *probabilistic polynomial-time algorithms* (Gen, Encaps, Decaps) *such that:*

1. *The* key-generation algorithm Gen *takes as input the security parameter* 1^n *and outputs a public-/private-key pair* (pk, sk). *We assume pk and sk each has length at least n, and that n can be determined from pk.*

2. *The* encapsulation algorithm Encaps *takes as input a public key pk (which implicitly defines n). It outputs a ciphertext c and a key* $k \in \{0,1\}^{\ell(n)}$ *where ℓ is the* key *length. We write this as* $(c, k) \leftarrow \text{Encaps}_{pk}(1^n)$.

3. *The deterministic* decapsulation algorithm Decaps *takes as input a private key sk and a ciphertext c, and outputs a key k or a special symbol $\perp$ denoting failure. We write this as* $k := \text{Decaps}_{sk}(c)$.

It is required that with all but negligible probability over the randomness of Gen *and* Encaps, *if* $\text{Encaps}_{pk}(1^n)$ *outputs* (c, k) *then* $\text{Decaps}_{sk}(c)$ *outputs k.*

In the definition we assume for simplicity that Encaps always outputs (a ciphertext c and) a key of some fixed length $\ell(n)$. One could also consider a more general definition in which Encaps takes 1^ℓ as an additional input and outputs a key of length ℓ.

Any public-key encryption scheme trivially gives a KEM by choosing a random key k and encrypting it. As we will see, however, dedicated constructions of KEMs can be more efficient.

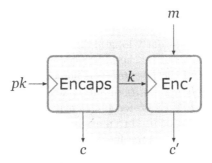

FIGURE 12.2: Hybrid encryption using the KEM/DEM approach.

Using a KEM (with key length n), we can implement hybrid encryption as in Figure 12.2. The sender runs $\text{Encaps}_{pk}(1^n)$ to obtain c along with a key k; it then uses a private-key encryption scheme to encrypt its message m, using k as the key. In this context, the private-key encryption scheme is called a *data-encapsulation mechanism* (DEM) for obvious reasons. The ciphertext sent to the receiver includes both c and the ciphertext c' from the private-key scheme. Construction 12.10 gives a formal specification.

What is the efficiency of the resulting hybrid encryption scheme Π^{hy}? For some fixed value of n, let α denote the cost of encapsulating an n-bit key

---

CONSTRUCTION 12.10

Let $\Pi = (\mathsf{Gen}, \mathsf{Encaps}, \mathsf{Decaps})$ be a KEM with key length n, and let $\Pi' = (\mathsf{Gen}', \mathsf{Enc}', \mathsf{Dec}')$ be a private-key encryption scheme. Construct a public-key encryption scheme $\Pi^{\mathsf{hy}} = (\mathsf{Gen}^{\mathsf{hy}}, \mathsf{Enc}^{\mathsf{hy}}, \mathsf{Dec}^{\mathsf{hy}})$ as follows:

- $\mathsf{Gen}^{\mathsf{hy}}$: on input 1^n run $\mathsf{Gen}(1^n)$ and use the public and private keys (pk, sk) that are output.

- $\mathsf{Enc}^{\mathsf{hy}}$: on input a public key pk and a message $m \in \{0,1\}^*$ do:

    1. Compute $(c, k) \leftarrow \mathsf{Encaps}_{pk}(1^n)$.
    2. Compute $c' \leftarrow \mathsf{Enc}'_k(m)$.
    3. Output the ciphertext $\langle c, c' \rangle$.

- $\mathsf{Dec}^{\mathsf{hy}}$: on input a private key sk and a ciphertext $\langle c, c' \rangle$ do:

    1. Compute $k := \mathsf{Decaps}_{sk}(c)$.
    2. Output the message $m := \mathsf{Dec}'_k(c')$.

Hybrid encryption using the KEM/DEM paradigm.

using Encaps, and let β denote the cost (per bit of plaintext) of encryption using Enc'. Assume $|m| > n$, which is the interesting case. Then the cost, per bit of plaintext, of encrypting a message m using Π^{hy} is

$$\frac{\alpha + \beta \cdot |m|}{|m|} = \frac{\alpha}{|m|} + \beta, \tag{12.8}$$

which approaches β for sufficiently long m. In the limit of very long messages, then, the cost per bit incurred by the *public-key* encryption scheme Π^{hy} is the same as the cost per bit of the *private-key* scheme Π'. Hybrid encryption thus allows us to achieve the *functionality* of public-key encryption at the *efficiency* of private-key encryption, at least for sufficiently long messages.

A similar calculation can be used to measure the effect of hybrid encryption on the ciphertext length. For some fixed value of n, let L denote the length of the ciphertext output by Encaps, and say the private-key encryption of a message m using Enc' results in a ciphertext of length $n + |m|$ (this can be achieved using one of the modes of encryption discussed in Section 3.6; actually, even ciphertext length $|m|$ is possible since, as we will see, Π' need not be CPA-secure). Then the total length of a ciphertext in scheme Π^{hy} is

$$L + n + |m|. \tag{12.9}$$

In contrast, when using block-by-block encryption as in Equation (12.1), and assuming that public-key encryption of an n-bit message using Enc results in a ciphertext of length L, encryption of a message m would result in a ciphertext of length $L \cdot \lceil |m|/n \rceil$. The ciphertext length given by Equation (12.9) is a significant improvement for sufficiently long m.

We can use some rough estimates to get a sense for what the above results mean in practice. (We stress that these numbers are only meant to give the reader a feel for the improvement; realistic values would depend on a variety of factors.) A typical value for the length of the key k might be $n = 128$. Furthermore, a "base" public-key encryption scheme might yield 256-bit ciphertexts when encrypting 128-bit messages; assume a KEM has ciphertexts of the same length when encapsulating a 128-bit key. Letting α, as before, denote the computational cost of public-key encryption/encapsulation of a 128-bit key, we see that block-by-block encryption as in Equation (12.1) would encrypt a 1 MB ($\approx 2^{20}$-bit) message with computational cost $\alpha \cdot \lceil 2^{20}/128 \rceil \approx 8200 \cdot \alpha$ and the ciphertext would be 2 MB long. Compare this to the efficiency of hybrid encryption. Letting β, as before, denote the per-bit computational cost of private-key encryption, a reasonable approximation is $\beta \approx \alpha/2^{11}$. Using Equation (12.8), we see that the overall computational cost for hybrid encryption for a 1 Mb message is

$$\alpha + 2^{20} \cdot \frac{\alpha}{2^{11}} \approx 512 \cdot \alpha,$$

and the ciphertext would be only slightly longer than 1 MB. Thus, hybrid encryption improves the computational efficiency in this case by a factor of 16, and the ciphertext length by a factor of 2.

It remains to analyze the security of Π^{hy}. This, of course, depends on the security of its underlying components Π and Π'. In the following sections we define notions of CPA-security and CCA-security for KEMs, and show:

- If Π is a CPA-secure KEM and the private-key scheme Π' is EAV-secure, then Π^{hy} is a CPA-secure public-key encryption scheme. Notice that it suffices for Π' to satisfy a weaker definition of security—which, recall, does *not* imply CPA-security in the private-key setting—in order for the hybrid scheme Π^{hy} to be CPA-secure. Intuitively, the reason is that a *fresh*, uniform key k is chosen each time a new message is encrypted. Since each key k is used only once, security of Π' for a single encryption suffices for CPA-security of the hybrid scheme Π^{hy}. This means that basic private-key encryption using a pseudorandom generator (or stream cipher), as in Construction 3.17, suffices.

- If Π is a CCA-secure KEM and Π' is a CCA-secure private-key encryption scheme, then Π^{hy} is a CCA-secure public-key encryption scheme.

12.3.1 CPA-Security

For simplicity, we assume in this and the next section a KEM with key length n. We define a notion of CPA-security for KEMs by analogy with Definition 12.2. As there, the adversary here eavesdrops on a single ciphertext c. Definition 12.2 requires that the attacker is unable to distinguish whether c is an encryption of some message m_0 or some other message m_1. With a KEM there is no message, and we require instead that the encapsulated key k is indistinguishable from a uniform key that is independent of the ciphertext c.

Let $\Pi = (\mathsf{Gen}, \mathsf{Encaps}, \mathsf{Decaps})$ be a KEM and $\mathcal{A}$ an arbitrary adversary.

The CPA indistinguishability experiment $\mathsf{KEM}^{\mathsf{cpa}}_{\mathcal{A},\Pi}(n)$:

1. $\mathsf{Gen}(1^n)$ is run to obtain keys (pk, sk). Then $\mathsf{Encaps}_{pk}(1^n)$ is run to generate (c, k) with $k \in \{0, 1\}^n$.

2. A uniform bit $b \in \{0, 1\}$ is chosen. If $b = 0$ set $\hat{k} := k$. If $b = 1$ then choose a uniform $\hat{k} \in \{0, 1\}^n$.

3. Give $(pk, c, \hat{k})$ to $\mathcal{A}$, who outputs a bit b'. The output of the experiment is defined to be 1 if $b' = b$, and 0 otherwise.

In the experiment, $\mathcal{A}$ is given the ciphertext c and either the actual key k corresponding to c, or an independent, uniform key. The KEM is CPA-secure if no efficient adversary can distinguish between these possibilities.

DEFINITION 12.11   *A key-encapsulation mechanism Π is* CPA-secure *if for all probabilistic polynomial-time adversaries $\mathcal{A}$ there exists a negligible function* negl *such that*

$$\Pr[\mathsf{KEM}^{\mathsf{cpa}}_{\mathcal{A},\Pi}(n) = 1] \leq \frac{1}{2} + \mathsf{negl}(n).$$

In the remainder of this section we prove the following theorem:

THEOREM 12.12   *If Π is a CPA-secure KEM and Π' is an EAV-secure private-key encryption scheme, then Π^{hy} as in Construction 12.10 is a CPA-secure public-key encryption scheme.*

Before proving the theorem formally, we give some intuition. Let the notation "$X \overset{c}{\equiv} Y$" mean that no polynomial-time adversary can distinguish between two distributions X and Y. (This concept is treated more formally in Section 8.8, although we do not rely on that section here.) For example, let $\mathsf{Encaps}^{(1)}_{pk}(1^n)$ (resp., $\mathsf{Encaps}^{(2)}_{pk}(1^n)$) denote the ciphertext (resp., key) output by Encaps. The fact that Π is CPA-secure means that

$$\left(pk, \mathsf{Encaps}_{pk}(1^n)\right) \overset{c}{\equiv} \left(pk, \mathsf{Encaps}^{(1)}_{pk}(1^n), k'\right),$$

where pk is generated by $\mathsf{Gen}(1^n)$ and k' is chosen independently and uniformly from $\{0, 1\}^n$. Similarly, the fact that Π' is EAV-secure means that for any m_0, m_1 output by $\mathcal{A}$ we have $\mathsf{Enc}'_k(m_0) \overset{c}{\equiv} \mathsf{Enc}'_k(m_1)$ if k is chosen uniformly at random.

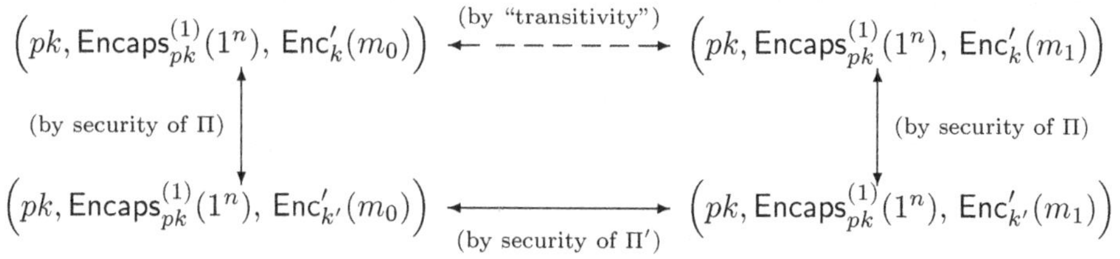

$$\left(pk, \mathsf{Encaps}_{pk}^{(1)}(1^n), \mathsf{Enc}_k'(m_0)\right) \xleftarrow{\quad\text{(by ``transitivity'')}\quad} \left(pk, \mathsf{Encaps}_{pk}^{(1)}(1^n), \mathsf{Enc}_k'(m_1)\right)$$

(by security of Π) ↑   ↑ (by security of Π)

$$\left(pk, \mathsf{Encaps}_{pk}^{(1)}(1^n), \mathsf{Enc}_{k'}'(m_0)\right) \xleftrightarrow{\qquad\qquad\qquad} \left(pk, \mathsf{Encaps}_{pk}^{(1)}(1^n), \mathsf{Enc}_{k'}'(m_1)\right)$$

(by security of Π′)

FIGURE 12.3: High-level structure of the proof of Theorem 12.12 (the arrows represent indistinguishability).

In order to prove CPA-security of Π^{hy} we need to show that

$$\left(pk, \mathsf{Encaps}_{pk}^{(1)}(1^n), \mathsf{Enc}_k'(m_0)\right) \stackrel{c}{\equiv} \left(pk, \mathsf{Encaps}_{pk}^{(1)}(1^n), \mathsf{Enc}_k'(m_1)\right) \quad (12.10)$$

for m_0, m_1 output by a PPT adversary $\mathcal{A}$, where $k = \mathsf{Encaps}_{pk}^{(2)}(1^n)$. (Equation (12.10) shows that Π^{hy} has indistinguishable encryptions in the presence of an eavesdropper; by Proposition 12.3 this implies that Π^{hy} is CPA-secure.)

The proof proceeds in three steps. (See Figure 12.3.) First we prove that

$$\left(pk, \mathsf{Encaps}_{pk}^{(1)}(1^n), \mathsf{Enc}_k'(m_0)\right) \stackrel{c}{\equiv} \left(pk, \mathsf{Encaps}_{pk}^{(1)}(1^n), \mathsf{Enc}_{k'}'(m_0)\right), \quad (12.11)$$

where on the left k is output by $\mathsf{Encaps}_{pk}^{(2)}(1^n)$, and on the right k' is an independent, uniform key. This follows via a fairly straightforward reduction, since CPA-security of Π means exactly that $\mathsf{Encaps}_{pk}^{(2)}(1^n)$ cannot be distinguished from a uniform key k' even given pk and $\mathsf{Encaps}_{pk}^{(1)}(1^n)$.

Next, we prove that

$$\left(pk, \mathsf{Encaps}_{pk}^{(1)}(1^n), \mathsf{Enc}_{k'}'(m_0)\right) \stackrel{c}{\equiv} \left(pk, \mathsf{Encaps}_{pk}^{(1)}(1^n), \mathsf{Enc}_{k'}'(m_1)\right). \quad (12.12)$$

Here the difference is between encrypting m_0 or m_1 using Π' *and a uniform, independent key* k'. Equation (12.12) follows since Π' is EAV-secure.

Exactly as in the case of Equation (12.11), we can also show that

$$\left(pk, \mathsf{Encaps}_{pk}^{(1)}(1^n), \mathsf{Enc}_k'(m_1)\right) \stackrel{c}{\equiv} \left(pk, \mathsf{Encaps}_{pk}^{(1)}(1^n), \mathsf{Enc}_{k'}'(m_1)\right), \quad (12.13)$$

(where, again, on the left k is output by $\mathsf{Encaps}_{pk}^{(2)}(1^n)$) using CPA-security of Π. Equations (12.11)–(12.13) imply, by transitivity, the desired result of Equation (12.10). (Transitivity will be implicit in the proof we give below.)

We now present the full proof.

PROOF (of Theorem 12.12) We prove that Π^{hy} has indistinguishable encryptions in the presence of an eavesdropper; by Proposition 12.3, this implies it is CPA-secure.

Fix an arbitrary PPT adversary $\mathcal{A}^{\mathsf{hy}}$, and consider experiment $\mathsf{PubK}^{\mathsf{eav}}_{\mathcal{A}^{\mathsf{hy}}, \Pi^{\mathsf{hy}}}(n)$. Our goal is to prove that there is a negligible function negl such that

$$\Pr[\mathsf{PubK}^{\mathsf{eav}}_{\mathcal{A}^{\mathsf{hy}}, \Pi^{\mathsf{hy}}}(n) = 1] \leq \frac{1}{2} + \mathsf{negl}(n).$$

By definition of the experiment, we have

$$\Pr[\mathsf{PubK}^{\mathsf{eav}}_{\mathcal{A}^{\mathsf{hy}}, \Pi^{\mathsf{hy}}}(n) = 1] \qquad (12.14)$$

$$= \frac{1}{2} \cdot \Pr[\mathcal{A}^{\mathsf{hy}}(pk, \mathsf{Encaps}^{(1)}_{pk}(1^n), \mathsf{Enc}'_k(m_0)) = 0]$$

$$+ \frac{1}{2} \cdot \Pr[\mathcal{A}^{\mathsf{hy}}(pk, \mathsf{Encaps}^{(1)}_{pk}(1^n), \mathsf{Enc}'_k(m_1)) = 1],$$

where in each case k equals $\mathsf{Encaps}^{(2)}_{pk}(1^n)$. Consider the following PPT adversary $\mathcal{A}_1$ attacking Π.

Adversary $\mathcal{A}_1$:

1. $\mathcal{A}_1$ is given $(pk, c, \hat{k})$.
2. $\mathcal{A}_1$ runs $\mathcal{A}^{\mathsf{hy}}(pk)$ to obtain two messages m_0, m_1. Then $\mathcal{A}_1$ computes $c' \leftarrow \mathsf{Enc}'_{\hat{k}}(m_0)$, gives ciphertext $\langle c, c' \rangle$ to $\mathcal{A}^{\mathsf{hy}}$, and outputs the bit b' that $\mathcal{A}^{\mathsf{hy}}$ outputs.

Consider the behavior of $\mathcal{A}_1$ when attacking Π in experiment $\mathsf{KEM}^{\mathsf{cpa}}_{\mathcal{A}_1, \Pi}(n)$. When $b = 0$ in that experiment, then $\mathcal{A}_1$ is given $(pk, c, \hat{k})$ where c and $\hat{k}$ were both output by $\mathsf{Encaps}_{pk}(1^n)$. This means that $\mathcal{A}^{\mathsf{hy}}$ is given a ciphertext of the form $\langle c, c' \rangle = \langle c, \mathsf{Enc}'_k(m_0) \rangle$, where k is the key encapsulated by c. So,

$$\Pr[\mathcal{A}_1 \text{ outputs } 0 \mid b = 0] = \Pr[\mathcal{A}^{\mathsf{hy}}(pk, \mathsf{Encaps}^{(1)}_{pk}(1^n), \mathsf{Enc}'_k(m_0)) = 0].$$

On the other hand, when $b = 1$ in experiment $\mathsf{KEM}^{\mathsf{cpa}}_{\mathcal{A}_1, \Pi}(n)$ then $\mathcal{A}_1$ is given $(pk, c, \hat{k})$ with $\hat{k}$ uniform and independent of c. If we denote such a key by k', this means $\mathcal{A}^{\mathsf{hy}}$ is given a ciphertext of the form $\langle c, \mathsf{Enc}'_{k'}(m_0) \rangle$, and

$$\Pr[\mathcal{A}_1 \text{ outputs } 1 \mid b = 1] = \Pr[\mathcal{A}^{\mathsf{hy}}(pk, \mathsf{Encaps}^{(1)}_{pk}(1^n), \mathsf{Enc}'_{k'}(m_0)) = 1].$$

Since Π is a CPA-secure KEM, there is a negligible function negl_1 such that

$$\frac{1}{2} + \mathsf{negl}_1(n) \geq \Pr[\mathsf{KEM}^{\mathsf{cpa}}_{\mathcal{A}_1, \Pi}(n) = 1] \qquad (12.15)$$

$$= \frac{1}{2} \cdot \Pr[\mathcal{A}_1 \text{ outputs } 0 \mid b = 0] + \frac{1}{2} \cdot \Pr[\mathcal{A}_1 \text{ outputs } 1 \mid b = 1]$$

$$= \frac{1}{2} \cdot \Pr[\mathcal{A}^{\mathsf{hy}}(pk, \mathsf{Encaps}^{(1)}_{pk}(1^n), \mathsf{Enc}'_k(m_0)) = 0]$$

$$+ \frac{1}{2} \cdot \Pr[\mathcal{A}^{\mathsf{hy}}(pk, \mathsf{Encaps}^{(1)}_{pk}(1^n), \mathsf{Enc}'_{k'}(m_0)) = 1]$$

where k is equal to $\mathsf{Encaps}_{pk}^{(2)}(1^n)$ and k' is a uniform and independent key.

Next, consider the following PPT adversary $\mathcal{A}'$ that eavesdrops on a message encrypted using the private-key scheme Π'.

Adversary $\mathcal{A}'$:

1. $\mathcal{A}'(1^n)$ runs $\mathsf{Gen}(1^n)$ on its own to generate keys (pk, sk). It also computes $c \leftarrow \mathsf{Encaps}_{pk}^{(1)}(1^n)$.

2. $\mathcal{A}'$ runs $\mathcal{A}^{hy}(pk)$ to obtain two messages m_0, m_1. These are output by $\mathcal{A}'$, and it is given in return a ciphertext c'.

3. $\mathcal{A}'$ gives the ciphertext $\langle c, c' \rangle$ to $\mathcal{A}^{hy}$, and outputs the bit b' that $\mathcal{A}^{hy}$ outputs.

When $b = 0$ in experiment $\mathsf{PrivK}_{\mathcal{A}',\Pi'}^{eav}(n)$, adversary $\mathcal{A}'$ is given a ciphertext c' which is an encryption of m_0 using a key k' that is uniform and independent of anything else. So $\mathcal{A}^{hy}$ is given a ciphertext of the form $\langle c, \mathsf{Enc}_{k'}'(m_0) \rangle$ where k' is uniform and independent of c, and

$$\Pr[\mathcal{A}' \text{ outputs } 0 \mid b = 0] = \Pr[\mathcal{A}^{hy}(pk, \mathsf{Encaps}_{pk}^{(1)}(1^n), \mathsf{Enc}_{k'}'(m_0)) = 0].$$

On the other hand, when $b = 1$ in experiment $\mathsf{PrivK}_{\mathcal{A}',\Pi'}^{eav}(n)$, then $\mathcal{A}'$ is given an encryption of m_1 using a uniform, independent key k'. This means $\mathcal{A}^{hy}$ is given a ciphertext of the form $\langle c, \mathsf{Enc}_{k'}'(m_1) \rangle$ and so

$$\Pr[\mathcal{A}' \text{ outputs } 1 \mid b = 1] = \Pr[\mathcal{A}^{hy}(pk, \mathsf{Encaps}_{pk}^{(1)}(1^n), \mathsf{Enc}_{k'}'(m_1)) = 1].$$

Since Π' is EAV-secure, there is a negligible function negl' such that

$$\frac{1}{2} + \mathsf{negl}'(n) \geq \Pr[\mathsf{PrivK}_{\mathcal{A}',\Pi'}^{eav}(n) = 1] \tag{12.16}$$

$$= \frac{1}{2} \cdot \Pr[\mathcal{A}' \text{ outputs } 0 \mid b = 0] + \frac{1}{2} \cdot \Pr[\mathcal{A}' \text{ outputs } 1 \mid b = 1]$$

$$= \frac{1}{2} \cdot \Pr[\mathcal{A}^{hy}(pk, \mathsf{Encaps}_{pk}^{(1)}(1^n), \mathsf{Enc}_{k'}'(m_0)) = 0]$$

$$+ \frac{1}{2} \cdot \Pr[\mathcal{A}^{hy}(pk, \mathsf{Encaps}_{pk}^{(1)}(1^n), \mathsf{Enc}_{k'}'(m_1)) = 1].$$

Proceeding exactly as we did to prove Equation (12.15), one can show there is a negligible function negl_2 such that

$$\frac{1}{2} + \mathsf{negl}_2(n) \geq \Pr[\mathsf{KEM}_{\mathcal{A}_2,\Pi}^{cpa}(n) = 1] \tag{12.17}$$

$$= \frac{1}{2} \cdot \Pr[\mathcal{A}_2 \text{ outputs } 0 \mid b = 0] + \frac{1}{2} \cdot \Pr[\mathcal{A}_2 \text{ outputs } 1 \mid b = 1]$$

$$= \frac{1}{2} \cdot \Pr[\mathcal{A}^{hy}(pk, \mathsf{Encaps}_{pk}^{(1)}(1^n), \mathsf{Enc}_k'(m_1)) = 1]$$

$$+ \frac{1}{2} \cdot \Pr[\mathcal{A}^{hy}(pk, \mathsf{Encaps}_{pk}^{(1)}(1^n), \mathsf{Enc}_{k'}'(m_1)) = 0].$$

Summing Equations (12.15)–(12.17) and using the fact that the sum of three negligible functions is negligible, we see there exists a negligible function negl such that

$$\frac{3}{2} + \mathsf{negl}(n) \geq$$

$$\frac{1}{2} \cdot \Big(\Pr[\mathcal{A}^{\mathsf{hy}}(pk, c, \mathsf{Enc}'_k(m_0)) = 0] + \Pr[\mathcal{A}^{\mathsf{hy}}(pk, c, \mathsf{Enc}'_{k'}(m_0)) = 1]$$
$$+ \Pr[\mathcal{A}^{\mathsf{hy}}(pk, c, \mathsf{Enc}'_{k'}(m_0)) = 0] + \Pr[\mathcal{A}^{\mathsf{hy}}(pk, c, \mathsf{Enc}'_{k'}(m_1)) = 1]$$
$$+ \Pr[\mathcal{A}^{\mathsf{hy}}(pk, c, \mathsf{Enc}'_k(m_1)) = 1] + \Pr[\mathcal{A}^{\mathsf{hy}}(pk, c, \mathsf{Enc}'_{k'}(m_1)) = 0] \Big),$$

where $c = \mathsf{Encaps}^{(1)}_{pk}(1^n)$ in all the above. Note that

$$\Pr[\mathcal{A}^{\mathsf{hy}}(pk, c, \mathsf{Enc}'_{k'}(m_0)) = 1] + \Pr[\mathcal{A}^{\mathsf{hy}}(pk, c, \mathsf{Enc}'_{k'}(m_0)) = 0] = 1,$$

since the probabilities of complementary events always sum to 1. Similarly,

$$\Pr[\mathcal{A}^{\mathsf{hy}}(pk, c, \mathsf{Enc}'_{k'}(m_1)) = 1] + \Pr[\mathcal{A}^{\mathsf{hy}}(pk, c, \mathsf{Enc}'_{k'}(m_1)) = 0] = 1.$$

Therefore,

$$\frac{1}{2} + \mathsf{negl}(n)$$

$$\geq \frac{1}{2} \cdot \Big(\Pr[\mathcal{A}^{\mathsf{hy}}(pk, c, \mathsf{Enc}'_k(m_0)) = 0] + \Pr[\mathcal{A}^{\mathsf{hy}}(pk, c, \mathsf{Enc}'_k(m_1)) = 1] \Big)$$
$$= \Pr[\mathsf{PubK}^{\mathsf{eav}}_{\mathcal{A}^{\mathsf{hy}}, \Pi^{\mathsf{hy}}}(n) = 1]$$

(using Equation (12.14) for the last equality), proving the theorem. ∎

12.3.2   CCA-Security

If the private-key encryption scheme Π' is not itself secure against chosen-ciphertext attacks, then (regardless of the KEM used) neither is the resulting hybrid encryption scheme Π^{hy}. As a simple, illustrative example, say we take Construction 3.17 as our private-key encryption scheme. Then, leaving the KEM unspecified, encryption of a message m by Π^{hy} is done by computing $(c, k) \leftarrow \mathsf{Encaps}_{pk}(1^n)$ and then outputting the ciphertext

$$\langle c, G(k) \oplus m \rangle,$$

where G is a pseudorandom generator. Given a ciphertext $\langle c, c' \rangle$, an attacker can simply flip the last bit of c' to obtain a modified ciphertext that is a valid encryption of m with its last bit flipped.

The natural way to fix this is to use a CCA-secure private-key encryption scheme. But this is clearly not enough if the KEM is susceptible to chosen-ciphertext attacks. Since we have not yet defined this notion, we do so now.

As in Definition 12.11, we require that an adversary given a ciphertext c cannot distinguish the key k encapsulated by that ciphertext from a uniform and independent key k'. Now, however, we additionally allow the attacker to request *decapsulation* of ciphertexts of its choice (as long as they are different from the challenge ciphertext).

Formally, let $\mathcal{A}$ be an adversary and let $\Pi = (\mathsf{Gen}, \mathsf{Encaps}, \mathsf{Decaps})$ be a KEM with key length n, and consider the following experiment:

The CCA indistinguishability experiment $\mathsf{KEM}^{\mathsf{cca}}_{\mathcal{A},\Pi}(n)$:

1. $\mathsf{Gen}(1^n)$ *is run to obtain keys* (pk, sk). *Then* $\mathsf{Encaps}_{pk}(1^n)$ *is run to generate* (c, k) *with* $k \in \{0,1\}^n$.

2. *Choose a uniform bit* $b \in \{0,1\}$. *If* $b = 0$ *set* $\hat{k} := k$. *If* $b = 1$ *then choose a uniform* $\hat{k} \in \{0,1\}^n$.

3. $\mathcal{A}$ *is given* $(pk, c, \hat{k})$ *and access to an oracle* $\mathsf{Decaps}_{sk}(\cdot)$, *but may not request decapsulation of* c *itself.*

4. $\mathcal{A}$ *outputs a bit* b'. *The output of the experiment is defined to be* 1 *if* $b' = b$, *and* 0 *otherwise.*

DEFINITION 12.13 *A key-encapsulation mechanism* Π *is* CCA-secure *if for all probabilistic polynomial-time adversaries* $\mathcal{A}$ *there is a negligible function* negl *such that*

$$\Pr[\mathsf{KEM}^{\mathsf{cca}}_{\mathcal{A},\Pi}(n) = 1] \leq \frac{1}{2} + \mathsf{negl}(n).$$

Using a CCA-secure KEM in combination with a CCA-secure private-key encryption scheme results in a CCA-secure public-key encryption scheme.

THEOREM 12.14 *If* Π *is a CCA-secure KEM and* Π' *is a CCA-secure private-key encryption scheme, then* Π^{hy} *as in Construction 12.10 is a CCA-secure public-key encryption scheme.*

A proof is obtained by suitable modification of the proof of Theorem 12.12.

12.4 CDH/DDH-Based Encryption

So far we have discussed public-key encryption abstractly, but have not yet seen any concrete examples of public-key encryption schemes (or KEMs). Here we explore some constructions based on the *Diffie–Hellman problems*. (The Diffie–Hellman problems are introduced in Section 9.3.2.)

12.4.1 El Gamal Encryption

In 1985, Taher El Gamal observed that the Diffie–Hellman key-exchange protocol (cf. Section 11.3) could be adapted to give a public-key encryption scheme. Recall that in the Diffie–Hellman protocol, Alice sends a message to Bob and then Bob responds with a message to Alice; based on these messages, Alice and Bob can derive a shared value k which is indistinguishable (to an eavesdropper) from a uniform element of some group $\mathbb{G}$. We could imagine Bob using that shared value to encrypt a message $m \in \mathbb{G}$ by simply sending $k \cdot m$ to Alice; Alice can clearly recover m using her knowledge of k, and we will argue below that an eavesdropper learns nothing about m.

In the *El Gamal encryption scheme* we simply change our perspective on the above interaction. We view Alice's initial message as her public key, and Bob's reply (both his initial response and $k \cdot m$) as a ciphertext. CPA-security based on the decisional Diffie–Hellman (DDH) assumption follows fairly easily from security of the Diffie–Hellman key-exchange protocol (Theorem 11.3).

In our formal treatment, we begin by stating and proving a simple lemma that underlies the El Gamal encryption scheme. Let $\mathbb{G}$ be a finite group, and let $m \in \mathbb{G}$ be an arbitrary element. The lemma states that multiplying m by a uniform group element k yields a uniformly distributed group element c. Importantly, the distribution of c is independent of m; this means that c *contains no information about m.*

LEMMA 12.15 *Let $\mathbb{G}$ be a finite group, and let $m \in \mathbb{G}$ be arbitrary. Then choosing uniform $k \in \mathbb{G}$ and setting $c := k \cdot m$ results in a uniformly distributed $c \in \mathbb{G}$. Put differently, for any $\hat{c} \in \mathbb{G}$, we have*

$$\Pr[k \cdot m = \hat{c}] = 1/|\mathbb{G}|,$$

where the probability is taken over uniform choice of $k \in \mathbb{G}$.

PROOF Let $\hat{c} \in \mathbb{G}$ be arbitrary. Then

$$\Pr[k \cdot m = \hat{c}] = \Pr[k = \hat{c} \cdot m^{-1}].$$

Since k is uniform, the probability that k is equal to the fixed element $\hat{c} \cdot m^{-1}$ is exactly $1/|\mathbb{G}|$. ∎

The above lemma suggests a way to construct a perfectly secret *private-key* encryption scheme with message space $\mathbb{G}$. The sender and receiver share as their secret key a uniform element $k \in \mathbb{G}$. To encrypt the message $m \in \mathbb{G}$, the sender computes the ciphertext $c := k \cdot m$. The receiver can recover the message from the ciphertext c by computing $m := c/k$. Perfect secrecy follows immediately from the lemma above. In fact, we have already seen this scheme in a different guise—the one-time pad encryption scheme is an instantiation of this approach, with the underlying group $\mathbb{G}$ being the set $\{0,1\}^{\ell}$ under the operation of bit-wise XOR.

We can adapt the above ideas to the public-key setting by providing the parties with a way to generate a shared, "random-looking" value k by interacting over a public channel. This should sound familiar since it is exactly what the Diffie–Hellman protocol achieves. We proceed with the details.

As in Section 9.3.2, let $\mathcal{G}$ be a polynomial-time algorithm that takes as input 1^n and (except possibly with negligible probability) outputs a description of a cyclic group $\mathbb{G}$, its order q (with $\|q\| = n$), and a generator g. The El Gamal encryption scheme is described in Construction 12.16.

CONSTRUCTION 12.16

Let $\mathcal{G}$ be as in the text. Define a public-key encryption scheme as follows:

- **Gen:** on input 1^n run $\mathcal{G}(1^n)$ to obtain $(\mathbb{G}, q, g)$. Then choose a uniform $x \in \mathbb{Z}_q$ and compute $h := g^x$. The public key is $\langle \mathbb{G}, q, g, h \rangle$ and the private key is $\langle \mathbb{G}, q, g, x \rangle$. The message space is $\mathbb{G}$.

- **Enc:** on input a public key $pk = \langle \mathbb{G}, q, g, h \rangle$ and a message $m \in \mathbb{G}$, choose a uniform $y \in \mathbb{Z}_q$ and output the ciphertext
$$\langle g^y, \ h^y \cdot m \rangle.$$

- **Dec:** on input a private key $sk = \langle \mathbb{G}, q, g, x \rangle$ and a ciphertext $\langle c_1, c_2 \rangle$, output
$$\hat{m} := c_2 / c_1^x.$$

The El Gamal encryption scheme.

To see that decryption succeeds, let $\langle c_1, c_2 \rangle = \langle g^y, \ h^y \cdot m \rangle$ with $h = g^x$. Then

$$\hat{m} = \frac{c_2}{c_1^x} = \frac{h^y \cdot m}{(g^y)^x} = \frac{(g^x)^y \cdot m}{g^{xy}} = \frac{g^{xy} \cdot m}{g^{xy}} = m.$$

Example 12.17

Let $q = 83$ and $p = 2q + 1 = 167$, and let $\mathbb{G}$ denote the group of *quadratic residues* (i.e., squares) modulo p. (Since p and q are prime, $\mathbb{G}$ is a subgroup of $\mathbb{Z}_p^*$ with order q. See Section 9.3.3.) Since the order of $\mathbb{G}$ is prime, any element of $\mathbb{G}$ except 1 is a generator; take $g = 2^2 = 4 \bmod 167$. Say the receiver chooses secret key $37 \in \mathbb{Z}_{83}$ and so the public key is

$$pk = \langle p, q, g, h \rangle = \langle 167, 83, 4, [4^{37} \bmod 167] \rangle = \langle 167, 83, 4, 76 \rangle,$$

where we use p to represent $\mathbb{G}$ (it is assumed that the receiver knows that the group is the set of quadratic residues modulo p).

Say a sender encrypts the message $m = 65 \in \mathbb{G}$ (note $65 = 30^2 \bmod 167$ and so 65 is an element of $\mathbb{G}$). If $y = 71$, the ciphertext is

$$\langle[4^{71} \bmod 167], [76^{71} \cdot 65 \bmod 167]\rangle = \langle132, 44\rangle.$$

To decrypt, the receiver first computes $124 = [132^{37} \bmod 167]$; then, since $66 = [124^{-1} \bmod 167]$, the receiver recovers $m = 65 = [44 \cdot 66 \bmod 167]$. $\Diamond$

We now prove security of the scheme. (The reader may want to compare the proof of the following to the proofs of Theorems 3.16 and 11.3.)

THEOREM 12.18   *If the DDH problem is hard relative to $\mathcal{G}$, then the El Gamal encryption scheme is CPA-secure.*

PROOF   Let Π denote the El Gamal encryption scheme. We prove that Π has indistinguishable encryptions in the presence of an eavesdropper; by Proposition 12.3, this implies it is CPA-secure.

Let $\mathcal{A}$ be a probabilistic polynomial-time adversary. We want to show that there is a negligible function negl such that

$$\Pr[\mathsf{PubK}^{\mathsf{eav}}_{\mathcal{A},\Pi}(n) = 1] \leq \frac{1}{2} + \mathsf{negl}(n).$$

Consider the modified "encryption scheme" $\widetilde{\Pi}$ where Gen is the same as in Π, but encryption of a message m with respect to the public key $\langle\mathbb{G}, q, g, h\rangle$ is done by choosing uniform $y, z \in \mathbb{Z}_q$ and outputting the ciphertext

$$\langle g^y, \ g^z \cdot m \rangle.$$

Although $\widetilde{\Pi}$ is not actually an encryption scheme (as there is no way for the receiver to decrypt), the experiment $\mathsf{PubK}^{\mathsf{eav}}_{\mathcal{A},\widetilde{\Pi}}(n)$ is still well-defined since that experiment depends only on the key-generation and encryption algorithms.

Lemma 12.15 and the discussion that immediately follows it imply that the second component of the ciphertext in scheme $\widetilde{\Pi}$ is a uniformly distributed group element and, in particular, is independent of the message m being encrypted. (Remember that g^z is a uniform element of $\mathbb{G}$ when z is chosen uniformly from $\mathbb{Z}_q$.) The first component of the ciphertext is trivially independent of m. Taken together, this means that the entire ciphertext contains no information about m. It follows that

$$\Pr[\mathsf{PubK}^{\mathsf{eav}}_{\mathcal{A},\widetilde{\Pi}}(n) = 1] = \frac{1}{2}.$$

Now consider the following PPT algorithm D that attempts to solve the DDH problem relative to $\mathcal{G}$. Recall that D receives $(\mathbb{G}, q, g, h_1, h_2, h_3)$ where $h_1 = g^x$, $h_2 = g^y$, and h_3 is either g^{xy} or g^z (for uniform x, y, z); the goal of D is to determine which is the case.

Algorithm D:
The algorithm is given $(\mathbb{G}, q, g, h_1, h_2, h_3)$ as input.

- Set $pk = \langle \mathbb{G}, q, g, h_1 \rangle$ and run $\mathcal{A}(pk)$ to obtain two messages $m_0, m_1 \in \mathbb{G}$.
- Choose a uniform bit b, and set $c_1 := h_2$ and $c_2 := h_3 \cdot m_b$.
- Give the ciphertext $\langle c_1, c_2 \rangle$ to $\mathcal{A}$ and obtain an output bit b'. If $b' = b$, output 1; otherwise, output 0.

Let us analyze the behavior of D. There are two cases to consider:

Case 1: Say the input to D is generated by running $\mathcal{G}(1^n)$ to obtain $(\mathbb{G}, q, g)$, then choosing uniform $x, y, z \in \mathbb{Z}_q$, and finally setting $h_1 := g^x$, $h_2 := g^y$, and $h_3 := g^z$. Then D runs $\mathcal{A}$ on a public key constructed as

$$pk = \langle \mathbb{G}, q, g, g^x \rangle$$

and a ciphertext constructed as

$$\langle c_1, c_2 \rangle = \langle g^y, \ g^z \cdot m_b \rangle.$$

We see that in this case the view of $\mathcal{A}$ when run as a subroutine by D is distributed identically to $\mathcal{A}$'s view in experiment $\mathsf{PubK}^{\mathsf{eav}}_{\mathcal{A}, \widetilde{\Pi}}(n)$. Since D outputs 1 exactly when the output b' of $\mathcal{A}$ is equal to b, we have that

$$\Pr[D(\mathbb{G}, q, g, g^x, g^y, g^z) = 1] = \Pr[\mathsf{PubK}^{\mathsf{eav}}_{\mathcal{A}, \widetilde{\Pi}}(n) = 1] = \frac{1}{2}.$$

Case 2: Say the input to D is generated by running $\mathcal{G}(1^n)$ to obtain $(\mathbb{G}, q, g)$, then choosing uniform $x, y \in \mathbb{Z}_q$, and finally setting $h_1 := g^x$, $h_2 := g^y$, and $h_3 := g^{xy}$. Then D runs $\mathcal{A}$ on a public key constructed as

$$pk = \langle \mathbb{G}, q, g, g^x \rangle$$

and a ciphertext constructed as

$$\langle c_1, c_2 \rangle = \langle g^y, \ g^{xy} \cdot m_b \rangle = \langle g^y, \ (g^x)^y \cdot m_b \rangle.$$

We see that in this case the view of $\mathcal{A}$ when run as a subroutine by D is distributed identically to $\mathcal{A}$'s view in experiment $\mathsf{PubK}^{\mathsf{eav}}_{\mathcal{A}, \Pi}(n)$. Since D outputs 1 exactly when the output b' of $\mathcal{A}$ is equal to b, we have that

$$\Pr[D(\mathbb{G}, q, g, g^x, g^y, g^{xy}) = 1] = \Pr[\mathsf{PubK}^{\mathsf{eav}}_{\mathcal{A}, \Pi}(n) = 1].$$

Under the assumption that the DDH problem is hard relative to $\mathcal{G}$, there is a negligible function negl such that

$$\mathsf{negl}(n) \geq \left| \Pr[D(\mathbb{G}, q, g, g^x, g^y, g^z) = 1] - \Pr[D(\mathbb{G}, q, g, g^x, g^y, g^{xy}) = 1] \right|$$

$$= \left| \frac{1}{2} - \Pr[\mathsf{PubK}^{\mathsf{eav}}_{\mathcal{A}, \Pi}(n) = 1] \right|.$$

This implies $\Pr[\mathsf{PubK}^{\mathsf{eav}}_{\mathcal{A}, \Pi}(n) = 1] \leq \frac{1}{2} + \mathsf{negl}(n)$, completing the proof. ∎

El Gamal Implementation Issues

We briefly discuss some practical issues related to El Gamal encryption.

Sharing public parameters. Our description of the El Gamal encryption scheme in Construction 12.16 requires the receiver to run $\mathcal{G}$ to generate $\mathbb{G}, q, g$. In practice, it is common for these parameters to be generated and fixed "once-and-for-all," and then shared by multiple receivers. (Of course, each receiver must choose their own secret value x and publish their own public key $h = g^x$.) For example, NIST has published a set of recommended parameters suitable for use in the El Gamal encryption scheme. Sharing parameters in this way does not impact security (assuming the parameters were generated correctly and honestly in the first place). Looking ahead, we remark that this is in contrast to the case of RSA, where parameters cannot safely be shared (see Section 12.5.1).

Choice of group. As discussed in Section 9.3.2, the group order q should be prime. As far as specific groups are concerned, elliptic curves are one increasingly popular choice; an alternative is to let $\mathbb{G}$ be a prime-order subgroup of $\mathbb{Z}_p^*$, for p prime. We refer to Section 10.4 for a tabulation of recommended key lengths for achieving different levels of security.

The message space. An inconvenient aspect of the El Gamal encryption scheme is that the message space is a group $\mathbb{G}$ rather than bit-strings of some specified length. For some choices of the group, it is possible to address this by defining a reversible encoding of bit-strings as group elements. In such cases, the sender can first encode their message $m \in \{0,1\}^\ell$ as a group element $\hat{m} \in \mathbb{G}$ and then apply El Gamal encryption to $\hat{m}$. The receiver can decrypt as in Construction 12.16 to obtain the encoded message $\hat{m}$, and then reverse the encoding to recover the original message m.

A simpler approach is to use (a variant of) El Gamal encryption as part of a hybrid encryption scheme. For example, the sender could choose a uniform group element $m \in \mathbb{G}$, encrypt this using the El Gamal encryption scheme, and then encrypt their actual message using a private-key encryption scheme and key $H(m)$, where $H : \mathbb{G} \to \{0,1\}^n$ is an appropriate key-derivation function (see the following section). In this case, it would be more efficient to use the DDH-based KEM that we describe next.

12.4.2 DDH-Based Key Encapsulation

At the end of the previous section we noted that El Gamal encryption can be used as part of a hybrid encryption scheme by simply encrypting a uniform group element m and using a hash of that element as a key. But this is wasteful! The proof of security for El Gamal encryption shows that c_1^x (where c_1 is the first component of the ciphertext, and x is the private key of the receiver) is already indistinguishable from a uniform group element, so the sender/receiver may as well use that to derive a key. Construction 12.19

illustrates the KEM that follows this approach. Note the resulting ciphertext consists of just a single group element. In contrast, if we were to use El Gamal encryption of a uniform group element, the ciphertext would contain two group elements.

CONSTRUCTION 12.19

Let $\mathcal{G}$ be as in the previous section. Define a KEM as follows:

- **Gen:** on input 1^n run $\mathcal{G}(1^n)$ to obtain $(\mathbb{G}, q, g)$. choose a uniform $x \in \mathbb{Z}_q$ and set $h := g^x$. Also specify a function $H : \mathbb{G} \to \{0,1\}^{\ell(n)}$ for some function ℓ (see text). The public key is $\langle \mathbb{G}, q, g, h, H \rangle$ and the private key is $\langle \mathbb{G}, q, g, x \rangle$.

- **Encaps:** on input a public key $pk = \langle \mathbb{G}, q, g, h, H \rangle$, choose a uniform $y \in \mathbb{Z}_q$ and output the ciphertext g^y and the key $H(h^y)$.

- **Decaps:** on input a private key $sk = \langle \mathbb{G}, q, g, x \rangle$ and a ciphertext $c \in \mathbb{G}$, output the key $H(c^x)$.

An "El Gamal-like" KEM.

The construction leaves the key-derivation function H unspecified, and there are several options for instantiating it. (See Section 6.6.4 for more on key derivation in general.) One possibility is to choose a function $H : \mathbb{G} \to \{0,1\}^\ell$ that is (close to) *regular*, meaning that for each possible key $k \in \{0,1\}^\ell$ the number of group elements that map to k is approximately the same. (Formally, we need a negligible function negl such that

$$\frac{1}{2} \cdot \textstyle\sum_{k \in \{0,1\}^{\ell(n)}} \left| \Pr[H(g) = k] - 2^{-\ell(n)} \right| \le \mathsf{negl}(n),$$

where the probability is taken over uniform $g \in \mathbb{G}$. This means the distribution of the key is statistically close to uniform.) Both the complexity of H, as well as the achievable key length ℓ, depend on the specific group $\mathbb{G}$ used.

A second possibility is to let H be a *keyed* function, where the (uniform) key for H is included as part of the receiver's public key. This works if H is a strong extractor, as mentioned briefly in Section 6.6.4. Appropriate choice of ℓ here (to ensure that the resulting key is statistically close to uniform) will depend on the size of $\mathbb{G}$.

In either of the above cases, a proof of CPA-security based on the decisional Diffie–Hellman (DDH) assumption follows easily by adapting the proof of security for the Diffie–Hellman key-exchange protocol (Theorem 11.3).

THEOREM 12.20 *If the DDH problem is hard relative to $\mathcal{G}$, and H is chosen as described, then Construction 12.19 is a CPA-secure KEM.*

If one is willing to model H as a *random oracle*, Construction 12.19 can be proven CPA-secure based on the (weaker) *computational* Diffie–Hellman (CDH) assumption. We discuss this in the following section.

12.4.3 *A CDH-Based KEM in the Random-Oracle Model

In this section, we show that if one is willing to model H as a *random oracle*, then Construction 12.19 can be proven CPA-secure based on the CDH assumption. (Readers may want to review Section 6.5 to remind themselves of the random-oracle model.) Intuitively, the CDH assumption implies that an attacker observing $h = g^x$ (from the public key) and the ciphertext $c = g^y$ cannot compute $\mathsf{DH}_g(h, c) = h^y$. In particular, then, an attacker cannot query h^y to the random oracle. But this means that the encapsulated key $H(h^y)$ is completely random from the attacker's point of view. This intuition is turned into a formal proof below

As indicated by the intuition above, the proof inherently relies on modeling H as a random oracle.[2] Specifically, the proof relies on the facts that (1) the only way to learn $H(h^y)$ is to explicitly query h^y to H, which would mean that the attacker has solved a CDH instance (this is called "extractability" in Section 6.5.1), and (2) if an attacker does *not* query h^y to H, then the value $H(h^y)$ is uniform from the attacker's point of view. These properties only hold—indeed, they only make sense—if H is modeled as a random oracle.

THEOREM 12.21 *If the CDH problem is hard relative to $\mathcal{G}$, and H is modeled as a random oracle, then Construction 12.19 is CPA-secure.*

PROOF Let Π denote Construction 12.19, and let $\mathcal{A}$ be a PPT adversary. We want to show that there is a negligible function negl such that

$$\Pr[\mathsf{KEM}^{\mathsf{cpa}}_{\mathcal{A},\Pi}(n) = 1] \leq \frac{1}{2} + \mathsf{negl}(n).$$

The above probability is also taken over uniform choice of the function H, to which $\mathcal{A}$ is given oracle access.

Consider an execution of experiment $\mathsf{KEM}^{\mathsf{cpa}}_{\mathcal{A},\Pi}(n)$ in which the public key is $\langle \mathbb{G}, q, g, h \rangle$ and the ciphertext is $c = g^y$, and let Query be the event that $\mathcal{A}$ queries $\mathsf{DH}_g(h, c) = h^y$ to H. We have

$$\Pr[\mathsf{KEM}^{\mathsf{cpa}}_{\mathcal{A},\Pi}(n) = 1] = \Pr[\mathsf{KEM}^{\mathsf{cpa}}_{\mathcal{A},\Pi}(n) = 1 \wedge \overline{\mathsf{Query}}]$$
$$+ \Pr[\mathsf{KEM}^{\mathsf{cpa}}_{\mathcal{A},\Pi}(n) = 1 \wedge \mathsf{Query}]$$
$$\leq \Pr[\mathsf{KEM}^{\mathsf{cpa}}_{\mathcal{A},\Pi}(n) = 1 \wedge \overline{\mathsf{Query}}] + \Pr[\mathsf{Query}]. \quad (12.18)$$

If $\Pr[\overline{\mathsf{Query}}] = 0$ then $\Pr[\mathsf{KEM}^{\mathsf{cpa}}_{\mathcal{A},\Pi}(n) = 1 \wedge \overline{\mathsf{Query}}] = 0$. Otherwise,

$$\Pr[\mathsf{KEM}^{\mathsf{cpa}}_{\mathcal{A},\Pi}(n) = 1 \wedge \overline{\mathsf{Query}}] = \Pr[\mathsf{KEM}^{\mathsf{cpa}}_{\mathcal{A},\Pi}(n) = 1 \mid \overline{\mathsf{Query}}] \cdot \Pr[\overline{\mathsf{Query}}]$$
$$\leq \Pr[\mathsf{KEM}^{\mathsf{cpa}}_{\mathcal{A},\Pi}(n) = 1 \mid \overline{\mathsf{Query}}].$$

---

[2]This is true as long as we wish to rely only on the CDH assumption. As noted earlier, a proof without random oracles is possible if we rely on the stronger DDH assumption.

In experiment $\mathsf{KEM}^{\mathsf{cpa}}_{\mathcal{A},\Pi}(n)$, the adversary $\mathcal{A}$ is given the public key and the ciphertext, plus either the encapsulated key $k \stackrel{\text{def}}{=} H(h^y)$ or a uniform key. If Query does not occur, then k is uniformly distributed from the perspective of the adversary, and so there is no way $\mathcal{A}$ can distinguish between these two possibilities. This means that

$$\Pr[\mathsf{KEM}^{\mathsf{cpa}}_{\mathcal{A},\Pi}(n) = 1 \mid \overline{\mathsf{Query}}] = \frac{1}{2}.$$

Returning to Equation (12.18), we thus have

$$\Pr[\mathsf{KEM}^{\mathsf{cpa}}_{\mathcal{A},\Pi}(n) = 1] \leq \frac{1}{2} + \Pr[\mathsf{Query}].$$

We next show that $\Pr[\mathsf{Query}]$ is negligible, completing the proof.

Let $t = t(n)$ be a (polynomial) upper bound on the number of queries that $\mathcal{A}$ makes to the random oracle H. Define the following PPT algorithm $\mathcal{A}'$ for the CDH problem relative to $\mathcal{G}$:

Algorithm $\mathcal{A}'$:
The algorithm is given $\mathbb{G}, q, g, h, c$ as input.

- Set $pk := \langle \mathbb{G}, q, g, h \rangle$ and choose a uniform $k \in \{0,1\}^\ell$.

- Run $\mathcal{A}(pk, c, k)$. When $\mathcal{A}$ makes a query to H, answer it by choosing a fresh, uniform ℓ-bit string.

- At the end of $\mathcal{A}$'s execution, let $y_1, \ldots, y_t$ be the list of queries that $\mathcal{A}$ has made to H. Choose a uniform index $i \in \{1, \ldots, t\}$ and output y_i.

We are interested in the probability with which $\mathcal{A}'$ solves the CDH problem, i.e., $\Pr[\mathcal{A}'(\mathbb{G}, q, g, h, c) = \mathsf{DH}_g(h, c)]$, where the probability is taken over $\mathbb{G}, q, g$ output by $\mathcal{G}(1^n)$, uniform $h, c \in \mathbb{G}$, and the randomness of $\mathcal{A}'$. To analyze this probability, note first that event Query is still well-defined in the execution of $\mathcal{A}'$, even though $\mathcal{A}'$ cannot detect whether it occurs. Moreover, the probability of event Query when $\mathcal{A}$ is run as a subroutine by $\mathcal{A}'$ is identical to the probability of Query in experiment $\mathsf{KEM}^{\mathsf{cpa}}_{\mathcal{A},\Pi}(n)$. This follows because the view of $\mathcal{A}$ is identical in both cases until event Query occurs: in each case, $\mathbb{G}, q, g$ are output by $\mathcal{G}(1^n)$; in each case, h and c are uniform elements of $\mathbb{G}$ and k is a uniform ℓ-bit string, and in each case queries to H *other than* $H(\mathsf{DH}_g(h,c))$ are answered with a uniform ℓ-bit string. (In $\mathsf{KEM}^{\mathsf{cpa}}_{\mathcal{A},\Pi}(n)$, the query $H(\mathsf{DH}_g(h,c))$ is answered with the actual encapsulated key, which is equal to k with probability $1/2$, whereas when $\mathcal{A}$ is run as a subroutine by $\mathcal{A}'$ the query $H(\mathsf{DH}_g(h,c))$ is answered with a uniform ℓ-bit string that is independent of k. But when this query is made, event Query occurs.)

Finally, observe that when Query occurs then $\mathsf{DH}_g(h,c) \in \{y_1, \ldots, y_t\}$ by definition, and so $\mathcal{A}'$ outputs the correct result $\mathsf{DH}_g(h,c)$ with probability at least $1/t$. We therefore conclude that

$$\Pr[\mathcal{A}'(\mathbb{G}, q, g, h, c) = \mathsf{DH}_g(h, c)] \geq \Pr[\mathsf{Query}]/t,$$

or $\Pr[\mathsf{Query}] \leq t \cdot \Pr[\mathcal{A}'(\mathbb{G}, q, g, h, c) = \mathsf{DH}_g(h, c)]$. Since the CDH problem is hard relative to $\mathcal{G}$ and t is polynomial, this implies that $\Pr[\mathsf{Query}]$ is negligible and completes the proof. ∎

In the next section we will see that Construction 12.19 can even be shown to be secure against chosen-*ciphertext* attacks based on a stronger variant of the CDH assumption (if we continue to model H as a random oracle).

12.4.4   *Chosen-Ciphertext Security and DHIES/ECIES

The El Gamal encryption scheme is vulnerable to chosen-ciphertext attacks. This follows from the fact that it is *malleable*. Recall that an encryption scheme is malleable, informally, if given a ciphertext c that is an encryption of some unknown message m, it is possible to generate a modified ciphertext c' that is an encryption of a message m' having some known relation to m. In the case of El Gamal encryption, consider an adversary $\mathcal{A}$ who intercepts a ciphertext $c = \langle c_1, c_2 \rangle$ encrypted using the public key $pk = \langle \mathbb{G}, q, g, h \rangle$, and who then constructs the modified ciphertext $c' = \langle c_1, c_2' \rangle$ where $c_2' = c_2 \cdot \alpha$ for some $\alpha \in \mathbb{G}$. If c is an encryption of a message $m \in \mathbb{G}$ (which may be unknown to $\mathcal{A}$), we have $c_1 = g^y$ and $c_2 = h^y \cdot m$ for some $y \in \mathbb{Z}_q$. But then

$$c_1 = g^y \quad \text{and} \quad c_2' = h^y \cdot (\alpha \cdot m),$$

and so c' is a valid encryption of the message $\alpha \cdot m$. In other words, $\mathcal{A}$ *can transform an encryption of the (unknown) message m into an encryption of the (unknown) message $\alpha \cdot m$.* As discussed in Scenario 3 in Section 12.2.3, this sort of attack can have serious consequences.

The KEM discussed in the previous section might also be malleable depending on the key-derivation function H being used. If H is modeled as a random oracle, however, then such attacks no longer seem possible. In fact, one can prove in this case that Construction 12.19 is CCA-secure (which, as we have noted, implies non-malleability) based on the *gap-CDH assumption*. Recall the CDH assumption is that given group elements g^x and g^y (for some generator g), it is infeasible to compute g^{xy}. The gap-CDH assumption says that this remains infeasible even given access to an oracle $\mathcal{O}_y$ such that $\mathcal{O}_y(U, V)$ returns 1 exactly when $V = U^y$. Stated differently, the gap-CDH assumption is that the CDH problem remains hard even given an oracle that solves the DDH problem. (We do not give a formal definition since we do not use the assumption in the rest of the book.) The gap-CDH assumption is believed to hold for all cryptographic groups in which the DDH assumption holds.

A proof of the following is very similar to the proof of Theorem 12.38.

THEOREM 12.22 *If the gap-CDH problem is hard relative to $\mathcal{G}$, and H is modeled as a random oracle, then Construction 12.19 is CCA-secure.*

It is interesting to observe that the same construction (namely, Construction 12.19) can be analyzed under different assumptions and in different models, yielding different results. Assuming only that the DDH problem is hard (and for H chosen appropriately), the scheme is CPA-secure. If we model H as a random oracle (which imposes more stringent requirements on H), then we obtain CPA-security under the weaker CDH assumption, and CCA-security under the stronger gap-CDH assumption.

CONSTRUCTION 12.23

Let $\mathcal{G}$ be as in the text. Let $\Pi_E = (\mathsf{Enc}', \mathsf{Dec}')$ be a private-key encryption scheme, and let $\Pi_M = (\mathsf{Mac}, \mathsf{Vrfy})$ be a message authentication code. Define a public-key encryption scheme as follows:

- **Gen:** On input 1^n run $\mathcal{G}(1^n)$ to obtain $(\mathbb{G}, q, g)$. Choose uniform $x \in \mathbb{Z}_q$, set $h := g^x$, and specify a function $H : \mathbb{G} \to \{0,1\}^{2n}$. The public key is $\langle \mathbb{G}, q, g, h, H \rangle$ and the private key is $\langle \mathbb{G}, q, g, x, H \rangle$.

- **Enc:** On input a public key $pk = \langle \mathbb{G}, q, g, h, H \rangle$, choose a uniform $y \in \mathbb{Z}_q$ and set $k_E \| k_M := H(h^y)$. Compute $c' \leftarrow \mathsf{Enc}'_{k_E}(m)$, and output the ciphertext $\langle g^y, c', \mathsf{Mac}_{k_M}(c') \rangle$.

- **Dec:** On input a private key $sk = \langle \mathbb{G}, q, g, x, H \rangle$ and a ciphertext $\langle c, c', t \rangle$, output $\perp$ if $c \notin \mathbb{G}$. Else, compute $k_E \| k_M := H(c^x)$. If $\mathsf{Vrfy}_{k_M}(c', t) \neq 1$ then output $\perp$; otherwise, output $\mathsf{Dec}'_{k_E}(c')$.

DHIES/ECIES.

CCA-secure encryption with Construction 12.19. Combining the KEM in Construction 12.19 with any CCA-secure private-key encryption scheme yields a CCA-secure public-key encryption scheme. (See Theorem 12.14.) Instantiating this approach using Construction 5.6 for the private-key component matches what is done in DHIES/ECIES, variants of which are included in the ISO/IEC 18033-2 standard for public-key encryption. (See Construction 12.23.) Encryption of a message m in these schemes takes the form

$$\langle g^y, \ \mathsf{Enc}'_{k_E}(m), \mathsf{Mac}_{k_M}(c') \rangle,$$

where Enc' denotes a CPA-secure private-key encryption scheme and c' denotes $\mathsf{Enc}'_{k_E}(m)$. DHIES, the *Diffie–Hellman Integrated Encryption Scheme*, can be used generically to refer to any scheme of this form, or to refer specifically to the case when the group $\mathbb{G}$ is a cyclic subgroup of a finite field. ECIES, the *Elliptic Curve Integrated Encryption Scheme*, refers to the case when $\mathbb{G}$ is an elliptic-curve group. We remark that in Construction 12.23 it is critical to check during decryption that c, the first component of the ciphertext, is in $\mathbb{G}$. Otherwise, an attacker might request decryption of a malformed ciphertext $\langle c, c', t \rangle$ in which $c \notin \mathbb{G}$; decrypting such a ciphertext (i.e., without returning $\bot$) might leak information about the private key.

By Theorem 5.7, encrypting a message and then applying a (strong) message authentication code yields a CCA-secure private-key encryption scheme. Combining this with Theorem 12.14, we conclude:

COROLLARY 12.24 *Let Π_E be a CPA-secure private-key encryption scheme, and let Π_M be a strongly secure message authentication code. If the gap-CDH problem is hard relative to $\mathcal{G}$, and H is modeled as a random oracle, then Construction 12.23 is a CCA-secure public-key encryption scheme.*

12.5 RSA-Based Encryption

In this section we turn our attention to encryption schemes based on the *RSA assumption* defined in Section 9.2.4. We remark that although RSA-based encryption is still in use, there is currently a gradual shift away from using RSA—and toward using CDH/DDH-based cryptosystems relying on elliptic-curve groups—because of the longer key lengths required for RSA-based schemes. We refer to Section 10.4 for further discussion.

12.5.1 Plain RSA Encryption

We begin by describing a simple encryption scheme based on the RSA problem. Although the scheme is insecure, it provides a useful starting point for the secure schemes that follow.

Let GenRSA be a PPT algorithm that, on input 1^n, outputs a modulus N that is the product of two n-bit primes, along with integers e, d satisfying $ed = 1 \bmod \phi(N)$. (As usual, the algorithm may fail with negligible probability but we ignore that here.) Recall from Section 9.2.4 that such an algorithm can be easily constructed from any algorithm GenModulus that outputs a composite modulus N along with its factorization; see Algorithm 12.25.

ALGORITHM 12.25
RSA key generation GenRSA

Input: Security parameter 1^n
Output: N, e, d as described in the text

$(N, p, q) \leftarrow \mathsf{GenModulus}(1^n)$
$\phi(N) := (p-1) \cdot (q-1)$
choose $e > 1$ such that $\gcd(e, \phi(N)) = 1$
compute $d := [e^{-1} \bmod \phi(N)]$
return N, e, d

Let N, e, d be as above, and let $c = m^e \bmod N$ for some $m \in \mathbb{Z}_N^*$. RSA encryption relies on the fact that someone who knows d can recover m from c by computing $[c^d \bmod N]$; this works because

$$c^d = (m^e)^d = m^{ed} = m \bmod N,$$

as discussed in Section 9.2.4. On the other hand, *without* knowledge of d— even if N and e are known—the RSA assumption (cf. Definition 9.46) implies that it is difficult to recover m from c, at least if m is chosen uniformly from $\mathbb{Z}_N^*$. This naturally suggests the public-key encryption scheme shown as Construction 12.26: The receiver runs GenRSA to obtain N, e, d; it publishes N and e as its public key, and keeps d in its private key. To encrypt a message $m \in \mathbb{Z}_N^*$, a sender computes the ciphertext $c := [m^e \bmod N]$. As we have just noted, the receiver—who knows d—can decrypt c and recover m.

CONSTRUCTION 12.26

Let GenRSA be as in the text. Define a public-key encryption scheme as follows:

- Gen: on input 1^n run GenRSA(1^n) to obtain N, e, and d. The public key is $\langle N, e \rangle$ and the private key is $\langle N, d \rangle$.

- Enc: on input a public key $pk = \langle N, e \rangle$ and a message $m \in \mathbb{Z}_N^*$, compute the ciphertext
$$c := [m^e \bmod N].$$

- Dec: on input a private key $sk = \langle N, d \rangle$ and a ciphertext $c \in \mathbb{Z}_N^*$, compute the message
$$m := [c^d \bmod N].$$

The plain RSA encryption scheme.

The following gives a worked example of the above (see also Example 9.49).

Example 12.27
Say GenRSA outputs $(N, e, d) = (391, 3, 235)$. (Note that $391 = 17 \cdot 23$ and so $\phi(391) = 16 \cdot 22 = 352$. Moreover, $3 \cdot 235 = 1 \bmod 352$.) So the public key is $\langle 391, 3 \rangle$ and the private key is $\langle 391, 235 \rangle$.

To encrypt the message $m = 158 \in \mathbb{Z}_{391}^*$ using the public key $(391, 3)$, we simply compute $c := [158^3 \bmod 391] = 295$; this is the ciphertext. To decrypt, the receiver computes $[295^{235} \bmod 391] = 158$. $\diamond$

Is the plain RSA encryption scheme secure? The factoring assumption implies that it is computationally infeasible for an attacker who is given the public key to derive the corresponding private key; see Section 9.2.5. This is necessary—but not sufficient—for a public-key encryption scheme to be secure. The RSA assumption implies that if the message m is *chosen uniformly from* $\mathbb{Z}_N^*$ then an eavesdropper given N, e, and c (namely, the public key and the ciphertext) *cannot recover* m *in its entirety*. But these are weak guarantees, and fall short of the level of security we want. In particular, they leave open the possibility that an attacker can recover the message when it is *not* chosen uniformly from $\mathbb{Z}_N^*$—and, indeed, when m is chosen from a small range it is easy to see that an attacker *can* compute m from the public key and ciphertext. In addition, it does not rule out the possibility that an attacker can learn *partial information* about the message, even when it is uniform. (In fact, this is known to be possible.) Moreover, plain RSA encryption is *deterministic* and so cannot be CPA-secure, as we have discussed in Section 12.2.1.

More Attacks on Plain RSA

We have already noted that plain RSA encryption is not CPA-secure. Nevertheless, there may be a temptation to use plain RSA for encrypting "random messages" and/or in situations where leaking a few bits of information about the message is acceptable. We warn against this in general, and provide here a few examples of what can go wrong. (Some of the attacks assume $e = 3$. In several cases the attacks can be extended, at least partially, to larger e; in any case, as noted in Section 9.2.4, setting $e = 3$ is often done in practice. The attacks should be taken as demonstrating that Construction 12.26 is inadequate, not as indicating that using $e = 3$ is a bad choice in general.)

A quadratic improvement in recovering m. Since plain RSA encryption is deterministic, if an attacker knows that $m < B$ then the attacker can determine m from the ciphertext $c = [m^e \bmod N]$ in time $\mathcal{O}(B)$ using the brute-force attack discussed in Section 12.2.1. One might hope, however, that plain RSA encryption can be used if B is large, i.e., if the message is chosen from a reasonably large set of values. One possible scenario where this might occur is in the context of hybrid encryption (cf. Section 12.3), where the "message" is a uniform n-bit key and so $B = 2^n$. Unfortunately, there is a clever attack that recovers m, with high probability, in time roughly $\mathcal{O}(\sqrt{B})$. This can make a significant difference in practice: a 2^{80}-time attack (say) is infeasible, but an attack running in time 2^{40} is relatively easy to carry out.

A description of the attack is given as Algorithm 12.28. In our description, we assume $B = 2^n$ and let $\alpha \in (\frac{1}{2}, 1)$ denote some fixed constant (see below). Binary search is used in the second-to-last line to check whether there exists an r with $x_r = [s^e \bmod N]$. The time for the attack is dominated by the time to perform $2T = \mathcal{O}(2^{\alpha n})$ exponentiations.

We now sketch why the attack recovers m with high probability. Let $c = m^e \bmod N$. For appropriate choice of $\alpha \approx \frac{1}{2}$, it can be shown that if m is a uniform n-bit integer then with high probability there exist r, s with $1 < r \leq s \leq 2^{\alpha n}$ for which $m = r \cdot s$. (For example, if $n = 64$ and so m is a uniform 64-bit string, then with probability 0.35 there exist r, s of length at most 34 bits such that $m = r \cdot s$. See the references at the end of the chapter

---

ALGORITHM 12.28
An attack on plain RSA encryption

Input: Public key $\langle N, e \rangle$; ciphertext c; bound 2^n
Output: $m < 2^n$ such that $m^e = c \bmod N$

set $T := 2^{\alpha n}$
for $r = 1$ to T:
$\quad x_r := [c/r^e \bmod N]$
sort the pairs $\{(r, x_r)\}_{r=1}^T$ by their second component
for $s = 1$ to T:
$\quad$ **if** $[s^e \bmod N] \stackrel{?}{=} x_r$ for some r
$\quad\quad$ **return** $[r \cdot s \bmod N]$

---

for details.) Assuming this to be the case, the above algorithm finds m since

$$c = m^e = (r \cdot s)^e = r^e \cdot s^e \bmod N,$$

and so $x_r = c/r^e = s^e \bmod N$ with $r, s \leq T$.

Encrypting short messages using small e. The previous attack shows how to recover a message m known to be smaller than some bound B in time roughly $\mathcal{O}(\sqrt{B})$. Here we show how to do the same thing in time $\mathsf{poly}(\|N\|)$ if $B \leq N^{1/e}$ (where this means the eth root of N as a real number).

The attack relies on the observation that when $m < N^{1/e}$, raising m to the eth power modulo N involves no modular reduction; i.e., $[m^e \bmod N]$ is equal to the integer m^e. This means that given the ciphertext $c = [m^e \bmod N]$, an attacker can determine m by computing $m := c^{1/e}$ *over the integers* (i.e., not modulo N); this can be done easily in time $\mathsf{poly}(\|c\|) = \mathsf{poly}(\|N\|)$ since finding eth roots is easy over the integers and hard only when working modulo N.

For small e this represents a serious weakness of plain RSA encryption. For example, if we take $e = 3$ and assume $\|N\| \approx 2048$ bits, then the attack works even when m is a uniform 256-bit string; this once again rules out security of plain RSA even when used as part of a hybrid encryption scheme.

Encrypting a partially known message. This attack assumes a sender who encrypts a message, part of which is known to the adversary (something that should not lead to an attack when using a secure scheme). We rely on a powerful theorem due to Coppersmith that we state without proof:

THEOREM 12.29 *Let $p(x)$ be a polynomial of degree e. Then in time* $\mathsf{poly}(\|N\|, e)$ *one can find all m such that $p(m) = 0 \bmod N$ and $|m| \leq N^{1/e}$.*

Due to the dependence of the running time on e, the attack is only practical for small e. In what follows we assume $e = 3$ for concreteness.

Assume a sender encrypts a message $m = m_1 \| m_2$, where m_1 is known but m_2 is not. Say m_2 is k bits long, so $m = 2^k \cdot m_1 + m_2$. Given the resulting ciphertext $c = [(m_1 \| m_2)^3 \bmod N]$, an eavesdropper can define $p(x) \stackrel{\text{def}}{=}$

$(2^k \cdot m_1 + x)^3 - c$, a cubic polynomial. This polynomial has m_2 as a root (modulo N), and $|m_2| < 2^k$. Theorem 12.29 thus implies that the attacker can compute m_2 efficiently as long as $2^k \leq N^{1/3}$. A similar attack works when m_2 is known but m_1 is not.

Encrypting related messages.[3] This attack assumes a sender who encrypts two related messages to the same receiver (something that should not result in an attack when using a secure encryption scheme). Assume the sender encrypts both m and $m+\delta$, where the offset δ is known but m is not. Given the two ciphertexts $c_1 = [m^e \bmod N]$ and $c_2 = [(m+\delta)^e \bmod N]$, an eavesdropper can define the two polynomials $f_1(x) \overset{\text{def}}{=} x^e - c_1$ and $f_2(x) \overset{\text{def}}{=} (x+\delta)^e - c_2$, each of degree e. Note that $x = m$ is a root (modulo N) of both polynomials, and so the linear term $(x - m)$ is a factor of both. Thus, if the greatest common divisor of $f_1(x)$ and $f_2(x)$ (modulo N) is linear, it will reveal m. The greatest common divisor can be computed in time $\mathsf{poly}(\|N\|, e)$ using an algorithm similar to the one in Appendix B.1.2; thus, this attack is feasible for small e.

Sending the same message to multiple receivers.[4] Our final attack assumes a sender who encrypts the same message to multiple receivers (something that, once again, should not result in an attack when using a secure encryption scheme). Let $e = 3$, and say the same message m is encrypted to three different parties holding public keys $pk_1 = \langle N_1, 3\rangle$, $pk_2 = \langle N_2, 3\rangle$, and $pk_3 = \langle N_3, 3\rangle$, respectively. Assume $\gcd(N_i, N_j) = 1$ for distinct i, j (if not, then at least one of the moduli can be factored immediately and the message m can be easily recovered). An eavesdropper sees

$$c_1 = [m^3 \bmod N_1], \quad c_2 = [m^3 \bmod N_2], \quad \text{and} \quad c_3 = [m^3 \bmod N_3].$$

Let $N^* = N_1 N_2 N_3$. An extended version of the Chinese remainder theorem says that there exists a unique non-negative integer $\hat{c} < N^*$ such that

$$\hat{c} = c_1 \bmod N_1$$
$$\hat{c} = c_2 \bmod N_2$$
$$\hat{c} = c_3 \bmod N_3.$$

Moreover, using techniques similar to those shown in Section 9.1.5 it is possible to compute $\hat{c}$ efficiently given the public keys and the above ciphertexts. Note finally that m^3 satisfies the above equations, and $m^3 < N^*$ since $m < \min\{N_1, N_2, N_3\}$. This means that $\hat{c} = m^3$ *over the integers* (i.e., with no modular reduction taking place), and so the message m can be recovered by computing the integer cube root of $\hat{c}$.

---

[3] This attack relies on some algebra slightly beyond what we have covered in this book.

[4] This attack relies on the Chinese remainder theorem presented in Section 9.1.5.

12.5.2 Padded RSA and PKCS #1 v1.5

Although plain RSA is insecure, it does suggest one general approach to public-key encryption based on the RSA problem: to encrypt a message m using public key $\langle N, e \rangle$, first map m to an element $\hat{m} \in \mathbb{Z}_N^*$; then compute the ciphertext $c := [\hat{m}^e \bmod N]$. To decrypt a ciphertext c, the receiver computes $\hat{m} := [c^d \bmod N]$ and then recovers the original message m. For the receiver to be able to recover the message, the mapping from messages to elements of $\mathbb{Z}_N^*$ must be (efficiently) *reversible*. For a scheme following this approach to have a hope of being CPA-secure, the mapping must be *randomized* so encryption is not deterministic. This is, of course, a necessary condition but not a sufficient one, and security of the encryption scheme depends critically on the specific mapping that is used.

One simple implementation of the above idea is to *randomly pad* the message before encrypting. That is, to map a message m (viewed as a bit-string) to an element of $\mathbb{Z}_N^*$, the sender chooses a uniform bit-string $r \in \{0, 1\}^\ell$ (for some appropriate ℓ) and sets $\hat{m} := r\|m$; the resulting value can naturally be interpreted as an integer in $\mathbb{Z}_N^*$. (This mapping is clearly reversible.) See Construction 12.30, where the bounds on $\ell(n)$ and the length of m ensure that the integer $\hat{m}$ is less than N.

CONSTRUCTION 12.30

Let GenRSA be as before, and let ℓ be a function with $\ell(n) < 2n$. Define a public-key encryption scheme as follows:

- **Gen:** on input 1^n, run GenRSA(1^n) to obtain (N, e, d). Output the public key $pk = \langle N, e \rangle$, and the private key $sk = \langle N, d \rangle$.

- **Enc:** on input a public key $pk = \langle N, e \rangle$ and a message $m \in \{0, 1\}^{\|N\| - \ell(n) - 1}$, choose a uniform string $r \in \{0, 1\}^{\ell(n)}$ and interpret $\hat{m} := r\|m$ as an element of $\mathbb{Z}_N^*$. Output the ciphertext

$$c := [\hat{m}^e \bmod N].$$

- **Dec:** on input a private key $sk = \langle N, d \rangle$ and a ciphertext $c \in \mathbb{Z}_N^*$, compute

$$\hat{m} := [c^d \bmod N],$$

and output the $\|N\| - \ell(n) - 1$ least-significant bits of $\hat{m}$.

The padded RSA encryption scheme.

The construction is parameterized by a value ℓ that determines the length of the random padding used. Security of the scheme depends on ℓ. There is an obvious brute-force attack on the scheme that runs in time 2^ℓ, so if ℓ is too short (in particular, if $\ell(n) = \mathcal{O}(\log n)$), the scheme is insecure. At the other extreme, the result we show in the following section shows that when the padding is as large as possible, and m is just a single bit, then it is possible to prove security based on the RSA assumption. In intermediate cases, the situation is less clear: for certain ℓ we cannot prove security based on the RSA assumption but no polynomial-time attacks are known either. We defer further discussion until after our treatment of PKCS #1 v1.5 next.

RSA PKCS #1 v1.5. The *RSA Laboratories Public-Key Cryptography Standard (PKCS) #1 version 1.5*, issued in 1993, utilizes a variant of padded RSA encryption. For a public key $pk = \langle N, e \rangle$ of the usual form, let k denote the length of N in bytes; i.e., k is the integer satisfying $2^{8(k-1)} \leq N < 2^{8k}$. Messages m to be encrypted are assumed to have length an integer number of bytes ranging from one to $k - 11$. Encryption of a D-byte message m is computed as

$$[(0x00\|0x02\|r\|0x00\|m)^e \bmod N],$$

where r is a randomly generated, $(k - D - 3)$-byte string with none of its bytes equal to 0x00. (This latter condition enables the message to be unambiguously recovered upon decryption.) Note that the maximum allowed length of m ensures that r is at least 8 bytes long.

Unfortunately, PKCS #1 v1.5 as specified is not CPA-secure because it allows using random padding that is too short. This is best illustrated by showing that an attacker can determine the initial portion of a message known to have many trailing 0s. For simplicity, say $m = b\|\underbrace{0\cdots0}_{L}$ where $b \in \{0, 1\}$ is unknown and m is as long as possible (so $L = 8 \cdot (k - 11) - 1$). Encryption of m gives a ciphertext c with

$$c = (0x00\|0x02\|r\|0x00\|b\|0\cdots0)^e \bmod N.$$

An attacker can compute $c' := c/(2^L)^e \bmod N$; note that

$$c' = \left(\frac{0x00\|0x02\|r\|0x00\|b\|0\cdots0}{10\cdots0}\right)^e = (0x02\|r\|0x00\|b)^e \bmod N.$$

The integer $0x02\|r\|0x00\|b$ is 75 bits long (note that $0x02 = 0000\,0010$, and all the high-order 0-bits don't count), and so an attacker can now apply the "short-message attack," or the attack based on encrypting a partially known message, from the previous section. To avoid these attacks we need to take r of length at least $\|N\|/e$. Even if e is large, however, the "quadratic-improvement attack" from the previous section shows that r can be recovered, with high probability, in time roughly $2^{\|r\|/2}$.

If we force r to be roughly half the length of N, and correspondingly reduce the maximum message length, then it is reasonable to conjecture that the encryption scheme in PKCS #1 v1.5 is CPA-secure. (We stress, however, that no proof of security based on the RSA assumption is known.) Nevertheless, because of a serious chosen-*ciphertext* attack on the scheme, described briefly in Section 12.5.5, newer versions of the PKCS #1 standard have been introduced and should be used instead.

12.5.3 *CPA-Secure Encryption without Random Oracles

In this section we show an encryption scheme that can be proven to be CPA-secure based on the RSA assumption. We begin by describing a specific hard-core predicate (see Section 8.1.3) for the RSA problem and then show how to use that hard-core predicate to encrypt a single bit. We then extend this scheme to give a KEM.

The schemes described in this section are mainly of theoretical interest and are not used in practice. This is because they are less efficient than alternative RSA-based constructions that can be proven secure in the random-oracle model (cf. Section 6.5). We will see examples of such encryption schemes in the sections that follow.

A hard-core predicate for the RSA problem. Loosely speaking, the RSA assumption says that given N, e, and $[x^e \bmod N]$ (for x chosen uniformly from $\mathbb{Z}_N^*$), it is infeasible to recover x. By itself, this says nothing about the computational difficulty of computing some specific information about x. Can we isolate some particular bit of information about x that is hard to compute from N, e and $[x^e \bmod N]$? The notion of a *hard-core predicate* captures exactly this requirement. (Hard-core predicates were introduced in Section 8.1.3. The fact that the RSA assumption gives a family of one-way permutations is discussed in Section 9.4.1. Our treatment here, however, is self-contained.) It is possible to show that the least-significant bit of x, denoted $\mathsf{lsb}(x)$, is a hard-core predicate for the RSA problem.

Define the following experiment for a given algorithm GenRSA (with the usual behavior) and algorithm $\mathcal{A}$:

The RSA hard-core predicate experiment RSA-lsb$_{\mathcal{A},\mathsf{GenRSA}}(1^n)$:

1. *Run* GenRSA(1^n) *to obtain* (N, e, d).
2. *Choose a uniform* $x \in \mathbb{Z}_N^*$ *and compute* $y := [x^e \bmod N]$.
3. $\mathcal{A}$ *is given* N, e, y, *and outputs a bit* b.
4. *The output of the experiment is* 1 *if and only if* $\mathsf{lsb}(x) = b$.

Observe that $\mathsf{lsb}(x)$ is a uniform bit when $x \in \mathbb{Z}_N^*$ is uniform. $\mathcal{A}$ can guess $\mathsf{lsb}(x)$ with probability $1/2$ by simply outputting a uniform bit b. The following theorem states that if the RSA problem is hard, then no efficient algorithm $\mathcal{A}$ can do significantly better than this; i.e., the least-significant bit is a hard-core predicate of the RSA permutation.

THEOREM 12.31 *If the RSA problem is hard relative to* GenRSA *then for all probabilistic polynomial-time algorithms* $\mathcal{A}$ *there is a negligible function* negl *such that* $\Pr[\textsf{RSA-lsb}_{\mathcal{A},\mathsf{GenRSA}}(n) = 1] \leq \frac{1}{2} + \mathsf{negl}(n)$.

A full proof of this theorem is beyond the scope of this book. However, we provide some intuition for the theorem by sketching a proof of a weaker result: that the RSA assumption implies $\Pr[\text{RSA-lsb}_{\mathcal{A},\text{GenRSA}}(n) = 1] < 1$ for all probabilistic polynomial-time $\mathcal{A}$. To prove this we show that an efficient algorithm that *always* correctly computes $\text{lsb}(r)$ from N, e, and $[r^e \bmod N]$ can be used to efficiently recover x (in its entirety) from N, e, and $[x^e \bmod N]$.

Fix N and e, and let $\mathcal{A}$ be an algorithm such that $\mathcal{A}([r^e \bmod N]) = \text{lsb}(r)$. Given N, e, and $y = [x^e \bmod N]$, we will recover the bits of x one-by-one, from least to most significant. To determine $\text{lsb}(x)$ we simply run $\mathcal{A}(y)$. There are now two cases:

Case 1: $\text{lsb}(x) = 0$. Note that $y/2^e = (x/2)^e \bmod N$, and because x is even (i.e., $\text{lsb}(x) = 0$), 2 divides the integer x. So $x/2$ is just the right-wise bit-shift of x, and $\text{lsb}(x/2)$ is equal to $2\text{sb}(x)$, the 2nd-least-significant bit of x. So we can obtain $2\text{sb}(x)$ by computing $y' := [y/2^e \bmod N]$ and then running $\mathcal{A}(y')$.

Case 2: $\text{lsb}(x) = 1$. Here $[x/2 \bmod N] = (x + N)/2$. So $\text{lsb}([x/2 \bmod N])$ is equal to $2\text{sb}(x + N)$; the latter is equal to $1 \oplus 2\text{sb}(N) \oplus 2\text{sb}(x)$ (we have a carry bit in the second position because both x and N are odd). So if we compute $y' := [y/2^e \bmod N]$, then $2\text{sb}(x) = \mathcal{A}(y') \oplus 1 \oplus 2\text{sb}(N)$.

Continuing in this way, we can recover all the bits of x.

Encrypting one bit. We can use the hard-core predicate identified above to encrypt a single bit. The idea is straightforward: to encrypt the message $m \in \{0, 1\}$, the sender chooses uniform $r \in \mathbb{Z}_N^*$ subject to the constraint that $\text{lsb}(r) = m$; the ciphertext is $c := [r^e \bmod N]$. See Construction 12.32.

CONSTRUCTION 12.32

Let GenRSA be as usual, and define a public-key encryption scheme as follows:

- Gen: on input 1^n, run GenRSA(1^n) to obtain (N, e, d). Output the public key $pk = \langle N, e \rangle$, and the private key $sk = \langle N, d \rangle$.

- Enc: on input a public key $pk = \langle N, e \rangle$ and a message $m \in \{0, 1\}$, choose a uniform $r \in \mathbb{Z}_N^*$ subject to the constraint that $\text{lsb}(r) = m$. Output the ciphertext $c := [r^e \bmod N]$.

- Dec: on input a private key $sk = \langle N, d \rangle$ and a ciphertext c, compute $r := [c^d \bmod N]$ and output $\text{lsb}(r)$.

Single-bit encryption using a hard-core predicate for RSA.

THEOREM 12.33 *If the RSA problem is hard relative to GenRSA then Construction 12.32 is CPA-secure.*

PROOF   Let Π denote Construction 12.32. We prove that Π has indistinguishable encryptions in the presence of an eavesdropper; by Proposition 12.3, this implies it is CPA-secure.

Let $\mathcal{A}$ be a probabilistic polynomial-time adversary. Without loss of generality, we may assume $m_0 = 0$ and $m_1 = 1$ in experiment $\mathsf{PubK}^{\mathsf{eav}}_{\mathcal{A},\Pi}(n)$. So

$$\Pr[\mathsf{PubK}^{\mathsf{eav}}_{\mathcal{A},\Pi}(n) = 1] = \frac{1}{2} \cdot \Pr[\mathcal{A}(N, e, c) = 0 \mid c \text{ is an encryption of } 0]$$
$$+ \frac{1}{2} \cdot \Pr[\mathcal{A}(N, e, c) = 1 \mid c \text{ is an encryption of } 1].$$

Consider running $\mathcal{A}$ in experiment RSA-lsb. By definition,

$$\Pr[\mathsf{RSA\text{-}lsb}_{\mathcal{A},\mathsf{GenRSA}}(n) = 1] = \Pr[\mathcal{A}\left(N, e, [r^e \bmod N]\right) = \mathsf{lsb}(r)],$$

where r is uniform in $\mathbb{Z}_N^*$. Since $\Pr[\mathsf{lsb}(r) = 1] = 1/2$, we have

$$\Pr[\mathsf{RSA\text{-}lsb}_{\mathcal{A},\mathsf{GenRSA}}(n) = 1] = \frac{1}{2} \cdot \Pr[\mathcal{A}\left(N, e, [r^e \bmod N]\right) = 0 \mid \mathsf{lsb}(r) = 0]$$
$$+ \frac{1}{2} \cdot \Pr[\mathcal{A}\left(N, e, [r^e \bmod N]\right) = 1 \mid \mathsf{lsb}(r) = 1].$$

Noting that encrypting $m \in \{0, 1\}$ corresponds exactly to choosing uniform r subject to the constraint that $\mathsf{lsb}(r) = m$, we see that

$$\Pr[\mathsf{PubK}^{\mathsf{eav}}_{\mathcal{A},\Pi}(n) = 1] = \Pr[\mathsf{RSA\text{-}lsb}_{\mathcal{A},\mathsf{GenRSA}}(n) = 1].$$

Theorem 12.31 thus implies that there is a negligible function negl such that

$$\Pr[\mathsf{PubK}^{\mathsf{eav}}_{\mathcal{A},\Pi}(n) = 1] \leq \frac{1}{2} + \mathsf{negl}(n),$$

as desired. ∎

Constructing a KEM. We now show how to extend Construction 12.32 so as to obtain a KEM with key length n. A naive way of doing this would be to simply choose a uniform, n-bit key k and then encrypt the bits of k one-by-one using n invocations of Construction 12.32. This would result in a rather long ciphertext consisting of n elements of $\mathbb{Z}_N^*$.

A better approach is for the sender to apply the RSA permutation (namely, raising to the eth power modulo N) repeatedly, starting from an initial, uniform value c_1. That is, the sender will successively compute c_1^e, followed by $(c_1^e)^e = c_1^{e^2}$, and so on, up to $c_1^{e^n}$ (all modulo N). The final value $[c_1^{e^n} \bmod N]$ will be the ciphertext, and the sequence of bits $\mathsf{lsb}(c_1), \mathsf{lsb}(c_1^e), \ldots, \mathsf{lsb}(c_1^{e^{n-1}})$ is the key. To decrypt a ciphertext c, the receiver simply reverses this process, successively computing $c^d, (c^d)^d = c^{d^2}$ up to c^{d^n} (again, all modulo N) to recover the initial value $c_1 = c^{d^n}$ used by the sender. Having recovered c_1, as well as the intermediate values $c_1^{e^n}, \ldots, c_1^e$, the receiver can compute the key.

It is possible to implement decryption more efficiently using the fact that the receiver knows the order of the group $\mathbb{Z}_N^*$. At key-generation time, the receiver can pre-compute $d' := [d^n \bmod \phi(N)]$ and store d' as part of its private key. To decrypt, the receiver can then directly compute $c_1 := [c^{d'} \bmod N]$, after which it can compute $c_1^e, \ldots, c_1^{e^n}$. (Exponentiations to the power e are more efficient than exponentiations to the power d since $e \ll d$ in practice.) This works, of course, since

$$c^{d^n} \bmod N = c^{[d^n \bmod \phi(N)]} = c^{d'} \bmod N.$$

The above is formally described as Construction 12.34.

CONSTRUCTION 12.34

Let GenRSA be as usual, and define a KEM as follows:

- Gen: on input 1^n, run GenRSA(1^n) to obtain (N, e, d). Then compute $d' := [d^n \bmod \phi(N)]$ (note that $\phi(N)$ can be computed from (N, e, d) or obtained during the course of running GenRSA). Output $pk = \langle N, e \rangle$ and $sk = \langle N, d' \rangle$.

- Encaps: on input $pk = \langle N, e \rangle$, choose a uniform $c_1 \in \mathbb{Z}_N^*$. Then for $i = 1, \ldots, n$ do:

  1. Compute $k_i := \mathsf{lsb}(c_i)$.
  2. Compute $c_{i+1} := [c_i^e \bmod N]$.

  Output the ciphertext c_{n+1} and the key $k = k_1 \cdots k_n$.

- Decaps: on input $sk = \langle N, d' \rangle$ and a ciphertext c, compute $c_1 := [c^{d'} \bmod N]$. Then for $i = 1, \ldots, n$ do:

  1. Compute $k_i := \mathsf{lsb}(c_i)$.
  2. Compute $c_{i+1} := [c_i^e \bmod N]$.

  Output the key $k = k_1 \cdots k_n$.

A KEM using a hard-core predicate for RSA.

The construction is reminiscent of the approach used to construct a pseudorandom generator from a one-way permutation toward the end of Section 8.4.2. If we let f denote the RSA permutation relative to some public key $\langle N, e \rangle$ (i.e., $f(x) \stackrel{\text{def}}{=} [x^e \bmod N]$), then CPA-security of Construction 12.34 is equivalent to pseudorandomness of $\mathsf{lsb}(f^{n-1}(c_1)), \ldots, \mathsf{lsb}(c_1)$ even conditioned on the value $c = f^n(c_1)$. This, in turn, can be proven using Theorem 12.31 and the techniques from Section 8.4.2. (The only difference is that in Section 8.4.2 the value $f^n(c_1)$ was itself a uniform n-bit string, whereas here it is a uniform element of $\mathbb{Z}_N^*$. Pseudorandomness of the successive hard-core predicates is independent of the domain of f.) Summarizing:

THEOREM 12.35 *If the RSA problem is hard relative to* GenRSA *then Construction 12.34 is a CPA-secure KEM.*

Efficiency. Construction 12.34 is reasonably efficient. To be concrete, assume that $n = 128$, the RSA modulus N is 2048 bits long, and the public exponent e is 3 so that exponentiation to the power e modulo N can be computed using two modular multiplications. (See Appendix B.2.3.) Encryption then requires $2n = 256$ modular multiplications. Decryption can be done with one full modular exponentiation (at the cost of approximately $1.5 \cdot 2048 = 3072$ modular multiplications) plus an additional 256 modular multiplications. The cost of decryption is thus only about 8% less efficient than for the plain RSA encryption scheme. Encryption is significantly more expensive than in plain RSA, but in many applications decryption time is more important (since it may be implemented by a server that is performing thousands of decryptions simultaneously).

12.5.4 OAEP and PKCS #1 v2

We now consider CCA-security for RSA-based encryption schemes. We begin by showing that all the RSA-based encryption schemes we have seen so far are vulnerable to chosen-ciphertext attacks.

Plain RSA encryption. Plain RSA is not even CPA-secure. But it does ensure that if $m \in \mathbb{Z}_N^*$ is uniform then an attacker who eavesdrops on the encryption $c = [m^e \bmod N]$ of m with respect to the public key $\langle N, e \rangle$ cannot recover m. Even this weak guarantee no longer holds in a setting where chosen-ciphertext attacks are possible. As in the case of El Gamal encryption, this is a consequence of the fact that plain RSA is *malleable*: given the encryption $c = [m^e \bmod N]$ of an unknown message m, it is easy to generate a ciphertext c' that is an encryption of $[2m \bmod N]$ by setting

$$c' := [2^e \cdot c \bmod N]$$
$$= 2^e \cdot m^e = (2m)^e \bmod N.$$

In fact, we have used this observation several times already.

RSA PKCS #1 v1.5. Padded RSA encryption, which is conjectured to be CPA-secure for the right setting of the parameters, is vulnerable to essentially the same attack as plain RSA encryption is. But there is also a more interesting chosen-ciphertext attack on PKCS #1 v1.5 encryption that, in contrast to an attack that exploits malleability, does *not* require full access to a decryption oracle; it only requires access to a "partial" decryption oracle that indicates whether or not decryption of some ciphertext returns an error. This makes the attack much more practical, as it can be carried out whenever an attacker can distinguish a decryption success from a decryption failure, as in the case of the padding-oracle attack discussed in Section 5.1.1.

Recall that the public-key encryption scheme defined in the PKCS #1 v1.5 standard uses a variant of padded RSA encryption where the padding is done in a specific way. In particular, the two high-order bytes of the padded message are always $\texttt{0x00}\|\texttt{0x02}$. When decrypting, the receiver is supposed to check that the two high-order bytes of the intermediate result match these values, and return an error if this is not the case. In 1998, Bleichenbacher developed a chosen-ciphertext attack that exploits the fact that this check is done. Roughly, given a ciphertext c that corresponds to an honest encryption of some unknown message m with respect to a public key $\langle N, e \rangle$, Bleichenbacher's attack repeatedly chooses uniform $s \in \mathbb{Z}_N^*$ and submits the ciphertext $c' := [s^e \cdot c \bmod N]$ to the receiver. Say $c = [\hat{m}^e \bmod N]$ where

$$\hat{m} = \texttt{0x00}\|\texttt{0x02}\|r\|\texttt{0x00}\|m,$$

as specified by PKCS #1 v1.5. Then decryption of c' will give the intermediate result $\hat{m}' = [s \cdot \hat{m} \bmod N]$, and the receiver will return an error unless the top two bytes of $\hat{m}'$ are exactly $\texttt{0x00}\|\texttt{0x02}$. (Other checks are done as well, but we ignore those for simplicity.) Thus, whenever decryption succeeds the attacker learns that the top two bytes of $s \cdot \hat{m} \bmod N$ are $\texttt{0x00}\|\texttt{0x02}$, where s is known. Sufficiently many equations of this type suffice for the attacker to learn $\hat{m}$ and recover all of the original message m.

The CPA-secure KEM. In Section 12.5.3 we showed a construction of a KEM that can be proven CPA-secure based on the RSA assumption. That construction is also insecure against a chosen-ciphertext attack; we leave the details as an exercise.

RSA-OAEP

We explore a construction of CCA-secure encryption from RSA using what is called *optimal asymmetric encryption padding* (OAEP). The resulting RSA-OAEP scheme follows the idea (used also in Section 12.5.2) of taking a message m, mapping it to an element $\hat{m} \in \mathbb{Z}_N^*$, and then letting $c = [\hat{m}^e \bmod N]$ be the ciphertext. The transformation here, however, is more complex than before. A version of RSA-OAEP has been standardized as part of RSA PKCS #1 since version 2.0.

Let $\ell(n), k(n)$ be integer-valued functions with $k(n) = \Theta(n)$, and such that $\ell(n) + 2k(n)$ is less than the bit-length of moduli output by $\mathsf{GenRSA}(1^n)$. Fix n, and let $\ell = \ell(n)$ and $k = k(n)$. Let $G : \{0,1\}^k \to \{0,1\}^{\ell+k}$ and $H : \{0,1\}^{\ell+k} \to \{0,1\}^k$ be two hash functions that will be modeled as independent random oracles. (Although using more than one random oracle was not discussed in Section 6.5.1, we can do so in the natural way.) The transformation defined by OAEP is based on a two-round Feistel network with G and H as round functions; see Figure 12.4. Mapping a message $m \in \{0,1\}^\ell$ to $\hat{m}$ is done as follows: first set $m' := m\|0^k$ and choose a uniform $r \in \{0,1\}^k$.

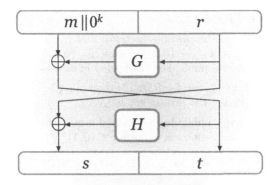

FIGURE 12.4: The OAEP transformation.

Then compute

$$t := m' \oplus G(r) \in \{0,1\}^{\ell+k}, \quad s := r \oplus H(t) \in \{0,1\}^k,$$

and set $\hat{m} := s\|t$. (The PKCS #1 standard differs from what we have described, but the differences are unimportant for our purposes.) To encrypt a message m with respect to the public key $\langle N, e \rangle$, the sender generates $\hat{m}$ as above and outputs the ciphertext $c := [\hat{m}^e \bmod N]$. (Note that $\hat{m}$, interpreted as an integer, is less than N because of the constraints on ℓ, k.)

To decrypt, the receiver computes $\hat{m} := [c^d \bmod N]$ and lets $s\|t := \hat{m}$ with s and t of the appropriate lengths. It then inverts the Feistel network by computing $r := H(t) \oplus s$ and $m' := G(r) \oplus t$. Importantly, the receiver then verifies that the trailing k bits of m' are all 0; if not, the ciphertext is rejected and an error message is returned. Otherwise, the k least-significant 0s of m' are discarded, and the remaining ℓ bits of m' are output as the message. This process is described in Construction 12.36.

RSA-OAEP can be proven to be CCA-secure based on the RSA assumption if G and H are modeled as random oracles. The proof is rather complicated, and we do not give it here; instead, we merely provide some intuition. First consider CPA-security. During encryption the sender computes

$$m' := m\|0^{k_1}, \quad t := m' \oplus G(r), \quad s := r \oplus H(t)$$

for uniform r; the ciphertext is $[(s\|t)^e \bmod N]$. If the attacker never queries r to G then, since we model G as a random function, the value $G(r)$ is uniform from the attacker's point of view and so m is masked with a uniform string just as in the one-time pad encryption scheme. Thus, if the attacker never queries r to G then no information about the message is leaked.

Can the attacker query r to G? The value of r is itself masked by $H(t)$. So the attacker has no information about r unless it first queries t to H. If the attacker does not query t to H then the attacker may get lucky and guess r anyway, but if r is sufficiently long the probability of doing so is negligible.

CONSTRUCTION 12.36

Let GenRSA be as in the previous sections, and ℓ, k be as described in the text. Let $G : \{0,1\}^k \rightarrow \{0,1\}^{\ell+k}$ and $H : \{0,1\}^{\ell+k} \rightarrow \{0,1\}^k$ be functions. Construct a public-key encryption scheme as follows:

- Gen: on input 1^n, run GenRSA(1^n) to obtain (N, e, d). The public key is $\langle N, e \rangle$ and the private key is $\langle N, d \rangle$.

- Enc: on input a public key $\langle N, e \rangle$ and a message $m \in \{0,1\}^\ell$, set $m' := m \| 0^k$ and choose a uniform $r \in \{0,1\}^k$. Then compute

$$t := m' \oplus G(r), \quad s := r \oplus H(t)$$

and set $\hat{m} := s \| t$. Output the ciphertext $c := [\hat{m}^e \bmod N]$.

- Dec: on input a private key $\langle N, d \rangle$ and a ciphertext $c \in \mathbb{Z}_N^*$, compute $\hat{m} := [c^d \bmod N]$. If $\|\hat{m}\| > \ell + 2k$, output $\bot$. Otherwise, parse $\hat{m}$ as $s \| t$ with $s \in \{0,1\}^{\ell+k}$ and $t \in \{0,1\}^k$. Compute $r := H(t) \oplus s$ and $m' := G(r) \oplus t$. If the k least-significant bits of m' are not all 0, output $\bot$. Otherwise, output the ℓ most-significant bits of m'.

The RSA-OAEP encryption scheme.

Can the attacker query t to H? Doing so would require the attacker to compute t from $[(s\|t)^e \bmod N]$. Note that doing so does not directly solve the RSA problem, which instead would require computing *both* s and t. Nevertheless, for the right settings of the parameters it is possible to show that recovering t is computationally infeasible if the RSA problem is hard.

Arguing CCA-security involves additional complications, but the basic idea is to show that every decryption-oracle query c made by the attacker falls into one of two categories: either the attacker obtained c by legally encrypting some message m itself (in which case the attacker learns nothing from the decryption query), or else decryption of c returns an error. This is a consequence of the fact that the receiver checks that the k least-significant bits of m' are 0 during decryption; if the attacker did not generate the ciphertext c using the prescribed encryption algorithm, the probability that this condition holds is negligible. The formal proof is complicated by the fact that the attacker's decryption-oracle queries must be answered correctly without knowledge of the private key, which means there must be an efficient way to determine whether to return an error or not and, if not, what message to return. This is accomplished by looking at the adversary's queries to the random oracles G, H.

Manger's chosen-ciphertext attack on PKCS #1 v2.0. In 2001, James Manger showed a chosen-ciphertext attack against certain implementations of the RSA encryption scheme specified in PKCS #1 v2.0—even though what was specified was a variant of RSA-OAEP! Since Construction 12.36 can be proven to be CCA-secure, how is this possible?

Examining the decryption algorithm in Construction 12.36, note that there are two ways an error can occur: either $\hat{m} \in \mathbb{Z}_N^*$ is too large, or $m' \in \{0,1\}^{\ell+k}$ does not have enough trailing 0s. In Construction 12.36, the receiver is supposed to return the *same* error (denoted $\perp$) in either case. In some implementations, however, the receiver would output *different* errors depending on which step failed. This single bit of additional information enabled a chosen-ciphertext attack that could recover a message m in its entirety from a corresponding ciphertext using only $\approx \|N\|$ queries to an oracle leaking the type of error upon decryption. This shows the importance of implementing cryptographic schemes exactly as specified, since the resulting proof and analysis may no longer apply if aspects of the scheme are changed.

Note that even if the same error is returned in both cases, an attacker might be able to determine where the error occurs if the *time* to return the error is different. (This is a great example of how an attacker is not limited to examining the inputs/outputs of an algorithm, but can use *side-channel information* to attack a scheme.) Implementations should ensure that the time to return an error is identical in either case.

12.5.5 *A CCA-Secure KEM in the Random-Oracle Model

We show here a construction of an RSA-based KEM that is CCA-secure in the random-oracle model; this scheme is included as part of the ISO/IEC 18033-2 standard for public-key encryption. (Recall from Theorem 12.14 that any such construction can be used in conjunction with any CCA-secure *private*-key encryption scheme to give a CCA-secure *public*-key encryption scheme.) As compared to the RSA-OAEP scheme from the previous section, the main advantage is the simplicity of both the construction and its proof of security. Its main disadvantage is that it results in longer ciphertexts when encrypting short messages since it requires the KEM/DEM paradigm whereas RSA-OAEP does not. For encrypting long messages, however, RSA-OAEP would also be used as part of a hybrid encryption scheme, and would result in an encryption scheme having similar efficiency to what would be obtained using the KEM shown here.

The public key of the scheme includes $\langle N, e \rangle$ as usual, as well as a specification of a function $H : \mathbb{Z}_N^* \to \{0,1\}^n$ that will be modeled as a random oracle in the analysis. (This function can be based on some underlying cryptographic hash function, as discussed in Section 6.5. We omit the details.) To encapsulate a key, the sender chooses uniform $r \in \mathbb{Z}_N^*$ and then computes the ciphertext $c := [r^e \bmod N]$ and the key $k := H(r)$. To decrypt a ciphertext c, the receiver simply recovers r in the usual way and then re-derives the same key $k := H(r)$. See Construction 12.37.

CPA-security of the scheme is immediate. Indeed, the ciphertext c is equal to $[r^e \bmod N]$ for uniform $r \in \mathbb{Z}_N^*$, and so the RSA assumption implies that an eavesdropper who observes c will be unable to compute r. This means, in turn, that (except with negligible probability) the eavesdropper will not

---

CONSTRUCTION 12.37

Let GenRSA be as usual, and construct a KEM as follows:

- Gen: on input 1^n, run $\mathsf{GenRSA}(1^n)$ to compute (N, e, d). The public key is $\langle N, e \rangle$, and the private key is $\langle N, d \rangle$.

  As part of key generation, a function $H : \mathbb{Z}_N^* \to \{0,1\}^n$ is specified, but we leave this implicit.

- Encaps: on input public key $\langle N, e \rangle$, choose a uniform $r \in \mathbb{Z}_N^*$. Output the ciphertext $c := [r^e \bmod N]$ and the key $k := H(r)$.

- Decaps: on input private key $\langle N, d \rangle$ and a ciphertext $c \in \mathbb{Z}_N^*$, compute $r := [c^d \bmod N]$ and output the key $k := H(r)$.

A CCA-secure KEM (in the random-oracle model).

query r to H, and thus the value of the key $k \overset{\text{def}}{=} H(r)$ will remain uniform from the attacker's point of view.

In fact, the above extends to show CCA-security as well. This is because answering a decapsulation-oracle query for any ciphertext $\tilde{c} \neq c$ only involves evaluating H at some input $[\tilde{c}^d \bmod N] = \tilde{r} \neq r$. Thus, the attacker's decapsulation-oracle queries do not reveal any additional information about the key $H(r)$ encapsulated by the challenge ciphertext. (A formal proof is slightly more involved since we must show how it is possible to *simulate* the answers to decapsulation-oracle queries without knowledge of the private key. Nevertheless, this turns out to be not very difficult.)

THEOREM 12.38   *If the RSA problem is hard relative to* GenRSA *and H is modeled as a random oracle, then Construction 12.37 is CCA-secure.*

PROOF   Let Π denote Construction 12.37, and let $\mathcal{A}$ be a probabilistic polynomial-time adversary. For convenience, and because this is the first proof where we use the full power of the random-oracle model, we explicitly describe the steps of experiment $\mathsf{KEM}_{\mathcal{A},\Pi}^{\mathsf{cca}}(n)$:

1. $\mathsf{GenRSA}(1^n)$ is run to obtain (N, e, d). In addition, a random function $H : \mathbb{Z}_N^* \to \{0,1\}^n$ is chosen.

2. Uniform $r \in \mathbb{Z}_N^*$ is chosen, and the ciphertext $c := [r^e \bmod N]$ and key $k := H(r)$ are computed.

3. A uniform bit $b \in \{0,1\}$ is chosen. If $b = 0$ set $\hat{k} := k$. If $b = 1$ then choose a uniform $\hat{k} \in \{0,1\}^n$.

4. $\mathcal{A}$ is given $pk = \langle N, e \rangle$, c, and $\hat{k}$, and may query $H(\cdot)$ (on any input) and the decapsulation oracle $\mathsf{Decaps}_{\langle N, d \rangle}(\cdot)$ on any ciphertext $\hat{c} \neq c$.

5. $\mathcal{A}$ outputs a bit b'. The output of the experiment is defined to be 1 if $b' = b$, and 0 otherwise.

In an execution of experiment $\mathsf{KEM}^{\mathsf{cca}}_{\mathcal{A},\Pi}(n)$, let Query be the event that, at any point during its execution, $\mathcal{A}$ queries r to the random oracle H. We let Success denote the event that $b' = b$ (i.e., the experiment outputs 1). Then

$$\Pr[\mathsf{Success}] = \Pr\left[\mathsf{Success} \wedge \overline{\mathsf{Query}}\right] + \Pr[\mathsf{Success} \wedge \mathsf{Query}]$$
$$\leq \Pr\left[\mathsf{Success} \wedge \overline{\mathsf{Query}}\right] + \Pr[\mathsf{Query}],$$

where all probabilities are taken over the randomness used in experiment $\mathsf{KEM}^{\mathsf{cca}}_{\mathcal{A},\Pi}(n)$. We show that $\Pr\left[\mathsf{Success} \wedge \overline{\mathsf{Query}}\right] \leq \frac{1}{2}$ and that $\Pr[\mathsf{Query}]$ is negligible. The theorem follows.

We first argue that $\Pr\left[\mathsf{Success} \wedge \overline{\mathsf{Query}}\right] \leq \frac{1}{2}$. If $\Pr[\overline{\mathsf{Query}}] = 0$ this is immediate. Otherwise, $\Pr\left[\mathsf{Success} \wedge \overline{\mathsf{Query}}\right] \leq \Pr\left[\mathsf{Success}|\overline{\mathsf{Query}}\right]$. Now, conditioned on $\overline{\mathsf{Query}}$, the value of the correct key $k = H(r)$ is uniform because H is a random function. Consider $\mathcal{A}$'s information about k in experiment $\mathsf{KEM}^{\mathsf{cca}}_{\mathcal{A},\Pi}(n)$. The public key pk and ciphertext c, by themselves, do not contain any information about k. (They do uniquely determine r, but since H is chosen independently of anything else, this gives no information about $H(r)$.) Queries that $\mathcal{A}$ makes to H also do not reveal any information about r, unless $\mathcal{A}$ queries r to H (in which case Query occurs); this, again, relies on the fact that H is a random function. Finally, queries that $\mathcal{A}$ makes to its decapsulation oracle only reveal $H(\tilde{r})$ for $\tilde{r} \neq r$. This follows from the fact that $\mathsf{Decaps}_{\langle N,d \rangle}(\tilde{c}) = H(\tilde{r})$ where $\tilde{r} = [\tilde{c}^d \bmod N]$, but $\tilde{c} \neq c$ implies $\tilde{r} \neq r$. Once again, this and the fact that H is a random function mean that no information about $H(r)$ is revealed unless Query occurs.

The above shows that, as long as Query does not occur, the value of the correct key k is uniform even given $\mathcal{A}$'s view of the public key, ciphertext, and the answers to all its oracle queries. In that case, then, there is no way $\mathcal{A}$ can distinguish (any better than random guessing) whether $\hat{k}$ is the correct key or a uniform, independent key. Therefore, $\Pr\left[\mathsf{Success}|\overline{\mathsf{Query}}\right] = \frac{1}{2}$.

We highlight that nowhere in the above argument did we rely on the fact that $\mathcal{A}$ is computationally bounded, and in fact $\Pr\left[\mathsf{Success} \wedge \overline{\mathsf{Query}}\right] \leq \frac{1}{2}$ even if no computational restrictions are placed on $\mathcal{A}$. This indicates part of the power of the random-oracle model.

To complete the proof of the theorem, we show

CLAIM 12.39 *If the RSA problem is hard relative to* GenRSA *and* H *is modeled as a random oracle, then* $\Pr[\mathsf{Query}]$ *is negligible.*

To prove this, we construct an algorithm $\mathcal{A}'$ that uses $\mathcal{A}$ as a subroutine. $\mathcal{A}'$ is given an instance N, e, c of the RSA problem, and its goal is to compute r for which $r^e = c \bmod N$. To do so, it will run $\mathcal{A}$, answering its queries to H and Decaps. Handling queries to H is simple, since $\mathcal{A}'$ can just return a random value. Queries to Decaps are trickier, however, since $\mathcal{A}'$ does not know the private key associated with the effective public key $\langle N, e \rangle$.

On further thought, however, decapsulation queries are also easy to answer since $\mathcal{A}'$ can just return a random value here as well. That is, although the query $\mathsf{Decaps}(\tilde{c})$ is supposed to be computed by first computing $\tilde{r}$ such that $\tilde{r}^e = \tilde{c} \bmod N$ and then evaluating $H(\tilde{r})$, the result is just a uniform value. Thus, $\mathcal{A}'$ can simply return a random value without performing the intermediate computation. The only "catch" is that $\mathcal{A}'$ must ensure consistency between its answers to H-queries and Decaps-queries; namely, it must ensure that for any $\tilde{r}, \tilde{c}$ with $\tilde{r}^e = \tilde{c} \bmod N$ it holds that $H(\tilde{r}) = \mathsf{Decaps}(\tilde{c})$. This is handled using simple bookkeeping and lists L_H and L_{Decaps} that keep track of the answers $\mathcal{A}'$ has given in response to the respective oracle queries. We now give the details.

Algorithm $\mathcal{A}'$:
The algorithm is given (N, e, c) as input.

1. Initialize empty lists L_H, L_{Decaps}. choose a uniform $k \in \{0,1\}^n$ and store (c, k) in L_{Decaps}.

2. Choose a uniform bit $b \in \{0,1\}$. If $b = 0$ set $\hat{k} := k$. If $b = 1$ then choose a uniform $\hat{k} \in \{0,1\}^n$. Run $\mathcal{A}$ on $\langle N, e \rangle$, c, and $\hat{k}$.

   When $\mathcal{A}$ makes a query $H(\tilde{r})$, answer it as follows:
   - If there is an entry in L_H of the form $(\tilde{r}, k)$ for some k, return k.
   - Otherwise, let $\tilde{c} := [\tilde{r}^e \bmod N]$. If there is an entry in L_{Decaps} of the form $(\tilde{c}, k)$ for some k, return k and store $(\tilde{r}, k)$ in L_H.
   - Otherwise, choose a uniform $k \in \{0,1\}^n$, return k, and store $(\tilde{r}, k)$ in L_H.

   When $\mathcal{A}$ makes a query $\mathsf{Decaps}(\tilde{c})$, answer it as follows:
   - If there is an entry in L_{Decaps} of the form $(\tilde{c}, k)$ for some k, return k.
   - Otherwise, for each entry $(\tilde{r}, k) \in L_H$, check if $\tilde{r}^e = \tilde{c} \bmod N$ and, if so, output k.
   - Otherwise, choose a uniform $k \in \{0,1\}^n$, return k, and store $(\tilde{c}, k)$ in L_{Decaps}.

3. At the end of $\mathcal{A}$'s execution, if there is an entry (r, k) in L_H for which $r^e = c \bmod N$ then return r.

Clearly $\mathcal{A}'$ runs in polynomial time, and the view of $\mathcal{A}$ when run as a subroutine by $\mathcal{A}'$ in experiment $\mathsf{RSA\text{-}inv}_{\mathcal{A}', \mathsf{GenRSA}}(n)$ is *identical* to the view of $\mathcal{A}$ in experiment $\mathsf{KEM}^{\mathsf{cca}}_{\mathcal{A}, \Pi}(n)$: the inputs given to $\mathcal{A}$ clearly have the right distribution, the answers to $\mathcal{A}$'s oracle queries are consistent, and the responses to all H-queries are uniform and independent. Finally, $\mathcal{A}'$ outputs the correct solution exactly when Query occurs. Hardness of the RSA problem relative to GenRSA thus implies that $\Pr[\text{Query}]$ is negligible, as required. ∎

It is worth remarking on the various properties of the random-oracle model (see Section 6.5.1) that are used in the above proof. First, we rely on the fact that the value $H(r)$ is uniform unless r is queried to H—even if H is queried on multiple other values $\tilde{r} \neq r$. We also, implicitly, use extractability to argue that the attacker cannot query r to H; otherwise, we could use this attacker to solve the RSA problem. Finally, the proof relies on programmability in order to simulate the adversary's decapsulation-oracle queries.

12.5.6 RSA Implementation Issues and Pitfalls

We close this section with a brief discussion of some issues related to the implementation of RSA-based schemes, and some pitfalls to be aware of.

Using Chinese remaindering. In implementations of RSA-based encryption, the receiver can use the Chinese remainder theorem (Section 9.1.5) to speed up computation of eth roots modulo N during decryption. Specifically, let $N = pq$ and say the receiver wishes to compute the eth root of some value y using $d = [e^{-1} \bmod \phi(N)]$. The receiver can use the correspondence $[y^d \bmod N] \leftrightarrow ([y^d \bmod p], [y^d \bmod q])$ to compute the partial results

$$x_p := [y^d \bmod p] = \left[y^{[d \bmod (p-1)]} \bmod p\right] \tag{12.19}$$

and

$$x_q := [y^d \bmod q] = \left[y^{[d \bmod (q-1)]} \bmod q\right], \tag{12.20}$$

and then combine these to obtain $x \leftrightarrow (x_p, x_q)$, as discussed in Section 9.1.5. Note that $[d \bmod (p-1)]$ and $[d \bmod (q-1)]$ could be pre-computed since they are independent of y.

Why is this better? Assume exponentiation modulo an ℓ-bit integer takes $\gamma \cdot \ell^3$ operations for some constant γ. If p, q are each n bits long, then naively computing $[y^d \bmod N]$ takes $\gamma \cdot (2n)^3 = 8\gamma \cdot n^3$ steps (because $\|N\| = 2n$). Using Chinese remaindering reduces this to roughly $2 \cdot (\gamma \cdot n^3)$ steps (because $\|p\| = \|q\| = n$), or roughly $1/4$ of the time.

Example 12.40

We revisit Example 9.49. Recall that $N = 143 = 11 \cdot 13$ and $d = 103$, and

$y = 64$ there. To calculate $[64^{103} \bmod 143]$ we compute

$$\left([64 \bmod 11],\ [64 \bmod 13]\right)^{103} = \left([(-2)^{103} \bmod 11],\ [(-1)^{103} \bmod 13]\right)$$
$$= \left([(-2)^{[103 \bmod 10]} \bmod 11],\ -1\right)$$
$$= \left([-8 \bmod 11],\ -1\right) = (3,\ -1).$$

We can compute $1_p = 78 \leftrightarrow (1,0)$ and $1_q = 66 \leftrightarrow (0,1)$, as discussed in Section 9.1.5. (Note these values can be pre-computed, as they are independent of y.) Then $(3,-1) \leftrightarrow 3 \cdot 1_p - 1_q = 3 \cdot 78 - 66 = 168 = 25 \bmod 143$, in agreement with the answer previously obtained. $\qquad\qquad\qquad\qquad\qquad \diamond$

A fault attack when using Chinese remaindering. When using Chinese remaindering as just described, one should be aware of a potential attack that can be carried out if *faults* occur (or can be induced to occur by an attacker, e.g., by hardware tampering) during the course of the computation.

Consider what happens if $[y^d \bmod N]$ is computed twice: the first time with no error (giving the correct result x), but the second time with an error during computation of Equation (12.20) but not Equation (12.19) (the same attack applies in the opposite case). The second computation yields an incorrect result x' for which $x' = x \bmod p$ but $x' \neq x \bmod q$. This means that $p \mid (x'-x)$ but $q \nmid (x' - x)$. But then $\gcd(x' - x, N) = p$, yielding the factorization of N.

One possible countermeasure is to verify correctness of the result before using it, by checking that $x^e = y \bmod N$. (Since $\|e\| \ll \|d\|$, using Chinese remaindering still gives better efficiency.) This is recommended in hardware implementations.

Dependent public keys I. When multiple receivers wish to utilize the same encryption scheme, they should use *independent* public keys. This and the following attack demonstrate what can go wrong when this is not done.

Imagine a company wants to use the same modulus N for each of its employees. Since it is not desirable for messages encrypted to one employee to be read by any other employee, the company issues different (e_i, d_i) pairs to each employee. That is, the public key of the ith employee is $pk_i = \langle N, e_i \rangle$ and their private key is $sk = \langle N, d_i \rangle$, where $e_i \cdot d_i = 1 \bmod \phi(N)$ for all i.

This approach is insecure and allows any employee to read messages encrypted to all other employees. The reason is that, as noted in Section 9.2.4, given N and e_i, d_i with $e_i \cdot d_i = 1 \bmod \phi(N)$, the factorization of N can be efficiently computed. Given the factorization of N, of course, it is possible to compute $d_j := e_j^{-1} \bmod \phi(N)$ for any j.

Dependent public keys II. The attack just shown allows any employee to decrypt messages sent to any other employee. This still leaves the possibility that sharing the modulus N is fine as long as all employees trust each other (or, alternatively, as long as confidentiality need only be preserved against outsiders but not against other members of the company). Here we show a scenario indicating that sharing a modulus is still a bad idea, at least when plain RSA encryption is used.

Say the same message m is encrypted and sent to two different (known) employees with public keys (N, e_1) and (N, e_2) where $e_1 \neq e_2$. Assume further that $\gcd(e_1, e_2) = 1$. Then an eavesdropper sees the two ciphertexts

$$c_1 = m^{e_1} \bmod N \quad \text{and} \quad c_2 = m^{e_2} \bmod N.$$

Since $\gcd(e_1, e_2) = 1$, there exist integers X, Y such that $Xe_1 + Ye_2 = 1$ by Proposition 9.2. Moreover, given the public exponents e_1 and e_2 it is possible to efficiently compute X and Y using the extended Euclidean algorithm (see Appendix B.1.2). We claim that $m = [c_1^X \cdot c_2^Y \bmod N]$, which can easily be calculated. This is true because

$$c_1^X \cdot c_2^Y = m^{Xe_1} m^{Ye_2} = m^{Xe_1 + Ye_2} = m^1 = m \bmod N.$$

A similar attack applies when using padded RSA or RSA-OAEP if the sender uses the same transformed message $\hat{m}$ when encrypting to two users.

Randomness quality in RSA key generation. Throughout this book, we always assume that honest parties have access to sufficient, high-quality randomness. When this assumption is violated then security may fail to hold. For example, if an ℓ-bit string is chosen from some set $S \subset \{0,1\}^\ell$ rather than uniformly from $\{0,1\}^\ell$, then an attacker can perform a brute-force search (in time $\mathcal{O}(|S|)$) to attack the system.

In some cases the situation may be even worse. Consider in particular the case of RSA key generation, where random bits r_p is used to choose the first prime p, and random bits r_q is used to generate the second prime q. Assume further that many public/private keys are generated using the same source of poor-quality randomness, in which r_p, r_q are each chosen uniformly from some set S of size 2^s. After generating roughly $2^{s/2}$ public keys (see Appendix A.4), we expect to obtain two different moduli N, N' that were generated using identical randomness $r_p = r_p'$ but different randomness $r_q \neq r_q'$. These two moduli share a prime factor which can be easily found by computing $\gcd(N, N')$. An attacker can attempt to exploit this by scraping the Internet for a large set of RSA public keys, computing their pairwise gcd's, and thus hoping to factor some subset of them. Although computing pairwise gcd's of $2^{s/2}$ moduli would naively take time $\mathcal{O}(2^s)$, it turns out that this can be significantly improved using a "divide-and-conquer" approach that is beyond the scope of this book. The upshot is that an attacker can factor a small number of public moduli in time less than 2^s. Note also that the attack works even if the set S is unknown to the attacker.

The above scenario was verified experimentally by two research teams working independently, who carried out exactly the above attack on public keys obtained over the Internet, and were able to successfully factor a significant fraction of the keys they found.

References and Additional Reading

The idea of public-key encryption was first proposed in the open literature by Diffie and Hellman [65]. Rivest, Shamir, and Adleman [171] introduced the RSA assumption and proposed a public-key encryption scheme based on this assumption. As pointed out in the previous chapter, other pioneers of public-key cryptography include Merkle and Rabin (in academic publications) and Ellis, Cocks, and Williamson (in classified publications).

Definition 12.2 is rooted in the seminal work of Goldwasser and Micali [87], who were also the first to recognize the necessity of *probabilistic* encryption for satisfying this definition. As noted in Chapter 4, chosen-ciphertext attacks were first formally defined by Naor and Yung [147] and Rackoff and Simon [168]. The expository article by Shoup [180] discusses the importance of security against chosen-ciphertext attacks. Bellare et al. give a unified, modern treatment of various security notions for public-key encryption [18].

A proof of CPA-security for hybrid encryption was first given by Blum and Goldwasser [40]. The case of CCA-security was treated in [63].

Somewhat amazingly, the El Gamal encryption scheme [77] was not suggested until 1984, even though it can be viewed as a direct transformation of the Diffie–Hellman key-exchange protocol (see Exercise 12.4). DHIES was introduced in [2]. The ISO/IEC 18033-2 standard for public-key encryption can be found at http://www.shoup.net/iso.

Plain RSA encryption corresponds to the original scheme introduced by Rivest, Shamir, and Adleman [171]. The attacks on plain RSA encryption described in Section 12.5.1 are due to [186, 62, 92, 55, 44]; see [137, Chapter 8] and [42] for additional attacks and further information. Proofs of Coppersmith's theorem can be found in the original work [54] or several subsequent expositions (e.g., [76, 135]).

The PKCS #1 standards are available as RFCs [107, 108, 145]. For progress toward proving security of the padded RSA encryption scheme, see the work of Smith and Zhang [189]. The chosen-plaintext attack on PKCS #1 v1.5 described here is due to Coron et al. [57]. A description of Bleichenbacher's chosen-ciphertext attack on PKCS #1 v1.5 can be found in the original paper [38]. See the work of Bardou et al. [13] for subsequent improvements.

Proofs of Theorem 12.31, and generalizations, can be found in [8, 94, 73, 7]. See Section 15.1.2 for a general treatment of schemes of this form. Construction 12.37 appears to have been introduced and first analyzed by Shoup [181]. OAEP was introduced by Bellare and Rogaway [25]. The original proof of security for OAEP was later found to be flawed; other proofs have since been given [43, 182, 75]. For details of Manger's chosen-ciphertext attack on implementations of PKCS #1 v2.0, see [133].

The pairwise-gcd attack described in Section 12.5.6 was carried out by Lenstra et al. [125] and Heninger et al. [96].

When using any encryption scheme in practice, the question arises as to what key length to use. This issue should not be taken lightly, and we refer the reader to Section 10.4 and references therein for an in-depth treatment.

The first efficient CCA-secure public-key encryption scheme not relying on the random-oracle model was shown by Cramer and Shoup [58] based on the DDH assumption. Subsequently, Hoffheinz and Kiltz have shown an efficient CCA-secure scheme without random oracles based on the RSA assumption [100].

Exercises

12.1 Assume a public-key encryption scheme for single-bit messages with no decryption error. Show that, given pk and a ciphertext c computed via $c \leftarrow \mathsf{Enc}_{pk}(m)$, it is possible for an unbounded adversary to determine m with probability 1.

12.2 Show that for any CPA-secure public-key encryption scheme for single-bit messages, the length of the ciphertext must be superlogarithmic in the security parameter.

    Hint: If not, the range of possible ciphertexts has polynomial size.

12.3 Say a public-key encryption scheme $(\mathsf{Gen}, \mathsf{Enc}, \mathsf{Dec})$ is *one-way* if any PPT adversary $\mathcal{A}$ has negligible probability of success in the following experiment:

- $\mathsf{Gen}(1^n)$ *is run to obtain keys* (pk, sk).
- *A uniform message* m *in the message space is chosen, and a ciphertext* $c \leftarrow \mathsf{Enc}_{pk}(m)$ *is computed.*
- $\mathcal{A}$ *is given* pk *and* c, *and outputs a message* m'. *We say* $\mathcal{A}$ *succeeds if* $m' = m$.

   (a) Construct a CPA-secure KEM in the random-oracle model based on a one-way public-key encryption scheme with message space $\{0,1\}^n$.

   (b) Can a *deterministic* public-key encryption scheme be one-way? If not, prove impossibility; if so, give a construction based on any of the assumptions introduced in this book.

12.4 Show that any two-round key-exchange protocol (that is, where each party sends a single message) satisfying Definition 11.1 can be converted into a CPA-secure public-key encryption scheme.

12.5 Show that Claim 12.7 does not hold in the setting of CCA-security.

12.6 Consider the following public-key encryption scheme. The public key is $(\mathbb{G}, q, g, h)$ and the private key is x, generated exactly as in the El Gamal encryption scheme. In order to encrypt a bit b, the sender does the following:

  (a) If $b = 0$ then choose a uniform $y \in \mathbb{Z}_q$ and compute $c_1 := g^y$ and $c_2 := h^y$. The ciphertext is $\langle c_1, c_2 \rangle$.

  (b) If $b = 1$ then choose independent uniform $y, z \in \mathbb{Z}_q$, compute $c_1 := g^y$ and $c_2 := g^z$, and set the ciphertext equal to $\langle c_1, c_2 \rangle$.

Show that it is possible to decrypt efficiently given knowledge of x. Prove that this encryption scheme is CPA-secure if the decisional Diffie–Hellman problem is hard relative to $\mathcal{G}$.

12.7 Consider the following variant of El Gamal encryption. Let $p = 2q + 1$, let $\mathbb{G}$ be the group of squares modulo p (so $\mathbb{G}$ is a subgroup of $\mathbb{Z}_p^*$ of order q), and let g be a generator of $\mathbb{G}$. The private key is $(\mathbb{G}, g, q, x)$ and the public key is $(\mathbb{G}, g, q, h)$, where $h = g^x$ and $x \in \mathbb{Z}_q$ is chosen uniformly. To encrypt a message $m \in \mathbb{Z}_p$, choose a uniform $r \in \mathbb{Z}_q$, compute $c_1 := g^r \bmod p$ and $c_2 := h^r + m \bmod p$, and let the ciphertext be $\langle c_1, c_2 \rangle$. Is this scheme CPA-secure? Prove your answer.

12.8 Consider the following protocol for two parties A and B to flip a fair coin (more complicated versions of this might be used for Internet gambling): (1) a trusted party T publishes her public key pk; (2) then A chooses a uniform bit b_A, encrypts it using pk, and announces the ciphertext c_A to B and T; (3) next, B acts symmetrically and announces a ciphertext $c_B \neq c_A$; (4) T decrypts both c_A and c_B, and the parties XOR the results to obtain the value of the coin.

  (a) Argue that even if A is dishonest (but B is honest), the final value of the coin is uniformly distributed.

  (b) Assume the parties use El Gamal encryption (where the bit b is encoded as the group element g^b before being encrypted—note that efficient decryption is still possible). Show how a dishonest B can bias the coin to any value he likes.

  (c) Suggest what type of encryption scheme would be appropriate to use here. Can you define an appropriate notion of security and prove that your suggestion achieves this definition?

12.9 Prove formally that the El Gamal encryption scheme is not CCA-secure.

12.10 In Section 12.4.4 we showed that El Gamal encryption is malleable, and specifically that given a ciphertext $\langle c_1, c_2 \rangle$ that is the encryption of some unknown message m, it is possible to produce a ciphertext $\langle c_1, c_2' \rangle$ that is the encryption of $\alpha \cdot m$ (for known α). A receiver who receives both these ciphertexts might be suspicious since both ciphertexts share the first component. Show that it is possible to generate $\langle c_1', c_2' \rangle$ that is the encryption of $\alpha \cdot m$, with $c_1' \neq c_1$ and $c_2' \neq c_2$.

12.11 Prove Theorem 12.22.

12.12 One of the attacks on plain RSA discussed in Section 12.5.1 involves a sender who encrypts two related messages using the same public key. Formulate an appropriate definition of security ruling out such attacks, and show that any CPA-secure public-key encryption scheme satisfies your definition.

12.13 One of the attacks on plain RSA discussed in Section 12.5.1 involves a sender who encrypts the same message to three different receivers. Formulate an appropriate definition of security ruling out such attacks, and show that any CPA-secure public-key encryption scheme satisfies your definition.

12.14 Consider the following modified version of padded RSA encryption: Assume messages to be encrypted have length exactly $\|N\|/2$. To encrypt, first compute $\hat{m} := \texttt{0x00}\|r\|\texttt{0x00}\|m$ where r is a uniform string of length $\|N\|/2 - 16$. Then compute the ciphertext $c := [\hat{m}^e \bmod N]$. When decrypting a ciphertext c, the receiver computes $\hat{m} := [c^d \bmod N]$ and returns an error if $\hat{m}$ does not consist of $\texttt{0x00}$ followed by $\|N\|/2 - 16$ arbitrary bits followed by $\texttt{0x00}$. Show that this scheme is not CCA-secure. Why is it easier to construct a chosen-ciphertext attack on this scheme than on PKCS #1 v1.5?

12.15 Consider the RSA-based encryption scheme in which a user encrypts a message $m \in \{0,1\}^\ell$ with respect to the public key $\langle N, e \rangle$ by computing $\hat{m} := H(m)\|m$ and outputting the ciphertext $[\hat{m}^e \bmod N]$. (Here, let $H : \{0,1\}^\ell \to \{0,1\}^n$ and assume $\ell + n < \|N\|$.) Is this scheme CPA-secure if H is modeled as a random oracle?

12.16 Show a chosen-ciphertext attack on Construction 12.34.

12.17 Say three users have RSA public keys $\langle N_1, 3 \rangle$, $\langle N_2, 3 \rangle$, and $\langle N_3, 3 \rangle$ (i.e., they all use $e = 3$), with $N_1 < N_2 < N_3$. Consider the following method for sending the same message $m \in \{0,1\}^\ell$ to each of these parties: choose a uniform $r \leftarrow \mathbb{Z}_{N_1}^*$, and send to everyone the same ciphertext

$$\langle [r^3 \bmod N_1], [r^3 \bmod N_2], [r^3 \bmod N_3], H(r) \oplus m \rangle,$$

where $H : \mathbb{Z}_{N_1}^* \to \{0,1\}^\ell$. Assume $\|N_1\| = \|N_2\| = \|N_3\| = n \ll \ell$.

(a) Show that this is not CPA-secure, and an adversary can recover m from the ciphertext even when H is modeled as a random oracle.
    Hint: See Section 12.5.1.

(b) Show a simple way to fix this and get a CPA-secure method that transmits a ciphertext of length $3\ell + \mathcal{O}(n)$.

(c) Show a better approach that is still CPA-secure but with a ciphertext of length $\ell + \mathcal{O}(n)$.

12.18 Let $\Pi = (\mathsf{Gen}, \mathsf{Enc}, \mathsf{Dec})$ be a CPA-secure public-key encryption scheme, and let $\Pi' = (\mathsf{Gen}', \mathsf{Enc}', \mathsf{Dec}')$ be a CCA-secure private-key encryption scheme. Consider the following construction:

CONSTRUCTION 12.41

Let $H : \{0,1\}^n \to \{0,1\}^n$ be a function. Construct a public-key encryption scheme as follows:

- Gen^*: on input 1^n, run $\mathsf{Gen}(1^n)$ to obtain (pk, sk). Output these as the public and private keys, respectively.

- Enc^*: on input a public key pk and a message $m \in \{0,1\}^n$, choose a uniform $r \in \{0,1\}^n$ and output the ciphertext

$$\langle \mathsf{Enc}_{pk}(r), \mathsf{Enc}'_{H(r)}(m) \rangle.$$

- Dec^*: on input a private key sk and a ciphertext $\langle c_1, c_2 \rangle$, compute $r := \mathsf{Dec}_{sk}(c_1)$ and set $k := H(r)$. Then output $\mathsf{Dec}'_k(c_2)$.

Does the above construction have indistinguishable encryptions under a chosen-*ciphertext* attack, if H is modeled as a random oracle? If yes, provide a proof. If not, where does the approach used to prove Theorem 12.38 break down?

12.19 Consider the following variant of Construction 12.32:

CONSTRUCTION 12.42

Let GenRSA be as usual, and define a public-key encryption scheme as follows:

- Gen: on input 1^n, run $\mathsf{GenRSA}(1^n)$ to obtain (N, e, d). Output the public key $pk = \langle N, e \rangle$, and the private key $sk = \langle N, d \rangle$.

- Enc: on input a public key $pk = \langle N, e \rangle$ and a message $m \in \{0,1\}$, choose a uniform $r \in \mathbb{Z}_N^*$. Output the ciphertext $\langle [r^e \bmod N], \mathsf{lsb}(r) \oplus m \rangle$.

- Dec: on input a private key $sk = \langle N, d \rangle$ and a ciphertext $\langle c, b \rangle$, compute $r := [c^d \bmod N]$ and output $\mathsf{lsb}(r) \oplus b$.

Prove that this scheme is CPA-secure.

12.20 Fix an RSA public key $\langle N, e \rangle$ and assume we have an algorithm $\mathcal{A}$ that always correctly computes $\mathsf{lsb}(x)$ given $[x^e \bmod N]$. Write full pseudocode for an algorithm $\mathcal{A}'$ that computes x from $[x^e \bmod N]$.

Chapter 13

Digital Signature Schemes

13.1 Digital Signatures – An Overview

In the previous chapter we explored how public-key encryption can be used to achieve *secrecy* in the public-key setting. *Integrity* (or *authenticity*) in the public-key setting is provided using *digital signature schemes*. These can be viewed as the public-key analogue of message authentication codes although, as we will see, there are several important differences between these primitives.

Signature schemes allow a *signer* S who has established a public key pk to "sign" a message using the associated private key sk in such a way that anyone who knows pk (and knows that this public key was established by S) can *verify* that the message originated from S and was not modified in transit. (Note that, in contrast to public-key encryption, in the context of digital signatures the owner of the public key acts as the *sender*.) As a prototypical application, consider a software company that wants to disseminate software updates in an authenticated manner; that is, when the company releases an update it should be possible for any of its clients to verify that the update is authentic, and a malicious third party should never be able to fool a client into accepting an update that was not actually released by the company. To do this, the company can generate a public key pk along with a private key sk, and then distribute pk in some reliable manner to its clients while keeping sk secret. (As in the case of public-key encryption, we assume that this initial distribution of the public key is carried out correctly so that all clients have a correct copy of pk. In the current example, pk could be bundled with the original software purchased by a client.) When releasing a software update m, the company computes a digital signature σ on m using its private key sk, and sends (m, σ) to every client. Each client can verify the authenticity of m by checking that σ is a correct signature on m with respect to the public key pk.

A malicious party might try to issue a fraudulent update by sending (m', σ') to a client, where m' represents an update that was never released by the company. This m' might be a modified version of some previous update, or it might be completely new and unrelated to any prior updates. If the signature scheme is "secure" (in a sense we will define more carefully soon), however, then when the client attempts to verify σ' it will find that this is an *invalid* signature on m' with respect to pk, and will therefore *reject* the signature. The client will reject even if m' is modified only slightly from a genuine update m.

The above is not just a theoretical application of digital signatures, but one that is in widespread use today for distributing software updates.

Comparison to Message Authentication Codes

Both message authentication codes and digital signature schemes are used to ensure the integrity of transmitted messages. Although the discussion in Chapter 11 comparing the public-key and private-key settings focused mainly on encryption, that discussion applies also to message integrity. Using digital signatures rather than message authentication codes simplifies key distribution and management, especially when a sender needs to communicate with multiple receivers as in the software-update example above. By using a digital signature scheme the sender avoids having to establish a distinct secret key with each potential receiver, and avoids having to compute a separate MAC tag with respect to each such key. Instead, the sender need only compute a single signature that can be verified by all recipients.

A *qualitative* advantage that digital signatures have as compared to message authentication codes is that signatures are *publicly verifiable*. This means that if a receiver verifies that a signature on a given message is legitimate, then all other parties who receive this signed message will also verify it as legitimate. This feature is not achieved by message authentication codes if the signer shares a separate key with each receiver: in such a setting a malicious sender might compute a correct MAC tag with respect to the key it shares with receiver A but an incorrect MAC tag with respect to the key it shares with a different user B. In this case, A knows that he received an authentic message from the sender but has no guarantee that B will agree.

Public verifiability implies that signatures are *transferable*: a signature σ on a message m by a signer S can be shown to a third party, who can then verify herself that σ is a legitimate signature on m with respect to S's public key (here, we assume this third party also knows S's public key). By making a copy of the signature, this third party can then show the signature to another party and convince *them* that S authenticated m, and so on. Public verifiability and transferability are essential for the application of digital signatures to certificates and public-key infrastructures, as we will discuss in Section 13.6.

Digital signature schemes also provide the very important property of *non-repudiation*. This means that once S signs a message he cannot later deny having done so (assuming the public key of S is widely publicized and distributed). This aspect of digital signatures is crucial for legal applications where a recipient may need to prove to a third party (say, a judge) that a signer did indeed "certify" a particular message (e.g., a contract): assuming S's public key is known to the judge, or is otherwise publicly available, a valid signature on a message serves as convincing evidence that S indeed signed that message. Message authentication codes simply cannot provide non-repudiation. To see this, say users S and R share a key k_{SR}, and S

sends a message m to R along with a (valid) MAC tag t computed using this key. Since the judge does *not* know k_{SR} (indeed, this key is kept secret by S and R), there is no way for the judge to determine whether t is valid or not. If R were to reveal the key k_{SR} to the judge, there would be no way for the judge to know whether this is the "actual" key that S and R shared, or whether it is some "fake" key manufactured by R. Finally, even if we assume the judge can somehow obtain the actual key k_{SR} shared by the parties, there is no way for the judge to distinguish whether S generated t or whether R did—this is because message authentication codes are a *symmetric*-key primitive; anything S can do, R can do also.

As in the case of private-key vs. public-key encryption, message authentication codes have the advantage of being shorter and roughly 2–3 orders of magnitude more efficient to generate/verify than digital signatures. Thus, in situations where public verifiability, transferability, and/or non-repudiation are not needed, and the sender communicates primarily with a single recipient (with whom it is able to share a secret key), message authentication codes should be used.

Relation to Public-Key Encryption

Digital signatures are often mistakenly viewed as the "inverse" of public-key encryption, with the roles of the sender and receiver interchanged. Historically,[1] in fact, it has been suggested that digital signatures can be obtained by "reversing" public-key encryption, i.e., signing a message m by decrypting it (using the private key) to obtain σ, and verifying a signature σ by encrypting it (using the corresponding public key) and checking whether the result is m. The suggestion to construct signature schemes in this way is *completely unfounded*: in most cases, it is simply inapplicable, and even in cases where it can be applied it results in signature schemes that are not secure.

13.2 Definitions

Digital signatures are the public-key counterpart of message authentication codes, and their syntax and security guarantees are analogous. The algorithm that the sender applies to a message is here denoted Sign (rather than Mac), and the output of this algorithm is now called a *signature* (rather than a tag).

---

[1]This view no doubt arose because, as we will see in Section 13.4.1, plain RSA signatures are the reverse of plain RSA encryption. However, neither plain RSA signatures nor plain RSA encryption meet even minimal notions of security.

The algorithm that the receiver applies to a message and a signature in order to check validity is still denoted Vrfy.

DEFINITION 13.1   *A* (digital) signature scheme *consists of three proba-bilistic polynomial-time algorithms* (Gen, Sign, Vrfy) *such that:*

1. *The* key-generation algorithm Gen *takes as input a security parameter* 1^n *and outputs a pair of keys* (pk, sk). *These are called the* public key *and the* private key, *respectively. We assume that pk and sk each has length at least n, and that n can be determined from pk or sk.*

2. *The* signing algorithm Sign *takes as input a private key sk and a mes-sage m from some message space (that may depend on pk). It outputs a signature* σ, *and we write this as* $\sigma \leftarrow \text{Sign}_{sk}(m)$.

3. *The deterministic* verification algorithm Vrfy *takes as input a public key pk, a message m, and a signature* σ. *It outputs a bit b, with* $b = 1$ *meaning* valid *and* $b = 0$ *meaning* invalid. *We write this as* $b := \text{Vrfy}_{pk}(m, \sigma)$.

It is required that except with negligible probability over (pk, sk) *output by* Gen(1^n), *it holds that* $\text{Vrfy}_{pk}(m, \text{Sign}_{sk}(m)) = 1$ *for every (legal) message m.*

If there is a function ℓ *such that for every* (pk, sk) *output by* Gen(1^n) *the message space is* $\{0,1\}^{\ell(n)}$, *then we say that* (Gen, Sign, Vrfy) *is a* signature scheme for messages of length $\ell(n)$.

We call σ a *valid signature* on a message m (with respect to some public key pk understood from the context) if $\text{Vrfy}_{pk}(m, \sigma) = 1$.

A signature scheme is used in the following way. One party S, who acts as the *sender*, runs Gen(1^n) to obtain keys (pk, sk). The public key pk is then publicized as belonging to S; e.g., S can put the public key on its webpage or place it in some public directory. As in the case of public-key encryption, we assume that any other party is able to obtain a legitimate copy of S's public key (see discussion below). When S wants to authenticate a message m, it computes the signature $\sigma \leftarrow \text{Sign}_{sk}(m)$ and sends (m, σ). Upon receipt of (m, σ), a receiver who knows pk can verify the authenticity of m by checking whether $\text{Vrfy}_{pk}(m, \sigma) \overset{?}{=} 1$. This establishes both that S sent m, and also that m was not modified in transit. As in the case of message authentication codes, however, it does not say anything about *when* m was sent, and replay attacks are still possible (see Section 4.2).

The assumption that parties are able to obtain a legitimate copy of S's public key implies that S is able to transmit at least one message (namely, pk itself) in a reliable and authenticated manner. If S is able to transmit messages reliably, however, then why does it need a signature scheme at all? The answer is that reliable distribution of pk may be difficult and expensive, but using a signature scheme means that such distribution need only be carried out *once*, after which an unlimited number of messages can subsequently be sent reliably. Furthermore, as we will discuss in Section 13.6, signature schemes themselves are used to ensure the reliable distribution of *other* public keys. They thus serve as a central tool for setting up a "public-key infrastructure" to address the key-distribution problem.

Security of signature schemes. For a fixed public key *pk* generated by a signer *S*, a *forgery* is a message *m* along with a valid signature σ, where *m* was not previously signed by *S*. Security of a signature scheme means that an adversary should be unable to output a forgery even if it obtains signatures on many other messages of its choice. This is the direct analogue of the definition of security for message authentication codes, and we refer the reader to Section 4.2 for motivation and further discussion.

The formal definition of security is essentially the same as Definition 4.2, with the main difference being that here the adversary is given a public key. Let $\Pi = (\mathsf{Gen}, \mathsf{Sign}, \mathsf{Vrfy})$ be a signature scheme, and consider the following experiment for an adversary $\mathcal{A}$ and parameter *n*:

The signature experiment Sig-forge$_{\mathcal{A},\Pi}(n)$:

1. $\mathsf{Gen}(1^n)$ *is run to obtain keys* (pk, sk).

2. *Adversary* $\mathcal{A}$ *is given pk and access to an oracle* $\mathsf{Sign}_{sk}(\cdot)$. *The adversary then outputs* (m, σ). *Let* $\mathcal{Q}$ *denote the set of all queries that* $\mathcal{A}$ *asked its oracle.*

3. $\mathcal{A}$ *succeeds if and only if (1)* $\mathsf{Vrfy}_{pk}(m, \sigma) = 1$ *and (2)* $m \notin \mathcal{Q}$. *In this case the output of the experiment is defined to be 1.*

DEFINITION 13.2 *A signature scheme* $\Pi = (\mathsf{Gen}, \mathsf{Sign}, \mathsf{Vrfy})$ *is* existentially unforgeable under an adaptive chosen-message attack, *or just* secure, *if for all probabilistic polynomial-time adversaries* $\mathcal{A}$, *there is a negligible function* negl *such that:*
$$\Pr[\text{Sig-forge}_{\mathcal{A},\Pi}(n) = 1] \leq \mathsf{negl}(n).$$

Strong security can be defined analogously to Definition 4.3.

13.3 The Hash-and-Sign Paradigm

As in the case of public-key vs. private-key encryption, "native" signature schemes are orders of magnitude less efficient than message authentication codes. Fortunately, as with hybrid encryption (see Section 12.3), it is possible to obtain the functionality of digital signatures at the asymptotic cost of a private-key operation, at least for sufficiently long messages. This can be done using the *hash-and-sign* approach, discussed next.

The intuition behind the hash-and-sign approach is straightforward. Say we have a signature scheme for messages of length ℓ, and wish to sign a (longer) message $m \in \{0,1\}^*$. Rather than sign m itself, we can instead use a hash function H to *hash* the message to a fixed-length *digest* $H(m)$ of length ℓ, and then sign the resulting digest. This approach is exactly analogous to the hash-and-MAC approach discussed in Section 6.3.1.

CONSTRUCTION 13.3

Let $\Pi = (\mathsf{Gen}, \mathsf{Sign}, \mathsf{Vrfy})$ be a signature scheme for messages of length $\ell(n)$, and let $\Pi_H = (\mathsf{Gen}_H, H)$ be a hash function with output length $\ell(n)$. Construct signature scheme $\Pi' = (\mathsf{Gen}', \mathsf{Sign}', \mathsf{Vrfy}')$ as follows:

- Gen': on input 1^n, run $\mathsf{Gen}(1^n)$ to obtain (pk, sk) and run $\mathsf{Gen}_H(1^n)$ to obtain s; the public key is $\langle pk, s \rangle$ and the private key is $\langle sk, s \rangle$.

- Sign': on input a private key $\langle sk, s \rangle$ and a message $m \in \{0,1\}^*$, output $\sigma \leftarrow \mathsf{Sign}_{sk}(H^s(m))$.

- Vrfy': on input a public key $\langle pk, s \rangle$, a message $m \in \{0,1\}^*$, and a signature σ, output 1 if and only if $\mathsf{Vrfy}_{pk}(H^s(m), \sigma) \stackrel{?}{=} 1$.

The hash-and-sign paradigm.

THEOREM 13.4   *If Π is a secure signature scheme for messages of length ℓ and Π_H is collision resistant, then Construction 13.3 is a secure signature scheme (for arbitrary-length messages).*

The proof of this theorem is almost identical to that of Theorem 6.6.

13.4 RSA-Based Signatures

We begin our consideration of concrete signature schemes with a discussion of schemes based on the RSA assumption.

13.4.1 Plain RSA Signatures

We first describe a simple, RSA-based signature scheme. Although the scheme is insecure, it serves as a useful starting point.

As usual, let GenRSA be a PPT algorithm that, on input 1^n, outputs a modulus N that is the product of two n-bit primes (except with negligible probability), along with integers e, d satisfying $ed = 1 \bmod \phi(N)$. Key generation in plain RSA involves simply running GenRSA, and outputting $\langle N, e \rangle$ as the public key and $\langle N, d \rangle$ as the private key. To sign a message $m \in \mathbb{Z}_N^*$, the signer computes $\sigma := [m^d \bmod N]$. Verification of a signature σ on a message m with respect to the public key $\langle N, e \rangle$ is carried out by checking whether $m \overset{?}{=} \sigma^e \bmod N$. See Construction 13.5.

CONSTRUCTION 13.5

Let GenRSA be as in the text. Define a signature scheme as follows:

- Gen: on input 1^n run GenRSA(1^n) to obtain (N, e, d). The public key is $\langle N, e \rangle$ and the private key is $\langle N, d \rangle$.

- Sign: on input a private key $sk = \langle N, d \rangle$ and a message $m \in \mathbb{Z}_N^*$, compute the signature

$$\sigma := [m^d \bmod N].$$

- Vrfy: on input a public key $pk = \langle N, e \rangle$, a message $m \in \mathbb{Z}_N^*$, and a signature $\sigma \in \mathbb{Z}_N^*$, output 1 if and only if

$$m \overset{?}{=} [\sigma^e \bmod N].$$

The plain RSA signature scheme.

It is easy to see that verification of a legitimately generated signature is always successful since

$$\sigma^e = (m^d)^e = m^{[ed \bmod \phi(N)]} = m^1 = m \bmod N.$$

One might expect this scheme to be secure since, for an adversary knowing only the public key $\langle N, e \rangle$, computing a valid signature on a message m seems to require solving the RSA problem (since the signature is exactly the eth root of m). Unfortunately, this reasoning is incorrect. For one thing, the RSA assumption only implies hardness of computing a signature (that is, computing an eth root) of a *uniform* message m; it says nothing about hardness of computing a signature on a nonuniform m or on some message m of the attacker's choice. Moreover, the RSA assumption says nothing about what an attacker might be able to do once it learns signatures on *other* messages. The following examples demonstrate that both of these observations lead to attacks on the plain RSA signature scheme.

A no-message attack. The first attack we describe generates a forgery using the public key alone, without obtaining any signatures from the legitimate signer. The attack works as follows: given a public key $pk = \langle N, e \rangle$, choose a uniform $\sigma \in \mathbb{Z}_N^*$ and compute $m := [\sigma^e \bmod N]$. Then output the forgery (m, σ). It is immediate that σ is a valid signature on m, and this is a forgery since no signatures at all were issued by the owner of the public key. We conclude that the plain RSA signature scheme does not satisfy Definition 13.2.

One might argue that this does not constitute a "realistic" attack since the adversary has "no control" over the message m for which it forges a valid signature. This is irrelevant as far as Definition 13.2 is concerned, and we have already discussed (in Chapter 4) why it is dangerous to assume any semantics for messages that are going to be authenticated using any cryptographic scheme. Moreover, the adversary does have *some* control over m: for example, by choosing multiple, uniform values of σ it can (with high probability) obtain an m with a few bits set in some desired way. By choosing σ in some specific manner, it may also be possible to influence the resulting message for which a forgery is output.

Forging a signature on an arbitrary message. A more damaging attack on the plain RSA signature scheme requires the adversary to obtain *two* signatures from the signer, but allows the adversary to output a forged signature on any message of its choice. Say the adversary wants to forge a signature on the message $m \in \mathbb{Z}_N^*$ with respect to the public key $pk = \langle N, e \rangle$. The adversary chooses arbitrary $m_1, m_2 \in \mathbb{Z}_N^*$ distinct from m such that $m = m_1 \cdot m_2 \bmod N$. It then obtains signatures σ_1, σ_2 on m_1, m_2, respectively. Finally, it outputs $\sigma := [\sigma_1 \cdot \sigma_2 \bmod N]$ as a valid signature on m. This works because

$$\sigma^e = (\sigma_1 \cdot \sigma_2)^e = (m_1^d \cdot m_2^d)^e = m_1^{ed} \cdot m_2^{ed} = m_1 \cdot m_2 = m \bmod N,$$

using the fact that σ_1, σ_2 are valid signatures on m_1, m_2.

Being able to forge a signature on an arbitrary message is devastating. Nevertheless, one might argue that this attack is unrealistic since an adversary will not be able to convince a signer to sign the exact messages m_1 and m_2. Once again, this is irrelevant as far as Definition 13.2 is concerned. Furthermore, it is dangerous to make assumptions about what messages the signer may or may not be willing to sign. For example, a client may use a signature scheme to authenticate to a server by signing a random challenge sent by the server. Here, a malicious server would be able to obtain a signature on any message(s) of its choice. More generally, it may be possible for the adversary to choose m_1 and m_2 as "legitimate" messages that the signer will agree to sign. Finally, note that the attack can be generalized: if an adversary obtains valid signatures on q arbitrary messages $M = \{m_1, \ldots, m_q\}$, then the adversary can output a valid signature on any of $2^q - q$ other messages obtained by taking products of subsets of M (of size different from 1).

13.4.2 RSA-FDH and PKCS #1 Standards

One can attempt to prevent the attacks from the previous section by applying some transformation to messages before signing them. That is, the signer will now specify as part of its public key a (deterministic) function H with certain cryptographic properties (described below) mapping messages to $\mathbb{Z}_N^*$; the signature on a message m will be $\sigma := [H(m)^d \bmod N]$, and verification of the signature σ on the message m will be done by checking whether $\sigma^e \stackrel{?}{=} H(m) \bmod N$. See Construction 13.6.

CONSTRUCTION 13.6

Let GenRSA be as in the previous sections, and construct a signature scheme as follows:

- Gen: on input 1^n, run GenRSA(1^n) to compute (N, e, d). The public key is $\langle N, e \rangle$ and the private key is $\langle N, d \rangle$.

  As part of key generation, a function $H : \{0,1\}^* \to \mathbb{Z}_N^*$ is specified, but we leave this implicit.

- Sign: on input a private key $\langle N, d \rangle$ and a message $m \in \{0,1\}^*$, compute
  $$\sigma := [H(m)^d \bmod N].$$

- Vrfy: on input a public key $\langle N, e \rangle$, a message m, and a signature σ, output 1 if and only if
  $$\sigma^e \stackrel{?}{=} H(m) \bmod N.$$

The RSA-FDH signature scheme.

What properties does H need in order for this construction to be secure? At a minimum, to prevent the no-message attack it should be infeasible for an attacker to start with σ, compute $\hat{m} := [\sigma^e \bmod N]$, and then find a message m such that $H(m) = \hat{m}$. This, in particular, means that H should be hard to invert in some sense. To prevent the second attack, we need an H that does not admit "multiplicative relations," that is, for which it is hard to find three messages m, m_1, m_2 with $H(m) = H(m_1) \cdot H(m_2) \bmod N$. Finally, it must be hard to find collisions in H: if $H(m_1) = H(m_2)$, then m_1 and m_2 have the same signature and forgery becomes trivial.

There is no known way to choose H so that Construction 13.6 can be proven secure. However, it *is* possible to prove security if H is modeled as a random oracle that maps its inputs uniformly onto $\mathbb{Z}_N^*$; the resulting scheme is called the *RSA full-domain hash* (RSA-FDH) signature scheme. One can check that a random function of this sort satisfies the requirements discussed in the previous paragraph: a random function (with large range) is hard to invert, does not have any easy-to-find multiplicative relations, and is collision resistant. Of course, this informal reasoning does not rule out all possible attacks, but the proof of security below does.

Before continuing, we stress that it is critical for the range of H to be (close to) *all* of $\mathbb{Z}_N^*$; in particular it does not suffice to simply let H be an "off-the-shelf" cryptographic hash function such as SHA-2. (The output length of SHA-2 is much smaller than the length of RSA moduli used in practice.) Indeed, *practical attacks* on Construction 13.6 are known if the output length of H is too small (e.g., if the output length is 256 bits as would be the case if a version of SHA-2 were used directly as H).

Before turning to the formal proof, we provide some intuition. Our goal is to prove that if the RSA problem is hard relative to GenRSA, then RSA-FDH is secure when H is modeled as a random oracle. We consider first security against a no-message attack, i.e., when the adversary $\mathcal{A}$ cannot request any signatures. Here the adversary is limited to making queries to the random oracle, and we assume without loss of generality that $\mathcal{A}$ always makes exactly q (distinct) queries to H and that if the adversary outputs a forgery (m, σ) then it had previously queried m to H.

Say there is an efficient adversary $\mathcal{A}$ that carries out a no-message attack and makes exactly q queries to H. We construct an efficient algorithm $\mathcal{A}'$ solving the RSA problem relative to GenRSA. Given input (N, e, y), algorithm $\mathcal{A}'$ runs $\mathcal{A}$ on the public key $pk = \langle N, e \rangle$. Let $m_1, \ldots, m_q$ denote the q (distinct) queries that $\mathcal{A}$ makes to H. Our algorithm $\mathcal{A}'$ answers these random-oracle queries of $\mathcal{A}$ with uniform elements of $\mathbb{Z}_N^*$ except for one query—say, the ith query, chosen uniformly from the oracle queries of $\mathcal{A}$—that is answered with y itself. Note that, from the point of view of $\mathcal{A}$, all its random-oracle queries are answered with uniform elements of $\mathbb{Z}_N^*$ (recall that y is uniform as well, although it is not chosen by $\mathcal{A}'$), and so $\mathcal{A}$ has no information about i. Moreover, the view of $\mathcal{A}$ when run as a subroutine by $\mathcal{A}'$ is identically distributed to the view of $\mathcal{A}$ when attacking the original signature scheme.

If $\mathcal{A}$ outputs a forgery (m, σ) then, because $m \in \{m_1, \ldots, m_q\}$, with probability $1/q$ we will have $m = m_i$. In that case,

$$\sigma^e = H(m) = H(m_i) = y \bmod N$$

and $\mathcal{A}'$ can output σ as the solution to its given RSA instance (N, e, y). We conclude that if $\mathcal{A}$ outputs a forgery with probability ε, then $\mathcal{A}'$ solves the RSA problem with probability ε/q. Since q is polynomial, we conclude that ε must be negligible if the RSA problem is hard relative to GenRSA.

Handling the case when the adversary is allowed to request signatures on messages of its choice is more difficult. The complication arises since our algorithm $\mathcal{A}'$ above does not know the decryption exponent d, yet now has to compute valid signatures on messages queried by $\mathcal{A}$ to its signing oracle. This seems impossible (and possibly even contradictory!) until we realize that $\mathcal{A}'$ *can* correctly compute a signature on a message m as long as it sets $H(m)$ to be equal to $[\sigma^e \bmod N]$ for a *known* value σ. (Here we are using the fact that the random oracle is "programmable.") If σ is uniform then $[\sigma^e \bmod N]$ is uniform as well, and so the random oracle is still emulated "properly" by $\mathcal{A}'$.

The above intuition is formalized in the proof of the following:

THEOREM 13.7   *If the RSA problem is hard relative to* GenRSA *and H is modeled as a random oracle, then Construction 13.6 is secure.*

PROOF   Let $\Pi = (\mathsf{Gen}, \mathsf{Sign}, \mathsf{Vrfy})$ denote Construction 13.6, and let $\mathcal{A}$ be a probabilistic polynomial-time adversary. We assume without loss of generality that if $\mathcal{A}$ requests a signature on a message m, or outputs a forgery (m, σ), then it previously queried m to H. Let $q(n)$ be a polynomial upper bound on the number of queries $\mathcal{A}$ makes to H on security parameter n; we assume without loss of generality that $\mathcal{A}$ makes exactly $q(n)$ distinct queries to H.

For convenience, we list the steps of experiment $\mathsf{Sig\text{-}forge}_{\mathcal{A}, \Pi}(n)$:

1. $\mathsf{GenRSA}(1^n)$ *is run to obtain* (N, e, d). *A random function* $H : \{0, 1\}^* \to \mathbb{Z}_N^*$ *is chosen.*

2. *The adversary* $\mathcal{A}$ *is given* $pk = \langle N, e \rangle$, *and may query* H *as well as a signing oracle* $\mathsf{Sign}_{\langle N, d \rangle}(\cdot)$ *that, on input a message* m, *returns* $\sigma := [H(m)^d \bmod N]$.

3. $\mathcal{A}$ *outputs* (m, σ), *where it had not previously requested a signature on* m. *The output of the experiment is* 1 *if and only if* $\sigma^e = H(m) \bmod N$.

We define a modified experiment $\mathsf{Sig\text{-}forge}'_{\mathcal{A}, \Pi}(n)$ in which a guess is made at the outset as to which message (from among the q messages that $\mathcal{A}$ queries to H) will correspond to the eventual forgery (if any) output by $\mathcal{A}$:

1. *Choose uniform* $j \in \{1, \dots, q\}$.

2. $\mathsf{GenRSA}(1^n)$ *is run to obtain* (N, e, d). *A random function* $H : \{0, 1\}^* \to \mathbb{Z}_N^*$ *is chosen.*

3. *The adversary* $\mathcal{A}$ *is given* $pk = \langle N, e \rangle$, *and may query* H *as well as a signing oracle* $\mathsf{Sign}_{\langle N, d \rangle}(\cdot)$ *that, on input a message* m, *returns* $\sigma := [H(m)^d \bmod N]$.

4. $\mathcal{A}$ *outputs* (m, σ), *where it had not previously requested a signature on* m. *Let* i *be such that* $m = m_i$.[2] *The output of the experiment is* 1 *if and only if* $\sigma^e = H(m) \bmod N$ *and* $j = i$.

Since j is uniform and independent of everything else, the probability that $j = i$ (even conditioned on the event that $\mathcal{A}$ outputs a forgery) is exactly $1/q$. Therefore $\Pr[\mathsf{Sig\text{-}forge}'_{\mathcal{A}, \Pi}(n) = 1] = \frac{1}{q(n)} \cdot \Pr[\mathsf{Sig\text{-}forge}_{\mathcal{A}, \Pi}(n) = 1]$.

Now consider the modified experiment $\mathsf{Sig\text{-}forge}''_{\mathcal{A}, \Pi}(n)$ in which the experiment is aborted if $\mathcal{A}$ ever requests a signature on the message m_j (where m_j denotes the jth message queried to H, and j is the uniform value chosen

---

[2] Here m_i denotes the ith query made to H. Recall, by assumption, that if $\mathcal{A}$ requests a signature on a message m, then it must have previously queried m to H.

at the outset). This does not change the probability that the output of the experiment is 1, since if $\mathcal{A}$ ever requests a signature on m_j then it cannot possible output a forgery on m_j. In words,

$$\Pr[\text{Sig-forge}''_{\mathcal{A},\Pi}(n) = 1] = \Pr[\text{Sig-forge}'_{\mathcal{A},\Pi}(n) = 1]$$
$$= \frac{\Pr[\text{Sig-forge}_{\mathcal{A},\Pi}(n) = 1]}{q(n)}. \qquad (13.1)$$

Finally, consider the following algorithm $\mathcal{A}'$ solving the RSA problem:

Algorithm $\mathcal{A}'$:
The algorithm is given (N, e, y) as input.

1. Choose uniform $j \in \{1, \ldots, q\}$.

2. Run $\mathcal{A}$ on input the public key $pk = \langle N, e \rangle$. Store triples $(\cdot, \cdot, \cdot)$ in a table, initially empty. An entry (m_i, σ_i, y_i) indicates that $\mathcal{A}'$ has set $H(m_i) = y_i$, and $\sigma_i^e = y_i \bmod N$.

3. When $\mathcal{A}$ makes its ith random-oracle query $H(m_i)$, answer it as follows:

   - If $i = j$, return y as the answer to the query.
   - Else choose uniform $\sigma_i \in \mathbb{Z}_N^*$, compute $y_i := [\sigma_i^e \bmod N]$, return y_i as the answer to the query, and store (m_i, σ_i, y_i) in the table.

   When $\mathcal{A}$ requests a signature on message m, let i be such that $m = m_i$ and answer the query as follows[3]

   - If $i = j$ then $\mathcal{A}'$ aborts.
   - If $i \neq j$ then there is an entry (m_i, σ_i, y_i) in the table. Return σ_i as the answer to the query.

4. At the end of $\mathcal{A}$'s execution, it outputs (m, σ). If $m = m_j$ and $\sigma^e = y \bmod N$, then output σ.

Clearly, $\mathcal{A}'$ runs in probabilistic polynomial time. Say the input (N, e, y) to $\mathcal{A}'$ is generated by running $\text{GenRSA}(1^n)$ to obtain (N, e, d), and then choosing uniform $y \in \mathbb{Z}_N^*$. The crucial observation is that the view of $\mathcal{A}$ when run as a subroutine by $\mathcal{A}'$ is identical to the view of $\mathcal{A}$ in experiment $\text{Sig-forge}''_{\mathcal{A},\Pi}(n)$. In particular, all Sign-oracle queries are answered correctly, and each of the random-oracle queries of $\mathcal{A}$ when run as a subroutine by $\mathcal{A}'$ is answered with a uniform element of $\mathbb{Z}_N^*$:

- The query $H(m_j)$ is answered with y, a uniform element of $\mathbb{Z}_N^*$.

---

[3] Here m_i denotes the ith query made to H. Recall, by assumption, that if $\mathcal{A}$ requests a signature on a message m, then it must have previously queried m to H.

- Queries $H(m_i)$ with $i \neq j$ are answered with $y_i = [\sigma_i^e \bmod N]$, where σ_i is uniform in $\mathbb{Z}_N^*$. Since exponentiation to the eth power is a one-to-one function, y_i is uniformly distributed as well.

Finally, observe that whenever experiment $\mathsf{Sig\text{-}forge}''_{\mathcal{A},\Pi}(n)$ would output 1, then $\mathcal{A}'$ outputs a correct solution to its given RSA instance. This follows since $\mathsf{Sig\text{-}forge}''_{\mathcal{A},\Pi}(n) = 1$ implies that $j = i$ and $\sigma^e = H(m_i) \bmod N$. Now, when $j = i$, algorithm $\mathcal{A}'$ does not abort and in addition $H(m_i) = y$. Thus, $\sigma^e = H(m_i) = y \bmod N$, and so σ is the desired inverse. Using Equation (13.1), this means that

$$\Pr[\mathsf{RSA\text{-}inv}_{\mathcal{A}',\mathsf{GenRSA}}(n) = 1] = \Pr[\mathsf{Sig\text{-}forge}''_{\mathcal{A},\Pi}(n) = 1]$$
$$= \frac{\Pr[\mathsf{Sig\text{-}forge}_{\mathcal{A},\Pi}(n) = 1]}{q(n)}. \tag{13.2}$$

If the RSA problem is hard relative to GenRSA, there is a negligible function negl such that $\Pr[\mathsf{RSA\text{-}inv}_{\mathcal{A}',\mathsf{GenRSA}}(n) = 1] \leq \mathsf{negl}(n)$. Since $q(n)$ is polynomial, we conclude from Equation (13.2) that $\Pr[\mathsf{Sig\text{-}forge}_{\mathcal{A},\Pi}(n) = 1]$ is negligible as well. This completes the proof. ∎

RSA PKCS #1 standards. RSA PKCS #1 v1.5 specifies a signature scheme that is very similar to RSA-FDH. A more-complex scheme that can be viewed as a randomized variant of RSA-FDH has been included in the PKCS #1 standard since version 2.1.

13.5 Signatures from the Discrete-Logarithm Problem

Signature schemes can be based on the discrete-logarithm assumption as well, although the assumption does not lend itself as readily to signatures as the RSA assumption does. In Sections 13.5.1 and 13.5.2 we describe the Schnorr signature scheme that can be proven secure in the random-oracle model. In Section 13.5.3 we describe the DSA and ECDSA signature schemes; these standardized schemes are widely used even though they have no full proof of security.

13.5.1 Identification Schemes and Signatures

The underlying intuition for the Schnorr signature scheme is best explained by taking a slight detour to discuss (public-key) *identification schemes*. We then describe the *Fiat–Shamir transform* that can be used to convert identification schemes to signature schemes in the random-oracle model. Finally, we present the Schnorr identification scheme—and corresponding signature scheme—based on the discrete-logarithm problem.

Identification Schemes

An identification scheme is an interactive protocol that allows one party to prove its identity (i.e., to *authenticate* itself) to another. This is a very natural notion, and it is common nowadays to authenticate oneself when logging in to a website. We call the party identifying herself (e.g., the user) the "prover," and the party verifying the identity (e.g., the web server) the "verifier." Here, we are interested in the public-key setting where the prover and verifier do not share any secret information (such as a password) in advance; instead, the verifier only knows the public key of the prover. Successful execution of the identification protocol convinces the verifier that it is communicating with the intended prover rather than an imposter.

We will only consider three-round identification protocols of a specific form, where the prover is specified by two algorithms $\mathcal{P}_1, \mathcal{P}_2$ and the verifier's side of the protocol is specified by an algorithm $\mathcal{V}$. The prover runs $\mathcal{P}_1(sk)$ using its private key sk to obtain an initial message I along with some state st, and initiates the protocol by sending I to the verifier. In response, the verifier sends a challenge r chosen uniformly from some set Ω_{pk} defined by the prover's public key pk. Next, the prover runs $\mathcal{P}_2(sk, \mathsf{st}, r)$ to compute a response s that it sends back to the verifier. Finally, the verifier computes $\mathcal{V}(pk, r, s)$ and accepts if and only if this results in the initial message I; see Figure 13.1. Of course, for correctness we require that if the legitimate prover executes the protocol correctly then the verifier should always accept.

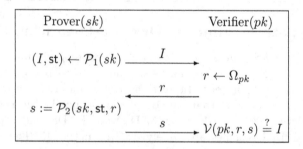

FIGURE 13.1: A three-round identification scheme.

For technical reasons, we assume identification schemes that are "non-degenerate," which intuitively means that there are many possible initial messages I, and none has a high probability of being sent. Formally, a scheme is *non-degenerate* if for every private key sk and any fixed initial message I, the probability that $\mathcal{P}_1(sk)$ outputs I is negligible. (Any identification scheme can be trivially modified to be non-degenerate by sending a uniform n-bit string along with the initial message.)

The basic security requirement of an identification scheme is that an adversary who does not know the prover's secret key should be unable to fool the verifier into accepting. This should hold even if the attacker is able to passively eavesdrop on multiple (honest) executions of the protocol between the prover and verifier. We formalize such eavesdropping via an oracle Trans_{sk} that, when called without any input, runs an honest execution of the protocol and returns to the adversary the entire transcript (I, r, s) of the interaction.

Let $\Pi = (\mathsf{Gen}, \mathcal{P}_1, \mathcal{P}_2, \mathcal{V})$ be an identification scheme, and consider the following experiment for an adversary $\mathcal{A}$ and parameter n:

The identification experiment $\mathsf{Ident}_{\mathcal{A},\Pi}(n)$:

1. $\mathsf{Gen}(1^n)$ *is run to obtain keys* (pk, sk).

2. *Adversary* $\mathcal{A}$ *is given* pk *and access to an oracle* Trans_{sk} *that it can query as often as it likes.*

3. *At any point during the experiment,* $\mathcal{A}$ *outputs a message* I. *A uniform challenge* $r \in \Omega_{pk}$ *is chosen and given to* $\mathcal{A}$, *who responds with some* s. *($\mathcal{A}$ may continue to query* Trans_{sk} *even after receiving* r.)

4. *The experiment outputs* 1 *if and only if* $\mathcal{V}(pk, r, s) \overset{?}{=} I$.

DEFINITION 13.8 *An identification scheme* $\Pi = (\mathsf{Gen}, \mathcal{P}_1, \mathcal{P}_2, \mathcal{V})$ *is* secure against a passive attack, *or just* secure, *if for all probabilistic polynomial-time adversaries* $\mathcal{A}$, *there exists a negligible function* negl *such that:*

$$\Pr[\mathsf{Ident}_{\mathcal{A},\Pi}(n) = 1] \leq \mathsf{negl}(n).$$

It is also possible to consider stronger notions of security, for example, where the adversary can also carry out *active* attacks on the protocol by impersonating a verifier and possibly sending maliciously chosen values r. We will not need this for our application to signature schemes.

From Identification Schemes to Signatures

The Fiat–Shamir transform (Construction 13.9) provides a way to convert any (interactive) identification scheme into a (non-interactive) signature scheme. The basic idea is for the signer to act as a prover, running the identification protocol *by itself.* That is, to sign a message m, the signer first computes I, and next generates the challenge r by applying some function H to I and m. It then derives the correct response s. The signature on m is (r, s), which can be verified by (1) recomputing $I := \mathcal{V}(pk, r, s)$ and then (2) checking that $H(I, m) \overset{?}{=} r$.

CONSTRUCTION 13.9

Let $(\mathsf{Gen}_{\mathsf{id}}, \mathcal{P}_1, \mathcal{P}_2, \mathcal{V})$ be an identification scheme, and construct a signature scheme as follows:

- Gen: on input 1^n, simply run $\mathsf{Gen}_{\mathsf{id}}(1^n)$ to obtain keys pk, sk.

  The public key pk specifies a set of challenges Ω_{pk}. As part of key generation, a function $H : \{0,1\}^* \to \Omega_{pk}$ is specified, but we leave this implicit.

- Sign: on input a private key sk and a message $m \in \{0,1\}^*$, do:

  1. Compute $(I, \mathsf{st}) \leftarrow \mathcal{P}_1(sk)$.
  2. Compute $r := H(I, m)$.
  3. Compute $s := \mathcal{P}_2(sk, \mathsf{st}, r)$.

  Output the signature (r, s).

- Vrfy: on input a public key pk, a message m, and a signature (r, s), compute $I := \mathcal{V}(pk, r, s)$ and output 1 if and only if

$$H(I, m) \stackrel{?}{=} r.$$

The Fiat–Shamir transform.

A signature (r, s) is "bound" to a specific message m because r is a function of both I and m; changing m thus results in a completely different r. If H is modeled as a random oracle mapping inputs uniformly onto Ω_{pk}, then the challenge r is uniform; intuitively, it will be just as difficult for an adversary (who does not know sk) to find a valid signature (r, s) on a message m as it would be to impersonate the prover in an honest execution of the protocol. This intuition is formalized in the proof of the following theorem.

THEOREM 13.10 *Let Π be an identification scheme, and let Π' be the signature scheme that results by applying the Fiat–Shamir transform to it. If Π is secure and H is modeled as a random oracle, then Π' is secure.*

PROOF Let $\mathcal{A}'$ be a probabilistic polynomial-time adversary attacking the signature scheme Π', with $q = q(n)$ an upper bound on the number of queries that $\mathcal{A}'$ makes to H. We make a number of simplifying assumptions without loss of generality. First, we assume that $\mathcal{A}'$ makes any given query to H only once. We also assume that after being given a signature (r, s) on a message m with $\mathcal{V}(pk, r, s) = I$, the adversary $\mathcal{A}'$ never queries $H(I, m)$ (since it knows the answer will be r). Finally, we assume that if $\mathcal{A}'$ outputs a forged signature (r, s) on a message m with $\mathcal{V}(pk, r, s) = I$, then $\mathcal{A}'$ had previously queried $H(I, m)$.

We construct an efficient adversary $\mathcal{A}$ that uses $\mathcal{A}'$ as a subroutine and attacks the identification scheme Π:

Algorithm $\mathcal{A}$:
The algorithm is given pk and access to an oracle Trans_{sk}.

1. Choose uniform $j \in \{1, \ldots, q\}$.
2. Run $\mathcal{A}'(pk)$. Answer its queries as follows:
   When $\mathcal{A}'$ makes its ith random-oracle query $H(I_i, m_i)$, answer it as follows:
   - If $i = j$, output I_j and receive in return a challenge r. Return r to $\mathcal{A}'$ as the answer to its query.
   - If $i \neq j$, choose a uniform $r \in \Omega_{pk}$ and return r as the answer to the query.

   When $\mathcal{A}'$ requests a signature on m, answer it as follows:
   (a) Query Trans_{sk} to obtain a transcript (I, r, s) of an honest execution of the protocol.
   (b) Return the signature (r, s).
3. If $\mathcal{A}'$ outputs a forged signature (r, s) on a message m, compute $I := \mathcal{V}(pk, r, s)$ and check whether $(I, m) \stackrel{?}{=} (I_j, m_j)$. If so, then output s. Otherwise, abort.

The view of $\mathcal{A}'$ when run as a subroutine by $\mathcal{A}$ in experiment $\mathsf{Ident}_{\mathcal{A},\Pi}(n)$ is *almost* identical to the view of $\mathcal{A}'$ in experiment $\mathsf{Sig\text{-}forge}_{\mathcal{A}',\Pi'}(n)$. Indeed, all the H-queries that $\mathcal{A}'$ makes are answered with a uniform value from Ω_{pk}, and all the signing queries that $\mathcal{A}'$ makes are answered with valid signatures having the correct distribution. The only difference between the views is that when $\mathcal{A}'$ is run as a subroutine by $\mathcal{A}$ it is possible for there to be an inconsistency in the answers $\mathcal{A}'$ receives from its queries to H: specifically, this happens if $\mathcal{A}$ ever answers a signing query for a message m using a transcript (I, r, s) for which $H(I, m)$ is already defined (that is, $\mathcal{A}'$ had previously queried (I, m) to H) and $H(I, m) \neq r$. However, if Π is non-degenerate then this only ever happens with negligible probability. Thus, the probability that $\mathcal{A}'$ outputs a forgery when run as a subroutine by $\mathcal{A}$ is $\Pr[\mathsf{Sig\text{-}forge}_{\mathcal{A}',\Pi'}(n) = 1] - \mathsf{negl}(n)$ for some negligible function negl.

Consider an execution of experiment $\mathsf{Ident}_{\mathcal{A},\Pi}(n)$ in which $\mathcal{A}'$ outputs a forged signature (r, s) on a message m, and let $I := \mathcal{V}(pk, r, s)$. Since j is uniform and independent of everything else, the probability that $(I, m) = (I_j, m_j)$ (even conditioned on the event that $\mathcal{A}'$ outputs a forgery) is exactly $1/q$. (Recall we assume that if $\mathcal{A}'$ outputs a forged signature (r, s) on a message m with $\mathcal{V}(pk, r, s) = I$, then $\mathcal{A}'$ had previously queried $H(I, m)$.) When both events happen, $\mathcal{A}$ successfully impersonates the prover. Indeed, $\mathcal{A}$ sends I_j as its initial message, receives in response a challenge r, and responds with s. But $H(I_j, m_j) = r$ and (since the forged signature is valid) $\mathcal{V}(pk, r, s) = I$. Putting everything together, we see that

$$\Pr[\mathsf{Ident}_{\mathcal{A},\Pi}(n) = 1] \geq \frac{1}{q(n)} \cdot \left(\Pr[\mathsf{Sig\text{-}forge}_{\mathcal{A}',\Pi'}(n) = 1] - \mathsf{negl}(n) \right)$$

or

$$\Pr[\mathsf{Sig\text{-}forge}_{\mathcal{A}',\Pi'}(n) = 1] \leq q(n) \cdot \Pr[\mathsf{Ident}_{\mathcal{A},\Pi}(n) = 1] + \mathsf{negl}(n).$$

If Π is secure then $\Pr[\mathsf{Ident}_{\mathcal{A},\Pi}(n) = 1]$ is negligible; since $q(n)$ is polynomial this implies that $\Pr[\mathsf{Sig\text{-}forge}_{\mathcal{A}',\Pi'}(n) = 1]$ is also negligible. Because $\mathcal{A}'$ was arbitrary, this means Π' is secure. ∎

13.5.2 The Schnorr Identification/Signature Schemes

The Schnorr identification scheme is based on hardness of the discrete-logarithm problem. Let $\mathcal{G}$ be a polynomial-time algorithm that takes as input 1^n and (except possibly with negligible probability) outputs a description of a cyclic group $\mathbb{G}$, its order q (with $\|q\| = n$), and a generator g. To generate its keys, the prover runs $\mathcal{G}(1^n)$ to obtain $(\mathbb{G}, q, g)$, chooses a uniform $x \in \mathbb{Z}_q$, and sets $y := g^x$; the public key is $\langle \mathbb{G}, q, g, y \rangle$ and the private key is x. To execute the protocol (see Figure 13.2), the prover begins by choosing a uniform $k \in \mathbb{Z}_q$ and setting $I := g^k$; it sends I as the initial message. The verifier chooses and sends a uniform challenge $r \in \mathbb{Z}_q$; in response, the prover computes $s := [rx + k \bmod q]$. The verifier accepts if and only if $g^s \cdot y^{-r} \stackrel{?}{=} I$. Correctness holds because

$$g^s \cdot y^{-r} = g^{rx+k} \cdot (g^x)^{-r} = g^k = I.$$

Note that I is uniform in $\mathbb{G}$, and so the scheme is non-degenerate.

Before giving the proof, we provide some high-level intuition. A first important observation is that passive eavesdropping is of no help to the attacker. The reason is that the attacker can *simulate* transcripts of honest executions on its own, based only on the public key and *without* knowledge of the private key. To do this, the attacker just reverses the order of the steps: it first chooses uniform and independent $r, s \in \mathbb{Z}_q$ and then sets $I := g^s \cdot y^{-r}$. In an honest transcript (I, r, s), the initial message I is a uniform element of $\mathbb{G}$, the

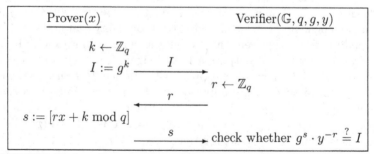

FIGURE 13.2: An execution of the Schnorr identification scheme.

challenge is an independent, uniform element of $\mathbb{Z}_q$, and s is then uniquely determined as $s = \log_g(I \cdot y^r)$. Simulated transcripts constructed by an attacker have the same distribution: $r \in \mathbb{Z}_q$ is uniform and, because s is uniform in $\mathbb{Z}_q$ and independent of r, we see that I is uniform in $\mathbb{G}$ and independent of r. Finally, s is uniquely determined as satisfying the same constraint as before. Due to this, we may effectively assume that when attacking the identification scheme, an attacker does not eavesdrop on honest executions at all.

So, we have reduced to an attacker who gets a public key y, sends an initial message I, is given in response a uniform challenge r, and then must send a response s for which $g^s \cdot y^{-r} = I$. Informally, if an attacker is able to do this with high probability then it must, in particular, be able to compute correct responses s_1, s_2 to at least two different challenges $r_1, r_2 \in \mathbb{Z}_q$. Note

$$g^{s_1} \cdot y^{-r_1} = I = g^{s_2} \cdot y^{-r_2},$$

and so $g^{s_1-s_2} = y^{r_1-r_2}$. But this implies that the attacker (who, recall, is able to generate s_1 in response to r_1, and s_2 in response to r_2) can implicitly compute the discrete logarithm

$$\log_g y = [(s_1 - s_2) \cdot (r_1 - r_2)^{-1} \bmod q],$$

contradicting the assumed hardness of the discrete-logarithm problem.

THEOREM 13.11 *If the discrete-logarithm problem is hard relative to $\mathcal{G}$, then the Schnorr identification scheme is secure.*

PROOF Let Π denote the Schnorr identification scheme, and let $\mathcal{A}$ be a PPT adversary attacking the scheme. We construct the following PPT algorithm $\mathcal{A}'$ solving the discrete-logarithm problem relative to $\mathcal{G}$:

Algorithm $\mathcal{A}'$:
The algorithm is given $\mathbb{G}, q, g, y$ as input.

1. Run $\mathcal{A}(pk)$, answering all its queries to Trans_{sk} as described in the intuition given previously.

2. When $\mathcal{A}$ outputs I, choose a uniform $r_1 \in \mathbb{Z}_q$ as the challenge. Give r_1 to $\mathcal{A}$, who responds with s_1.

3. Run $\mathcal{A}(pk)$ a second time (from the beginning), using the same randomness as before except for uniform and independent $r_2 \in \mathbb{Z}_q$. Eventually, $\mathcal{A}$ responds with s_2.

4. If $g^{s_1} \cdot h^{-r_1} = I$ and $g^{s_2} \cdot h^{-r_2} = I$ and $r_1 \neq r_2$ then output $[(s_1 - s_2) \cdot (r_1 - r_2)^{-1} \bmod q]$. Else, output nothing.

Considering a single run of $\mathcal{A}$ as a subroutine of $\mathcal{A}'$, let ω denote the randomness used in that execution except for the challenge itself. So, ω comprises any randomness used by $\mathcal{G}$, the choice of (unknown) private key x, any randomness used by $\mathcal{A}$ itself, and the randomness used by $\mathcal{A}'$ when answering queries to Trans_{sk}. Define $V(\omega, r)$ to be equal to 1 if and only if $\mathcal{A}$ correctly responds to challenge r when randomness ω is used in the rest of the execution. For any fixed ω, define $\delta_\omega \stackrel{\mathrm{def}}{=} \mathrm{Pr}_r[V(\omega, r) = 1]$; having fixed ω, this is the probability over choice of the challenge r that $\mathcal{A}$ responds correctly.

Define $\delta(n) \stackrel{\mathrm{def}}{=} \mathrm{Pr}[\mathsf{Ident}_{\mathcal{A},\Pi}(n) = 1]$. Since the simulation of the Trans_{sk} oracle is perfect, we have

$$\delta(n) = \mathrm{Pr}_{\omega,r}[V(\omega, r) = 1] = \sum_\omega \mathrm{Pr}[\omega] \cdot \delta_\omega.$$

Moreover, the intuition preceding the proof shows that $\mathcal{A}'$ correctly computes the discrete logarithm of y whenever $\mathcal{A}$ succeeds twice and $r_1 \neq r_2$. Thus:

$$\begin{aligned}
\mathrm{Pr}[\mathsf{DLog}_{\mathcal{A}',\mathcal{G}}(n) = 1] &= \mathrm{Pr}_{\omega,r_1,r_2}[V(\omega, r_1) \wedge V(\omega, r_2) \wedge r_1 \neq r_2] \\
&\geq \mathrm{Pr}_{\omega,r_1,r_2}[V(\omega, r_1) \wedge V(\omega, r_2)] - \mathrm{Pr}_{\omega,r_1,r_2}[r_1 = r_2] \\
&= \sum_\omega \mathrm{Pr}[\omega] \cdot (\delta_\omega)^2 - 1/q \\
&\geq \left(\sum_\omega \mathrm{Pr}[\omega] \cdot \delta_\omega \right)^2 - 1/q \\
&= \delta(n)^2 - 1/q,
\end{aligned}$$

using Jensen's inequality in the second-to-last step. (Jensen's inequality says that $\sum_i a_i \cdot b_i^2 \geq (\sum_i a_i)^{-1} \cdot (\sum_i a_i \cdot b_i)^2$ for positive $\{a_i\}$.) If the discrete-logarithm problem is hard relative to $\mathcal{G}$ then $\mathrm{Pr}[\mathsf{DLog}_{\mathcal{A}',\mathcal{G}}(n) = 1]$ is negligible. Since $1/q$ is negligible (because $\|q\| = n$), this implies that $\delta(n)$ is also negligible, and so Π is a secure identification scheme. ∎

The Schnorr signature scheme is obtained by applying the Fiat–Shamir transform to the Schnorr identification scheme. See Construction 13.12.

CONSTRUCTION 13.12

Let $\mathcal{G}$ be as described in the text.

- **Gen:** run $\mathcal{G}(1^n)$ to obtain $(\mathbb{G}, q, g)$. Choose a uniform $x \in \mathbb{Z}_q$ and set $y := g^x$. The private key is x and the public key is $(\mathbb{G}, q, g, y)$. As part of key generation, a function $H : \{0,1\}^* \to \mathbb{Z}_q$ is specified, but we leave this implicit.

- **Sign:** on input a private key x and a message $m \in \{0,1\}^*$, choose uniform $k \in \mathbb{Z}_q$ and set $I := g^k$. Then compute $r := H(I, m)$, followed by $s := [rx + k \bmod q]$. Output the signature (r, s).

- **Vrfy:** on input a public key $(\mathbb{G}, q, g, y)$, a message m, and a signature (r, s), compute $I := g^s \cdot y^{-r}$ and output 1 if $H(I, m) \stackrel{?}{=} r$.

The Schnorr signature scheme.

EdDSA is an efficient, standardized version of Schnorr signatures that uses a specific elliptic-curve group.

13.5.3 DSA and ECDSA

The *Digital Signature Algorithm* (DSA) and *Elliptic Curve Digital Signature Algorithm* (ECDSA) are based on the discrete-logarithm problem in different classes of groups. They have been around in some form since 1991, and are both included in the current *Digital Signature Standard* (DSS) issued by NIST (although in 2019 NIST proposed to deprecate DSA).

Both schemes follow a common template and can be viewed as being constructed from an underlying identification scheme (see the previous section). Let $\mathbb{G}$ be a cyclic group of prime order q with generator g. Consider the following identification scheme in which the prover's private key is x and public key is $(\mathbb{G}, q, g, y)$ with $y = g^x$:

1. The prover chooses uniform $k \in \mathbb{Z}_q^*$ and sends $I := g^k$.

2. The verifier chooses and sends uniform $\alpha, r \in \mathbb{Z}_q$ as the challenge.

3. The prover sends $s := [k^{-1} \cdot (\alpha + xr) \bmod q]$ as the response.

4. The verifier accepts if $s \neq 0$ and $g^{\alpha s^{-1}} \cdot y^{rs^{-1}} \overset{?}{=} I$.

Note $s \neq 0$ unless $\alpha = -xr \bmod q$, which occurs with negligible probability. Assuming $s \neq 0$, the inverse $s^{-1} \bmod q$ exists and

$$g^{\alpha s^{-1}} \cdot y^{rs^{-1}} = g^{\alpha s^{-1}} \cdot g^{xrs^{-1}} = g^{(\alpha + xr) \cdot s^{-1}} = g^{(\alpha + xr) \cdot k \cdot (\alpha + xr)^{-1}} = I.$$

We thus see that correctness holds with all but negligible probability.

One can show that this identification scheme is secure if the discrete-logarithm problem is hard relative to $\mathcal{G}$. We merely sketch the argument, assuming familiarity with the results of the previous section. First of all, transcripts of honest executions can be simulated: to do so, simply choose uniform $\alpha, r \in \mathbb{Z}_q$ and $s \in \mathbb{Z}_q^*$, and then set $I := g^{\alpha s^{-1}} \cdot y^{rs^{-1}}$. (This no longer gives a *perfect* simulation, but it is close enough.) Moreover, if an attacker outputs an initial message I for which it can give correct responses $s_1, s_2 \in \mathbb{Z}_q^*$ to distinct challenges $(\alpha, r_1), (\alpha, r_2)$ then

$$g^{\alpha s_1^{-1}} \cdot y^{r_1 s_1^{-1}} = I = g^{\alpha s_2^{-1}} \cdot y^{r_2 s_2^{-1}},$$

and so $g^{\alpha(s_1^{-1} - s_2^{-1})} = y^{r_1 s_1^{-1} - r_2 s_2^{-1}}$ and $\log_g y$ can be computed as in the previous section. The same holds if the attacker gives correct responses to distinct challenges $(\alpha_1, r), (\alpha_2, r)$.

The DSA/ECDSA signature schemes are constructed by "collapsing" the above identification scheme into a non-interactive algorithm run by the signer. In contrast to the Fiat–Shamir transform, however, the transformation here is carried out as follows (see Construction 13.13):

- Set $\alpha := H(m)$, where m is the message being signed and H is a cryptographic hash function.

- Set $r := F(I)$ for a (specified) function $F : \mathbb{G} \to \mathbb{Z}_q$. Here, F is a "simple" function that is *not* intended to act like a random oracle.

The function F depends on the group $\mathbb{G}$, which in turn depends on the scheme. In DSA, $\mathbb{G}$ is taken to be an order-q subgroup of $\mathbb{Z}_p^*$, for p prime (cf. Section 9.3.3), and $F(I) \overset{\text{def}}{=} [I \bmod q]$. In ECDSA, $\mathbb{G}$ is an order-q subgroup of an elliptic-curve group $E(\mathbb{Z}_p)$, for p prime.[4] Recall from Section 9.3.4 that any element of such a group can be represented as a pair $(x, y) \in \mathbb{Z}_p \times \mathbb{Z}_p$. The function F in this case is defined as $F((x, y)) \overset{\text{def}}{=} [x \bmod q]$.

CONSTRUCTION 13.13

Let $\mathcal{G}$ be as in the text.

- Gen: on input 1^n, run $\mathcal{G}(1^n)$ to obtain $(\mathbb{G}, q, g)$. Choose uniform $x \in \mathbb{Z}_q$ and set $y := g^x$. The public key is $\langle \mathbb{G}, q, g, y \rangle$ and the private key is x.

  As part of key generation, two functions $H : \{0,1\}^* \to \mathbb{Z}_q$ and $F : \mathbb{G} \to \mathbb{Z}_q$ are specified, but we leave this implicit.

- Sign: on input the private key x and a message $m \in \{0,1\}^*$, choose uniform $k \in \mathbb{Z}_q^*$ and set $r := F(g^k)$. Then compute $s := [k^{-1} \cdot (H(m) + xr) \bmod q]$. (If $r = 0$ or $s = 0$ then start again with a fresh choice of k.) Output the signature (r, s).

- Vrfy: on input a public key $\langle \mathbb{G}, q, g, y \rangle$, a message $m \in \{0,1\}^*$, and a signature (r, s) with $r, s \neq 0 \bmod q$, output 1 if and only if

$$r \overset{?}{=} F\left(g^{H(m) \cdot s^{-1}} y^{r \cdot s^{-1}}\right).$$

DSA and ECDSA—abstractly.

Assuming hardness of the discrete-logarithm problem, DSA and ECDSA can be proven secure if H and F are modeled as random oracles. As we have discussed above, however, while the random-oracle model may be reasonable for H, it is *not* an appropriate model for F. No proofs of security are known for the specific choices of F in the standard. Nevertheless, DSA and ECDSA have been used and studied for decades without any attacks being found.

Proper generation of k. The DSA/ECDSA schemes specify that the signer should choose a uniform $k \in \mathbb{Z}_q^*$ when computing a signature. Failure to

---

[4]ECDSA also allows elliptic curves over other fields, but we have only covered the case of prime fields in Section 9.3.4.

choose k properly (e.g., due to poor random-number generation) can lead to catastrophic results. For starters, if an attacker can predict the value of k used to compute a signature (r, s) on a message m, then they can compute the signer's private key. This is true because $s = k^{-1} \cdot (H(m) + xr) \bmod q$, and if k is known then the only unknown is the private key x.

Even if k is unpredictable, the attacker can compute the signer's private key if the *same* k is ever used to generate two different signatures. The attacker can easily tell when this happens because then r repeats as well. Say (r, s_1) and (r, s_2) are signatures on messages m_1 and m_2, respectively. Then

$$s_1 = k^{-1} \cdot (H(m_1) + xr) \bmod q$$
$$s_2 = k^{-1} \cdot (H(m_2) + xr) \bmod q.$$

Subtracting gives $s_1 - s_2 = k^{-1} (H(m_1) - H(m_2)) \bmod q$, from which k can be computed; given k, the attacker can determine the private key x as in the previous paragraph. This very attack was used by hackers to extract the master private key from the Sony PlayStation (PS3) in 2010.

13.6 Certificates and Public-Key Infrastructures

In this section we briefly discuss one of the primary applications of digital signatures: the secure distribution of public keys. This brings us full circle in our discussion of public-key cryptography. In this and the previous chapter we have seen how to *use* public-key cryptography once public keys are securely distributed. Now we show how public-key cryptography itself can be used to securely distribute public keys. This may sound circular, but it is not. What we will show is that once a *single* public key, belonging to a trusted party, is distributed in a secure fashion, that key can be used to "bootstrap" the secure distribution of arbitrarily many other public keys. Thus, at least in principle, the problem of secure key distribution need only be solved *once*.

The key notion here is a *digital certificate*, which is simply a signature binding an entity to some public key. To be concrete, say a party Charlie has generated keys (pk_C, sk_C) for a secure digital signature scheme (in this section, we will only be concerned with signature schemes satisfying Definition 13.2). Assume further that another party Bob has also generated keys (pk_B, sk_B) (in the present discussion, these may be keys for either a signature scheme or a public-key encryption scheme), and that Charlie *knows* that pk_B is Bob's public key. Then Charlie can compute the signature

$$\mathsf{cert}_{C \rightarrow B} \overset{\text{def}}{=} \mathsf{Sign}_{sk_C}(\text{'Bob's key is } pk_B\text{'})$$

and give this signature to Bob. We call $\mathsf{cert}_{C \rightarrow B}$ a *certificate* for Bob's key issued by Charlie. In practice a certificate should unambiguously identify the party holding a particular public key and so a more uniquely descriptive term than "Bob" would be used, for example, Bob's full name and email address, or the URL of Bob's website.

Now say Bob wants to communicate with some other party Alice who already knows pk_C. Bob can send $(pk_B, \mathsf{cert}_{C \to B})$ to Alice, who can then verify that $\mathsf{cert}_{C \to B}$ is indeed a valid signature on the message 'Bob's key is pk_B' with respect to pk_C. Assuming verification succeeds, Alice now knows that Charlie has signed the indicated message. If Alice trusts Charlie, she can accept pk_B as Bob's legitimate public key.

All communication between Bob and Alice can occur over an *insecure* and *unauthenticated* channel. If an active adversary interferes with the transmission of $(pk_B, \mathsf{cert}_{C \to B})$ from Bob to Alice, that adversary will be unable to generate a valid certificate linking Bob to any *other* public key pk'_B unless Charlie had previously signed some other certificate linking Bob with pk'_B (in which case this is anyway not much of an attack). This all assumes that Charlie is not dishonest and that his private key has not been compromised.

We have omitted many details in the above description. Most prominently, we have not discussed how Alice learns pk_C in the first place; how Charlie can be sure that pk_B is Bob's public key; and how Alice decides whether to trust Charlie. Fully specifying such details (and others) defines a *public-key infrastructure* (PKI) that enables the widespread distribution of public keys. A variety of different PKI models have been suggested, and we mention a few of the more popular ones now. Our treatment here will be kept at a relatively high level, and the reader interested in further details is advised to consult the references at the end of this chapter.

A single certificate authority. The simplest PKI assumes a single *certificate authority* (CA) who is completely trusted by everybody and who issues certificates for everyone's public key. A certificate authority would not typically be a person, but would more likely be a company whose business it is to certify public keys, a government agency, or perhaps a department within an organization (although in this latter case the CA would likely only be used by people within the organization). Anyone who wants to rely on the services of the CA would have to obtain a legitimate copy of the CA's public key pk_{CA}. Clearly, this step must be carried out in a secure fashion since if some party obtains an incorrect version of pk_{CA} then that party may not be able to obtain an authentic copy of anyone else's public key. This means that pk_{CA} must be distributed over an *authenticated* channel. The easiest way of doing this is via physical means: for example, if the CA is within an organization then any employee can obtain an authentic copy of pk_{CA} directly from the CA on their first day of work. If the CA is a company, then other users would have to go to this company at some point and, say, pick up a USB stick that contains the CA's public key. This inconvenient step need only be carried out once.

A common way for a CA to distribute its public key in practice is to "bundle" this public key with some other software. For example, this occurs today in many popular web browsers: a CA's public key is provided together with the browser, and the browser is programmed to automatically verify certificates as they arrive. (Actually, modern web browsers have public keys of *multiple* CAs hard-wired into their code, and so more accurately fall into the "multiple CA" model discussed below.)

The mechanism by which a CA issues a certificate to some party Bob must also be very carefully controlled, although the details may vary from CA to CA. As one example, Bob may have to show up in person with a copy of his public key pk_B along with identification proving that his name (or his email address) is what he claims. Only then would the CA issue the certificate.

In the model where there is a single CA, parties completely trust this CA to issue certificates only when appropriate; this is why it is crucial that a detailed verification process be used before a certificate is issued. As a consequence, if Alice receives a certificate $\mathsf{cert}_{CA \to B}$ certifying that pk_B is Bob's public key, Alice will accept this assertion as valid, and use pk_B as Bob's public key.

Multiple certificate authorities. While the model in which there is only one CA is simple and appealing, it is not very practical. For one thing, outside of a single organization it is unlikely for *everyone* to trust the same CA. This need not imply that anyone thinks the CA is corrupt; it could simply be the case that someone finds the CA's verification process to be insufficient (say, the CA asks for only one form of identification when generating a certificate but Alice would prefer that two be used instead). Moreover, the CA is a single point of failure for the entire system. If the CA is corrupt, or can be bribed, or even if the CA is merely lax with the way it protects its private key, the legitimacy of issued certificates may be called into question. It is also inconvenient for all parties who want certificates to have to contact this CA.

One approach to alleviating these issues is to rely on multiple CAs. A party Bob who wants to obtain a certificate on his public key can choose which CA(s) it wants to issue a certificate, and a party Alice who is presented with a certificate, or even multiple certificates issued by different CAs, can choose which CA's certificates she trusts. There is no harm in having Bob obtain a certificate from more than one CA (apart from some inconvenience and expense for Bob), but Alice must be more careful since the security of her communication is ultimately only as good as the least-secure CA that she trusts. That is, say Alice trusts two CAs CA_1 and CA_2, and CA_2 is corrupted by an adversary. Then, although this adversary will not be able to forge certificates issued by CA_1, it will be able to issue fake certificates in the name of CA_2 for any identity/public key of its choice. This is a real problem in current systems. As mentioned earlier, operating systems/web browsers typically come pre-configured with many CAs' public keys, and the default setting is for all these CAs to be treated as equally trustworthy. Essentially any company willing to pay, however, can be included as a CA. So the list of pre-configured CAs includes some reputable, well-established companies along with other, newer companies whose trustworthiness cannot be easily established. It is left to the user to manually configure their settings so as to only accept certificates from CAs the user trusts.

Delegation and certificate chains. Another approach which alleviates some of the burden on a single CA (but does not address the security concerns of having a single point of failure) is to use *certificate chains*. We present the idea for certificate chains of length 2, although it is easy to see that everything we say generalizes to chains of arbitrary length.

Say Charlie, acting as a CA, issues a certificate for Bob as in our original discussion. Assume further that Bob's key pk_B is a public key for a signature scheme. Bob, in turn, can issue his own certificates for other parties. For example, Bob may issue a certificate for Alice of the form

$$\mathsf{cert}_{B \to A} \stackrel{\text{def}}{=} \mathsf{Sign}_{sk_B}(\texttt{'Alice's key is } pk_A \texttt{'}).$$

Now, if Alice wants to communicate with some fourth party Dave who knows Charlie's public key (but not Bob's), then Alice can send

$$pk_A, \ \mathsf{cert}_{B \to A}, \ pk_B, \ \mathsf{cert}_{C \to B},$$

to Dave. What can Dave deduce from this? Well, he can first verify that Charlie, whom he trusts and whose public key is already in his possession, has signed a certificate $\mathsf{cert}_{C \to B}$ indicating that pk_B indeed belongs to someone named Bob. Dave can also verify that this person named Bob has signed a certificate $\mathsf{cert}_{B \to A}$ indicating that pk_A indeed belongs to Alice. If Dave trusts Charlie to issue certificates only to trustworthy people, then Dave may accept pk_A as being the authentic key of Alice.

We highlight that in this example stronger semantics are associated with a certificate $\mathsf{cert}_{C \to B}$. In our prior discussion, a certificate of this form was only an assertion that Bob holds public key pk_B. Now, a certificate asserts that Bob holds public key pk_B *and Bob is trusted to issue other certificates*. When Charlie signs a certificate for Bob having these stronger semantics, Charlie is, in effect, *delegating* his ability to issue certificates to Bob. Bob can now act as a proxy for Charlie, issuing certificates on Charlie's behalf.

Coming back to a CA-based PKI, we can imagine one "root" CA and n "second-level" CAs $CA_1, \ldots, CA_n$. The root CA can issue certificates for each of the second-level CAs, who can then in turn issue certificates for other principles holding public keys. This eases the burden on the root CA, and also makes it more convenient for parties to obtain certificates (since they may now contact the second-level CA who is closest to them, for example). On the other hand, managing these second-level CAs may be difficult, and their presence means that there are now more points of attack in the system.

The "web of trust" model. The last example of a PKI we will discuss is a fully distributed model, with no central points of trust, called the "web of trust." A variant of this model is used by the PGP ("Pretty Good Privacy") email-encryption software for distribution of public keys.

In the "web of trust" model, anyone can issue certificates to anyone else and each user has to make their own decision about how much trust to place in certificates issued by other users. As an example of how this might work, say a user Alice is already in possession of public keys pk_1, pk_2, pk_3 for some users C_1, C_2, C_3. (We discuss below how these public keys might initially be obtained by Alice.) Another user Bob who wants to communicate with Alice might have certificates $\text{cert}_{C_1 \to B}$, $\text{cert}_{C_3 \to B}$, and $\text{cert}_{C_4 \to B}$, and will send these certificates (along with his public key pk_B) to Alice. Alice cannot verify $\text{cert}_{C_4 \to B}$ (since she doesn't have C_4's public key), but she can verify the other two certificates. Now she has to decide how much trust she places in C_1 and C_3. She may decide to accept pk_B if she unequivocally trusts C_1, or if she trusts both C_1 and C_3 to a lesser extent. (She may, for example, consider it likely that either C_1 or C_3 is corrupt, but consider it unlikely for them *both* to be corrupt.)

In this model, as described, users are expected to collect both public keys of other parties, as well as certificates on their own public key. In the context of PGP, this used to be done at "key-signing parties" where PGP users got together (say, at a conference), gave each other authentic copies of their public keys, and issued certificates for each other. In general the users at a key-signing party may not know each other, but they can check a driver's license, say, before accepting or issuing a certificate for someone's public key.

Public keys and certificates can also be stored in a central database, and this is done for PGP (see `http://pgp.mit.edu`). When Alice wants to send an encrypted message to Bob, she can search for Bob's public key in this database; along with Bob's public key, the database will return a list of all certificates it holds that have been issued for Bob's public key. It is also possible that multiple public keys for Bob will be found in the database, and each of these public keys may be certified by certificates issued by a different set of parties. Once again, Alice then needs to decide how much trust to place in any of these public keys before using them.

The web of trust model is attractive because it does not require trust in any central authority. On the other hand, while it may work well for the average user encrypting their email, it does not seem appropriate for settings where security is more critical, or for the distribution of organizational public keys (e.g., for e-commerce on the web). If a user wants to communicate with his bank, for example, it is unlikely that he would trust people he met at a conference to certify his bank's public key, and also unlikely that a bank representative will go to a key-signing party to get the bank's key certified.

Invalidating Certificates

One important issue we have not yet touched upon at all is the fact that certificates should generally not be valid indefinitely. An employee may leave a company, in which case he or she is no longer allowed to receive encrypted communication from others within the company; a user's private key might also be stolen, at which point the user (assuming they know about the theft) will want to generate a new set of public/private keys and remove the old public key from circulation. In either of these scenarios, we need a way to render previously issued certificates invalid.

Approaches for handling these issues are varied and complex, and we will only mention two relatively simple ideas that, in some sense, represent opposite extremes. (Improving these methods is an active area of real-world network-security research.)

Expiration. One method for preventing certificates from being used indefinitely is to include an *expiry date* as part of the certificate. A certificate issued by a CA Charlie for Bob's public key might now have the form

$$\mathsf{cert}_{C \to B} \stackrel{\text{def}}{=} \mathsf{Sign}_{sk_C}(\texttt{`Bob's key is } pk_B\texttt{'}, \texttt{date}),$$

where `date` is some date in the future at which point the certificate becomes invalid. (For example, one year from the day the certificate is issued.) When another user verifies this certificate, they need to know not only pk_B but also the expiry date, and they now need to check not only that the signature is valid, but also that the expiry date has not passed. A user who holds a certificate must contact the CA to get a new certificate issued whenever their current one expires; at this point, the CA verifies the identity/credentials of the user again before issuing another certificate.

Using expiry dates provides a very coarse-grained solution to the problems mentioned earlier. If an employee leaves a company the day after getting a certificate, and the certificate expires one year after its issuance date, then this employee can use his or her public key illegitimately for an entire year until the expiry date passes. For this reason, this approach is typically used in conjunction with other methods such as the one we describe next.

Revocation. When an employee leaves an organization, or a user's private key is stolen, we would like the certificates that have been issued for their public keys to become invalid immediately, or at least as soon as possible. This can be achieved by having the CA explicitly *revoke* the certificate. For simplicity we assume a single CA, but everything we say applies more generally if the user had certificates issued by multiple CAs.

There are many different ways revocation can be handled. One possibility (the only one we will discuss) is for the CA to include a serial number in every certificate it issues; that is, a certificate will now have the form

$$\mathsf{cert}_{C \to B} \stackrel{\text{def}}{=} \mathsf{Sign}_{sk_C}(\texttt{`Bob's key is } pk_B\texttt{'}, \texttt{\#\#\#}),$$

where "###" represents the serial number of this certificate. Each certificate should have a unique serial number, and the CA will store the information $(\mathsf{Bob}, pk_B, \texttt{\#\#\#})$ for each certificate it generates.

If a user Bob's private key corresponding to a public key pk_B is stolen, Bob can alert the CA to this fact. (The CA must verify Bob's identity here, to prevent another user from falsely revoking a certificate issued to Bob.) The CA will then search its database to find the serial number associated with the certificate issued for Bob and pk_B. At the end of each day, say, the CA will then generate a *certificate revocation list* (CRL) with the serial numbers of all revoked certificates, and sign the CRL and the current date. The signed CRL is then widely distributed or otherwise made available to potential verifiers. Verification of a certificate now requires checking that the signature in the certificate is valid, checking that the serial number does not appear on the most current revocation list, and verifying the CA's signature on the revocation list itself.

In this approach the way we have described it, there is a gap of at most one day before a certificate becomes invalid. This offers more flexibility than an approach based only on expiry dates.

13.7 Putting It All Together – TLS

The *Transport Layer Security* (TLS) protocol is used by your web browser every time you securely connect to a website using `https`. TLS is a standardized protocol based on a precursor called SSL (or *Secure Sockets Layer*) that was developed by Netscape in the mid-1990s. TLS version 1.0 was released in 1999, and then updated to version 1.1 in 2006, version 1.2 in 2008, and version 1.3 (the current version) in 2018. In this section, we describe the "cryptographic core" of the TLS protocol; this serves as a nice culmination of everything we have covered in the book so far, and also demonstrates the real-world applicability of what we have learned. Our description corresponds roughly to TLS 1.3 but, as usual, we have slightly simplified and abstracted parts of the protocol in order to convey the main point, and our description should not be relied upon for an implementation. (The actual protocol is more complex, and also includes several other interesting features that are outside the scope of this book.) We do not formally define or prove security of the protocol; this is a topic of active research.

The TLS protocol allows a client (e.g., a web browser) and a server (e.g., a website) to agree on a set of shared keys and then use those keys to encrypt and authenticate their subsequent communication. It consists of two parts: a *handshake protocol* that performs (authenticated) key exchange to establish the shared keys, and a *record-layer protocol* that uses those shared keys to encrypt/authenticate the parties' communication. Although TLS allows for clients to authenticate to servers, it is primarily used only for authentication of servers to clients because typically only servers have certificates. (After a TLS session is established, client-to-server authentication—if desired—can be done at the application layer by, e.g., having the client send a password.)

The handshake protocol. We describe the basic flow of the handshake protocol in the most typical case. At the outset, the client C holds a set of CAs' public keys $\{pk_1, \ldots, pk_n\}$, and the server S holds keys (pk_S, sk_S) for a digital signature scheme along with a certificate $\mathsf{cert}_{i \to S}$ on pk_S issued by one of the CAs whose public key C knows. The parties run the following steps.

1. C begins by sending to S the initial message of the Diffie–Hellman key-exchange protocol (cf. Section 11.3). This message includes a specification of the underlying group $\mathbb{G}$ being used by the client (along with the group order q and a generator g), as well as the value g^x for a random secret value x chosen by the client. The underlying group is selected by the client from a set of standardized options, and can be either a prime-order subgroup of $\mathbb{Z}_p^*$ for some prime p or an elliptic-curve group. The client also sends a uniform value (a "nonce") $N_C \in \{0,1\}^n$.

   This message from C also includes information about which cryptographic algorithms (or *ciphersuites*) are supported by the client.

2. S completes the Diffie–Hellman key exchange by sending a message to the client containing g^y for a random secret value y chosen by the server. The server also includes its own uniform value $N_S \in \{0,1\}^n$.

   At this point, S can compute a shared secret $K = g^{xy}$. It applies a key-derivation function (cf. Section 6.6.4) to K to derive keys k_S', k_C', k_S, k_C for an authenticated encryption (AE) scheme. Supported AE schemes include GCM, CCM, and ChaCha20–Poly1305 (cf. Section 5.3.2).

   Finally, S sends its public key pk_S and its certificate $\mathsf{cert}_{i \to S}$, along with a signature σ computed by the server (using its long-term key sk_S) on the handshake messages exchanged thus far. These values sent by the server are all encrypted using k_S'.

3. C computes K from the server's response, and also derives the keys k_S', k_C', k_S, and k_C. It uses k_S' to recover pk_S and the associated certificate, as well as the signature σ. The client checks whether one of the CA's public keys that it holds matches the CA who issued S's certificate. If so, C verifies the certificate (and also checks that it has not expired or been revoked) and, if this was successful, learns that pk_S is indeed S's public key. C then verifies the signature σ on the handshake messages with respect to pk_S, and aborts if verification fails.

   Finally, C computes a MAC of the handshake messages exchanged thus far using k_C'. It sends the result back to S, who verifies the tag before proceeding to the record-layer protocol.

At the end of the handshake protocol, C and S share *session keys* k_C and k_S that they can use to encrypt and authenticate their subsequent communication. (The keys k_C', k_S' are only used for the handshake.)

As some intuition for why the handshake protocol is secure, note first that since C verifies the certificate, it knows that pk_S is the correct public key of the intended server. If the signature σ is valid, then C knows it must be communicating with the server because only someone with knowledge of the associated secret key sk_S could have generated a valid signature. (It is important here that the handshake messages being signed have high entropy, so as to prevent a replay attack. This is why the client includes a random nonce N_C as part of its initial message.) Moreover, since the server signs all the messages of the Diffie–Hellman key-exchange protocol, C knows that none of those values were modified in transit as would be the case if an active adversary were carrying out a man-in-the-middle attack (see Section 11.3). Of course, the Diffie–Hellman protocol itself ensures that a passive eavesdropper learns nothing about K (and hence nothing about the derived keys) from the messages exchanged. In summary, then, by the end of the handshake phase C knows that it shares keys k_C, k_S with the legitimate S, and that no adversary could have learned anything about those keys.

TLS version 1.2 provided a variant that allowed C and S to agree on shared keys using public-key encryption instead of Diffie–Hellman key exchange. In that variant, the server's long-term keys (pk_S, sk_S) corresponded to a public-key encryption scheme, and the client simply chose a key K and encrypted it using pk_S. (Several other aspects of the protocol were also different, and in particular the client verified the certificate on the server's public key before encryption was done.) This variant was purposefully eliminated in version 1.3 due to the desire to ensure *forward secrecy*, i.e., secrecy of previous session keys in the event of a server compromise. Diffie–Hellman key exchange provides forward secrecy since the server's "ephemeral" secret value y used in the handshake protocol can be erased once the handshake is finished; without y an eavesdropper has no way to recover K. On the other hand, using public-key encryption as just described does not provide forward secrecy since the server's long-term secret key sk_S cannot be erased; if an adversary obtains it, then it can decrypt ciphertexts from past executions of the handshake protocol and recover the session keys used by the parties involved.

The record-layer protocol. Once keys have been agreed upon by C and S, the parties use those keys to encrypt and authenticate all their subsequent communication using an AE scheme. C uses k_C for the messages it sends to S, whereas S uses k_S for the messages it sends to C. Sequence numbers are used to prevent replay attacks, as discussed in Section 5.4.

13.8   *Signcryption

To close this chapter, we briefly and informally discuss the issue of joint secrecy and integrity in the public-key setting. While this parallels our treatment from Section 5.2, the fact that we are now in the public-key setting introduces several additional complications.

We consider a setting in which all relevant parties have public/private keys for *both* encrypting and signing. We let (ek, dk) denote a (public) encryption key and (private) decryption key, and use (vk, sk) for a (public) verification key and (private) signing key. We assume all parties know all public keys.

Informally, our goal is to design a mechanism that allows a sender S to send a message m to a receiver R while ensuring that (1) no other party in the network can learn any information about m (i.e., secrecy) and (2) R is assured that the message came from S (i.e., integrity). We consider both of these security properties even against active (e.g., chosen-ciphertext) attacks by other parties in the system.

Following our discussion in Section 5.2, a natural idea is to use an "encrypt-then-authenticate" approach in which S sends $\langle S, c, \mathsf{Sign}_{sk_S}(c) \rangle$ to R, where c is an encryption of m using R's encryption key ek_R. (We explicitly include the sender's identity here for convenience.) However, there is a clever chosen-ciphertext attack here regardless of the encryption scheme used. Having observed a transmission as above, another (adversarial) party A can strip off S's signature and replace it with its own, sending $\langle A, c, \mathsf{Sign}_{sk_A}(c) \rangle$ to R. In this case, R would not detect anything wrong, and would mistakenly think that A has sent it the message m. If R replies to A, or otherwise behaves toward A in a way that depends on the contents of the message, then A can potentially learn the unknown message m.

(Another problem with this scheme, although somewhat independent of our discussion here, is that it no longer provides *non-repudiation*. That is, R cannot easily prove to a third party that S has signed the message m, at least not without divulging its own decryption key dk_R.)

One could instead try an "authenticate-then-encrypt" approach. Here, S would first compute a signature $\sigma \leftarrow \mathsf{Sign}_{sk_S}(m)$ and then send

$$\langle S, \mathsf{Enc}_{ek_R}(m\|\sigma) \rangle.$$

(Note that this solves the non-repudiation issue mentioned above.) If the encryption scheme is only CPA-secure then problems just like those mentioned in Section 5.2 apply, so let us assume a CCA-secure encryption scheme is used instead. Even then, there is an attack that can be carried out by a malicious R. Upon receiving $\langle S, \mathsf{Enc}_{ek_R}(m\|\sigma) \rangle$ from S, a malicious R can decrypt to obtain $m\|\sigma$, and then re-encrypt and send $\langle S, \mathsf{Enc}_{ek_{R'}}(m\|\sigma) \rangle$ to another receiver R'. This (honest) receiver R' will then think that S sent it the message m. This can have serious consequences, e.g., if m is the message "I owe you \$100."

These attacks can be prevented if parties are more careful about how they handle identifiers. When encrypting, a sender should encrypt its own identity along with the message; when signing, a party should sign the identity of the intended recipient along with what is being signed. For example, the second approach would be modified so that S first computes $\sigma \leftarrow \mathsf{Sign}_{sk_S}(m\|R)$, and then sends $\langle S, \mathsf{Enc}_{ek_R}(S\|m\|\sigma)\rangle$ to R. When decrypting, the receiver should check that the decrypted value includes the (purported) sender's identity; when verifying, the receiver should check that what was signed incorporates its own identity. When including identities in this way, both authenticate-then-encrypt and encrypt-then-authenticate are secure if a CCA-secure encryption scheme and a strongly secure signature scheme are used.

References and Additional Reading

Notable early work on signatures includes that of Diffie and Hellman [65], Rabin [165, 166], Rivest, Shamir, and Adleman [171], and Goldwasser, Micali, and Yao [89]. For an extensive treatment of signature schemes beyond what is covered here, see the monograph by Katz [109].

Goldwasser, Micali, and Rivest [88] defined the notion of existential unforgeability under an adaptive chosen-message attack, and also gave the first construction of a stateful signature scheme satisfying this definition.

Plain RSA signatures date to the original RSA paper [171]. RSA-FDH was proposed by Bellare and Rogaway in their paper introducing the random-oracle model [24], although the idea (without proof) of using a cryptographic hash function to prevent algebraic attacks can be traced back to Rabin [166]. A later improvement of RSA-FDH [26] was standardized in PKCS #1 v2.1.

The Fiat–Shamir transform [72] and the Schnorr signature scheme [175] both date to the late-1980s. The proof of Theorem 13.10 is due to Abdalla et al. [1] and the proof of Theorem 13.11 is inspired by Bellare and Neven [22]. The DSA and ECDSA standards are described in [150, 151].

The notion of certificates was first described by Kohnfelder [118] in his undergraduate thesis. Public-key infrastructures are discussed in greater detail in [113, Chapter 15]; see also [3, 69]. The TLS version 1.3 standard is available as an RFC [170]. A formal treatment of combined secrecy and integrity in the public-key setting is given by An et al. [10].

Exercises

13.1 Show that Construction 4.7 for constructing a variable-length MAC from any fixed-length MAC can also be used (with appropriate modifications) to construct a signature scheme for arbitrary-length messages from any signature scheme for messages of fixed length $\ell(n) \geq n$.

13.2 In Section 13.4.1 we showed an attack on the plain RSA signature scheme in which an attacker forges a signature on an arbitrary message using two signing queries. Show how an attacker can forge a signature on an arbitrary message using a *single* signing query.

13.3 Assume the RSA problem is hard. Show that the plain RSA signature scheme satisfies the following weak definition of security: an attacker is given the public key $\langle N, e \rangle$ and a uniform message $m \in \mathbb{Z}_N^*$. The adversary succeeds if it can output a valid signature on m without making any signing queries.

13.4 Consider a "padded RSA" signature scheme where the public key is $\langle N, e \rangle$ as usual, and a signature on a message $m \in \{0,1\}^\ell$ is computed by choosing uniform $r \in \{0,1\}^{2n-\ell-1}$ and outputting $[(r\|m)^d \bmod N]$.

   (a) How can verification be done for this scheme?

   (b) Show that this scheme is insecure.

13.5 Another approach (besides hashing) that has been explored to construct secure RSA-based signatures is to *encode* the message before applying the RSA permutation. Here the signer fixes a public encoding function enc : $\{0,1\}^\ell \to \mathbb{Z}_N^*$ as part of its public key, and the signature on a message m is $\sigma := [\mathsf{enc}(m)^d \bmod N]$.

   (a) How is verification performed in such a scheme?

   (b) Suggest an appropriate encoding function for $\ell \ll \|N\|$ that heuristically prevents the "no-message attack" described in Section 13.4.1.

   (c) Show that encoded RSA is insecure if $\mathsf{enc}(m) = m\|0^{\kappa/10}$ (where $\kappa \stackrel{\text{def}}{=} \|N\|$, $|m| \stackrel{\text{def}}{=} 4\kappa/5$, and m is not the all-0 message). Assume $e = 3$.

   (d) Show that encoded RSA is insecure for $\mathsf{enc}(m) = m\|0\|m$ (where $|m| \stackrel{\text{def}}{=} (\|N\|-1)/2$ and m is not the all-0 message). Assume $e = 3$.

   (e) Show attacks in parts (c) and (d) for arbitrary e.

13.6 Consider a variant of the Fiat–Shamir transform in which the signature is (I, s) rather than (r, s) and verification is changed in the natural way. Show that if the underlying identification scheme is secure, then this variant signature scheme is secure as well.

13.7 Show that ECDSA is not strongly secure. Specifically, show that if (r, s) is a valid signature on a message m, then so is $(r, -s)$.

   Hint: You will need to consider the representation of elliptic-curve points.

13.8 Consider a variant of DSA in which the message space is $\mathbb{Z}_q$ and H is omitted. (So the second component of the signature is now $s := [k^{-1} \cdot (m + xr) \bmod q]$.) Show that this variant is not secure.

13.9 Assume revocation of certificates is handled in the following way: when a user Bob claims that the private key corresponding to his public key pk_B has been stolen, the user sends to the CA a statement of this fact *signed with respect to pk_B*. Upon receiving such a signed message, the CA revokes the appropriate certificate.

Explain why it is not necessary for the CA to check Bob's identity in this case. In particular, explain why it is of no concern that an adversary who has stolen Bob's private key can forge signatures with respect to pk_B.

Chapter 14

*Post-Quantum Cryptography

So far in this book (cf. Section 3.1.2), we have equated the notion of "efficient adversaries" with adversarial algorithms running in (probabilistic) polynomial-time *on a classical computer*. Thus, when evaluating the security of our schemes we only considered efficient *classical* attacks. We did not, however, consider the potential impact of *quantum computers*—that is, computers that rely in an essential way on the principles of quantum mechanics. As we will see here, quantum algorithms can in some cases be faster than classical algorithms—possibly much faster—and thus quantum computers can have a dramatic impact on the security of cryptosystems.

While the theoretical impact of quantum computing on cryptography has been recognized since the mid-1990s, its potential impact in practice is currently unclear. As of this writing, no large-scale, general-purpose quantum computer has been built, and the timeframe for developing such a computer is uncertain due to the numerous engineering difficulties involved. Even if such computers are one day built, the true cost in time or money of executing quantum algorithms on those computers (as distinguished from the theoretical analysis of the number of steps those algorithms take in theory) is not well understood. Nevertheless, the current consensus is that there is a strong chance that well-funded attackers (e.g., government agencies) will be able to build quantum computers capable of attacking currently deployed cryptosystems in the next 10–15 years. Assuming this to be the case, we *cannot* wait 10–15 years to worry about the problem: standardizing and deploying new cryptographic algorithms takes time, and there may be messages encrypted now that must remain secret for more than a decade.

The above concerns have motivated an intense research effort over the past several years aimed at designing, analyzing, and developing "post-quantum" cryptosystems that would remain secure even against (polynomial-time) quantum algorithms. This work accelerated in 2017, when NIST announced an effort to evaluate and (eventually) standardize quantum-resistant public-key schemes. As in the case of the earlier AES and SHA-3 competitions, NIST solicited proposals for public-key encryption schemes and signature schemes from cryptographers around the world, eventually receiving 69 candidates; 26 of those made it to the second round in early 2019. In contrast to the AES and SHA-3 process, NIST is not expected to choose a single "winner" in each category; instead, the idea is to identify multiple schemes judged to be secure. NIST is expected to issue a set of draft standards for such schemes by 2024.

The goal of this chapter is to describe the impact of quantum algorithms on the schemes used today, and to offer a glimpse of some schemes offering plausible post-quantum security. We do not assume any background in quantum mechanics or quantum computing, and will not present any quantum algorithms in detail. Rather, we explain what existing quantum algorithms can do (without describing in detail how they do it) and otherwise treat them as "black boxes." The post-quantum cryptosystems we describe are similar to current leading candidates in the NIST post-quantum standardization effort, but we have simplified them for pedagogical purposes.

Post-quantum cryptography vs. quantum cryptography. *Quantum cryptography* is related to, but distinct from, post-quantum cryptography as we use the term here. Quantum cryptography refers to cryptosystems that are implemented using quantum computers, quantum-mechanical phenomena, and quantum communication channels; for this reason, they would be difficult to deploy widely over the existing Internet. Post-quantum cryptosystems, on the other hand, are entirely classical—but are intended to ensure security even if an attacker has access to a quantum computer.

Interestingly, quantum cryptosystems can in some cases be proven secure *unconditionally* (i.e., without any computational assumptions), even against quantum attackers. In contrast, post-quantum cryptosystems—as with the rest of the schemes in this book—rely on assumptions regarding the hardness of certain mathematical problems even for quantum algorithms.

14.1 Post-Quantum Symmetric-Key Cryptography

We begin by exploring the impact of quantum computers on symmetric-key cryptography. While there are known quantum attacks that can outperform classical attacks in this setting, the net result is only a polynomial speedup and so the overall impact on symmetric-key cryptography is relatively minor.

14.1.1 Grover's Algorithm and Symmetric-Key Lengths

Consider the following abstract problem: Given oracle access to a function $f : D \to \{0, 1\}$, find an input x for which $f(x) = 1$. If there is only one such input, chosen uniformly in D, then it is not hard to show that any classical algorithm for this problem requires $\mathcal{O}(|D|)$ evaluations of f; this effectively corresponds to exhaustive search over the domain D of the function.

In a surprising result published in 1996, Lov Grover showed that *quantum* algorithms can do better. Specifically, he gave an algorithm that finds x as above using only $\mathcal{O}(|D|^{1/2})$ evaluations of f—a quadratic speedup. It was later shown that this is optimal, i.e., no quantum algorithm can do better.

Let us explore the impact this has on the required key length for symmetric-key cryptosystems. For concreteness, consider the case of a block cipher $F : \{0,1\}^n \times \{0,1\}^\ell \to \{0,1\}^\ell$ for which exhaustive search is the best attack, and an attacker whose goal is to determine the key $k \in \{0,1\}^n$ given a constant number of input/output pairs $\{(x_i, y_i)\}$ with $y_i = F_k(x_i)$. Say we want security against attacks running in time 2^κ. Classically, it suffices to set $n = \kappa$ (since exhaustive search for k requires time $\approx 2^n$). But if we define

$$f(k) = 1 \Leftrightarrow F_k(x_i) = y_i \text{ for all } i,$$

then Grover's algorithm allows an attacker to find the key using $\mathcal{O}(2^{n/2})$ evaluations of f, or equivalently $\mathcal{O}(2^{n/2})$ evaluations of F. Thus, to achieve the desired level of security we must set $n = 2\kappa$. Summarizing:

> *To ensure equivalent security against exhaustive-search attacks in the quantum setting, symmetric-key cryptosystems must use keys that are* **double** *the length of keys used in the classical setting.*

We stress that the above applies only if exhaustive-search attacks are the best possible; in other cases quantum algorithms may give even larger speedups.

14.1.2 Collision-Finding Algorithms and Hash Functions

Consider next the problem of finding a collision in some hash function $H : \{0,1\}^m \to \{0,1\}^n$ (with $m > n$). As we have seen already in Section 6.4.1, this can be done classically via a "birthday attack" using $\mathcal{O}(2^{n/2})$ evaluations of H. Is it possible to do better using a quantum algorithm?

It is indeed possible to do better via clever use of Grover's algorithm. For simplicity in the analysis we will model H as a random function (as we did in Section 6.4.1); the collision-finding algorithm we describe can be adapted for arbitrary H as well. The approach is as follows. Let $\ell \ll 2^n$ be a parameter that we will set later. Let C, D be disjoint subsets of $\{0,1\}^m$ with $|C| = \ell$ and $|D| = \ell^2$; for example, we can let C be the set of all strings whose first $\log \ell$ bits are all 0 and take D to be the set of all strings whose first $2 \log \ell$ bits are all 1. For $x_i \in C$, set $y_i := H(x_i)$ using ℓ evaluations of H; define $C' = \{y_i\}$. If $y_i = y_j$ for some $i \neq j$ then a collision has already been found. Otherwise, define the function $f : D \to \{0,1\}$ as

$$f(x) = 1 \Leftrightarrow H(x) \in C'.$$

If there is any x with $f(x) = 1$, then we can use Grover's algorithm to find such an x using $\mathcal{O}(|D|^{1/2})$ evaluations of f, or equivalently $\mathcal{O}(|D|^{1/2})$ evaluations of H. The overall number of evaluations of H, then, is $\mathcal{O}(\ell + \sqrt{\ell^2}) = \mathcal{O}(\ell)$.

What is the probability that such an x exists? We only run the second stage of the algorithm if all the $\{y_i\}$ are distinct. Since we model H as a random function, for any particular $x \in D$ the probability that $H(x) \in C'$ is $\frac{\ell}{2^n}$, and so the probability that the hash of some element of D lies in C' is

$$1 - \left(1 - \frac{\ell}{2^n}\right)^{\ell^2} \geq 1 - e^{-\ell^3/2^n}$$

(using Proposition A.3). We thus see that taking $\ell = \Theta(2^{n/3})$ gives a constant probability of finding a collision using only $\mathcal{O}(2^{n/3})$ evaluations of H.

Consider the impact this has on the required output length of a hash function $H : \{0,1\}^m \to \{0,1\}^n$ in order to achieve some desired level of security. As in the previous section, say we want security (i.e., inability to find collisions) against attackers running in time 2^κ, and assume there are no structural weaknesses in H so generic attacks are the best possible. Classically, it suffices to set $n = 2\kappa$ (since a birthday attack would then require time $\mathcal{O}(2^{n/2}) = \mathcal{O}(2^\kappa)$). But achieving the same level of security in the quantum setting requires $n = 3\kappa$. Summarizing:

> *To ensure equivalent security against generic collision-finding attacks in the quantum setting, the output length of a hash function must be* **50% larger** *than the output length in the classical setting.*

14.2 Shor's Algorithm and its Impact on Cryptography

In the previous section we have seen quantum algorithms that offer a polynomial speedup as compared to the best classical algorithms for the same problems. These improved algorithms necessitate changes in the underlying parameters of symmetric-key schemes, but do not fundamentally render those schemes insecure. Here, in contrast, we discuss quantum algorithms that result in *exponential* speedups for solving certain number-theoretic problems—in particular, we show polynomial-time quantum algorithms for factoring and computing discrete logarithms. The existence of such algorithms means that all the public-key schemes we have discussed so far in this book are insecure (at least asymptotically) against a quantum attacker.

We begin by discussing an abstract mathematical problem with no explicit connection to cryptography. Let $f : \mathbb{H} \to R$ be a function whose domain $\mathbb{H}$ is an abelian group. (For now, R can be arbitrary.) Assume further that f is *periodic*, i.e., there is a $\delta \in \mathbb{H}$ (not equal to the identity) called the *period* such that for all $x \in \mathbb{H}$

$$f(x) = f(x + \delta).$$

(Note that if δ is a period then so is 2δ, etc., and so the period is not unique.) The *period-finding problem* is to find a period, given oracle access to f.

Classically, it is not clear how to solve this problem efficiently; even verifying that a given δ is a period seems difficult given only oracle access to f. In

1994, Peter Shor stunned researchers by showing a polynomial-time *quantum* algorithm for this problem for certain groups $\mathbb{H}$. His result was subsequently generalized by others to handle larger classes of groups. The details of Shor's algorithm lie outside our scope, but we discuss the cryptographic implications of Shor's algorithm below.

Implications for factoring and computing discrete logarithms. Period finding is a powerful tool: in particular, it can be used to factor and compute discrete logarithms! All we need to do is carefully choose the function whose period gives us the solution we are looking for.

First consider the problem of factoring. Fix a composite number N that is the product of two distinct primes. Taking any $x \in \mathbb{Z}_N^*$, define the function $f_{x,N} : \mathbb{Z} \to \mathbb{Z}_N^*$ by

$$f_{x,N}(r) = [x^r \bmod N].$$

The key observation is that this function has period $\phi(N)$ since

$$f_{x,N}(r + \phi(N)) = [x^{r+\phi(N)} \bmod N] = [x^r \cdot x^{\phi(N)} \bmod N] = [x^r \bmod N]$$

for any r. Thus, for any $x \in \mathbb{Z}_N^*$ of our choice we can run Shor's algorithm to obtain some period of $f_{x,N}$, i.e., a nonzero integer k such that $x^k = 1 \bmod N$. Theorem 9.50 shows that this enables us to factor N using polynomially many calls to Shor's algorithm and polynomial-time classical computation. (Shor's algorithm in this case runs in time polynomial in the logarithm of the smallest period—which is at most $\phi(N)$—and so this gives a quantum algorithm running in polynomial time overall.)

Period finding can also be used to compute discrete logarithms. Fix some cyclic group $\mathbb{G}$ of prime order q with generator g, and say we are given some element $h \in \mathbb{G}$. Consider the function $f_{g,h} : \mathbb{Z}_q \times \mathbb{Z}_q \to \mathbb{G}$ given by

$$f(a,b) = g^a \cdot h^{-b}.$$

If we let $x = \log_g h$, then $f_{g,h}$ has period $(x,1)$ since

$$f_{g,h}(a+x, b+1) = g^{a+x}h^{-b-1} = g^a g^x h^{-b} h^{-1} = g^a h^{-b}$$

for any a, b. Moreover, for any period (x', y') we have $g^{x'} h^{-y'} = 1 = g^0 h^0$. Lemma 9.65 thus shows that we can use any period to compute $\log_g h$ using classical polynomial-time computation. A quantum polynomial-time algorithm for computing $\log_g h$ follows from the fact that the running time of the period-finding algorithm in this case is polynomial in $\log q$.

Since the hardness of factoring and computing discrete logarithms underlies *all* the public-key cryptosystems we have seen so far in the book (and, indeed, all public-key algorithms in wide use today), we conclude that

> *all public-key cryptosystems we have covered thus far can be broken in polynomial time by a quantum computer.*

This stark fact highlights the importance of post-quantum cryptography.

14.3   Post-Quantum Public-Key Encryption

As noted at the end of the previous section, both the factoring and discrete-logarithm problems become "easy" given a quantum computer. To have any hope of constructing public-key schemes with post-quantum security, then, we need to look for mathematical problems that are computationally hard even for quantum computers. As in the classical case, we generally cannot *prove* unconditionally that a specific problem is hard for quantum algorithms; all we can do is rely on plausible *conjectures* about the (quantum) hardness of certain problems. One notable difference from the classical setting is that the problems being considered for post-quantum cryptography have, on the whole, not been studied as long as the factoring and discrete-logarithm problems; thus, in some sense, we have less confidence that they are truly hard.

In this section we introduce one computational problem that has received a lot of attention, and is widely believed to be hard even for quantum algorithms. We then show how to construct a public-key encryption scheme based on the assumed hardness of that problem. We stress that our goal here is merely to provide a taste of recent work on post-quantum cryptography; in particular, we describe the scheme somewhat loosely without including every detail. For pedagogical purposes we also focus on a simple encryption scheme without attempting to optimize its efficiency.

The remainder of this section assumes a very basic knowledge of linear algebra, but can be appreciated even without this background if the reader is willing to accept certain facts on faith.

Throughout this section we let q be an odd prime. We let $\lfloor \cdot \rfloor$ denote the standard "floor" function, so $\lfloor x \rfloor$ is the largest integer less than or equal to x. In this section we also change our view of $\mathbb{Z}_q$, equating it with the set

$$\{-\lfloor (q-1)/2 \rfloor, \ldots, 0, \ldots, \lfloor q/2 \rfloor\}$$

(as opposed to $\{0, \ldots, q-1\}$ as we have done until now). This viewpoint is better suited to the present context, where we will say that an element of $\mathbb{Z}_q$ is "small" if it is "close" to 0.

The LWE assumption. Consider the following problem: A matrix $\mathbf{B} \in \mathbb{Z}_q^{m \times n}$ is chosen, along with a vector[1] $\mathbf{s} \in \mathbb{Z}_q^n$. We are then given $\mathbf{B}$ and $\mathbf{t} := [\mathbf{B} \cdot \mathbf{s} \bmod q]$ (i.e., all operations are done modulo q); the goal is to find any value $\mathbf{s}' \in \mathbb{Z}_q^n$ such that $\mathbf{B}\mathbf{s}' = \mathbf{t} \bmod q$. This problem is easy and can be solved used standard (efficient) linear-algebraic techniques.

Consider next the following variant of the problem. Choose $\mathbf{B}$ and $\mathbf{s}$ as before, but now also choose a short "error vector" $\mathbf{e} \in \mathbb{Z}_q^m$. (We use the standard Euclidean norm to define the length of vectors. That is, the length of a

---

[1]By default, our vectors are column vectors and so we write, e.g., $\mathbf{s}^T$ (the transpose of $\mathbf{s}$) to denote a row vector.

vector $\mathbf{e} = [e_1, \ldots, e_m]^T$, denoted $\|\mathbf{e}\|$, is simply $\sqrt{\sum_i e_i^2}$. At the moment, we do not quantify what we mean by "short.") The value $\mathbf{t}$ is now computed as $\mathbf{t} := [\mathbf{Bs} + \mathbf{e} \bmod q]$ and the goal, given $\mathbf{B}$ and $\mathbf{t}$, is to find any $\mathbf{s}' \in \mathbb{Z}_q^n$ such that $[\mathbf{t} - \mathbf{Bs}' \bmod q]$ is short. For historical reasons, this is called the *learning with errors (LWE) problem*. When parameters are chosen appropriately, this problem appears to be significantly more difficult than the previous problem (when there are no errors), and efficient algorithms for solving it—even allowing for quantum algorithms—are not known.

For our purposes it is useful to consider a different version of the above called the *decisional LWE* problem. Here, roughly speaking, the goal is to distinguish whether $\mathbf{t}$ was generated by the process described above, or whether $\mathbf{t}$ was sampled *uniformly* from $\mathbb{Z}_q^m$. It is possible to show (for certain settings of the parameters) that this problem is hard if and only if the the LWE problem itself is hard.

We now formalize the above discussion. Let m, q be deterministic functions of the security parameter n with $m > n$; we leave the dependence on n implicit. Let ψ be an efficient randomized algorithm that takes as input 1^n and outputs an integer; this ψ represents the distribution on the errors, and we also leave its dependence on n implicit. The following defines what it means for the decisional LWE problem to be (quantum-)hard for some m, q, ψ.

DEFINITION 14.1 *We say the decisional* $\mathsf{LWE}_{m,q,\psi}$ *problem is quantum-hard if for all quantum polynomial-time algorithms $\mathcal{A}$ there is a negligible function* negl *such that*

$$\Big| \Pr[\mathbf{B} \leftarrow \mathbb{Z}_q^{m \times n}; \mathbf{s} \leftarrow \psi^n; \mathbf{e} \leftarrow \psi^m : \mathcal{A}\left(\mathbf{B}, [\mathbf{Bs} + \mathbf{e} \bmod q]\right) = 1]$$
$$- \Pr[\mathbf{B} \leftarrow \mathbb{Z}_q^{m \times n}; \mathbf{t} \leftarrow \mathbb{Z}_q^m : \mathcal{A}(\mathbf{B}, \mathbf{t}) = 1] \Big| \leq \mathsf{negl}(n).$$

(Note that we also choose $\mathbf{s}$ to be short.) Clearly, if the decisional $\mathsf{LWE}_{m,q,\psi}$ problem is hard then so is the decisional $\mathsf{LWE}_{m',q,\psi}$ problem for any $m' \leq m$. It is only slightly more difficult to show that increasing the length of $\mathbf{s}$ can only make the problem harder. We leave the following as an exercise.

LEMMA 14.2 *If the decisional* $\mathsf{LWE}_{m,q,\psi}$ *problem is quantum-hard, then for all quantum polynomial-time algorithms $\mathcal{A}$ and all functions m', ℓ with $m'(n) \leq m(n)$ and $\ell(n) \geq n$ there is a negligible function* negl *such that*

$$\Big| \Pr[\mathbf{B} \leftarrow \mathbb{Z}_q^{m' \times \ell}; \mathbf{s} \leftarrow \psi^\ell; \mathbf{e} \leftarrow \psi^{m'} : \mathcal{A}\left(\mathbf{B}, [\mathbf{Bs} + \mathbf{e} \bmod q]\right) = 1]$$
$$- \Pr[\mathbf{B} \leftarrow \mathbb{Z}_q^{m' \times \ell}; \mathbf{t} \leftarrow \mathbb{Z}_q^{m'} : \mathcal{A}(\mathbf{B}, \mathbf{t}) = 1] \Big| \leq \mathsf{negl}(n).$$

LWE-Based encryption. We motivate the construction of an encryption scheme from the decisional LWE problem by first describing an insecure key-exchange protocol that can be viewed as a linear-algebraic version of Diffie-Hellman key exchange. Fix n, q, ψ, and $m > n$, and consider the following protocol run between two parties Alice and Bob. Alice begins by generating a uniform $\mathbf{B} \in \mathbb{Z}_q^{m \times n}$ and choosing $\mathbf{s} \leftarrow \psi^n$; she then sends $(\mathbf{B}, \mathbf{t}_A := [\mathbf{B} \cdot \mathbf{s} \bmod q])$. Bob chooses $\hat{\mathbf{s}} \leftarrow \psi^m$ and replies with $\mathbf{t}_B^T := [\hat{\mathbf{s}}^T \cdot \mathbf{B} \bmod q]$. Finally, Alice computes $k_A := [\mathbf{t}_B^T \cdot \mathbf{s} \bmod q]$ and Bob computes $k_B := [\hat{\mathbf{s}}^T \cdot \mathbf{t}_A \bmod q]$. Note that

$$k_A = \mathbf{t}_B^T \cdot \mathbf{s} = \hat{\mathbf{s}}^T \cdot \mathbf{B} \cdot \mathbf{s} = \hat{\mathbf{s}}^T \cdot \mathbf{t}_A = k_B$$

(where all calculations above are done modulo q), and so Alice and Bob have agreed on a shared key!

Of course, the protocol above is not secure since an eavesdropper can use linear algebra to recover $\mathbf{s}$, $\hat{\mathbf{s}}$, or both, and thus compute the key as well. By judiciously adding noise, however (and under the assumption that the decisional LWE problem is hard), it is possible for Alice and Bob to agree on a key while preventing an adversary from learning it. Adapting the resulting protocol to give an encryption scheme (in the same way the Diffie-Hellman protocol is adapted to give El Gamal encryption), we obtain Construction 14.3.

CONSTRUCTION 14.3

Let m, q, ψ be as in the text. Define a public-key encryption scheme as follows:

- Gen: on input 1^n choose uniform $\mathbf{B} \leftarrow \mathbb{Z}_q^{m \times n}$ as well as $\mathbf{s} \leftarrow \psi^n$ and $\mathbf{e} \leftarrow \psi^m$. Set $\mathbf{t} := [\mathbf{B} \cdot \mathbf{s} + \mathbf{e} \bmod q]$. The public key is $\langle \mathbf{B}, \mathbf{t} \rangle$ and the private key is $\mathbf{s}$.

- Enc: on input a public key $pk = \langle \mathbf{B}, \mathbf{t} \rangle$ and a bit b, choose $\hat{\mathbf{s}} \leftarrow \psi^m$ and $\hat{\mathbf{e}} \leftarrow \psi^{n+1}$, and output the ciphertext

$$\mathbf{c}^T := \left[\hat{\mathbf{s}}^T \cdot [\mathbf{B} \mid \mathbf{t}] + \hat{\mathbf{e}}^T + \underbrace{[0, \ldots, 0, b \cdot \lfloor \tfrac{q}{2} \rfloor]}_{n+1} \bmod q \right].$$

- Dec: on input a private key $\mathbf{s}$ and a ciphertext $\mathbf{c}^T$, first compute $k := \left[\mathbf{c}^T \cdot \begin{bmatrix} -\mathbf{s} \\ 1 \end{bmatrix} \bmod q \right]$. Then output 1 if k is closer to $\lfloor \tfrac{q}{2} \rfloor$ than to 0 (see text), and 0 otherwise.

An encryption scheme based on the decisional LWE problem.

During decryption, "closeness" of k to $\lfloor \tfrac{q}{2} \rfloor$ is determined by looking at the absolute value of $[k - \lfloor \tfrac{q}{2} \rfloor \bmod q]$. Here it is important that we use the particular representation of $\mathbb{Z}_q$ described at the beginning of this section.

The construction is somewhat complicated, so it is worth stepping through the process of encryption and decryption to verify that the scheme is correct (at least with high probability) when parameters are set appropriately. Let

$\mathbf{c}^T$ be an honestly generated ciphertext, so

$$\mathbf{c}^T = \hat{\mathbf{s}}^T \cdot [\mathbf{B} \mid \mathbf{t}] + \hat{\mathbf{e}}^T + \mathbf{b}^T,$$

where we let $\mathbf{b}^T = [0, \ldots, 0, b \cdot \lfloor \frac{q}{2} \rfloor]$. (By default from now on, all operations are performed modulo q.) During decryption, the receiver computes

$$
\begin{aligned}
k = \mathbf{c}^T \cdot \begin{bmatrix} -\mathbf{s} \\ 1 \end{bmatrix} \\
= (\hat{\mathbf{s}}^T \cdot [\mathbf{B} \mid \mathbf{t}] + \hat{\mathbf{e}}^T + \mathbf{b}^T) \cdot \begin{bmatrix} -\mathbf{s} \\ 1 \end{bmatrix} \\
= -\hat{\mathbf{s}}^T \mathbf{B} \mathbf{s} + \hat{\mathbf{s}}^T \mathbf{t} + \hat{\mathbf{e}}^T \cdot \begin{bmatrix} -\mathbf{s} \\ 1 \end{bmatrix} + b \cdot \lfloor \frac{q}{2} \rfloor \\
= \hat{\mathbf{s}}^T \mathbf{e} + \hat{\mathbf{e}}^T \cdot \begin{bmatrix} -\mathbf{s} \\ 1 \end{bmatrix} + b \cdot \lfloor \frac{q}{2} \rfloor,
\end{aligned}
$$

using the fact that $\mathbf{t} = \mathbf{B} \cdot \mathbf{s} + \mathbf{e}$. At this point, it is unclear that the receiver recovers the correct bit. However, simple algebra shows that as long as

$$\left| \hat{\mathbf{s}}^T \mathbf{e} + \hat{\mathbf{e}}^T \cdot \begin{bmatrix} -\mathbf{s} \\ 1 \end{bmatrix} \right| < (q-1)/4 \tag{14.1}$$

the receiver will output the same bit b used by the sender. Note that if we let $\hat{\mathbf{s}}^T = [\hat{s}_1, \ldots, \hat{s}_m]$, and similarly for $\mathbf{e}, \hat{\mathbf{e}}$, and $\mathbf{s}$, then we may write Equation (14.1) as

$$\left| \sum_{i=1}^m \hat{s}_i e_i - \sum_{i=1}^n \hat{e}_i s_i + e_{n+1} \right| < (q-1)/4,$$

so the left-hand side is a sum of products of integers output by ψ. Thus, if the distribution ψ is chosen appropriately—specifically, so that it outputs integers sufficiently small so that Equation (14.1) holds (at least with overwhelming probability)—then correctness of the encryption scheme follows.

We now prove that Construction 14.3 is CPA-secure[2] (even for quantum adversaries) if the decisional $\mathsf{LWE}_{m,q,\psi}$ problem is quantum-hard.

THEOREM 14.4 *If the $\mathsf{LWE}_{m,q,\psi}$ problem is quantum-hard, then Construction 14.3 is CPA-secure (even for quantum adversaries).*

---

[2]One can easily define a notion of CPA-security for quantum adversaries by simply replacing "probabilistic polynomial-time" with "quantum polynomial-time" in Definition 12.2. In doing so, we continue to assume the adversary has only *classical* access to the encryption oracle in experiment, i.e., it can only request the encryption of classical messages. It is possible to consider stronger notions of security where the attacker is given *quantum* access to the encryption oracle; this is beyond the scope of our book.

PROOF Let Π denote Construction 14.3. We prove that Π has indistinguishable encryptions in the presence of an eavesdropper even for quantum adversaries; as in the classical case, this implies that Π is CPA-secure even for quantum adversaries).

Let $\mathcal{A}$ be a quantum polynomial-time adversary. Consider a modified encryption scheme $\widetilde{\Pi}$ in which key generation is done by choosing $\mathbf{B}$ as before, but where $\mathbf{t}$ is chosen uniformly from $\mathbb{Z}_q^m$. (Encryption is done as in Π.) Although $\widetilde{\Pi}$ is not actually an encryption scheme (as there is no way for the receiver to decrypt), the experiment $\mathsf{PubK}_{\mathcal{A},\widetilde{\Pi}}^{\mathsf{eav}}(n)$ is still well-defined since that experiment depends only on the key-generation and encryption algorithms.

CLAIM 14.5 $\left| \Pr[\mathsf{PubK}_{\mathcal{A},\Pi}^{\mathsf{eav}}(n) = 1] - \Pr[\mathsf{PubK}_{\mathcal{A},\widetilde{\Pi}}^{\mathsf{eav}}(n) = 1] \right|$ *is negligible.*

PROOF The proof is by a direct reduction to the decisional LWE problem as specified in Definition 14.1. Consider the following algorithm D that attempts to solve the decisional $\mathsf{LWE}_{m,q,\psi}$ problem:

Algorithm D:
The algorithm is given $\mathbf{B} \in \mathbb{Z}_q^{m \times n}$ and $\mathbf{t} \in \mathbb{Z}_q^m$ as input.

- Set $pk := \langle \mathbf{B}, \mathbf{t} \rangle$ and run $\mathcal{A}(pk)$ to obtain $m_0, m_1 \in \{0,1\}$.
- Choose a uniform bit b, and set
$$\mathbf{c}^T := \left[\hat{\mathbf{s}}^T \cdot [\mathbf{B} \mid \mathbf{t}] + \hat{\mathbf{e}}^T + [0,\ldots,0, m_b \cdot \lfloor \tfrac{q}{2} \rfloor] \mod q \right].$$
- Give the ciphertext $\mathbf{c}^T$ to $\mathcal{A}$ and obtain an output bit b'. If $b' = b$, output 1; otherwise, output 0.

Note that D is a quantum polynomial-time algorithm since $\mathcal{A}$ is.

It is immediate that
$$\Pr[\mathbf{B} \leftarrow \mathbb{Z}_q^{m \times n}; \mathbf{s} \leftarrow \psi^n; \mathbf{e} \leftarrow \psi^m : D\left(\mathbf{B}, [\mathbf{Bs} + \mathbf{e} \mod q]\right) = 1]$$
$$= \Pr[\mathsf{PubK}_{\mathcal{A},\Pi}^{\mathsf{eav}}(n) = 1]$$

and
$$\Pr[\mathbf{B} \leftarrow \mathbb{Z}_q^{m \times n}; \mathbf{t} \leftarrow \mathbb{Z}_q^m : D(\mathbf{B}, \mathbf{t}) = 1] = \Pr[\mathsf{PubK}_{\mathcal{A},\widetilde{\Pi}}^{\mathsf{eav}}(n) = 1].$$

Quantum hardness of the $\mathsf{LWE}_{m,q,\psi}$ problem implies the claim. ∎

Consider now a second modified encryption scheme $\widetilde{\Pi}'$ in which key generation is done as in $\widetilde{\Pi}$, but encryption of a bit b is done by choosing a uniform $\hat{\mathbf{t}} \in \mathbb{Z}_q^{n+1}$ and outputting the ciphertext
$$\mathbf{c}^T := \hat{\mathbf{t}}^T + [0,\ldots,0, b \cdot \lfloor \tfrac{q}{2} \rfloor].$$

CLAIM 14.6 $\left| \Pr[\mathsf{PubK}^{\mathsf{eav}}_{\mathcal{A},\widetilde{\Pi}}(n) = 1] - \Pr[\mathsf{PubK}^{\mathsf{eav}}_{\mathcal{A},\widetilde{\Pi}'}(n) = 1] \right|$ *is negligible.*

PROOF We begin by rewriting the way encryption is done in $\widetilde{\Pi}$. Fixing some public key $\langle \mathbf{B}, \mathbf{t} \rangle$, define $\hat{\mathbf{B}} = [\mathbf{B} \mid \mathbf{t}]^T \in \mathbb{Z}_q^{(n+1) \times m}$. Encrypting a bit b in $\widetilde{\Pi}$ is then equivalent to choosing $\hat{\mathbf{s}} \leftarrow \psi^m$ and $\hat{\mathbf{e}} \leftarrow \psi^{n+1}$, computing $\hat{\mathbf{t}} := \hat{\mathbf{B}}\hat{\mathbf{s}} + \hat{\mathbf{e}}$, and then outputting the ciphertext

$$\mathbf{c}^T := \hat{\mathbf{t}}^T + [0, \ldots, 0, b \cdot \lfloor \tfrac{q}{2} \rfloor].$$

The crucial observation is that $\hat{\mathbf{t}}$ is computed exactly as in the decisional LWE assumption, though with different parameters (namely, $\hat{\mathbf{B}} \in \mathbb{Z}_q^{(n+1) \times m}$ instead of $\mathbf{B} \in \mathbb{Z}_q^{m \times n}$). However, since $m > n$, and hence also $n + 1 \leq m$, Lemma 14.2 shows that the decisional LWE problem is hard for this setting of the parameters as well. The claim can thus be proved similarly to the previous claim. ∎

CLAIM 14.7 $\Pr[\mathsf{PubK}^{\mathsf{eav}}_{\mathcal{A},\widetilde{\Pi}'}(n) = 1] = \tfrac{1}{2}$.

PROOF This follows from the fact that, in $\widetilde{\Pi}'$, the "message vector" $[0, \ldots, 0, b \cdot \lfloor \tfrac{q}{2} \rfloor]$ is added to a uniform vector $\hat{\mathbf{t}}^T \in \mathbb{Z}_q^{n+1}$ modulo q. ∎

The preceding three claims prove the theorem. ∎

14.4 Post-Quantum Signatures

Security of all the signature schemes presented in Chapter 13 required either the hardness of factoring or the hardness of computing discrete logarithms. As we have discussed, constructions from alternate assumptions are needed if we want security in a post-quantum world. While it is possible to construct signature schemes from the LWE assumption introduced in the previous section, such schemes are complex and we explore a different approach here

Somewhat surprisingly, and in contrast to the case of public-key encryption, it is possible to construct signature schemes based on hash functions, a *symmetric-key* primitve. Since existing cryptographic hash functions such as SHA-3 are believed to be secure even against quantum algorithms (subject to the increase in parameters discussed in Section 14.1), this provides a promising approach to constructing post-quantum signatures.

Signatures based on hash functions are interesting for several other reasons, as well. First, it is amazing (and perhaps counterintuitive) that signatures can be constructed without any number-theoretic assumptions, unlike public-key encryption schemes. Moreover, as we will see, the ideas developed here can be used to construct signature schemes from the minimal assumption that *one-way functions* exist. It is also worth noting that the schemes we present here do not rely on random oracles, as opposed to all the constructions we saw in Chapter 13. Finally, signatures based on hash functions can be more efficient than those relying on number-theoretic assumptions.

In the rest of this section, we no longer mention quantum attacks explicitly. However, all security claims hold against such attacks so long as the hash function used is quantum-secure (in the appropriate sense).

14.4.1 Lamport's Signature Scheme

We initiate our study of signature schemes based on hash functions by considering the relatively weak notion of *one-time signature schemes*. Informally, such schemes are "secure" as long as a given private key is used to sign only a *single* message. Schemes satisfying this notion of security may be appropriate for some applications, and also serve as useful building blocks for achieving stronger notions of security, as we will see in the following section.

Let $\Pi = (\mathsf{Gen}, \mathsf{Sign}, \mathsf{Vrfy})$ be a signature scheme, and consider the following experiment for an adversary $\mathcal{A}$ and parameter n:

The one-time signature experiment $\mathsf{Sig\text{-}forge}_{\mathcal{A},\Pi}^{\text{1-time}}(n)$:

1. $\mathsf{Gen}(1^n)$ *is run to obtain keys* (pk, sk).

2. *Adversary* $\mathcal{A}$ *is given* pk *and asks a* single *query* m' *to its oracle* $\mathsf{Sign}_{sk}(\cdot)$. $\mathcal{A}$ *then outputs* (m, σ) *with* $m \neq m'$.

3. *The output of the experiment is defined to be 1 if and only if* $\mathsf{Vrfy}_{pk}(m, \sigma) = 1$.

DEFINITION 14.8 *Signature scheme* $\Pi = (\mathsf{Gen}, \mathsf{Sign}, \mathsf{Vrfy})$ *is existentially unforgeable under a single-message attack, or is a one-time signature scheme, if for all probabilistic polynomial-time adversaries* $\mathcal{A}$, *there exists a negligible function* negl *such that:*

$$\Pr\left[\mathsf{Sig\text{-}forge}_{\mathcal{A},\Pi}^{\text{1-time}}(n) = 1\right] \leq \mathsf{negl}(n).$$

Leslie Lamport gave a construction of a one-time signature scheme in 1979. We illustrate the idea for the case of signing 3-bit messages. Let H be a cryptographic hash function. A private key consists of six uniform values $x_{1,0}$, $x_{1,1}$, $x_{2,0}$, $x_{2,1}$, $x_{3,0}$, $x_{3,1} \in \{0,1\}^n$, and the corresponding public key contains the results obtained by applying H to each of these elements. These keys can be visualized as two-dimensional arrays:

$$pk = \begin{pmatrix} y_{1,0} & y_{2,0} & y_{3,0} \\ y_{1,1} & y_{2,1} & y_{3,1} \end{pmatrix} \quad sk = \begin{pmatrix} x_{1,0} & x_{2,0} & x_{3,0} \\ x_{1,1} & x_{2,1} & x_{3,1} \end{pmatrix}.$$

To sign a message $m = m_1 m_2 m_3$ (where $m_i \in \{0, 1\}$), the signer releases the appropriate preimage x_{i,m_i} for each bit of the message; the signature σ consists of the three values $(x_{1,m_1}, x_{2,m_2}, x_{3,m_3})$. Verification is carried out in the natural way: presented with the candidate signature (x_1, x_2, x_3) on the message $m = m_1 m_2 m_3$, accept if and only if $H(x_i) \overset{?}{=} y_{i,m_i}$ for $1 \le i \le 3$. This is shown graphically in Figure 14.1, and the general case—for messages of any length ℓ—is described formally in Construction 14.9.

Signing $m = 011$:

$$sk = \begin{pmatrix} \boxed{x_{1,0}} & x_{2,0} & x_{3,0} \\ x_{1,1} & \boxed{x_{2,1}} & \boxed{x_{3,1}} \end{pmatrix} \Rightarrow \sigma = (x_{1,0}, x_{2,1}, x_{3,1})$$

Verifying for $m = 011$ and $\sigma = (x_1, x_2, x_3)$:

$$pk = \begin{pmatrix} \boxed{y_{1,0}} & y_{2,0} & y_{3,0} \\ y_{1,1} & \boxed{y_{2,1}} & \boxed{y_{3,1}} \end{pmatrix} \Bigg\} \Rightarrow \begin{array}{l} H(x_1) \overset{?}{=} y_{1,0} \\ H(x_2) \overset{?}{=} y_{2,1} \\ H(x_3) \overset{?}{=} y_{3,1} \end{array}$$

FIGURE 14.1: The Lamport scheme used to sign the message $m = 011$.

After observing a signature on a message, an attacker who wishes to forge a signature on any *other* message must find a preimage of one of the three "unused" elements in the public key. If H is *one-way* (see Definition 9.73), then finding any such preimage is computationally difficult.

THEOREM 14.10 *Let ℓ be any polynomial. If H is a one-way function, then Construction 14.9 is a one-time signature scheme.*

PROOF Let $\ell = \ell(n)$ throughout. As noted above, the key observation is this: say an attacker $\mathcal{A}$ requests a signature on a message m', and consider any other message $m \ne m'$. There must be at least one position $i^* \in \{1, \dots, \ell\}$ on which m and m' differ. Say $m_{i^*} = b \ne m'_{i^*}$. Then forging a signature on m requires, at least, finding a preimage (under H) of element y_{i^*, b^*} of the public key. Since H is one-way, this is infeasible. We now formalize this intuition.

Let Π denote the Lamport scheme, and let $\mathcal{A}$ be a probabilistic polynomial-time adversary. In a particular execution of $\mathsf{Sig\text{-}forge}^{\text{1-time}}_{\mathcal{A}, \Pi}(n)$, let m' denote the message whose signature is requested by $\mathcal{A}$ (we assume without loss of generality that $\mathcal{A}$ always requests a signature on a message), and let (m, σ) be the

---

CONSTRUCTION 14.9

Let $H : \{0,1\}^* \to \{0,1\}^*$ be a function. Construct a signature scheme
for messages of length $\ell = \ell(n)$ as follows:

- Gen: on input 1^n, proceed as follows for $i \in \{1,\dots,\ell\}$:

  1. Choose uniform $x_{i,0}, x_{i,1} \in \{0,1\}^n$.

  2. Compute $y_{i,0} := H(x_{i,0})$ and $y_{i,1} := H(x_{i,1})$.

  The public key pk and the private key sk are

  $$pk = \begin{pmatrix} y_{1,0} & y_{2,0} & \cdots & y_{\ell,0} \\ y_{1,1} & y_{2,1} & \cdots & y_{\ell,1} \end{pmatrix} \quad sk = \begin{pmatrix} x_{1,0} & x_{2,0} & \cdots & x_{\ell,0} \\ x_{1,1} & x_{2,1} & \cdots & x_{\ell,1} \end{pmatrix}.$$

- Sign: on input a private key sk as above and a message $m \in \{0,1\}^\ell$
  with $m = m_1 \cdots m_\ell$, output the signature $(x_{1,m_1}, \dots, x_{\ell,m_\ell})$.

- Vrfy: on input a public key pk as above, a message $m \in \{0,1\}^\ell$
  with $m = m_1 \cdots m_\ell$, and a signature $\sigma = (x_1, \dots, x_\ell)$, output 1 if
  and only if $H(x_i) = y_{i,m_i}$ for all $1 \le i \le \ell$.

The Lamport signature scheme.

final output of $\mathcal{A}$. We say that $\mathcal{A}$ *outputs a forgery at* (i,b) if $\mathsf{Vrfy}_{pk}(m,\sigma) = 1$
and furthermore $m_i \ne m_i'$ (i.e., messages m and m' differ on their ith posi-
tion) and $m_i = b \ne m_i'$. Note that whenever $\mathcal{A}$ outputs a forgery, it outputs
a forgery at *some* (i,b).

Consider the following PPT algorithm $\mathcal{I}$ attempting to invert H:

Algorithm $\mathcal{I}$:
The algorithm is given 1^n and y as input.

1. Choose uniform $i^* \in \{1,\dots,\ell\}$ and $b^* \in \{0,1\}$. Set $y_{i^*,b^*} := y$.

2. For all $i \in \{1,\dots,\ell\}$ and $b \in \{0,1\}$ with $(i,b) \ne (i^*,b^*)$:

   - Choose uniform $x_{i,b} \in \{0,1\}^n$ and set $y_{i,b} := H(x_{i,b})$.

3. Run $\mathcal{A}$ on input $pk := \begin{pmatrix} y_{1,0} & y_{2,0} & \cdots & y_{\ell,0} \\ y_{1,1} & y_{2,1} & \cdots & y_{\ell,1} \end{pmatrix}$.

4. When $\mathcal{A}$ requests a signature on the message m':

   - If $m_{i^*}' = b^*$, then $\mathcal{I}$ aborts the execution.

   - Otherwise, $\mathcal{I}$ returns the signature $\sigma = (x_{1,m_1'}, \dots, x_{\ell,m_\ell'})$.

5. When $\mathcal{A}$ outputs (m,σ) with $\sigma = (x_1,\dots,x_\ell)$:

   - If $\mathcal{A}$ outputs a forgery at (i^*,b^*), then output x_{i^*}.

Whenever $\mathcal{A}$ outputs a forgery at (i^*,b^*), algorithm $\mathcal{I}$ succeeds in inverting
its given input y. We are interested in the probability that this occurs when
the input to $\mathcal{I}$ is generated by choosing uniform $x \in \{0,1\}^n$ and setting
$y := H(x)$ (cf. Definition 9.73). Imagine a "mental experiment" in which $\mathcal{I}$ is

given x at the outset, sets $x_{i^*, b^*} := x$, and then always returns a signature to $\mathcal{A}$ in step 4 (i.e., even if $m'_{i^*} = b^*$). The view of $\mathcal{A}$ when run as a subroutine by $\mathcal{I}$ in this mental experiment is distributed identically to the view of $\mathcal{A}$ in experiment $\mathsf{Sig\text{-}forge}_{\mathcal{A},\Pi}^{\text{1-time}}(n)$. Because (i^*, b^*) was chosen uniformly at the beginning of the experiment, and the view of $\mathcal{A}$ is independent of this choice, the probability that $\mathcal{A}$ outputs a forgery at (i^*, b^*), conditioned on the fact that $\mathcal{A}$ outputs a forgery at all, is at least $1/2\ell$. (The easiest way to see this is to simply consider deferring the choice of (i^*, b^*) to the end of the experiment.) We conclude that, in this mental experiment, the probability that $\mathcal{A}$ outputs a forgery at (i^*, b^*) is at least $\frac{1}{2\ell} \cdot \Pr[\mathsf{Sig\text{-}forge}_{\mathcal{A},\Pi}^{\text{1-time}}(n) = 1]$.

Returning to the real experiment involving $\mathcal{I}$ as initially described, the key point is that *the probability that $\mathcal{A}$ outputs a forgery at (i^*, b^*) is unchanged.* This is because the mental experiment and the real experiment coincide if $\mathcal{A}$ outputs a forgery at (i^*, b^*). That is, the experiments only differ if $m'_{i^*} = b^*$, but if this happens then it is impossible (by definition) for $\mathcal{A}$ to subsequently output a forgery at (i^*, b^*). So the probability that $\mathcal{A}$ outputs a forgery at (i^*, b^*) is still at least $\frac{1}{2\ell} \cdot \Pr[\mathsf{Sig\text{-}forge}_{\mathcal{A},\Pi}^{\text{1-time}}(n) = 1]$. In other words,

$$\Pr[\mathsf{Invert}_{\mathcal{I},H}(n) = 1] \geq \frac{1}{2\ell} \cdot \Pr[\mathsf{Sig\text{-}forge}_{\mathcal{A},\Pi}^{\text{1-time}}(n) = 1].$$

Because H is a one-way function, there is a negligible function negl such that

$$\mathsf{negl}(n) \geq \Pr[\mathsf{Invert}_{\mathcal{I},H}(n) = 1].$$

Since ℓ is polynomial this implies that $\Pr[\mathsf{Sig\text{-}forge}_{\mathcal{A},\Pi}^{\text{1-time}}(n) = 1]$ is negligible, completing the proof. ∎

COROLLARY 14.11 *If one-way functions exist, then for any polynomial ℓ there is a one-time signature scheme for messages of length ℓ.*

14.4.2 Chain-Based Signatures

Being able to sign only a single message with a given private key is obviously a significant drawback. We show here an approach based on collision-resistant hash functions that allows a signer to sign *arbitrarily many* messages, at the expense of maintaining *state* that must be updated after each signature is generated. In Section 14.4.3 we discuss a more efficient variant of this approach (that still requires state), and then describe how that construction can be made *stateless*. The result shows that full-fledged signature schemes satisfying Definition 13.2 can be constructed from collision-resistant hash functions.

We first define signature schemes that allow the signer to maintain *state* that is updated after every signature is produced.

DEFINITION 14.12 *A stateful signature scheme is a tuple of probabilistic polynomial-time algorithms* (Gen, Sign, Vrfy) *satisfying the following:*

1. *The* key-generation algorithm Gen *takes as input a security parameter* 1^n *and outputs* (pk, sk, s_0). *These are called the* public key, private key, *and* initial state, *respectively. We assume* pk *and* sk *each has length at least* n, *and that* n *can be determined from* pk, sk.

2. *The* signing algorithm Sign *takes as input a private key* sk, *a value* s_{i-1}, *and a message* $m \in \{0,1\}^*$. *It outputs a signature* σ *and a value* s_i.

3. *The deterministic* verification algorithm Vrfy *takes as input a public key* pk, *a message* m, *and a signature* σ. *It outputs a bit* b.

We require that for every n, *every* (pk, sk, s_0) *output by* Gen(1^n), *and any messages* $m_1, \ldots, m_t \in \{0,1\}^*$, *if we iteratively compute* $(\sigma_i, s_i) \leftarrow \mathsf{Sign}_{sk,s_{i-1}}(m_i)$ *for* $i = 1, \ldots, t$, *then for every* $i \in \{1, \ldots, t\}$, *it holds that* $\mathsf{Vrfy}_{pk}(m_i, \sigma_i) = 1$.

We emphasize that the verifier does not need to know the signer's state in order to verify a signature; in fact, in some schemes the state must be kept secret by the signer in order for security to hold. Signature schemes that do not maintain state (as in Definition 13.1) are called *stateless* to distinguish them from stateful schemes. Clearly, stateless schemes are preferable (although stateful schemes can still potentially be useful). We introduce stateful signatures as a stepping stone to an eventual stateless construction.

Security for stateful signatures schemes is exactly analogous to Definition 13.2, with the only subtleties being that the signing oracle returns only the signature (and *not* the state), and that the signing oracle updates the state each time it is invoked.

For any polynomial $t = t(n)$, we can easily construct a stateful "t-time-secure" signature scheme. (The definition of security here would be the obvious generalization of Definition 14.8.) We can do this by simply letting the public key (resp., private key) consist of t independently generated public keys (resp., private keys) for any one-time signature scheme; i.e., set $pk := \langle pk_1, \ldots, pk_t \rangle$ and $sk := \langle sk_1, \ldots, sk_t \rangle$ where each (pk_i, sk_i) is an independently generated key-pair for some one-time signature scheme. The state is a counter i initially set to 1. To sign a message m using the private key sk and current state $i \leq t$, compute $\sigma \leftarrow \mathsf{Sign}_{sk_i}(m)$ (that is, generate a signature on m using the private key sk_i) and output (σ, i); the state is updated to $i := i + 1$. Since the state starts at 1, this means the ith message is signed using sk_i. Verification of a signature (σ, i) on a message m is done by checking whether σ is a valid signature on m with respect to pk_i. This scheme is secure if used to sign t messages since each private key of the underlying one-time scheme is used to sign only a *single* message.

As described, signatures have constant length (i.e., independent of t), but the public key has length *linear* in t. It is possible to trade off the length of the public key and signatures by having the signer compute a Merkle tree $h := \mathcal{MT}_t(pk_1, \ldots, pk_t)$ (see Section 6.6.2) over the t underlying public keys from the one-time scheme. That is, the public key will now be $\langle t, h \rangle$, and the signature on the ith message will include (σ, i), as before, along with the ith value pk_i and a proof π_i that this is the correct value corresponding to h. (Verification is done in the natural way.) The public key now has constant size, and the signature length grows only logarithmically with t.

Since t can be an arbitrary polynomial, why don't the previous schemes give us the solution we are looking for? The main drawback is that they require the upper bound t on the number of messages that can be signed *to be fixed in advance*, at the time of key generation. This is a potentially severe limitation since once the upper bound is reached a new public key would have to be generated and distributed. We would prefer instead to have a single, fixed public key that can be used to sign an *unbounded* number of messages.

Let $\Pi = (\mathsf{Gen}, \mathsf{Sign}, \mathsf{Vrfy})$ be a one-time signature scheme. In the scheme we have just described (ignoring the Merkle-tree optimization), the signer runs t invocations of Gen to obtain public keys $pk_1, \ldots, pk_t$, and includes each of these in its actual public key pk. The signer is then restricted to signing at most t messages. We can do better by using a "chain-based" scheme in which the signer generates additional public keys *on-the-fly*, as needed.

In the chain-based scheme, the public key consists of just a single public key pk_1 generated using Gen, and the private key is just the associated private key sk_1. To sign the first message m_1, the signer first generates a new key-pair (pk_2, sk_2) using Gen, and then signs both m_1 and pk_2 using sk_1 to obtain $\sigma_1 \leftarrow \mathsf{Sign}_{sk_1}(m_1 \| pk_2)$. The signature that is output includes both pk_2 and σ_1, and the signer adds $(m_1, pk_2, sk_2, \sigma_1)$ to its current state. In general, when it comes time to sign the ith message the signer will have stored $\{(m_j, pk_{j+1}, sk_{j+1}, \sigma_j)\}_{j=1}^{i-1}$ as part of its state. To sign the ith message m_i, the signer first generates a new key-pair (pk_{i+1}, sk_{i+1}) using Gen, and then signs m_i and pk_{i+1} using sk_i to obtain a signature $\sigma_i \leftarrow \mathsf{Sign}_{sk_i}(m_i \| pk_{i+1})$. The actual signature that is output includes pk_{i+1}, σ_i, and also the values $\{m_j, pk_{j+1}, \sigma_j\}_{j=1}^{i-1}$. The signer then adds $(m_i, pk_{i+1}, sk_{i+1}, \sigma_i)$ to its state. See Figure 14.2 for a graphical depiction of this process.

To verify a signature $(pk_{i+1}, \sigma_i, \{m_j, pk_{j+1}, \sigma_j\}_{j=1}^{i-1})$ on a message $m = m_i$ with respect to public key pk_1, the receiver verifies each link between a public key pk_j and the next public key pk_{j+1} in the chain, as well as the link between the last public key pk_i and m. That is, verification outputs 1 if and only if $\mathsf{Vrfy}_{pk_j}(m_j \| pk_{j+1}, \sigma_j) \stackrel{?}{=} 1$ for all $j \in \{1, \ldots, i\}$. (Refer to Figure 14.2.)

It is not hard to be convinced—at least on an intuitive level—that this signature scheme is existentially unforgeable under an adaptive chosen-message attack (regardless of how many messages are signed). Informally, this is once again due to the fact that each key-pair (pk_i, sk_i) is used to sign only a single "message," where in this case the "message" is actually a message/public-key pair $m_i \| pk_{i+1}$. Since we will prove security of a more efficient scheme in the

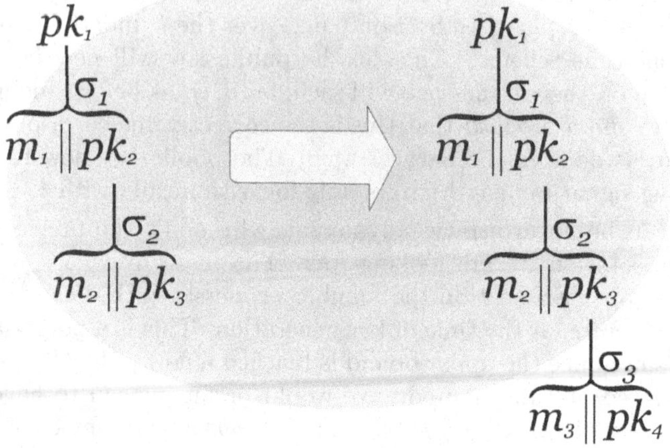

FIGURE 14.2: Chain-based signatures: the situation before and after signing the third message m_3.

next section, we do not prove security for the chain-based scheme here.

In the chain-based scheme, each public key pk_i is used to sign both a message and another public key. Thus, it is essential that the underlying one-time signature scheme Π *is capable of signing messages longer than the public key.* The Lamport scheme presented in Section 14.4.1 does *not* have this property. However, if we apply the hash-and-sign paradigm from Section 13.3 to the Lamport scheme, we *do* obtain a one-time signature scheme that can sign messages of arbitrary length. (Although Theorem 13.4 was stated only with regard to signature schemes satisfying Definition 13.2, it is not hard to see that an identical proof works for one-time signature schemes.) Because this result is crucial for the next section, we state it formally. (Note that the existence of collision-resistant hash functions implies the existence of one-way functions; see Exercise 8.4.)

LEMMA 14.13 *If collision-resistant hash functions exist, then there exists a one-time signature scheme (for messages of arbitrary length).*

The chain-based signature scheme is a stateful signature scheme that is existentially unforgeable under an adaptive chosen-message attack. It has a number of disadvantages, though. For one, there is no immediate way to eliminate the state (recall that our ultimate goal is a stateless scheme satisfying Definition 13.2). It is also not very efficient, in that the signature length, size of the state, and verification time are all linear in the number of messages that have been signed. Finally, each signature reveals all previously signed messages, and this may be undesirable in some contexts.

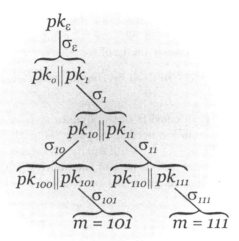

FIGURE 14.3:   Tree-based signatures (conceptually).

14.4.3   Tree-Based Signatures

The signer in the chain-based scheme of the previous section can be viewed as maintaining a *tree* of degree 1, rooted at the public key pk_1, and with depth equal to the number of messages signed so far (cf. Figure 14.2). A natural way to improve efficiency is to use a *binary* tree in which each node has degree 2. As before, a signature will correspond to a "signed" path in the tree from a leaf to the root; as long as the tree has polynomial depth (even if it has exponential size!), verification can be done in polynomial time.

Concretely, to sign messages of length n we will work with a binary tree of depth n having 2^n leaves. As before, the signer will add nodes to the tree "on-the-fly," as needed. In contrast to the chain-based scheme, however, only leaves (and not internal nodes) will be used for signing messages. Each leaf of the tree will correspond to one of the possible messages of length n.

In more detail, we imagine a binary tree of depth n where the root is labeled by ε (i.e., the empty string), and a node that is labeled with the binary string w (of length less than n) has left-child labeled $w0$ and right-child labeled $w1$. This tree is never constructed in its entirety (note that it has exponential size), but is instead built up by the signer as needed.

For every node w, we associate a pair of keys pk_w, sk_w for a one-time signature scheme Π. The public key of the root, pk_ε, is the actual public key of the signer. To sign a message $m \in \{0,1\}^n$, the signer does the following:

1. It first generates keys (as needed) for all nodes on the path from the root to the leaf labeled m. (Some of these public keys may have been generated in the process of signing previous messages, and in that case are not generated again.)

2. Next, it "certifies" the path from the root to the leaf labeled m by computing a signature on $pk_{w0}\|pk_{w1}$, using private key sk_w, for each string w that is a proper prefix of m.

3. Finally, it "certifies" m itself by computing a signature on m using the private key sk_m.

The final signature on m consists of the signature on m with respect to pk_m, as well as all the information needed to verify the path from the leaf labeled m to the root; see Figure 14.3. Additionally, the signer updates its state by storing all the keys generated as part of the above signing process. A formal description of this scheme is given as Construction 14.14.

CONSTRUCTION 14.14

Let $\Pi = (\mathsf{Gen}, \mathsf{Sign}, \mathsf{Vrfy})$ be a signature scheme. For a binary string m, let $m|_i \overset{\text{def}}{=} m_1 \cdots m_i$ denote the i-bit prefix of m (with $m|_0 \overset{\text{def}}{=} \varepsilon$, the empty string). Construct the scheme $\Pi^* = (\mathsf{Gen}^*, \mathsf{Sign}^*, \mathsf{Vrfy}^*)$ as follows:

- Gen^*: on input 1^n, compute $(pk_\varepsilon, sk_\varepsilon) \leftarrow \mathsf{Gen}(1^n)$ and output the public key pk_ε. The private key and initial state are sk_ε.

- Sign^*: on input a message $m \in \{0,1\}^n$, carry out the following.

  1. For $i = 0$ to $n - 1$:
     - If $pk_{m|_i0}, pk_{m|_i1}$, and $\sigma_{m|_i}$ are not in the state, compute $(pk_{m|_i0}, sk_{m|_i0}) \leftarrow \mathsf{Gen}(1^n)$, $(pk_{m|_i1}, sk_{m|_i1}) \leftarrow \mathsf{Gen}(1^n)$, and $\sigma_{m|_i} \leftarrow \mathsf{Sign}_{sk_{m|_i}}(pk_{m|_i0}\,\|\,pk_{m|_i1})$. In addition, add all of these values to the state.

  2. If σ_m is not yet included in the state, compute $\sigma_m \leftarrow \mathsf{Sign}_{sk_m}(m)$ and store it as part of the state.

  3. Output the signature $\left(\{\sigma_{m|_i}, pk_{m|_i0}, pk_{m|_i1}\}_{i=0}^{n-1},\ \sigma_m\right)$.

- Vrfy^*: on input a public key pk_ε, message m, and signature $\left(\{\sigma_{m|_i}, pk_{m|_i0}, pk_{m|_i1}\}_{i=0}^{n-1},\ \sigma_m\right)$, output 1 if and only if:

  1. $\mathsf{Vrfy}_{pk_{m|_i}}(pk_{m|_i0}\,\|\,pk_{m|_i1}, \sigma_{m|_i}) \overset{?}{=} 1$ for all $i \in \{0, \ldots, n-1\}$.

  2. $\mathsf{Vrfy}_{pk_m}(m, \sigma_m) \overset{?}{=} 1$.

A "tree-based" signature scheme.

Notice that each of the underlying keys in this scheme is used to sign only a *single* "message." Each key associated with an internal node signs a pair of public keys, and a key at a leaf is used to sign only a single message. Since each key is used to sign a *pair* of other keys, we again need the one-time signature scheme Π to be capable of signing messages longer than the public key. Lemma 14.13 shows that such schemes can be constructed based on collision-resistant hash functions.

Before proving security of this tree-based approach, note that it improves on the chain-based scheme in a number of respects. It still allows for signing an unbounded number of messages. (Although there are only 2^n leaves, the message space contains only 2^n messages. In any case, 2^n is eventually larger than any polynomial function of n.) In terms of efficiency, the signature length and verification time are now proportional to the message length n but are *independent* of the number of messages signed. The scheme is still stateful, but we will see how this can be avoided after we prove the following result.

THEOREM 14.15 *Let Π be a one-time signature scheme. Then Construction 14.14 is a secure signature scheme.*

PROOF Let Π^* denote Construction 14.14. Let $\mathcal{A}^*$ be a probabilistic polynomial time adversary, let $\ell^* = \ell^*(n)$ be a (polynomial) upper bound on the number of signing queries made by $\mathcal{A}^*$, and set $\ell = \ell(n) \stackrel{\text{def}}{=} 2n\ell^*(n) + 1$. Note that ℓ upper bounds the number of public keys from Π that are needed to generate ℓ^* signatures using Π^*. This is because each signature in Π^* requires at most $2n$ new keys from Π (in the worst case), and one additional key from Π is used as the actual public key pk_ε.

Consider the following PPT adversary $\mathcal{A}$ attacking the one-time signature scheme Π:

Adversary $\mathcal{A}$:
$\mathcal{A}$ is given as input a public key pk (the security parameter n is implicit).

- Choose a uniform index $i^* \in \{1, \ldots, \ell\}$. Construct a list $pk^1, \ldots, pk^\ell$ of keys as follows:

    - Set $pk^{i^*} := pk$.
    - For $i \neq i^*$, compute $(pk^i, sk^i) \leftarrow \mathsf{Gen}(1^n)$.

- Run $\mathcal{A}^*$ on input public key $pk_\varepsilon = pk^1$. When $\mathcal{A}^*$ requests a signature on a message m do:

    1. For $i = 0$ to $n - 1$:
        - If the values $pk_{m|_i 0}, pk_{m|_i 1}$, and $\sigma_{m|_i}$ have not yet been defined, then set $pk_{m|_i 0}$ and $pk_{m|_i 1}$ equal to the next two unused public keys pk^j and pk^{j+1}, and compute a signature $\sigma_{m|_i}$ on $pk_{m|_i 0} \,\|\, pk_{m|_i 1}$ with respect to $pk_{m|_i}$.[3]

---

[3] If $i \neq i^*$ then $\mathcal{A}$ can compute a signature with respect to pk^i by itself. $\mathcal{A}$ can also obtain a (single) signature with respect to pk^{i^*} by making the appropriate query to its signing oracle. This is what is meant here.

2. If σ_m is not yet defined, compute a signature σ_m on m with respect to pk_m (see footnote 3).

3. Give $\left(\left\{ \sigma_{m|_i}, pk_{m|_i 0}, pk_{m|_i 1} \right\}_{i=0}^{n-1}, \sigma_m \right)$ to $\mathcal{A}^*$.

- Say $\mathcal{A}^*$ outputs a message m (for which it had not previously requested a signature) and a signature $\left(\left\{ \sigma'_{m|_i}, pk'_{m|_i 0}, pk'_{m|_i 1} \right\}_{i=0}^{n-1}, \sigma'_m \right)$. If this is a valid signature on m, then:

Case 1: Say there exists a $j \in \{0, \ldots, n-1\}$ for which $pk'_{m|_j 0} \neq pk_{m|_j 0}$ or $pk'_{m|_j 1} \neq pk_{m|_j 1}$; this includes the case when $pk_{m|_j 0}$ or $pk_{m|_j 1}$ were never defined by $\mathcal{A}$. Take the minimal such j, and let i be such that $pk^i = pk_{m|_j} = pk'_{m|_j}$ (such an i exists by the minimality of j). If $i = i^*$, output $(pk'_{m|_j 0} \| pk'_{m|_j 1}, \sigma'_{m|_j})$.

Case 2: If case 1 does not hold, then $pk'_m = pk_m$. Let i be such that $pk^i = pk_m$. If $i = i^*$, output (m, σ'_m).

In experiment $\mathsf{Sig\text{-}forge}_{\mathcal{A},\Pi}^{\text{1-time}}(n)$, the view of $\mathcal{A}^*$ being run as a subroutine by $\mathcal{A}$ is distributed identically to the view of $\mathcal{A}^*$ in experiment $\mathsf{Sig\text{-}forge}_{\mathcal{A}^*,\Pi^*}(n)$.[4] Thus, the probability that $\mathcal{A}^*$ outputs a forgery is exactly $\delta(n)$ when it is run as a subroutine by $\mathcal{A}$ in this experiment. Given that $\mathcal{A}^*$ outputs a forgery, consider each of the two possible cases described above:

Case 1: Since i^* is uniform and independent of the view of $\mathcal{A}^*$, the probability that $i = i^*$ is exactly $1/\ell$. If $i = i^*$, then $\mathcal{A}$ requested a signature on the message $pk_{m|_j 0} \| pk_{m|_j 1}$ with respect to the public key $pk = pk^{i^*} = pk_{m|_j}$ that it was given (and requested no other signatures). Moreover,

$$pk'_{m|_j 0} \| pk'_{m|_j 1} \neq pk_{m|_j 0} \| pk_{m|_j 1}$$

and yet $\sigma'_{m|_j}$ is a valid signature on $pk'_{m|_j 0} \| pk'_{m|_j 1}$ with respect to pk. Thus, $\mathcal{A}$ outputs a forgery in this case.

Case 2: Again, since i^* was chosen uniformly at random and is independent of the view of $\mathcal{A}^*$, the probability that $i = i^*$ is exactly $1/\ell$. If $i = i^*$, then $\mathcal{A}$ did not request any signatures with respect to the public key $pk = pk^i = pk_m$ and yet σ'_m is a valid signature on m with respect to pk.

We see that, conditioned on $\mathcal{A}^*$ outputting a forgery, $\mathcal{A}$ outputs a forgery with probability exactly $1/\ell$. This means that

$$\Pr[\mathsf{Sig\text{-}forge}_{\mathcal{A},\Pi}^{\text{1-time}}(n) = 1] = \Pr[\mathsf{Sig\text{-}forge}_{\mathcal{A}^*,\Pi^*}(n) = 1]/\ell(n).$$

---

[4]As we have mentioned, $\mathcal{A}$ never "runs out" of public keys. A signing query of $\mathcal{A}^*$ uses $2n$ public keys; thus, even if new public keys were required to answer *every* signing query of $\mathcal{A}^*$ (which will in general not be the case), only $2n\ell^*(n)$ public keys would be needed by $\mathcal{A}$ in addition to the "root" public key pk_ε.

Because Π is a one-time signature scheme, there is a negligible function negl for which

$$\Pr[\text{Sig-forge}_{\mathcal{A},\Pi}^{\text{1-time}}(n) = 1] \leq \text{negl}(n).$$

Since ℓ is polynomial, this means $\Pr[\text{Sig-forge}_{\mathcal{A}^*,\Pi^*}(n) = 1]$ is negligible. ∎

A Stateless Solution

As described, the signer generates state on-the-fly as needed. However, we can imagine having the signer generate the necessary information for all the nodes in the entire tree *in advance*, at the time of key generation. (That is, at the time of key generation the signer could generate the keys $\{(pk_w, sk_w)\}$ and the signatures $\{\sigma_w\}$ for all binary strings w of length at most n.) If key generation were done in this way, the signer would not have to update its state at all; these values could all be stored as part of a (huge) private key, and we would obtain a stateless scheme. The problem with this approach, of course, is that generating all these values would require *exponential* time, and storing them all would require exponential memory.

An alternative is to store some *randomness* that can be used to generate the values $\{(pk_w, sk_w)\}$ and $\{\sigma_w\}$, as needed, rather than storing the values themselves. That is, the signer could store a random string r_w for each w, and whenever the values pk_w, sk_w are needed the signer can compute $(pk_w, sk_w) :=$ $\text{Gen}(1^n; r_w)$, where this denotes the generation of a length-n key using random coins r_w. Similarly, if the signing procedure is probabilistic, the signer can store r'_w and then set $\sigma_w := \text{Sign}_{sk_w}(pk_{w0}\|pk_{w1}; r'_w)$ (assuming here that $|w| < n$). Generating and storing sufficiently many random strings, however, still requires exponential time and memory.

A simple modification of this alternative gives a polynomial-time solution. Instead of storing random r_w and r'_w as suggested above, the signer can store two keys k, k' for a pseudorandom function F. When needed, the values pk_w, sk_w can now be generated by the following two-step process:

1. Compute $r_w := F_k(w)$.[5]

2. Compute $(pk_w, sk_w) := \text{Gen}(1^n; r_w)$ (as before).

In addition, the key k' is used to generate the value r'_w that is used to compute the signature σ_w. This gives a *stateless* scheme in which key generation (as well as signing and verifying) can be done in polynomial time. Intuitively, this is secure because storing a random function is equivalent to storing all the r_w and r'_w values that are needed, and storing a pseudorandom function is "just as good." We leave it as an exercise to give a formal proof that this modified scheme remains secure.

---

[5]We assume that the output length of F is sufficiently long, and that w is padded to some fixed-length string in a one-to-one fashion. We ignore these technicalities here.

Since the existence of collision-resistant hash functions implies the existence of one-way functions (cf. Exercise 8.4), and the latter implies the existence of pseudorandom functions (see Chapter 8), we have:

THEOREM 14.16 *If collision-resistant hash functions exist, then there exists a (stateless) secure signature scheme.*

We remark that it is possible to construct signature schemes satisfying Definition 13.2 from the (minimal) assumption that one-way functions exist; a proof of this result is beyond the scope of this book.

References and Additional Reading

Quantum computing is covered in the text by Nielsen and Chuang [153], which also describes Grover's algorithm [90] and Shor's algorithm [178]. The collision-finding algorithm in Section 14.1.2 is due to Brassard et al. [46].

For details of the NIST post-quantum cryptography standardization effort, see https://csrc.nist.gov/projects/post-quantum-cryptography. The LWE problem originated in the work of Regev [169]. Several of the candidate public-key encryption schemes submitted to NIST can be viewed as following the approach of the LWE-based scheme presented here (which is also due to Regev [169]), with the most similar being Frodo (see https://frodokem.org).

Lamport's signature scheme was published in 1979 [124], although it was already described by Diffie and Hellman [65]. A tree-based construction similar in spirit to Construction 14.14 was suggested by Merkle [138, 139], and a tree-based approach was also used in other schemes [88]. Goldreich [81] suggested a way to make the Goldwasser–Micali–Rivest scheme [88] stateless, and we have adapted his ideas in Section 14.4.3. Naor and Yung [146] showed that one-way permutations suffice for constructing one-time signatures that can sign messages of arbitrary length, and this was improved by Rompel [174], who showed that one-way functions are sufficient. (See also [110].) As we have seen in Section 14.4.3, one-time signatures of this sort can be used to construct secure signature schemes, implying that one-way functions suffice for the existence of (stateless) secure signatures. SPHINCS+ (see https://sphincs.org) is a hash-based signature scheme submitted to the NIST post-quantum cryptography standardization effort.

Exercises

14.1 Prove Lemma 14.2.

14.2 Prove that the existence of a one-time signature scheme for 1-bit messages implies the existence of one-way functions.

14.3 Let f be a one-way permutation. Consider the following signature scheme for messages in the set $\{1, \ldots, \ell\}$:

- To generate keys, choose uniform $x \in \{0, 1\}^n$ and set $y := f^{(\ell)}(x)$ (where $f^{(i)}(\cdot)$ refers to i-fold iteration of f, and $f^{(0)}(x) \overset{\text{def}}{=} x$). The public key is y and the private key is x.
- To sign message $i \in \{1, \ldots, \ell\}$, output $f^{(\ell-i)}(x)$.
- To verify signature σ on message i with respect to public key y, check whether $y \overset{?}{=} f^{(i)}(\sigma)$.

(a) Show that the above is not a one-time signature scheme. Given a signature on a message i, for what messages j can an adversary output a forgery?

(b) Prove that no PPT adversary given a signature on i can output a forgery on any message $j > i$ except with negligible probability.

(c) Suggest how to modify the scheme so as to obtain a one-time signature scheme.
   Hint: Include two values y, y' in the public key.

14.4 A *strong* one-time signature scheme satisfies the following (informally): given a signature σ' on a message m', it is infeasible to output $(m, \sigma) \neq (m', \sigma')$ for which σ is a valid signature on m (note that $m = m'$ is allowed).

(a) Give a formal definition of strong one-time signatures.

(b) Assuming the existence of one-way functions, show a one-way function for which Lamport's scheme is *not* a strong one-time signature scheme.

(c) Construct a strong one-time signature scheme based on any assumption used in this book.
   Hint: Use a particular one-way function in Lamport's scheme.

14.5 Show an adversary attacking the Lamport scheme who obtains signatures on *two* messages of its choice and can then forge signatures on any message it likes.

14.6 The Lamport scheme uses 2ℓ values in the public key to sign messages of length ℓ. Consider the variant in which the private key contains 2ℓ values $x_1, \ldots, x_{2\ell}$ and the public key contains the values $y_1, \ldots, y_{2\ell}$ with $y_i := f(x_i)$. A message $m \in \{0,1\}^{\ell'}$ is mapped in a one-to-one fashion to a subset $S_m \subset \{1, \ldots, 2\ell\}$ of size ℓ. To sign m, the signer reveals $\{x_i\}_{i \in S_m}$. Prove that this gives a one-time signature scheme. What is the maximum message length ℓ' that this scheme supports?

14.7 At the end of Section 14.4.3, we show how a pseudorandom function can be used to make Construction 14.14 stateless. Does a similar approach work for the chain-based scheme described in Section 14.4.2? If so, sketch a construction and proof. If not, explain why and modify the scheme to obtain a stateless variant.

14.8 Prove Theorem 14.16.

Chapter 15

*Advanced Topics in Public-Key Encryption

In Chapter 12 we saw several examples of public-key encryption schemes used in practice. Here, we explore some schemes that are currently more of theoretical interest—although in some cases it is possible that these schemes (or variants thereof) will be used more widely in the future.

We begin with a treatment of *trapdoor permutations*, a generalization of one-way permutations, and show how to use them to construct public-key encryption schemes. Trapdoor permutations neatly encapsulate the key characteristics of the RSA permutation that make it so useful. As such, they often provide a useful abstraction for designing new cryptosystems.

Next, we present three schemes based on problems related to factoring:

- The *Paillier encryption scheme* is an example of an encryption scheme that is *homomorphic*. This property turns out to be useful for constructing more-complex cryptographic protocols, something we touch on briefly in Section 15.3.

- The *Goldwasser–Micali encryption scheme* is of historical interest as the first scheme to be proven CPA-secure. It is also homomorphic, and uses some interesting number theory that can be applied in other contexts.

- Finally, we discuss the *Rabin trapdoor permutation*, which can be used to construct a public-key encryption scheme. Although superficially similar to the RSA trapdoor permutation, the Rabin trapdoor permutation is distinguished by the fact that its security is based *directly* on the hardness of factoring. (Recall from Section 9.2.5 that hardness of the RSA problem appears to be a stronger assumption.)

15.1   Encryption from Trapdoor Permutations

In Section 12.5.3 we saw how to construct a CPA-secure public-key encryption scheme based on the RSA assumption. By distilling those properties of

RSA that are used in the construction, and defining an abstract notion that encapsulates those properties, we obtain a general *template* for constructing secure encryption schemes based on any primitive satisfying the same set of properties. *Trapdoor permutations* turn out to be the "right" abstraction here.

In the following section we define (families of) trapdoor permutations and observe that the RSA family of one-way permutations (Construction 9.77) satisfies the additional requirements needed to be a family of *trapdoor* permutations. In Section 15.1.2 we generalize the construction from Section 12.5.3 and show that public-key encryption can be constructed from any trapdoor permutation. These results will be used again in Section 15.5, where we show a second example of a trapdoor permutation, this time based directly on the factoring assumption.

In this section we rely on the material from Section 9.4.1 or, alternately, Chapter 8.

15.1.1   Trapdoor Permutations

Recall the definitions of families of functions and families of one-way permutations from Section 9.4.1. In that section, we showed that the RSA assumption naturally gives rise to a family of one-way permutations. The astute reader may have noticed that the construction we gave (Construction 9.77) has a special property that was not remarked upon there: namely, the parameter-generation algorithm Gen outputs some additional information along with I that *enables efficient inversion of* f_I. We refer to such additional information as a *trapdoor*, and call families of one-way permutations with this additional property *families of trapdoor permutations*. A formal definition follows.

DEFINITION 15.1   *A tuple of polynomial-time algorithms* (Gen, Samp, f, Inv) *is a* family of trapdoor permutations (*or a* trapdoor permutation) *if:*

- *The probabilistic* parameter-generation algorithm Gen, *on input* 1^n, *outputs* (I, td) *with* $|I| \geq n$. *Each value of* I *defines a set* $\mathcal{D}_I$ *that constitutes the domain and range of a permutation (i.e., bijection)* $f_I : \mathcal{D}_I \to \mathcal{D}_I$.

- *Let* Gen_1 *denote the algorithm that results by running* Gen *and outputting only* I. *Then* $(\mathsf{Gen}_1, \mathsf{Samp}, f)$ *is a family of one-way permutations.*

- *Let* (I, td) *be an output of* $\mathsf{Gen}(1^n)$. *The deterministic* inverting algorithm Inv, *on input* td *and* $y \in \mathcal{D}_I$, *outputs* $x \in \mathcal{D}_I$. *We denote this by* $x := \mathsf{Inv}_{\mathsf{td}}(y)$. *It is required that with all but negligible probability over* (I, td) *output by* $\mathsf{Gen}(1^n)$ *and uniform choice of* $x \in \mathcal{D}_I$, *we have*

$$\mathsf{Inv}_{\mathsf{td}}(f_I(x)) = x.$$

As shorthand, we drop explicit mention of Samp and simply refer to trapdoor permutation (Gen, f, Inv). For (I, td) output by Gen we write $x \leftarrow \mathcal{D}_I$ to denote uniform selection of $x \in \mathcal{D}_I$ (with the understanding that this is done by algorithm Samp).

The second condition above implies that f_I cannot be efficiently inverted without td, but the final condition means that f_I *can* be efficiently inverted with td. It is immediate that Construction 9.77 can be modified to give a family of trapdoor permutations if the RSA problem is hard relative to GenRSA, and so we refer to that construction as the *RSA trapdoor permutation*.

15.1.2 Public-Key Encryption from Trapdoor Permutations

We now sketch how a public-key encryption scheme can be constructed from an arbitrary family of trapdoor permutations. The construction is simply a generalization of what was already done for the specific RSA trapdoor permutation in Section 12.5.3.

We begin by (re-)introducing the notion of a hard-core predicate. This is the natural adaptation of Definition 8.4 to our context, and also generalizes our previous discussion of one specific hard-core predicate for the RSA trapdoor permutation in Section 12.5.3.

DEFINITION 15.2 *Let* $\Pi = (\mathsf{Gen}, f, \mathsf{Inv})$ *be a family of trapdoor permutations, and let* hc *be a deterministic polynomial-time algorithm that, on input* I *and* $x \in \mathcal{D}_I$*, outputs a single bit* $\mathsf{hc}_I(x)$*. We say that* hc *is a hard-core* predicate *of* Π *if for every probabilistic polynomial-time algorithm* $\mathcal{A}$ *there is a negligible function* negl *such that*

$$\Pr[\mathcal{A}(I, f_I(x)) = \mathsf{hc}_I(x)] \leq \frac{1}{2} + \mathsf{negl}(n),$$

where the probability is taken over the experiment in which $\mathsf{Gen}(1^n)$ *is run to generate* (I, td) *and then* x *is chosen uniformly from* $\mathcal{D}_I$*.*

The asymmetry provided by trapdoor permutations implies that anyone who knows the trapdoor td associated with I can recover x from $f_I(x)$ and thus compute $\mathsf{hc}_I(x)$ from $f_I(x)$. But given only I, it is infeasible to compute $\mathsf{hc}_I(x)$ from $f_I(x)$ for a uniform x.

The following can be proved by a suitable modification of Theorem 8.5:

THEOREM 15.3 *Given a family of trapdoor permutations* Π*, there is a family of trapdoor permutations* $\widehat{\Pi}$ *with a hard-core predicate* hc *for* $\widehat{\Pi}$*.*

Given a family of trapdoor permutations $\widehat{\Pi} = (\widehat{\mathsf{Gen}}, f, \mathsf{Inv})$ with hard-core predicate hc, we can construct a single-bit encryption scheme via the following approach (see Construction 15.4 below, and compare to Construction 12.32): To generate keys, run $\widehat{\mathsf{Gen}}(1^n)$ to obtain (I, td); the public key is I and the private key is td. Given a public key I, encryption of a message $m \in \{0,1\}$ works by choosing uniform $r \in \mathcal{D}_I$ subject to the constraint that $hc_I(r) = m$, and then setting the ciphertext equal to $f_I(r)$. In order to decrypt, the receiver uses td to recover r from $f_I(r)$ and then outputs the message $m := hc_I(r)$.

CONSTRUCTION 15.4

Let $\widehat{\Pi} = (\widehat{\mathsf{Gen}}, f, \mathsf{Inv})$ be a family of trapdoor permutations with hard-core predicate hc. Define a public-key encryption scheme as follows:

- Gen: on input 1^n, run $\widehat{\mathsf{Gen}}(1^n)$ to obtain (I, td). Output the public key I and the private key td.

- Enc: on input a public key I and a message $m \in \{0,1\}$, choose a uniform $r \in \mathcal{D}_I$ subject to the constraint that $hc_I(r) = m$. Output the ciphertext $c := f_I(r)$.

- Dec: on input a private key td and a ciphertext c, compute the value $r := \mathsf{Inv}_{\mathsf{td}}(c)$ and output the message $hc_I(r)$.

Public-key encryption from any family of trapdoor permutations.

A proof of security follows along the lines of the proof of Theorem 12.33.

THEOREM 15.5 *If $\widehat{\Pi}$ is a family of trapdoor permutations with hard-core predicate hc, then Construction 15.4 is CPA-secure.*

PROOF Let Π denote Construction 15.4. We prove that Π has indistinguishable encryptions in the presence of an eavesdropper; by Proposition 12.3, this implies it is CPA-secure.

We first observe that hc must be *unbiased* in the following sense. Let

$$\delta_0(n) \overset{\text{def}}{=} \Pr_{(I,\mathsf{td}) \leftarrow \widehat{\mathsf{Gen}}(1^n); x \leftarrow \mathcal{D}_I} [hc_I(x) = 0]$$

and

$$\delta_1(n) \overset{\text{def}}{=} \Pr_{(I,\mathsf{td}) \leftarrow \widehat{\mathsf{Gen}}(1^n); x \leftarrow \mathcal{D}_I} [hc_I(x) = 1].$$

Then there is a negligible function negl such that

$$\delta_0(n), \delta_1(n) \geq \frac{1}{2} - \mathsf{negl}(n);$$

if not, then an attacker who simply outputs the more frequently occurring bit would violate Definition 15.2.

Now let $\mathcal{A}$ be a probabilistic polynomial-time adversary. Without loss of generality, we may assume $m_0 = 0$ and $m_1 = 1$ in experiment $\mathsf{PubK}^{\mathsf{eav}}_{\mathcal{A},\Pi}(n)$.

We then have

$$\Pr[\mathsf{PubK}^{\mathsf{eav}}_{\mathcal{A},\Pi}(n) = 1] = \frac{1}{2} \cdot \Pr[\mathcal{A}(pk, c) = 0 \mid c \text{ is an encryption of } 0]$$
$$+ \frac{1}{2} \cdot \Pr[\mathcal{A}(pk, c) = 1 \mid c \text{ is an encryption of } 1].$$

But then

$$\Pr[\mathcal{A}(I, f_I(x)) = \mathsf{hc}_I(x)]$$
$$= \delta_0(n) \cdot \Pr[\mathcal{A}(I, f_I(x)) = 0 \mid \mathsf{hc}_I(x) = 0]$$
$$+ \delta_1(n) \cdot \Pr[\mathcal{A}(I, f_I(x)) = 1 \mid \mathsf{hc}_I(x) = 1]$$
$$\geq \left(\frac{1}{2} - \mathsf{negl}(n)\right) \cdot \Pr[\mathcal{A}(I, f_I(x)) = 0 \mid \mathsf{hc}_I(x) = 0]$$
$$+ \left(\frac{1}{2} - \mathsf{negl}(n)\right) \cdot \Pr[\mathcal{A}(I, f_I(x)) = 1 \mid \mathsf{hc}_I(1) = 1]$$
$$\geq \frac{1}{2} \cdot \Pr[\mathcal{A}(I, f_I(x)) = 0 \mid \mathsf{hc}_I(x) = 0]$$
$$+ \frac{1}{2} \cdot \Pr[\mathcal{A}(I, f_I(x)) = 1 \mid \mathsf{hc}_I(1) = 1] - 2 \cdot \mathsf{negl}(n)$$
$$= \Pr[\mathsf{PubK}^{\mathsf{eav}}_{\mathcal{A},\Pi}(n) = 1] - 2 \cdot \mathsf{negl}(n).$$

Since hc is a hard-core predicate for $\widehat{\Pi}$, there is a negligible function negl' such that $\mathsf{negl}'(n) \geq \Pr[\mathcal{A}(I, f_I(x)) = \mathsf{hc}_I(x)]$; this means that

$$\Pr[\mathsf{PubK}^{\mathsf{eav}}_{\mathcal{A},\Pi}(n) = 1] \leq \mathsf{negl}'(n) + 2 \cdot \mathsf{negl}(n),$$

completing the proof. ∎

Encrypting longer messages. Using Claim 12.7, we know that we can extend Construction 15.4 to encrypt ℓ-bit messages using ciphertexts ℓ times as long. Better efficiency can be obtained by constructing a KEM, following along the lines of Construction 12.34. We leave the details as an exercise.

15.2 The Paillier Encryption Scheme

In this section we describe the *Paillier encryption scheme*, a public-key encryption scheme whose security is based on an assumption related (but not known to be equivalent) to the hardness of factoring. This encryption scheme is particularly interesting because it possesses some nice *homomorphic* properties, as we will discuss further in Section 15.2.3.

The Paillier encryption scheme utilizes the group $\mathbb{Z}_{N^2}^*$, the multiplicative group of elements in the range $\{1, \ldots, N^2\}$ that are relatively prime to N, for N a product of two distinct primes. To understand the scheme it is helpful to first understand the structure of $\mathbb{Z}_{N^2}^*$. A useful characterization of this group is given by the following proposition, which says, among other things, that $\mathbb{Z}_{N^2}^*$ is isomorphic to $\mathbb{Z}_N \times \mathbb{Z}_N^*$ (cf. Definition 9.23) for N of the form we will be interested in. We prove the proposition in the next section. (The reader willing to accept the proposition on faith can skip to Section 15.2.2.)

PROPOSITION 15.6   *Let $N = pq$, where p, q are distinct odd primes of equal length. Then:*

1. $\gcd(N, \phi(N)) = 1$.

2. *For any integer $a \geq 0$, we have $(1+N)^a = (1+aN) \bmod N^2$.*

   As a consequence, the order of $(1+N)$ in $\mathbb{Z}_{N^2}^*$ is N. That is, $(1+N)^N = 1 \bmod N^2$ and $(1+N)^a \neq 1 \bmod N^2$ for any $1 \leq a < N$.

3. $\mathbb{Z}_N \times \mathbb{Z}_N^*$ *is isomorphic to $\mathbb{Z}_{N^2}^*$, with isomorphism $f \colon \mathbb{Z}_N \times \mathbb{Z}_N^* \to \mathbb{Z}_{N^2}^*$ given by*
$$f(a, b) = \left[(1+N)^a \cdot b^N \bmod N^2 \right].$$

In light of the last part of the above proposition, we introduce some convenient notation. With N understood, and $x \in \mathbb{Z}_{N^2}^*$, $a \in \mathbb{Z}_N$, $b \in \mathbb{Z}_N^*$, we write $x \leftrightarrow (a, b)$ if $f(a, b) = x$ where f is the isomorphism from the proposition above. One way to think about this notation is that it means "x in $\mathbb{Z}_{N^2}^*$ *corresponds to* (a, b) in $\mathbb{Z}_N \times \mathbb{Z}_N^*$." We have used the same notation in this book with regard to the isomorphism $\mathbb{Z}_N^* \simeq \mathbb{Z}_p^* \times \mathbb{Z}_q^*$ given by the Chinese remainder theorem; we keep the notation because in both cases it refers to an isomorphism of groups. Nevertheless, there should be no confusion since the group $\mathbb{Z}_{N^2}^*$ and the above proposition are only used in this section. We remark that here the isomorphism—but not its inverse—is efficiently computable even without the factorization of N.

15.2.1   The Structure of $\mathbb{Z}_{N^2}^*$

This section is devoted to a proof of Proposition 15.6. Throughout, we let N, p, q be as in the proposition.

CLAIM 15.7 $\gcd(N, \phi(N)) = 1$.

PROOF   Recall that $\phi(N) = (p-1)(q-1)$. Assume $p > q$ without loss of generality. Since p is prime and $p > p - 1 > q - 1$, clearly $\gcd(p, \phi(N)) = 1$. Similarly, $\gcd(q, q-1) = 1$. Now, if $\gcd(q, p-1) \neq 1$ then $\gcd(q, p-1) = q$

since q is prime. But then $(p-1)/q \geq 2$, contradicting the assumption that p and q have the same length. ∎

CLAIM 15.8 *For $a \geq 0$ an integer, we have $(1+N)^a = 1 + aN \bmod N^2$. Thus, the order of $(1+N)$ in $\mathbb{Z}_{N^2}^*$ is N.*

PROOF Using the binomial expansion theorem (Theorem A.1):

$$(1+N)^a = \sum_{i=0}^{a} \binom{a}{i} N^i.$$

Reducing the right-hand side modulo N^2, all terms with $i \geq 2$ become 0 and so $(1+N)^a = 1 + aN \bmod N^2$. The smallest nonzero a such that $(1+N)^a = 1 \bmod N^2$ is therefore $a = N$. ∎

CLAIM 15.9 *The group $\mathbb{Z}_N \times \mathbb{Z}_N^*$ is isomorphic to the group $\mathbb{Z}_{N^2}^*$, with isomorphism $f \colon \mathbb{Z}_N \times \mathbb{Z}_N^* \to \mathbb{Z}_{N^2}^*$ given by $f(a,b) = [(1+N)^a \cdot b^N \bmod N^2]$.*

PROOF Note that $(1+N)^a \cdot b^N$ does not have a factor in common with N^2 since $\gcd((1+N), N^2) = 1$ and $\gcd(b, N^2) = 1$ (because $b \in \mathbb{Z}_N^*$). So $[(1+N)^a \cdot b^N \bmod N^2]$ lies in $\mathbb{Z}_{N^2}^*$. We now prove that f is an isomorphism.

We first show that f is a bijection. Since

$$|\mathbb{Z}_{N^2}^*| = \phi(N^2) = p \cdot (p-1) \cdot q \cdot (q-1) = pq \cdot (p-1)(q-1)$$
$$= |\mathbb{Z}_N| \cdot |\mathbb{Z}_N^*| = |\mathbb{Z}_N \times \mathbb{Z}_N^*|$$

(see Theorem 9.19 for the second equality), it suffices to show that f is one-to-one. Say $a_1, a_2 \in \mathbb{Z}_N$ and $b_1, b_2 \in \mathbb{Z}_N^*$ are such that $f(a_1, b_1) = f(a_2, b_2)$. Then:

$$(1+N)^{a_1 - a_2} \cdot (b_1/b_2)^N = 1 \bmod N^2. \tag{15.1}$$

(Note that $b_2 \in \mathbb{Z}_N^*$ and thus $b_2 \in \mathbb{Z}_{N^2}^*$, and so b_2 has a multiplicative inverse modulo N^2.) Raising both sides to the power $\phi(N)$ and using the fact that the order of $\mathbb{Z}_{N^2}^*$ is $\phi(N^2) = N \cdot \phi(N)$ we obtain

$$(1+N)^{(a_1 - a_2) \cdot \phi(N)} \cdot (b_1/b_2)^{N \cdot \phi(N)} = 1 \bmod N^2$$
$$\Rightarrow (1+N)^{(a_1 - a_2) \cdot \phi(N)} = 1 \bmod N^2.$$

By Claim 15.8, $(1+N)$ has order N modulo N^2. Applying Proposition 9.54, we see that $(a_1 - a_2) \cdot \phi(N) = 0 \bmod N$ and so N divides $(a_1 - a_2) \cdot \phi(N)$. Since $\gcd(N, \phi(N)) = 1$ by Claim 15.7, it follows that $N \mid (a_1 - a_2)$. Since $a_1, a_2 \in \mathbb{Z}_N$, this can only occur if $a_1 = a_2$.

Returning to Equation (15.1) and setting $a_1 = a_2$, we thus have $b_1^N = b_2^N \bmod N^2$. This implies $b_1^N = b_2^N \bmod N$. Since N is relatively prime to $\phi(N)$, the order of $\mathbb{Z}_N^*$, exponentiation to the power N is a bijection in $\mathbb{Z}_N^*$ (cf. Corollary 9.17). This means that $b_1 = b_2 \bmod N$; since $b_1, b_2 \in \mathbb{Z}_N^*$, we have $b_1 = b_2$. We conclude that f is one-to-one, and hence a bijection.

To show that f is an isomorphism, we show that $f(a_1, b_1) \cdot f(a_2, b_2) = f(a_1 + a_2, b_1 \cdot b_2)$. (Note that multiplication on the left-hand side of the equality takes place modulo N^2, while addition/multiplication on the right-hand side takes place modulo N.) We have:

$$f(a_1, b_1) \cdot f(a_2, b_2) = \big((1+N)^{a_1} \cdot b_1^N\big) \cdot \big((1+N)^{a_2} \cdot b_2^N\big) \bmod N^2$$
$$= (1+N)^{a_1+a_2} \cdot (b_1 b_2)^N \bmod N^2.$$

Since $(1+N)$ has order N modulo N^2 (by Claim 15.8), we can apply Proposition 9.53 and obtain

$$f(a_1, b_1) \cdot f(a_2, b_2) = (1+N)^{a_1+a_2} \cdot (b_1 b_2)^N \bmod N^2$$
$$= (1+N)^{[a_1+a_2 \bmod N]} \cdot (b_1 b_2)^N \bmod N^2. \quad (15.2)$$

We are not yet done, since $b_1 b_2$ in Equation (15.2) represents multiplication modulo N^2 whereas we would like it to be modulo N. Let $b_1 b_2 = r + \gamma N$, where γ, r are integers with $1 \le r < N$ (r cannot be 0 since $b_1, b_2 \in \mathbb{Z}_N^*$ and so their product cannot be divisible by N). Note that $r = b_1 b_2 \bmod N$. We also have

$$(b_1 b_2)^N = (r + \gamma N)^N \bmod N^2$$
$$= \sum_{k=0}^{N} \binom{N}{k} r^{N-k} (\gamma N)^k \bmod N^2$$
$$= r^N + N \cdot r^{N-1} \cdot (\gamma N) = r^N = ([b_1 b_2 \bmod N])^N \bmod N^2,$$

using the binomial expansion theorem as in Claim 15.8. Plugging this in to Equation (15.2) we get the desired result:

$$f(a_1, b_1) \cdot f(a_2, b_2) = (1+N)^{[a_1+a_2 \bmod N]} \cdot (b_1 b_2 \bmod N)^N \bmod N^2$$
$$= f(a_1 + a_2, b_1 b_2),$$

proving that f is an isomorphism from $\mathbb{Z}_N \times \mathbb{Z}_N^*$ to $\mathbb{Z}_{N^2}^*$. ∎

15.2.2 The Paillier Encryption Scheme

Let $N = pq$ be a product of two distinct primes of equal length. Proposition 15.6 says that $\mathbb{Z}_N \times \mathbb{Z}_N^*$ is isomorphic to $\mathbb{Z}_{N^2}^*$, with isomorphism given by $f(a, b) = [(1+N)^a \cdot b^N \bmod N^2]$. A consequence is that a uniform element

$y \in \mathbb{Z}_{N^2}^*$ corresponds to a uniform element $(a,b) \in \mathbb{Z}_N \times \mathbb{Z}_N^*$ or, in other words, an element (a,b) with uniform $a \in \mathbb{Z}_N$ and uniform $b \in \mathbb{Z}_N^*$.

Call $y \in \mathbb{Z}_{N^2}^*$ an *Nth residue modulo* N^2 if y is an Nth power, that is, if there exists an $x \in \mathbb{Z}_{N^2}^*$ with $y = x^N \bmod N^2$. We denote the set of Nth residues modulo N^2 by $\mathsf{Res}(N^2)$. Let us characterize the Nth residues in $\mathbb{Z}_{N^2}^*$. Taking any $x \in \mathbb{Z}_{N^2}^*$ with $x \leftrightarrow (a,b)$ and raising it to the Nth power gives:

$$[x^N \bmod N^2] \leftrightarrow (a,b)^N = (N \cdot a \bmod N, b^N \bmod N) = (0, b^N \bmod N).$$

(Recall that the group operation in $\mathbb{Z}_N \times \mathbb{Z}_N^*$ is addition modulo N in the first component and multiplication modulo N in the second component.) Moreover, we claim that any element y with $y \leftrightarrow (0,b)$ is an Nth residue. To see this, recall that $\gcd(N, \phi(N)) = 1$ and so $d \stackrel{\text{def}}{=} [N^{-1} \bmod \phi(N)]$ exists. So

$$(a, [b^d \bmod N])^N = (Na \bmod N, [b^{dN} \bmod N]) = (0,b) \leftrightarrow y$$

for any $a \in \mathbb{Z}_N$. We have thus shown that $\mathsf{Res}(N^2)$ corresponds to the set

$$\left\{ (0,b) \mid b \in \mathbb{Z}_N^* \right\}.$$

The above also demonstrates that the number of Nth roots of any $y \in \mathsf{Res}(N^2)$ is exactly N, and so computing Nth powers is an N-to-1 function. As such, if $r \in \mathbb{Z}_{N^2}^*$ is uniform then $[r^N \bmod N^2]$ is a uniform element of $\mathsf{Res}(N^2)$.

The *decisional composite residuosity problem*, roughly speaking, is to distinguish a uniform element of $\mathbb{Z}_{N^2}^*$ from a uniform element of $\mathsf{Res}(N^2)$. Formally, let $\mathsf{GenModulus}$ be a polynomial-time algorithm that, on input 1^n, outputs (N,p,q) where $N = pq$, and p and q are n-bit primes (except with probability negligible in n). Then:

DEFINITION 15.10 *The* decisional composite residuosity problem is hard relative to $\mathsf{GenModulus}$ *if for all probabilistic polynomial-time algorithms* D *there is a negligible function* negl *such that*

$$\left| \Pr[D(N, [r^N \bmod N^2]) = 1] - \Pr[D(N,r) = 1] \right| \leq \mathsf{negl}(n),$$

where in each case the probabilities are taken over the experiment in which $\mathsf{GenModulus}(1^n)$ *outputs* (N,p,q), *and then a uniform* $r \in \mathbb{Z}_{N^2}^*$ *is chosen.* (*Recall that* $[r^N \bmod N^2]$ *is a uniform element of* $\mathsf{Res}(N^2)$.)

The *decisional composite residuosity (DCR) assumption* is the assumption that there is a $\mathsf{GenModulus}$ relative to which the decisional composite residuosity problem is hard.

As we have discussed, elements of $\mathbb{Z}_{N^2}^*$ have the form (r',r) with r' and r arbitrary (in the appropriate groups), whereas Nth residues have the form $(0,r)$ with $r \in \mathbb{Z}_N^*$ arbitrary. The DCR assumption is that it is hard to distinguish uniform elements of the first type from uniform elements of the second type. This suggests the following abstract way to encrypt a message $m \in \mathbb{Z}_N$ with respect to a public key N: choose a uniform Nth residue $(0, r)$ and set the ciphertext equal to

$$c \leftrightarrow (m, 1) \cdot (0, r) = (m + 0, \, 1 \cdot r) = (m, r).$$

Without worrying for now how this can be carried out efficiently by the sender, or how the receiver can decrypt, let us simply convince ourselves (on an intuitive level) that this is secure. Since a uniform Nth residue $(0, r)$ cannot be distinguished from a uniform element (r', r), the ciphertext as constructed above is indistinguishable (from the point of an eavesdropper who does not know the factorization of N) from the ciphertext

$$c' \leftrightarrow (m, 1) \cdot (r', r) = ([m + r' \bmod N], r)$$

for uniform $r' \in \mathbb{Z}_N$ and $r \in \mathbb{Z}_N^*$. Lemma 12.15 shows that $[m + r' \bmod N]$ is uniformly distributed in $\mathbb{Z}_N$ and so, in particular, this ciphertext c' is independent of the message m. CPA-security follows. A formal proof that proceeds exactly along these lines is given further below.

Before turning to the formal description and proof of security, we show how encryption and decryption can be performed efficiently.

Encryption. We have described encryption above as though it is taking place in $\mathbb{Z}_N \times \mathbb{Z}_N^*$. In fact it takes place in the isomorphic group $\mathbb{Z}_{N^2}^*$. That is, the sender generates a ciphertext $c \in \mathbb{Z}_{N^2}^*$ by choosing a uniform[1] $r \in \mathbb{Z}_N^*$ and then computing

$$c := [(1 + N)^m \cdot r^N \bmod N^2].$$

Observe that

$$c = \left((1 + N)^m \cdot 1^N \right) \cdot \left((1 + N)^0 \cdot r^N \right) \bmod N^2 \leftrightarrow (m, 1) \cdot (0, r),$$

and so $c \leftrightarrow (m, r)$ as desired.

Decryption. We now describe how decryption can be performed efficiently given the factorization of N. For c constructed as above, we claim that m is recovered by the following steps:

- Set $\hat{c} := [c^{\phi(N)} \bmod N^2]$.

- Set $\hat{m} := (\hat{c} - 1)/N$. (Note that this is carried out over the integers.)

- Set $m := \left[\hat{m} \cdot \phi(N)^{-1} \bmod N \right]$.

---

[1] We remark that it does not make any difference whether the sender chooses uniform $r \in \mathbb{Z}_N^*$ or uniform $r \in \mathbb{Z}_{N^2}^*$, since in either case the distribution of $[r^N \bmod N^2]$ is the same (as can be verified by looking at what happens in the isomorphic group $\mathbb{Z}_N \times \mathbb{Z}_N^*$).

To see why this works, let $c \leftrightarrow (m, r)$ for an arbitrary $r \in \mathbb{Z}_N^*$. Then

$$\hat{c} \stackrel{\text{def}}{=} [c^{\phi(N)} \bmod N^2]$$
$$\leftrightarrow (m, r)^{\phi(N)}$$
$$= \left([m \cdot \phi(N) \bmod N], [r^{\phi(N)} \bmod N]\right)$$
$$= ([m \cdot \phi(N) \bmod N], 1).$$

By Proposition 15.6(3), this means that $\hat{c} = (1 + N)^{[m \cdot \phi(N) \bmod N]} \bmod N^2$. Using Proposition 15.6(2), we know that

$$\hat{c} = (1 + N)^{[m \cdot \phi(N) \bmod N]} = (1 + [m \cdot \phi(N) \bmod N] \cdot N) \bmod N^2.$$

Since $1 + [m \cdot \phi(N) \bmod N] \cdot N$ is always less that N^2 we can drop the $\bmod N^2$ at the end and view the above as an equality over the integers. Thus, $\hat{m} \stackrel{\text{def}}{=} (\hat{c} - 1)/N = [m \cdot \phi(N) \bmod N]$ and, finally,

$$m = [\hat{m} \cdot \phi(N)^{-1} \bmod N],$$

as required. (Note that $\phi(N)$ is invertible modulo N since $\gcd(N, \phi(N)) = 1$.)

We give a complete description of the Paillier encryption scheme, followed by an example of the above calculations.

CONSTRUCTION 15.11

Let GenModulus be a polynomial-time algorithm that, on input 1^n, outputs (N, p, q) where $N = pq$ and p and q are n-bit primes (except with probability negligible in n). Define the following encryption scheme:

- Gen: on input 1^n run GenModulus(1^n) to obtain (N, p, q). The public key is N, and the private key is $\langle N, \phi(N) \rangle$.

- Enc: on input a public key N and a message $m \in \mathbb{Z}_N$, choose a uniform $r \leftarrow \mathbb{Z}_N^*$ and output the ciphertext

$$c := [(1 + N)^m \cdot r^N \bmod N^2].$$

- Dec: on input a private key $\langle N, \phi(N) \rangle$ and a ciphertext c, compute

$$m := \left[\frac{[c^{\phi(N)} \bmod N^2] - 1}{N} \cdot \phi(N)^{-1} \bmod N \right].$$

The Paillier encryption scheme.

Example 15.12

Let $N = 11 \cdot 17 = 187$ (and so $N^2 = 34969$), and consider encrypting the message $m = 175$ and then decrypting the corresponding ciphertext. Choosing

$r = 83 \in \mathbb{Z}_{187}^*$, we compute the ciphertext

$$c := [(1 + 187)^{175} \cdot 83^{187} \bmod 34969] = 23911$$

corresponding to $(175, 83)$. To decrypt, note that $\phi(N) = 160$. So we first compute $\hat{c} := [23911^{160} \bmod 34969] = 25620$. Subtracting 1 and dividing by 187 gives $\hat{m} := (25620 - 1)/187 = 137$; since $90 = [160^{-1} \bmod 187]$, the message is recovered as $m := [137 \cdot 90 \bmod 187] = 175$.   ◇

THEOREM 15.13   *If the decisional composite residuosity problem is hard relative to GenModulus, then the Paillier encryption scheme is CPA-secure.*

PROOF   Let Π denote the Paillier encryption scheme. We prove that Π has indistinguishable encryptions in the presence of an eavesdropper; by Theorem 12.6 this implies that it is CPA-secure.

Let $\mathcal{A}$ be an arbitrary probabilistic polynomial-time adversary. Consider the following PPT algorithm D that attempts to solve the decisional composite residuosity problem relative to GenModulus:

> **Algorithm D:**
> The algorithm is given N, y as input.
>
> - Set $pk = N$ and run $\mathcal{A}(pk)$ to obtain two messages m_0, m_1.
> - Choose a uniform bit b and set $c := [(1 + N)^{m_b} \cdot y \bmod N^2]$.
> - Give the ciphertext c to $\mathcal{A}$ and obtain an output bit b'. If $b' = b$, output 1; otherwise, output 0.

Let us analyze the behavior of D. There are two cases to consider:

Case 1: Say the input to D was generated by running GenModulus(1^n) to obtain (N, p, q), choosing uniform $r \in \mathbb{Z}_{N^2}^*$, and setting $y := [r^N \bmod N^2]$. (That is, y is a uniform element of Res(N^2).) In this case,

$$c = [(1 + N)^{m_b} \cdot r^N \bmod N^2]$$

for uniform $r \in \mathbb{Z}_{N^2}^*$. Recalling that the distribution on $[r^N \bmod N^2]$ is the same whether r is chosen uniformly from $\mathbb{Z}_N^*$ or from $\mathbb{Z}_{N^2}^*$, we see that in this case the view of $\mathcal{A}$ when run as a subroutine by D is distributed identically to $\mathcal{A}$'s view in experiment $\mathsf{PubK}_{\mathcal{A},\Pi}^{\mathsf{eav}}(n)$. Since D outputs 1 exactly when the output b' of $\mathcal{A}$ is equal to b, we have

$$\Pr\left[D(N, [r^N \bmod N^2]) = 1\right] = \Pr[\mathsf{PubK}_{\mathcal{A},\Pi}^{\mathsf{eav}}(n) = 1],$$

where the first probability is taken over the experiment as in Definition 15.10.

Case 2: Say the input to D was generated by running GenModulus(1^n) to obtain (N, p, q) and choosing uniform $y \in \mathbb{Z}_{N^2}^*$. We claim that the view of $\mathcal{A}$

in this case is *independent* of the bit b. This follows because y is a uniform element of the group $\mathbb{Z}_{N^2}^*$, and so the ciphertext c is uniformly distributed in $\mathbb{Z}_{N^2}^*$ (see Lemma 12.15), independent of m. Thus, the probability that $b' = b$ in this case is exactly $\frac{1}{2}$. That is,

$$\Pr[D(N,r) = 1] = \frac{1}{2},$$

where the probability is taken over the experiment as in Definition 15.10.

Combining the above, we see that

$$\left| \Pr\left[D(N, [r^N \bmod N^2]) = 1\right] - \Pr[D(N,r) = 1] \right|$$
$$= \left| \Pr[\mathsf{PubK}^{\mathsf{eav}}_{\mathcal{A},\Pi}(n) = 1] - \tfrac{1}{2} \right|.$$

By the assumption that the decisional composite residuosity problem is hard relative to GenModulus, there is a negligible function negl such that

$$\left| \Pr[\mathsf{PubK}^{\mathsf{eav}}_{\mathcal{A},\Pi}(n) = 1] - \tfrac{1}{2} \right| \leq \mathsf{negl}(n).$$

Thus $\Pr[\mathsf{PubK}^{\mathsf{eav}}_{\mathcal{A},\Pi}(n) = 1] \leq \frac{1}{2} + \mathsf{negl}(n)$, completing the proof. ∎

15.2.3 Homomorphic Encryption

The Paillier encryption scheme is useful in a number of settings because it is *homomorphic*. Roughly, a homomorphic encryption scheme enables (certain) computations to be performed on encrypted data, yielding a ciphertext containing the encrypted result. In the case of Paillier encryption, the computation that can be performed is (modular) *addition*. Specifically, fix a public key $pk = N$. Then the Paillier scheme has the property that multiplying an encryption of m_1 and an encryption of m_2 (with multiplication done modulo N^2) results in an encryption of $[m_1 + m_2 \bmod N]$; this is because

$$\big((1+N)^{m_1} \cdot r_1^N\big) \cdot \big((1+N)^{m_2} \cdot r_2^N\big)$$
$$= (1+N)^{[m_1+m_2 \bmod N]} \cdot (r_1 r_2)^N \bmod N^2.$$

Although the ability to add encrypted values may not seem very useful, it suffices for several interesting applications including voting, discussed below.

We present a general definition, of which Pailler encryption is a special case.

DEFINITION 15.14 *A public-key encryption scheme* (Gen, Enc, Dec) *is homomorphic if for all n and all (pk, sk) output by* Gen(1^n), *it is possible to define groups $\mathcal{M}, \mathcal{C}$ (depending on pk only) such that:*

- *The message space is $\mathcal{M}$, and all ciphertexts output by* Enc$_{pk}$ *are elements of $\mathcal{C}$. For notational convenience, we write $\mathcal{M}$ as an additive group and $\mathcal{C}$ as a multiplicative group.*

- *For any $m_1, m_2 \in \mathcal{M}$, any c_1 output by $\mathsf{Enc}_{pk}(m_1)$, and any c_2 output by $\mathsf{Enc}_{pk}(m_2)$, it holds that*

$$\mathsf{Dec}_{sk}(c_1 \cdot c_2) = m_1 + m_2.$$

Moreover, the distribution on ciphertexts obtained by encrypting m_1, encrypting m_2, and then multiplying the results is identical to the distribution on ciphertexts obtained by encrypting $m_1 + m_2$.

The last part of the definition ensures that if ciphertexts $c_1 \leftarrow \mathsf{Enc}_{pk}(m_1)$ and $c_2 \leftarrow \mathsf{Enc}_{pk}(m_2)$ are generated and the result $c_3 := c_1 \cdot c_2$ is computed, then the resulting ciphertext c_3 contains no more information about m_1 or m_2 than the sum m_3.

The Paillier encryption scheme with $pk = N$ is homomorphic with $\mathcal{M} = \mathbb{Z}_N$ and $\mathcal{C} = \mathbb{Z}_{N^2}^*$. This is not the first example of a homomorphic encryption scheme we have seen; El Gamal encryption is also homomorphic. Specifically, for public key $pk = \langle \mathbb{G}, q, g, h \rangle$ we can take $\mathcal{M} = \mathbb{G}$ and $\mathcal{C} = \mathbb{G} \times \mathbb{G}$; then

$$\langle g^{y_1}, h^{y_1} \cdot m_1 \rangle \cdot \langle g^{y_2}, h^{y_2} \cdot m_2 \rangle = \langle g^{y_1+y_2}, h^{y_1+y_2} \cdot m_1 m_2 \rangle,$$

where multiplication of ciphertexts is component-wise. The Goldwasser–Micali encryption scheme we will see later is also homomorphic (see Exercise 15.11).

A nice feature of Paillier encryption is that it is homomorphic over a large additive group (namely, $\mathbb{Z}_N$). To see an application of this, consider the following distributed voting scheme, where ℓ voters can vote "no" or "yes" and the goal is to tabulate the number of "yes" votes:

1. A voting authority generates a public key N for the Paillier encryption scheme and publicizes N.

2. Let 0 stand for a "no," and let 1 stand for a "yes." Each voter casts their vote by encrypting it. That is, voter i casts her vote v_i by computing $c_i := [(1 + N)^{v_i} \cdot (r_i)^N \bmod N^2]$ for a uniform $r_i \in \mathbb{Z}_N^*$.

3. Each voter broadcasts their vote c_i. These votes are then publicly *aggregated* by computing

$$c_{\text{total}} := \left[\prod_{i=1}^{\ell} c_i \bmod N^2 \right].$$

4. The authority is given c_{total}. (We assume the authority has not been able to observe what goes on until now.) By decrypting it, the authority obtains the vote total

$$v_{\text{total}} \stackrel{\text{def}}{=} \sum_{i=1}^{\ell} v_i \bmod N.$$

If ℓ is small (so that $v_{\text{total}} \ll N$), there is no wrap-around modulo N and $v_{\text{total}} = \sum_{i=1}^{\ell} v_i$.

Key features of the above are that *no voter learns anyone else's vote*, and calculation of the total is *publicly verifiable* if the authority is trusted to correctly compute v_{total} from c_{total}. Also, the authority obtains the correct total *without learning any individual votes*. (Here, we assume the authority cannot see voters' ciphertexts. In Section 15.3.3 we show a protocol in which votes are kept hidden from authorities even if they see all the communication.) We assume all voters act honestly (and only try to learn others' votes based on information they observe); an entire research area of cryptography is dedicated to addressing potential threats from participants who might be malicious and not follow the protocol.

15.3 Secret Sharing and Threshold Encryption

Motivated by the discussion of distributed voting in the previous section, we briefly consider *secure (interactive) protocols*. Such protocols can be significantly more complicated than the basic cryptographic primitives (e.g., encryption and signature schemes) we have focused on until now, both because they can involve multiple parties exchanging several rounds of messages, as well as because they are intended to realize more-complex security requirements.

The goal of this section is mainly to give the reader a taste of this fascinating area, and no attempt is made at being comprehensive or complete. Although the protocols presented here can be proven secure (with respect to appropriate definitions), we omit formal definitions, details, and proofs and instead rely on informal discussion.

15.3.1 Secret Sharing

Consider the following problem. A *dealer* holds a *secret* $s \in \{0,1\}^{\ell}$— say, a nuclear-launch code—that it wishes to share among some set of N users $P_1, \ldots, P_N$ by giving each user a *share*. Any t users should be able to pool their shares and reconstruct the secret, but no coalition of fewer than t users should get *any* information about s from their collective shares (beyond whatever information they had about s already). We refer to such a sharing mechanism as a (t, N)-*threshold secret-sharing scheme*. Such a scheme ensures that s is not revealed without sufficient authorization, while also guaranteeing availability of s when needed (since any t users can reconstruct it). Beyond their direct application, secret-sharing schemes are also a building block of many cryptographic protocols.

There is a simple solution for the case $t = N$. The dealer chooses uniform $s_1, \ldots, s_{N-1} \in \{0,1\}^{\ell}$ and sets $s_N := s \oplus \left(\bigoplus_{i=1}^{N-1} s_i \right)$; the share of user P_i is s_i. Since $\bigoplus_{i=1}^{N} s_i = s$ by construction, clearly all the users together can recover s. However, the shares of any coalition of $N-1$ users are (jointly) uniform and independent of s, and thus reveal no information about s. This is clear when the coalition is $P_1, \ldots, P_{N-1}$. In the general case, when the coalition includes everyone except for P_j ($j \neq N$), this is true because $s_1, \ldots, s_{j-1}, s_{j+1}, \ldots, s_{N-1}$ are uniform and independent of s by construction, and

$$s_N = s \oplus \left(\bigoplus_{i < N, i \neq j} s_i \right) \oplus s_j;$$

thus, even conditioned on some fixed values for s and the shares of the other members of the coalition, the share s_N of user P_N is uniform because s_j is uniform and independent of s.

We can extend this to obtain a solution for $t < N$. The basic idea is to replicate the above scheme for each subset $T \subset N$ of size t. That is, for each such subset $T = \{P_{i_1}, \ldots, P_{i_t}\}$, we choose uniform shares $s_{T,i_1}, \ldots, s_{T,i_t}$ subject to the constraint that $\oplus_{j=1}^{t} s_{T,i_j} = s$, and give s_{T,i_j} to user P_{i_j}. It is not hard to see that this satisfies the requirements.

Unfortunately, this extension of the original scheme is not *efficient*. Each user now stores a share $s_{T,i}$ for *every* subset T of which she is a member. For each user there are $\binom{N-1}{t-1}$ such subsets, which is exponential in N if $t \approx N/2$.

Shamir's scheme. Fortunately, it is possible to do significantly better using a secret-sharing scheme introduced by Adi Shamir (of RSA fame). This scheme is based on polynomials[2] over a finite field $\mathbb{F}$, where $\mathbb{F}$ is chosen so that $s \in \mathbb{F}$ and $|\mathbb{F}| > N$. (See Appendix A.5 for a brief discussion of finite fields.) Before describing the scheme, we briefly review some background related to polynomials over a field $\mathbb{F}$.

A value $x \in \mathbb{F}$ is a *root* of a polynomial p if $p(x) = 0$. We use the well-known fact that any nonzero, degree-t polynomial over a field has at most t roots. This implies:

COROLLARY 15.15 *Any two distinct degree-t polynomials p and q agree on at most t points.*

PROOF If not, then the nonzero, degree-t polynomial $p - q$ would have more than t roots. ∎

Shamir's scheme relies on the fact that for any t pairs of elements (x_1, y_1), $\ldots, (x_t, y_t)$ from $\mathbb{F}$ (with the $\{x_i\}$ distinct), there is a *unique* polynomial p of degree $t - 1$ such that $p(x_i) = y_i$ for $1 \leq i \leq t$. We can prove this quite easily. The fact that there exists such a p uses standard polynomial interpolation.

---

[2] A degree-t polynomial p over $\mathbb{F}$ is given by $p(X) = \sum_{i=0}^{t} a_i X^i$, where $a_i \in \mathbb{F}$ and X is a formal variable. (Note that we allow $a_t = 0$ and so we really mean a polynomial of degree *at most t*.) Any such polynomial naturally defines a function mapping $\mathbb{F}$ to itself, given by evaluating the polynomial on its input.

In detail: for $i = 1, \ldots, t$, define the degree-$(t - 1)$ polynomial

$$\delta_i(X) \overset{\text{def}}{=} \frac{\prod_{j=1, j \neq i}^{t}(X - x_j)}{\prod_{j=1, j \neq i}^{t}(x_i - x_j)}.$$

Note that $\delta_i(x_j) = 0$ for any $j \neq i$, and $\delta_i(x_i) = 1$. So $p(X) \overset{\text{def}}{=} \sum_{i=1}^{t} \delta_i(X) \cdot y_i$ is a polynomial of degree $(t - 1)$ with $p(x_i) = y_i$ for $1 \leq i \leq t$. (We remark that this, in fact, demonstrates that the desired polynomial p can be found *efficiently*.) Uniqueness follows from Corollary 15.15.

We now describe Shamir's (t, N)-threshold secret-sharing scheme. Let $\mathbb{F}$ be a finite field that contains the domain of possible secrets, and with $|\mathbb{F}| > N$. Let $x_1, \ldots, x_N \in \mathbb{F}$ be distinct, nonzero elements that are fixed and publicly known. (Such elements exist since $|\mathbb{F}| > N$.) The scheme works as follows:

Sharing: Given a secret $s \in \mathbb{F}$, the dealer chooses uniform $a_1, \ldots, a_{t-1} \in \mathbb{F}$ and defines the polynomial $p(X) \overset{\text{def}}{=} s + \sum_{i=1}^{t-1} a_i X^i$. This is a uniform degree-$(t - 1)$ polynomial with constant term s. The share of user P_i is $s_i := p(x_i) \in \mathbb{F}$.

Reconstruction: Say t users $P_{i_1}, \ldots, P_{i_t}$ pool their shares $s_{i_1}, \ldots, s_{i_t}$. Using polynomial interpolation, they compute the unique degree-$(t - 1)$ polynomial p' for which $p'(x_{i_j}) = s_{i_j}$ for $1 \leq j \leq t$. The secret is $p'(0)$.

It is clear that reconstruction works since $p' = p$ and $p(0) = s$.

It remains to show that any $t - 1$ users learn nothing about the secret s from their shares. By symmetry, it suffices to consider the shares of users $P_1, \ldots, P_{t-1}$. We claim that for *any* secret s, the shares $s_1, \ldots, s_{t-1}$ are (jointly) uniform. Since the dealer chooses $a_1, \ldots, a_{t-1}$ uniformly, this follows if we show that there is a one-to-one correspondence between the polynomial p chosen by the dealer and the shares $s_1, \ldots, s_{t-1}$. But this is a direct consequence of Corollary 15.15.

15.3.2 Verifiable Secret Sharing

So far we have considered *passive* attacks in which $t - 1$ users may try to use their shares to learn information about the secret. But we may also be concerned about *active*, malicious behavior. Here there are two separate concerns: First, a corrupted *dealer* may give inconsistent shares to the users, i.e., such that different secrets are recovered depending on which t users pool their shares. Second, in the reconstruction phase a malicious user may present a *different* share from the one given to them by the dealer, and thus affect the recovered secret. (While this could be addressed by having the dealer sign the shares, this does not work when the dealer itself may be dishonest.) *Verifiable secret-sharing (VSS) schemes* prevent both these attacks.

More formally, we allow any $t - 1$ users to be corrupted and to collude with each other and, possibly, the dealer. We require: (1a) at the end of the sharing phase, a secret s is defined such that any set of users that includes t uncorrupted users will successfully recover s in the reconstruction phase; moreover, (1b) if the dealer is honest, then s corresponds to the dealer's secret. In addition, (2) when the dealer is honest then, as before, any set of $t - 1$ corrupted users learns nothing about the secret from their shares and any public information the dealer publishes. Since we want there to be t uncorrupted users even if $t-1$ users are corrupted, we require $N \geq t+(t-1)$ or $N > 2t$; in other words, we assume a majority of the users remain uncorrupted.

We describe a VSS scheme due to Feldman that relies on an algorithm $\mathcal{G}$ relative to which the discrete-logarithm problem is hard. For simplicity, we describe it in the random-oracle model and let H denote a function to be modeled as a random oracle. We also assume that some trusted parameters $(\mathbb{G}, q, g)$, generated using $\mathcal{G}(1^n)$, are published in advance, where q is prime and so $\mathbb{Z}_q$ is a field. Finally, we assume that all users have access to a broadcast channel, such that a message broadcast by any user is heard by everyone.

The sharing phase now involves the N users running an interactive protocol with the dealer that proceeds as follows:

1. To share a secret s, the dealer chooses uniform $a_0 \in \mathbb{Z}_q$ and then shares a_0 as in Shamir's scheme. That is, the dealer chooses uniform $a_1, \ldots, a_{t-1} \in \mathbb{Z}_q$ and defines the polynomial $p(X) \overset{\text{def}}{=} \sum_{j=0}^{t-1} a_j X^j$. The dealer sends the share $s_i := p(i) = \sum_{j=0}^{t-1} a_j \cdot i^j$ to user P_i.[3]

   In addition, the dealer publicly broadcasts the values $A_0 := g^{a_0}$, $\ldots$, $A_{t-1} := g^{a_{t-1}}$, and the "masked secret" $c := H(a_0) \oplus s$.

2. Each user P_i verifies that its share s_i satisfies

$$g^{s_i} \overset{?}{=} \prod_{j=0}^{t-1} (A_j)^{i^j}. \tag{15.3}$$

   If not, P_i publicly broadcasts a complaint.

   Note that if the dealer is honest, we have

$$\prod_{j=0}^{t-1} (A_j)^{i^j} = \prod_{j=0}^{t-1} (g^{a_j})^{i^j} = g^{\sum_{j=0}^{t-1} a_j \cdot i^j} = g^{p(i)} = g^{s_i},$$

   and so no honest user will complain. Since there are at most $t - 1$ corrupted users, there are at most $t-1$ complaints if the dealer is honest.

3. If more than $t - 1$ users complain, the dealer is disqualified and the protocol is aborted. Otherwise, the dealer responds to a complaint from P_i by broadcasting s_i. If this share does not satisfy Equation (15.3) (or if the dealer refuses to respond to a complaint at all), the dealer is disqualified and the protocol is aborted. Otherwise, P_i uses the broadcast value (rather than the value it received in the first round) as its share.

---

[3]Note that we are now setting $x_i = i$, which is fine since we are using the field $\mathbb{Z}_q$.

In the reconstruction phase, say a group of users (that includes at least t uncorrupted users) pool their shares. A share s_i provided by a user P_i is discarded if it does not satisfy Equation (15.3). Among the remaining shares, any t of them are used to recover a_0 exactly as in Shamir's scheme. The original secret is then computed as $s := c \oplus H(a_0)$.

We now argue that this protocol meets the desired security requirements. We first show that, assuming the dealer is not disqualified, the value recovered in the reconstruction phase is uniquely determined by the public information; specifically, the recovered value is $c \oplus H(\log_g A_0)$. (Combined with the fact that an honest dealer is never disqualified, this proves that conditions (1a) and (1b) hold.) Define $a_i := \log_g A_i$ for $0 \le i \le t - 1$; the $\{a_i\}$ cannot be computed efficiently if the discrete-logarithm problem is hard, but they are still well-defined. Define the polynomial $p(X) \stackrel{\text{def}}{=} \sum_{i=0}^{t-1} a_i X^i$. Any share s_i, contributed by party P_i, that is not discarded during the reconstruction phase must satisfy Equation (15.3), and hence satisfies $s_i = p(i)$. It follows that, regardless of which shares are used, the parties will reconstruct polynomial p, compute $a_0 = p(0)$, and then recover $s = c \oplus H(a_0)$.

It is also possible to show that condition (2) holds for computationally bounded adversaries if the discrete-logarithm problem is hard for $\mathcal{G}$. (In contrast to Shamir's secret-sharing scheme, secrecy here is no longer unconditional. Unconditionally secure VSS schemes are possible, but are beyond the scope of our treatment.) Intuitively, this is because the secret s is masked by the random value $H(a_0)$, and the information given to any $t - 1$ users in the sharing phase—namely, their shares and the public values $\{A_i\}$—reveals only g^{a_0}, from which it is hard to compute a_0. This intuition can be made rigorous, but we do not do so here.

15.3.3 Threshold Encryption and Electronic Voting

In Section 15.2.3 we introduced the notion of homomorphic encryption schemes and gave the Paillier encryption scheme as an example. Here we show a different homomorphic encryption scheme that is a variant of El Gamal encryption. Specifically, given a public key $pk = \langle \mathbb{G}, q, g, h \rangle$ as in regular El Gamal encryption, we now encrypt a message $m \in \mathbb{Z}_q$ by setting $M := g^m$, choosing a uniform $y \in \mathbb{Z}_q$, and sending the ciphertext $c := \langle g^y, h^y \cdot M \rangle$. To decrypt, the receiver recovers M as in standard El Gamal decryption and then computes $m := \log_g M$. Although this is not efficient if m comes from a large domain, if m is from a small domain—as it will be in our application—then the receiver can compute $\log_g M$ efficiently using exhaustive search. The advantage of this variant scheme is that it is homomorphic with respect to *addition* in $\mathbb{Z}_q$. That is,

$$\langle g^{y_1}, h^{y_1} \cdot g^{m_1} \rangle \cdot \langle g^{y_2}, h^{y_2} \cdot g^{m_2} \rangle = \langle g^{y_1+y_2}, h^{y_1+y_2} \cdot g^{m_1+m_2} \rangle.$$

Recall that the basic approach to electronic voting using homomorphic encryption has each voter i encrypt her vote $v_i \in \{0, 1\}$ to obtain a ciphertext c_i. Once everyone has voted, the ciphertexts are multiplied to obtain an encryption of the sum $v_{total} \stackrel{\text{def}}{=} \sum_i v_i \bmod q = \sum_i v_i$. (The value q is, in practice, large enough so that no wrap-around modulo q occurs.) Since $0 \leq v_{total} \leq \ell$, where ℓ is the total number of voters, an authority with the private key can efficiently decrypt the final ciphertext and recover v_{total}.

A drawback of this approach is that the authority is trusted, both to (correctly) decrypt the final ciphertext as well as not to decrypt any of the individual voters' ciphertexts. (In Section 15.2.3 we assumed the authority could not see the individual voters' ciphertexts.) We might instead prefer to *distribute* trust among a set of N authorities, such that any set of t authorities is able to jointly decrypt an agreed-upon ciphertext (this ensures availability even if some authorities are down or unwilling to help decrypt), but no collection of $t - 1$ authorities is able to decrypt any ciphertext on their own (this ensures privacy as long as fewer than t authorities are corrupted).

At first glance, it may seem that secret sharing solves the problem. If we share the private key among the N authorities, then no set of $t-1$ authorities learns the private key and so they cannot decrypt. On the other hand, any t authorities can pool their shares, recover the private key, and then decrypt any desired ciphertext.

A little thought shows that this does not quite work. If the authorities reconstruct the private key in order to decrypt some ciphertext, then as part of this process *all the authorities learn the private key*! Thus, afterward, any authority could decrypt any ciphertext of its choice, on its own.

We need instead a modified approach in which the "secret" (namely, the private key) is never reconstructed in the clear, yet is implicitly reconstructed only enough to enable decryption of one, agreed-upon ciphertext. We can achieve this for the specific case of El Gamal encryption in the following way. Fix a public key $pk = \langle \mathbb{G}, q, g, h \rangle$, and let $x \in \mathbb{Z}_q$ be the private key, i.e., $g^x = h$. Each authority is given a share $x_i \in \mathbb{Z}_q$ exactly as in Shamir's secret-sharing scheme. That is, a uniform degree-$(t-1)$ polynomial p with $p(0) = x$ is chosen, and the ith authority is given $x_i := p(i)$. (We assume a trusted dealer who knows x and securely deletes it once it is shared. It is possible to eliminate the dealer entirely, but this is beyond our present scope.)

Now, say some t authorities $i_1, \ldots, i_t$ wish to jointly decrypt a ciphertext $\langle c_1, c_2 \rangle$. To do so, authority i_j first publishes the value $w_j := c_1^{x_{i_j}}$. Recall from the previous section that there exist publicly computable polynomials $\{\delta_j(X)\}$ (that depend on the identities of these t authorities) such that $p(X) \stackrel{\text{def}}{=} \sum_{j=1}^t \delta_j(X) \cdot x_{i_j}$. Setting $\delta_j \stackrel{\text{def}}{=} \delta_j(0)$, we see that there exist publicly computable values $\delta_1, \ldots, \delta_t \in \mathbb{Z}_q$ for which $x = p(0) = \sum_{j=1}^t \delta_j \cdot x_{i_j}$. Any authority can then compute

$$M' := \frac{c_2}{\prod_{j=1}^t w_j^{\delta_j}}.$$

(They can then each compute $\log_g M$, if desired.) To see that this correctly recovers the message, say $c_1 = g^y$ and $c_2 = h^y \cdot M$. Then

$$\prod_{j=1}^{t} w_j^{\delta_j} = \prod_{j=1}^{t} c_1^{x_{i_j}\delta_j} = c_1^{\sum_{j=1}^{t} x_{i_j}\delta_j} = c_1^{p(0)} = c_1^x,$$

and so

$$M' \stackrel{\text{def}}{=} \frac{c_2}{\prod_{j=1}^{t} w_j^{\delta_j}} = \frac{h^y \cdot M}{c_1^x} = \frac{(g^x)^y \cdot M}{(g^y)^x} = M.$$

Note that any set of $t - 1$ corrupted authorities learns nothing about the private key x from their shares. Moreover, it is possible to show that they learn nothing from the decryption process beyond the recovered value M.

Malicious (active) adversaries. Our treatment above assumes that the authorities decrypting some ciphertext all behave correctly. (If they do not, it would be easy for any of them to cause an incorrect result by publishing an arbitrary value w_j.) We also assume that voters behave honestly, and encrypt a vote of either 0 or 1. (Note that a voter could unfairly sway the election by encrypting a large value or a negative value.) Potential malicious behavior of this sort can be prevented using techniques beyond the scope of this book.

15.4   The Goldwasser–Micali Encryption Scheme

Before we present the Goldwasser–Micali encryption scheme, we need to develop a better understanding of *quadratic residues*. We first explore the easier case of quadratic residues modulo a prime p, and then look at the slightly more complicated case of quadratic residues modulo a composite N.

Throughout this section, p and q denote odd primes, and $N = pq$ denotes a product of two distinct, odd primes.

15.4.1   Quadratic Residues Modulo a Prime

In a group $\mathbb{G}$, an element $y \in \mathbb{G}$ is a *quadratic residue* if there exists an $x \in \mathbb{G}$ with $x^2 = y$. In this case, we call x a *square root* of y. An element that is not a quadratic residue is called a *quadratic non-residue*. In an abelian group, the set of quadratic residues forms a subgroup.

In the specific case of $\mathbb{Z}_p^*$, we have that y is a quadratic residue if there exists an x with $x^2 = y \bmod p$. We begin with an easy observation.

PROPOSITION 15.16   *Let $p > 2$ be prime. Every quadratic residue in $\mathbb{Z}_p^*$ has exactly two square roots.*

PROOF   This follows from Theorem 9.66, but we give a direct proof here. Let $y \in \mathbb{Z}_p^*$ be a quadratic residue. Then there exists an $x \in \mathbb{Z}_p^*$ such that $x^2 = y \bmod p$. Clearly, $(-x)^2 = x^2 = y \bmod p$. Furthermore, $-x \neq x \bmod p$: if $-x = x \bmod p$ then $2x = 0 \bmod p$, which implies $p \mid 2x$. Since p is prime, this would mean that either $p \mid 2$ (which is impossible since $p > 2$) or $p \mid x$ (which is impossible since $0 < x < p$). So, $[x \bmod p]$ and $[-x \bmod p]$ are distinct elements of $\mathbb{Z}_p^*$, and y has at least two square roots.

Let $x' \in \mathbb{Z}_p^*$ be a square root of y. Then $x^2 = y = (x')^2 \bmod p$, implying that $x^2 - (x')^2 = 0 \bmod p$. Factoring the left-hand side we obtain

$$(x - x')(x + x') = 0 \bmod p,$$

so that (by Proposition 9.3) either $p \mid (x - x')$ or $p \mid (x + x')$. In the first case, $x' = x \bmod p$ and in the second case $x' = -x \bmod p$, showing that y indeed has only $[\pm x \bmod p]$ as square roots.   ∎

Let $\mathsf{sq}_p : \mathbb{Z}_p^* \to \mathbb{Z}_p^*$ be the function $\mathsf{sq}_p(x) \stackrel{\mathrm{def}}{=} [x^2 \bmod p]$. The above shows that sq_p is a two-to-one function when $p > 2$ is prime. This immediately implies that *exactly half the elements of $\mathbb{Z}_p^*$ are quadratic residues*. We denote the set of quadratic residues modulo p by $\mathcal{QR}_p$, and the set of quadratic non-residues by $\mathcal{QNR}_p$. We have just seen that for $p > 2$ prime

$$|\mathcal{QR}_p| = |\mathcal{QNR}_p| = \frac{|\mathbb{Z}_p^*|}{2} = \frac{p-1}{2}.$$

Define $\mathcal{J}_p(x)$, the *Jacobi symbol of x modulo p*, as follows.[4] Let $p > 2$ be prime, and $x \in \mathbb{Z}_p^*$. Then

$$\mathcal{J}_p(x) \stackrel{\mathrm{def}}{=} \begin{cases} +1 \text{ if } x \text{ is a quadratic residue modulo } p \\ -1 \text{ if } x \text{ is } not \text{ a quadratic residue modulo } p. \end{cases}$$

The notation can be extended in the natural way for any x relatively prime to p by setting $\mathcal{J}_p(x) \stackrel{\mathrm{def}}{=} \mathcal{J}_p([x \bmod p])$.

Can we characterize the quadratic residues in $\mathbb{Z}_p^*$? We begin with the fact that $\mathbb{Z}_p^*$ is a cyclic group of order $p-1$ (see Theorem 9.57). Let g be a generator of $\mathbb{Z}_p^*$. This means that

$$\mathbb{Z}_p^* = \{g^0, g^1, g^2, \dots, g^{\frac{p-1}{2}-1}, g^{\frac{p-1}{2}}, g^{\frac{p-1}{2}+1}, \dots, g^{p-2}\}$$

(recall that p is odd, so $p-1$ is even). Squaring each element in this list and reducing modulo $p-1$ in the exponent (cf. Corollary 9.15) yields a list of all the quadratic residues in $\mathbb{Z}_p^*$:

$$\mathcal{QR}_p = \{g^0, g^2, g^4, \dots, g^{p-3}, g^0, g^2, \dots, g^{p-3}\}.$$

---

[4]For p prime, $\mathcal{J}_p(x)$ is also sometimes called the *Legendre symbol* of x and denoted by $\mathcal{L}_p(x)$; we have chosen our notation to be consistent with notation introduced later.

Each quadratic residue appears twice in this list. Therefore, the quadratic residues in $\mathbb{Z}_p^*$ are exactly those elements that can be written as g^i with $i \in \{0, \ldots, p-2\}$ an *even* integer.

The above characterization leads to a simple way to compute the Jacobi symbol and thus tell whether an element $x \in \mathbb{Z}_p^*$ is a quadratic residue or not.

PROPOSITION 15.17 *Let $p > 2$ be a prime. Then $\mathcal{J}_p(x) = x^{\frac{p-1}{2}} \bmod p$.*

PROOF Let g be an arbitrary generator of $\mathbb{Z}_p^*$. If x is a quadratic residue modulo p, our earlier discussion shows that $x = g^i$ for some even integer i. Writing $i = 2j$ with j an integer we then have

$$x^{\frac{p-1}{2}} = \left(g^{2j}\right)^{\frac{p-1}{2}} = g^{(p-1)j} = \left(g^{p-1}\right)^j = 1^j = 1 \bmod p,$$

and so $x^{\frac{p-1}{2}} = +1 = \mathcal{J}_p(x) \bmod p$ as claimed.

On the other hand, if x is not a quadratic residue then $x = g^i$ for some odd integer i. Writing $i = 2j + 1$ with j an integer, we have

$$x^{\frac{p-1}{2}} = \left(g^{2j+1}\right)^{\frac{p-1}{2}} = \left(g^{2j}\right)^{\frac{p-1}{2}} \cdot g^{\frac{p-1}{2}} = 1 \cdot g^{\frac{p-1}{2}} = g^{\frac{p-1}{2}} \bmod p.$$

Now,

$$\left(g^{\frac{p-1}{2}}\right)^2 = g^{p-1} = 1 \bmod p,$$

and so $g^{\frac{p-1}{2}} = \pm 1 \bmod p$ since $[\pm 1 \bmod p]$ are the two square roots of 1 (cf. Proposition 15.16). Since g is a generator, it has order $p - 1$ and so $g^{\frac{p-1}{2}} \neq 1 \bmod p$. It follows that $x^{\frac{p-1}{2}} = -1 = \mathcal{J}_p(x) \bmod p$. ∎

Proposition 15.17 directly gives a polynomial-time algorithm (cf. Algorithm 15.18) for testing whether an element $x \in \mathbb{Z}_p^*$ is a quadratic residue.

ALGORITHM 15.18
Deciding quadratic residuosity modulo a prime

Input: A prime p; an element $x \in \mathbb{Z}_p^*$
Output: $\mathcal{J}_p(x)$ (or, equivalently, whether x is a quadratic residue or quadratic non-residue)

$b := \left[x^{\frac{p-1}{2}} \bmod p\right]$
if $b = 1$ **return** "quadratic residue"
else return "quadratic non-residue"

We conclude this section by noting a nice multiplicative property of quadratic residues and non-residues modulo p.

PROPOSITION 15.19   *Let $p > 2$ be a prime, and $x, y \in \mathbb{Z}_p^*$. Then*

$$\mathcal{J}_p(xy) = \mathcal{J}_p(x) \cdot \mathcal{J}_p(y).$$

PROOF   Using the previous proposition,

$$\mathcal{J}_p(xy) = (xy)^{\frac{p-1}{2}} = x^{\frac{p-1}{2}} \cdot y^{\frac{p-1}{2}} = \mathcal{J}_p(x) \cdot \mathcal{J}_p(y) \bmod p.$$

Since $\mathcal{J}_p(xy), \mathcal{J}_p(x), \mathcal{J}_p(y) = \pm 1$, equality holds over the integers as well. ∎

COROLLARY 15.20   *Let $p > 2$ be prime, and say $x, x' \in \mathcal{QR}_p$ and $y, y' \in \mathcal{QNR}_p$. Then:*

1. *$[xx' \bmod p] \in \mathcal{QR}_p$.*

2. *$[yy' \bmod p] \in \mathcal{QR}_p$.*

3. *$[xy \bmod p] \in \mathcal{QNR}_p$.*

15.4.2   Quadratic Residues Modulo a Composite

We now turn our attention to quadratic residues in the group $\mathbb{Z}_N^*$, where $N = pq$. Characterizing the quadratic residues modulo N is easy if we use the results of the previous section in conjunction with the Chinese remainder theorem. Recall that the Chinese remainder theorem says that $\mathbb{Z}_N^* \simeq \mathbb{Z}_p^* \times \mathbb{Z}_q^*$, and we let $y \leftrightarrow (y_p, y_q)$ denote the correspondence guaranteed by the theorem (i.e., $y_p = [y \bmod p]$ and $y_q = [y \bmod q]$). The key observation is:

PROPOSITION 15.21   *Let $N = pq$ with p, q distinct primes, and $y \in \mathbb{Z}_N^*$ with $y \leftrightarrow (y_p, y_q)$. Then y is a quadratic residue modulo N if and only if y_p is a quadratic residue modulo p and y_q is a quadratic residue modulo q.*

PROOF   If y is a quadratic residue modulo N then, by definition, there exists an $x \in \mathbb{Z}_N^*$ such that $x^2 = y \bmod N$. Let $x \leftrightarrow (x_p, x_q)$. Then

$$(y_p, y_q) \leftrightarrow y = x^2 \leftrightarrow (x_p, x_q)^2 = ([x_p^2 \bmod p], [x_q^2 \bmod q]),$$

where $(x_p, x_q)^2$ is simply the square of the element (x_p, x_q) in the group $\mathbb{Z}_p^* \times \mathbb{Z}_q^*$. We have thus shown that:

$$y_p = x_p^2 \bmod p \quad \text{and} \quad y_q = x_q^2 \bmod q \qquad (15.4)$$

and y_p, y_q are quadratic residues (with respect to the appropriate moduli).

Conversely, if $y \leftrightarrow (y_p, y_q)$ and y_p, y_q are quadratic residues modulo p and q, respectively, then there exist $x_p \in \mathbb{Z}_p^*$ and $x_q \in \mathbb{Z}_q^*$ such that Equation (15.4) holds. Let $x \in \mathbb{Z}_N^*$ be such that $x \leftrightarrow (x_p, x_q)$. Reversing the above steps shows that x is a square root of y modulo N. ∎

The above proposition characterizes the quadratic residues modulo N. A careful examination of the proof yields another important observation: each quadratic residue $y \in \mathbb{Z}_N^*$ has exactly *four* square roots. To see this, let $y \leftrightarrow (y_p, y_q)$ be a quadratic residue modulo N and let x_p, x_q be square roots of y_p and y_q modulo p and q, respectively. Then the four square roots of y are given by the elements in $\mathbb{Z}_N^*$ corresponding to:

$$(x_p,\ x_q),\quad (-x_p,\ x_q),\quad (x_p,\ -x_q),\quad (-x_p,\ -x_q). \tag{15.5}$$

Each of these is a square root of y since

$$(\pm x_p,\ \pm x_q)^2 = \left([(\pm x_p)^2 \bmod p],\ [(\pm x_q)^2 \bmod q] \right)$$
$$= ([x_p^2 \bmod p],\ [x_q^2 \bmod q]) = (y_p,\ y_q) \leftrightarrow y$$

(where again the notation $(\cdot, \cdot)^2$ refers to squaring in the group $\mathbb{Z}_p \times \mathbb{Z}_q$). The Chinese remainder theorem guarantees that the four elements in Equation (15.5) correspond to *distinct* elements of $\mathbb{Z}_N^*$, since x_p and $-x_p$ are unique modulo p (and similarly for x_q and $-x_q$ modulo q).

Example 15.22
Consider $\mathbb{Z}_{15}^*$ (the correspondence given by the Chinese remainder theorem is tabulated in Example 9.25). Element 4 is a quadratic residue modulo 15 with square root 2. Since $2 \leftrightarrow (2,2)$, the other square roots of 4 are given by

- $(2, [-2 \bmod 3]) = (2,1) \leftrightarrow 7$;

- $([-2 \bmod 5], 2) = (3,2) \leftrightarrow 8$; and

- $([-2 \bmod 5], [-2 \bmod 3]) = (3,1) \leftrightarrow 13$.

One can verify that $7^2 = 8^2 = 13^2 = 4 \bmod 15$. ◇

Let $\mathcal{QR}_N$ denote the set of quadratic residues modulo N. Since squaring modulo N is a four-to-one function, we see that exactly $1/4$ of the elements of $\mathbb{Z}_N^*$ are quadratic residues. Alternately, we could note that since $y \in \mathbb{Z}_N^*$ is a quadratic residue if and only if y_p, y_q are quadratic residues, there is a one-to-one correspondence between $\mathcal{QR}_N$ and $\mathcal{QR}_p \times \mathcal{QR}_q$. Thus, the fraction of quadratic residues modulo N is,

$$\frac{|\mathcal{QR}_N|}{|\mathbb{Z}_N^*|} = \frac{|\mathcal{QR}_p| \cdot |\mathcal{QR}_q|}{|\mathbb{Z}_N^*|} = \frac{\frac{p-1}{2} \cdot \frac{q-1}{2}}{(p-1)(q-1)} = \frac{1}{4},$$

in agreement with the above.

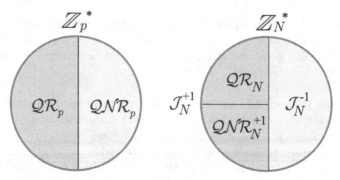

FIGURE 15.1:   The structure of $\mathbb{Z}_p^*$ and $\mathbb{Z}_N^*$.

In the previous section, we defined the Jacobi symbol $\mathcal{J}_p(x)$ for $p > 2$ prime. We extend the definition to the case of N a product of distinct, odd primes p and q as follows. For any x relatively prime to $N = pq$,

$$\mathcal{J}_N(x) \stackrel{\text{def}}{=} \mathcal{J}_p(x) \cdot \mathcal{J}_q(x)$$
$$= \mathcal{J}_p([x \bmod p]) \cdot \mathcal{J}_q([x \bmod q]).$$

We define $\mathcal{J}_N^{+1}$ as the set of elements in $\mathbb{Z}_N^*$ having Jacobi symbol $+1$, and define $\mathcal{J}_N^{-1}$ analogously.

We know from Proposition 15.21 that if x is a quadratic residue modulo N, then $[x \bmod p]$ and $[x \bmod q]$ are quadratic residues modulo p and q, respectively; that is, $\mathcal{J}_p(x) = \mathcal{J}_q(x) = +1$. So $\mathcal{J}_N(x) = +1$ and we see that:

If x is a quadratic residue modulo N, then $\mathcal{J}_N(x) = +1$.

However, $\mathcal{J}_N(x) = +1$ can also occur when $\mathcal{J}_p(x) = \mathcal{J}_q(x) = -1$, that is, when *both* $[x \bmod p]$ and $[x \bmod q]$ are *not* quadratic residues modulo p and q (and so x is not a quadratic residue modulo N). This turns out to be useful for the Goldwasser–Micali encryption scheme, and we therefore introduce the notation $\mathcal{QNR}_N^{+1}$ for the set of elements of this type. That is,

$$\mathcal{QNR}_N^{+1} \stackrel{\text{def}}{=} \left\{ x \in \mathbb{Z}_N^* \;\middle|\; \begin{array}{l} x \text{ is not a quadratic residue modulo } N, \\ \text{but } \mathcal{J}_N(x) = +1 \end{array} \right\}.$$

It is now easy to prove the following (see Figure 15.1):

PROPOSITION 15.23   *Let $N = pq$ with p, q distinct, odd primes. Then:*

1. *Exactly half the elements of $\mathbb{Z}_N^*$ are in $\mathcal{J}_N^{+1}$.*

2. *$\mathcal{QR}_N$ is contained in $\mathcal{J}_N^{+1}$.*

3. *Exactly half the elements of $\mathcal{J}_N^{+1}$ are in $\mathcal{QR}_N$ (the other half are in $\mathcal{QNR}_N^{+1}$).*

PROOF We know that $\mathcal{J}_N(x) = +1$ if either $\mathcal{J}_p(x) = \mathcal{J}_q(x) = +1$ or $\mathcal{J}_p(x) = \mathcal{J}_q(x) = -1$. We also know (from the previous section) that exactly half the elements of $\mathbb{Z}_p^*$ have Jacobi symbol $+1$, and half have Jacobi symbol -1 (and similarly for $\mathbb{Z}_q^*$). Defining $\mathcal{J}_p^{+1}$, $\mathcal{J}_p^{-1}$, $\mathcal{J}_q^{+1}$, and $\mathcal{J}_q^{-1}$ in the natural way, we thus have

$$
\begin{aligned}
\left|\mathcal{J}_N^{+1}\right| &= \left|\mathcal{J}_p^{+1} \times \mathcal{J}_q^{+1}\right| + \left|\mathcal{J}_p^{-1} \times \mathcal{J}_q^{-1}\right| \\
&= \left|\mathcal{J}_p^{+1}\right| \cdot \left|\mathcal{J}_q^{+1}\right| + \left|\mathcal{J}_p^{-1}\right| \cdot \left|\mathcal{J}_q^{-1}\right| \\
&= \frac{(p-1)}{2}\frac{(q-1)}{2} + \frac{(p-1)}{2}\frac{(q-1)}{2} = \frac{\phi(N)}{2}.
\end{aligned}
$$

So $\left|\mathcal{J}_N^{+1}\right| = \left|\mathbb{Z}_N^*\right|/2$, proving that half the elements of $\mathbb{Z}_N^*$ are in $\mathcal{J}_N^{+1}$.

We have noted earlier that all quadratic residues modulo N have Jacobi symbol $+1$, showing that $\mathcal{QR}_N \subseteq \mathcal{J}_N^{+1}$.

Since $x \in \mathcal{QR}_N$ if and only if $\mathcal{J}_p(x) = \mathcal{J}_q(x) = +1$, we have:

$$
|\mathcal{QR}_N| = \left|\mathcal{J}_p^{+1} \times \mathcal{J}_q^{+1}\right| = \frac{(p-1)}{2}\frac{(q-1)}{2} = \frac{\phi(N)}{4},
$$

and so $|\mathcal{QR}_N| = \left|\mathcal{J}_N^{+1}\right|/2$. Since $\mathcal{QR}_N$ is a subset of $\mathcal{J}_N^{+1}$, this proves that half the elements of $\mathcal{J}_N^{+1}$ are in $\mathcal{QR}_N$. ∎

The next two results are analogues of Proposition 15.19 and Corollary 15.20.

PROPOSITION 15.24 *Let $N = pq$ be a product of distinct, odd primes, and $x, y \in \mathbb{Z}_N^*$. Then $\mathcal{J}_N(xy) = \mathcal{J}_N(x) \cdot \mathcal{J}_N(y)$.*

PROOF Using the definition of $\mathcal{J}_N(\cdot)$ and Proposition 15.19:

$$
\begin{aligned}
\mathcal{J}_N(xy) = \mathcal{J}_p(xy) \cdot \mathcal{J}_q(xy) &= \mathcal{J}_p(x) \cdot \mathcal{J}_p(y) \cdot \mathcal{J}_q(x) \cdot \mathcal{J}_q(y) \\
&= \mathcal{J}_p(x) \cdot \mathcal{J}_q(x) \cdot \mathcal{J}_p(y) \cdot \mathcal{J}_q(y) = \mathcal{J}_N(x) \cdot \mathcal{J}_N(y).
\end{aligned}
$$

∎

COROLLARY 15.25 *Let $N = pq$ be a product of distinct, odd primes, and say $x, x' \in \mathcal{QR}_N$ and $y, y' \in \mathcal{QNR}_N^{+1}$. Then:*

1. *$[xx' \bmod N] \in \mathcal{QR}_N$.*

2. *$[yy' \bmod N] \in \mathcal{QR}_N$.*

3. *$[xy \bmod N] \in \mathcal{QNR}_N^{+1}$.*

PROOF   We prove the final claim; proofs of the others are similar. Since $x \in \mathcal{QR}_N$, we have $\mathcal{J}_p(x) = \mathcal{J}_q(x) = +1$. Since $y \in \mathcal{QNR}_N^{+1}$, we have $\mathcal{J}_p(y) = \mathcal{J}_q(y) = -1$. Using Proposition 15.19,

$$\mathcal{J}_p(xy) = \mathcal{J}_p(x) \cdot \mathcal{J}_p(y) = -1 \quad \text{and} \quad \mathcal{J}_q(xy) = \mathcal{J}_q(x) \cdot \mathcal{J}_q(y) = -1,$$

and so $\mathcal{J}_N(xy) = +1$. But xy is not a quadratic residue modulo N, since $\mathcal{J}_p(xy) = -1$ and so $[xy \bmod p]$ is not a quadratic residue modulo p. We conclude that $xy \in \mathcal{QNR}_N^{+1}$. ∎

In contrast to Corollary 15.20, it is *not* true that $y, y' \in \mathcal{QNR}_N$ implies $yy' \in \mathcal{QR}_N$. (Instead, as indicated in the corollary, this is only guaranteed if $y, y' \in \mathcal{QNR}_N^{+1}$.) For example, we could have $\mathcal{J}_p(y) = +1$, $\mathcal{J}_q(y) = -1$ and $\mathcal{J}_p(y') = -1$, $\mathcal{J}_q(y') = +1$, so $\mathcal{J}_p(yy') = \mathcal{J}_q(yy') = -1$ and yy' is not a quadratic residue even though $\mathcal{J}_N(yy') = +1$.

15.4.3   The Quadratic Residuosity Assumption

In Section 15.4.1, we showed an efficient algorithm for deciding whether an input x is a quadratic residue modulo a prime p. Can we adapt the algorithm to work modulo a composite number N? Proposition 15.21 gives an easy solution to this problem *provided the factorization of N is known*. See Algorithm 15.26.

ALGORITHM 15.26
Deciding quadratic residuosity modulo a composite of known factorization

Input: Composite $N = pq$; the factors p and q; element $x \in \mathbb{Z}_N^*$
Output: A decision as to whether $x \in \mathcal{QR}_N$

compute $\mathcal{J}_p(x)$ and $\mathcal{J}_q(x)$
if $\mathcal{J}_p(x) = \mathcal{J}_q(x) = +1$ **return** "quadratic residue"
else return "quadratic non-residue"

(As always, we assume the factors of N are distinct odd primes.) A simple modification of the above algorithm allows for computing $\mathcal{J}_N(x)$ when the factorization of N is known.

When the factorization of N is *unknown*, however, there is no known polynomial-time algorithm for deciding whether a given x is a quadratic residue modulo N or not. Somewhat surprisingly, a polynomial-time algorithm *is* known for computing $\mathcal{J}_N(x)$ without the factorization of N. (Although the algorithm itself is not that complicated, its proof of correctness is beyond the scope of this book and we therefore do not present the algorithm at all. The interested reader can refer to the references listed at the end of this chapter.) This leads to a partial test of quadratic residuosity: if, for a given input x, it holds that $\mathcal{J}_N(x) = -1$, then x cannot possibly be a quadratic residue. (See Proposition 15.23.) This test says nothing when $\mathcal{J}_N(x) = +1$, and there is *no* known polynomial-time algorithm for deciding quadratic residuosity in that case (that does better than random guessing).

We now formalize the assumption that this problem is hard. Let GenModulus be a polynomial-time algorithm that, on input 1^n, outputs (N, p, q) where $N = pq$, and p and q are n-bit primes except with probability negligible in n.

DEFINITION 15.27 *We say* deciding quadratic residuosity is hard relative to GenModulus *if for all probabilistic polynomial-time algorithms D there exists a negligible function* negl *such that*

$$\Big| \Pr[D(N, \mathsf{qr}) = 1] - \Pr[D(N, \mathsf{qnr}) = 1] \Big| \leq \mathsf{negl}(n),$$

where in each case the probabilities are taken over the experiment in which GenModulus(1^n) *is run to give (N, p, q),* qr *is chosen uniformly from $\mathcal{QR}_N$, and* qnr *is chosen uniformly from $\mathcal{QNR}_N^{+1}$.*

It is crucial in the above that qnr is chosen from $\mathcal{QNR}_N^{+1}$ rather than $\mathcal{QNR}_N$; if qnr were chosen from $\mathcal{QNR}_N$ then with probability $2/3$ it would be the case that $\mathcal{J}_N(x) = -1$ and so distinguishing qnr from a uniform quadratic residue would be easy. (Recall that $\mathcal{J}_N(x)$ can be computed efficiently even without the factorization of N.)

The quadratic residuosity assumption is simply the assumption that there exists a GenModulus relative to which deciding quadratic residuosity is hard. It is easy to see that if deciding quadratic residuosity is hard relative to GenModulus, then factoring must be hard relative to GenModulus as well.

15.4.4 The Goldwasser–Micali Encryption Scheme

The preceding section immediately suggests a public-key encryption scheme for single-bit messages based on the quadratic residuosity assumption:

- The public key is a modulus N, and the private key is its factorization.

- To encrypt a '0,' send a uniform quadratic residue; to encrypt a '1,' send a uniform quadratic non-residue with Jacobi symbol $+1$.

- The receiver can decrypt a ciphertext c with its private key by using the factorization of N to decide whether c is a quadratic residue or not.

CPA-security of this scheme follows almost trivially from the hardness of the quadratic residuosity problem as formalized in Definition 15.27.

One thing missing from the above description is a specification of how the sender, who does not know the factorization of N, can choose a uniform element of $\mathcal{QR}_N$ (to encrypt a 0) or a uniform element of $\mathcal{QNR}_N^{+1}$ (to encrypt a 1). The first of these is easy, while the second requires some ingenuity.

Choosing a uniform quadratic residue. Choosing a uniform element $y \in \mathcal{QR}_N$ is easy: simply pick a uniform $x \in \mathbb{Z}_N^*$ (see Appendix B.2.5) and set $y := x^2 \bmod N$. Clearly $y \in \mathcal{QR}_N$. The fact that y is uniformly distributed in $\mathcal{QR}_N$ follows from the facts that squaring modulo N is a 4-to-1 function (see Section 15.4.2) and that x is chosen uniformly from $\mathbb{Z}_N^*$. In more detail, fix any $\hat{y} \in \mathcal{QR}_N$ and let us compute the probability that $y = \hat{y}$ after the above procedure. Denote the four square roots of $\hat{y}$ by $\pm\hat{x}, \pm\hat{x}'$. Then:

$$\Pr[y = \hat{y}] = \Pr[x \text{ is a square root of } \hat{y}]$$
$$= \Pr[x \in \{\pm\hat{x}, \pm\hat{x}'\}]$$
$$= \frac{4}{|\mathbb{Z}_N^*|} = \frac{1}{|\mathcal{QR}_N|}.$$

Since the above holds for every $\hat{y} \in \mathcal{QR}_N$, we see that y is distributed uniformly in $\mathcal{QR}_N$.

Choosing a uniform element of $\mathcal{QNR}_N^{+1}$. In general, it is not known how to choose a uniform element of $\mathcal{QNR}_N^{+1}$ if the factorization of N is unknown. What saves us in the present context is that *the receiver can help* by including certain information in the public key. Specifically, we modify the scheme so that the receiver additionally chooses a uniform $z \in \mathcal{QNR}_N^{+1}$ and includes z as part of its public key. (This is easy for the receiver to do since it knows the factorization of N; see Exercise 15.7.) The sender can choose a uniform element $y \in \mathcal{QNR}_N^{+1}$ by choosing a uniform $x \in \mathbb{Z}_N^*$ (as above) and setting $y := [z \cdot x^2 \bmod N]$. It follows from Corollary 15.25 that $y \in \mathcal{QNR}_N^{+1}$. We leave it as an exercise to show that y is uniformly distributed in $\mathcal{QNR}_N^{+1}$; we do not use this fact directly in the proof of security given below.

We give a complete description of the Goldwasser–Micali encryption scheme, implementing the above ideas, in Construction 15.28.

THEOREM 15.29 *If the quadratic residuosity problem is hard relative to* GenModulus, *then the Goldwasser–Micali encryption scheme is CPA-secure.*

PROOF Let Π denote the Goldwasser–Micali encryption scheme. We prove that Π has indistinguishable encryptions in the presence of an eavesdropper; by Theorem 12.6 this implies that it is CPA-secure.

Let $\mathcal{A}$ be an arbitrary probabilistic polynomial-time adversary. Consider the following PPT adversary D that attempts to solve the quadratic residuosity problem relative to GenModulus:

> **Algorithm D:**
> The algorithm is given N and z as input, and its goal is to determine if $z \in \mathcal{QR}_N$ or $z \in \mathcal{QNR}_N^{+1}$.

CONSTRUCTION 15.28

Let GenModulus be as usual. Construct a public-key encryption scheme as follows:

- **Gen:** on input 1^n, run GenModulus(1^n) to obtain (N, p, q), and choose a uniform $z \in \mathcal{QNR}_N^{+1}$. The public key is $pk = \langle N, z \rangle$ and the private key is $sk = \langle p, q \rangle$.

- **Enc:** on input a public key $pk = \langle N, z \rangle$ and a message $m \in \{0, 1\}$, choose a uniform $x \in \mathbb{Z}_N^*$ and output the ciphertext

$$c := [z^m \cdot x^2 \bmod N].$$

- **Dec:** on input a private key sk and a ciphertext c, determine whether c is a quadratic residue modulo N using, e.g., Algorithm 15.26. If yes, output 0; otherwise, output 1.

The Goldwasser–Micali encryption scheme.

- Set $pk = \langle N, z \rangle$ and run $\mathcal{A}(pk)$ to obtain two single-bit messages m_0, m_1.
- Choose a uniform bit b and a uniform $x \in \mathbb{Z}_N^*$, and then set $c := [z^{m_b} \cdot x^2 \bmod N]$.
- Give the ciphertext c to $\mathcal{A}$, who in turn outputs a bit b'. If $b' = b$, output 1; otherwise, output 0.

Let us analyze the behavior of D. There are two cases to consider:

Case 1: Say the input to D was generated by running GenModulus(1^n) to obtain (N, p, q) and then choosing a uniform $z \in \mathcal{QNR}_N^{+1}$. Then D runs $\mathcal{A}$ on a public key constructed exactly as in Π, and we see that in this case the view of $\mathcal{A}$ when run as a subroutine by D is distributed identically to $\mathcal{A}$'s view in experiment $\mathsf{PubK}_{\mathcal{A},\Pi}^{\mathsf{eav}}(n)$. Since D outputs 1 exactly when the output b' of $\mathcal{A}$ is equal to b, we have

$$\Pr[D(N, \mathsf{qnr}) = 1] = \Pr[\mathsf{PubK}_{\mathcal{A},\Pi}^{\mathsf{eav}}(n) = 1],$$

where qnr represents a uniform element of $\mathcal{QNR}_N^{+1}$ as in Definition 15.27.

Case 2: Say the input to D was generated by running GenModulus(1^n) to obtain (N, p, q) and then choosing a uniform $z \in \mathcal{QR}_N$. We claim that the view of $\mathcal{A}$ in this case is *independent* of the bit b. To see this, note that the ciphertext c given to $\mathcal{A}$ is a uniform quadratic residue regardless of whether a 0 or a 1 is encrypted:

- When a 0 is encrypted, $c = [x^2 \bmod N]$ for a uniform $x \in \mathbb{Z}_N^*$, and so c is a uniform quadratic residue.

- When a 1 is encrypted, $c = [z \cdot x^2 \bmod N]$ for a uniform $x \in \mathbb{Z}_N^*$. Let $\hat{x} \overset{\text{def}}{=} [x^2 \bmod N]$, and note that $\hat{x}$ is a uniformly distributed element of the group $\mathcal{QR}_N$. Since $z \in \mathcal{QR}_N$, we can apply Lemma 12.15 to conclude that c is uniformly distributed in $\mathcal{QR}_N$ as well.

Since $\mathcal{A}$'s view is independent of b, the probability that $b' = b$ in this case is exactly $\frac{1}{2}$. That is,

$$\Pr[D(N, \mathsf{qr}) = 1] = \frac{1}{2},$$

where qr represents a uniform element of $\mathcal{QR}_N$ as in Definition 15.27.

Thus,

$$\left| \Pr[D(N, \mathsf{qr}) = 1] - \Pr[D(N, \mathsf{qnr}) = 1] \right| = \left| \Pr[\mathsf{PubK}^{\mathsf{eav}}_{\mathcal{A}, \Pi}(n) = 1] - \frac{1}{2} \right|.$$

By the assumption that the quadratic residuosity problem is hard relative to GenModulus, there is a negligible function negl such that

$$\left| \varepsilon(n) - \frac{1}{2} \right| \leq \mathsf{negl}(n);$$

thus, $\varepsilon(n) \leq \frac{1}{2} + \mathsf{negl}(n)$. This completes the proof. ∎

15.5   The Rabin Encryption Scheme

As mentioned at the beginning of this chapter, the Rabin encryption scheme is attractive because its security is *equivalent* to the assumption that factoring is hard. An analogous result is *not* known for RSA-based encryption, and the RSA problem may potentially be easier than factoring. (The same is true of the Goldwasser–Micali encryption scheme, and it is possible that deciding quadratic residuosity modulo N is easier than factoring N.)

Interestingly, the Rabin encryption scheme is (superficially, at least) very similar to the RSA encryption scheme yet has the advantage of being based on a potentially weaker assumption. The fact that RSA is more widely used than the former seems to be due more to historical factors than technical ones; we discuss this further at the end of this section.

We begin with some preliminaries about computing modular square roots. We then introduce a trapdoor permutation that can be based directly on the assumption that factoring is hard. The Rabin encryption scheme (or, at least, one instantiation of it) is then obtained by applying the results from Section 15.1. Throughout this section, we continue to let p and q denote odd primes, and let $N = pq$ denote a product of two distinct, odd primes.

15.5.1   Computing Modular Square Roots

The Rabin encryption scheme requires the receiver to compute modular square roots, and so in this section we explore the algorithmic complexity of this problem. We first show an efficient algorithm for computing square roots modulo a prime p, and then extend this algorithm to enable computation of square roots modulo a composite N of *known* factorization. The reader willing to accept the existence of these algorithms on faith can skip to the following section, where we show that computing square roots modulo a composite N with *unknown* factorization is equivalent to factoring N.

Let p be an odd prime. Computing square roots modulo p is relatively simple when $p = 3 \bmod 4$, but much more involved when $p = 1 \bmod 4$. (The easier case is all we need for the Rabin encryption scheme as presented in Section 15.5.3; we include the second case for completeness.) In both cases, we show how to compute one of the square roots of a quadratic residue $a \in \mathbb{Z}_p^*$. Note that if x is one of the square roots of a, then $[-x \bmod p]$ is the other.

We tackle the easier case first. Say $p = 3 \bmod 4$, meaning we can write $p = 4i + 3$ for some integer i. Since $a \in \mathbb{Z}_p^*$ is a quadratic residue, we have $\mathcal{J}_p(a) = 1 = a^{\frac{p-1}{2}} \bmod p$ (see Proposition 15.17). Multiplying both sides by a we obtain:

$$a = a^{\frac{p-1}{2}+1} = a^{2i+2} = \left(a^{i+1}\right)^2 \bmod p \,,$$

and so $a^{i+1} = a^{\frac{p+1}{4}} \bmod p$ is a square root of a. That is, we obtain a square root of a modulo p by simply computing $x := [a^{\frac{p+1}{4}} \bmod p]$.

It is crucial above that $(p+1)/2$ is *even* because this ensures that $(p+1)/4$ is an *integer* (this is necessary in order for $a^{\frac{p+1}{4}} \bmod p$ to be well-defined; recall that the exponent must be an integer). This approach does not succeed when $p = 1 \bmod 4$, in which case $p+1$ is an integer that is *not* divisible by 4.

When $p = 1 \bmod 4$ we proceed slightly differently. Motivated by the above approach, we might hope to find an odd integer r for which it holds that $a^r = 1 \bmod p$. Then, as above, $a^{r+1} = a \bmod p$ and $a^{\frac{r+1}{2}} \bmod p$ would be a square root of a with $(r+1)/2$ an integer. Although we will not be able to do this, we *can* do something just as good: we will find an odd integer r along with an element $b \in \mathbb{Z}_p^*$ and an *even* integer r' such that

$$a^r \cdot b^{r'} = 1 \bmod p.$$

Then $a^{r+1} \cdot b^{r'} = a \bmod p$ and $a^{\frac{r+1}{2}} \cdot b^{\frac{r'}{2}} \bmod p$ is a square root of a (with the exponents $(r+1)/2$ and $r'/2$ being integers).

We now describe the general approach to finding r, b, and r' with the stated properties. Let $\frac{p-1}{2} = 2^\ell \cdot m$ where ℓ, m are integers with $\ell \geq 1$ and m odd.[5] Since a is a quadratic residue, we know that

$$a^{2^\ell m} = a^{\frac{p-1}{2}} = 1 \bmod p. \tag{15.6}$$

This means that $a^{2^\ell m/2} = a^{2^{\ell-1}m} \bmod p$ is a square root of 1. The square roots of 1 modulo p are $\pm 1 \bmod p$, so $a^{2^{\ell-1}m} = \pm 1 \bmod p$. If $a^{2^{\ell-1}m} = 1 \bmod p$, we

---

[5] The integers ℓ and m can be computed easily by taking out factors of 2 from $(p-1)/2$.

are in the same situation as in Equation (15.6) except that the exponent of a is now divisible by a smaller power of 2. This is progress in the right direction: if we can get to the point where the exponent of a is not divisible by *any* power of 2 (as would be the case here if $\ell = 1$), then the exponent of a is *odd* and we can compute a square root as discussed earlier. We give an example, and discuss in a moment how to deal with the case when $a^{2^{\ell-1}m} = -1 \bmod p$.

Example 15.30
Take $p = 29$ and $a = 7$. Since 7 is a quadratic residue modulo 29, we have $7^{14} \bmod 29 = 1$ and we know that $7^7 \bmod 29$ is a square root of 1. In fact,

$$7^7 = 1 \bmod 29,$$

and the exponent 7 is odd. So $7^{(7+1)/2} = 7^4 = 23 \bmod 29$ is a square root of 7 modulo 29.   ◇

To summarize the algorithm so far: we begin with $a^{2^\ell m} = 1 \bmod p$ and we pull out factors of 2 from the exponent until one of two things happen: either $a^m = 1 \bmod p$, or $a^{2^{\ell'}m} = -1 \bmod p$ for some $\ell' < \ell$. In the first case, since m is odd we can immediately compute a square root of a as in Example 15.30. In the second case, we will "restore" the $+1$ on the right-hand side of the equation by multiplying each side of the equation by $-1 \bmod p$. However, as motivated at the beginning of this discussion, we want to achieve this by multiplying the left-hand side of the equation by some element b raised to an *even* power. If we have available a quadratic *non*-residue $b \in \mathbb{Z}_p^*$, this is easy: since $b^{2^\ell m} = b^{\frac{p-1}{2}} = -1 \bmod p$, we have

$$a^{2^{\ell'}m} \cdot b^{2^\ell m} = (-1)(-1) = +1 \bmod p.$$

With this we can proceed as before, taking a square root of the left-hand side to reduce the largest power of 2 dividing the exponent of a, and multiplying by $b^{2^\ell m}$ (as needed) so the right-hand side is always $+1$. Observe that the exponent of b is always divisible by a larger power of 2 than the exponent of a (and so we can indeed take square roots by dividing by 2 in both exponents). We continue performing these steps until the exponent of a is odd, and can then compute a square root of a as described earlier. Pseudocode for this algorithm, which gives another way of viewing what is going on, is given below in Algorithm 15.31. It can be verified that the algorithm runs in polynomial time given a quadratic non-residue b since the number of iterations of the inner loop is $\ell = \mathcal{O}(\log p)$.

One point we have not yet addressed is how to find b in the first place. In fact, no *deterministic* polynomial-time algorithm for finding a quadratic non-residue modulo p is known. Fortunately, it is easy to find a quadratic non-residue probabilistically: simply choose uniform elements of $\mathbb{Z}_p^*$ until a

ALGORITHM 15.31
Computing square roots modulo a prime

Input: Prime p; quadratic residue $a \in \mathbb{Z}_p^*$
Output: A square root of a

case $p = 3 \bmod 4$:
$\quad$ **return** $[a^{\frac{p+1}{4}} \bmod p]$
case $p = 1 \bmod 4$:
$\quad$ **let** b be a quadratic non-residue modulo p
$\quad$ **compute** $\ell \geq 1$ and odd m with $2^\ell \cdot m = \frac{p-1}{2}$
$\quad r := 2^\ell \cdot m,\ r' := 0$
$\quad$ **for** $i = \ell$ **to** 1 {
$\quad\quad$ // maintain the invariant $a^r \cdot b^{r'} = 1 \bmod p$
$\quad\quad r := r/2,\ r' := r'/2$
$\quad\quad$ **if** $a^r \cdot b^{r'} = -1 \bmod p$
$\quad\quad\quad r' := r' + 2^\ell \cdot m$
$\quad$ }
$\quad$ // now $r = m$, r' is even, and $a^r \cdot b^{r'} = 1 \bmod p$
$\quad$ **return** $\left[a^{\frac{r+1}{2}} \cdot b^{\frac{r'}{2}} \bmod p\right]$

quadratic non-residue is found. This works because exactly half the elements of $\mathbb{Z}_p^*$ are quadratic non-residues, and because a polynomial-time algorithm for deciding quadratic residuosity modulo a prime is known (see Section 15.4.1 for proofs of both these statements). This means that the algorithm we have shown is actually randomized when $p = 1 \bmod 4$; a deterministic polynomial-time algorithm for computing square roots in this case is not known.

Example 15.32

Here we consider the "worst case," when taking a square root always gives -1. Let $a \in \mathbb{Z}_p^*$ be the element whose square root we are trying to compute; let $b \in \mathbb{Z}_p^*$ be a quadratic non-residue; and let $\frac{p-1}{2} = 2^3 \cdot m$ where m is odd.

In the first step, we have $a^{2^3 m} = 1 \bmod p$. Since $a^{2^3 m} = \left(a^{2^2 m}\right)^2$ and the square roots of 1 are ± 1, this means that $a^{2^2 m} = \pm 1 \bmod p$; assuming the worst case, $a^{2^2 m} = -1 \bmod p$. So, we multiply by $b^{\frac{p-1}{2}} = b^{2^3 m} = -1 \bmod p$ to obtain

$$a^{2^2 m} \cdot b^{2^3 m} = 1 \bmod p.$$

In the second step, we observe that $a^{2m} \cdot b^{2^2 m}$ is a square root of 1; again assuming the worst case, we thus have $a^{2m} \cdot b^{2^2 m} = -1 \bmod p$. Multiplying by $b^{2^3 m}$ to "correct" this gives

$$a^{2m} \cdot b^{2^2 m} \cdot b^{2^3 m} = 1 \bmod p.$$

In the third step, taking square roots and assuming the worst case (as above) we obtain $a^m \cdot b^{2m} \cdot b^{2^2 m} = -1 \bmod p$; multiplying by the "correction factor" $b^{2^3 m}$ we get

$$a^m \cdot b^{2m} \cdot b^{2^2 m} \cdot b^{2^3 m} = 1 \bmod p.$$

We are now where we want to be. To conclude the algorithm, multiply both sides by a to obtain

$$a^{m+1} \cdot b^{2m + 2^2 m + 2^3 m} = a \bmod p.$$

Since m is odd, $(m+1)/2$ is an integer and $a^{\frac{m+1}{2}} \cdot b^{m + 2m + 2^2 m} \bmod p$ is a square root of a.   ◇

Example 15.33
Here we work out a concrete example. Let $p = 17$, $a = 2$, and $b = 3$. Note that here $(p-1)/2 = 2^3$ and $m = 1$.

We begin with $2^{2^3} = 1 \bmod 17$. So 2^{2^2} should be equal to $\pm 1 \bmod 17$; by calculation, $2^{2^2} = -1 \bmod 17$. Multiplying by 3^{2^3} gives $2^{2^2} \cdot 3^{2^3} = 1 \bmod 17$.

Continuing, we know that $2^2 \cdot 3^{2^2}$ is a square root of 1 and so must be equal to $\pm 1 \bmod 17$; calculation gives $2^2 \cdot 3^{2^2} = 1 \bmod 17$. So no correction term is needed here.

Halving the exponents again we find that $2 \cdot 3^2 = 1 \bmod 17$. We are now almost done: multiplying both sides by 2 gives $2^2 \cdot 3^2 = 2 \bmod 17$, and so $2 \cdot 3 = 6 \bmod 17$ is a square root of 2.   ◇

Computing Square Roots Modulo N

It is not hard to see that the algorithm we have shown for computing square roots modulo a prime extends easily to the case of computing square roots modulo a composite $N = pq$ of known factorization. Specifically, let $a \in \mathbb{Z}_N^*$ be a quadratic residue with $a \leftrightarrow (a_p, a_q)$ via the Chinese remainder theorem. Computing the square roots x_p, x_q of a_p, a_q modulo p and q, respectively, gives a square root (x_p, x_q) of a (see Section 15.4.2). Given x_p and x_q, the representation x corresponding to (x_p, x_q) can be recovered as discussed in Section 9.1.5. That is, to compute a square root of a modulo an integer $N = pq$ of known factorization:

- Compute $a_p := [a \bmod p]$ and $a_q := [a \bmod q]$.

- Using Algorithm 15.31, compute a square root x_p of a_p modulo p and a square root x_q of a_q modulo q.

- Convert from the representation $(x_p, x_q) \in \mathbb{Z}_p^* \times \mathbb{Z}_q^*$ to $x \in \mathbb{Z}_N^*$ with $x \leftrightarrow (x_p, x_q)$. Output x, which is a square root of a modulo N.

It is easy to modify the algorithm so that it returns all four square roots of a.

15.5.2 A Trapdoor Permutation Based on Factoring

We have seen that computing square roots modulo N can be carried out in polynomial time if the factorization of N is known. We show here that, in contrast, computing square roots modulo a composite N of *unknown* factorization is as hard as factoring N.

More formally, let GenModulus be a polynomial-time algorithm that, on input 1^n, outputs (N, p, q) where $N = pq$ and p and q are n-bit primes except with probability negligible in n. Consider the following experiment for a given algorithm $\mathcal{A}$ and parameter n:

The square-root computation experiment $\mathsf{SQR}_{\mathcal{A},\mathsf{GenModulus}}(n)$:

1. *Run* $\mathsf{GenModulus}(1^n)$ *to obtain output* N, p, q.
2. *Choose a uniform* $y \in \mathcal{QR}_N$.
3. $\mathcal{A}$ *is given* (N, y), *and outputs* $x \in \mathbb{Z}_N^*$.
4. *The output of the experiment is defined to be 1 if* $x^2 = y \bmod N$, *and 0 otherwise.*

DEFINITION 15.34 *We say that* computing square roots is hard relative to GenModulus *if for all probabilistic polynomial-time algorithms* $\mathcal{A}$ *there exists a negligible function* negl *such that*

$$\Pr[\mathsf{SQR}_{\mathcal{A},\mathsf{GenModulus}}(n) = 1] \leq \mathsf{negl}(n).$$

It is easy to see that if computing square roots is hard relative to GenModulus then factoring must be hard relative to GenModulus too: if moduli N output by GenModulus could be factored easily, then it would be easy to compute square roots modulo N by first factoring N and then applying the algorithm discussed in the previous section. Our goal now is to show the converse: that if factoring is *hard* relative to GenModulus then so is the problem of computing square roots. We emphasize again that an analogous result is not known for the RSA problem or the problem of deciding quadratic residuosity.

The key is the following lemma, which says that two "unrelated" square roots of any element in $\mathbb{Z}_N^*$ can be used to factor N.

LEMMA 15.35 *Let* $N = pq$ *with* p, q *distinct, odd primes. Given* $x, \hat{x}$ *such that* $x^2 = y = \hat{x}^2 \bmod N$ *but* $x \neq \pm\hat{x} \bmod N$, *it is possible to factor* N *in time polynomial in* $\|N\|$.

PROOF We claim that either $\gcd(N, x + \hat{x})$ or $\gcd(N, x - \hat{x})$ is equal to one of the prime factors of N.[6] Since gcd computations can be carried out in polynomial time (see Appendix B.1.2), this proves the lemma.

---

[6] In fact, both of these are equal to one of the prime factors of N.

If $x^2 = \hat{x}^2 \bmod N$ then

$$0 = x^2 - \hat{x}^2 = (x - \hat{x}) \cdot (x + \hat{x}) \bmod N,$$

and so $N \mid (x - \hat{x})(x + \hat{x})$. Then $p \mid (x - \hat{x})(x + \hat{x})$ and so p divides one of these terms. Say $p \mid (x + \hat{x})$ (the proof proceeds similarly if $p \mid (x - \hat{x})$). If $q \mid (x + \hat{x})$ then $N \mid (x + \hat{x})$, but this cannot be the case since $x \neq -\hat{x} \bmod N$. So $q \nmid (x + \hat{x})$ and $\gcd(N, x + \hat{x}) = p$. ∎

An alternative way of proving the above is to look at what happens in the Chinese remaindering representation. Say $x \leftrightarrow (x_p, x_q)$. Then, because x and $\hat{x}$ are square roots of the same value y, we know that $\hat{x}$ corresponds to either $(-x_p, x_q)$ or $(x_p, -x_q)$. (It cannot correspond to (x_p, x_q) or $(-x_p, -x_q)$ since the first corresponds to x while the second corresponds to $[-x \bmod N]$, and both possibilities are ruled out by the assumption of the lemma.) Say $\hat{x} \leftrightarrow (-x_p, x_q)$. Then

$$[x + \hat{x} \bmod N] \leftrightarrow (x_p, x_q) + (-x_p, x_q) = (0, [2x_q \bmod q]),$$

and we see that $x + \hat{x} = 0 \bmod p$ while $x + \hat{x} \neq 0 \bmod q$. It follows that $\gcd(N, x + \hat{x}) = p$, a factor of N.

We can now prove the main result of this section.

THEOREM 15.36 *If factoring is hard relative to* GenModulus, *then computing square roots is hard relative to* GenModulus.

PROOF Let $\mathcal{A}$ be a probabilistic polynomial-time algorithm computing square roots (as in Definition 15.34). Consider the following probabilistic polynomial-time algorithm $\mathcal{A}_{\mathsf{fact}}$ for factoring moduli output by GenModulus:

> **Algorithm $\mathcal{A}_{\mathsf{fact}}$:**
> The algorithm is given a modulus N as input.
>
> - Choose a uniform $x \in \mathbb{Z}_N^*$ and compute $y := [x^2 \bmod N]$.
> - Run $\mathcal{A}(N, y)$ to obtain output $\hat{x}$.
> - If $\hat{x}^2 = y \bmod N$ and $\hat{x} \neq \pm x \bmod N$, then factor N using Lemma 15.35.

By Lemma 15.35, we know that $\mathcal{A}_{\mathsf{fact}}$ succeeds in factoring N exactly when $\hat{x} \neq \pm x \bmod N$ and $\hat{x}^2 = y \bmod N$. That is,

$$\Pr[\mathsf{Factor}_{\mathcal{A}_{\mathsf{fact}}, \mathsf{GenModulus}}(n) = 1]$$
$$= \Pr\left[\hat{x} \neq \pm x \bmod N \wedge \hat{x}^2 = y \bmod N\right]$$
$$= \Pr\left[\hat{x} \neq \pm x \bmod N \mid \hat{x}^2 = y \bmod N\right] \cdot \Pr\left[\hat{x}^2 = y \bmod N\right], \quad (15.7)$$

where the above probabilities all refer to experiment $\mathsf{Factor}_{\mathcal{A}_{\mathsf{fact}},\mathsf{GenModulus}}(n)$ (see Section 9.2.3 for a description of this experiment). In the experiment, the modulus N given as input to $\mathcal{A}_{\mathsf{fact}}$ is generated by $\mathsf{GenModulus}(1^n)$, and y is a uniform quadratic residue modulo N since x was chosen uniformly from $\mathbb{Z}_N^*$ (see Section 15.4.4). So the view of $\mathcal{A}$ when run as a subroutine by $\mathcal{A}_{\mathsf{fact}}$ is distributed exactly as $\mathcal{A}$'s view in experiment $\mathsf{SQR}_{\mathcal{A},\mathsf{GenModulus}}(n)$. Therefore,

$$\Pr\left[\hat{x}^2 = y \bmod N\right] = \Pr\left[\mathsf{SQR}_{\mathcal{A},\mathsf{GenModulus}}(n) = 1\right]. \tag{15.8}$$

Conditioned on the value of the quadratic residue y used in experiment $\mathsf{Factor}_{\mathcal{A}_{\mathsf{fact}},\mathsf{GenModulus}}(n)$, the value x is equally likely to be any of the four possible square roots of y. This means that from the point of view of algorithm $\mathcal{A}$ (being run as a subroutine by $\mathcal{A}_{\mathsf{fact}}$), x is equally likely to be each of the four square roots of y. This in turn means that, conditioned on $\mathcal{A}$ outputting *some* square root $\hat{x}$ of y, the probability that $\hat{x} = \pm x \bmod N$ is exactly $1/2$. (We stress that we do not make any assumption about how $\hat{x}$ is distributed among the square roots of y, and in particular are not assuming here that $\mathcal{A}$ outputs a uniform square root of y. Rather we are using the fact that x is uniformly distributed among the square roots of y.) That is,

$$\Pr\left[\hat{x} \neq \pm x \bmod N \mid \hat{x}^2 = y \bmod N\right] = \frac{1}{2}. \tag{15.9}$$

Combining Equations (15.7)–(15.9), we see that

$$\Pr\left[\mathsf{Factor}_{\mathcal{A}_{\mathsf{fact}},\mathsf{GenModulus}}(n) = 1\right] = \frac{1}{2} \cdot \Pr\left[\mathsf{SQR}_{\mathcal{A},\mathsf{GenModulus}}(n) = 1\right].$$

Since factoring is hard relative to $\mathsf{GenModulus}$, there is a negligible function negl such that

$$\Pr[\mathsf{Factor}_{\mathcal{A}_{\mathsf{fact}},\mathsf{GenModulus}}(n) = 1] \leq \mathsf{negl}(n),$$

which implies $\Pr\left[\mathsf{SQR}_{\mathcal{A},\mathsf{GenModulus}}(n) = 1\right] \leq 2{\cdot}\mathsf{negl}(n)$. Since $\mathcal{A}$ was arbitrary, this completes the proof. ∎

The previous theorem leads directly to a family of one-way functions (see Definition 9.76) based on any $\mathsf{GenModulus}$ relative to which factoring is hard:

- Algorithm Gen, on input 1^n, runs $\mathsf{GenModulus}(1^n)$ to obtain (N, p, q) and outputs $I = N$. The domain $\mathcal{D}_I$ is $\mathbb{Z}_N^*$ and the range $\mathcal{R}_I$ is $\mathcal{QR}_N$.

- Algorithm Samp, on input N, chooses a uniform element $x \in \mathbb{Z}_N^*$.

- Algorithm f, on input N and $x \in \mathbb{Z}_N^*$, outputs $[x^2 \bmod N]$.

The preceding theorem shows that this family is one-way if factoring is hard relative to $\mathsf{GenModulus}$.

We can turn this into a family of one-way *permutations* by using moduli N of a special form and letting $\mathcal{D}_I$ be a subset of $\mathbb{Z}_N^*$. (See Exercise 15.20 for another way to make this a permutation.) Call $N = pq$ a *Blum integer* if p and q are distinct primes with $p = q = 3 \bmod 4$. The key to building a permutation is the following proposition.

PROPOSITION 15.37 *Let N be a Blum integer. Then every quadratic residue modulo N has exactly one square root that is also a quadratic residue.*

PROOF Say $N = pq$ with $p = q = 3 \bmod 4$. Using Proposition 15.17, we see that -1 is not a quadratic residue modulo p or q. This is because for $p = 3 \bmod 4$ it holds that $p = 4i + 3$ for some i and so

$$(-1)^{\frac{p-1}{2}} = (-1)^{2i+1} = -1 \bmod p$$

(because $2i+1$ is odd). Now let $y \leftrightarrow (y_p, y_q)$ be an arbitrary quadratic residue modulo N with the four square roots

$$(x_p, x_q), \quad (-x_p, x_q), \quad (x_p, -x_q), \quad (-x_p, -x_q).$$

We claim that exactly one of these is a quadratic residue modulo N. To see this, assume $\mathcal{J}_p(x_p) = +1$ and $\mathcal{J}_q(x_q) = -1$ (the proof proceeds similarly in any other case). Using Proposition 15.19, we have

$$\mathcal{J}_q(-x_q) = \mathcal{J}_q(-1) \cdot \mathcal{J}_q(x_q) = +1,$$

and so $(x_p, -x_q)$ corresponds to a quadratic residue modulo N (using Proposition 15.21). Similarly, $\mathcal{J}_p(-x_p) = -1$ and so none of the other square roots of y are quadratic residues modulo N. ∎

Expressed differently, the above proposition says that when N is a Blum integer, the function $f_N : \mathcal{QR}_N \to \mathcal{QR}_N$ given by $f_N(x) = [x^2 \bmod N]$ is a permutation over $\mathcal{QR}_N$. Modifying the sampling algorithm Samp above to choose a uniform $x \in \mathcal{QR}_N$ (which, as we have already seen, can be done easily by choosing uniform $r \in \mathbb{Z}_N^*$ and setting $x := [r^2 \bmod N]$) gives a family of one-way *permutations*. Finally, because square roots modulo N can be computed in polynomial time given the factorization of N, a straightforward modification yields a family of *trapdoor* permutations based on any GenModulus relative to which factoring is hard. This is sometimes called the *Rabin* family of trapdoor permutations. In summary:

THEOREM 15.38 *Let GenModulus be an algorithm that, on input 1^n, outputs (N, p, q) where $N = pq$ and p and q are distinct primes (except possibly with negligible probability) with $p = q = 3 \bmod 4$. If factoring is hard relative to GenModulus, then there exists a family of trapdoor permutations.*

15.5.3   The Rabin Encryption Scheme

We can apply the results of Section 15.1.2 to the Rabin trapdoor permutation to obtain a public-key encryption scheme whose security is based on factoring. To do this, we first need to identify a hard-core predicate for this trapdoor permutation. Although we could appeal to Theorem 15.3, which states that a suitable hard-core predicate always exists, it turns out that the least significant bit lsb is a hard-core predicate for the Rabin trapdoor permutation just as it is for the case of RSA (see Section 12.5.3). Using this as our hard-core predicate, we obtain the scheme of Construction 15.39.

CONSTRUCTION 15.39

Let GenModulus be a polynomial-time algorithm that, on input 1^n, outputs (N, p, q) where $N = pq$ and p and q are n-bit primes (except with probability negligible in n) with $p = q = 3 \bmod 4$. Construct a public-key encryption scheme as follows:

- Gen: on input 1^n run GenModulus(1^n) to obtain (N, p, q). The public key is N, and the private key is $\langle p, q \rangle$.

- Enc: on input a public-key N and message $m \in \{0, 1\}$, choose a uniform $x \in \mathcal{QR}_N$ subject to the constraint that $\mathsf{lsb}(x) = m$. Output the ciphertext $c := [x^2 \bmod N]$.

- Dec: on input a private key $\langle p, q \rangle$ and a ciphertext c, compute the unique $x \in \mathcal{QR}_N$ such that $x^2 = c \bmod N$, and output $\mathsf{lsb}(x)$.

The Rabin encryption scheme.

Theorems 15.5 and 15.38 imply the following result.

THEOREM 15.40   *If factoring is hard relative to* GenModulus, *then Construction 15.39 is CPA-secure.*

Rabin Encryption vs. RSA Encryption

It is worthwhile to remark on the similarities and differences between the Rabin and RSA cryptosystems. (The discussion here applies to any cryptographic construction—not necessarily a public-key encryption scheme—based on the Rabin or RSA trapdoor permutations.)

The RSA and Rabin trapdoor permutations appear quite similar, with squaring in the case of Rabin corresponding to taking $e = 2$ in the case of RSA. (Of course, 2 is *not* relatively prime to $\phi(N)$ and so Rabin is not a special case of RSA.) In terms of the security offered by each construction, hardness of computing modular square roots is equivalent to hardness of factoring, whereas hardness of solving the RSA problem is not known to be implied by the hardness of factoring. The Rabin trapdoor permutation is thus based on a potentially *weaker* assumption: it is theoretically possible that someone might develop an efficient algorithm for solving the RSA problem, yet computing square roots will remain hard. Or, someone may develop an algorithm that solves the RSA problem faster than known factoring algorithms. Lemma 15.35 ensures, however, that computing square roots modulo N can never be much faster than the best available algorithm for factoring N.

In terms of their efficiency, the RSA and Rabin permutations are essentially the same. Actually, if a large exponent e is used in the case of RSA then computing eth powers (as in RSA) is slightly slower than squaring (as in Rabin). On the other hand, a bit more care is required when working with the Rabin permutation since it is only a permutation over a *subset* of $\mathbb{Z}_N^*$, in contrast to RSA, which gives a permutation over all of $\mathbb{Z}_N^*$.

A "plain Rabin" encryption scheme, constructed in a manner analogous to plain RSA encryption, is vulnerable to a chosen-ciphertext attack that enables an adversary to learn the entire private key (see Exercise 15.18). Although plain RSA is not CCA-secure either, known chosen-ciphertext attacks on plain RSA are less damaging since they recover the message but not the private key. Perhaps the existence of such an attack on "plain Rabin" influenced cryptographers, early on, to reject the use of Rabin encryption entirely.

In summary, the RSA permutation is much more widely used in practice than the Rabin permutation, but in light of the above this appears to be due more to historical accident than to any compelling technical justification.

References and Additional Reading

The existence of public-key encryption based on arbitrary trapdoor permutations was shown by Yao [205], and the efficiency improvement discussed at the end of Section 15.1.2 is due to Blum and Goldwasser [40].

Childs [51] and Shoup [183] provide further coverage of the (computational) number theory used in this chapter. A good description of the algorithm for computing the Jacobi symbol modulo a composite of unknown factorization, along with a proof of correctness, is given in [64].

The Paillier encryption scheme was introduced in [157]. Shoup [183, Section 7.5] gives a characterization of $\mathbb{Z}_{N^e}^*$ for arbitrary integers N, e (and not just $N = pq$, $e = 2$ as done here).

The problem of deciding quadratic residuosity modulo a composite of unknown factorization goes back to Gauss [78] and is related to other (conjectured) hard number-theoretic problems. The Goldwasser–Micali encryption scheme [87], introduced in 1982, was the first public-key encryption scheme with a proof of security.

Rabin [166] showed that computing square roots modulo a composite is equivalent to factoring. The results of Section 15.5.2 are due to Blum [39]. Hard-core predicates for the Rabin trapdoor permutation are discussed in [8, 94, 7] and references therein.

Exercises

15.1 Construct and prove CPA-security for a KEM based on any trapdoor permutation by suitably generalizing Construction 12.34.

15.2 Show that the isomorphism of Proposition 15.6 can be efficiently inverted when the factorization of N is known.

15.3 Generalize the Paillier encryption scheme so $(1 + N)$ is replaced by any $g \in \mathbb{Z}_{N^2}^*$ of order N. I.e., the public key now includes g, and encryption of m is done by computing the ciphertext $c := [g^m \cdot r^N \bmod N^2]$.

    (a) Show how decryption can be done.

    (b) Prove CPA-security under the same assumption as in Theorem 15.13.

15.4 Let $\Psi(N^2)$ denote the set $\{(a, 1) \mid a \in \mathbb{Z}_N\} \subset \mathbb{Z}_{N^2}^*$. Show that it is *not* hard to decide whether a given element $y \in \mathbb{Z}_{N^2}^*$ is in $\Psi(N^2)$.

15.5 Let $\mathbb{G}$ be an abelian group. Show that the set of quadratic residues in $\mathbb{G}$ forms a subgroup.

15.6 This question concerns the quadratic residues in the additive group $\mathbb{Z}_N$. (An element $y \in \mathbb{Z}_N$ is a quadratic residue if and only if there exists an $x \in \mathbb{Z}_N$ with $2x = y \bmod N$.)

    (a) Let p be an odd prime. How many elements of $\mathbb{Z}_p$ are quadratic residues?

    (b) Let $N = pq$ be a product of two odd primes p and q. How many elements of $\mathbb{Z}_N$ are quadratic residues?

    (c) Let N be an even integer. How many elements of $\mathbb{Z}_N$ are quadratic residues?

15.7 Let $N = pq$ with p, q distinct, odd primes. Show a PPT algorithm for choosing a uniform element of $\mathcal{QNR}_N^{+1}$ when the factorization of N is known. (Your algorithm can have failure probability negligible in $\|N\|$.)

15.8 Let $N = pq$ with p, q distinct, odd primes. Prove that if $x \in \mathcal{QR}_N$ then $[x^{-1} \bmod N] \in \mathcal{QR}_N$, and if $x \in \mathcal{QNR}_N^{+1}$ then $[x^{-1} \bmod N] \in \mathcal{QNR}_N^{+1}$.

15.9 Let $N = pq$ with p, q distinct, odd primes, and fix $z \in \mathcal{QNR}_N^{+1}$. Show that choosing uniform $x \in \mathcal{QR}_N$ and setting $y := [z \cdot x \bmod N]$ gives a y that is uniformly distributed in $\mathcal{QNR}_N^{+1}$. That is, for any $\hat{y} \in \mathcal{QNR}_N^{+1}$

$$\Pr[z \cdot x = \hat{y} \bmod N] = 1/|\mathcal{QNR}_N^{+1}|,$$

where the probability is taken over uniform choice of $x \in \mathcal{QR}_N$.

   Hint: Use the previous exercise.

15.10 Let N be the product of 5 distinct, odd primes. If $y \in \mathbb{Z}_N^*$ is a quadratic residue, how many solutions are there to the equation $x^2 = y \bmod N$?

15.11 Show that the Goldwasser–Micali encryption scheme is homomorphic if the message space $\{0, 1\}$ is viewed as the group $\mathbb{Z}_2$.

15.12 Consider the following variation of the Goldwasser–Micali encryption scheme: GenModulus(1^n) is run to obtain (N, p, q) where $N = pq$ and $p = q = 3 \bmod 4$, (i.e., N is a Blum integer.) The public key is N and the private key is $\langle p, q \rangle$. To encrypt $m \in \{0, 1\}$, the sender chooses uniform $x \in \mathbb{Z}_N$ and computes the ciphertext $c := [(-1)^m \cdot x^2 \bmod N]$.

   (a) Prove that for N of the stated form, $[-1 \bmod N] \in \mathcal{QNR}_N^{+1}$.

   (b) Prove that the scheme described has indistinguishable encryptions under a chosen-plaintext attack if deciding quadratic residuosity is hard relative to GenModulus.

15.13 Assume deciding quadratic residuosity is hard for GenModulus. Show that this implies the hardness of distinguishing a uniform element of $\mathcal{QR}_N$ from a uniform element of $\mathcal{J}_N^{+1}$.

15.14 Show that plain RSA encryption of a message m leaks $\mathcal{J}_N(m)$.

15.15 Consider the following variation of the Goldwasser–Micali encryption scheme: GenModulus(1^n) is run to obtain (N, p, q). The public key is N and the private key is $\langle p, q \rangle$. To encrypt a 0, the sender chooses n uniform elements $c_1, \ldots, c_n \in \mathcal{QR}_N$. To encrypt a 1, the sender chooses n uniform elements $c_1, \ldots, c_n \in \mathcal{J}_N^{+1}$. In each case, the resulting ciphertext is $c^* = \langle c_1, \ldots, c_n \rangle$.

   (a) Show how the sender can generate a uniform element of $\mathcal{J}_N^{+1}$ in polynomial time, where failing with negligible probability.

   (b) Suggest a way for the receiver to decrypt efficiently, although with negligible error probability.

   (c) Prove that if deciding quadratic residuosity is hard relative to GenModulus, this scheme is CPA-secure.

      Hint: Use the previous exercise.

15.16 Let $\mathcal{G}$ be a polynomial-time algorithm that, on input 1^n, outputs a prime p with $\|p\| = n$ and a generator g of $\mathbb{Z}_p^*$. Prove that the DDH problem is *not* hard relative to $\mathcal{G}$.

> **Hint:** Use the fact that quadratic residuosity can be decided efficiently modulo a prime.

15.17 The discrete logarithm problem is believed to be hard for $\mathcal{G}$ as in the previous exercise. This means that the function (family) $f_{p,g}$ where $f_{p,g}(x) \overset{\text{def}}{=} [g^x \bmod p]$ is one-way. Let $\mathsf{lsb}(x)$ denote the least-significant bit of x. Show that lsb is *not* a hard-core predicate for $f_{p,g}$.

15.18 Consider the plain Rabin encryption scheme in which a message $m \in \mathcal{QR}_N$ is encrypted relative to a public key N (where N is a Blum integer) by computing the ciphertext $c := [m^2 \bmod N]$. Show a chosen-ciphertext attack on this scheme that recovers the entire private key.

15.19 The plain Rabin signature scheme is like the plain RSA signature scheme, except using the Rabin trapdoor permutation. Show an attack on plain Rabin signatures by which the attacker learns the signer's private key.

15.20 Let N be a Blum integer.

(a) Define the set $S \overset{\text{def}}{=} \{x \in \mathbb{Z}_N^* \mid x < N/2 \text{ and } \mathcal{J}_N(x) = +1\}$. Define the function $f_N : S \to \mathbb{Z}_N^*$ by:

$$f_N(x) = \begin{cases} [x^2 \bmod N] & \text{if } [x^2 \bmod N] < N/2 \\ [-x^2 \bmod N] & \text{if } [x^2 \bmod N] > N/2 \end{cases}$$

Show that f_N is a permutation over S.

(b) Define a family of trapdoor permutations based on factoring using f_N as defined above.

15.21 Let N be a Blum integer. Define the function $\mathsf{half}_N : \mathbb{Z}_N^* \to \{0,1\}$ as

$$\mathsf{half}_N(x) = \begin{cases} -1 \text{ if } x < N/2 \\ +1 \text{ if } x > N/2 \end{cases}$$

Show that the function $f : \mathbb{Z}_N^* \to \mathcal{QR}_N \times \{-1,+1\}^2$ defined as

$$f(x) = \left([x^2 \bmod N], \mathcal{J}_N(x), \mathsf{half}_N(x)\right)$$

is one-to-one.

Index of Common Notation

General notation:

- $:=$ refers to deterministic assignment

- If S is a set, then $x \leftarrow S$ denotes that x is chosen uniformly from S

- If A is a randomized algorithm, then $y \leftarrow A(x)$ denotes running A on input x with a uniform random tape and assigning the output to y. We write $y := A(x;r)$ to denote running A on input x using random tape r and assigning the output to y

- $\wedge$ denotes Boolean conjunction (the AND operator)

- $\vee$ denotes Boolean disjunction (the OR operator)

- $\oplus$ denotes the exclusive-or (XOR) operator; this operator can be applied to single bits or entire strings (in the latter case, the XOR is bitwise)

- $\{0,1\}^n$ is the set of all bit-strings of length n

- $\{0,1\}^{\leq n}$ is the set of all bit-strings of length at most n

- $\{0,1\}^*$ is the set of all finite bit-strings; $\{0,1\}^+$ is the set of all non-empty, finite bit-strings

- 0^n (resp., 1^n) denotes the string comprised of n zeroes (resp., n ones)

- $\|x\|$ denotes the length of the binary representation of the (positive) integer x, written with leading bit 1. Note that $\log x < \|x\| \leq \log x + 1$

- $|x|$ denotes the length of the binary string x (which may have leading 0s), or the absolute value of the real number x

- $\mathcal{O}(\cdot), \Theta(\cdot), \Omega(\cdot), \omega(\cdot)$ are used for asymptotic running times; see Appendix A.2

- $\mathtt{0x}$ denotes that digits are being represented in hexadecimal

- $x\|y$ and (x,y) are used interchangeably to denote concatenation of the strings x and y

- $\Pr[X]$ denotes the probability of event X

- $\log x$ denotes the base-2 logarithm of x

Cryptographic notation:

- n is the security parameter

- PPT stands for "probabilistic polynomial time"

- $\mathcal{A}^{\mathcal{O}(\cdot)}$ denotes the algorithm $\mathcal{A}$ with oracle access to $\mathcal{O}$

- k typically denotes a secret key (as in private-key encryption and MACs)

- (pk, sk) denotes a public/private key pair (for public-key encryption and digital signatures)

- $\perp$ denotes a generic error

- $\mathsf{negl}(n)$ denotes a negligible function; see Definition 3.4

- $\mathsf{poly}(n)$ denotes an arbitrary polynomial

- Func_n denotes the set of functions mapping n-bit strings to n-bit strings

- Perm_n denotes the set of bijections on n-bit strings

- IV denotes an initialization vector

Algorithms and procedures:

- G denotes a pseudorandom generator

- F denotes a keyed function that is typically a pseudorandom function or permutation

- $(\mathsf{Gen}, \mathsf{Enc}, \mathsf{Dec})$ denote the key-generation, encryption, and decryption procedures, respectively, for both private- and public-key encryption. For the case of private-key encryption, when Gen is unspecified then $\mathsf{Gen}(1^n)$ outputs a uniform $k \in \{0,1\}^n$

- $(\mathsf{Gen}, \mathsf{Mac}, \mathsf{Vrfy})$ denote the key-generation, tag-generation, and verification procedures, respectively, for a message authentication code. When Gen is unspecified then $\mathsf{Gen}(1^n)$ outputs a uniform $k \in \{0,1\}^n$

- $(\mathsf{Gen}, \mathsf{Sign}, \mathsf{Vrfy})$ denote the key-generation, signature-generation, and verification procedures, respectively, for a digital signature scheme

- $\mathsf{GenPrime}$ denotes a PPT algorithm that, on input 1^n, outputs an n-bit prime except with probability negligible in n

- $\mathsf{GenModulus}$ denotes a PPT algorithm that, on input 1^n, outputs (N, p, q) where $N = pq$ and (except with negligible probability) p and q are n-bit primes

- GenRSA denotes a PPT algorithm that, on input 1^n, outputs (except with negligible probability) a modulus N, an integer $e > 0$ with $\gcd(e, \phi(N)) = 1$, and an integer d satisfying $ed = 1 \bmod \phi(N)$

- $\mathcal{G}$ denotes a PPT algorithm that, on input 1^n, outputs (except with negligible probability) a description of a cyclic group $\mathbb{G}$, the group order q (with $\|q\| = n$), and a generator $g \in \mathbb{G}$.

Number theory:

- $\mathbb{Z}$ denotes the set of integers

- $a \mid b$ means a divides b

- $a \nmid b$ means that a does not divide b

- $\gcd(a, b)$ denotes the greatest common divisor of a and b

- $[a \bmod b]$ denotes the remainder of a when divided by b

- $x_1 = x_2 = \cdots = x_n \bmod N$ means that $x_1, \ldots, x_n$ are all congruent modulo N

  Note: $x = y \bmod N$ means that x and y are congruent modulo N, whereas $x = [y \bmod N]$ means that x is equal to the remainder of y when divided by N

- $\mathbb{Z}_N$ denotes the additive group of integers modulo N as well as the set $\{0, \ldots, N-1\}$. Note: in Section 14.3 only, we let $\mathbb{Z}_N$ also refer to the set $\{-\lfloor (N-1)/2 \rfloor, \ldots, 0, \ldots, \lfloor N/2 \rfloor\}$

- $\mathbb{Z}_N^*$ denotes the multiplicative group of invertible integers modulo N (i.e., those that are relatively prime to N)

- $\phi(N)$ denotes the size of $\mathbb{Z}_N^*$

- $\mathbb{G}$ and $\mathbb{H}$ denote groups

- $\mathbb{G}_1 \simeq \mathbb{G}_2$ means that groups $\mathbb{G}_1$ and $\mathbb{G}_2$ are isomorphic. If this isomorphism is given by f and $f(x_1) = x_2$ then we write $x_1 \leftrightarrow x_2$

- g is typically a generator of a group

- $\log_g h$ denotes the discrete logarithm of h to the base g

- $\langle g \rangle$ denotes the group generated by g

- p and q usually denote primes

- N typically denotes the product of two distinct primes p and q of equal length

- $\mathcal{QR}_p$ is the set of quadratic residues modulo p

- $\mathcal{QNR}_p$ is the set of quadratic non-residues modulo p

- $\mathcal{J}_p(x)$ is the Jacobi symbol of x modulo p

- $\mathcal{J}_N^{+1}$ is the set of elements with Jacobi symbol $+1$ modulo N

- $\mathcal{J}_N^{-1}$ is the set of elements with Jacobi symbol -1 modulo N

- $\mathcal{QNR}_N^{+1}$ is the set of quadratic non-residues modulo N having Jacobi symbol $+1$

Appendix A

Mathematical Background

A.1 Identities and Inequalities

We list some standard identities and inequalities that are used at various points throughout the text.

THEOREM A.1 (Binomial expansion theorem) *Let x, y be real numbers, and let n be a positive integer. Then*

$$(x + y)^n = \sum_{i=0}^{n} \binom{n}{i} x^i \, y^{n-i}.$$

PROPOSITION A.2 *For all $x \geq 1$ it holds that $(1 - 1/x)^x \leq e^{-1}$.*

PROPOSITION A.3 *For all x it holds that $1 - x \leq e^{-x}$.*

PROPOSITION A.4 *For all x with $0 \leq x \leq 1$ it holds that*

$$e^{-x} \leq 1 - \left(1 - \frac{1}{e}\right) \cdot x \leq 1 - \frac{x}{2}.$$

A.2 Asymptotic Notation

We use standard notation for expressing asymptotic behavior of functions.

DEFINITION A.5 *Let $f(n), g(n)$ be functions from non-negative integers to non-negative reals. Then:*

- *$f(n) = \mathcal{O}(g(n))$ means that there exist positive integers c and n' such that for all $n > n'$ it holds that $f(n) \leq c \cdot g(n)$.*

- $f(n) = \Omega(g(n))$ *means that there exist positive integers c and n' such that for all $n > n'$ it holds that $f(n) \geq c \cdot g(n)$.*

- $f(n) = \Theta(g(n))$ *means that there exist positive integers $c_1, c_2,$ and n' such that for all $n > n'$ it holds that $c_1 \cdot g(n) \leq f(n) \leq c_2 \cdot g(n)$.*

- $f(n) = o(g(n))$ *means that $\lim_{n \to \infty} \frac{f(n)}{g(n)} = 0$.*

- $f(n) = \omega(g(n))$ *means that $\lim_{n \to \infty} \frac{f(n)}{g(n)} = \infty$.*

Example A.6
Let $f(n) = n^4 + 3n + 500$. Then:

- $f(n) = \mathcal{O}(n^4)$.

- $f(n) = \mathcal{O}(n^5)$. In fact, $f(n) = o(n^5)$.

- $f(n) = \Omega(n^3 \log n)$. In fact, $f(n) = \omega(n^3 \log n)$.

- $f(n) = \Theta(n^4)$.

$\diamondsuit$

A.3 Basic Probability

We assume the reader is familiar with basic probability theory, on the level of what is covered in a typical undergraduate course on discrete mathematics. Here we simply remind the reader of some notation and basic facts.

If E is an event, then $\bar{E}$ denotes the complement of that event; i.e., $\bar{E}$ is the event that E does *not* occur. By definition, $\Pr[E] = 1 - \Pr[\bar{E}]$. If E_1 and E_2 are events, then $E_1 \wedge E_2$ denotes their conjunction; i.e., $E_1 \wedge E_2$ is the event that *both* E_1 and E_2 occur. By definition, $\Pr[E_1 \wedge E_2] \leq \Pr[E_1]$. Events E_1 and E_2 are said to be *independent* if $\Pr[E_1 \wedge E_2] = \Pr[E_1] \cdot \Pr[E_2]$.

If E_1 and E_2 are events, then $E_1 \vee E_2$ denotes the disjunction of E_1 and E_2; that is, $E_1 \vee E_2$ is the event that *either* E_1 or E_2 occurs. It follows from the definition that $\Pr[E_1 \vee E_2] \geq \Pr[E_1]$. The *union bound* is often a very useful upper bound of this quantity.

PROPOSITION A.7 (Union Bound)

$$\Pr[E_1 \vee E_2] \leq \Pr[E_1] + \Pr[E_2].$$

Repeated application of the union bound for any events $E_1, \ldots, E_k$ gives

$$\Pr\left[\bigvee_{i=1}^k E_i\right] \le \sum_{i=1}^k \Pr[E_i].$$

The *conditional probability of E_1 given E_2*, denoted $\Pr[E_1 \mid E_2]$, is defined as

$$\Pr[E_1 \mid E_2] \stackrel{\text{def}}{=} \frac{\Pr[E_1 \wedge E_2]}{\Pr[E_2]}$$

as long as $\Pr[E_2] \ne 0$. (If $\Pr[E_2] = 0$ then $\Pr[E_1 \mid E_2]$ is undefined.) This represents the probability that event E_1 occurs, given that event E_2 has occurred. It follows immediately from the definition that

$$\Pr[E_1 \wedge E_2] = \Pr[E_1 \mid E_2] \cdot \Pr[E_2]\,;$$

equality holds even if $\Pr[E_2] = 0$ as long as we interpret multiplication by zero on the right-hand side in the obvious way.

We can now easily derive Bayes' theorem.

THEOREM A.8 (Bayes' Theorem) *If $\Pr[E_2] \ne 0$ then*

$$\Pr[E_1 \mid E_2] = \frac{\Pr[E_2 \mid E_1] \cdot \Pr[E_1]}{\Pr[E_2]}.$$

PROOF This follows because

$$\Pr[E_1 \mid E_2] = \frac{\Pr[E_1 \wedge E_2]}{\Pr[E_2]} = \frac{\Pr[E_2 \wedge E_1]}{\Pr[E_2]} = \frac{\Pr[E_2 \mid E_1] \cdot \Pr[E_1]}{\Pr[E_2]}.$$

■

Let $E_1, \ldots, E_n$ be disjoint events, so that $\Pr[E_i \wedge E_j] = 0$ for all $i \ne j$. That is, at most one of the $\{E_i\}$ occur. Assume further than $\Pr[E_i] > 0$ for all i. Then for any event F

$$\Pr[F] \le \sum_{i=1}^n \Pr[F \wedge E_i]$$
$$= \sum_{i=1}^n \Pr[F \mid E_i] \cdot \Pr[E_i],$$

with equality when $\Pr[E_1 \vee \cdots \vee E_n] = 1$. A special case is when we take E_1 and $\bar{E}_1$ as our disjoint events. Taking $F = E_1 \vee E_2$ for any event E_2, we get a potentially tighter version of the union bound:

$$\Pr[E_1 \vee E_2] = \Pr[E_1 \vee E_2 \mid E_1] \cdot \Pr[E_1] + \Pr[E_1 \vee E_2 \mid \bar{E}_1] \cdot \Pr[\bar{E}_1]$$
$$\le \Pr[E_1] + \Pr[E_2 \mid \bar{E}_1].$$

Extending this to n events we obtain

PROPOSITION A.9

$$\Pr\left[\bigvee_{i=1}^{n} E_i\right] \leq \Pr[E_1] + \sum_{i=2}^{n} \Pr[E_i \mid \bar{E}_1 \wedge \cdots \wedge \bar{E}_{i-1}].$$

*Useful Probability Bounds

We review some terminology and state probability bounds that are standard, but may not be encountered in a basic discrete mathematics course. The material here is used only in Section 8.3.

A (discrete, real-valued) random variable X is a variable whose value is assigned probabilistically from some finite set S of real numbers. X is nonnegative if it does not take negative values; it is a $0/1$-random variable if $S = \{0, 1\}$. The $0/1$-random variables $X_1, \ldots, X_k$ are *independent* if for all $b_1, \ldots, b_k$ it holds that $\Pr[X_1 = b_1 \wedge \cdots \wedge X_k = b_k] = \prod_{i=1}^{k} \Pr[X_i = b_i]$.

We let $\mathsf{Exp}[X]$ denote the expectation of a random variable X; if X takes values in a set S then $\mathsf{Exp}[X] \stackrel{\text{def}}{=} \sum_{s \in S} s \cdot \Pr[X = s]$. One of the most important facts is that expectation is *linear*; for random variables $X_1, \ldots, X_k$ (with arbitrary dependencies) we have $\mathsf{Exp}[\sum_i X_i] = \sum_i \mathsf{Exp}[X_i]$. If X_1, X_2 are independent, then $\mathsf{Exp}[X_i \cdot X_j] = \mathsf{Exp}[X_i] \cdot \mathsf{Exp}[X_j]$.

Markov's inequality is useful when little is known about X.

PROPOSITION A.10 (Markov's inequality)   *Let X be a non-negative random variable and $v > 0$. Then $\Pr[X \geq v] \leq \mathsf{Exp}[X]/v$.*

PROOF   Say X takes values in a set S. We have

$$\mathsf{Exp}[X] = \sum_{s \in S} s \cdot \Pr[X = s]$$

$$\geq \sum_{x \in S,\ x < v} \Pr[X = s] \cdot 0 + \sum_{x \in S,\ x \geq v} v \cdot \Pr[X = s]$$

$$= v \cdot \Pr[X \geq v].$$

The desired result follows.   ∎

The variance of X, denoted $\mathsf{Var}[X]$, measures how much X deviates from its expectation. We have $\mathsf{Var}[X] \stackrel{\text{def}}{=} \mathsf{Exp}[(X - \mathsf{Exp}[X])^2] = \mathsf{Exp}[X^2] - \mathsf{Exp}[X]^2$, and one can easily show that $\mathsf{Var}[aX + b] = a^2 \mathsf{Var}[X]$. For a $0/1$-random variable X_i, we have $\mathsf{Var}[X_i] \leq 1/4$ because in this case $\mathsf{Exp}[X_i] = \mathsf{Exp}[X_i^2]$ and so $\mathsf{Var}[X_i] = \mathsf{Exp}[X_i](1 - \mathsf{Exp}[X_i])$, which is maximized when $\mathsf{Exp}[X_i] = \frac{1}{2}$.

PROPOSITION A.11 (Chebyshev's inequality)   *Let X be a random variable and $\delta > 0$. Then:*

$$\Pr[|X - \mathsf{Exp}[X]| \geq \delta] \leq \frac{\mathsf{Var}[X]}{\delta^2}.$$

PROOF   Define the non-negative random variable $Y \stackrel{\text{def}}{=} (X - \mathsf{Exp}[X])^2$ and then apply Markov's inequality. So,

$$\Pr[|X - \mathsf{Exp}[X]| \geq \delta] = \Pr[(X - \mathsf{Exp}[X])^2 \geq \delta^2]$$
$$\leq \frac{\mathsf{Exp}[(X - \mathsf{Exp}[X])^2]}{\delta^2} = \frac{\mathsf{Var}[X]}{\delta^2}.$$

∎

The $0/1$-random variables $X_1, \ldots, X_m$ are *pairwise independent* if for every $i \neq j$ and every $b_i, b_j \in \{0, 1\}$ it holds that

$$\Pr[X_i = b_i \ \wedge \ X_j = b_j] = \Pr[X_i = b_i] \cdot \Pr[X_j = h_j].$$

If $X_1, \ldots, X_m$ are pairwise independent then $\mathsf{Var}[\sum_{i=1}^m X_i] = \sum_{i=1}^m \mathsf{Var}[X_i]$. (This follows since $\mathsf{Exp}[X_i \cdot X_j] = \mathsf{Exp}[X_i] \cdot \mathsf{Exp}[X_j]$ when $i \neq j$, using pairwise independence.) An important corollary of Chebyshev's inequality follows.

COROLLARY A.12   *Let $X_1, \ldots, X_m$ be pairwise-independent random variables with the same expectation μ and variance σ^2. Then for every $\delta > 0$,*

$$\Pr\left[\left|\frac{\sum_{i=1}^m X_i}{m} - \mu\right| \geq \delta\right] \leq \frac{\sigma^2}{\delta^2 m}.$$

PROOF   By linearity of expectation, $\mathsf{Exp}[\sum_{i=1}^m X_i / m] = \mu$. Applying Chebyshev's inequality to the random variable $\sum_{i=1}^m X_i / m$, we have

$$\Pr\left[\left|\frac{\sum_{i=1}^m X_i}{m} - \mu\right| \geq \delta\right] \leq \frac{\mathsf{Var}\left[\frac{1}{m} \cdot \sum_{i=1}^m X_i\right]}{\delta^2}.$$

Using pairwise independence, it follows that

$$\mathsf{Var}\left[\frac{1}{m} \cdot \sum_{i=1}^m X_i\right] = \frac{1}{m^2} \sum_{i=1}^m \mathsf{Var}[X_i] = \frac{1}{m^2} \sum_{i=1}^m \sigma^2 = \frac{\sigma^2}{m}.$$

The inequality is obtained by combining the above two equations.   ∎

Say $0/1$-random variables $X_1, \ldots, X_m$ each provides an estimate of some fixed (unknown) bit b. That is, $\Pr[X_i = b] \geq 1/2 + \varepsilon$ for all i, where $\varepsilon > 0$.

We can estimate b by looking at the value of X_1; this estimate will be correct with probability $\Pr[X_1 = b]$. A better estimate can be obtained by looking at the values of $X_1, \ldots, X_m$ and taking the value that occurs the majority of the time. The following allows us to analyze how well this does when $X_1, \ldots, X_m$ are pairwise independent.

PROPOSITION A.13 *Fix $\varepsilon > 0$ and $b \in \{0, 1\}$, and let $\{X_i\}$ be pairwise-independent, 0/1-random variables for which $\Pr[X_i = b] \geq \frac{1}{2} + \varepsilon$ for all i. Consider the process in which m values $X_1, \ldots, X_m$ are recorded and X is set to the value that occurs a strict majority of the time. Then*

$$\Pr[X \neq b] \leq \frac{1}{4 \cdot \varepsilon^2 \cdot m}.$$

PROOF By symmetry, we may assume $b = 1$. Then $\mathsf{Exp}[X_i] \geq \frac{1}{2} + \varepsilon$; we assume $\mathsf{Exp}[X_i] = \frac{1}{2} + \varepsilon$ as that is the worst case. Let X denote the strict majority of the $\{X_i\}$, and note that $X \neq 1$ if and only if $\sum_{i=1}^{m} X_i \leq m/2$. So

$$\begin{aligned}
\Pr[X \neq 1] &= \Pr\left[\sum_{i=1}^{m} X_i \leq m/2\right] \\
&= \Pr\left[\frac{\sum_{i=1}^{m} X_i}{m} - \frac{1}{2} \leq 0\right] \\
&= \Pr\left[\frac{\sum_{i=1}^{m} X_i}{m} - \left(\frac{1}{2} + \varepsilon\right) \leq -\varepsilon\right] \\
&\leq \Pr\left[\left|\frac{\sum_{i=1}^{m} X_i}{m} - \left(\frac{1}{2} + \varepsilon\right)\right| \geq \varepsilon\right].
\end{aligned}$$

Since $\mathsf{Var}[X_i] \leq 1/4$ for all i, applying the previous corollary shows that $\Pr[X \neq 1] \leq \frac{1}{4\varepsilon^2 m}$ as claimed. ∎

A better bound is possible if the $\{X_i\}$ are independent:

PROPOSITION A.14 (Chernoff bound) *Fix $\varepsilon > 0$ and $b \in \{0, 1\}$, and let $\{X_i\}$ be independent 0/1-random variables with $\Pr[X_i = b] = \frac{1}{2} + \varepsilon$ for all i. The probability that their majority value is not b is at most $e^{-\varepsilon^2 m/2}$.*

A.4 The "Birthday" Problem

If we choose q elements $y_1, \ldots, y_q$ uniformly from a set of size N, what is the probability that there exist distinct i, j with $y_i = y_j$? We refer to the stated event as a *collision*, and let $\mathsf{coll}(q, N)$ denote the probability of this event. This problem is related to the so-called *birthday problem*, which asks what size group of people we need such that with probability $1/2$ some pair of people in the group share a birthday. To see the relationship, let y_i denote the birthday of the ith person in the group. If there are q people in the group then we have q values $y_1, \ldots, y_q$ chosen uniformly from $\{1, \ldots, 365\}$, making the simplifying assumption that birthdays are uniformly and independently distributed among the 365 days of a non-leap year. Furthermore, matching birthdays correspond to a collision, i.e., distinct i, j with $y_i = y_j$. So the desired solution to the birthday problem is given by the minimal (integer) value of q for which $\mathsf{coll}(q, 365) \geq 1/2$. (The answer may surprise you—taking $q = 23$ people suffices!)

The following shows that when $q \leq \sqrt{2N}$, the probability of a collision is $\Theta(q^2/N)$; alternately, for $q = \Theta(\sqrt{N})$ the probability of a collision is constant.

LEMMA A.15 *Fix a positive integer N, and say $q \leq \sqrt{2N}$ elements $y_1, \ldots, y_q$ are chosen uniformly and independently from a set of size N. Then*

$$\frac{q \cdot (q-1)}{4N} \leq 1 - e^{-q(q-1)/2N} \leq \mathsf{coll}(q, N) \leq \frac{q \cdot (q-1)}{2N}.$$

PROOF The upper bound, which holds for arbitrary q, can be proven by a simple application of the union bound (Proposition A.7). Recall that a *collision* means that there exist distinct i, j with $y_i = y_j$. Let Coll denote the event of a collision, and let $\mathsf{Coll}_{i,j}$ denote the event that $y_i = y_j$. It is immediate that $\Pr[\mathsf{Coll}_{i,j}] = 1/N$ for any distinct i, j. Furthermore, $\mathsf{Coll} = \bigvee_{i \neq j} \mathsf{Coll}_{i,j}$ and so repeated application of the union bound implies that

$$\Pr[\mathsf{Coll}] = \Pr\left[\bigvee_{i \neq j} \mathsf{Coll}_{i,j}\right]$$

$$\leq \sum_{i \neq j} \Pr[\mathsf{Coll}_{i,j}] = \binom{q}{2} \cdot \frac{1}{N}.$$

For the lower bound, let NoColl_i be the event that there is *no* collision among $y_1, \ldots, y_i$; that is, $y_j \neq y_k$ for all $j < k \leq i$. Then $\mathsf{NoColl}_q = \overline{\mathsf{Coll}}$ is the event that there is no collision at all. If NoColl_q occurs then NoColl_i must also have occurred for all $i \leq q$. Thus,

$$\Pr[\mathsf{NoColl}_q] = \Pr[\mathsf{NoColl}_1] \cdot \Pr[\mathsf{NoColl}_2 \mid \mathsf{NoColl}_1] \cdots \Pr[\mathsf{NoColl}_q \mid \mathsf{NoColl}_{q-1}].$$

Now, $\Pr[\mathsf{NoColl}_1] = 1$ since y_1 cannot collide with itself. Furthermore, if event NoColl_i occurs then $\{y_1, \ldots, y_i\}$ contains i distinct values; so, the probability that y_{i+1} collides with one of these values is $\frac{i}{N}$ and hence the probability that y_{i+1} does *not* collide with any of these values is $1 - \frac{i}{N}$. This means

$$\Pr[\mathsf{NoColl}_{i+1} \mid \mathsf{NoColl}_i] = 1 - \frac{i}{N},$$

and so

$$\Pr[\mathsf{NoColl}_q] = \prod_{i=1}^{q-1} \left(1 - \frac{i}{N}\right).$$

Since $i/N < 1$ for all i, we have $1 - \frac{i}{N} \le e^{-i/N}$ (by Inequality A.3) and so

$$\Pr[\mathsf{NoColl}_q] \le \prod_{i=1}^{q-1} e^{-i/N} = e^{-\sum_{i=1}^{q-1}(i/N)} = e^{-q(q-1)/2N}.$$

We conclude that

$$\Pr[\mathsf{Coll}] = 1 - \Pr[\mathsf{NoColl}_q] \ge 1 - e^{-q(q-1)/2N} \ge \frac{q(q-1)}{4N},$$

using Inequality A.4 in the last step (note that $q(q-1)/2N < 1$). ∎

As a simple application of Lemma A.15, we show that any pseudorandom permutation is also a pseudorandom function (cf. Proposition 3.26). Recall that a pseudorandom permutation has $\ell_{in} = \ell_{out}$, meaning that its input and output lengths are equal. Our proof here is adapted from [27].

PROPOSITION A.16 *If F is a pseudorandom permutation and furthermore $\ell_{out}(n) \ge n$, then F is also a pseudorandom function.*

PROOF For simplicity of notation, we assume $\ell_{in} = \ell_{out} = n$. The crux of the proof is to show that a random permutation is indistinguishable (using polynomially many queries) from a random function. Let D be an algorithm, and let $q = q(n)$ be the number of queries that D makes to its oracle. (We assume without loss of generality that D always makes exactly q queries, and that it never repeats a query.) We will allow D to be all-powerful (and hence may assume it is deterministic), but will assume that the number of queries q that it makes is polynomial. We show

$$\left| \Pr_{f \leftarrow \mathsf{Func}_n}[D^{f(\cdot)}(1^n) = 1] - \Pr_{f \leftarrow \mathsf{Perm}_n}[D^{f(\cdot)}(1^n) = 1] \right| < \frac{q^2}{2^{n+1}}. \tag{A.1}$$

The intuition for this is that the only way D can tell that its oracle f is *not* a permutation is by observing a collision, i.e., two distinct inputs that map to the same output. However, the probability of finding such a collision when querying a random function q times is at most $\text{coll}(q, 2^n) \leq q^2/2^n$, which is negligible for any polynomial q.

Formally, let Coll be the event that two queries by D to its oracle return the same result. We claim first that

$$\Pr_{f \leftarrow \text{Func}_n}[D^{f(\cdot)}(1^n) = 1 \mid \overline{\text{Coll}}] = \Pr_{f \leftarrow \text{Perm}_n}[D^{f(\cdot)}(1^n) = 1]. \qquad (A.2)$$

To see this, observe that the behavior of D is completely characterized by the set $S \subseteq (\{0,1\}^n)^q$ of q-tuples such that $\vec{a} = (a_1, \ldots, a_q) \in S$ iff D outputs 1 when it receives a_i as the response to its ith oracle query for all i. Let distinct $\subset (\{0,1\}^n)^q$ denote the set of q-tuples where each entry is distinct. When f is a permutation, then each $\vec{a} \in$ distinct is equally likely and $\vec{a} \notin$ distinct cannot occur; thus

$$\Pr_{f \leftarrow \text{Perm}_n}[D^{f(\cdot)}(1^n) = 1] = \frac{|S \cap \text{distinct}|}{|\text{distinct}|}.$$

When f is a random function, each q-tuple in $(\{0,1\}^n)^q$ occurs with probability 2^{-nq}. So, using Bayes' theorem

$$\Pr_{f \leftarrow \text{Func}_n}[D^{f(\cdot)}(1^n) = 1 \mid \overline{\text{Coll}}] = \frac{\Pr_{f \leftarrow \text{Func}_n}[D^{f(\cdot)}(1^n) = 1 \wedge \overline{\text{Coll}}]}{\Pr_{f \leftarrow \text{Func}_n}[\overline{\text{Coll}}]}$$

$$= \frac{2^{-nq} \cdot |S \cap \text{distinct}|}{2^{-nq} \cdot |\text{distinct}|}.$$

Equation (A.2) follows.

As a consequence,

$$\left| \Pr_{f \leftarrow \text{Func}_n}[D^{f(\cdot)}(1^n) = 1] - \Pr_{f \leftarrow \text{Perm}_n}[D^{f(\cdot)}(1^n) = 1] \right|$$

$$= \left| \Pr_{f \leftarrow \text{Func}_n}[D^{f(\cdot)}(1^n) = 1 \mid \overline{\text{Coll}}] \cdot \Pr[\overline{\text{Coll}}] \right.$$

$$\left. + \Pr_{f \leftarrow \text{Func}_n}[D^{f(\cdot)}(1^n) = 1 \mid \text{Coll}] \cdot \Pr[\text{Coll}] - \Pr_{f \leftarrow \text{Perm}_n}[D^{f(\cdot)}(1^n) = 1] \right|$$

$$= \left| \Pr_{f \leftarrow \text{Func}_n}[D^{f(\cdot)}(1^n) = 1 \mid \text{Coll}] \cdot \Pr[\text{Coll}] - \Pr_{f \leftarrow \text{Perm}_n}[D^{f(\cdot)}(1^n) = 1] \cdot \Pr[\text{Coll}] \right|$$

$$\leq \Pr[\text{Coll}].$$

With Lemma A.15, this implies Equation (A.1) and completes the proof. ∎

While the above shows that a pseudorandom permutation (PRP) is asymptotically also a pseudorandom function (PRF), it does also indicate a concrete-security gap: namely, a PRP *can* be distinguished from a PRF with probability $\mathcal{O}(q^2/2^{\ell_{out}(n)})$ using q queries. This is important to keep in mind when using a block cipher and treating it in the analysis as a PRF.

A.5 *Finite Fields

We use finite fields only sparingly in the book, but we include a definition and some basic facts for completeness. Further details can be found in any textbook on abstract algebra.

DEFINITION A.17 *A (finite) field is a (finite) set $\mathbb{F}$ along with two binary operations $+, \cdot$ for which the following hold:*

- *$\mathbb{F}$ is an abelian group with respect to the operation '$+$.' We let 0 denote the identity element of this group.*

- *$\mathbb{F} \setminus \{0\}$ is an abelian group with respect to the operation '$\cdot$.' We let 1 denote the identity element of this group.*

  As usual, we often write ab in place of $a \cdot b$.

- *(**Distributivity:**) For all $a, b, c \in \mathbb{F}$, we have $a \cdot (b + c) = ab + ac$.*

The *additive inverse* of $a \in \mathbb{F}$, denoted by $-a$, is the unique element satisfying $a + (-a) = 0$; we write $b - a$ in place of $b + (-a)$. The *multiplicative inverse* of $a \in \mathbb{F} \setminus \{0\}$, denoted by a^{-1}, is the unique element satisfying $aa^{-1} = 1$; we often write b/a in place of ba^{-1}.

Example A.18
It follows from the results of Section 9.1.4 that for any prime p the set $\{0, \ldots, p-1\}$ is a finite field with respect to addition and multiplication modulo p. We denote this field by $\mathbb{F}_p$. $\diamond$

Finite fields have a rich theory. For our purposes, we need only a few basic facts. The *order* of $\mathbb{F}$ is the number of elements in $\mathbb{F}$ (assuming it is finite). Recall also that q is a prime power if $q = p^r$ for some prime p and integer $r \geq 1$.

THEOREM A.19 *If $\mathbb{F}$ is a finite field, then the order of $\mathbb{F}$ is a prime power. Conversely, for every prime power q there is a finite field of order q, which is moreover the unique such field (up to relabeling of the elements).*

For $q = p^r$ with p prime, we let $\mathbb{F}_q$ denote the (unique) field of order q. We call p the *characteristic* of $\mathbb{F}_q$.

As in the case of groups, if n is a positive integer and $a \in \mathbb{F}$ then

$$n \cdot a \overset{\text{def}}{=} \underbrace{a + \cdots + a}_{n \text{ times}} \quad \text{and} \quad a^n \overset{\text{def}}{=} \underbrace{a \cdots a}_{n \text{ times}}.$$

The notation is extended for $n \leq 0$ in the natural way.

THEOREM A.20 Let $\mathbb{F}_q$ be a finite field of characteristic p. Then for all $a \in \mathbb{F}_q$ we have $p \cdot a = 0$.

Let $q = p^r$ with p prime. For $r = 1$, we have seen in Example A.18 that $\mathbb{F}_q = \mathbb{F}_p$ can be taken to be the set $\{0, \dots, p-1\}$ under addition and multiplication modulo p. We caution, however, that for $r > 1$ the set $\{0, \dots, q-1\}$ is *not* a field under addition and multiplication modulo q. For example, if we take $q = 3^2 = 9$ then the element 3 does not have a multiplicative inverse modulo 9.

Finite fields of characteristic p can be represented using polynomials over $\mathbb{F}_p$. We give an example to demonstrate the flavor of the construction, without discussing why the construction works or describing the general case. We construct the field $\mathbb{F}_4$ by working with polynomials over $\mathbb{F}_2$. Fix the polynomial $r(x) = x^2 + x + 1$, and note that $r(x)$ has no roots over $\mathbb{F}_2$ since $r(0) = r(1) = 1$ (recall that we are working in $\mathbb{F}_2$, which means that all operations are carried out modulo 2). In the same way that we can introduce the imaginary number i to be a root of $x^2 + 1$ over the reals, we can introduce a value ω to be a root of $r(x)$ over $\mathbb{F}_2$; that is, $\omega^2 = -\omega - 1$. We then define $\mathbb{F}_4$ to be the set of all degree-1 polynomials in ω over $\mathbb{F}_2$; that is, $\mathbb{F}_4 = \{0, 1, \omega, \omega + 1\}$. Addition in $\mathbb{F}_4$ will just be regular polynomial addition, remembering that operations on the coefficients are done in $\mathbb{F}_2$ (that is, modulo 2). Multiplication in $\mathbb{F}_4$ will be polynomial multiplication (again, with operations on the coefficients carried out modulo 2) followed by the substitution $\omega^2 = -\omega - 1$; this also ensures that the result lies in $\mathbb{F}_4$. So, for example,

$$\omega + (\omega + 1) = 2\omega + 1 = 1$$

and

$$(\omega + 1) \cdot (\omega + 1) = \omega^2 + 2\omega + 1 = (-\omega - 1) + 1 = -\omega = \omega.$$

Although not obvious, one can check that this is a field; the only difficult condition to verify is that every nonzero element has a multiplicative inverse. We need only one other result.

THEOREM A.21 Let $\mathbb{F}_q$ be a finite field of order q. Then the abelian group $\mathbb{F}_q \setminus \{0\}$ with respect to '$\cdot$' is a **cyclic** group of order $q - 1$.

Appendix B

Basic Algorithmic Number Theory

For the cryptographic constructions given in this book to be efficient (i.e., to run in time polynomial in the lengths of their inputs), it is necessary for these constructions to utilize efficient (that is, polynomial-time) algorithms for performing basic number-theoretic operations. Although in some cases there exist "trivial" algorithms that would work, it is still worthwhile to carefully consider their efficiency since for cryptographic applications it is not uncommon to use integers that are *thousands* of bits long. In other cases obtaining any polynomial-time algorithm requires a bit of cleverness, and an analysis of their performance may rely on non-trivial group-theoretic results.

In Appendix B.1 we describe basic algorithms for integer arithmetic. Here we cover the familiar algorithms for addition, subtraction, etc., as well as the Euclidean algorithm for computing greatest common divisors. We also discuss the extended Euclidean algorithm, assuming there that the reader has covered the material in Section 9.1.1.

In Appendix B.2 we show various algorithms for modular arithmetic. In addition to a brief discussion of basic modular operations (i.e., modular reduction, addition, multiplication, and inversion), we also describe *Montgomery multiplication*, which can greatly simplify (and speed up) implementations of modular arithmetic. We then discuss algorithms for problems that are less common outside the field of cryptography: exponentiation modulo N (as well as in arbitrary groups) and choosing a uniform element of $\mathbb{Z}_N$ or $\mathbb{Z}_N^*$ (or in an arbitrary group). This section assumes familiarity with the basic group theory covered in Section 9.1.

The material above is used implicitly throughout the second half of the book, although it is not absolutely necessary to read this material in order to follow the book. (In particular, the reader willing to accept the results of this Appendix without proof can simply read the summary of those results in the theorems below.) Appendix B.3, which discusses finding generators in cyclic groups (when the factorization of the group order is known) and assumes the results of Section 9.3.1, contains material that is hardly used at all; it is included for completeness and reference.

Since our goal is only to establish that certain problems can be solved in polynomial time, we have opted for *simplicity* rather than *efficiency* in our selection of algorithms and their descriptions (as long as the algorithms run in polynomial time). For this reason, we generally will not be interested in the exact running times of the algorithms we present beyond establishing that they indeed run in polynomial time. The reader who is seriously interested in implementing these algorithms is forewarned to look at other sources for more efficient alternatives as well as various techniques for speeding up the necessary computations.

The results in this Appendix are summarized by the theorems that follow. Throughout, we assume that any integer a provided as input is written using exactly $\|a\|$ bits; i.e., the high-order bit is 1. In Appendix B.1 we show:

THEOREM B.1 (Integer operations)   *Given integers a and b, it is possible to perform the following operations in time polynomial in $\|a\|$ and $\|b\|$:*

1. *Computing the sum $a + b$ and the difference $a - b$;*

2. *Computing the product ab;*

3. *Computing positive integers q and $r < b$ such that $a = qb + r$ (i.e.,* computing division with remainder*);*

4. *Computing the greatest common divisor of a and b, $\gcd(a, b)$;*

5. *Computing integers X, Y with $Xa + Yb = \gcd(a, b)$.*

The following results are proved in Appendix B.2:

THEOREM B.2 (Modular operations)   *Given integers $N > 1$, a, and b, it is possible to perform the following operations in time polynomial in $\|a\|$, $\|b\|$, and $\|N\|$:*

1. *Computing the modular reduction $[a \bmod N]$;*

2. *Computing the sum $[(a + b) \bmod N]$, the difference $[(a - b) \bmod N]$, and the product $[ab \bmod N]$;*

3. *Determining whether a is invertible modulo N and, if so, computing the multiplicative inverse $[a^{-1} \bmod N]$;*

4. *Computing the exponentiation $[a^b \bmod N]$.*

The following generalizes Theorem B.2(5) to arbitrary groups:

THEOREM B.3 (Group exponentiation)   *Let $\mathbb{G}$ be a group, written multiplicatively. Let g be an element of the group and let b be a non-negative integer. Then g^b can be computed using $\mathsf{poly}(\|b\|)$ group operations.*

THEOREM B.4 (Choosing uniform elements) *There exists a randomized algorithm with the following properties: on input N,*

- *The algorithm runs in time polynomial in $\|N\|$;*

- *The algorithm outputs* fail *with probability negligible in $\|N\|$; and*

- *Conditioned on not outputting* fail, *the algorithm outputs a uniformly distributed element of $\mathbb{Z}_N$.*

An algorithm with analogous properties exists for $\mathbb{Z}_N^*$ as well.

Since the probability that either algorithm referenced in the above theorem outputs fail is negligible, we ignore this possibility (and instead leave it implicit). In Appendix B.2 we also discuss generalizations of the above to the case of selecting a uniform element from any finite group (subject to certain requirements on the representation of group elements).

A proof of the following is in Appendix B.3:

THEOREM B.5 (Testing and finding generators) *Let $\mathbb{G}$ be a cyclic group of order q, and assume that the group operation and selection of a uniform group element can be carried out in unit time.*

1. *There is an algorithm that on input q, the prime factorization of q, and an element $g \in \mathbb{G}$, runs in $\mathsf{poly}(\|q\|)$ time and decides whether g is a generator of $\mathbb{G}$.*

2. *There is a randomized algorithm that on input q and the prime factorization of q, runs in $\mathsf{poly}(\|q\|)$ time and outputs a generator of $\mathbb{G}$ except with probability negligible in $\|q\|$. Conditioned on the output being a generator, it is uniformly distributed among the generators of $\mathbb{G}$.*

B.1 Integer Arithmetic

B.1.1 Basic Operations

We begin our exploration of algorithmic number theory with a discussion of integer addition/subtraction, multiplication, and division with remainder. A little thought shows that all these operations can be carried out in time polynomial in the input length using the standard "grade-school" algorithms for these problems. For example, addition of two positive integers a and b with $a > b$ can be done in time linear in $\|a\|$ by stepping one-by-one through the bits of a and b, starting with the low-order bits, and computing the corresponding output bit and a "carry bit" at each step. (Details are omitted.) Multiplication of two n-bit integers a and b, to take another example, can be done by first generating a list of n integers of length at most $2n$ (each of which is equal to $a \cdot 2^{i-1} \cdot b_i$, where b_i is the ith bit of b) and then adding these n integers together to obtain the final result. (Division with remainder is trickier to implement efficiently, but can also be done.)

Although these grade-school algorithms suffice to demonstrate that the aforementioned problems can be solved in polynomial time, it is interesting to note that these algorithms are in some cases not the best ones available. As an example, the simple algorithm for multiplication given above multiplies two n-bit numbers in time $\mathcal{O}(n^2)$, but there exists a better algorithm running in time $\mathcal{O}(n^{\log_2 3})$ (and even that is not the best possible). While the difference is insignificant for numbers of the size we encounter daily, it becomes noticeable when the numbers are large. In cryptographic applications it is not uncommon to use integers that are thousands of bits long (i.e., $n > 1000$), and a judicious choice of which algorithms to use then becomes critical.

B.1.2　The Euclidean and Extended Euclidean Algorithms

Recall from Section 9.1 that $\gcd(a, b)$, the *greatest common divisor* of two integers a and b, is the largest integer d that divides both a and b. We state an easy proposition regarding the greatest common divisor, and then show how this leads to an efficient algorithm for computing gcd's.

PROPOSITION B.6　*Let $a, b > 1$ with $b \nmid a$. Then*

$$\gcd(a, b) = \gcd(b, [a \bmod b]).$$

PROOF　If $b > a$ the claim is immediate; so, assume $a > b$. Write $a = qb + r$ for q, r positive integers and $r < b$ (cf. Proposition 9.1); note $r > 0$ because $b \nmid a$. Since $r = [a \bmod b]$, it remains to show that $\gcd(a, b) = \gcd(b, r)$.

The claim follows since for any positive integer d we have

$$d \mid a \text{ and } d \mid b \iff d \mid (a - qb) \text{ and } d \mid b,$$

and $r = a - qb$.　　　　　　　　　　　　　　　　　　　　　　　　　　■

The above suggests the recursive *Euclidean algorithm* (Algorithm B.7) for computing the greatest common divisor $\gcd(a, b)$ of two integers a and b. Correctness of the algorithm follows readily from Proposition B.6. As for its running time, we show below that on input (a, b) the algorithm makes fewer than $2 \cdot \|b\|$ recursive calls. Since checking whether b divides a and computing $[a \bmod b]$ can both be done in time polynomial in $\|a\|$ and $\|b\|$, this implies that the entire algorithm runs in polynomial time.

PROPOSITION B.8　*Consider an execution of $\mathsf{GCD}(a_0, b_0)$ where it holds*

> **ALGORITHM B.7**
> **The Euclidean algorithm** GCD
>
> **Input:** Integers a, b with $a \geq b > 0$
> **Output:** The greatest common divisor of a and b
>
> **if** b divides a
>   **return** b
> **else return** GCD$(b, [a \bmod b])$

that $a_0 \geq b_0 > 0$, and let a_i, b_i (for $i = 1, \ldots, \ell$) denote the arguments to the ith recursive call of GCD. Then $b_{i+2} \leq b_i/2$ for $0 \leq i \leq \ell - 2$.

PROOF First note that for any $a > b$ we have $[a \bmod b] < a/2$. To see this, consider the two cases: If $b \leq a/2$ then $[a \bmod b] < b \leq a/2$ is immediate. On the other hand, if $b > a/2$ then $[a \bmod b] = a - b < a/2$.

Now fix arbitrary i with $0 \leq i \leq \ell - 2$. Then $b_{i+2} = [a_{i+1} \bmod b_{i+1}] < a_{i+1}/2 = b_i/2$. ∎

COROLLARY B.9 *In an execution of algorithm GCD(a, b), there are at most $2\|b\| - 2$ recursive calls to GCD.*

PROOF Let a_i, b_i (for $i = 1, \ldots, \ell$) denote the arguments to the ith recursive call of GCD. The $\{b_i\}$ are always greater than zero, and the algorithm makes no further recursive calls if it ever happens that $b_i = 1$ (since then $b_i \mid a_i$). The previous proposition indicates that the $\{b_i\}$ decrease by a multiplicative factor of (at least) 2 in every two iterations. It follows that the number of recursive calls to GCD is at most $2 \cdot (\|b\| - 1)$. ∎

By Proposition 9.2, we know that for positive integers a, b there exist integers X, Y with $Xa + Yb = \gcd(a, b)$. A simple modification of the Euclidean algorithm, called the *extended Euclidean algorithm*, can be used to find X, Y in addition to computing $\gcd(a, b)$; see Algorithm B.10. You are asked to show correctness of the extended Euclidean algorithm in Exercise B.1, and to prove that the algorithm runs in polynomial time in Exercise B.2.

B.2 Modular Arithmetic

We now turn our attention to basic arithmetic operations modulo $N > 1$. We will use $\mathbb{Z}_N$ to refer both to the set $\{0, \ldots, N - 1\}$ as well as to the group

---

ALGORITHM B.10
The extended Euclidean algorithm eGCD

Input: Integers a, b with $a \geq b > 0$
Output: (d, X, Y) with $d = \gcd(a, b)$ and $Xa + Yb = d$

if b divides a
    return $(b, 0, 1)$
else
    Compute integers q, r with $a = qb + r$ and $0 < r < b$
    $(d, X, Y) := \text{eGCD}(b, r)$ // note that $Xb + Yr = d$
    return $(d, Y, X - Yq)$

---

that results by considering addition modulo N among the elements of this set.

B.2.1 Basic Operations

Efficient algorithms for the basic arithmetic operations over the integers immediately imply efficient algorithms for the corresponding arithmetic operations modulo N. For example, computing the modular reduction $[a \bmod N]$ can be done in time polynomial in $\|a\|$ and $\|N\|$ by computing division-with-remainder over the integers. Next consider modular operations on two elements $a, b \in \mathbb{Z}_N$ where $\|N\| = n$. (Note that a, b have length at most n. Actually, it is convenient to simply assume that all elements of $\mathbb{Z}_N$ have length *exactly* n, padding to the left with 0s if necessary.) Addition of a and b modulo N can be done by first computing $a + b$, an integer of length at most $n + 1$, and then reducing this intermediate result modulo N. Similarly, multiplication modulo N can be performed by first computing the integer ab of length at most $2n$ and then reducing the result modulo N. Since addition, multiplication, and division-with-remainder can all be done in polynomial time, these give polynomial-time algorithms for addition and multiplication modulo N.

B.2.2 Computing Modular Inverses

Our discussion thus far has shown how to add, subtract, and multiply modulo N. One operation we are missing is "division" or, equivalently, computing multiplicative inverses modulo N. Recall from Section 9.1.2 that the multiplicative inverse (modulo N) of an element $a \in \mathbb{Z}_N$ is an element $a^{-1} \in \mathbb{Z}_N$ such that $a \cdot a^{-1} = 1 \bmod N$. Proposition 9.7 shows that a has an inverse if and only if $\gcd(a, N) = 1$, i.e., if and only if $a \in \mathbb{Z}_N^*$. Thus, using the Euclidean algorithm we can easily determine whether a given element a has a multiplicative inverse modulo N.

Given N and $a \in \mathbb{Z}_N$ with $\gcd(a, N) = 1$, Proposition 9.2 tells us that there exist integers X, Y with $Xa + YN = 1$. This means that $[X \bmod N]$ is the multiplicative inverse of a. Integers X and Y satisfying $Xa + YN = 1$ can be found efficiently using the extended Euclidean algorithm eGCD shown in

---

ALGORITHM B.11
Computing modular inverses

Input: Modulus N; element a
Output: $[a^{-1} \bmod N]$ (if it exists)

$(d, X, Y) := \mathsf{eGCD}(a, N)$ // note that $Xa + YN = \gcd(a, N)$
if $d \neq 1$ **return** "a is not invertible modulo N"
else return $[X \bmod N]$

Section B.1.2. This leads to a polynomial-time algorithm (Algorithm B.11) for computing multiplicative inverses.

B.2.3 Modular Exponentiation

A more challenging task is that of *exponentiation* modulo N, that is, computing $[a^b \bmod N]$ for base $a \in \mathbb{Z}_N$ and integer exponent $b > 0$. (When $b = 0$ the problem is easy. When $b < 0$ and $a \in \mathbb{Z}_N^*$ then $a^b = (a^{-1})^{-b} \bmod N$ and the problem is reduced to the case of exponentiation with a positive exponent given that we can compute inverses, as discussed in the previous section.) Notice that the basic approach used in the case of addition and multiplication (i.e., computing the integer a^b and then reducing this intermediate result modulo N) does *not* work here: the integer a^b has length $\|a^b\| = \Theta(\log a^b) = \Theta(b \cdot \|a\|)$, and so even storing the intermediate result a^b would require time exponential in $\|b\| = \Theta(\log b)$.

We can address this problem by reducing modulo N at all intermediate steps of the computation, rather than only reducing modulo N at the end. This has the effect of keeping the intermediate results "small" throughout the computation. Even with this important initial observation, it is still nontrivial to design a polynomial-time algorithm for modular exponentiation. Consider the naïve approach of Algorithm B.12, which simply performs b multiplications by a. This still runs in time that is *exponential* in $\|b\|$.

ALGORITHM B.12
A naïve algorithm for modular exponentiation

Input: Modulus N; base $a \in \mathbb{Z}_N$; integer exponent $b > 0$
Output: $[a^b \bmod N]$

$x := 1$
for $i = 1$ **to** b:
    $x := [x \cdot a \bmod N]$
return x

This naïve algorithm can be viewed as relying on the following recurrence:

$$[a^b \bmod N] = [a \cdot a^{b-1} \bmod N] = [a \cdot a \cdot a^{b-2} \bmod N] = \cdots$$

Any algorithm based on this relationship will require $\Theta(b)$ time. We can do better by relying on the following recurrence:

$$[a^b \bmod N] = \begin{cases} \left[\left(a^{\frac{b}{2}} \right)^2 \bmod N \right] & \text{when } b \text{ is even} \\ \left[a \cdot \left(a^{\frac{b-1}{2}} \right)^2 \bmod N \right] & \text{when } b \text{ is odd.} \end{cases}$$

Doing so leads to an algorithm—called, for obvious reasons, "square-and-multiply" (or "repeated squaring")—that requires only $\mathcal{O}(\log b) = \mathcal{O}(\|b\|)$ modular squarings/multiplications; see Algorithm B.13. In this algorithm, the length of b decreases by 1 in each iteration; it follows that the number of iterations is $\|b\|$, and so the overall algorithm runs in time polynomial in $\|a\|$, $\|b\|$, and $\|N\|$. More precisely, the number of modular squarings is exactly $\|b\|$, and the number of additional modular multiplications is exactly the Hamming weight of b (i.e., the number of 1s in the binary representation of b). This explains the preference, discussed in Section 9.2.4, for choosing the public RSA exponent e to have small length/Hamming weight.

ALGORITHM B.13
Algorithm ModExp for efficient modular exponentiation

Input: Modulus N; base $a \in \mathbb{Z}_N$; integer exponent $b > 0$
Output: $[a^b \bmod N]$

$x := a$
$t := 1$
// maintain the invariant that the answer is $[t \cdot x^b \bmod N]$
while $b > 0$ **do:**
    if b is odd
        $t := [t \cdot x \bmod N]$, $b := b - 1$
    $x := [x^2 \bmod N]$, $b := b/2$
return t

Fix a and N and consider the modular exponentiation function given by $f_{a,N}(b) = [a^b \bmod N]$. We have just seen that computing $f_{a,N}$ is easy. In contrast, computing the *inverse* of this function—that is, computing b given a, N, and $[a^b \bmod N]$—is believed to be hard for appropriate choice of a and N. Inverting this function requires solving the *discrete-logarithm problem*, something we discuss in detail in Section 9.3.2.

Using precomputation. If the base a is known in advance, and there is a bound on the length of the exponent b, then one can use precomputation and a small amount of memory to speed up computation of $[a^b \bmod N]$. Say $\|b\| \leq n$. Then we precompute and store the n values

$$x_0 := a, \quad x_1 := [a^2 \bmod N], \quad \ldots, \quad x_{n-1} := [a^{2^{n-1}} \bmod N].$$

Given exponent b with binary representation $b_{n-1} \cdots b_0$ (written from most to least significant bit), we then have

$$a^b = a^{\sum_{i=0}^{n-1} 2^i \cdot b_i} = \prod_{i=0}^{n-1} x_i^{b_i} \bmod N.$$

Since $b_i \in \{0, 1\}$, the number of multiplications needed to compute the result is exactly one less than the Hamming weight of b.

Exponentiation in Arbitrary Groups

The efficient modular exponentiation algorithm given above carries over in a straightforward way to enable efficient exponentiation in any group, as long as the underlying group operation can be performed efficiently. Specifically, if $\mathbb{G}$ is a group and g is an element of $\mathbb{G}$, then g^b can be computed using at most $2 \cdot \|b\|$ applications of the underlying group operation. Precomputation could also be used, exactly as described above.

If the order q of $\mathbb{G}$ is known, then $a^b = a^{[b \bmod q]}$ (cf. Proposition 9.53) and this can be used to speed up the computation by reducing b modulo q first.

Considering the (additive) group $\mathbb{Z}_N$, the group exponentiation algorithm just described gives a method for computing the "exponentiation"

$$[b \cdot g \bmod N] \stackrel{\text{def}}{=} [\underbrace{g + \cdots + g}_{b \text{ times}} \bmod N]$$

that differs from the method discussed earlier that relies on standard integer multiplication followed by a modular reduction. In comparing the two approaches to solving the same problem, note that the original algorithm uses specific information about $\mathbb{Z}_N$; in particular, it (essentially) treats the "exponent" b as an element of $\mathbb{Z}_N$ (possibly by reducing b modulo N first). In contrast, the "square-and-multiply" algorithm just presented treats $\mathbb{Z}_N$ only as an abstract group. (Of course, the group operation of addition modulo N relies on the specifics of $\mathbb{Z}_N$.) The point of this discussion is merely to illustrate that some group algorithms are *generic* (i.e., they apply equally well to all groups) while some group algorithms rely on specific properties of a *particular* group or class of groups. We saw some examples of this phenomenon in Chapter 10.

B.2.4   *Montgomery Multiplication

Although division over the integers (and hence modular reduction) can be done in polynomial time, algorithms for integer division are slow in compari-

Introduction to Modern Cryptography

son to, say, algorithms for integer multiplication. Montgomery multiplication provides a way to perform modular multiplication *without* carrying out any expensive modular reductions. Since pre- and postprocessing is required, the method is advantageous only when several modular multiplications will be done in sequence as, e.g., when computing a modular exponentiation.

Fix an odd modulus N with respect to which modular operations are to be done. Let $R > N$ be a power of two, say $R = 2^w$, and note that $\gcd(R, N) = 1$. The key property we will exploit is that division by R is fast: the quotient of x upon division by R is obtained by simply shifting x to the right w positions, and $[x \bmod R]$ is just the w least-significant bits of x.

Define the *Montgomery representation* of $x \in \mathbb{Z}_N^*$ by $\bar{x} \overset{\text{def}}{=} [xR \bmod N]$. Montgomery multiplication of $\bar{x}, \bar{y} \in \mathbb{Z}_N^*$ is defined as

$$\mathsf{Mont}(\bar{x}, \bar{y}) \overset{\text{def}}{=} [\bar{x}\bar{y}R^{-1} \bmod N].$$

(We show below how this can be computed without any expensive modular reductions.) Note that

$$\mathsf{Mont}(\bar{x}, \bar{y}) = \bar{x}\bar{y}R^{-1} = (xR)(yR)R^{-1} = (xy)R = \overline{xy} \bmod N.$$

This means we can multiply several values in $\mathbb{Z}_N$ by (1) converting to the Montgomery representation, (2) carrying out all multiplications using Montgomery multiplication to obtain the final result, and then (3) converting the result from Montgomery representation back to the standard representation.

Let $\alpha \overset{\text{def}}{=} [-N^{-1} \bmod R]$, a value which can be precomputed. (Computation of α, and conversion to/from Montgomery representation, can also be done without any expensive modular reductions; details are beyond our scope.) To compute $c \overset{\text{def}}{=} \mathsf{Mont}(x, y)$ without any expensive modular reductions do:

1. Let $z := x \cdot y$ (over the integers).

2. Set $c' := (z + [z\alpha \bmod R] \cdot N) / R$.

3. If $c' < N$ then set $c := c'$; else set $c := c' - N$.

To see that this works, we first need to verify that step 2 is well-defined, namely, that the numerator is divisible by R. This follows because

$$z + [z\alpha \bmod R] \cdot N = z + z\alpha N = z - zN^{-1}N = 0 \bmod R.$$

Next, note that $c' = z/R \bmod N$ after step 2; moreover, since $z < N^2 < RN$ we have $0 < c' < (z + RN)/R < 2RN/R = 2N$. But then $[c' \bmod N] = c'$ if $c' < N$, and $[c' \bmod N] = c' - N$ if $c' > N$. We conclude that

$$c = [c' \bmod N] = [z/R \bmod N] = [xyR^{-1} \bmod N],$$

as desired.

B.2.5 Choosing a Uniform Group Element

For cryptographic applications, it is often necessary to choose a uniform element of a group $\mathbb{G}$. We first treat the problem in an abstract setting, and then focus specifically on the cases of $\mathbb{Z}_N$ and $\mathbb{Z}_N^*$.

Note that if $\mathbb{G}$ is a cyclic group of order q, and a generator $g \in \mathbb{G}$ is known, then choosing a uniform element $h \in \mathbb{G}$ reduces to choosing a uniform integer $x \in \mathbb{Z}_q$ and setting $h := g^x$. In what follows we make no assumptions on $\mathbb{G}$.

Elements of a group $\mathbb{G}$ must be specified using some *representation* of these elements as bit-strings, where we assume without any real loss of generality that all elements are represented using strings of the same length. (It is also crucial that there is a *unique* string representing each group element.) For example, if $\|N\| = n$ then elements of $\mathbb{Z}_N$ can all be represented as strings of length n, where the integer $a \in \mathbb{Z}_N$ is padded to the left with 0s if $\|a\| < n$.

We do not focus much on the issue of representation, since for all the groups considered in this text the representation can simply be taken to be the "natural" one (as in the case of $\mathbb{Z}_N$, above). Note, however, that different representations of the same group can affect the complexity of performing various computations, and so choosing the "right" representation for a given group is often important in practice. Since our goal is only to show *polynomial-time* algorithms for each of the operations we need (and not to show the most efficient algorithms known), the exact representation used is less important for our purposes. Moreover, most of the "higher-level" algorithms we present use the group operation in a "black-box" manner, so that as long as the group operation can be performed in polynomial time (in some parameter), the resulting algorithm will run in polynomial time as well.

Given a group $\mathbb{G}$ where elements are represented by strings of length ℓ, a uniform group element can be selected by choosing uniform ℓ-bit strings until the first string that corresponds to a group element is found. (Note this assumes that testing group membership can be done efficiently.) To obtain an algorithm with bounded running time, we introduce a parameter t bounding the maximum number of times this process is repeated; if all t iterations fail to find an element of $\mathbb{G}$, then the algorithm outputs fail. (An alternative is to output an arbitrary element of $\mathbb{G}$.) That is:

ALGORITHM B.14
Choosing a uniform group element

Input: A (description of a) group $\mathbb{G}$; length-parameter ℓ;
  parameter t
Output: A uniform element of $\mathbb{G}$

for $i = 1$ to t:
  Choose uniform $x \in \{0,1\}^\ell$
  if $x \in \mathbb{G}$ return x
return "fail"

It is clear that whenever the above algorithm does *not* output fail, it outputs a uniformly distributed element of $\mathbb{G}$. This is simply because each element of $\mathbb{G}$ is equally likely to be chosen in any iteration. Formally, if we let Fail be the event that the algorithm outputs fail, then for any element $g \in \mathbb{G}$ we have

$$\Pr\left[\text{output of the algorithm equals } g \mid \overline{\text{Fail}}\right] = \frac{1}{|\mathbb{G}|}.$$

What is the probability that the algorithm outputs fail? In any iteration the probability that $x \in \mathbb{G}$ is exactly $|\mathbb{G}|/2^\ell$, and so the probability that x does *not* lie in $\mathbb{G}$ in any of the t iterations is

$$\left(1 - \frac{|\mathbb{G}|}{2^\ell}\right)^t. \tag{B.1}$$

There is a trade-off between the running time of Algorithm B.14 and the probability that the algorithm outputs fail: increasing t decreases the probability of failure but increases the worst-case running time. For cryptographic applications we need an algorithm where the worst-case running time is polynomial in the security parameter n, while the failure probability is negligible in n. Let $K \stackrel{\text{def}}{=} 2^\ell/|\mathbb{G}|$. If we set $t := K \cdot n$ then the probability that the algorithm outputs fail is:

$$\left(1 - \frac{1}{K}\right)^{K \cdot n} = \left(\left(1 - \frac{1}{K}\right)^K\right)^n \leq \left(e^{-1}\right)^n = e^{-n},$$

using Proposition A.2. Thus, if $K = \mathsf{poly}(n)$ (we assume some group-generation algorithm that depends on the security parameter n, and so both $|\mathbb{G}|$ and ℓ are functions of n), we obtain an algorithm with the desired properties.

The case of $\mathbb{Z}_N$. Consider the group $\mathbb{Z}_N$, with $n = \|N\|$. Checking whether an n-bit string x (interpreted as a positive integer of length at most n) is an element of $\mathbb{Z}_N$ simply requires checking whether $x < N$. Furthermore,

$$\frac{2^n}{|\mathbb{Z}_N|} = \frac{2^n}{N} \leq \frac{2^n}{2^{n-1}} = 2,$$

and so we can sample a uniform element of $\mathbb{Z}_N$ in $\mathsf{poly}(n)$ time and with failure probability negligible in n.

The case of $\mathbb{Z}_N^*$. Consider next the group $\mathbb{Z}_N^*$, with $n = \|N\|$ as before. Determining whether an n-bit string x is an element of $\mathbb{Z}_N^*$ is also easy (see the exercises). Moreover,

$$\frac{2^n}{|\mathbb{Z}_N^*|} = \frac{2^n}{\phi(N)} = \frac{2^n}{N} \cdot \frac{N}{\phi(N)} \leq 2 \cdot \frac{N}{\phi(N)}.$$

A $\mathsf{poly}(n)$ upper-bound is a consequence of the following theorem.

THEOREM B.15 *For $N \geq 3$ of length n, we have $\frac{N}{\phi(N)} < 2n$.*

(Stronger bounds are known, but the above suffices for our purpose.) The theorem can be proved using Bertrand's Postulate (Theorem 9.32), but we content ourselves with a proof in two special cases: when N is prime and when N is a product of two equal-length (distinct) primes.

The analysis is easy when N is an odd prime. Here $\phi(N) = N - 1$ and so

$$\frac{N}{\phi(N)} \leq \frac{2^n}{\phi(N)} = \frac{2^n}{N-1} \leq \frac{2^n}{2^{n-1}} = 2$$

(using the fact that N is odd for the second inequality). Consider next the case of $N = pq$ for p and q distinct, odd primes. Then

$$\frac{N}{\phi(N)} = \frac{pq}{(p-1)(q-1)} = \frac{p}{p-1} \cdot \frac{q}{q-1} < \left(\frac{3}{2}\right) \cdot \left(\frac{5}{4}\right) < 2.$$

We conclude that when N is prime or the product of two distinct, odd primes, there is an algorithm for generating a uniform element of $\mathbb{Z}_N^*$ that runs in time polynomial in $n = \|N\|$ and outputs fail with probability negligible in n.

Throughout this book, when we speak of sampling a uniform element of $\mathbb{Z}_N$ or $\mathbb{Z}_N^*$ we simply ignore the negligible probability of outputting fail with the understanding that this has no significant effect on the analysis.

B.3   *Finding a Generator of a Cyclic Group

In this section we address the problem of finding a generator of an arbitrary cyclic group $\mathbb{G}$ of order q. Here, q does not necessarily denote a prime number; indeed, finding a generator when q is prime is trivial by Corollary 9.56.

We actually show how to sample a *uniform* generator, proceeding in a manner very similar to that of Section B.2.5. Here, we repeatedly sample uniform elements of $\mathbb{G}$ until we find an element that is a generator. As in Section B.2.5, an analysis of this method requires understanding two things:

- How to efficiently test whether a given element is a generator; and

- the fraction of group elements that are generators.

In order to understand these issues, we first develop a bit of additional group-theoretic background.

B.3.1   Group-Theoretic Background

We tackle the second issue first. Recall that the order of an element h is the smallest positive integer i for which $h^i = 1$. Let g be a generator of a group $\mathbb{G}$ of order $q > 1$; this means the order of g is q. Consider an element $h \in \mathbb{G}$ that is not the identity (the identity cannot be a generator of $\mathbb{G}$), and let us ask whether h might also be a generator of $\mathbb{G}$. Since g generates $\mathbb{G}$, we can write $h = g^x$ for some $x \in \{1, \ldots, q-1\}$ (note $x \neq 0$ since h is not the identity). Consider two cases:

Case 1: $\gcd(x, q) = r > 1$. Write $x = \alpha \cdot r$ and $q = \beta \cdot r$ with α, β non-zero integers less than q. Then:

$$h^\beta = (g^x)^\beta = g^{\alpha r \beta} = (g^q)^\alpha = 1.$$

So the order of h is at most $\beta < q$, and h cannot be a generator of $\mathbb{G}$.

Case 2: $\gcd(x, q) = 1$. Let $i \leq q$ be the order of h. Then

$$g^0 = 1 = h^i = (g^x)^i = g^{xi},$$

implying $xi = 0 \bmod q$ by Proposition 9.54. This means that $q \mid xi$. Since $\gcd(x, q) = 1$, however, Proposition 9.3 shows that $q \mid i$ and so $i = q$. We conclude that h is a generator of $\mathbb{G}$.

Summarizing the above, we see that for $x \in \{1, \ldots, q-1\}$ the element $h = g^x$ is a generator of $\mathbb{G}$ exactly when $\gcd(x, q) = 1$. We have thus proved the following:

THEOREM B.16   *Let $\mathbb{G}$ be a cyclic group of order $q > 1$ with generator g. There are $\phi(q)$ generators of $\mathbb{G}$, and these are exactly given by $\{g^x \mid x \in \mathbb{Z}_q^*\}$.*

In particular, if $\mathbb{G}$ is a group of prime order q, then it has $\phi(q) = q - 1$ generators—exactly in agreement with Corollary 9.56.

We turn next to the first issue, that of deciding whether a given element h is a generator of $\mathbb{G}$. Of course, one way to check whether h generates $\mathbb{G}$ is to enumerate $\{h^0, h^1, \ldots, h^{q-1}\}$ and see whether this list includes every element of $\mathbb{G}$. This requires time linear in q (i.e., exponential in $\|q\|$) and is therefore unacceptable for our purposes. Another approach, if we already know a generator g, is to compute the discrete logarithm $x = \log_g h$ and then apply the previous theorem; in general, however, we may not have such a g, and anyway computing the discrete logarithm may itself be a hard problem.

If we know the factorization of q, we can do better.

PROPOSITION B.17   *Let $\mathbb{G}$ be a group of order q, and let $q = \prod_{i=1}^{k} p_i^{e_i}$ be the prime factorization of q, where the $\{p_i\}$ are distinct primes and $e_i \geq 1$. Set $q_i = q/p_i$. Then $h \in \mathbb{G}$ is a generator of $\mathbb{G}$ if and only if*

$$h^{q_i} \neq 1 \quad for\ i = 1, \ldots, k.$$

PROOF   One direction is easy. Say $h^{q_i} = 1$ for some i. Then the order of h is at most $q_i < q$, and so h cannot be a generator.

Conversely, say h is not a generator but instead has order $q' < q$. By Proposition 9.55, we know $q' \mid q$. This implies that q' can be written as $q' = \prod_{i=1}^{k} p_i^{e'_i}$, where $e'_i \geq 0$ and for at least one index j we have $e'_j < e_j$. But then q' divides $q_j = p_j^{e_j-1} \cdot \prod_{i \neq j} p_i^{e_i}$, and so (using Proposition 9.54) $h^{q_j} = h^{[q_j \bmod q']} = h^0 = 1$. ∎

The proposition does not require $\mathbb{G}$ to be cyclic; if $\mathbb{G}$ is not cyclic then every element $h \in \mathbb{G}$ will satisfy $h^{q_i} = 1$ for some i and there are no generators.

B.3.2   Efficient Algorithms

Armed with the results of the previous section, we show how to efficiently *test* whether a given element is a generator, as well as how to efficiently *find* a generator in an arbitrary group.

Testing if an element is a generator. Proposition B.17 immediately suggests an efficient algorithm for deciding whether a given element h is a generator or not.

ALGORITHM B.18
Testing whether an element is a generator

Input: Group order q; prime factors $\{p_i\}_{i=1}^{k}$ of q; element $h \in \mathbb{G}$
Output: A decision as to whether h is a generator of $\mathbb{G}$

for $i = 1$ to k:
  if $h^{q/p_i} = 1$ return "h is not a generator"
return "h is a generator"

Correctness of the algorithm is evident from Proposition B.17. We now show that the algorithm terminates in time polynomial in $\|q\|$. Since, in each iteration, h^{q/p_i} can be computed in polynomial time, we need only show that the number of iterations k is polynomial. This is the case since an integer q can have no more than $\log_2 q = \mathcal{O}(\|q\|)$ prime factors; this is because

$$q = \prod_{i=1}^{k} p_i^{e_i} \geq \prod_{i=1}^{k} p_i \geq \prod_{i=1}^{k} 2 = 2^k$$

and so $k \leq \log_2 q$.

Algorithm B.18 requires the prime factors of the group order q to be provided as input. Interestingly, there is no known efficient algorithm for testing whether an element of an arbitrary group is a generator when the factors of the group order are *not* known.

The fraction of elements that are generators. As shown in Theorem B.16, the fraction of elements of a group $\mathbb{G}$ of order q that are generators is $\phi(q)/q$. Theorem B.15 says that $\phi(q)/q = \Omega(1/\|q\|)$. The fraction of elements that are generators is thus sufficiently high to ensure that sampling a polynomial number of elements from the group will yield a generator with all but negligible probability. (The analysis is the same as in Section B.2.5.)

Concrete examples in $\mathbb{Z}_p^*$. Putting everything together, we see there is an efficient probabilistic algorithm for finding a generator of a group $\mathbb{G}$ *as long as the factorization of the group order is known.* When selecting a group for cryptographic applications, it is therefore important that the group is chosen in such a way that this holds. This explains again the preference, discussed extensively in Section 9.3.2, for working in an appropriate prime-order subgroup of $\mathbb{Z}_p^*$. Another possibility is to use $\mathbb{G} = \mathbb{Z}_p^*$ for p a strong prime (i.e., $p = 2q+1$ with q also prime), in which case the prime factorization of the group order $p-1$ is known. One final possibility is to generate a prime p in such a way that the factorization of $p-1$ is known. Further details are beyond the scope of this book.

References and Additional Reading

The book by Shoup [183] is highly recommended for those seeking to explore the topics of this chapter in further detail. In particular, bounds on $\phi(N)/N$ (and an asymptotic version of Theorem B.15) can be found in [183, Chapter 5]. Hankerson et al. [91] also provide extensive detail on the implementation of number-theoretic algorithms for cryptography.

Exercises

B.1 Prove correctness of the extended Euclidean algorithm.

B.2 Prove that the extended Euclidean algorithm runs in time polynomial in the lengths of its inputs.

      Hint: First prove a proposition analogous to Proposition B.8.

B.3 Prove that, on input integers $a \geq b > 0$, the extended Euclidean algorithm outputs (d, X, Y) with $|X| \leq b$ and $|Y| \leq a$.

      Hint: Use induction on the recursive call.

B.4 Show how to determine that an n-bit string is in $\mathbb{Z}_N^*$ in polynomial time.

References

[1] M. Abdalla, J.H. An, M. Bellare, and C. Namprempre. From identification to signatures via the Fiat-Shamir transform: Necessary and sufficient conditions for security and forward-security. *IEEE Trans. Information Theory*, 54(8):3631–3646, 2008.

[2] M. Abdalla, M. Bellare, and P. Rogaway. The oracle Diffie-Hellman assumptions and an analysis of DHIES. In *Cryptographers' Track— RSA 2001*, volume 2020 of *LNCS*, pages 143–158. Springer, 2001. See http://cseweb.ucsd.edu/~mihir/papers/dhies.html.

[3] C. Adams and S. Lloyd. *Understanding PKI: Concepts, Standards, and Deployment Considerations.* Addison Wesley, 2nd edition, 2002.

[4] L.M. Adleman. A subexponential algorithm for the discrete logarithm problem with applications to cryptography. In *20th Annual Symposium on Foundations of Computer Science*, pages 55–60. IEEE, 1979.

[5] M. Agrawal, N. Kayal, and N. Saxena. PRIMES is in P. *Annals of Mathematics*, 160(2):781–793, 2004.

[6] W. Aiello, M. Bellare, G. Di Crescenzo, and R. Venkatesan. Security amplification by composition: The case of doubly-iterated, ideal ciphers. In *Advances in Cryptology—Crypto '98*, volume 1462 of *LNCS*, pages 390–407. Springer, 1998.

[7] A. Akavia, S. Goldwasser, and S. Safra. Proving hard-core predicates using list decoding. In *Proc. 44th Annual Symposium on Foundations of Computer Science*, pages 146–157. IEEE, 2003.

[8] W. Alexi, B. Chor, O. Goldreich, and C.P. Schnorr. RSA and Rabin functions: Certain parts are as hard as the whole. *SIAM Journal on Computing*, 17(2):194–209, 1988.

[9] N.J. AlFardan, D.J. Bernstein, K.G. Paterson, B. Poettering, and J.C.N. Schuldt. On the security of RC4 in TLS and WPA. In *USENIX Security Symposium*, 2013.

[10] J.H. An, Y. Dodis, and T. Rabin. On the security of joint signature and encryption. In *Advances in Cryptology—Eurocrypt 2002*, volume 2332 of *LNCS*, pages 83–107. Springer, 2002.

[11] ANSI X9.9. American national standard for financial institution message authentication (wholesale), 1981.

[12] R. Barbulescu, P. Gaudry, A. Joux, and E. Thomé. A heuristic quasi-polynomial algorithm for discrete logarithm in finite fields of small characteristic. In *Advances in Cryptology—Eurocrypt 2014*, volume 8441 of *LNCS*, pages 1–16. Springer, 2014.

[13] R. Bardou, R. Focardi, Y. Kawamoto, L. Simionato, G. Steel, and J.-K. Tsay. Efficient padding oracle attacks on cryptographic hardware. In *Advances in Cryptology—Crypto 2012*, volume 7417 of *LNCS*, pages 608–625. Springer, 2012.

[14] E. Barker. Recommendation for key management—general, January 2016. NIST Special Publication 800-57, Part 1, Revision 4.

[15] E. Barker, L. Chen, A. Roginsky, A. Vassilev, and R. Davis. Recommendations for pair-wise key-establishment schemes using discrete logarithm cryptography, April 2018. NIST Special Publication 800-56A, Revision 3; available at https://doi.org/10.6028/NIST.SP.800-56Ar3.

[16] M. Bellare, R. Canetti, and H. Krawczyk. Keying hash functions for message authentication. In *Advances in Cryptology—Crypto '96*, volume 1109 of *LNCS*, pages 1–15. Springer, 1996.

[17] M. Bellare, A. Desai, E. Jokipii, and P. Rogaway. A concrete security treatment of symmetric encryption. In *Proc. 38th Annual Symposium on Foundations of Computer Science*, pages 394–403. IEEE, 1997.

[18] M. Bellare, A. Desai, D. Pointcheval, and P. Rogaway. Relations among notions of security for public-key encryption schemes. In *Advances in Cryptology—Crypto '98*, volume 1462 of *LNCS*, pages 26–45. Springer, 1998.

[19] M. Bellare, O. Goldreich, and A. Mityagin. The power of verification queries in message authentication and authenticated encryption. Available at https://eprint.iacr.org/2004/309.

[20] M. Bellare, J. Kilian, and P. Rogaway. The security of the cipher block chaining message authentication code. *Journal of Computer and System Sciences*, 61(3):362–399, 2000.

[21] M. Bellare and C. Namprempre. Authenticated encryption: Relations among notions and analysis of the generic composition paradigm. In *Advances in Cryptology—Asiacrypt 2000*, volume 1976 of *LNCS*, pages 531–545. Springer, 2000.

[22] M. Bellare and G. Neven. Multi-signatures in the plain public-key model and a general forking lemma. In *13th ACM Conf. on Computer and Communications Security*, pages 390–399. ACM Press, 2006.

[23] M. Bellare, K. Pietrzak, and P. Rogaway. Improved security analyses for CBC MACs. In *Advances in Cryptology—Crypto 2005*, volume 3621 of *LNCS*, pages 527–545. Springer, 2005.

[24] M. Bellare and P. Rogaway. Random oracles are practical: A paradigm for designing efficient protocols. In *1st ACM Conf. on Computer and Communications Security*, pages 62–73. ACM, 1993.

[25] M. Bellare and P. Rogaway. Optimal asymmetric encryption. In *Advances in Cryptology—Eurocrypt '94*, volume 950 of *LNCS*, pages 92–111. Springer, 1994.

[26] M. Bellare and P. Rogaway. The exact security of digital signatures: How to sign with RSA and Rabin. In *Advances in Cryptology—Eurocrypt '96*, volume 1070 of *LNCS*, pages 399–416. Springer, 1996.

[27] M. Bellare and P. Rogaway. The security of triple encryption and a framework for code-based game-playing proofs. In *Advances in Cryptology—Eurocrypt 2006*, volume 4004 of *LNCS*, pages 409–426. Springer, 2006. Available at https://eprint.iacr.org/2004/331.

[28] S.M. Bellovin. Frank Miller: Inventor of the one-time pad. *Cryptologia*, 35(3):203–222, 2011.

[29] D. Bernstein. ChaCha, a variant of Salsa20, 2008. Available at http://cr.yp.to/chacha/chacha-20080128.pdf.

[30] D.J. Bernstein. A short proof of the unpredictability of cipher block chaining. Available at http://cr.yp.to/papers.html#easycbc.

[31] D.J. Bernstein. The Poly1305-AES message-authentication code. In *Fast Software Encryption—FSE 2005*, volume 3557 of *LNCS*, pages 32–49. Springer, 2005.

[32] G. Bertoni, J. Daemen, M. Peeters, and G. Van Assche. On the indifferentiability of the sponge construction. In *Advances in Cryptology—Eurocrypt 2008*, volume 4965 of *LNCS*, pages 181–197. Springer, 2008.

[33] K. Bhargavan and G. Leurent. On the practical (in-)security of 64-bit block ciphers: Collision attacks on HTTP over TLS and OpenVPN. In *Proc. 2016 ACM Conf. on Computer and Communications Security*, pages 456–467. ACM, 2016.

[34] E. Biham and A. Shamir. Differential cryptanalysis of DES-like cryptosystems. *Journal of Cryptology*, 4(1):3–72, 1991.

[35] E. Biham and A. Shamir. *Differential Cryptanalysis of the Data Encryption Standard.* Springer, 1993.

[36] J. Black and P. Rogaway. CBC MACs for arbitrary-length messages: The three-key constructions. *Journal of Cryptology,* 18(2):111–131, 2005.

[37] J. Black, P. Rogaway, T. Shrimpton, and M. Stam. An analysis of the blockcipher-based hash functions from PGV. *J. Cryptology,* 23(4):519–545, 2010.

[38] D. Bleichenbacher. Chosen ciphertext attacks against protocols based on the RSA encryption standard PKCS#1. In *Advances in Cryptology—Crypto '98,* volume 1462 of *Lecture Notes in Computer Science,* pages 1–12. Springer, 1998.

[39] M. Blum. Coin flipping by telephone. In *Proc. IEEE COMPCOM,* pages 133–137, 1982.

[40] M. Blum and S. Goldwasser. An efficient probabilistic public-key encryption scheme which hides all partial information. In *Advances in Cryptology—Crypto '84,* volume 196 of *Lecture Notes in Computer Science,* pages 289–302. Springer, 1985.

[41] M. Blum and S. Micali. How to generate cryptographically strong sequences of pseudo-random bits. *SIAM Journal on Computing,* 13(4):850–864, 1984.

[42] D. Boneh. Twenty years of attacks on the RSA cryptosystem. *Notices of the American Mathematical Society,* 46(2):203–213, 1999.

[43] D. Boneh. Simplified OAEP for the RSA and Rabin functions. In *Advances in Cryptology—Crypto 2001,* volume 2139 of *LNCS,* pages 275–291. Springer, 2001.

[44] D. Boneh, A. Joux, and P.Q. Nguyen. Why textbook ElGamal and RSA encryption are insecure. In *Advances in Cryptology—Asiacrypt 2000,* volume 1976 of *LNCS,* pages 30–43. Springer, 2000.

[45] F. Boudot, P. Gaudry, A. Guillevic, N. Heninger, E. Thomé, and P. Zimmermann. Comparing the difficulty of factorization and discrete logarithm: A 240-digit experiment, 2020. Available at https://eprint.iacr.org/2020/697.

[46] G. Brassard, P. Høyer, and A. Tapp. Quantum algorithm for the collision problem, 1997. Available at https://arxiv.org/abs/quant-ph/9705002.

[47] R. Canetti, O. Goldreich, and S. Halevi. The random oracle methodology, revisited. *Journal of the ACM*, 51(4):557–594, 2004.

[48] J.L. Carter and M.N. Wegman. Universal classes of hash functions. *J. Computer and System Sciences*, 18(2):143–154, 1979.

[49] D. Chaum, E. van Heijst, and B. Pfitzmann. Cryptographically strong undeniable signatures, unconditionally secure for the signer. In *Advances in Cryptology—Crypto '91*, volume 576 of *LNCS*, pages 470–484. Springer, 1992.

[50] L. Chen, D. Moody, A. Regenscheid, and K. Randall. Recommendations for discrete logarithm-based cryptography: Elliptic curve domain parameters, October 2019. Draft NIST Special Publication 800-186; available at https://doi.org/10.6028/NIST.SP.800-186-draft.

[51] L.N. Childs. *A Concrete Introduction to Higher Algebra*. Undergraduate Texts in Mathematics. Springer, 2nd edition, 2000.

[52] B. Cogliati, Y. Dodis, J. Katz, J. Lee, J. Steinberger, A. Thiruvengadam, and Z. Zhang. Provable security of (tweakable) block ciphers based on substitution-permutation networks. In *Advances in Cryptology—Crypto 2018, Part I*, volume 10991 of *LNCS*, pages 722–753. Springer, 2018.

[53] D. Coppersmith. The Data Encryption Standard (DES) and its strength against attacks. *IBM Journal of Research and Development*, 38(3):243–250, 1994.

[54] D. Coppersmith. Small solutions to polynomial equations, and low exponent RSA vulnerabilities. *Journal of Cryptology*, 10(4):233–260, 1997.

[55] D. Coppersmith, M.K. Franklin, J. Patarin, and M.K. Reiter. Low-exponent RSA with related messages. In *Advances in Cryptology—Eurocrypt '96*, volume 1070 of *LNCS*, pages 1–9. Springer, 1996.

[56] J.-S. Coron, Y. Dodis, C. Malinaud, and P. Puniya. Merkle–Damgård revisited: How to construct a hash function. In *Advances in Cryptology—Crypto 2005*, volume 3621 of *LNCS*, pages 430–448. Springer, 2005.

[57] J.-S. Coron, M. Joye, D. Naccache, and P. Paillier. New attacks on PKCS #1 v1.5 encryption. In *Advances in Cryptology—Eurocrypt 2000*, volume 1807 of *LNCS*, pages 369–381. Springer, 2000.

[58] R. Cramer and V. Shoup. Design and analysis of practical public-key encryption schemes secure against adaptive chosen ciphertext attack. *SIAM Journal on Computing*, 33(1):167–226, 2003.

[59] R. Crandall and C. Pomerance. *Prime Numbers: A Computational Perspective*. Springer, 2nd edition, 2005.

[60] I. Damgård. Collision free hash functions and public key signature schemes. In *Advances in Cryptology—Eurocrypt '87*, volume 304 of *LNCS*, pages 203–216. Springer, 1988.

[61] I. Damgård. A design principle for hash functions. In *Advances in Cryptology—Crypto '89*, volume 435 of *LNCS*, pages 416–427. Springer, 1990.

[62] J. DeLaurentis. A further weakness in the common modulus protocol for the RSA cryptoalgorithm. *Cryptologia*, 8:253–259, 1984.

[63] G. Di Crescenzo, J. Katz, R. Ostrovsky, and A. Smith. Efficient and non-interactive non-malleable commitment. In *Advances in Cryptology—Eurocrypt 2001*, volume 2045 of *LNCS*, pages 40–59. Springer, 2001.

[64] M. Dietzfelbinger. *Primality Testing in Polynomial Time*. Springer, 2004.

[65] W. Diffie and M. Hellman. New directions in cryptography. *IEEE Transactions on Information Theory*, 22(6):644–654, 1976.

[66] W. Diffie and M. Hellman. Exhaustive cryptanalysis of the NBS data encryption standard. *Computer*, pages 74–84, June 1977.

[67] J.D. Dixon. Asymptotically fast factorization of integers. *Mathematics of Computation*, 36:255–260, 1981.

[68] D. Dolev, C. Dwork, and M. Naor. Non-malleable cryptography. *SIAM J. Computing*, 30(2):391–437, 2000. Preliminary version in STOC '91.

[69] C. Ellison and B. Schneier. Ten risks of PKI: What you're not being told about public key infrastructure. *Computer Security Journal*, 16(1):1–7, 2000.

[70] S. Even and Y. Mansour. A construction of a cipher from a single pseudorandom permutation. *J. Cryptology*, 10(3):151–162, 1997.

[71] H. Feistel. Cryptography and computer privacy. *Scientific American*, 228(5):15–23, 1973.

[72] A. Fiat and A. Shamir. How to prove yourself: Practical solutions to identification and signature problems. In *Advances in Cryptology—Crypto '86*, volume 263 of *LNCS*, pages 186–194. Springer, 1987.

[73] R. Fischlin and C.-P. Schnorr. Stronger security proofs for RSA and Rabin bits. *Journal of Cryptology*, 13(2):221–244, 2000.

[74] J.B. Fraleigh. *A First Course in Abstract Algebra*. Addison Wesley, 7th edition, 2002.

[75] E. Fujisaki, T. Okamoto, D. Pointcheval, and J. Stern. RSA-OAEP is secure under the RSA assumption. *Journal of Cryptology*, 17(2):81–104, 2004.

[76] S.D. Galbraith. *The Mathematics of Public Key Cryptography*. Cambridge University Press, 2012.

[77] T. El Gamal. A public-key cryptosystem and a signature scheme based on discrete logarithms. *IEEE Trans. Info. Theory*, 31(4):469–472, 1985.

[78] C.F. Gauss. *Disquisitiones Arithmeticae*. Springer, 1986. (English edition).

[79] R. Gennaro, Y. Gertner, and J. Katz. Lower bounds on the efficiency of encryption and digital signature schemes. In *35th Annual ACM Symposium on Theory of Computing*, pages 417–425. ACM Press, 2003.

[80] E.N. Gilbert, F.J. MacWilliams, and N.J.A. Sloane. Codes which detect deception. *Bell Systems Technical Journal*, 53(3):405–424, 1974.

[81] O. Goldreich. Two remarks concerning the Goldwasser-Micali-Rivest signature scheme. In *Advances in Cryptology—Crypto '86*, volume 263 of *LNCS*, pages 104–110. Springer, 1987.

[82] O. Goldreich. *Foundations of Cryptography, vol. 1: Basic Tools*. Cambridge University Press, 2001.

[83] O. Goldreich. *Foundations of Cryptography, vol. 2: Basic Applications*. Cambridge University Press, 2004.

[84] O. Goldreich, S. Goldwasser, and S. Micali. On the cryptographic applications of random functions. In *Advances in Cryptology—Crypto '84*, volume 196 of *LNCS*, pages 276–288. Springer, 1985.

[85] O. Goldreich, S. Goldwasser, and S. Micali. How to construct random functions. *Journal of the ACM*, 33(4):792–807, 1986.

[86] O. Goldreich and L.A. Levin. A hard-core predicate for all one-way functions. In *21st Annual ACM Symposium on Theory of Computing*, pages 25–32. ACM Press, 1989.

[87] S. Goldwasser and S. Micali. Probabilistic encryption. *Journal of Computer and System Sciences*, 28(2):270–299, 1984.

[88] S. Goldwasser, S. Micali, and R. Rivest. A digital signature scheme secure against adaptive chosen-message attacks. *SIAM J. Computing*, 17(2):281–308, 1988.

[89] S. Goldwasser, S. Micali, and A.C.-C. Yao. Strong signature schemes. In *Proc. 15th Annual ACM Symposium on Theory of Computing*, pages 431–439. ACM, 1983.

[90] L.K. Grover. A fast quantum mechanical algorithm for database search. In *28th Annual ACM Symposium on Theory of Computing*, pages 212–219. ACM Press, 1996. Available at https://arxiv.org/abs/quant-ph/9605043.

[91] D. Hankerson, A.J. Menezes, and S.A. Vanstone. *Guide to Elliptic Curve Cryptography*. Springer, 2004.

[92] J. Håstad. Solving simultaneous modular equations of low degree. *SIAM Journal on Computing*, 17(2):336–341, 1988.

[93] J. Håstad, R. Impagliazzo, L. Levin, and M. Luby. A pseudorandom generator from any one-way function. *SIAM Journal on Computing*, 28(4):1364–1396, 1999.

[94] J. Håstad and M. Näslund. The security of all RSA and discrete log bits. *Journal of the ACM*, 51(2):187–230, 2004.

[95] M. Hellman. A cryptanalytic time-memory trade-off. *IEEE Trans. Information Theory*, 26(4):401–406, 1980.

[96] N. Heninger, Z. Durumeric, E. Wustrow, and J.A. Halderman. Mining your Ps and Qs: Detection of widespread weak keys in network devices. In *Proc. 21st USENIX Security Symposium*, 2012.

[97] I.N. Herstein. *Abstract Algebra*. Wiley, 3rd edition, 1996.

[98] H.M. Heys. *The Design of Substitution-Permutation Network Ciphers Resistant to Cryptanalysis*. PhD thesis, Queen's University, 1994.

[99] H.M. Heys. A tutorial on linear and differential cryptanalysis. *Cryptologia*, 26(3):189–221, 2002. Also available at http://www.engr.mun.ca/~howard/Research/Papers/.

[100] D. Hofheinz and E. Kiltz. Practical chosen ciphertext secure encryption from factoring. In *Advances in Cryptology—Eurocrypt 2009*, volume 5479 of *LNCS*, pages 313–332. Springer, 2009.

[101] R. Impagliazzo and M. Luby. One-way functions are essential for complexity-based cryptography. In *30th Annual Symposium on Foundations of Computer Science*, pages 230–235. IEEE, 1989.

[102] ISO/IEC 9797. Data cryptographic techniques—data integrity mechanism using a cryptographic check function employing a block cipher algorithm, 1989.

[103] T. Iwata and K. Kurosawa. OMAC: One-key CBC MAC. In *Fast Software Encryption—FSE 2003*, volume 2887 of *LNCS*, pages 129–153. Springer, 2003.

[104] J. Jonsson. On the security of CTR + CBC-MAC. In *9th Annual International Workshop on Selected Areas in Cryptography—SAC 2002*, volume 2595 of *LNCS*, pages 76–93. Springer, 2003.

[105] A. Joux. *Algorithmic Cryptanalysis*. Chapman & Hall/CRC Press, 2009.

[106] D. Kahn. *The Codebreakers: The Comprehensive History of Secret Communication from Ancient Times to the Internet*. Scribner, 1996.

[107] B. Kaliski. PKCS #1: RSA encryption, version 1.5, 1998. RFC 2313, available at https://tools.ietf.org/html/rfc2313.

[108] B. Kaliski and J. Staddon. PKCS #1: RSA cryptography specifications, version 2.0, 1998. RFC 2437, available at https://tools.ietf.org/html/rfc2437.

[109] J. Katz. *Digital Signatures*. Springer, 2010.

[110] J. Katz and C.-Y. Koo. On constructing universal one-way hash functions from arbitrary one-way functions. *J. Cryptology*, to appear. Available at https://eprint.iacr.org/2005/328.

[111] J. Katz and M. Yung. Unforgeable encryption and chosen ciphertext secure modes of operation. In *Fast Software Encryption—FSE 2000*, volume 1978 of *LNCS*, pages 284–299. Springer, 2000.

[112] J. Katz and M. Yung. Characterization of security notions for probabilistic private-key encryption. *Journal of Cryptology*, 19(1):67–96, 2006.

[113] C. Kaufman, R. Perlman, and M. Speciner. *Network Security: Private Communication in a Public World*. Prentice Hall, 2nd edition, 2002.

[114] A. Kerckhoffs. La cryptographie militaire. *Journal des Sciences Militaires*, IX:5–38, January 1883. A copy of the paper is available at http://www.petitcolas.net/fabien/kerckhoffs.

[115] A. Kerckhoffs. La cryptographie militaire. *Journal des Sciences Militaires*, IX:161–191, February 1883. A copy of the paper is available at http://www.petitcolas.net/fabien/kerckhoffs.

[116] T. Kleinjung and B. Wesolowski. Discrete logarithms in quasi-polynomial time in finite fields of fixed characteristic, 2019. Available at https://eprint.iacr.org/2019/751.

[117] L. Knudsen and M.J.B. Robshaw. *The Block Cipher Companion*. Springer, 2011.

[118] L.M. Kohnfelder. Towards a practical public-key cryptosystem, 1978. Undergraduate thesis, MIT.

[119] T. Kohno, J. Viega, and D. Whiting. CWC: A high-performance conventional authenticated encryption mode. In *Fast Software Encryption— FSE 2004*, volume 3017 of *LNCS*, pages 408–426. Springer, 2004.

[120] H. Krawczyk. LFSR-based hashing and authentication. In *Advances in Cryptology—Crypto '94*, volume 839 of *LNCS*, pages 129–139. Springer, 1994.

[121] H. Krawczyk. New hash functions for message authentication. In *Advances in Cryptology—Eurocrypt '95*, volume 921 of *LNCS*, pages 301–310. Springer, 1995.

[122] H. Krawczyk. The order of encryption and authentication for protecting communications (or: How secure is SSL?). In *Advances in Cryptology— Crypto 2001*, volume 2139 of *LNCS*, pages 310–331. Springer, 2001.

[123] H. Krawczyk. Cryptographic extraction and key derivation: The HKDF scheme. In *Advances in Cryptology—Crypto 2010*, volume 6223 of *LNCS*, pages 631–648. Springer, 2010.

[124] L. Lamport. Constructing digital signatures from a one-way function. Technical Report CSL-98, SRI International, 1978.

[125] A.K. Lenstra, J.P. Hughes, M. Augier, J.W. Bos, T. Kleinjung, and C. Wachter. Public keys. In *Advances in Cryptology—Crypto 2012*, volume 7417 of *LNCS*, pages 626–642. Springer, 2012.

[126] A.K. Lenstra and E.R. Verheul. Selecting cryptographic key sizes. *Journal of Cryptology*, 14(4):255–293, 2001.

[127] G. Leurent and T. Peyrin. From collisions to chosen-prefix collisions— application to full SHA-1. In *Advances in Cryptology—Eurocrypt 2019, Part III*, volume 11478 of *LNCS*, pages 527–555. Springer, 2019. Available at https://eprint.iacr.org/2019/459.

[128] G. Leurent and T. Peyrin. SHA-1 is a shambles: First chosen-prefix collision on SHA-1 and application to the PGP web of trust, 2020. Available at https://eprint.iacr.org/2020/014.

[129] S. Levy. *Crypto: How the Code Rebels Beat the Government—Saving Privacy in the Digital Age*. Viking, 2001.

[130] R. Lidl and H. Niederreiter. *Introduction to Finite Fields and Their Applications*. Cambridge University Press, 2nd edition, 1994.

[131] M. Luby. *Pseudorandomness and Cryptographic Applications*. Princeton University Press, 1996.

[132] M. Luby and C. Rackoff. How to construct pseudorandom permutations from pseudorandom functions. *SIAM J. Computing*, 17(2):373–386, 1988.

[133] J. Manger. A chosen ciphertext attack on RSA optimal asymmetric encryption padding (OAEP) as standardized in PKCS #1 v2.0. In *Advances in Cryptology—Crypto 2001*, volume 2139 of *LNCS*, pages 230–238. Springer, 2001.

[134] M. Matsui. Linear cryptanalysis method for DES cipher. In *Advances in Cryptology—Eurocrypt '93*, volume 765 of *LNCS*, pages 386–397. Springer, 1993.

[135] A. May. *New RSA Vulnerabilities Using Lattice Reduction Methods*. PhD thesis, University of Paderborn, 2003.

[136] D.A. McGrew and J. Viega. The security and performance of the Galois/counter mode (GCM) of operation. In *Progress in Cryptology—Indocrypt 2004*, volume 3348 of *LNCS*, pages 343–355. Springer, 2004.

[137] A.J. Menezes, P.C. van Oorschot, and S.A. Vanstone. *Handbook of Applied Cryptography*. CRS Press, 1997.

[138] R.C. Merkle. A digital signature based on a conventional encryption function. In *Advances in Cryptology—Crypto '87*, volume 293 of *LNCS*, pages 369–378. Springer, 1988.

[139] R.C. Merkle. A certified digital signature. In *Advances in Cryptology—Crypto '89*, volume 435 of *LNCS*, pages 218–238. Springer, 1990.

[140] R.C. Merkle. One way hash functions and DES. In *Advances in Cryptology—Crypto '89*, volume 435 of *LNCS*, pages 428–446. Springer, 1990.

[141] R.C. Merkle and M. Hellman. On the security of multiple encryption. *Communications of the ACM*, 24(7):465–467, 1981.

[142] S. Micali, C. Rackoff, and B. Sloan. The notion of security for probabilistic cryptosystems. *SIAM J. Computing*, 17(2):412–426, 1988.

[143] G.L. Miller. Riemann's hypothesis and tests for primality. *Journal of Computer and System Sciences*, 13(3):300–317, 1976.

[144] C.J. Mitchell. On the security of 2-key triple DES, 2016. Available at https://arxiv.org/pdf/1602.06229v2.pdf.

[145] K. Moriarty, B. Kaliski, J. Jonsson, and A. Rusch. PKCS #1: RSA cryptography specifications, version 2.2, 2016. RFC 8017, available at https://tools.ietf.org/html/rfc8017.

[146] M. Naor and M. Yung. Universal one-way hash functions and their cryptographic applications. In *Proc. 21st Annual ACM Symposium on Theory of Computing*, pages 33–43. ACM, 1989.

[147] M. Naor and M. Yung. Public-key cryptosystems provably secure against chosen ciphertext attacks. In *22nd Annual ACM Symposium on Theory of Computing*, pages 427–437. ACM Press, 1990.

[148] National Bureau of Standards. Federal information processing standard publication 81: DES modes of operation, 1980.

[149] National Institute of Standards and Technology. Federal information processing standard publication 198-1: The keyed-hash message authentication code (HMAC), July 2008.

[150] National Institute of Standards and Technology. Federal information processing standards publication 186-4: Digital signature standard (DSS), July 2013.

[151] National Institute of Standards and Technology. Federal information processing standards publication 186-5 (draft): Digital signature standard (DSS), October 2019. Available at https://doi.org/10.6028/NIST.FIPS.186-5-draft.

[152] V.I. Nechaev. Complexity of a determinate algorithm for the discrete logarithm. *Mathematical Notes*, 55(2):165–172, 1994.

[153] M.A. Nielsen and I.L. Chuang. *Quantum Computation and Quantum Information: 10th Anniversary Edition*. Cambridge University Press, 2011.

[154] Y. Nir and A. Langley. ChaCha20 and Poly1305 for IETF protocols, 2018. RFC 8439, available at https://tools.ietf.org/html/rfc8439.

[155] P. Oechslin. Making a faster cryptanalytic time-memory trade-off. In *Advances in Cryptology—Crypto 2003*, volume 2729 of *LNCS*, pages 617–630. Springer, 2003.

[156] C. Paar and J. Pelzl. *Understanding Cryptography*. Springer, 2010.

[157] P. Paillier. Public-key cryptosystems based on composite degree residuosity classes. In *Advances in Cryptology—Eurocrypt '99*, volume 1592 of *LNCS*, pages 223–238. Springer, 1999.

[158] E. Petrank and C. Rackoff. CBC MAC for real-time data sources. *Journal of Cryptology*, 13(3):315–338, 2000.

[159] S. Pohlig and M. Hellman. An improved algorithm for computing logarithms over GF(p) and its cryptographic significance. *IEEE Trans. Information Theory*, 24(1):106–110, 1978.

[160] J.M. Pollard. Theorems of factorization and primality testing. *Proc. Cambridge Philosophical Society*, 76:521–528, 1974.

[161] J.M. Pollard. A Monte Carlo method for factorization. *BIT Numerical Mathematics*, 15(3):331–334, 1975.

[162] J.M. Pollard. Monte Carlo methods for index computation (mod p). *Mathematics of Computation*, 32(143):918–924, 1978.

[163] C. Pomerance. The quadratic sieve factoring algorithm. In *Advances in Cryptology—Eurocrypt '84*, volume 209 of *LNCS*, pages 169–182. Springer, 1985.

[164] B. Preneel, R. Govaerts, and J. Vandewalle. Hash functions based on block ciphers: A synthetic approach. In *Advances in Cryptology—Crypto '93*, volume 773 of *LNCS*, pages 368–378. Springer, 1994.

[165] M.O. Rabin. Digitalized signatures. In R.A. DeMillo, D.P. Dobkin, A.K. Jones, and R.J. Lipton, editors, *Foundations of Secure Computation*, pages 155–168. Academic Press, 1978.

[166] M.O. Rabin. Digitalized signatures as intractable as factorization. Technical Report TR-212, MIT/LCS, 1979.

[167] M.O. Rabin. Probabilistic algorithm for testing primality. *Journal of Number Theory*, 12(1):128–138, 1980.

[168] C. Rackoff and D.R. Simon. Non-interactive zero-knowledge proof of knowledge and chosen ciphertext attack. In *Advances in Cryptology—Crypto '91*, volume 576 of *LNCS*, pages 433–444. Springer, 1992.

[169] O. Regev. On lattices, learning with errors, random linear codes, and cryptography. *J. ACM*, 56(6):34:1–34:40, 2009.

[170] E. Rescorla. The transport layer security (TLS) protocol version 1.3, 2018. RFC 8446, available at https://tools.ietf.org/html/rfc8446.

[171] R. Rivest, A. Shamir, and L. Adleman. A method for obtaining digital signatures and public-key cryptosystems. *Communications of the ACM*, 21(2):120–126, 1978.

[172] P. Rogaway. Nonce-based symmetric encryption. In *Fast Software Encryption—FSE 2004*, volume 3017 of *LNCS*, pages 348–359. Springer, 2004.

[173] P. Rogaway and T. Shrimpton. Cryptographic hash-function basics: Definitions, implications, and separations for preimage resistance, second-preimage resistance, and collision resistance. In *Fast Software Encryption—FSE 2004*, volume 3017 of *LNCS*, pages 371–388. Springer, 2004.

[174] J. Rompel. One-way functions are necessary and sufficient for secure signatures. In *Proc. 22nd Annual ACM Symposium on Theory of Computing*, pages 387–394. ACM, 1990.

[175] C.-P. Schnorr. Efficient identification and signatures for smart cards. In *Advances in Cryptology—Crypto '89*, volume 435 of *LNCS*, pages 239–252. Springer, 1990.

[176] D. Shanks. Class number, a theory of factorization, and genera. In *Proc. Symposia in Pure Mathematics 20*, pages 415–440. American Mathematical Society, 1971.

[177] C.E. Shannon. Communication theory of secrecy systems. *Bell Systems Technical Journal*, 28(4):656–715, 1949.

[178] P.W. Shor. Polynomial-time algorithms for prime factorization and discrete logarithms on a quantum computer. *SIAM Journal on Computing*, 26(5):1484–1509, 1997. Available at https://arxiv.org/abs/quant-ph/9508027.

[179] V. Shoup. Lower bounds for discrete logarithms and related problems. In *Advances in Cryptology—Eurocrypt '97*, volume 1233 of *LNCS*, pages 256–266. Springer, 1997.

[180] V. Shoup. Why chosen ciphertext security matters. Technical Report RZ 3076, IBM Zurich, November 1998. Available at http://shoup.net/papers/expo.pdf.

[181] V. Shoup. A proposal for an ISO standard for public key encryption, 2001. Available at https://eprint.iacr.org/201/112.

[182] V. Shoup. OAEP reconsidered. *Journal of Cryptology*, 15(4):223–249, 2002.

[183] V. Shoup. *A Computational Introduction to Number Theory and Algebra*. Cambridge University Press, 2nd edition, 2009. Also available at http://www.shoup.net/ntb.

[184] T. Shrimpton. A characterization of authenticated encryption as a form of chosen-ciphertext security, 2004. Available at eprint.iacr.org/2004/272.

[185] J.H. Silverman and J. Tate. *Rational Points on Elliptic Curves.* Undergraduate Texts in Mathematics. Springer, 1994.

[186] G. Simmons. A "weak" privacy protocol using the RSA crypto algorithm. *Cryptologia,* 7:180–182, 1983.

[187] G. Simmons. A survey of information authentication. In G. Simmons, editor, *Contemporary Cryptology: The Science of Information Integrity,* pages 379–419. IEEE Press, 1992.

[188] S. Singh. *The Code Book: The Science of Secrecy from Ancient Egypt to Quantum Cryptography.* Anchor Books, 2000.

[189] A. Smith and Y. Zhang. On the regularity of lossy RSA: Improved bounds and applications to padding-based encryption. In *12th Theory of Cryptography Conference — TCC 2015,* volume 9014 of *LNCS,* pages 609–628. Springer, 2015.

[190] R. Solovay and V. Strassen. A fast Monte-Carlo test for primality. *SIAM Journal on Computing,* 6(1):84–85, 1977.

[191] J. Song, R. Poovendran, J. Lee, and T. Iwata. The AES-CMAC algorithm, 2006. RFC 4493, available at https://tools.ietf.org/html/rfc4493.

[192] M. Stevens, E. Bursztein, P. Karpman, A. Albertini, and Y. Markov. The first collision for full SHA-1. In *Advances in Cryptology— Crypto 2017, Part I,* volume 10401 of *LNCS,* pages 570–596. Springer, 2017.

[193] D.R. Stinson. Universal hashing and authentication codes. *Designs, Codes, and Cryptography,* 4(4):369–380, 1994.

[194] D.R. Stinson. *Cryptography: Theory and Practice.* Chapman & Hall/ CRC Press, 1st edition, 1995.

[195] D.R. Stinson. *Cryptography: Theory and Practice.* Chapman & Hall/ CRC Press, 3rd edition, 2005.

[196] W. Trappe and L. Washington. *Introduction to Cryptography with Coding Theory.* Prentice Hall, 2nd edition, 2005.

[197] P.C. van Oorschot and M.J. Wiener. A known plaintext attack on two-key triple encryption. In *Advances in Cryptology—Eurocrypt '90,* volume 473 of *LNCS,* pages 318–325. Springer, 1990.

[198] P.C. van Oorschot and M.J. Wiener. Parallel collision search with cryptanalytic applications. *Journal of Cryptology,* 12(1):1–28, 1999.

[199] S. Vaudenay. Security flaws induced by CBC padding—applications to SSL, IPSEC, WTLS, ... In *Advances in Cryptology—Eurocrypt 2002*, volume 2332 of *LNCS*, pages 534–546. Springer, 2002.

[200] G.S. Vernam. Cipher printing telegraph systems for secret wire and radio telegraphic communications. *Journal of the American Institute for Electrical Engineers*, 55:109–115, 1926.

[201] S.S. Wagstaff, Jr. *Cryptanalysis of Number Theoretic Ciphers*. Chapman & Hall/CRC Press, 2003.

[202] L. Washington. *Elliptic Curves: Number Theory and Cryptography*. Chapman & Hall/CRC Press, 2003.

[203] M.N. Wegman and J.L. Carter. New hash functions and their use in authentication and set equality. *J. Computer and System Sciences*, 22(3):265–279, 1981.

[204] D. Whiting, R. Housley, and N. Ferguson. Counter with CBC-MAC (CCM), 2003. RFC 3610, available at https://tools.ietf.org/html/rfc3610.

[205] A.C.-C. Yao. Theory and applications of trapdoor functions. In *23rd Annual Symposium on Foundations of Computer Science*, pages 80–91. IEEE, 1982.

[206] G. Yuval. How to swindle Rabin. *Cryptologia*, 3:187–189, 1979.

Index

Printed in the United States
By Bookmasters